T0189331

Lecture Notes in Computer Science   **14188**

The series Lecture Notes in Computer Science (LNCS), including its subseries Lecture Notes in Artificial Intelligence (LNAI) and Lecture Notes in Bioinformatics (LNBI), has established itself as a medium for the publication of new developments in computer science and information technology research, teaching, and education.

LNCS enjoys close cooperation with the computer science R & D community, the series counts many renowned academics among its volume editors and paper authors, and collaborates with prestigious societies. Its mission is to serve this international community by providing an invaluable service, mainly focused on the publication of conference and workshop proceedings and postproceedings. LNCS commenced publication in 1973.

Gernot A. Fink · Rajiv Jain · Koichi Kise ·
Richard Zanibbi
Editors

Document Analysis and Recognition – ICDAR 2023

17th International Conference
San José, CA, USA, August 21–26, 2023
Proceedings, Part II

 Springer

Editors
Gernot A. Fink
TU Dortmund University
Dortmund, Germany

Rajiv Jain
Adobe
College Park, MN, USA

Koichi Kise
Osaka Metropolitan University
Osaka, Japan

Richard Zanibbi
Rochester Institute of Technology
Rochester, NY, USA

ISSN 0302-9743   ISSN 1611-3349 (electronic)
Lecture Notes in Computer Science
ISBN 978-3-031-41678-1   ISBN 978-3-031-41679-8 (eBook)
https://doi.org/10.1007/978-3-031-41679-8

This Springer imprint is published by the registered company Springer Nature Switzerland AG
The registered company address is: Gewerbestrasse 11, 6330 Cham, Switzerland

Foreword

We are delighted to welcome you to the proceedings of ICDAR 2023, the 17th IAPR International Conference on Document Analysis and Recognition, which was held in San Jose, in the heart of Silicon Valley in the United States. With the worst of the pandemic behind us, we hoped that ICDAR 2023 would be a fully in-person event. However, challenges such as difficulties in obtaining visas also necessitated the partial use of hybrid technologies for ICDAR 2023. The oral papers being presented remotely were synchronous to ensure that conference attendees interacted live with the presenters and the limited hybridization still resulted in an enjoyable conference with fruitful interactions.

ICDAR 2023 was the 17th edition of a longstanding conference series sponsored by the International Association of Pattern Recognition (IAPR). It is the premier international event for scientists and practitioners in document analysis and recognition. This field continues to play an important role in transitioning to digital documents. The IAPR-TC 10/11 technical committees endorse the conference. The very first ICDAR was held in St Malo, France in 1991, followed by Tsukuba, Japan (1993), Montreal, Canada (1995), Ulm, Germany (1997), Bangalore, India (1999), Seattle, USA (2001), Edinburgh, UK (2003), Seoul, South Korea (2005), Curitiba, Brazil (2007), Barcelona, Spain (2009), Beijing, China (2011), Washington, DC, USA (2013), Nancy, France (2015), Kyoto, Japan (2017), Sydney, Australia (2019) and Lausanne, Switzerland (2021).

Keeping with its tradition from past years, ICDAR 2023 featured a three-day main conference, including several competitions to challenge the field and a post-conference slate of workshops, tutorials, and a doctoral consortium. The conference was held at the San Jose Marriott on August 21–23, 2023, and the post-conference tracks at the Adobe World Headquarters in San Jose on August 24–26, 2023.

We thank our executive co-chairs, Venu Govindaraju and Tong Sun, for their support and valuable advice in organizing the conference. We are particularly grateful to Tong for her efforts in facilitating the organization of the post-conference in Adobe Headquarters and for Adobe's generous sponsorship.

The highlights of the conference include keynote talks by the recipient of the IAPR/ICDAR Outstanding Achievements Award, and distinguished speakers Marti Hearst, UC Berkeley School of Information; Vlad Morariu, Adobe Research; and Seiichi Uchida, Kyushu University, Japan.

A total of 316 papers were submitted to the main conference (plus 33 papers to the ICDAR-IJDAR journal track), with 53 papers accepted for oral presentation (plus 13 IJDAR track papers) and 101 for poster presentation. We would like to express our deepest gratitude to our Program Committee Chairs, featuring three distinguished researchers from academia, Gernot A. Fink, Koichi Kise, and Richard Zanibbi, and one from industry, Rajiv Jain, who did a phenomenal job in overseeing a comprehensive reviewing process and who worked tirelessly to put together a very thoughtful and interesting technical program for the main conference. We are also very grateful to the members of the Program Committee for their high-quality peer reviews. Thank you to our competition chairs, Kenny Davila, Chris Tensmeyer, and Dimosthenis Karatzas, for overseeing the competitions.

The post-conference featured 8 excellent workshops, four value-filled tutorials, and the doctoral consortium. We would like to thank Mickael Coustaty and Alicia Fornes, the workshop chairs, Elisa Barney-Smith and Laurence Likforman-Sulem, the tutorial chairs, and Jean-Christophe Burie and Andreas Fischer, the doctoral consortium chairs, for their efforts in putting together a wonderful post-conference program.

We would like to thank and acknowledge the hard work put in by our Publication Chairs, Anurag Bhardwaj and Utkarsh Porwal, who worked diligently to compile the camera-ready versions of all the papers and organize the conference proceedings with Springer. Many thanks are also due to our sponsorship, awards, industry, and publicity chairs for their support of the conference.

The organization of this conference was only possible with the tireless behind-the-scenes contributions of our webmaster and tech wizard, Edward Sobczak, and our secretariat, ably managed by Carol Doermann. We convey our heartfelt appreciation for their efforts.

Finally, we would like to thank for their support our many financial sponsors and the conference attendees and authors, for helping make this conference a success. We sincerely hope those who attended had an enjoyable conference, a wonderful stay in San Jose, and fruitful academic exchanges with colleagues.

August 2023

David Doermann
Srirangaraj (Ranga) Setlur

Preface

Welcome to the proceedings of the 17th International Conference on Document Analysis and Recognition (ICDAR) 2023. ICDAR is the premier international event for scientists and practitioners involved in document analysis and recognition.

This year, we received 316 conference paper submissions with authors from 42 different countries. In order to create a high-quality scientific program for the conference, we recruited 211 regular and 38 senior program committee (PC) members. Regular PC members provided a total of 913 reviews for the submitted papers (an average of 2.89 per paper). Senior PC members who oversaw the review phase for typically 8 submissions took care of consolidating reviews and suggested paper decisions in their meta-reviews. Based on the information provided in both the reviews and the prepared meta-reviews we PC Chairs then selected 154 submissions (48.7%) for inclusion into the scientific program of ICDAR 2023. From the accepted papers, 53 were selected for oral presentation, and 101 for poster presentation.

In addition to the papers submitted directly to ICDAR 2023, we continued the tradition of teaming up with the International Journal of Document Analysis and Recognition (IJDAR) and organized a special journal track. The journal track submissions underwent the same rigorous review process as regular IJDAR submissions. The ICDAR PC Chairs served as Guest Editors and oversaw the review process. From the 33 manuscripts submitted to the journal track, 13 were accepted and were published in a Special Issue of IJDAR entitled "Advanced Topics of Document Analysis and Recognition." In addition, all papers accepted in the journal track were included as oral presentations in the conference program.

A very prominent topic represented in both the submissions from the journal track as well as in the direct submissions to ICDAR 2023 was handwriting recognition. Therefore, we organized a Special Track on Frontiers in Handwriting Recognition. This also served to keep alive the tradition of the International Conference on Frontiers in Handwriting Recognition (ICFHR) that the TC-11 community decided to no longer organize as an independent conference during ICFHR 2022 held in Hyderabad, India. The handwriting track included oral sessions covering handwriting recognition for historical documents, synthesis of handwritten documents, as well as a subsection of one of the poster sessions. Additional presentation tracks at ICDAR 2023 featured Graphics Recognition, Natural Language Processing for Documents (D-NLP), Applications (including for medical, legal, and business documents), additional Document Analysis and Recognition topics (DAR), and a session highlighting featured competitions that were run for ICDAR 2023 (Competitions). Two poster presentation sessions were held at ICDAR 2023.

As ICDAR 2023 was held with in-person attendance, all papers were presented by their authors during the conference. Exceptions were only made for authors who could not attend the conference for unavoidable reasons. Such oral presentations were then provided by synchronous video presentations. Posters of authors that could not attend were presented by recorded teaser videos, in addition to the physical posters.

Three keynote talks were given by Marti Hearst (UC Berkeley), Vlad Morariu (Adobe Research), and Seichi Uchida (Kyushu University). We thank them for the valuable insights and inspiration that their talks provided for participants.

Finally, we would like to thank everyone who contributed to the preparation of the scientific program of ICDAR 2023, namely the authors of the scientific papers submitted to the journal track and directly to the conference, reviewers for journal-track papers, and both our regular and senior PC members. We also thank Ed Sobczak for helping with the conference web pages, and the ICDAR 2023 Publications Chairs Anurag Bharadwaj and Utkarsh Porwal, who oversaw the creation of this proceedings.

August 2023

Gernot A. Fink
Rajiv Jain
Koichi Kise
Richard Zanibbi

Organization

General Chairs

David Doermann   University at Buffalo, The State University of New York, USA

Srirangaraj Setlur   University at Buffalo, The State University of New York, USA

Executive Co-chairs

Venu Govindaraju   University at Buffalo, The State University of New York, USA

Tong Sun   Adobe Research, USA

PC Chairs

Gernot A. Fink   Technische Universität Dortmund, Germany (Europe)

Rajiv Jain   Adobe Research, USA (Industry)

Koichi Kise   Osaka Metropolitan University, Japan (Asia)

Richard Zanibbi   Rochester Institute of Technology, USA (Americas)

Workshop Chairs

Mickael Coustaty   La Rochelle University, France

Alicia Fornes   Universitat Autònoma de Barcelona, Spain

Tutorial Chairs

Elisa Barney-Smith   Luleå University of Technology, Sweden

Laurence Likforman-Sulem   Télécom ParisTech, France

Competitions Chairs

Kenny Davila   Universidad Tecnológica Centroamericana,
                                UNITEC, Honduras
Dimosthenis Karatzas   Universitat Autònoma de Barcelona, Spain
Chris Tensmeyer   Adobe Research, USA

Doctoral Consortium Chairs

Andreas Fischer   University of Applied Sciences and Arts Western
                                Switzerland
Veronica Romero   University of Valencia, Spain

Publications Chairs

Anurag Bharadwaj   Northeastern University, USA
Utkarsh Porwal   Walmart, USA

Posters/Demo Chair

Palaiahnakote Shivakumara   University of Malaya, Malaysia

Awards Chair

Santanu Chaudhury   IIT Jodhpur, India

Sponsorship Chairs

Wael Abd-Almageed   Information Sciences Institute USC, USA
Cheng-Lin Liu   Chinese Academy of Sciences, China
Masaki Nakagawa   Tokyo University of Agriculture and Technology,
                                Japan

Industry Chairs

Andreas Dengel   DFKI, Germany
Véronique Eglin   Institut National des Sciences Appliquées (INSA)
                                de Lyon, France
Nandakishore Kambhatla   Adobe Research, India

Publicity Chairs

Sukalpa Chanda   Østfold University College, Norway
Simone Marinai   University of Florence, Italy
Safwan Wshah   University of Vermont, USA

Technical Chair

Edward Sobczak   University at Buffalo, The State University of
                          New York, USA

Conference Secretariat

University at Buffalo, The State University of New York, USA

Program Committee

Senior Program Committee Members

Srirangaraj Setlur   Apostolos Antonacopoulos
Richard Zanibbi   Lianwen Jin
Koichi Kise   Nicholas Howe
Gernot Fink   Marc-Peter Schambach
David Doermann   Marcal Rossinyol
Rajiv Jain   Wataru Ohyama
Rolf Ingold   Nicole Vincent
Andreas Fischer   Faisal Shafait
Marcus Liwicki   Simone Marinai
Seiichi Uchida   Bertrand Couasnon
Daniel Lopresti   Masaki Nakagawa
Josep Llados   Anurag Bhardwaj
Elisa Barney Smith   Dimosthenis Karatzas
Umapada Pal   Masakazu Iwamura
Alicia Fornes   Tong Sun
Jean-Marc Ogier   Laurence Likforman-Sulem
C. V. Jawahar   Michael Blumenstein
Xiang Bai   Cheng-Lin Liu
Liangrui Peng   Luiz Oliveira
Jean-Christophe Burie   Robert Sabourin
Andreas Dengel   R. Manmatha
Robert Sablatnig   Angelo Marcelli
Basilis Gatos   Utkarsh Porwal

Program Committee Members

Harold Mouchere
Foteini Simistira Liwicki
Vernonique Eglin
Aurelie Lemaitre
Qiu-Feng Wang
Jorge Calvo-Zaragoza
Yuchen Zheng
Guangwei Zhang
Xu-Cheng Yin
Kengo Terasawa
Yasuhisa Fujii
Yu Zhou
Irina Rabaev
Anna Zhu
Soo-Hyung Kim
Liangcai Gao
Anders Hast
Minghui Liao
Guoqiang Zhong
Carlos Mello
Thierry Paquet
Mingkun Yang
Laurent Heutte
Antoine Doucet
Jean Hennebert
Cristina Carmona-Duarte
Fei Yin
Yue Lu
Maroua Mehri
Ryohei Tanaka
Adel M. M. Alimi
Heng Zhang
Gurpreet Lehal
Ergina Kavallieratou
Petra Gomez-Kramer
Anh Le Duc
Frederic Rayar
Muhammad Imran Malik
Vincent Christlein
Khurram Khurshid
Bart Lamiroy
Ernest Valveny
Antonio Parziale

Jean-Yves Ramel
Haikal El Abed
Alireza Alaei
Xiaoqing Lu
Sheng He
Abdel Belaid
Joan Puigcerver
Zhouhui Lian
Francesco Fontanella
Daniel Stoekl Ben Ezra
Byron Bezerra
Szilard Vajda
Irfan Ahmad
Imran Siddiqi
Nina S. T. Hirata
Momina Moetesum
Vassilis Katsouros
Fadoua Drira
Ekta Vats
Ruben Tolosana
Steven Simske
Christophe Rigaud
Claudio De Stefano
Henry A. Rowley
Pramod Kompalli
Siyang Qin
Alejandro Toselli
Slim Kanoun
Rafael Lins
Shinichiro Omachi
Kenny Davila
Qiang Huo
Da-Han Wang
Hung Tuan Nguyen
Ujjwal Bhattacharya
Jin Chen
Cuong Tuan Nguyen
Ruben Vera-Rodriguez
Yousri Kessentini
Salvatore Tabbone
Suresh Sundaram
Tonghua Su
Sukalpa Chanda

Mickael Coustaty
Donato Impedovo
Alceu Britto
Bidyut B. Chaudhuri
Swapan Kr. Parui
Eduardo Vellasques
Sounak Dey
Sheraz Ahmed
Julian Fierrez
Ioannis Pratikakis
Mehdi Hamdani
Florence Cloppet
Amina Serir
Mauricio Villegas
Joan Andreu Sanchez
Eric Anquetil
Majid Ziaratban
Baihua Xiao
Christopher Kermorvant
K. C. Santosh
Tomo Miyazaki
Florian Kleber
Carlos David Martinez Hinarejos
Muhammad Muzzamil Luqman
Badarinath T.
Christopher Tensmeyer
Musab Al-Ghadi
Ehtesham Hassan
Journet Nicholas
Romain Giot
Jonathan Fabrizio
Sriganesh Madhvanath
Volkmar Frinken
Akio Fujiyoshi
Srikar Appalaraju
Oriol Ramos-Terrades
Christian Viard-Gaudin
Chawki Djeddi
Nibal Nayef
Nam Ik Cho
Nicolas Sidere
Mohamed Cheriet
Mark Clement
Shivakumara Palaiahnakote
Shangxuan Tian

Ravi Kiran Sarvadevabhatla
Gaurav Harit
Iuliia Tkachenko
Christian Clausner
Vernonica Romero
Mathias Seuret
Vincent Poulain D'Andecy
Joseph Chazalon
Kaspar Riesen
Lambert Schomaker
Mounim El Yacoubi
Berrin Yanikoglu
Lluis Gomez
Brian Kenji Iwana
Ehsanollah Kabir
Najoua Essoukri Ben Amara
Volker Sorge
Clemens Neudecker
Praveen Krishnan
Abhisek Dey
Xiao Tu
Mohammad Tanvir Parvez
Sukhdeep Singh
Munish Kumar
Qi Zeng
Puneet Mathur
Clement Chatelain
Jihad El-Sana
Ayush Kumar Shah
Peter Staar
Stephen Rawls
David Etter
Ying Sheng
Jiuxiang Gu
Thomas Breuel
Antonio Jimeno
Karim Kalti
Enrique Vidal
Kazem Taghva
Evangelos Milios
Kaizhu Huang
Pierre Heroux
Guoxin Wang
Sandeep Tata
Youssouf Chherawala

Reeve Ingle
Aashi Jain
Carlos M. Travieso-Gonzales
Lesly Miculicich
Curtis Wigington
Andrea Gemelli
Martin Schall
Yanming Zhang
Dezhi Peng
Chongyu Liu
Huy Quang Ung
Marco Peer
Nam Tuan Ly
Jobin K. V.
Rina Buoy
Xiao-Hui Li
Maham Jahangir
Muhammad Naseer Bajwa

Oliver Tueselmann
Yang Xue
Kai Brandenbusch
Ajoy Mondal
Daichi Haraguchi
Junaid Younas
Ruddy Theodose
Rohit Saluja
Beat Wolf
Jean-Luc Bloechle
Anna Scius-Bertrand
Claudiu Musat
Linda Studer
Andrii Maksai
Oussama Zayene
Lars Voegtlin
Michael Jungo

Program Committee Subreviewers

Li Mingfeng
Houcemeddine Filali
Kai Hu
Yejing Xie
Tushar Karayil
Xu Chen
Benjamin Deguerre
Andrey Guzhov
Estanislau Lima
Hossein Naftchi
Giorgos Sfikas
Chandranath Adak
Yakn Li
Solenn Tual
Kai Labusch
Ahmed Cheikh Rouhou
Lingxiao Fei
Yunxue Shao
Yi Sun
Stephane Bres
Mohamed Mhiri
Zhengmi Tang
Fuxiang Yang
Saifullah Saifullah

Paolo Giglio
Wang Jiawei
Maksym Taranukhin
Menghan Wang
Nancy Girdhar
Xudong Xie
Ray Ding
Mélodie Boillet
Nabeel Khalid
Yan Shu
Moises Diaz
Biyi Fang
Adolfo Santoro
Glen Pouliquen
Ahmed Hamdi
Florian Kordon
Yan Zhang
Gerasimos Matidis
Khadiravana Belagavi
Xingbiao Zhao
Xiaotong Ji
Yan Zheng
M. Balakrishnan
Florian Kowarsch

Mohamed Ali Souibgui
Xuewen Wang
Djedjiga Belhadj
Omar Krichen
Agostino Accardo
Erika Griechisch
Vincenzo Gattulli
Thibault Lelore
Zacarias Curi
Xiaomeng Yang
Mariano Maisonnave
Xiaobo Jin
Corina Masanti
Panagiotis Kaddas
Karl Löwenmark
Jiahao Lv
Narayanan C. Krishnan
Simon Corbillé
Benjamin Fankhauser
Tiziana D'Alessandro
Francisco J. Castellanos
Souhail Bakkali
Caio Dias
Giuseppe De Gregorio
Hugo Romat
Alessandra Scotto di Freca
Christophe Gisler
Nicole Dalia Cilia
Aurélie Joseph
Gangyan Zeng
Elmokhtar Mohamed Moussa
Zhong Zhuoyao
Oluwatosin Adewumi
Sima Rezaei
Anuj Rai
Aristides Milios
Shreeganesh Ramanan
Wenbo Hu

Arthur Flor de Sousa Neto
Rayson Laroca
Sourour Ammar
Gianfranco Semeraro
Andre Hochuli
Saddok Kebairi
Shoma Iwai
Cleber Zanchettin
Ansgar Bernardi
Vivek Venugopal
Abderrhamne Rahiche
Wenwen Yu
Abhishek Baghel
Mathias Fuchs
Yael Iseli
Xiaowei Zhou
Yuan Panli
Minghui Xia
Zening Lin
Konstantinos Palaiologos
Loann Giovannangeli
Yuanyuan Ren
Shubhang Desai
Yann Soullard
Ling Fu
Juan Antonio Ramirez-Orta
Chixiang Ma
Truong Thanh-Nghia
Nathalie Girard
Kalyan Ram Ayyalasomayajula
Talles Viana
Francesco Castro
Anthony Gillioz
Huawen Shen
Sanket Biswas
Haisong Ding
Solène Tarride

Contents – Part II

Document Analysis and Recognition 2: Camera + Scene Text

Frontiers in Handwriting Recognition 3 (Synthesis)

Competition

Graphics Rec. 2: Tables and Charts

A Study on Reproducibility and Replicability of Table Structure Recognition Methods

Kehinde Ajayi[1]([✉]), Muntabir Hasan Choudhury[1], Sarah M. Rajtmajer[2], and Jian Wu[1]

[1] Computer Science, Old Dominion University, Norfolk, VA, USA
{kajay001,mchou001,j1wu}@odu.edu
[2] IST, Pennsylvania State University, University Park, PA, USA
smr48@psu.edu

Abstract. Concerns about reproducibility in artificial intelligence (AI) have emerged, as researchers have reported unsuccessful attempts to directly reproduce published findings in the field. Replicability, the ability to affirm a finding using the same procedures on new data, has not been well studied. In this paper, we examine both reproducibility and replicability of a corpus of 16 papers on table structure recognition (TSR), an AI task aimed at identifying cell locations of tables in digital documents. We attempt to reproduce published results using codes and datasets provided by the original authors. We then examine replicability using a dataset similar to the original as well as a new dataset, GenTSR, consisting of 386 annotated tables extracted from scientific papers. Out of 16 papers studied, we reproduce results consistent with the original in only four. Two of the four papers are identified as replicable using the similar dataset under certain IoU values. No paper is identified as replicable using the new dataset. We offer observations on the causes of irreproducibility and irreplicability. All code and data are available on Codeocean at https://codeocean.com/capsule/6680116/tree.

Keywords: reproducibility · replicability · generalizability · table structure recognition · artificial intelligence · science of science

1 Introduction

Concerns about reproducibility, replicability, and generalizability (RR&G) of findings in the social and behavioral sciences are now well-established [2,3,6]. More recently, RR&G concerns have been raised in the field of artificial intelligence (AI), e.g., [23,30]. There has been inconsistent use of these terms across the literature. Here, we adopt definitions from Goodman et al. [11]. By *reproducibility*, we refer to computational repeatability – obtaining consistent computational results using the same data, methods, code, and conditions of analysis. While, *replicability* is obtaining consistent results on a different but similar dataset using the same methods [1,11,22,25]. *Generalizability* refers to obtaining consistent results in settings outside of the experimental framework [11]. Each concept

G. A. Fink et al. (Eds.): ICDAR 2023, LNCS 14188, pp. 3–19, 2023.
https://doi.org/10.1007/978-3-031-41679-8_1

sets an incrementally higher standard than the previous one with reproducibility being the most basic requirement of fundamental science.

Existing studies of AI reproducibility have focused on empirical and computational AI, in which datasets, codes, and environments are essential conditions for reproduction. Some papers have examined the availability of certain information assumed critical to reproducibility. For example, Gunderson et al. [12] studied reproducibility of AI research by investigating whether research papers include adequate metadata, i.e., detailed documentation of methodology. Others have investigated the availability of open-access datasets and software [32] and the executability of source codes [24]. Directly reproducing results provides the most convincing evidence of reproducibility but usually requires more time, effort, and domain knowledge. Raff [30] conducted direct reproduction of AI papers. However, little effort has been put into the *replicability* of AI papers.

In this work, we investigate the reproducibility and replicability of methods for table structure recognition, an AI task aimed at parsing tables in digital documents and automatically identifying rows, columns, and cell positions in a detected table image within a document [26]. This task is different from a related task called table detection, automatically locating tables in document images [9]. Earlier methods attempted to solve these two tasks separately [33]. Recently, several end-to-end solutions based on neural networks have been proposed [7, 17,26]. The input of the TSR task is a table image and the output is usually an XML or JSON file containing coordinates of detected cells (row and column numbers and pixels of cell bounding boxes). The content of cells is not identified. Figure 1 illustrates the TSR problem. To the best of our knowledge, our work takes the first step to assess the replicability (based on the definition given above) of published research in the domain of document analysis and pattern recognition.

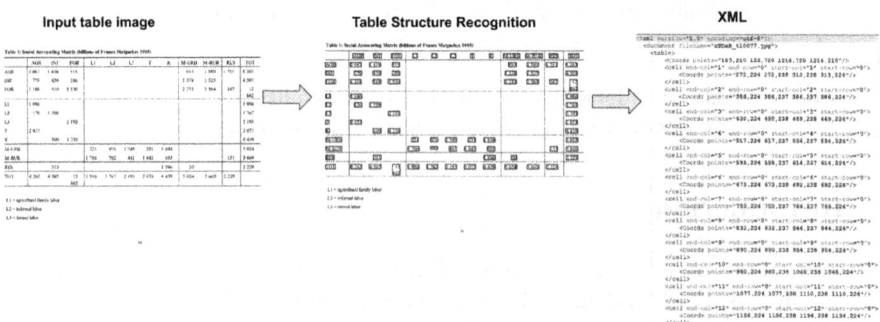

Fig. 1. An illustration of the TSR problem.

The goal of our work is twofold. First, we test *reproducibility* of published findings by examining whether results reported by the original authors can be reproduced. Second, we test the *replicability* of executable codes on two datasets: a dataset similar to the one used in the original paper and a new dataset built by manually annotating tables in six scientific domains. Our main contributions are summarized as follows:

1. We perform a study on AI reproducibility and replicability based on state-of-the-art TSR methods and identify reproducible and replicable papers under certain conditions.
2. We build a new, manually-annotated dataset, GenTSR, representing digital tables in papers from six scientific domains and demonstrate that the dataset is more challenging than widely adopted benchmarks such as ICDAR 2013 and ICDAR 2019, on the TSR task.
3. We observe possible causes of irreproducibility and irreplicability of AI papers based on our experiments.

2   Related Work

Concerns about reproducibility in computer science have been studied in the context of computer systems, e.g., [4], software engineering, e.g., [19], and recently on AI. e.g., [12,30]. Recent efforts have characterized the reproducibility of AI papers using automatic verification or meta-level information. Pimentel et al. [24] conducted an extensive study on over 1 million Jupyter notebooks from GitHub and found that only 24.11% executed without errors and only 4.03% produced the same results. Kamphius et al. conducted a large-scale study of reproducibility on BM25 scoring function variants [15]. Seibold et al. [34] investigated the reproducibility of analyses of longitudinal data associated with 11 articles published in PLOS ONE after contacting original authors. Prenkaj [27] compared several deep methods for trajectory forecasting on different datasets to provide insight into the actual novelty, reliability, and applicability of available methods. Salsabil et al. [32] proposed a hybrid classifier to automatically extract open-access datasets and software from scientific papers. Gunderson et al. [12] investigated 400 AI papers, and found that none contain documentation of published experiments, methods, and data altogether.

Although the work referenced above have highlighted the importance of including codes and data alongside published findings, studies were limited to meta-level indicators. There are a handful of studies that directly compare reproduced results of AI algorithms with published results. For example, Olorisade et al. [23] attempted to directly reproduce 6 AI papers using the codes and datasets from the original papers, reporting inconsistent results with published findings. Raff [30] directly compared the reported results of 255 AI papers without using the code from the original papers, and found out that 162 papers were at least 75% consistent with reported results while 93 were not.

Lack of transparency and reproducibility is particularly critical given the standard for AI papers to evaluate performance of proposed methods against baselines. Such problems have been found in AI research on machine learning analyses on clinical research [36] and deep metric learning [21]. *Replication* studies that provide side-by-side comparisons between AI papers addressing the same topic are rarely conducted. Our work fills this gap by directly comparing the implementations of TSR methods with reported results, and testing the replicability of these methods on new datasets. We chose TSR because recently, many learning-based methods on this task have been proposed and reportedly achieved high performance but not all of them were evaluated on the same datasets. Standard benchmarks are available in open competitions, i.e., ICDAR 2013 [10] and ICDAR 2019 (Task B2) [8]. Although the datasets were created 6 years apart, both resemble *generic* tables in a variety of documents, including government documents, scientific journals, forms, and financial statements. Scientific tables, on the other hand, usually contain precise measurements of experimental or analytical results. Compared with other types of tables, scientific tables are more heterogenous, with complex and freestyle structures. Therefore, for our replication study, we built a separate evaluation benchmark using tables extracted from six scientific domains to challenge existing runnable TSR algorithms. Our work is different from a typical survey paper in that our focus is not on outlining the proposed algorithms but on testing the reproducibility and replicability of state-of-the-art TSR algorithms.

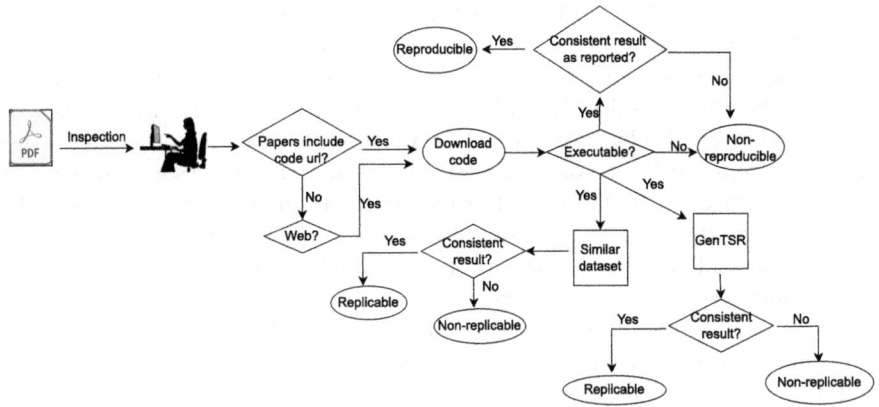

Fig. 2. Workflow of our study.

3   Methods

Prior work by Tatman et al. [37] proposed a taxonomy of reproducibility for AI, namely, "low", "medium", and "high" based on the availability of code, data, and adequate documentation of the experimental environment. This taxonomy is not directly applicable to our work because we not only verify the availability of code and data but also execute the code and compare with reported results. Figure 2 illustrates the workflow of our study. The process is summarized below.

1. **Sample Selection.** TSR papers were selected by searching "table structure recognition" as keywords on Google Scholar. Results were filtered to include papers published after 2017, in which the proposed methods accepted documents or table images as input. We downloaded 25 TSR papers from the conference websites, and we selected 16 candidate papers which use deep-learning based methods as our final sample.

2. **Meta-level Study.** We conducted meta-level study by inspecting each paper and determining whether the authors included URLs linking to source codes and datasets. If no URLs were found, we attempted to find open access codes and datasets by searching author names and framework names on Google. We bookmarked code and data repositories.

3. **Local Deployment.** We downloaded the data and source codes and deployed them in local computers by following the instructions in the original paper or on the code repositories.

4. **Reproducibility Tests.** We attempted to execute the codes using default settings and labeled each paper into one of three categories.
   (a) *Reproducible:* The source code was executed without errors and the results were consistent with the reported results within a deviation of an absolute F1-score of 10% below or above the reported results.
   (b) *Partially-reproducible:* The source code was executed without errors and the results were better than the reported results by more than an absolute F1-score of 10% deviation from the paper.
   (c) *Non-reproducible:* Otherwise.

5. **Replicability Tests on a Similar Dataset.** We tested executable TSR methods on a different but similar benchmark dataset. If results are consistent with the reported results within a deviation of an absolute F1-score of 10% under certain conditions, the paper was labeled as "conditionally replicable" with respect to this dataset.

6. **Replicability Tests on a New Dataset.** We tested executable TSR methods on a new dataset. If results are consistent within a deviation of 10% absolute F1-score, the paper was labeled as "conditionally replicable" with respect to the new dataset.

We create a separate virtual environment for each TSR method to avoid incompatibility issues. Then, we execute the code of each TSR algorithm to reproduce or replicate the reported F1 scores. Because we focus on reproducing results presented in original papers, we used the pre-trained models released by the authors. One important parameter that may affect the TSR result is Intersection over Union (IoU), defined as the percentage overlap between object regions provided by the ground truth and predicted by the model. In practice, IoU is measured by dividing the number of common pixels between the ground-truth bounding boxes and predicted bounding boxes by the total number of pixels across both bounding boxes. The IoU threshold defines the criterion of whether the predicted bounding boxes match the ground truth bounding boxes. The matching between two bounding boxes is counted if the predicted IoU is larger than the threshold.

For our reproducibility tests, we evaluate an executable model on the same datasets used in the original paper, either ICDAR 2013 or ICDAR 2019. If a paper used both ICDAR-2013 and ICDAR 2019 datasets, then we chose ICDAR-2019 because it contains more challenging tables. If a paper used neither dataset, then we used the dataset used in the original paper. For the replicability tests on a similar dataset, we evaluated each executable model on the alternate dataset of the two benchmarks, e.g., if ICDAR 2013 was used in the original paper, we use ICDAR 2019. For the second replicability test, we evaluated the model on a new dataset called GenTSR (introduced below). We compute F-scores at five IoU thresholds 0.5, 0.6, 0.7, 0.8, and 0.9.

For reproducibility tests, we define the discrepancy Δ as the absolute difference between the F1-score obtained by our reproduction $F1(R_0)$ and the F1-score reported in the original paper $F1(O)$, i.e., $\Delta_0 = F1(R_0) - F1(O)$. For replicability tests, discrepancy Δ is defined as F1-score of our replication $F1(R_x)$ and F1-score of our reproduction $F1(R_0)$, i.e., $\Delta_x = F1(R_x) - F1(R_0)$. We do not compare replicability against the original F1-score to ensure that compared results are obtained in exactly the same setting.

4   Data

We use two standard benchmarks and GenTSR, our manually-annotated dataset.

ICDAR 2013. This dataset, released for the table competition by ICDAR 2013, was used in 8 papers out of 16 papers (Table 1). ICDAR 2013 consists of 238 document pages in PDF format crawled from European and US government websites, out of which 128 documents include tables. We did not use the original ICDAR 2013 data as our ground truth because it consists of born-digital PDFs and all the TSR models we surveyed accept either documents or table images as input. Therefore, we cropped the tables from the PDFs based on the labeled coordinates preserving resolutions and adjust its annotations accordingly.

ICDAR 2019. This dataset, released for the Competition on Table Detection and Recognition (cTDaR) organized by ICDAR 2019, was used in 6 out of 16 papers we surveyed. The cTDaR competition includes two datasets including the modern and historical tables, respectively. The modern dataset contains 100 samples from scientific papers, forms, and financial documents, and the historical dataset includes images from hand-written accounting ledgers, and train schedules. Our experiments adopt the modern table dataset used in Track B2 (TB2).

GenTSR. This dataset consists of 386 table images obtained from research papers in six scientific domains, including three STEM (Chemistry, Biology, and Materials Science) and three non-STEM domains (Economics, Developmental Studies, and Business). The format of GenTSR is consistent with ICDAR 2019. The numbers of tables in each domain are 30 (Chemistry), 43 (Economics), 7 (Developmental Studies), 68 (Biology), and 208 (Materials Science). These tables were manually annotated by two graduate students independently using

Table 1. A summary of TSR papers used in our study, their properties, and direct reproducibility labels ("Rep" column). The columns labeled "Data" and "Code" indicate whether datasets and codes are publicly available at the time of writing. A dash ("-") means the resource is not available. Notes: NR: Not reproducible. R: Reproducible. PR: Partially-reproducible. CMDD: Chinese Medical Document Dataset. Rep: Reproducibility.

Reference	Year	Model Name	Venue	Training data	Original Eval. data	Data	Code	Rep
Schreiber et al. [33]	2017	DeepDeSRT	ICDAR	Marmot	Marmot	-	-	NR
Siddiqui et al. [35]	2019	DeeptabSTR	ICDAR	TabStructDB	ICDAR 2013	-	-	NR
Xue et al. [39][a]	2019	Res2TIM	ICDAR	CMDD + ICDAR 2013	ICDAR 2013	✓	✓	R
Qasim. [28][b]	2019	TIES-2.0	ICDAR	Synthetic data	Synthetic data	✓	✓	NR
Tensmeyer et al. [38][d]	2019	SPLERGE	ICDAR	Web-screaped PDFs + ICDAR 2013	ICDAR 2013	-	✓	NR
Prasad et al. [26][e]	2019	Cascade TabNet	CVPR	Marmot + ICDAR 2019	ICDAR 2019 Track-B2	✓	✓	PR
Hashmi et al. [13]	2019	No name	CVPR	TabStructDB	ICDAR 2013	-	-	NR
Khan et al. [16][c]	2020	No name	ICDAR	UNLV	ICDAR 2013	-	✓	NR
Raja et al. [31][f]	2020	TabStruct Net	ECCV	SciTSR	UNLV	✓	✓	NR
Fischer et al. [7][g]	2021	Multi-Type-TD-TSR	KI	ICDAR 2019	ICDAR 2019 Track-B2	✓	✓	R
Xue et al. [40][h]	2021	TGRNet	ICCV	TableGraph	ICDAR 2019	✓	✓	R
Qiao et al. [29][i]	2021	LGPMA	ICDAR	PubTabNet + SciTSR + ICDAR 2013	PubTabNet	✓	✓	NR
Lee et al. [17][j]	2021	Graph-based-TSR	MTA	ICDAR 2019	ICDAR 2019	✓	✓	R
Zheng et al. [41]	2022	GTE	WACV	PubTabNet	ICDAR 2013 + ICDAR 2019	-	-	NR
Jain et al. [14][k]	2022	TSR-DSAW	ESANN	PubTabNet	ICDAR 2013	-	-	NR
Li et al. [18][l]	2021	No name	ICDAR	PubTabNet	ICDAR 2019 + unlv	-	✓	NR

[a]https://github.com/xuewenyuan/ReS2TIM/
[b]https://github.com/shahrukhqasim/TIES-2.0
[c]https://github.com/saqib22/Table-Structure_Extraction-Bi-directional-GRU/
[d]https://github.com/pyxploiter/deep-splerge/
[e]https://github.com/DevashishPrasad/CascadeTabNet/
[f]https://github.com/sachinraja13/TabStructNet
[g]https://github.com/Psarpei/Multi-Type-TD-TSR/
[h]https://github.com/xuewenyuan/TGRNet/
[i]https://github.com/arushijain45/TSR-DSAW/
[j]https://github.com/hikopensource/DAVAR-Lab-OCR/tree/main/demo/table_recognition/lgpma/
[k]https://github.com/ejlee95/Graph-based-TSR/
[l]https://github.com/L597383845/row-col-table-recognition/

the VGG Image Annotator (VIA) [5]. VIA is open-source software for annotating images, videos, and audio. We drew rectangular bounding boxes around text content in a table cell and provided properties including "start-row", "start-col", "end-row", and "end-col" as labels. We followed the same schema as ICDAR 2019 dataset. We obtained a Cohan's $\kappa = 0.73$, indicating a substantial agreement between the two annotators [20]. The two annotators then discussed until they agreed on the remaining table cells that they initially did not agree with each other (Table 2).

Table 2. Three datasets used in our study.

Data	# tables	# cells	# row ranges	# column ranges
ICDAR-2013	158	14,278	2–58	2–13
ICDAR-2019	100	5,132	2–39	1–15
GenTSR	386	19,914	2–62	1–16

5   Experiment Results

We performed all experiments using two computing environments namely, a Linux server with an Intel Silver CPU, Nvidia GTX 2080 Ti and Google Colaboratory platform with P100 PCIE GPU of 16 GB GPU memory. It took approximately 12 h to reproduce the 5 papers that made their codes and data available. Specifically, it took about 6 h to reproduce TGRNet and Res2TIM, and approximately 2 h each for CascadeTabNet, Graph-based-TSR, and Multi-Type-TSR. The replication experiment took about 18 h (excluding the time to create the GenTSR dataset) even though all necessary packages used for each method had already been installed when conducting reproducibility experiment. This was due to the relatively large size of replication data and multiple replication attempts to cross-check results.

We answer the following research questions (RQs) using meta-level survey results and reproducibility and replicability experiment results. The reproducibility results are tabulated in Table 3. The replicability results are illustrated in Fig. 3 and Fig. 4.

RQ1: What is the data and code accessibility of TSR papers we sampled? Out of 16 papers we surveyed, 8 papers made their source code and data publicly available, 3 papers made only the codes available, and 5 papers did not provide either codes or data, making it difficult to validate the results of these methods without private communication with the original authors (Table 1).

RQ2: Are the accessible methods executable without contacting the original authors? Out of 11 papers with accessible data or codes, the codes of 5 papers were executable without contacting the original authors. The source codes of 6 papers were not executable [14, 16, 18, 28, 29, 31]. The code of one paper (TGRNet; [40]) was executable after we contacted the original authors. The reason was that the absolute paths to the evaluation data files in the original code were hard-coded. Therefore, the program could not find data files after they are transferred to a different environment. To evaluate the models, we wrote a script to replace the paths and the source codes could be executed. The source codes of the 6 papers were not executable due to multiple reasons such as dependency issues, errors in code, pretrained models not being released, or the absence of implementation in the authors' GitHub directory.

RQ3: What is the status of reproducibility based on our criteria? The status of reproducibility varies significantly depending on many factors. As shown above, most papers were labeled irreproducible because they do not provide datasets, codes, or executable codes. However, most papers with executable codes were labeled reproducible under our criteria. Specifically, 4 out of the 6 executable TSR methods were labeled reproducible, 1 paper was labeled partially-reproducible, and 1 paper was labeled not-reproducible. The case studies are below.

- Lee et al. [17] used only 19 document images with border lines from the ICDAR 2019 TB2 dataset to evaluate the Graph-based-TSR method. Therefore, we evaluated this method on the same 19 images. Table 3 indicates that the reproduction results are in general consistent with the reported results, with discrepancies $0.087 \leq \Delta_0 \leq 0.130$ depending on the IoU.
- The CascadeTabNet method was originally evaluated on 100 modern tables in ICDAR 2019. Surprisingly, our experiment on CascadeTabNet obtained higher F1-scores than the reported results by up to 0.682 at IoU = 0.9.
- The Multi-Type-TD-TSR method was originally evaluated on 162 tables from ICDAR-2019 TB2. The experiment results are consistent with the reported results with a discrepancy $\Delta_0 \leq 0.013$.
- TGRNet and ReS2TIM are both consistent with the reported results with a discrepancy $\Delta_0 \leq 0.003$. The authors of these two methods used only IoU = 0.5 to allow more cell boxes to be predicted.
- The SPLERGE method was originally evaluated on 34 randomly selected tables using ICDAR 2013 dataset but this dataset was not made publicly available. Thus, the SPLERGE method was marked "Non-reproducible".

RQ4: What reasons caused results to be not reproducible? We identified several major reasons that caused the results to be irreproducible.

- **Data and code availability:** This is the top reason that caused most papers to be irreproducible. However, most papers *with executable codes* are identified as reproducible.
  Several irreproducible papers have authors affiliated with the industry, which may impose intellectual property restrictions, e.g., [13, 16, 28, 33, 35, 41].

Table 3. The *reproducibility test* results of executable TSR models at different IoU thresholds. Data: the *original dataset*. SPLERGE does not provide evaluation data.

TSR Model	Data	IoU	$F1(O)$	$F1(R_0)$	Δ_0
CascadeTabNet	ICDAR 2019	0.6	0.438	0.770	0.332
CascadeTabNet	ICDAR 2019	0.7	0.354	0.760	0.406
CascadeTabNet	ICDAR 2019	0.8	0.190	0.745	0.555
CascadeTabNet	ICDAR 2019	0.9	0.036	0.718	0.682
Multi-Type-TD-TSR	ICDAR 2019	0.6	0.589	0.593	0.004
Multi-Type-TD-TSR	ICDAR 2019	0.7	0.404	0.397	−0.007
Multi-Type-TD-TSR	ICDAR 2019	0.8	0.137	0.124	−0.013
Multi-Type-TD-TSR	ICDAR 2019	0.9	0.015	0.012	−0.003
Graph-based-TSR	ICDAR 2019	0.6	0.966	0.879	−0.087
Graph-based-TSR	ICDAR 2019	0.7	0.966	0.868	−0.098
Graph-based-TSR	ICDAR 2019	0.8	0.966	0.856	−0.110
Graph-based-TSR	ICDAR 2019	0.9	0.828	0.815	−0.130
TGRNet	ICDAR 2013	0.5	0.667	0.670	0.003
ReS2TIM	ICDAR 2013	0.5	0.174	0.174	0.000
SPLERGE	ICDAR 2013	0.5	0.953	-	-

- **Portability:** Agile software engineering may develop software packages that are not portable when transferred to other platforms, e.g., [31].
- **Documentation:** This occurs when researchers do not provide detailed instructions or explanations to execute their codes, e.g., [14].
- **Dependency and compatibility issues:** Software that relies on out- dated dependencies can become prohibitive obstacles to reproducing reported results. Certain software did not provide the dependency version, making it extremely difficult or even impossible to find and install the right dependency, e.g., [29].
- **Data and code durability:** This occurs when the data and codes used in the original paper are updated after published results. Thus, a better result than what was reported may be obtained after executing the updated codes and data, thereby making it difficult to validate original reported results [26].

RQ5: What is the status of replicability with respect to a similar dataset? To answer this question, we evaluated each executable TSR method on a similar dataset. Here, we compare against the reproducibility experiment results instead of the original results to ensure the results to be compared are obtained in exactly the same setting.

Table 4 indicates the F1-scores of most methods were reduced by various lev- els depending on the IoU thresholds. In particular, the F1 of Graph-based-TSR decreases by 0.337. The F1 of Multi-Type-TD-TSR decreases by 0.009 to 0.586. The performance of CascadeTabNet decreases marginally, exhibiting better repli-

Table 4. The *replicability test* results of executable TSR models at different IoU thresholds. Data: the *similar dataset*. TGRNet and ReS2TIM do not allow inference on a custom dataset.

TSR Model	Data	IoU	$F1(R_0)$	$F1(R_1)$	Δ_1
CascadeTabNet	ICDAR 2013	0.6	0.770	0.690	−0.080
CascadeTabNet	ICDAR 2013	0.7	0.760	0.678	−0.082
CascadeTabNet	ICDAR 2013	0.8	0.745	0.661	−0.084
CascadeTabNet	ICDAR 2013	0.9	0.718	0.621	−0.097
Multi-Type-TD-TSR	ICDAR 2013	0.6	0.593	0.007	−0.586
Multi-Type-TD-TSR	ICDAR 2013	0.7	0.397	0.005	−0.392
Multi-Type-TD-TSR	ICDAR 2013	0.8	0.124	0.004	−0.120
Multi-Type-TD-TSR	ICDAR 2013	0.9	0.012	0.003	−0.009
Graph-based-TSR	ICDAR 2013	0.6	0.879	0.542	−0.337
Graph-based-TSR	ICDAR 2013	0.7	0.868	0.504	−0.364
Graph-based-TSR	ICDAR 2013	0.8	0.856	0.444	−0.412
Graph-based-TSR	ICDAR 2013	0.9	0.815	0.373	−0.442
TGRNet	ICDAR 2019	0.5	0.670	-	-
ReS2TIM	ICDAR 2019	0.5	0.174	-	-
SPLERGE	ICDAR 2019	0.5	-	0.121	-

cability. We could not replicate the results of ReS2TIM and TGRNet because they do not allow inference on an alternative dataset. We did not obtain the discrepancy Δ_1 for the SPLERGE method since it was not reproducible. Thus, out of the 6 methods that were either executable or reproducible, only 2 papers (CascadeTabNet and MUlti-Type-TD-TSR) were replicable under certain IoUs (CascadeTabNet on IoU from 0.6 to 0.9; Multi-Type-TD-TSR on IoU $= 0.9$).

RQ6: What is the status of replicability with respect to the new dataset? We test the replicability of each executable TSR model using GenTSR containing tables in six scientific domains. Similar to *RQ5.*, we compare against the reproducibility experiment results using the 10% threshold defined above. The results shown in Fig. 3 and Fig. 4 indicate that none of the 4 methods that allow inference on custom data [7,17,26,38] was replicable with respect to the GenTSR dataset, under a threshold of 10% absolute F1-score. Figure 4 also demonstrates that the performance of these methods varies significantly in scientific domains. Specifically, the CascadeTabNet achieved much higher F1-scores on five domains than biology. SPLERGE achieves comparable F1-scores in all domains. Graph-based-TSR performs remarkably well in Material Science but poorly in all other domains.

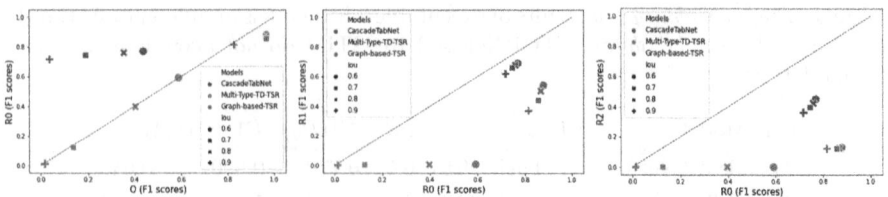

Fig. 3. The comparison of the original (O), reproducibility (R0), replicability on similar data (R1), and replicability on GenTSR (R2). The F1-scores of R2 are obtained by averaging the F1-scores across all domains for each IoU. SPLERGE was excluded because its results were not reproducible.

6   Discussion

Reproducibility. In reproducibility experiments, we were unable to produce exactly the same results reported in most papers. The discrepancies may be due to random factors, e.g., initialization, but certain discrepancies are too large to be explained by random factors. Investigating the reasons is beyond the scope of this paper, but the results suggest we define reproducibility using quantifiable criteria associated with thresholds. The exact criteria will differ depending on the results and the reproducer's needs. In addition, we obtained an interesting result in which the reproduced results were significantly better than the reported result (Table 3). Assuming both original and reproduced results are correct, the improvement could be attributed to the new versions of the codes and/or data. If that is the case, this poses another question of how long reproducibility can be preserved.

Replicability is More Challenging and Data Dependent. One requirement of replicability is that the original code is not only executable but also configurable, allowing users to test on different datasets. In our experiments, two methods did not allow inference on different datasets. In addition, the replicability performance could change the ranks of methods. For example, the Graph-based-TSR was the best in terms of the original and reproduced F1-scores, but it underperformed CascadeTabNet in two replicability tests (Table 4 and Fig. 3). This is likely to be caused by the nature of the model and the training process. The exact reason requires detailed ablation analysis and model surgery. In addition, only CascadeTabNet obtained reasonably consistent results using a similar dataset (Table 4). Using the new dataset, CascadeTabNet achieved a lower $F1(R_2)$ compared with $F1(O)$ and $F1(R_1)$. Figure 4 indicates that the performance exhibits a strong domain dependency. The decreased performance as seen in the new dataset indicates that TSR on scientific tables is still an unsolved problem and state-of-the-art methods still have a large space to improve.

Potentially Irreplicability Causes: Evaluation Bias. The replicability test results also indicate that the evaluation data of several TSR models may not be diverse enough. For example, the Graph-based-TSR was evaluated on only tables with

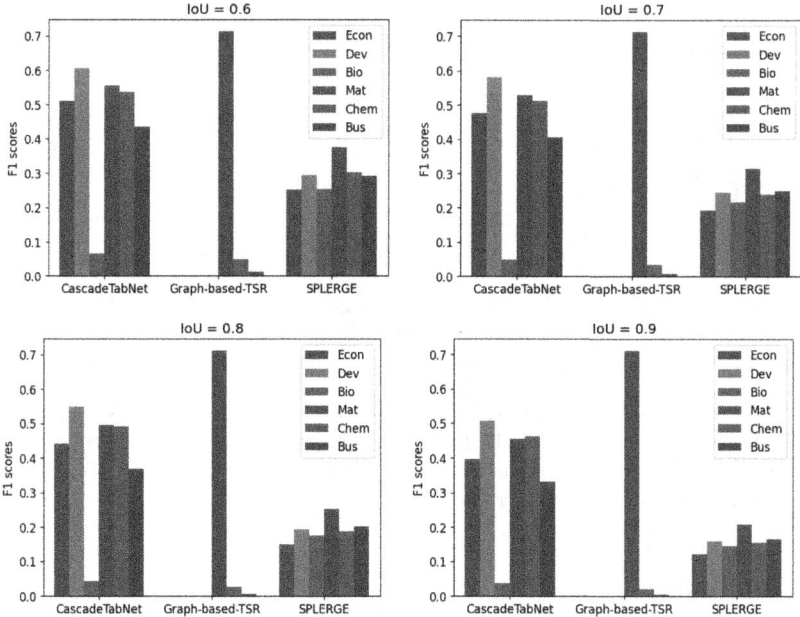

Fig. 4. The *replicability test* results of executable TSR model with respect to individual domains in GenTSR. Econ: Economics, Dev: Developmental Studies, Bio: Biology, Mat: Material Science, Chem: Chemistry, Bus: Business. Multi-Type-TD-TSR (not shown) obtains recall scores of zero across all the domains.

borders. In contrast, models that exhibit better robustness tend to be evaluated originally on more challenging datasets. For example, CascadeTabNet, which obtained the best replicability results was evaluated on ICDAR 2019 Task B2 which was more challenging and diverse compared to ICDAR 2013 and other small benchmarks.

Reproducibility and Venue Ranking. We inspected the relationship between reproducibility and venue ranking, which is characterized by the h5-index obtained in December, 2022. Created by Google Scholar, the calculation of h5-index is similar to h-index. H5-index is defined as the largest number h such that h articles published in the past 5 years have at least h citations each. Figure 5 shows the papers we studied and color-coded by reproducibility. It indicates that reproducibility is not necessarily associated with venue ranking.

Limitations. One limitation of our study is the relatively small sample size. Therefore the conclusions we draw may not directly be applicable to other AI tasks and domains. However, the way we selected the papers allowed us to focus on more papers on one topic and perform a side-to-side comparison between different methods. Our threshold of 10% absolute F1-score is also a very lenient threshold. Under a more strict threshold, fewer papers would be identified as reproducible.

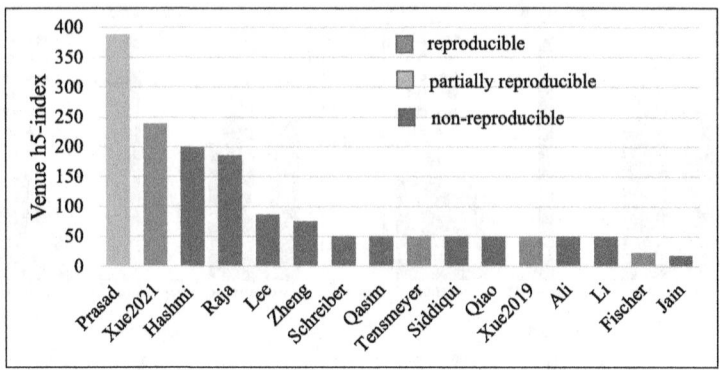

Fig. 5. The relationship between reproducibility and venue ranking, characterized by h5-index from Google Scholar at the time of writing. The h5-indices of ESANN (Jain) and KI (Fischer) are not available, so we used h-indices as surrogates, obtained from research.com and resurchify.com, respectively.

7   Conclusion

This work presents a study of reproducibility and replicability considering 16 recently-published papers on table structure recognition. We attempted to directly reproduce the results reported in the original papers. We then tested executable methods on an alternate benchmark dataset similar to the one used in the original paper as well as a new dataset of modern scientific tables extracted from six domains.

Under our criteria, most (12 out of 16) papers we examined were not fully reproducible. Only 2 papers [7, 26] were identified as conditionally replicable on a similar dataset, and none of the papers was identified as replicable with respect to the new dataset. Using a relatively small but focused dataset, our study reveals several challenges of reproducing and replicating methods proposed for the TSR task. Our study suggests that reproducibility should be defined under certain criteria with quantifiable thresholds and replicability is data-dependent. We also found that papers published in high-tier venues (characterized by h5-index) are not necessarily reproducible. The new dataset GenTSR can be used as ground truth for building more robust TSR models. Future work will investigate replicability at the model level. We suggest that this work provides evidence that infrastructure is needed for researchers to report RR&G of experiments.

Acknowledgment. This work was partially supported by the Defense Advanced Research Projects Agency (DARPA) under cooperative agreement No. W911NF-19-2-0272. The content of the information does not necessarily reflect the position or the policy of the Government, and no official endorsement should be inferred.

References

1. National Academies: Reproducibility and Replicability in Science. National Academies Press (2019). https://doi.org/10.17226/25303
2. Baker, M.: 1,500 scientists lift the lid on reproducibility. Nature **533**(7604), 452–454 (2016). https://doi.org/10.1038/533452a
3. Camerer, C.F., et al.: Evaluating the replicability of social science experiments in nature and science between 2010 and 2015. Nat. Hum. Behav. **2**(9), 637–644 (2018). https://doi.org/10.1038/s41562-018-0399-z
4. Collberg, C., Proebsting, T.A.: Repeatability in computer systems research. Commun. ACM **59**(3), 62–69 (2016). https://doi.org/10.1145/2812803
5. Dutta, A., Zisserman, A.: The via annotation software for images, audio and video. In: Proceedings of the 27th ACM International Conference on Multimedia, pp. 2276–2279 (2019)
6. Fanelli, D.: Opinion: is science really facing a reproducibility crisis, and do we need it to? Proc. Natl. Acad. Sci. **115**(11), 2628–2631 (2018). https://doi.org/10.1073/pnas.1708272114
7. Fischer, P., Smajic, A., Abrami, G., Mehler, A.: Multi-type-TD-TSR – extracting tables from document images using a multi-stage pipeline for table detection and table structure recognition: from OCR to structured table representations. In: Edelkamp, S., Möller, R., Rueckert, E. (eds.) KI 2021. LNCS (LNAI), vol. 12873, pp. 95–108. Springer, Cham (2021). https://doi.org/10.1007/978-3-030-87626-5_8
8. Gao, L., et al.: ICDAR 2019 competition on table detection and recognition (CTDAR). In: 2019 International Conference on Document Analysis and Recognition (ICDAR), pp. 1510–1515 (2019). https://doi.org/10.1109/ICDAR.2019.00243
9. Gatos, B., Danatsas, D., Pratikakis, I., Perantonis, S.J.: Automatic table detection in document images. In: Singh, S., Singh, M., Apte, C., Perner, P. (eds.) ICAPR 2005. LNCS, vol. 3686, pp. 609–618. Springer, Heidelberg (2005). https://doi.org/10.1007/11551188_67
10. Göbel, M., Hassan, T., Oro, E., Orsi, G.: ICDAR 2013 table competition. In: 2013 12th International Conference on Document Analysis and Recognition, pp. 1449–1453 (2013). https://doi.org/10.1109/ICDAR.2013.292
11. Goodman, S.N., Fanelli, D., Ioannidis, J.P.: What does research reproducibility mean? Sci. Transl. Med. **8**(341), 341ps12-341ps12 (2016)
12. Gundersen, O.E., Kjensmo, S.: State of the art: reproducibility in artificial intelligence. In: Proceedings of the AAAI Conference on Artificial Intelligence, vol. 32 (2018)
13. Hashmi, K.A., Stricker, D., Liwicki, M., Afzal, M.N., Afzal, M.Z.: Guided table structure recognition through anchor optimization. IEEE Access **9**, 113521–113534 (2021)
14. Jain, A., Paliwal, S., Sharma, M., Vig, L.: TSR-DSAW: table structure recognition via deep spatial association of words. In: 29th European Symposium on Artificial Neural Networks, Computational Intelligence and Machine Learning, ESANN 2021, Online event (Bruges, Belgium), 6–8 October 2021 (2021). https://doi.org/10.14428/esann/2021.ES2021-109
15. Kamphuis, C., de Vries, A.P., Boytsov, L., Lin, J.: Which BM25 do you mean? A large-scale reproducibility study of scoring variants. In: Jose, J.M., et al. (eds.) ECIR 2020. LNCS, vol. 12036, pp. 28–34. Springer, Cham (2020). https://doi.org/10.1007/978-3-030-45442-5_4

16. Khan, S.A., Khalid, S.M.D., Shahzad, M.A., Shafait, F.: Table structure extraction with bi-directional gated recurrent unit networks. In: 2019 International Conference on Document Analysis and Recognition (ICDAR), pp. 1366–1371. IEEE (2019)

17. Lee, E., Park, J., Koo, H.I., Cho, N.I.: Deep-learning and graph-based approach to table structure recognition. Multimedia Tools Appl. **81**(4), 5827–5848 (2022)

18. Li, Y., et al.: Rethinking table structure recognition using sequence labeling methods. In: Lladós, J., Lopresti, D., Uchida, S. (eds.) ICDAR 2021. LNCS, vol. 12822, pp. 541–553. Springer, Cham (2021). https://doi.org/10.1007/978-3-030-86331-9_35

19. Liu, C., Gao, C., Xia, X., Lo, D., Grundy, J.C., Yang, X.: On the reproducibility and replicability of deep learning in software engineering. ACM Trans. Softw. Eng. Methodol. **31**(1), 15:1–15:46 (2022). https://doi.org/10.1145/3477535

20. McHugh, M.L.: Interrater reliability: the kappa statistic. Biochem. Med. **22**(3), 276–82 (2012)

21. Musgrave, K., Belongie, S., Lim, S.-N.: A metric learning reality check. In: Vedaldi, A., Bischof, H., Brox, T., Frahm, J.-M. (eds.) ECCV 2020. LNCS, vol. 12370, pp. 681–699. Springer, Cham (2020). https://doi.org/10.1007/978-3-030-58595-2_41

22. Nosek, B.A., et al.: Replicability, robustness, and reproducibility in psychological science. Ann. Rev. Psychol. **73**(1), 719–748 (2022). https://doi.org/10.1146/annurev-psych-020821-114157. pMID: 34665669

23. Olorisade, B.K., Brereton, P., Andras, P.: Reproducibility of studies on text mining for citation screening in systematic reviews: evaluation and checklist. J. Biomed. Inform. **73**, 1–13 (2017). https://doi.org/10.1016/j.jbi.2017.07.010

24. Pimentel, J.F., Murta, L., Braganholo, V., Freire, J.: A large-scale study about quality and reproducibility of jupyter notebooks. In: 2019 IEEE/ACM 16th International Conference on Mining Software Repositories (MSR), pp. 507–517. IEEE (2019). https://doi.org/10.1109/MSR.2019.00077

25. Pineau, J., et al.: Improving reproducibility in machine learning research. J. Mach. Learn. Res. **22**, 7459–7478 (2021)

26. Prasad, D., Gadpal, A., Kapadni, K., Visave, M., Sultanpure, K.: Cascadetabnet: an approach for end to end table detection and structure recognition from image-based documents. In: Proceedings of the IEEE/CVF Conference on Computer Vision and Pattern Recognition Workshops, pp. 572–573 (2020)

27. Prenkaj, B., Velardi, P., Distante, D., Faralli, S.: A reproducibility study of deep and surface machine learning methods for human-related trajectory prediction. In: Proceedings of the 29th ACM International Conference on Information & Knowledge Management, CIKM 2020, pp. 2169–2172. Association for Computing Machinery, New York (2020). https://doi.org/10.1145/3340531.3412088

28. Qasim, S.R., Mahmood, H., Shafait, F.: Rethinking table recognition using graph neural networks. In: 2019 International Conference on Document Analysis and Recognition (ICDAR), pp. 142–147. IEEE (2019)

29. Qiao, L., et al.: LGPMA: complicated table structure recognition with local and global pyramid mask alignment. In: Lladós, J., Lopresti, D., Uchida, S. (eds.) ICDAR 2021. LNCS, vol. 12821, pp. 99–114. Springer, Cham (2021). https://doi.org/10.1007/978-3-030-86549-8_7

30. Raff, E.: A step toward quantifying independently reproducible machine learning research. In: Advances in Neural Information Processing Systems, vol. 32 (2019)

31. Raja, S., Mondal, A., Jawahar, C.V.: Table structure recognition using top-down and bottom-up cues. In: Vedaldi, A., Bischof, H., Brox, T., Frahm, J.-M. (eds.) ECCV 2020. LNCS, vol. 12373, pp. 70–86. Springer, Cham (2020). https://doi.org/10.1007/978-3-030-58604-1_5

32. Salsabil, L., et al.: A study of computational reproducibility using URLs linking to open access datasets and software. In: Companion Proceedings of the Web Conference 2022, WWW 2022, pp. 784–788. Association for Computing Machinery, New York (2022). https://doi.org/10.1145/3487553.3524658

33. Schreiber, S., Agne, S., Wolf, I., Dengel, A., Ahmed, S.: Deepdesrt: deep learning for detection and structure recognition of tables in document images. In: 2017 14th IAPR International Conference on Document Analysis and Recognition (ICDAR), vol. 1, pp. 1162–1167 (2017). https://doi.org/10.1109/ICDAR.2017.192

34. Seibold, H., et al.: A computational reproducibility study of PLOS ONE articles featuring longitudinal data analyses. PLoS ONE **16**(6), 1–15 (2021). https://doi.org/10.1371/journal.pone.0251194

35. Siddiqui, S.A., Fateh, I.A., Rizvi, S.T.R., Dengel, A., Ahmed, S.: Deeptabstr: deep learning based table structure recognition. In: 2019 International Conference on Document Analysis and Recognition (ICDAR), pp. 1403–1409. IEEE (2019)

36. Stevens, L.M., Mortazavi, B.J., Deo, R.C., Curtis, L., Kao, D.P.: Recommendations for reporting machine learning analyses in clinical research. Circ. Cardiovasc. Qual. Outcomes **13**(10), e006556 (2020). https://doi.org/10.1161/CIRCOUTCOMES.120.006556

37. Tatman, R., VanderPlas, J., Dane, S.: A practical taxonomy of reproducibility for machine learning research (2018)

38. Tensmeyer, C., Morariu, V.I., Price, B., Cohen, S., Martinez, T.: Deep splitting and merging for table structure decomposition. In: 2019 International Conference on Document Analysis and Recognition (ICDAR), pp. 114–121 (2019). https://doi.org/10.1109/ICDAR.2019.00027

39. Xue, W., Li, Q., Tao, D.: Res2tim: reconstruct syntactic structures from table images. In: 2019 International Conference on Document Analysis and Recognition (ICDAR), pp. 749–755 (2019). https://doi.org/10.1109/ICDAR.2019.00125

40. Xue, W., Yu, B., Wang, W., Tao, D., Li, Q.: TGRNet: a table graph reconstruction network for table structure recognition. In: Proceedings of the IEEE/CVF International Conference on Computer Vision, pp. 1295–1304 (2021)

41. Zheng, X., Burdick, D., Popa, L., Zhong, X., Wang, N.X.R.: Global table extractor (GTE): a framework for joint table identification and cell structure recognition using visual context. In: Proceedings of the IEEE/CVF Winter Conference on Applications of Computer Vision, pp. 697–706 (2021)

An End-to-End Local Attention Based Model for Table Recognition

Nam Tuan Ly[(✉)] [iD] and Atsuhiro Takasu [iD]

National Institute of Informatics (NII), Tokyo, Japan
namlytuan@gmail.com, takasu@nii.ac.jp

Abstract. Recently, due to the rapid development of deep learning, especially Transformer, many Transformer-based methods have been studied and proven to be very powerful for table recognition. However, Transformer-based models usually struggle to process big tables due to the limitation of their global attention mechanism. In this paper, we propose a local attention mechanism to address the limitation of the global attention mechanism. We also present an end-to-end local attention-based model for recognizing both table structure and table cell content from a table image. The proposed model consists of four main components: 1) an encoder for feature extraction; 2) the three decoders for the three sub-tasks of the table recognition problem. In the experiments, we evaluate the performance of the proposed model and the effectiveness of the local attention mechanism on the two large-scale datasets: PubTabNet and FinTabNet. The experiment results show that the proposed model outperforms the state-of-the-art methods on all benchmark datasets. Furthermore, we demonstrate the effectiveness of the local attention mechanism for table recognition, especially for big table recognition.

Keywords: Local Attention · Self-Attention · Table Recognition · End-to-End

1 Introduction

Table recognition is an important part of the document understanding system which aims to recognize the table structure and the text content of each table cell from an input table image, and to represent them in a machine-readable format such as HTML code [1–3] and LaTeX code [4, 5]. This task is a big challenging problem due to the diversity of table styles and the complexity of table structures and has been receiving much attention from numerous researchers around the world. Early works [6–8] of the table recognition primarily depend on hand-crafted features and heuristic rules are mainly applied to simple table structure or pre-defined table formats. In recent years, inspired by the success of deep learning-based object detection and semantic segmentation methods, many deep learning-based table recognition methods [9–12] have been proposed and proven to be powerful models. However, these systems rely on training datasets containing rich annotation information and are also difficult to be maintained and inferred due to consisting of multiple separate components.

© The Author(s), under exclusive license to Springer Nature Switzerland AG 2023
G. A. Fink et al. (Eds.): ICDAR 2023, LNCS 14188, pp. 20–36, 2023.
https://doi.org/10.1007/978-3-031-41679-8_2

In more recent years, inspired by the success of Transformer [13] in a wide range of machine learning tasks such as natural language processing and computer vision, many Transformer-based methods for table recognition [14–17] have been studied and achieved competitive results on the recent large-scale table image datasets. The success of these models is mostly due to the global attention mechanism in Transformer. However, the global attention mechanism requires a large amount of training data and usually struggles to work on the very long input sequence. Several recent studies [18–21] suggest that incorporating more focused attention on important local regions in the input sequence with an explicit bias could be more beneficial. Inspired by these suggestions, we propose the local attention mechanism in the decoder components to address this limitation of the global attention mechanism in Transformer-based table recognition models. As far as we know, this is the first study of the local attention mechanism in the decoder component for the document analysis problem.

In this paper, we also present an end-to-end local attention-based model for recognizing both table structure and the text content of each table cell from a table image. The proposed model consists of four main components: the CNN encoder, and the three decoders for three sub-tasks of the table recognition problems: table structure recognition; table cell detection; and table cell content recognition. The whole system can be trained in an end-to-end manner by stochastic gradient descent algorithms. The extensive experiments on the two large-scale table image datasets show that the proposed model achieves the state-of-the-art on the two large-scale datasets. The experiments also demonstrate the effectiveness of the local attention mechanism in the decoders for table recognition, especially for big table recognition.

In summary, the main contributions of this paper are as follows:

- We propose the local attention mechanism in the decoders to address the limitation of the global attention mechanism in the Transformer-based table recognition methods.
- We present a novel local attention-based model for table recognition. The proposed model can be easily trained and inferred in an end-to-end approach.
- Extensive experiments demonstrate the effectiveness of the local attention mechanism as well as the proposed model for table recognition.
- Across all the benchmark datasets, the proposed model outperforms the state-of-the-art methods.

The rest of this paper is organized as follows. In Sect. 2, we give a brief review of the related works. In Sect. 3, we introduce the overview of the local attention mechanism as well as the proposed model. In Sect. 4, we report the experimental results and analysis. Finally, we draw conclusions in Sect. 5.

2 Related Works

Most of the previous methods for table recognition are based on two-step approaches that divide the table recognition problem into two steps: table structure recognition and table cell content recognition. Due to the simplicity of the table cell content recognition, which can be easily overcome by the standard OCR methods [22, 23], many works only focused on the table structure recognition problem [6–11]. Early works [6–8] of table structure recognition primarily depend on hand-crafted features and heuristic rules and are mainly applied to simple table structure or pre-defined table formats. In recent years, inspired by the success of the deep learning, especially in object detection and semantic segmentation, many deep learning-based table structure recognition methods [9–11] have been proposed and proven to be powerful models. S. Schreiber et al. [11] presented two-fold system named DeepDeSRT that employs Faster RCNN [24] and FCN [25] for both table detection and table structure recognition. S. Raja et al. [10] proposed a table structure recognition model named TabStruct-Net that predicts the aligned cell regions and the localized cell relations in a joint manner.

Recently, some works [12, 14, 15, 26] focused on both table structure recognition and table cell content recognition to build a complete table recognition system. J. Ye et al. [15] proposed the TableMASTER model which consists of two decoders for predicting table structure and table cell location, and then combined it with a text line detector to detect text lines in each table cell. Finally, they used a text line recognizer based on MASTER [22] to recognize detected text lines in each table cell to build the complete table recognition system which achieved the third place in the ICDAR2021 competition [1]. Similar to [15], A. Nassar et al. [14] presented TableFormer which recognizes table structure and table cell location, and then uses the text content extracted from PDF to build the whole table recognition system. L. Qiao et al. [12] proposed a table structure recognition system named LGPMA which is built based on the Mask-RCNN. Then they combined LGPMA with a single line text detection and recognition models to get the OCR information. Their model achieved the first place in the ICDAR2021 competition [1].

Recently, due to the rapid development of deep learning and the availability of large-scale table image datasets, some researchers try to focus on end-to-end approaches [2, 4]. Y. Deng et al. [4] formulated table recognition as the image-to-LaTeX problem and directly employed IM2TEX model [27] for image-based table recognition. X. Zhong et al. [2] proposed an encoder-dual-decoder (EDD) model that consists of two decoders for predicting both table structure and text content of each table cell. However, their performance is still mediocre compared to the non-end-to-end approaches. Most recently, based on the global attention mechanism in Transformer, N. T. Ly et al. [16, 17] proposed the end-to-end multi-task learning methods for image-based table recognition, and then the weakly supervised learning method which reduces the annotation costs of preparing the training data. These methods achieved competitive accuracies compared to the non-end-to-end approaches; however, these methods do not work well on big table recognition due to the limitation of the global attention mechanism.

In 2021, IBM Research in conjunction with ICDAR2021 committee held the ICDAR2021 competition on scientific literature parsing (ICDAR2021 competition in short) [1]. The competition consists of two tasks: Task A: Document layout recognition and Task B: Table recognition which converts table images into HTML representations. In the Task B: Table recognition, the committee provided the PubTabNet dataset [2] for the participants to use to train and test their methods before evaluating in the final evaluation validation set. There are 30 submissions from 30 teams for the final evaluation phase and most of the top 10 solutions are non-end-to-end approaches and employ additional annotation information as well as ensemble techniques to improve their methods.

3   The Proposed Method

3.1   Local Attention Mechanism

Inspired by the work in [18], we employ the fixed-size window attention pattern in our local attention mechanism. Difference from the work in [18] is that our local attention mechanism is implemented in the decoder component and each token in the input sequence pays attention to w tokens on the backward side (instead of both sides in [18]). The local attention mechanism restricts the decoder to focus on important local (neighborhood) information. So, it helps the decoder easily learn to focus on the essential local features of very long input sequences. We will describe the details of our local attention mechanism as follows.

Local Attention. Let $X \in R^{l \times d_x}$ denote the input sequence of the local attention mechanism. First, the input sequence is linearly projected to get the *queries Q*, *keys K*, and *values V*. The attention weights are calculated from the *queries Q* and the keys K as follows:

$$Q = XW^Q \quad K = XW^K \quad V = XW^V \tag{1}$$

$$\text{AttnWeights} = \text{softmax}\left(\frac{QK^T}{\sqrt{d_k}}\right) \tag{2}$$

where l is the length of the input sequence, the projections are parameter matrices W^Q, $W^k \in \mathbb{R}^{d_x \times d_k}$ and $W^V \in \mathbb{R}^{d_x \times d_v}$. d_x, d_v, d_k are the dimensions of the input sequence, values and keys, respectively.

In order to implement the local attention mechanism, we define the mask matrix $M \in \mathbb{R}^{l \times l}$ that indicates the position in the *keys K* over which the *queries Q* should attend. In other words, the attentions are activated on the positions where the mask matrix element is *1*; otherwise, the position with *0* will be canceled out. The mask matrix M is defined as follows:

$$M_{ij} = \begin{cases} 1, & \text{if } 0 \le i - j \le w \\ 0, & \text{otherwise} \end{cases} \tag{3}$$

where w denotes the window size of the local attention mechanism.

Finally, the final outputs of the local attention function are calculated from the attention weights, the mask matrix M and the *values V* as follows:

$$\text{Attention}(X) = (\text{AttnWeights} \odot M)V \tag{4}$$

where $\odot$ denotes element-wise multiplication.

Masked Multi-head Local Attention. The masked multi-head local attention mechanism is the extension of the local attention mechanism, which allows the model to jointly attend to information from different representation subspaces at different positions. First, the masked multi-head local attention mechanism linearly projects the queries, keys, and values h times with different linear projections to obtain h different representations of

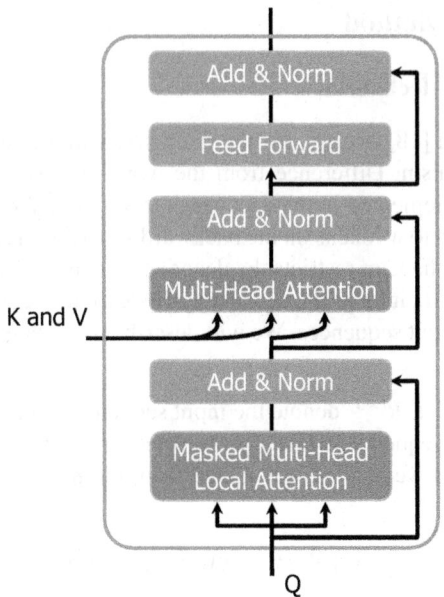

Fig. 1. The architecture of the local attention-based decoder layer.

(Q, K, V), and then performs the local attention function in parallel to get h heads output values. Finally, h heads' output values are concatenated and once again projected, resulting in the final output values, as follows:

$$\text{head}_i = \text{LocalAttention}\left(QW_i^Q, KW_i^K, VW_i^V\right) \tag{5}$$

$$\text{MultiHeadLocal}(Q, K, V) = \text{Concat}(\text{head}_1 \ldots \text{head}_h)W^O \tag{6}$$

where h is the number of heads, W_i^Q, $W_i^K \in R^{d_x \times d_k}$, $W_i^V \in R^{d_x \times d_v}$, and $W^O \in R^{d_x \times d_x}$ are parameter matrices of the linear projections, d_x is the dimension of the input sequence, $d_K = d_V = d_x/h$.

Local Attention-Based Decoder Layer. The local attention-based decoder layer consists of three sub-layers: a masked multi-head local attention layer, a multi-head attention layer, and a position-wise fully connected feed-forward layer as shown in Fig. 1. Each sub-layer is followed by a residual connection and a layer normalization similar to the global attention-based decoder layer in the original Transformer.

3.2 The Proposed Model

In this work, we proposed a local attention-based end-to-end model for table recognition. The proposed model is composed of four main components: a CNN-based encoder for feature extraction, and three decoders for predicting three sub-tasks of the table recognition problem: table structure recognition; table cell detection; and table cell content

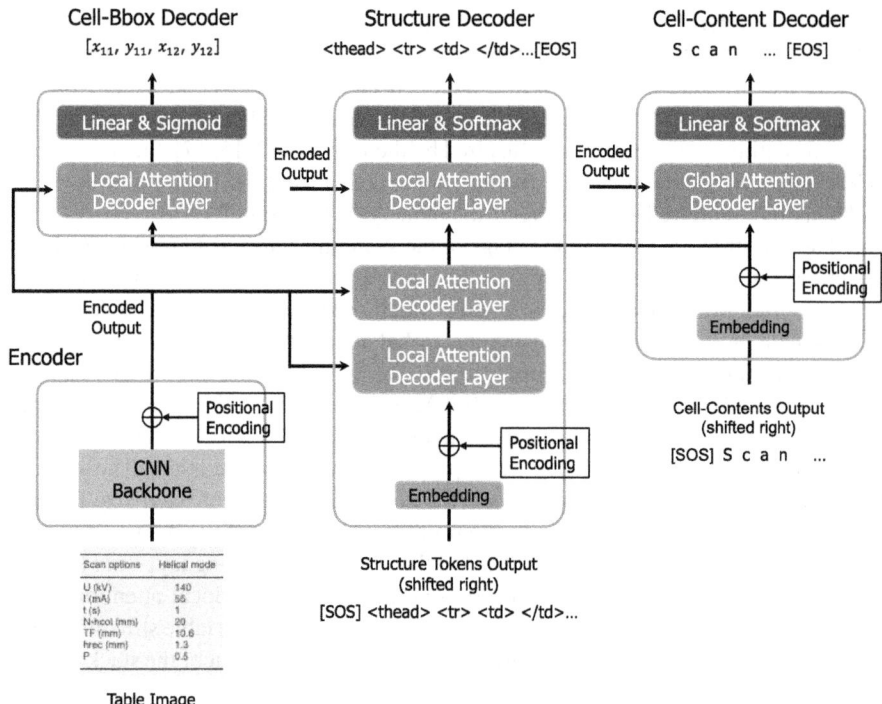

Fig. 2. The network architecture of the proposed model.

recognition as shown in Fig. 2. Given a table image, the CNN-based encoder extracts the features and encodes them as a sequence of features. The sequence of features is passed into the structure decoder to predict the sequence of structure tokens that represent the structure of the table. When the structure decoder produces the structure token that represents a table cell, the cell-bbox decoder and the cell-content decoder are triggered and use the hidden states of the structure decoder to predict the bounding box coordinates and the text content of that cell. Finally, the text content of each cell is inserted into its corresponding cell of the sequence of structure tokens to produce the final representations of the table. We describe the details of four components in the following sections.

Encoder. At the bottom of the proposed model, the encoder extracts the features from an input table image and then encodes them as a sequence of fixed-size features. This work employs a CNN backbone network followed by a positional encoding layer to build the encoder. Given an input table image of the size $\{h, w, c\}$, the CNN backbone network extracts a feature grid F of size $\{h', w', c'\}$, where c' is the number of the feature map in the last convolutional layer, and h' and w' depend on the h and w of the input image and the number of pooling layers in the CNN network. The feature grid F is unfolded into a sequence of features (column by column from left to right in each feature map), then fed into the positional encoding layer to get the encoded sequence of fixed-size features.

Structure Decoder. In this work, we formulate the table structure recognition problem as the sequence prediction problem that predicts a sequence of HTML table tags representing the structure information of a table. At the top of the encoder, the structure decoder uses the encoded sequence of fixed-size features to predict a sequence of HTML tags representing the table structure. Inspired by the works of [2, 15, 17], the sequence of HTML tags representing the table structure are tokenized at the HTML tag level except for the tag of a table cell. For the tag of the table cell, the form of '< td > < /td >' is treated as one structure token and the tags of the spanning cells ('< td rowspan/colspan ="number" >') are broken down into '< td', 'rowspan/colspan ="number"', and ' >'. Thus, the structure token of '< td > < /td >' and '< td' represent a new table cell.

As shown in Fig. 2, instead of using the global attention-based decoder layer of the original Transformer model, we employ a stack of three local attention-based decoder layers followed by a linear layer and a SoftMax layer to build the structure decoder. The local attention mechanism helps the structure decoder easily focus on the important local features (on neighbor tokens) when predicting one structure token. The stack of local attention-based decoder layers takes the encoded sequence of fixed-size features as the input value and key vectors. In the training phase, the right-shifted sequence of structure tokens (table structure target) is fed through the embedding layer and then the positional encoding layer before being passed into the stack of local attention-based decoder layers as the input query vector. In the inference phase, the table structure target is replaced by the output of the structure decoder. Finally, the output of the stack of local attention-based decoder layers are fed into the linear layer, and then the SoftMax layer to generate the sequence of structure tokens representing the table structure.

Cell-Bbox Decoder. At the top of the encoder, the cell-bbox decoder selects the hidden states of the structure decoder corresponding to the structure token representing a new table cell and uses them to predict the bounding box coordinates of table cells. Similar to the structure decoder, we also use the local attention-based decoder layer in the cell-bbox decoder to help this decoder focus on the important local features (on neighbor tokens) when predicting the location of one table cell. Specifically, the cell-bbox decoder consists of one local attention-based decoder layer followed by a linear layer and a sigmoid layer as shown in Fig. 2. When the structure decoder produces a new cell, the cell-bbox decoder is triggered. Then the local attention-based decoder layer takes the hidden states of the structure decoder as the query vectors, and the encoded sequence of fixed-size features as the value and key vectors. Finally, the output of the local attention-based decoder layer is fed into the linear layer and then the sigmoid layer to predict the four coordinates of this cell bounding box.

Cell-Content Decoder. Similar to the cell-bbox decoder, the cell-content decoder selects the hidden states of the structure decoder corresponding to the structure token representing a new table cell and uses them to predict the text content of table cells. In this work, the text contents of table cells are tokenized at character level and the cell-content decoder can be considered as a text recognizer.

As shown in Fig. 2, the cell-content decoder consists of one embedding layer, one positional encoding layer, and one global attention-based decoder layer followed by a linear layer and a SoftMax layer. Here, we do not employ the local attention mechanism in the cell-content decoder. The reason is that the number of tokens in the cell content is usually much smaller than that in the table structure, then the global attention mechanism can easily learn to focus on the important features. The following is the decoding mechanism of the cell-content decoder. When the structure decoder produces a new cell, the cell-content decoder is triggered, and takes the output of the encoder as the input value and key vectors of the identical layer. The right-shifted text content target of this table cell (the right-shirted output of the cell-content decoder in the inference phase) is fed into the embedded and the positional encoding layers, and then added to the hidden states of the structure decoder referring to this table cell before being passed into the identical layer as the query vectors. Finally, the output of the identical layer is fed through the linear layer and then SoftMax layer to predict the text content of this table cell.

Network Training. The whole model can be trained in an end-to-end manner by stochastic gradient descent algorithms on pairs of table images and their annotations of the table structure, the text content and its bounding box of each table cell. The overall loss of the proposed model is defined as follows:

$$\mathcal{L} = \lambda_1 \mathcal{L}_{\text{struc.}} + \lambda_2 \mathcal{L}_{\text{cont.}} + \lambda_3 \mathcal{L}_{\text{bbox}} \tag{7}$$

where $\mathcal{L}_{\text{struc.}}$ and $\mathcal{L}_{\text{cont.}}$ are the table structure recognition loss and the cell-content prediction loss, respectively that are implemented in Cross-Entropy loss, whereas $\mathcal{L}_{\text{bbox}}$ is the cell-bbox regression loss which is optimized by L1 loss. λ_1, λ_2, and λ_3 are weight hyperparameters.

4   Experiments

To evaluate the effectiveness of the local attention mechanism and the performance of the proposed model, we conducted experiments on the two large-scale table image datasets: PubTabNet and FinTabNet. The information of the datasets is given in Sect. 4.1; the implementation details are described in Sect. 4.2; the results of the experiments are shown in Sect. 4.3; and the visualization results are presented in Sect. 4.4.

4.1   Datasets

PubTabNet is a large-scale table image dataset published by X. Zhong et al. [28]. This dataset is created by collecting scientific articles from PubMed Central Open Access Subset (PMCOA) and consists of over 568k table images with corresponding annotations of table structure, the text content and bounding box of all non-empty table cells. PubTabNet is divided into 500,777 training samples; 9,115 validation samples; and 9,064 final evaluation samples used in the training phase, the development phase, and the Final Evaluation Phase of the ICDAR2021 competition, respectively.

PubTabNet250. We select the table images from PubTabNet which have more than 250 structure tokens in the table structure to create a dataset named PubTabNet250. This subset is used to verify the effectiveness of the local attention mechanism and consists of 114,111 table images for training, and 2,161 table images for validation.

FinTabNet. X. Zheng et al. [28] presented the large-scale table image dataset named FinTabNet which is collected from the annual reports of the S&P 500 companies with detailed annotations of the table HTML and the cell bounding boxes like PubtabNet. This dataset consists of about 112k table images which are divided into training, testing and validation sets with a ratio of 81%:9.5%:9.5%.

The statistics of these datasets are shown in Table 1.

Table 1. The statistics of the datasets.

Datasets	Training	Valid	Testing
PubTabNet	500,777	9,115	9,064
PubTabNet250	114,111	2,161	–
FinTabNet	91,596	10,635	10,656

4.2 Implementation Details

In the encoder, inspired by the works in [22], we employ the ResNet-31 backbone network [29] with the Multi-Aspect Global Context Attention (GCAttention) after each residual block to build the CNN backbone network. All the input images are resized to 520*520 pixels before being fed into the encoder and the size of the feature grid F outputted from the CNN backbone network is 65*65.

In the structure and cell-bbox decoders, all local attention-based decoder layers have the same architecture with 8 heads, the input feature size and the numbers of nodes of the feed-forward layer are 512 and 2048, respectively. The window size in the local attention mechanism is set as 300 in all the local attention-based decoder layers. The architecture of the global attention decoder layer in the cell-content decoder is the same as the local attention-based decoder layer except that the local attention mechanism is replaced by the global attention mechanism. The maximum sequence length of the structure tokens and the cell tokens (in the decoding process) are 600 and 150, respectively. The weight hyperparameters are set as $\lambda_1 = \lambda_2 = \lambda_3 = 1$.

The proposed model is implemented in the MMCV library [30] and trained on two NVIDIA A100 80G with a batch size of 8. The learning rate is initialized at 0.001 for the first 12 epochs and reduced to 0.0001 until the model converges.

4.3 Experiment Results

To evaluate the performance of the proposed model for table recognition, we employ the Tree-Edit-Distance-Based Similarity (TEDS) metric [2]. It represents the prediction and the ground-truth as a tree structure of HTML tags, and is calculated as follows:

$$\text{TEDS}(T_a, T_b) = 1 - \frac{\text{EditDist}(T_a, T_b)}{\max(|T_a|, |T_b|)} \tag{8}$$

where T_a and T_b represent tables in a tree structured HTML format, EditDist denotes the tree-edit distance, and $|T|$ represents the number of nodes in T.

We also denote TEDS-struc. as the TEDS score between two tables when considering only the table structure information.

4.3.1 Table Recognition

FinTabNet. The first experiment evaluates the performance of the proposed model on the FinTabNet dataset in terms of TEDS (Table recognition) and TEDS-struc. (Table structure recognition). To fairly compare with the previous methods, we do not use any data augmentation as well as ensemble techniques. Table 2 compares the TEDS and TEDS-struc. Scores between the proposed model and the previous methods on the test set of the FinTabNet dataset. With the TEDS of 95.74% and TEDS-struc. of 98.85%, the proposed model outperforms the best model in [16] and obtains state-of-the-art results on both table recognition and table structure recognition on the test set of the FinTabNet dataset.

Table 2. Table recognition results (%) on FinTabNet test set.

Models	TEDS-struc	TEDS
EDD [2]	90.60	–
GTE [28]	87.14	–
GTE$^{(PT)}$ [28]	91.02	–
TableFormer [14]	96.80	–
WSTabNet [16]	98.72	95.32
The Proposed Model	**98.85**	**95.74**

PubTabNet. In this experiment, we evaluate the performance of the proposed model for table recognition on the validation set of the PubTabNet dataset. To fairly compare with the previous methods, we also do not employ any data augmentation as well as ensemble techniques. Table 3 shows table recognition results by the proposed model in comparison with the previous methods on the validation set of the PubTabNet dataset. As shown in Table 3, the proposed model outperforms all the previous methods which do not employ any data augmentation as well as ensemble techniques. Furthermore, with TEDS of 98.07% on simple tables and TEDS of 95.42% on complex tables, the proposed model also achieves state-of-the-art results on both simple and complex table images.

Although, we do not use any data augmentation as well as ensemble techniques, the proposed method achieves competitive results compared to VCGroup + ME [15] which requires additional annotations of text-line bounding boxes of the text content in each table cell and employs three model ensembles in both table structure recognition and text line recognition models.

Table 3. Table recognition results on PubTabNet validation set.

Models	TEDS		
	Simple	Complex	All
EDD [2]	91.20	85.40	88.30
TabStruct-Net [10]	–	–	90.10
GTE [28]	–	–	93.00
TableFormer [14]	95.40	90.10	93.60
SEM [3] [26]	94.80	92.50	93.70
LGPMA + OCR [1] [12]	–	–	94.60
VCGoup [2] [15]	–	–	96.26
WSTabNet [16]	97.89	95.02	96.48
Multi-Tasks Model [17]	97.92	95.36	96.67
The Proposed Model	**98.07**	**95.42**	**96.77**
VCGoup + ME [2] [15]	–	–	96.84

Table 4. Table recognition results on PubTabNet final evaluation set.

Team Name	TEDS (%)		
	Simple	Complex	All
Davar-Lab-OCR	97.88	94.78	**96.36**
VCGroup [15]	**97.90**	94.68	96.32
XM [26]	97.60	**94.89**	96.27
The Proposed Model	97.77	94.58	96.21
YG	97.38	94.79	96.11
DBJ	97.39	93.87	95.66
TAL	97.30	93.93	95.65
PaodingAI	97.35	93.79	95.61
anyone	96.95	93.43	95.23
LTIAYN	97.18	92.40	94.84

We also evaluate the performance of the proposed model on the final evaluation set of the PubTabNet dataset in comparison with the top 10 solutions in the ICDAR2021 competition. As shown in Table 4, although we used neither any additional training data nor ensemble techniques, the proposed model outperforms the 4th ranking solution named YG and achieves competitive recognition results compared to the top 3 solutions in the final evaluation set of the ICDAR2021 competition. Note that the top 10 solutions are non-end-to-end approaches and most of them use additional training data or additional annotation information for training as well as ensemble techniques to improve the final recognition results.

4.3.2   Ablation Studies

We also conducted additional experiments on the PubTabNet250 dataset with ablation consideration.

The Effectiveness of the Local Attention Mechanism. In this study, we verify the effectiveness of the local attention mechanism in comparison with the traditional global attention mechanism. To do that, we prepared one variation, which is the same as the proposed model except for using the global attention mechanism instead of the local attention mechanism in the structure and cell-bbox decoders. We named this variation as Global Attention and the proposed model as Local Attention. Table 5 shows their table recognition results on the validation set of the PubTabNet250 dataset in different sequence lengths of the table structure tokens.

Table 5. Table recognition results (TEDS) with different attention mechanisms.

Models	TEDS (%)			
	All	len > 500	len > 600	len > 700
Local Attention	94.28	92.99	91.29	89.61
Global Attention	93.86	91.16	90.63	88.65

len: number of table structure tokens.

As shown in Table 5, Local Attention slightly outperforms Global Attention on the validation set of the PubTabNet250 dataset, and Local Attention is even more dominant on the tables having a large number of table structure tokens. The results imply that the local attention mechanism outperforms the global attention mechanism for table recognition, especially for big table recognition.

Table 6. TEDS scores with respect to window sizes.

Window Size	TEDS (%)			
	All	len > 500	len > 600	len > 700
100	92.88	89.10	86.76	83.10
200	93.74	91.67	89.82	88.01
300	94.28	92.99	91.29	89.61
400	94.25	92.27	90.86	88.95
500	94.07	92.42	90.72	88.69

len: number of structure tokens.

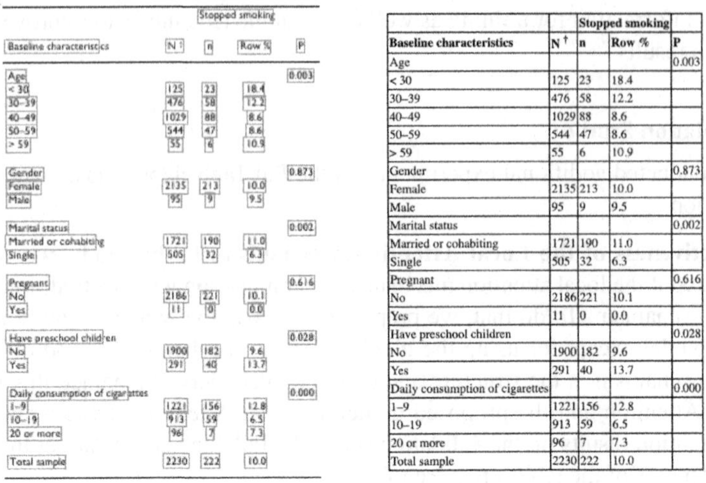

Baseline characteristics	N †	Stopped smoking n	Row %	P
Age				0.003
< 30	125	23	18.4	
30–39	476	58	12.2	
40–49	1029	88	8.6	
50–59	544	47	8.6	
> 59	55	6	10.9	
Gender				0.873
Female	2135	213	10.0	
Male	95	9	9.5	
Marital status				0.002
Married or cohabiting	1721	190	11.0	
Single	505	32	6.3	
Pregnant				0.616
No	2186	221	10.1	
Yes	11	0	0.0	
Have preschool children				0.028
No	1900	182	9.6	
Yes	291	40	13.7	
Daily consumption of cigarettes				0.000
1–9	1221	156	12.8	
10–19	913	59	6.5	
20 or more	96	7	7.3	
Total sample	2230	222	10.0	

Fig. 3. Visualization results of the proposed model on PubTabNet.

Window Size. In this study, we evaluate the effect of the window size on the performance of the local attention mechanism for table recognition. Table 6 shows the TEDS scores with respect to window sizes. We observe that the performance of the proposed model improves when the window size increases from 100 to 300. The results suggest that the local attention mechanism with the short window size does not work well on table recognition. The reason seems to be that the local attention mechanism with the too-short window size might lose information on the potential features. We also observe that the proposed model with a window size of 300 obtained the highest TEDS score and decreases slightly as the window size increases from 300 to 500. The results suggest that the too-long window size might decrease the performance of the local attention mechanism.

4.4 Visualization Results

In this section, we show visualization results of the proposed model on PubTabNet. As shown in Fig. 3, the proposed model is able to predict complex table structure as well as bounding boxes and contents for all table cells, even for the empty cells or cells that span multiple rows/columns. We also show the visualization results in comparison between the local attention and the global attention mechanisms. As shown in Fig. 4, the local attention mechanism can work more efficiently than the global attention mechanism on the big table image.

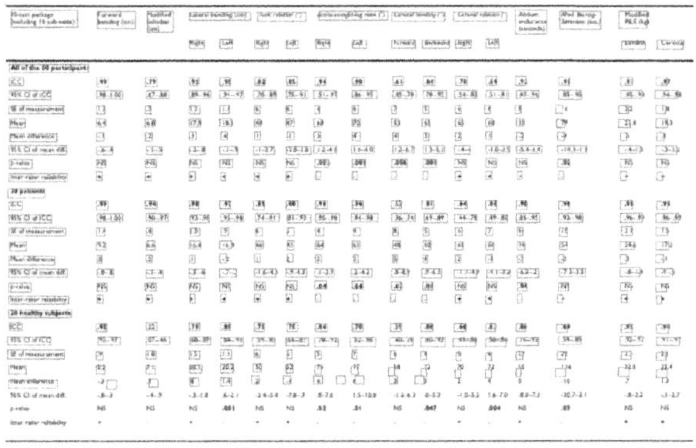

Global Attention

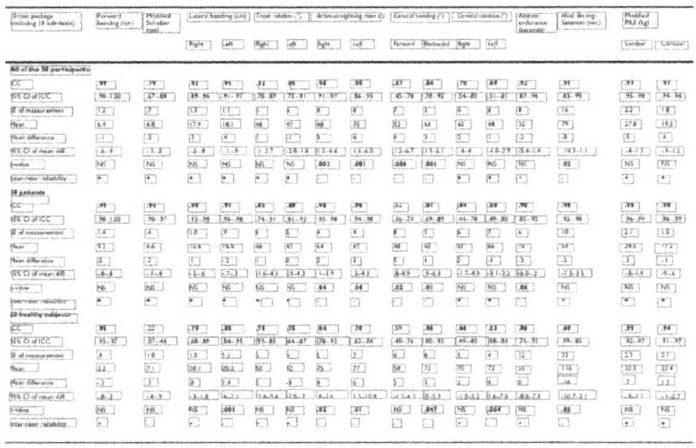

Local Attention

Fig. 4. Visualization results in comparison between the local and global attention.

5 Conclusion

In this paper, we present the local attention mechanism to address the limitation of the global attention mechanism and propose the end-to-end local attention-based model for image-based table recognition. The proposed model consists of four main components: a CNN encoder for feature extraction, and three decoders for three sub-tasks of table recognition: table structure recognition, table cell detection, and cell-content recognition. Extensive experiments on the two large-scale table image datasets demonstrate the proposed model outperforms the previous methods as well as the effectiveness of the local attention mechanism for table recognition, especially for big table recognition.

In the future, we will conduct the experiments of the proposed model on the other table image dataset. We also plan to explore the effectiveness of the local attention mechanism on other document analysis tasks such as text recognition and handwritten mathematical expression recognition.

Acknowledgments. This work was supported by the Cross-ministerial Strategic Innovation Promotion Program (SIP) Second Phase and "Big-data and AI-enabled Cyberspace Technologies" by New Energy and Industrial Technology Development Organization (NEDO) and partially supported by ROIS NII Open Collaborative Research 2023(23S0102).

References

1. Jimeno Yepes, A., Zhong, P., Burdick, D.: ICDAR 2021 competition on scientific literature parsing. In: Lladós, J., Lopresti, D., Uchida, S. (eds.) Document Analysis and Recognition - ICDAR 2021. Lecture Notes in Computer Science, vol. 12824, pp. 605–617. Springer, Cham (2021). https://doi.org/10.1007/978-3-030-86337-1_40

2. Zhong, X., ShafieiBavani, E., Jimeno Yepes, A.: Image-based table recognition: data, model, and evaluation. In: Vedaldi, A., Bischof, H., Brox, T., Frahm, JM. (eds.) Computer Vision – ECCV 2020. ECCV 2020. Lecture Notes in Computer Science(), vol. 12366, pp. 564–580. Springer, Cham (2020). https://doi.org/10.1007/978-3-030-58589-1_34/TABLES/3

3. Li, M., Cui, L., Huang, S., Wei, F., Zhou, M., Li, Z.: TableBank: a benchmark dataset for table detection and recognition (2019). https://doi.org/10.48550/arxiv.1903.01949

4. Deng, Y., Rosenberg, D., Mann, G.: Challenges in end-to-end neural scientific table recognition. In: Proceedings of the International Conference on Document Analysis and Recognition, ICDAR, pp. 894–901 (2019). https://doi.org/10.1109/ICDAR.2019.00148

5. Kayal, P., Anand, M., Desai, H., Singh, M.: ICDAR 2021 competition on scientific table image recognition to LaTeX. In: Lladós, J., Lopresti, D., Uchida, S. (eds.) Document Analysis and Recognition - ICDAR 2021. Lecture Notes in Computer Science, vol. 12824, pp. 754–766. Springer, Cham (2021). https://doi.org/10.1007/978-3-030-86337-1_50

6. Itonori, K.: Table structure recognition based on textblock arrangement and ruled line position. In: Proceedings of 2nd International Conference on Document Analysis and Recognition (ICDAR '93), pp. 765–768 (1993). https://doi.org/10.1109/ICDAR.1993.395625

7. Kieninger, T.G.: Table structure recognition based on robust block segmentation, vol. 3305, pp. 22–32 (1998). https://doi.org/10.1117/12.304642

8. Wang, Y., Phillips, I.T., Haralick, R.M.: Table structure understanding and its performance evaluation. Pattern Recognit. **37**, 1479–1497 (2004). https://doi.org/10.1016/J.PATCOG.2004.01.012

9. Prasad, D., Gadpal, A., Kapadni, K., Visave, M., Sultanpure, K.: CascadeTabNet: an approach for end to end table detection and structure recognition from image-based documents. In: IEEE Computer Society Conference on Computer Vision and Pattern Recognition Workshops, June 2020, pp. 2439–2447 (2020). https://doi.org/10.48550/arxiv.2004.12629

10. Raja, S., Mondal, A., Jawahar, C.V.: Table structure recognition using top-down and bottom-Up Cues. In: Vedaldi, A., Bischof, H., Brox, T., Frahm, JM. (eds.) Computer Vision – ECCV 2020. ECCV 2020. Lecture Notes in Computer Science(), vol. 12373, pp. 70–86. Springer, Cham (2020). https://doi.org/10.1007/978-3-030-58604-1_5/FIGURES/8

11. Schreiber, S., Agne, S., Wolf, I., Dengel, A., Ahmed, S.: DeepDeSRT: deep learning for detection and structure recognition of tables in document images. In: Proceedings of the International Conference on Document Analysis and Recognition, ICDAR, vol. 1, pp. 1162–1167 (2017). https://doi.org/10.1109/ICDAR.2017.192

12. Qiao, L., et al.: LGPMA: complicated table structure recognition with local and global pyramid mask alignment. In: Lladós, J., Lopresti, D., Uchida, S. (eds.) Document Analysis and Recognition – ICDAR 2021. ICDAR 2021. Lecture Notes in Computer Science(), vol. 12821, pp. 99–114. Springer, Cham (2021). https://doi.org/10.1007/978-3-030-86549-8_7/TABLES/4

13. Vaswani, A., et al.: Attention is all you need. In: Advances in Neural Information Processing Systems. Neural information processing systems foundation, pp. 5999–6009 (2017)
14. Nassar, A., Livathinos, N., Lysak, M., Staar, P.: TableFormer: table structure understanding with transformers (2022). https://doi.org/10.48550/arxiv.2203.01017
15. Ye, J., et al.: PingAn-VCGroup's solution for ICDAR 2021 competition on scientific literature parsing task B: table recognition to HTML (2021). https://doi.org/10.48550/arxiv.2105.01848
16. Ly, N.T., Takasu, A., Nguyen, P., Takeda, H.: Rethinking image-based table recognition using weakly supervised methods. In: In Proceedings of the 12th International Conference on Pattern Recognition Applications and Methods (ICPRAM 2023), Lisbon, Portugal (2023)
17. Ly, N.T., Takasu, A.: An end-to-end multi-task learning model for image-based table recognition. In: In Proceedings of the 18th International Joint Conference on Computer Vision, Imaging and Computer Graphics Theory and Applications (VISIGRAPP 2023) (2023)
18. Beltagy, I., Peters, M.E., Cohan, A.: Longformer: the long-document transformer. arXiv (2020)
19. Kovaleva, O., Romanov, A., Rogers, A., Rumshisky, A.: Revealing the dark secrets of BERT. In: EMNLP-IJCNLP 2019 - 2019 Conference on Empirical Methods in Natural Language Processing and 9th International Joint Conference on Natural Language Processing, Proceedings of the Conference, pp. 4365–4374 (2019). https://doi.org/10.18653/V1/D19-1445
20. Sperber, M., Niehues, J., Neubig, G., Stüker, S., Waibel, A.: Self-attentional acoustic models. In: Proceedings of the Annual Conference of the International Speech Communication Association, INTERSPEECH, September 2018, pp. 3723–3727 (2018). https://doi.org/10.21437/INTERSPEECH.2018-1910
21. Shaw, P., Uszkoreit, J., Vaswani, A.: Self-Attention with relative position representations. In: NAACL HLT 2018 - 2018 Conference of the North American Chapter of the Association for Computational Linguistics: Human Language Technologies - Proceedings of the Conference, vol. 2, pp. 464–468 (2018). https://doi.org/10.18653/V1/N18-2074
22. Lu, N., et al.: MASTER: multi-aspect non-local network for scene text recognition. Pattern Recognit. **117**, 107980 (2021). https://doi.org/10.1016/J.PATCOG.2021.107980
23. Ly, N.T., Nguyen, H.T., Nakagawa, M.: 2D self-attention convolutional recurrent network for offline handwritten text recognition. In: Lladós, J., Lopresti, D., Uchida, S. (eds.) Document Analysis and Recognition – ICDAR 2021. ICDAR 2021. Lecture Notes in Computer Science(), vol. 12821, pp. 191–204. Springer, Cham (2021). https://doi.org/10.1007/978-3-030-86549-8_13/COVER
24. Ren, S., He, K., Girshick, R., Sun, J.: Faster R-CNN: towards real-time object detection with region proposal networks. IEEE Trans. Pattern Anal. Mach. Intell. **39**, 1137–1149 (2015). https://doi.org/10.48550/arxiv.1506.01497
25. Long, J., Shelhamer, E., Darrell, T.: Fully convolutional networks for semantic segmentation. In: 2015 IEEE Conference on Computer Vision and Pattern Recognition (CVPR), 07–12 June 2015, pp. 3431–3440 (2015). https://doi.org/10.1109/CVPR.2015.7298965
26. Zhang, Z., Zhang, J., Du, J., Wang, F.: Split, embed and merge: an accurate table structure recognizer. Pattern Recognit. **126**, 108565 (2022). https://doi.org/10.1016/J.PATCOG.2022.108565
27. Deng, Y., Kanervisto, A., Ling, J., Rush, A.M.: Image-to-markup generation with coarse-to-fine attention. In: 34th International Conference on Machine Learning, ICML 2017, vol. 3, pp. 1631–1640 (2016). https://doi.org/10.48550/arxiv.1609.04938
28. Zheng, X., Burdick, D., Popa, L., Zhong, X., Wang, N.X.R.: Global table extractor (GTE): a framework for joint table identification and cell structure recognition using visual context. In: 2021 IEEE Winter Conference on Applications of Computer Vision (WACV), pp. 697–706 (2021). https://doi.org/10.1109/WACV48630.2021.00074

29. He, K., Zhang, X., Ren, S., Sun, J.: Deep residual learning for image recognition. In: 2016 IEEE Conference on Computer Vision and Pattern Recognition (CVPR), pp. 770–778. IEEE (2016). https://doi.org/10.1109/CVPR.2016.90
30. MMCV Contributors: {MMCV: OpenMMLab} Computer Vision Foundation (2018). https://github.com/open-mmlab/mmcv

Optimized Table Tokenization for Table Structure Recognition

Maksym Lysak[✉][iD], Ahmed Nassar[iD], Nikolaos Livathinos[iD],
Christoph Auer[iD], and Peter Staar[iD]

IBM Research, Zurich, Switzerland
{mly,ahn,nli,cau,taa}@zurich.ibm.com

Abstract. Extracting tables from documents is a crucial task in any document conversion pipeline. Recently, transformer-based models have demonstrated that table-structure can be recognized with impressive accuracy using Image-to-Markup-Sequence (Im2Seq) approaches. Taking only the image of a table, such models predict a sequence of tokens (e.g. in HTML, LaTeX) which represent the structure of the table. Since the token representation of the table structure has a significant impact on the accuracy and run-time performance of any Im2Seq model, we investigate in this paper how table-structure representation can be optimised. We propose a new, optimised table-structure language (OTSL) with a minimized vocabulary and specific rules. The benefits of OTSL are that it reduces the number of tokens to 5 (HTML needs 28+) and shortens the sequence length to half of HTML on average. Consequently, model accuracy improves significantly, inference time is halved compared to HTML-based models, and the predicted table structures are always syntactically correct. This in turn eliminates most post-processing needs. Popular table structure data-sets will be published in OTSL format to the community.

Keywords: Table Structure Recognition · Data Representation · Transformers · Optimization

1 Introduction

Tables are ubiquitous in documents such as scientific papers, patents, reports, manuals, specification sheets or marketing material. They often encode highly valuable information and therefore need to be extracted with high accuracy. Unfortunately, tables appear in documents in various sizes, styling and structure, making it difficult to recover their correct structure with simple analytical methods. Therefore, accurate table extraction is achieved these days with machine-learning based methods.

In modern document understanding systems [1,15], table extraction is typically a two-step process. Firstly, every table on a page is located with a bounding box, and secondly, their logical row and column structure is recognized. As of

G. A. Fink et al. (Eds.): ICDAR 2023, LNCS 14188, pp. 37–50, 2023.
https://doi.org/10.1007/978-3-031-41679-8_3

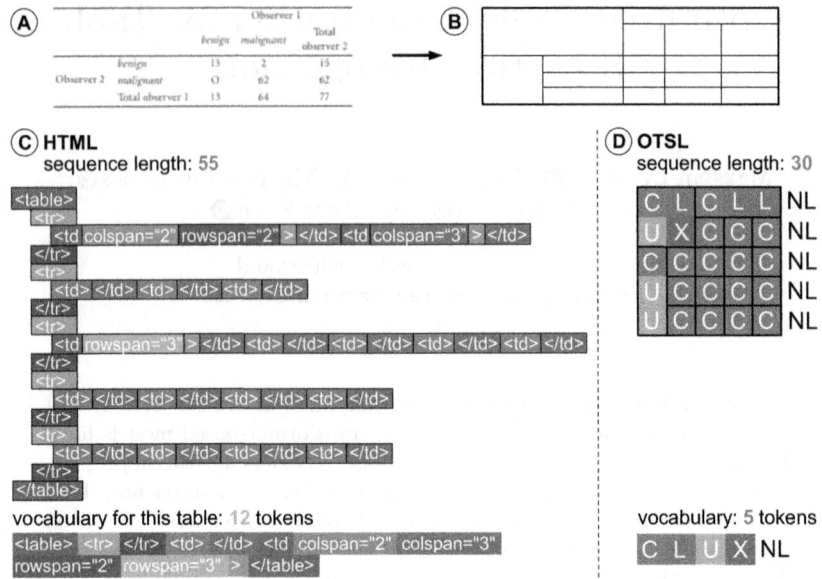

Fig. 1. Comparison between HTML and OTSL table structure representation: (A) table-example with complex row and column headers, including a 2D empty span, (B) minimal graphical representation of table structure using rectangular layout, (C) HTML representation, (D) OTSL representation. This example demonstrates many of the key-features of OTSL, namely its reduced vocabulary size (12 versus 5 in this case), its reduced sequence length (55 versus 30) and a enhanced internal structure (variable token sequence length per row in HTML versus a fixed length of rows in OTSL).

today, *table detection* in documents is a well understood problem, and the latest state-of-the-art (SOTA) object detection methods provide an accuracy comparable to human observers [7,8,10,14,23]. On the other hand, the problem of table structure recognition (TSR) is a lot more challenging and remains a very active area of research, in which many novel machine learning algorithms are being explored [3–5,9,11–14,17,18,21,22].

Recently emerging SOTA methods for table structure recognition employ transformer-based models, in which an image of the table is provided to the network in order to predict the structure of the table as a sequence of tokens. These image-to-sequence (Im2Seq) models are extremely powerful, since they allow for a purely data-driven solution. The tokens of the sequence typically belong to a markup language such as HTML, Latex or Markdown, which allow to describe table structure as rows, columns and spanning cells in various configurations. In Fig. 1, we illustrate how HTML is used to represent the table-structure of a particular example table. Public table-structure data sets such as PubTabNet [22], and FinTabNet [21], which were created in a semi-automated way from paired PDF and HTML sources (e.g. PubMed Central), popularized primarily the use of HTML as ground-truth representation format for TSR.

While the majority of research in TSR is currently focused on the development and application of novel neural model architectures, the table structure representation language (e.g. HTML in PubTabNet and FinTabNet) is usually adopted *as is* for the sequence tokenization in Im2Seq models. In this paper, we aim for the opposite and investigate the impact of the table structure representation language with an otherwise unmodified Im2Seq transformer-based architecture. Since the current state-of-the-art Im2Seq model is TableFormer [9], we select this model to perform our experiments.

The main contribution of this paper is the introduction of a new optimised table structure language (OTSL), specifically designed to describe table-structure in an compact and structured way for Im2Seq models. OTSL has a number of key features, which make it very attractive to use in Im2Seq models. Specifically, compared to other languages such as HTML, OTSL has a minimized vocabulary which yields short sequence length, strong inherent structure (e.g. strict rectangular layout) and a strict syntax with rules that only look backwards. The latter allows for syntax validation during inference and ensures a syntactically correct table-structure. These OTSL features are illustrated in Fig. 1, in comparison to HTML.

The paper is structured as follows. In Sect. 2, we give an overview of the latest developments in table-structure reconstruction. In Sect. 3 we review the current HTML table encoding (popularised by PubTabNet and FinTabNet) and discuss its flaws. Subsequently, we introduce OTSL in Sect. 4, which includes the language definition, syntax rules and error-correction procedures. In Sect. 5, we apply OTSL on the TableFormer architecture, compare it to TableFormer models trained on HTML and ultimately demonstrate the advantages of using OTSL. Finally, in Sect. 6 we conclude our work and outline next potential steps.

2   Related Work

Approaches to formalize the logical structure and layout of tables in electronic documents date back more than two decades [16]. In the recent past, a wide variety of computer vision methods have been explored to tackle the problem of table structure recognition, i.e. the correct identification of columns, rows and spanning cells in a given table. Broadly speaking, the current deep-learning based approaches fall into three categories: object detection (OD) methods, Graph-Neural-Network (GNN) methods and Image-to-Markup-Sequence (Im2Seq) methods. Object-detection based methods [11–14,21] rely on table-structure annotation using (overlapping) bounding boxes for training, and produce bounding-box predictions to define table cells, rows, and columns on a table image. Graph Neural Network (GNN) based methods [3,6,17,18], as the name suggests, represent tables as graph structures. The graph nodes represent the content of each table cell, an embedding vector from the table image, or geometric coordinates of the table cell. The edges of the graph define the relationship between the nodes, e.g. if they belong to the same column, row, or table cell. Other work [20] aims at predicting a grid for each table and deciding which cells must be merged using an attention network. Im2Seq methods cast the problem as a sequence generation task [4,5,9,22], and therefore need an internal table-structure representation language, which is often implemented with standard markup languages (e.g. HTML, LaTeX, Markdown). In theory, Im2Seq methods have a natural advantage over the OD and GNN methods by virtue of directly predicting the table-structure. As such, no post-processing or rules are needed in order to obtain the table-structure, which is necessary with OD and GNN approaches. In practice, this is not entirely true, because a predicted sequence of table-structure markup does not necessarily have to be syntactically correct. Hence, depending on the quality of the predicted sequence, some post-processing needs to be performed to ensure a syntactically valid (let alone correct) sequence.

Within the Im2Seq method, we find several popular models, namely the encoder-dual-decoder model (EDD) [22], TableFormer [9], Tabsplitter [2] and Ye et al. [19]. EDD uses two consecutive long short-term memory (LSTM) decoders to predict a table in HTML representation. The *tag decoder* predicts a sequence of HTML tags. For each decoded table cell ($<td>$), the attention is passed to the *cell decoder* to predict the content with an embedded OCR approach. The latter makes it susceptible to transcription errors in the cell content of the table. TableFormer address this reliance on OCR and uses two transformer decoders for HTML structure and cell bounding box prediction in an end-to-end architecture. The predicted cell bounding box is then used to extract text tokens from an originating (digital) PDF page, circumventing any need for OCR. TabSplitter [2] proposes a compact double-matrix representation of table rows and columns to do error detection and error correction of HTML structure sequences based on predictions from [19]. This compact double-matrix representation can not be used directly by the Img2seq model training, so the model uses HTML as an intermediate form. Chi et al. [4] introduce a data set and a baseline method using bidirectional LSTMs to predict LaTeX code. Kayal [5] introduces Gated ResNet transformers to predict LaTeX code, and a separate OCR module to extract content.

Im2Seq approaches have shown to be well-suited for the TSR task and allow a full end-to-end network design that can output the final table structure without pre- or post-processing logic. Furthermore, Im2Seq models have demonstrated to deliver state-of-the-art prediction accuracy [9]. This motivated the authors to investigate if the performance (both in accuracy and inference time) can be further improved by optimising the table structure representation language. We believe this is a necessary step before further improving neural network architectures for this task.

3   Problem Statement

All known Im2Seq based models for TSR fundamentally work in similar ways. Given an image of a table, the Im2Seq model predicts the structure of the table by generating a sequence of tokens. These tokens originate from a finite vocabulary and can be interpreted as a table structure. For example, with the HTML tokens $<table>$, $</table>$, $<tr>$, $</tr>$, $<td>$ and $</td>$, one can construct simple table structures without any spanning cells. In reality though, one needs at least 28 HTML tokens to describe the most common complex tables observed in real-world documents [21,22], due to a variety of spanning cells definitions in the HTML token vocabulary.

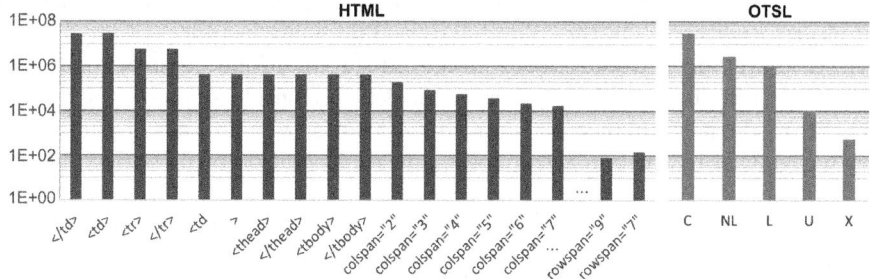

Fig. 2. Frequency of tokens in HTML and OTSL as they appear in PubTabNet.

Obviously, HTML and other general-purpose markup languages were not designed for Im2Seq models. As such, they have some serious drawbacks. First, the token vocabulary needs to be artificially large in order to describe all plausible tabular structures. Since most Im2Seq models use an autoregressive approach, they generate the sequence token by token. Therefore, to reduce inference time, a shorter sequence length is critical. Every table-cell is represented by at least two tokens ($<td>$ and $</td>$). Furthermore, when tokenizing the HTML structure, one needs to explicitly enumerate possible column-spans and row-spans as words. In practice, this ends up requiring 28 different HTML tokens (when including column- and row-spans up to 10 cells) just to describe every table in the PubTabNet dataset. Clearly, not every token is equally represented, as is depicted in Fig. 2. This skewed distribution of tokens in combination with variable token row-length makes it challenging for models to learn the HTML structure.

Additionally, it would be desirable if the representation would easily allow an early detection of invalid sequences on-the-go, before the prediction of the entire table structure is completed. HTML is not well-suited for this purpose as the verification of incomplete sequences is non-trivial or even impossible.

In a valid HTML table, the token sequence must describe a 2D grid of table cells, serialised in row-major ordering, where each row and each column have the same length (while considering row- and column-spans). Furthermore, every opening tag in HTML needs to be matched by a closing tag in a correct hierarchical manner. Since the number of tokens for each table row and column can vary significantly, especially for large tables with many row- and column-spans, it is complex to verify the consistency of predicted structures during sequence generation. Implicitly, this also means that Im2Seq models need to learn these complex syntax rules, simply to deliver valid output.

In practice, we observe two major issues with prediction quality when training Im2Seq models on HTML table structure generation from images. On the one hand, we find that on large tables, the visual attention of the model often starts to drift and is not accurately moving forward cell by cell anymore. This manifests itself in either in an increasing *location drift* for proposed table-cells in later rows on the same column or even complete loss of vertical alignment, as illustrated in Fig. 5. Addressing this with post-processing is partially possible, but clearly undesired. On the other hand, we find many instances of predictions with structural inconsistencies or plain invalid HTML output, as shown in Fig. 6, which are nearly impossible to properly correct. Both problems seriously impact the TSR model performance, since they reflect not only in the task of pure structure recognition but also in the equally crucial recognition or matching of table cell content.

4   Optimised Table Structure Language

To mitigate the issues with HTML in Im2Seq-based TSR models laid out before, we propose here our Optimised Table Structure Language (OTSL). OTSL is designed to express table structure with a minimized vocabulary and a simple set of rules, which are both significantly reduced compared to HTML. At the same time, OTSL enables easy error detection and correction during sequence generation. We further demonstrate how the compact structure representation and minimized sequence length improves prediction accuracy and inference time in the TableFormer architecture.

4.1   Language Definition

In Fig. 3, we illustrate how the OTSL is defined. In essence, the OTSL defines only 5 tokens that directly describe a tabular structure based on an atomic 2D grid.

The OTSL vocabulary is comprised of the following tokens:

– "C" cell - *a new table cell* that either has or does not have cell content
– "L" cell - *left-looking cell*, merging with the left neighbor cell to create a span
– "U" cell - *up-looking cell*, merging with the upper neighbor cell to create a span
– "X" cell - *cross cell*, to merge with both left and upper neighbor cells
– "NL" - *new-line*, switch to the next row.

A notable attribute of OTSL is that it has the capability of achieving lossless conversion to HTML.

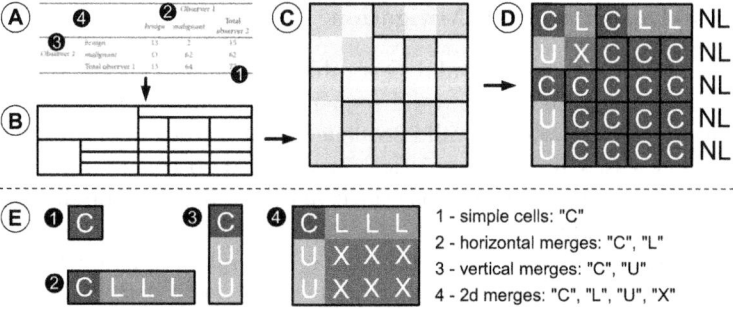

Fig. 3. OTSL description of table structure: A - table example; B - graphical representation of table structure; C - mapping structure on a grid; D - OTSL structure encoding; E - explanation on cell encoding

4.2   Language Syntax

The OTSL representation follows these syntax rules:

1. **Left-looking cell rule**: The left neighbour of an "L" cell must be either another "L" cell or a "C" cell.
2. **Up-looking cell rule**: The upper neighbour of a "U" cell must be either another "U" cell or a "C" cell.
3. **Cross cell rule**:
   The left neighbour of an "X" cell must be either another "X" cell or a "U" cell, and the upper neighbour of an "X" cell must be either another "X" cell or an "L" cell.
4. **First row rule**: Only "L" cells and "C" cells are allowed in the first row.
5. **First column rule**: Only "U" cells and "C" cells are allowed in the first column.
6. **Rectangular rule**: The table representation is always rectangular - all rows must have an equal number of tokens, terminated with "NL" token.

The application of these rules gives OTSL a set of unique properties. First of all, the OTSL enforces a strictly rectangular structure representation, where every new-line token starts a new row. As a consequence, all rows and all columns have exactly the same number of tokens, irrespective of cell spans. Secondly, the OTSL representation is unambiguous: Every table structure is represented in one way. In this representation every table cell corresponds to a "C"-cell token, which in case of spans is always located in the top-left corner of the table cell definition. Third, OTSL syntax rules are only backward-looking. As a consequence, every predicted token can be validated straight during sequence generation by looking at the previously predicted sequence. As such, OTSL can guarantee that every predicted sequence is syntactically valid.

These characteristics can be easily learned by sequence generator networks, as we demonstrate further below. We find strong indications that this pattern reduces significantly the column drift seen in the HTML based models (see Fig. 5).

4.3   Error-Detection and -Mitigation

The design of OTSL allows to validate a table structure easily on an unfinished sequence. The detection of an invalid sequence token is a clear indication of a prediction mistake, however a valid sequence by itself does not guarantee prediction correctness. Different heuristics can be used to correct token errors in an invalid sequence and thus increase the chances for accurate predictions. Such heuristics can be applied either after the prediction of each token, or at the end on the entire predicted sequence. For example a simple heuristic which can correct the predicted OTSL sequence on-the-fly is to verify if the token with the highest prediction confidence invalidates the predicted sequence, and replace it by the token with the next highest confidence until OTSL rules are satisfied.

5   Experiments

To evaluate the impact of OTSL on prediction accuracy and inference times, we conducted a series of experiments based on the TableFormer model (Fig. 4) with two objectives: Firstly we evaluate the prediction quality and performance of OTSL vs. HTML after performing Hyper Parameter Optimization (HPO) on the *canonical* PubTabNet data set. Secondly we pick the best hyper-parameters found in the first step and evaluate how OTSL impacts the performance of TableFormer after training on other publicly available data sets (FinTabNet, PubTables-1M [14]). The ground truth (GT) from all data sets has been converted into OTSL format for this purpose, and will be made publicly available.

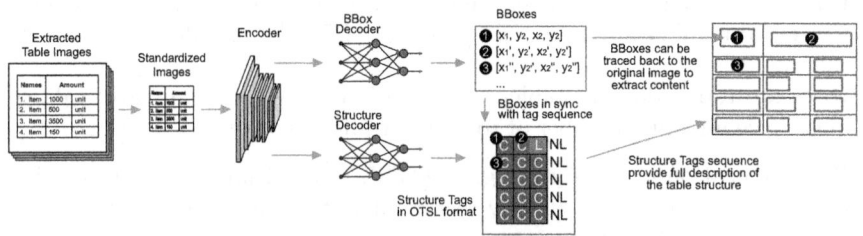

Fig. 4. Architecture sketch of the TableFormer model, which is a representative for the Im2Seq approach.

We rely on standard metrics such as Tree Edit Distance score (TEDs) for table structure prediction, and Mean Average Precision (mAP) with 0.75 Intersection Over Union (IOU) threshold for the bounding-box predictions of table cells. The predicted OTSL structures were converted back to HTML format in order to compute the TED score. Inference timing results for all experiments were obtained from the same machine on a single core with AMD EPYC 7763 CPU @2.45 GHz.

5.1 Hyper Parameter Optimization

We have chosen the PubTabNet data set to perform HPO, since it includes a highly diverse set of tables. Also we report TED scores separately for simple and complex tables (tables with cell spans). Results are presented in Table 1. It is evident that with OTSL, our model achieves the same TED score and slightly better mAP scores in comparison to HTML. However OTSL yields a *2x speed up* in the inference runtime over HTML.

Table 1. HPO performed in OTSL and HTML representation on the same transformer-based TableFormer [9] architecture, trained only on PubTabNet [22]. Effects of reducing the # of layers in encoder and decoder stages of the model show that smaller models trained on OTSL perform better, especially in recognizing complex table structures, and maintain a much higher mAP score than the HTML counterpart.

#enc-layers	#dec-layers	Language	TEDs			mAP (0.75)	Inference time (secs)
			simple	complex	all		
6	6	OTSL	0.965	0.934	0.955	**0.88**	**2.73**
		HTML	0.969	0.927	0.955	0.857	5.39
4	4	OTSL	0.938	0.904	0.927	**0.853**	**1.97**
		HTML	0.952	0.909	**0.938**	0.843	3.77
2	4	OTSL	0.923	0.897	0.915	**0.859**	**1.91**
		HTML	0.945	0.901	**0.931**	0.834	3.81
4	2	OTSL	0.952	0.92	**0.942**	**0.857**	**1.22**
		HTML	0.944	0.903	0.931	0.824	2

5.2 Quantitative Results

We picked the model parameter configuration that produced the best prediction quality (enc = 6, dec = 6, heads = 8) with PubTabNet alone, then independently trained and evaluated it on three publicly available data sets: PubTabNet (395k samples), FinTabNet (113k samples) and PubTables-1M (about 1M samples). Performance results are presented in Table 2. It is clearly evident that the model trained on OTSL outperforms HTML across the board, keeping high TEDs and mAP scores even on difficult financial tables (FinTabNet) that contain sparse and large tables.

Additionally, the results show that OTSL has an advantage over HTML when applied on a bigger data set like PubTables-1M and achieves significantly improved scores. Finally, OTSL achieves faster inference due to fewer decoding steps which is a result of the reduced sequence representation.

Table 2. TSR and cell detection results compared between OTSL and HTML on the PubTabNet [22], FinTabNet [21] and PubTables-1M [14] data sets using Table-Former [9] (with enc = 6, dec = 6, heads = 8).

Data set	Language	TEDs			mAP(0.75)	Inference time (secs)
		simple	complex	all		
PubTabNet	OTSL	0.965	0.934	0.955	**0.88**	**2.73**
	HTML	0.969	0.927	0.955	0.857	5.39
FinTabNet	OTSL	0.955	0.961	**0.959**	**0.862**	**1.85**
	HTML	0.917	0.922	0.92	0.722	3.26
PubTables-1M	OTSL	0.987	0.964	**0.977**	**0.896**	**1.79**
	HTML	0.983	0.944	0.966	0.889	3.26

5.3   Qualitative Results

To illustrate the qualitative differences between OTSL and HTML, Fig. 5 demonstrates less overlap and more accurate bounding boxes with OTSL. In Fig. 6, OTSL proves to be more effective in handling tables with longer token sequences, resulting in even more precise structure prediction and bounding boxes.

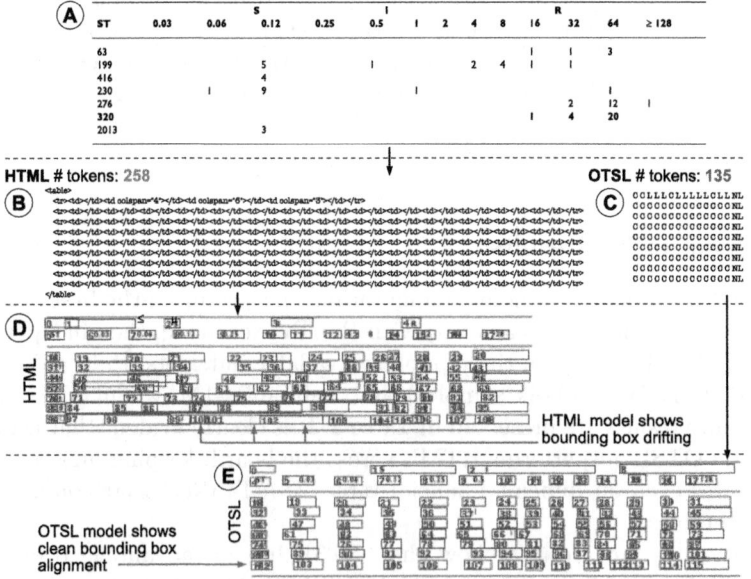

Fig. 5. The OTSL model produces more accurate bounding boxes with less overlap (E) than the HTML model (D), when predicting the structure of a sparse table (A), at twice the inference speed because of shorter sequence length (B), (C). "PMC2807444_006_00.png" PubTabNet.

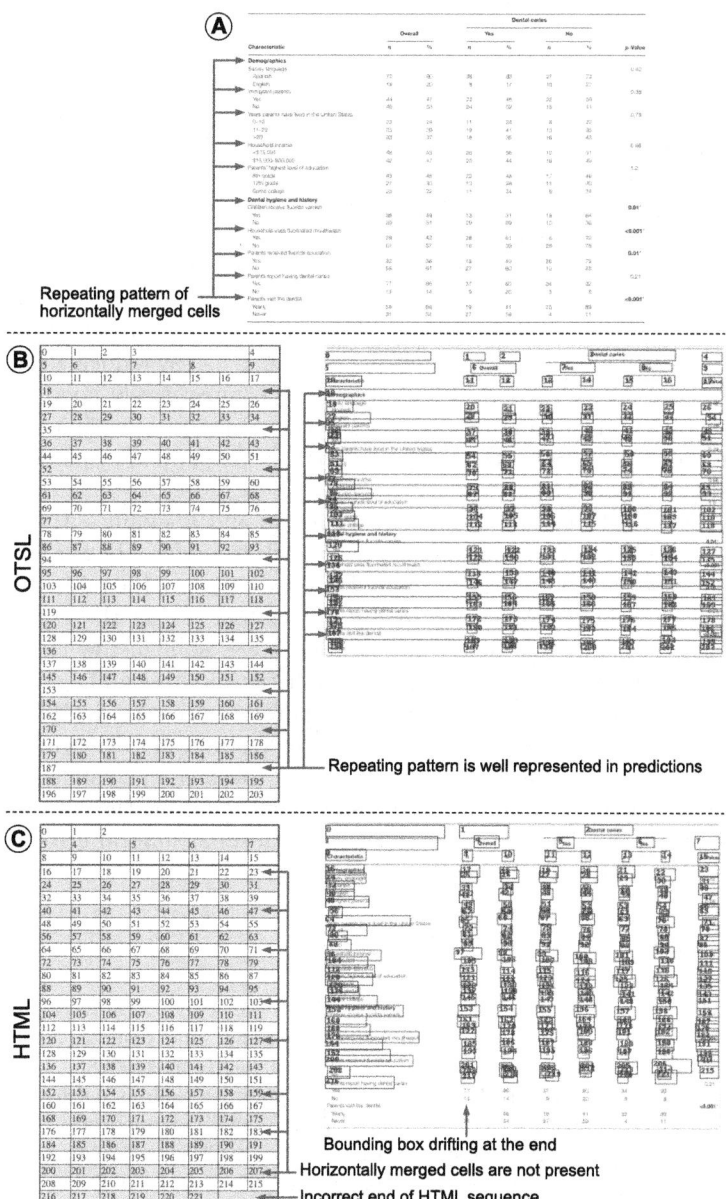

Fig. 6. Visualization of predicted structure and detected bounding boxes on a complex table with many rows. The OTSL model (B) captured repeating pattern of horizontally merged cells from the GT (A), unlike the HTML model (C). The HTML model also didn't complete the HTML sequence correctly and displayed a lot more of drift and overlap of bounding boxes. "PMC5406406_003_01.png" PubTabNet.

6   Conclusion

We demonstrated that representing tables in HTML for the task of table structure recognition with Im2Seq models is ill-suited and has serious limitations. Furthermore, we presented in this paper an Optimized Table Structure Language (OTSL) which, when compared to commonly used general purpose languages, has several key benefits.

First and foremost, given the same network configuration, inference time for a table-structure prediction is about 2 times faster compared to the conventional HTML approach. This is primarily owed to the shorter sequence length of the OTSL representation. Additional performance benefits can be obtained with HPO (hyper parameter optimization). As we demonstrate in our experiments, models trained on OTSL can be significantly smaller, e.g. by reducing the number of encoder and decoder layers, while preserving comparatively good prediction quality. This can further improve inference performance, yielding 5–6 times faster inference speed in OTSL with prediction quality comparable to models trained on HTML (see Table 1).

Secondly, OTSL has more inherent structure and a significantly restricted vocabulary size. This allows autoregressive models to perform better in the TED metric, but especially with regards to prediction accuracy of the table-cell bounding boxes (see Table 2). As shown in Fig. 5, we observe that the OTSL drastically reduces the drift for table cell bounding boxes at high row count and in sparse tables. This leads to more accurate predictions and a significant reduction in post-processing complexity, which is an undesired necessity in HTML-based Im2Seq models. Significant novelty lies in OTSL syntactical rules, which are few, simple and always backwards looking. Each new token can be validated only by analyzing the sequence of previous tokens, without requiring the entire sequence to detect mistakes. This in return allows to perform structural error detection and correction on-the-fly during sequence generation.

References

1. Auer, C., Dolfi, M., Carvalho, A., Ramis, C.B., Staar, P.W.J.: Delivering document conversion as a cloud service with high throughput and responsiveness. CoRR abs/2206.00785 (2022). https://doi.org/10.48550/arXiv.2206.00785
2. Chen, B., Peng, D., Zhang, J., Ren, Y., Jin, L.: Complex table structure recognition in the wild using transformer and identity matrix-based augmentation. In: Porwal, U., Fornés, A., Shafait, F. (eds.) ICFHR 2022. LNCS, vol. 13639, pp. 545–561. Springer, Cham (2022). https://doi.org/10.1007/978-3-031-21648-0_37
3. Chi, Z., Huang, H., Xu, H.D., Yu, H., Yin, W., Mao, X.L.: Complicated table structure recognition. arXiv preprint arXiv:1908.04729 (2019)
4. Deng, Y., Rosenberg, D., Mann, G.: Challenges in end-to-end neural scientific table recognition. In: 2019 International Conference on Document Analysis and Recognition (ICDAR), pp. 894–901. IEEE (2019)
5. Kayal, P., Anand, M., Desai, H., Singh, M.: Tables to LaTeX: structure and content extraction from scientific tables. Int. J. Doc. Anal. Recognit. (IJDAR) **26**, 1–10 (2022)

6. Lee, E., et al.: Table structure recognition based on grid shape graph. In: 2022 Asia-Pacific Signal and Information Processing Association Annual Summit and Conference (APSIPA ASC), pp. 1868–1873. IEEE (2022)

7. Li, M., Cui, L., Huang, S., Wei, F., Zhou, M., Li, Z.: Tablebank: a benchmark dataset for table detection and recognition (2019)

8. Livathinos, N., et al.: Robust pdf document conversion using recurrent neural networks. In: Proceedings of the AAAI Conference on Artificial Intelligence, vol. 35, no. 17, pp. 15137–15145 (2021). http://ojs.aaai.org/index.php/AAAI/article/view/17777

9. Nassar, A., Livathinos, N., Lysak, M., Staar, P.: Tableformer: table structure understanding with transformers. In: Proceedings of the IEEE/CVF Conference on Computer Vision and Pattern Recognition (CVPR), pp. 4614–4623 (2022)

10. Pfitzmann, B., Auer, C., Dolfi, M., Nassar, A.S., Staar, P.W.J.: Doclaynet: a large human-annotated dataset for document-layout segmentation. In: Zhang, A., Rangwala, H. (eds.) KDD 2022: The 28th ACM SIGKDD Conference on Knowledge Discovery and Data Mining, Washington, DC, USA, 14–18 August 2022, pp. 3743–3751. ACM (2022). https://doi.org/10.1145/3534678.3539043

11. Prasad, D., Gadpal, A., Kapadni, K., Visave, M., Sultanpure, K.: Cascadetabnet: an approach for end to end table detection and structure recognition from image-based documents. In: Proceedings of the IEEE/CVF Conference on Computer Vision and Pattern Recognition Workshops, pp. 572–573 (2020)

12. Schreiber, S., Agne, S., Wolf, I., Dengel, A., Ahmed, S.: Deepdesrt: deep learning for detection and structure recognition of tables in document images. In: 2017 14th IAPR International Conference on Document Analysis and Recognition (ICDAR), vol. 1, pp. 1162–1167. IEEE (2017)

13. Siddiqui, S.A., Fateh, I.A., Rizvi, S.T.R., Dengel, A., Ahmed, S.: Deeptabstr: deep learning based table structure recognition. In: 2019 International Conference on Document Analysis and Recognition (ICDAR), pp. 1403–1409 (2019). https://doi.org/10.1109/ICDAR.2019.00226

14. Smock, B., Pesala, R., Abraham, R.: PubTables-1M: towards comprehensive table extraction from unstructured documents. In: Proceedings of the IEEE/CVF Conference on Computer Vision and Pattern Recognition (CVPR), pp. 4634–4642 (2022)

15. Staar, P.W.J., Dolfi, M., Auer, C., Bekas, C.: Corpus conversion service: a machine learning platform to ingest documents at scale. In: Proceedings of the 24th ACM SIGKDD International Conference on Knowledge Discovery & Data Mining, KDD 2018, pp. 774–782. Association for Computing Machinery, New York (2018). https://doi.org/10.1145/3219819.3219834

16. Wang, X.: Tabular abstraction, editing, and formatting. Ph.D. thesis, CAN (1996). aAINN09397

17. Xue, W., Li, Q., Tao, D.: Res2tim: reconstruct syntactic structures from table images. In: 2019 International Conference on Document Analysis and Recognition (ICDAR), pp. 749–755. IEEE (2019)

18. Xue, W., Yu, B., Wang, W., Tao, D., Li, Q.: Tgrnet: a table graph reconstruction network for table structure recognition. In: Proceedings of the IEEE/CVF International Conference on Computer Vision, pp. 1295–1304 (2021)

19. Ye, J., et al.: PingAN-VCGroup's solution for ICDAR 2021 competition on scientific literature parsing Task B: table recognition to HTML (2021). https://doi.org/10.48550/ARXIV.2105.01848. http://arxiv.org/abs/2105.01848

20. Zhang, Z., Zhang, J., Du, J., Wang, F.: Split, embed and merge: an accurate table structure recognizer. Pattern Recogn. **126**, 108565 (2022)

21. Zheng, X., Burdick, D., Popa, L., Zhong, X., Wang, N.X.R.: Global table extractor (GTE): a framework for joint table identification and cell structure recognition using visual context. In: 2021 IEEE Winter Conference on Applications of Computer Vision (WACV), pp. 697–706 (2021). https://doi.org/10.1109/WACV48630.2021.00074
22. Zhong, X., ShafieiBavani, E., Jimeno Yepes, A.: Image-based table recognition: data, model, and evaluation. In: Vedaldi, A., Bischof, H., Brox, T., Frahm, J.-M. (eds.) ECCV 2020. LNCS, vol. 12366, pp. 564–580. Springer, Cham (2020). https://doi.org/10.1007/978-3-030-58589-1_34
23. Zhong, X., Tang, J., Yepes, A.J.: Publaynet: largest dataset ever for document layout analysis. In: 2019 International Conference on Document Analysis and Recognition (ICDAR), pp. 1015–1022. IEEE (2019)

Towards End-to-End Semi-Supervised Table Detection with Deformable Transformer

Tahira Shehzadi[1,2,4(✉)] , Khurram Azeem Hashmi[1,2,4] , Didier Stricker[1,2,4],
Marcus Liwicki[3], and Muhammad Zeshan Afzal[1,2,4]

[1] Department of Computer Science, Technical University of Kaiserslautern, 67663
Kaiserslautern, Germany
[2] Mindgarage, Technical University of Kaiserslautern, 67663 Kaiserslautern, Germany
[3] Department of Computer Science, Luleå University of Technology, 971 87 Luleå,
Sweden
[4] German Research Institute for Artificial Intelligence (DFKI), 67663 Kaiserslautern,
Germany
tahira.shehzadi@dfki.de

Abstract. Table detection is the task of classifying and localizing table objects within document images. With the recent development in deep learning methods, we observe remarkable success in table detection. However, a significant amount of labeled data is required to train these models effectively. Many semi-supervised approaches are introduced to mitigate the need for a substantial amount of label data. These approaches use CNN-based detectors that rely on anchor proposals and post-processing stages such as NMS. To tackle these limitations, this paper presents a novel end-to-end semi-supervised table detection method that employs the deformable transformer for detecting table objects. We evaluate our semi-supervised method on PubLayNet, DocBank, ICADR-19 and TableBank datasets, and it achieves superior performance compared to previous methods. It outperforms the fully supervised method (Deformable transformer) by +3.4 points on 10% labels of TableBank-both dataset and the previous CNN-based semi-supervised approach (Soft Teacher) by +1.8 points on 10% labels of PubLayNet dataset. We hope this work opens new possibilities towards semi-supervised and unsupervised table detection methods.

Keywords: Semi-Supervised Learning · Deformable Transformer · Table Analysis · Table Detection

1 Introduction

A visual summary is the main aspect of different applications in document analysis, such as recognizing graphical components in the visualization pipeline and summarizing the content of a document. As a result, localizing and detecting

G. A. Fink et al. (Eds.): ICDAR 2023, LNCS 14188, pp. 51–76, 2023.
https://doi.org/10.1007/978-3-031-41679-8_4

graphical items such as tables will be an important action in the analysis and summary of the document. Due to the increase in the number of documents, manually retrieving the table data is no longer practical. Automated processes offer efficient, reliable, and successful solutions for manual tasks. Previously, optical character recognition [1,2] and rule-based [3–5] table detection approaches were used to identify and locate them. Then, some automated methods [6–8] have been suggested to detect tables. However, these approaches are often rule-based because the documents have a set structure or dimension [9]. Moreover, they cannot generalize to a new table structure, such as borderless tables. Later on, deep learning methods were used by researchers to identify them [10–13], and shows that machine-learning approaches are more effective than traditional methods [14].

Deep learning approaches [15–20] do not rely on rules and can accurately generalize the problem. However, deep learning models take a considerable quantity of labeled data for training. These supervised methods achieve impressive results on public benchmarks, and their performance cannot be translated into industrial applications unless similar large-scale annotated datasets exist in that domain. It is potentially error-prone and time-consuming to generate this data manually or via other pre-processing approaches. Therefore, it is important to develop a semi-supervised approach due to concerns about the availability of labeled training data, which shifts the problem from a supervised to a semi-supervised setting. Recently, semi-supervised learning-based methods are introduced in computer vision containing two detectors. The first detector provides pseudo labels for unlabeled data. The second detector trains using pseudo labels generated by the first detector and a small percentage of label data and provides final predictions. Both detectors update each other during training. This approach has been described in several works, including [21–24]. In most cases, the first detector is not strong enough, which can negatively impact the pseudo-labeling process. Moreover, previous semi-supervised approaches used CNN-based networks [11] that depend on anchors to generate region proposals and post-processing stages such as Non-Maximal suppression (NMS) to reduce the number of overlapping predictions.

To address these limitations, this paper proposes a semi-supervised table detection approach that employs the deformable transformer [25]. It generates pseudo-labels for unlabeled data and then trains the detector using them and a small quantity of label data in each iteration. This approach aims to improve the pseudo-label generation procedure by iteratively refining the pseudo-labels and the detector. It involves training in two modules. The teacher module contains a pseudo-labeling framework. The student module is the final detection network that uses pseudo-labels and a small quantity of label data. The teacher module is simply an Exponential Moving-Average (EMA) of the student module, which ensures that the pseudo-label generation and detection modules are constantly updating each other. Unlike other pseudo-labeling methods, we propose the idea of employing the deformable transformer that allows completing the pseudo-labeling process without needing object proposals and post-processing steps as

Non-maximal suppression (NMS). Another benefit is having a dynamic effective receptive field to adapt fot tables of different sizes and scales in the input image. This allows the network to effectively detect tables of varying sizes and orientations, making it more robust and versatile. Additionally, this framework has a reinforcing effect, providing that the Teacher model consistently monitors the Student model. In this paper, we show through empirical evidence that this semi-supervised table detection approach that uses a deformable transformer can produce results comparable to CNN-based approaches without needing object proposals and post-processing steps such as Non-maximal suppression (NMS). In summary, the main contributions of the paper are as follows:

- We present an end-to-end semi-supervised table detection method that employs the deformable transformer and allows completing the pseudo-labeling process without needing object proposals and post-processing steps such as Non-maximal suppression (NMS).
- We formulate the problem of table detection as an object detection problem and leverage the potential of deformable detection transformer for this task. To the best of our knowledge, this work is the first that exploits the transformer-based method in a semi-supervised setting.
- We perform an exhaustive evaluation on four different datasets, PubLayNet, DocBank, ICDAR-19 and TableBank, and produce results comparable to CNN-based semi-supervised approaches without needing object proposals process and post-processing steps such as NMS.

2   Related Work

Table detection is an essential task for document image analysis. Many researchers have proposed different approaches for detecting tables containing arbitrary structures in document images. Previously, most presented approaches used custom rules or relied on extra meta-data input to deal with table detection tasks [26–29]. Recently, researchers employed statistical methods [30] and deep learning approaches to make the table detection systems more generalizable [15, 31–33]. This section gives a detailed summary of these techniques and an overview of the CNN-based semi-supervised object detection methods.

2.1   Rule-Based Approaches

To the best of our knowledge, Itonori et al. [26] presented a table detection approach for the first time on document images. This method represents the table as a text block that uses specified rules. Later, [28] introduced a table detection approach that works on horizontal and vertical lines. Pyreddy et al. [34] proposed a procedure that extracts tabular regions from the text using custom heuristics. Pivk et al. [35] presented a system that transforms HTML format table documents into logical forms. It introduces an appropriate tabular layout employed for extracting tables. Hu et al. [36] presented a table detection approach that relies on white regions and vertically connected elements in document images. Readers can find a complete overview of these rule-based methods in [3–5,37,38]. Though rule-based approaches perform fine on document images with matching table formats, these methods can not provide generic solutions. Therefore, systems with more generalizable abilities are needed to solve table detection tasks on document data.

2.2 Learning-Based Approaches

Cesarini et al. [39] presented a supervised learning system for detecting table objects in document images. It converts a document image into an MXY tree model and labels the blocks as tables confined in horizontal and vertical lines. Hidden Markov Models [40,41] and the SVM classifier with traditional heuristics [42] are applied to document images for table detection. Though these machine learning approaches performed better than ruled-based approaches on documents, these methods need additional information, such as ruling lines. Deep Learning-based approaches outperformed traditional approaches in accuracy and efficiency. These methods are categorised into object detection, semantic segmentation, and bottom-up approaches.

Semantic segmentation-based Approaches. These approaches [43–46] consider table detection a segmentation task and apply available semantic segmentation networks to generate segmentation masks on the pixel level and then combine table regions to provide final table detection. These methods performed better than traditional approaches on several benchmark datasets [47–53]. Yang et al. [43] presented a fully convolutional network (FCN) [54] for page object segmentation, which combines linguistic and visual features to enhance segmentation results for table and other page object detection. He et al. [44] presented a multi-scale FCN that provides segmentation masks table/text areas and their related contours and then refined to get final table blocks.

Bottom-up Approaches. These approaches consider table detection as a graph-labeling task and define graph nodes as page objects and graph edges connection between page objects. Li et al. [55] extracted line areas using the classic layout analysis approach, then used two CNN-CRF networks to categorise them into four categories: text, figure, formula and table and then provided a prediction of the corresponding cluster for pair of line areas. Holecek et al. [56] and Riba et al. [57] considered text areas as nodes, formed a graph to determine the design per document and then employed graph-neural networks for node-edge classification. These approaches rely on specific assumptions, such as text line boxes as an extra input.

Object Detection-based Approaches. Detecting tables from document images can be represented as an object detection task, with table objects treated as natural objects. Hao et al. [58] and Yi et al. [59] applied R-CNN for detecting tables, but the performance of these approaches still relies on heuristic rules as in previous methods. Later, more efficient single-stage object detectors like

RetinaNet [60] and YOLO [61] and two-stage object detectors like Fast R-CNN [10], Faster R-CNN [11], Mask R-CNN [62], and Cascade Mask R-CNN [63] were applied for other document objects such as figures and formulas detection in document images [9,15–17,64–69]. Furthermore, [65,68,70] applied different image transformation approaches, such as coloration and dilation, to improve the results further. Siddiqui et al. [31] combined deformable-convolution and RoI-Pooling [71] into Faster R-CNN to provide a more efficient network for geometrical modifications. Agarwal et al. [69] used a composite network [72] as a backbone with deformable convolution to increase the performance of two-stage Cascade R-CNN. These CNN-based object detectors have a few heuristic stages, like proposals generating step and post-processing steps such as non-maximal suppression (NMS). Our semi-supervised approach considers detection a set prediction task, eliminating the anchor generation and post-processing stages such as NMS and providing a simpler and more efficient detection pipeline.

2.3 Semi-supervised Object Detection

Semi-supervised learning approaches in object detection are divided into two types: consistency-based approaches [22,73] and pseudo-label generation-based approaches [21,74–79]. Our method falls into the pseudo-label type. Previous work [74,75] combined prediction results from varied data augmentation techniques to produce pseudo-labels for unlabeled data, while [76] trained a SelectiveNet to generate the pseudo-labels. In [76], a box from unlabeled data was placed onto labeled data and evaluated localization consistency on the labeled images. However, this method requires a very complex detection procedure due to the modification of the image. STAC [79] presented to perform strong augmentation on the data for pseudo-label generation and weak augmentation for model training. We propose an end-to-end semi-supervised table detection method that employs the deformable transformer. Similar to other pseudo-label generation approaches [21,74–76,79], it follows a multi-level training mechanism. It effectively avoids the need for anchors generation stage and post-processing steps such as Non-Maximal suppression (NMS).

3 Methodology

First, we revisit Deformable DETR, a modern transformer-based object detector, in Sect. 3.1. Later, we explain the proposed semi-supervised learning mechanism and its pseudo-label generation module in Sects. 3.2.

3.1 Revisiting Deformable DETR

Deformable DETR [25] contains a Transformer encoder-decoder network that considers object detection as a set-predictions task. It uses Hungarian loss and avoids overlapped predictions for ground-truth bounding boxes through bipartite matching. It eliminates the need for hand-crafted elements such as anchors and post-processing stages such as Non-maximal suppression (NMS) used in CNN-based object detectors. Deformable DETR is an extension of the DETR [80] architecture that addresses some of the limitations of DETR, such as slow training convergence and poor performance on small objects. Deformable DETR introduces deformable convolutions into the architecture, which allows for more flexible modeling of object shapes and better handling of objects of varying scales. This can lead to improved performance, particularly on small objects, and faster convergence during training. Here, we provide an overview of the encoder-decoder network, Multi-scale Feature processing and attention mechanism of deformable DETR. Figure 1 shows all modules of the deformable transformer, including multi-scale features and encoder-decoder network.

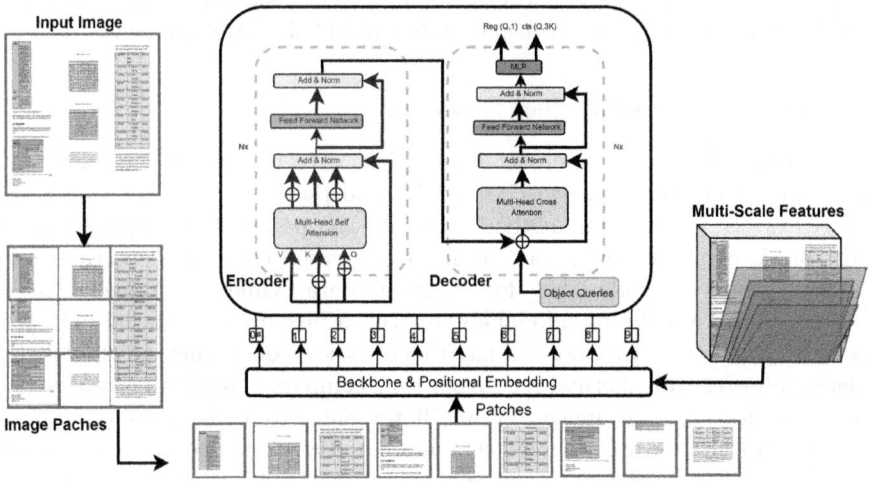

Fig. 1. Illustration of the deformable transformer employed in semi-supervised table detection method. We split the input image into small equal-sized patches, add position embeddings, and feed the resulting patches along with input multi-scale features to the transformer encoder. In the decoder, We use object queries as reference points and provide bounding boxes predictions and class labels as the final output.

Transformer Encoder. The CNN backbone (ResNet-50) extracts the input feature maps $f_m \in \mathbb{R}^{h_i \times w_i \times c_i}$. The spatial dimensional feature maps are converted into one-dimensional $z_m \in \mathbb{R}^{h_i \times w_i \times d_1}$ feature maps as the transformer encoder network takes input as a sequence. This one-dimensional vector is fed as input along with positional embeddings [81,82] to the transformer encoder network, which further transforms them into features for object queries. Every layer of the encoder module contains an attention network and a feed-forward network (FFN) where query and key values are the pixels of feature maps. Readers can refer to [83] for a detailed explanation of transformer.

Transformer Decoder. The decoder network takes the output of the encoder features and N number of object queries as input. It contains two attention types self-attention and cross-attention. The self-attention module finds the connection between object queries. Here both key and query matrics contain object queries. The cross-attention module extracts feature using object queries from the input feature map. Here key matrix contains the feature maps provided by the encoder module, and the query matrix is the object queries fed as input to the decoder. After the attention modules, feed-forward networks (FFN) and linear projection layers are added as the prediction head. The linear projection layer predicts class labels, while FFN provides final bounding-box coordinate values.

Deformable Attention Module. The attention module in the DETR network considers all spatial locations of the input feature map, which makes the training convergence slower. However, a deformable DETR can solve this issue using the deformable convolution-based [71, 84] attention network and multiscale input features [85, 86]. It considers only a few sample pixels near a reference pixel, whatever the size of input features, as illustrated in Fig. 2. The query matrix takes only a small set of keys, which resolves the slow training convergence issue of DETR. Readers can refer to [25] for a detailed explanation of Deformable DETR.

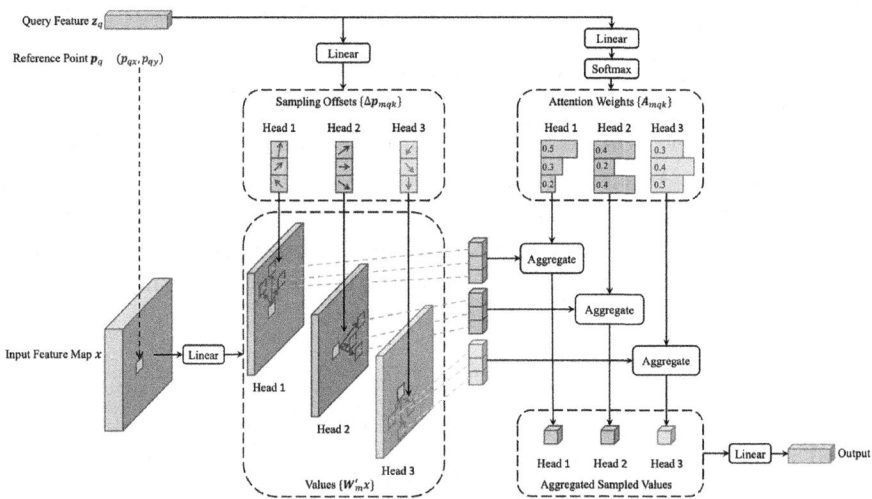

Fig. 2. Deformable Attention network. It considers only a few sample pixels near a reference pixel, whatever the size of input features. The query matrix takes only a small set of keys, which resolves the slow training convergence issue of DETR. (image from [25]).

3.2   Semi-Supervised Deformable DETR

In this subsection, we describe the learning mechanism of our proposed semi-supervised approach that employs the Deformable transformer and then explain the pseudo-labeling strategy. Semi-supervised Deformable-DETR is a unified learning approach that uses fully labeled and unlabeled data for object detection. It contains two modules a student module and a teacher module. The training data has two data types label data and unlabeled data. The student module takes both labeled and unlabeled images as input where strong augmentation is applied on unlabeled data while both (strong and weak augmentation) is applied on label data. The student module is trained using detection losses of labeled and unlabeled data through pseudo-boxes. The unlabeled data contains two groups of pseudo boxes for providing class labels and their bounding boxes. The teacher module only takes unlabeled images as input after applying weak augmentation. Figure 3 presents a summary of proposed pipeline. The teacher module feeds prediction results to the pseudo-labeling framework to get pseudo-labels. Then, the student module uses these pseudo-labels for supervised training. Here, weak augmentation on unlabeled data is used for the teacher module to generate more precise pseudo-labels. Strong augmentation on unlabeled data is used for the student module to have more challenging learning. The student module also takes a small percentage of labeled images with strong and weak augmentation as input. The student module s_m is optimized with the total loss as follows:

$$\mathcal{L}^{s_m} = \sum_n \mathcal{L}(x_j^{l,s_a}, y_j^{l,s_a}) + \mathcal{L}(x_j^{l,w_a}, y_j^{l,w_a}) + \sum_n \mathcal{L}(x_j^{u,s_a}, y_j^{t_m}) \tag{1}$$

where s_a represents strong augmentation, w_a represents weak augmentation. x_j^{l,s_a} is the strong augmented input image and its corresponding label is y_j^{l,s_a}. The term x_j^{l,w_a} is the weak augmented input image and its corresponding label is y_j^{l,w_a}. For the labeled images, strong and weak augmentations are also applied for learning, and are fed to the student module. The term x_j^{u,s_a} represents unlabeled strong augmented image fed to student module and the term $y_j^{t_m}$ is the pseudo-label from teacher module. Here, $\mathcal{L}$ is the weighted sum of classification (class labels) and regression (bounding box) loss as follows:

$$\mathcal{L} = \alpha_1 \mathcal{L}^{reg} + \alpha_2 \mathcal{L}^{cls} \tag{2}$$

where α_1 and α_2 are the weight values, the teacher-student modules are initialized randomly at the start of training. During training, the student module continuously updates the teacher module with an Exponential Moving-Average (EMA) [87] strategy. Pseudo-label generation for image classification tasks is easy, considering probability distribution as Pseudo-labels. In contrast, object detection tasks are more complicated as an image may include numerous objects, and annotation contains object location and class label. The CNN-based object detectors use anchors as object proposals and remove redundant boxes by post-processing steps such as non-maximal suppression (NMS). In contrast, transformers use attention mechanisms and object queries. Figure 4 shows sample points and attention weights from multi-scale deformable attention feature maps for both student and teacher networks. Its training complexity is $O(N_q c_i^2 + min(h_i w_i c_i^2, N_q k c_i^2) + 5N_q k c_i + 3N_q c_i p_s k)$. This takes into account the computation of the sampling coordinate offsets and attention weights, as well as the bilinear interpolation and weighted sum in the attention mechanism. N_q is the number of query elements, c_i is the channel dimension, k is the kernel size, p_s is the number of sampling points, and $h_i w_i$ is the height and width of the feature map. In our experiments, $p_s = 8, k \leq 4$ and $c_i = 256$ by default, thus $5k + 3p_s k < c_i$ and the complexity is of $O(2N_q c_i^2 + min(h_i w_i c_i^2, N_q k c_i^2))$. When used in the DETR encoder with $N_q = h_i w_i$, the complexity of the deformable attention module is $O(h_i w_i c_i^2)$, which scales linearly with the spatial size. When used in the DETR decoder with $N_q = N$ (the number of object queries), the complexity becomes $O(N k c_i^2)$, which is independent of the spatial size as attention is focused on the object queries.

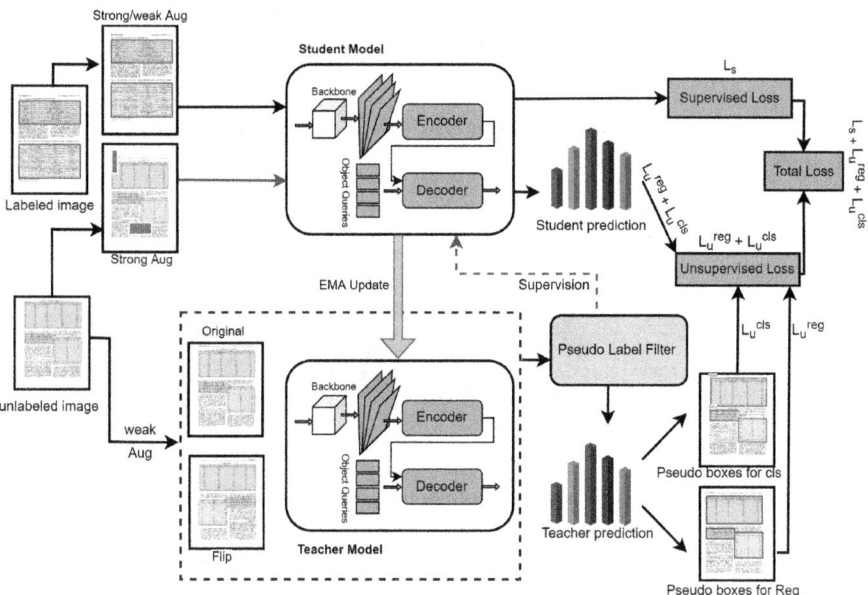

Fig. 3. Our proposed semi-supervised approach that employs Deformable transformer [25]. (1) The training data has two data types label data and unlabeled data. (2) It contains two modules a student module and a teacher module. (3) The teacher module only takes unlabeled images as input after applying weak augmentation. (4) After applying strong augmentation on unlabeled data type, the student module takes both labeled and unlabeled images as input. (5) During training, the student module continuously updates the teacher module with an Exponential Moving-Average (EMA) [87] strategy.

Training. The semi-supervised network is trained in two steps: a) train the student module independently on labeled data and generate pseudo-labels by teacher module; b) combine training of both modules to provide final prediction results.

Pseudo-Labeling Framework. We used a simple framework to provide pseudo-labels for unlabeled data at the output of the teacher module, as applied in SSOD [88]. Usually, object detectors give confidence score vector $s_k \in [0,1]^{C_i}$ for every provided bounding box b_k. A simple approach to provide pseudo-labels is to just thresholding these scores. In a simple pseudo-labeling filter, pseudo-labels can be formed by providing a threshold to the confidence value $s_k^{c_k}$ of the ground-truth class c_k. If the prediction value is not greater than the confidence value for a ground-truth class, the highest prediction value is considered the pseudo-label. Inspired by DETR [80], we develop the pseudo-label assignment task as a bipartite matching task between the teacher module predictions and the generated semi-labels. Specifically, the permutation of K elements is as follows:

$$\hat{\sigma} = \arg\min_{\sigma \in N} \sum_k^{N_i} \mathcal{L}_{match}(y_k, \hat{y}(k)), \tag{3}$$

where $\mathcal{L}_{match}(y_k, \hat{y}(k))$ is the match-cost between teacher labels and ground-truth semi-labels as follows:

$$\mathcal{L}_{match}(y_k, \hat{y}(k)) = -\mathbb{1}_{\{c_k \neq \phi\}} \hat{p}_{\sigma(k)}(c_k) + \mathbb{1}_{\{c_k \neq \phi\}} \mathcal{L}_{bbox}(b_k, \hat{b}_{\hat{\sigma}}(k)) \tag{4}$$

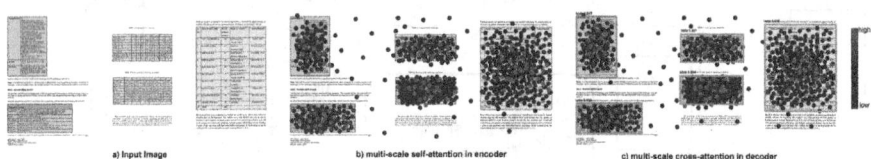

Fig. 4. Visualization of the sample points and attention weights from multi-scale deformable attention feature maps. Each sample point is denoted as a circle whose color represents its relative attention weight value. The reference points are the object queries taken as input in the encoder, represented by the green plus sign. In the decoder, the final bounding boxes are represented as green rectangles, and the class label and its confidence value are shown on the upper side in black text. (Color figure online)

The Pseudo-Labeling framework is applied to the predictions of teacher module $\hat{y}(k)$ where $\hat{y}(k) = \{\hat{y}^{class}, \hat{y}^{bbox}\}$ is the prediction, with $\hat{y}^{class}$ and $\hat{y}^{bbox}$ represent the class and box values, respectively. Here, $\hat{y}^{cls} = [v_1, ..., v_N]^T \in \mathbb{R}^{N \times C_i}$ and $\hat{y}^{box} = [\hat{b}_1, ..., \hat{b}_N]^T \in \mathbb{R}^{N \times 4}$, where v_N is the output vector (before the softmax) , $\hat{b}_N$ the related bounding-box prediction, and N is the object queries provided as input to the transformer decoder. y_k represents pseudo-labels generated from confidence-score. The optimal selection is allowed with the Hungarian match mechanism [80,89], giving pseudo-labels $\{(b_k, c_k)\}$. This approach to select matching between the teacher module's prediction and semi-labels generated by providing threshold works in the same way as the heuristic selection rules used for matching proposals [11] or anchors [85] with ground-truth objects in CNN-based object detectors. The main difference is that it determines one-to-one matching without duplicates. The second stage calculates the loss function, the Hungarian loss for all pair matching in the last stage. We define the loss similar to the previous object detector's losses as a linear combination of a negative log-likelihood for class label and a bounding box as follows:

$$\mathcal{L}_H(y, \hat{y}) = \sum_{i=1}^{N}[-log\hat{p}_{\hat{\sigma}(k)}(c_k) + \mathbb{1}_{\{c_k \neq \phi\}}\mathcal{L}_{box}(b_k, \hat{b}_{\hat{\sigma}}(k))] \quad (5)$$

Here, $b_k \in \mathbb{R}^4$ is the pseudo-bounding box, and c_k is the pseudo-class label. $\hat{\sigma}$ is the matching determined in the previous stage. In training, we reduce the weight of log probability by ten times when c_k for class imbalance. This mechanism is similar to the Faster R-CNN training strategy to balance proposals by subsampling [11].

4   Experimental Setup

4.1   Datasets

TableBank: TableBank [52] is the second-largest dataset in the document analysis domain for the table recognition problem. The dataset has 417,000 document images annotated through the arXiv database crawling procedure. The dataset features tables from three categories of document images: LaTeX images (253,817), Word images (163,417), and a combination of both (417,234). It also includes a dataset for recognizing the structures of the table. In our experiment, We only used the dataset for table detection from TableBank.

PubLayNet: PubLayNet [48] is a large public dataset with 335,703 images in the training set, 11,240 in the validation set, and 11,405 in the test set. It includes annotations such as polygonal segmentation and bounding boxes of figures, lists titles, tables, and text of images from research papers and articles. The dataset was evaluated using the coco analytic technique [90]. In our experiment, we only used 102,514 of the 86,460 table annotations.

DocBank: DocBank [91] is a large dataset of over 5,000 annotated document images from various sources designed to train and evaluate tasks such as text classification, entity recognition, and relation extraction. It includes annotations of title, author name, affiliation, abstract, body text, etc.

ICDAR-19: The competition for Table Detection and Recognition (cTDaR) [47] is organized at ICDAR in 2019. For the table detection task (TRACK A), two new datasets (modern and historical) are introduced in the competition. For direct comparison against the prior state-of-the-art [68], we provide results on the modern datasets with an IoU threshold ranging from 0.5-0.9.

4.2   Evaluation Criteria

We use some evaluation metrics to analyze the performance of our semi-supervised table detection approach that employs the deformable transformer. This section defines the employed evaluation metrics as precision, Recall, and F1-score. The Precision [92] is the fraction of actual instances as True Positives among the predicted instances as False Positives and True Positives). The Recall [92] is the fraction of actual instances as True Positives that were retrieved (True Positives + False Negatives). The F1-score [92] is the harmonic mean of Precision and Recall. We compute the intersection over union(IoU) by performing the intersection divided by the union for the region of the ground-truth box A_g and the formed bounding box A_p.

$$IoU = \frac{area(A_g \cap A_p)}{area(A_g \cup A_p)} \tag{6}$$

IoU estimates that either a detected table object is a false positive or a true positive. We find the average precision(AP) by a precision-recall (PR) curve following the context of MS COCO [90] evaluation. It is the area under the PR curve, calculated using the following equation:

$$AP = \sum_{k=1}^{N}(Re_{k+1} - Re_k)P_{intr}(Re_{k+1}) \tag{7}$$

where Re1, Re2, . . . , Re_k represent the recall parameter. The mean average precision (mAP) is often used to evaluate the performance of detection methods. It is calculated by taking the mean of average precision for all classes in a dataset. The mAP can be affected by changes in the performance of individual classes due to class mapping, which is a limitation of this metric. We set the intersection over union (IoU) threshold values at 0.5 and 0.95. The mAP is calculated as follows:

$$mAP = \frac{1}{S}\sum_{s=1}^{S}AP_s \tag{8}$$

where S represents total classes.

4.3   Implementation Details

We use the Deformable DETR [25] with a ResNet-50 [93] backbone pre-trained on the ImageNet [94] dataset as our detection framework for evaluating the usefulness of the semi-supervised approach. We perform training on PubLayNet, ICDAR-19, DocBank and all three splits of the TableBank dataset. We use 10%, 30% and 50% of labeled data and the rest as unlabeled data. The threshold value for pseudo-labeling is set at 0.7. We set the training epochs to 150 for all experiments with the learning rate reduced by a factor of 0.1 at the 120th epoch. We follow [25,88] to apply strong augmentation as horizontal flip, resize, remove patches, crop, grayscale and Gaussian blur. We use horizontal flipping to apply weak augmentation. The value N for the number of queries to the input of the decoder of Deformable DETR is set to 30 as it gives the best results. Unless otherwise specified, we evaluated the results using the mAP (AP50:95) metrics. All models are trained with a batch size of 16, using the same hyperparameters as Deformable DETR [25]. The weight α_1 is 2 and α_2 is 5 to balance the classification loss (L_{cls}) and regression loss (L_{box}). To make the training faster, we set the height and width of the input image to 600 pixels. We employ the standard size of 800 pixels for comparison with other approaches.

5   Results and Discussion

5.1   TableBank

In this subsection, we provide the experimental results on all splits of the TableBank dataset on different percentages of label data. We also compare the transformer-based semi-supervised approach with previous deep learning-based supervised and semi-supervised approaches. Furthermore, we give results on 10% TableBank-both data split for all IoU threshold values. Table 1 provides the results of semi-supervised approach that employs deformable transformer for TableBank-latex, TableBank-word, and TableBank-both data splits on 10%, 30% and 50% label data and the rest as unlabeled data. It shows that the TableBank-both data split has the highest AP_{50} value of 95.8%, TableBank-word has 93.5%, and TableBank-both has 92.5% at 10% label data.

Table 1. Performance of the semi-supervised approach that employs deformable transformer for TableBank-latex, TableBank-word, and TableBank-both data splits on different percentages of label data. Here, mAP represents mean AP at the IoU threshold range of (50:95), AP_{50} indicates AP at the IoU threshold of 0.5, and AP_{75} denotes AP at the IoU threshold of 0.75. AR_L indicates average recall for large objects.

Dataset	Label-percent	mAP	AP^{50}	AP^{75}	AR_L
TableBank-word	10%	80.5	92.5	87.7	87.1
	30%	88.3	95.7	93.1	92.1
	50%	91.5	96.7	95.2	94.5
TableBank-latex	10%	63.7	93.5	71.6	74.3
	30%	82.8	96.4	93.4	89.0
	50%	85.3	96.2	94.4	91.4
TableBank-both	10%	84.2	95.8	93.1	90.1
	30%	86.8	97.0	94.1	91.5
	50%	91.8	96.9	95.6	95.3

The qualitative analysis of semi-supervised learning for the TableBank-both data split is shown in Fig. 5. Part (b) of Fig. 5 has a matrix with a similar

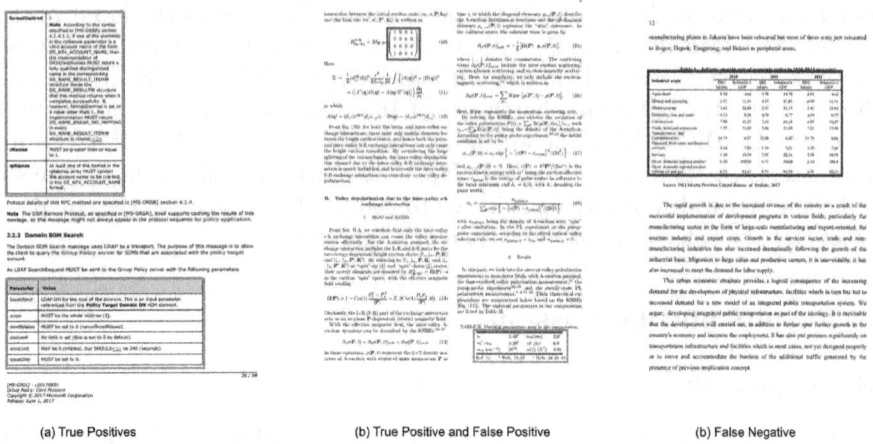

(a) True Positives   (b) True Positive and False Positive   (b) False Negative

Fig. 5. Semi-supervised table detection results that employs deformable transformer on TableBank-both data split. Green color represents true positives, blue denotes false negatives and red shows false positives. Here, (a) indicates true positive detection results, (b) shows true positive and false positive detection results, and (c) gives false negative detection results. (Color figure online)

structure as rows and columns, and the network detects the matrix as a table giving false positive detection results. Here, incorrect detection results indicate where the network fails to provide correct detection of table regions. Table 2 gives the results of this semi-supervised approach on different IoU threshold values for all splits of the TableBank dataset on 10% label data and the rest as unlabeled data. A visual comparison of Precision, Recall and F1-Score of semi-supervised network that employs deformable transformer with ResNet-50 backbone on different IoU threshold values on 10% labeled dataset of TableBank-both data split is shown in Fig. 6.

Table 2. The performance comparison of semi-supervised network that employs deformable transformer with ResNet-50 backbone on different IoU threshold values on 10% labeled dataset of TableBank-both data split.

Method	IoU	Precision	Recall	F1-score
Semi-Supervised	0.5	95.8	90.5	93.1
Deformable-DETR+ResNet-50	0.6	94.6	90.5	92.5
10% labels	0.7	93.3	90.3	91.8
	0.8	91.8	89.8	90.8
	0.9	89.1	87.2	88.1

Comparisons with Previous Supervised and Semi-supervised Approaches. Table 3 compares the deep learning-based supervised and semi-supervised networks on the ResNet-50 backbone. We also compare supervised deformable-DETR trained on 10%, 30% and 50% TableBank-both data split label data with our semi-supervised approach that employs deformable transformer. It shows that our attention mechanism-based semi-supervised approach provides comparable results without using proposal generation process and post-processing steps such as Non-maximal suppression (NMS).

Table 3. Performance comparison of previous supervised and semi-supervised approaches. Supervised Deformable-DETR and Faster R-CNN network trained on just 10%, 30% and 50% data of TableBank-both dataset while semi-supervised networks used 10%, 30% and 50% TableBank-both dataset as labeled and rest as unlabeled data using ResNet-50 backbone. Here, all results are represented on $mAP(0.5:0.95)$. The best threshold values are shown in bold.

Method	Approach	Detector	10%	30 %	50 %
Ren et al. [11]	supervised	Faster R-CNN	80.1	80.6	83.3
Zhu et al. [25]	supervised	Deformable DETR	80.8	82.6	86.9
STAC [79]	semi-supervised	Faster R-CNN	82.4	83.8	87.1
Unbiased Teacher [88]	semi-supervised	Faster R-CNN	83.9	86.4	88.5
Humble Teacher [95]	semi-supervised	Faster R-CNN	83.4	86.2	87.9
Soft Teacher [96]	semi-supervised	Faster R-CNN	83.6	**86.8**	89.6
Our	semi-supervised	Deformable DETR	**84.2**	**86.8**	**91.8**

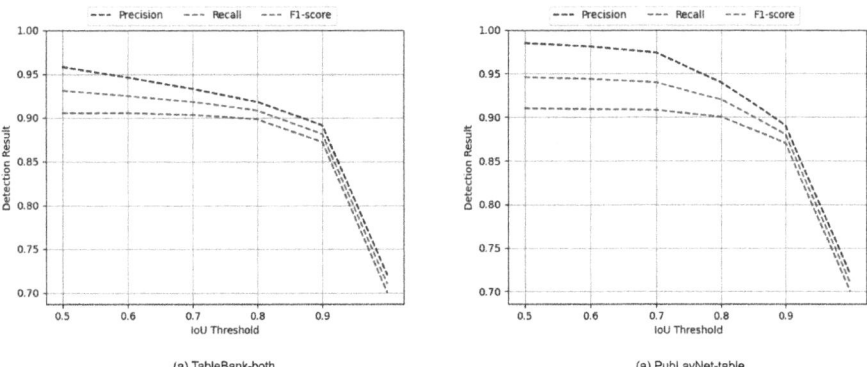

(a) TableBank-both   (a) PubLayNet-table

Fig. 6. A visual comparison of Precision, Recall and F1-Score of semi-supervised network that employs deformable transformer with ResNet-50 backbone on different IoU threshold values on 10% labeled dataset of TableBank-both data split and PubLayNet table class dataset. Here, blue indicates precision results on different IoU threshold values, red shows recall results on different IoU threshold values, and green represents F1-score results on different IoU threshold values. (Color figure online)

5.2   PubLayNet

In this subsection, we discuss the experimental results on PubLayNet table class dataset on different percentages of label data. We also compare the transformer-based semi-supervised approach with previous deep learning-based supervised and semi-supervised approaches. Furthermore, we give results on 10% Pub-LayNet dataset for all IoU threshold values. Table 4 provides the results of the semi-supervised approach that employs deformable transformer for PubLayNet table class on the different percentages of label data and rest as unlabeled data. Here, AP_{50} value is 98.5%, 98.8%, and 98.8% for 10%, 30% and 50% label data, respectively.

Table 4. Performance results for PubLayNet table class dataset. Here, mAP represents mean AP at the IoU threshold range of (50:95), AP_{50} indicates AP at the IoU threshold of 0.5 and AP_{75} denotes AP at the IoU threshold of 0.75. AR_L indicates average recall for large objects.

Dataset	Label-percent	mAP	AP^{50}	AP^{75}	AR_L
PubLayNet	10%	88.4	98.5	97.3	91.0
	30%	90.3	98.8	97.5	93.2
	50%	92.8	98.8	97.3	96.0

Table 5. The performance comparison of semi-supervised network that employs deformable transformer with ResNet-50 backbone on different IoU threshold values on 10% PubLayNet labeled Dataset.

Method	IoU	Precision	Recall	F1-score
Semi-Supervised	0.5	98.5	91.0	94.6
Deformable-DETR	0.6	98.1	90.9	94.4
10% labels	0.7	97.4	90.8	94.0
	0.8	94.0	90.0	92.0
	0.9	89.0	87.0	88.0

Furthermore, our semi-supervised network is trained on different IoU threshold values on 10% of labeled PubLayNet Dataset. Table 5 gives the results of the semi-supervised approach on different IoU threshold values for PubLayNet table class on 10% label data and the rest as unlabeled data. A visual comparison of Precision, Recall and F1-score of the semi-supervised network that employs the deformable transformer network with ResNet-50 backbone on different IoU threshold values on 10% labeled dataset of PubLayNet table class is shown in Fig. 6. Here, blue indicates precision results on different IoU threshold values

Table 6. Performance comparison of previous supervised and semi-supervised approaches. Deformable-DETR and Faster R-CNN trained on just 10%, 30% and 50% table data while semi-supervised networks used 10%, 30% and 50% PubLayNet dataset as labeled and rest as unlabeled data. Here, all results are represented on AP_{50} at the IoU threshold of 0.5. The best threshold values are shown in bold.

Method	Approach	Detector	10 %	30 %	50 %
Ren et al. [11]	supervised	Faster R-CNN	93.6	95.6	95.9
Zhu et al. [25]	supervised	Deformable DETR	93.9	96.2	97.1
STAC [79]	semi-supervised	Faster R-CNN	95.8	96.9	97.8
Unbiased Teacher [88]	semi-supervised	Faster R-CNN	96.1	97.4	98.1
Humble Teacher [95]	semi-supervised	Faster R-CNN	96.7	97.9	98.0
Soft Teacher [96]	semi-supervised	Faster R-CNN	96.5	98.1	98.5
Our	semi-supervised	Deformable DETR	**98.5**	**98.8**	**98.8**

on different IoU threshold values, red shows recall results, and green represents F1-score results on different IoU threshold values.

Comparisons with Previous Supervised and Semi-supervised Approaches. Table 6 compares the deep learning-based supervised and semi-supervised networks on PubLayNet table class using ResNet-50 backbone. We also compare supervised deformable-DETR trained on 10%, 30% and 50% PubLayNet table class label data with our semi-supervised approach that employs the deformable transformer. It shows that our semi-supervised approach provides comparable results without using proposal and post-processing steps such as Non-maximal suppression (NMS).

5.3 DocBank

In this subsection, we discuss the experimental results on DocBank dataset on different percentages of label data. We compare the transformer-based semi-supervised approach with previous CNN-based semi-supervised approach in Table 7.

Table 7. Performance comparison of previous semi-supervised approach and our Deformable-DETR based semi-supervised approach on DocBank dataset. Here, all results are represented on $mAP(0.5:0.95)$.

Method	Approach	Detector	10 %	30 %	50 %
Soft Teacher [96]	semi-supervised	Faster R-CNN	72.3	74.4	81.5
Our	semi-supervised	Deformable DETR	82.5	84.9	87.1

Furthermore, we also compare our semi-supervised approach on different percentages of label data with previous table detection and document analysis approaches for different datasets TableBank, PubLayNet, and DocBank in Table 8. Although we cannot directly compare our semi-supervised approach with previous supervised document analysis approaches. However, we can observe that even with 50% label data, we achieve comparable results with previous supervise approaches.

Table 8. Performance comparison of previous supervised approaches for document analysis. Our semi-supervised network uses 10%, 30% and 50% label data and rest as unlabeled data. Here, all results are represented on $mAP(0.5:0.95)$.

Method	Approach	Labels	TableBank	PubLayNet	DocBank
CDeC-Net [69]	supervised	100%	96.5	97.8	-
CasTabDetectoRS [32]	supervised	100%	95.3	-	-
Faster R-CNN [48]	supervised	100%	-	90	86.3
VSR [97]	supervised	100%	-	95.69	87.6
Our	semi-supervised	10%	84.2	88.4	82.5
Our	semi-supervised	30%	86.8	90.3	84.9
Our	semi-supervised	50%	91.8	92.8	87.1

5.4   ICDAR-19

We also evaluate our method for table detection on the Modern Track A portion of the table detection dataset from the cTDaR competition at ICDAR 2019. We summarize the quantitative results of our approach at different percentages of label data and compare it with previously supervised table detection approaches in Table 9. We evaluate results at higher IoU thresholds of 0.8 and 0.9. For a direct comparison with previous table detection approaches, we also evaluate our approach on 100% label data. Our approach achieved a precision of 92.6% and a recall of 91.3% on the IoU threshold of 0.9 on 100% label data.

Table 9. Performance comparison between the proposed semi-supervised approach and previous state-of-the-art results on the dataset of ICDAR 19 Track A (Modern).

Method	Approach	IoU = 0.8			IoU = 0.9		
		Recall	Precision	F1-Score	Recall	Precision	F1-Score
TableRadar [47]	supervised	94.0	95.0	94.5	89.0	90.0	89.5
NLPR-PAL [47]	supervised	93.0	93.0	93.0	86.0	86.0	86.0
Lenovo Ocean [47]	supervised	86.0	88.0	87.0	81.0	82.0	81.5
CascadeTabNet [68]	supervised	-	-	92.5	-	-	90.1
CDeC-Net [69]	supervised	93.4	95.3	94.4	90.4	92.2	91.3
HybridTabNet [33]	supervised	93.3	92.0	92.8	90.5	89.5	90.2
Our	semi-supervised (50%)	71.1	82.3	76.3	66.3	76.8	71.2
Our	supervised (100%)	92.1	94.9	93.5	91.3	92.6	91.9

5.5   Ablation Study

In this section, we validate the key design elements. Unless otherwise stated, all the ablation studies are conducted using a ResNet-50 backbone with 30% labeled images from the PubLayNet dataset.

Pseudo-Labeling Confidence Threshold. In Sect. 3.2, the threshold value (referred to as the confidence threshold) plays an important role in determining the balance between the accuracy and quantity of the generated pseudo-labels. As this threshold value increases, fewer examples will pass the filter, but they will be of higher quality. Conversely, a smaller threshold value will result in more examples passing but with a higher likelihood of false positives. The impact of various threshold values, ranging from 0.5 to 0.9, is presented in Table 10. The optimal threshold value was determined to be 0.7 based on the results.

Table 10. Performance comparison using different Pseudo-labeling confidence threshold values. The best threshold values are shown in bold.

Threshold	AP	AP^{50}	AP^{75}
0.5	86.9	91.6	90.1
0.6	89.5	98.1	95.7
0.7	**90.3**	**98.8**	**97.5**
0.8	89.4	97.2	95.3
0.9	87.9	96.3	94.5

Table 11. Performance comparison using different numbers of learnable queries to the decoder input. Here, best performance results are shown in bold.

N	AP	AP^{50}	AP^{75}
3	61.4	69.7	62.6
30	**90.3**	**98.8**	**97.5**
50	89.4	90.3	85.4
100	88.4	89.7	83.9
300	78.5	94.7	90.2

Influence of Learnable Queries Quantity. In our analysis, we investigate the impact of varying the number of queries fed as input in the decoder of deformable DETR. Figure 7 compares prediction results by varying the number of object queries fed as input in the decoder of deformable DETR. The optimal performance is attained when the number of queries N is set to 30; deviating

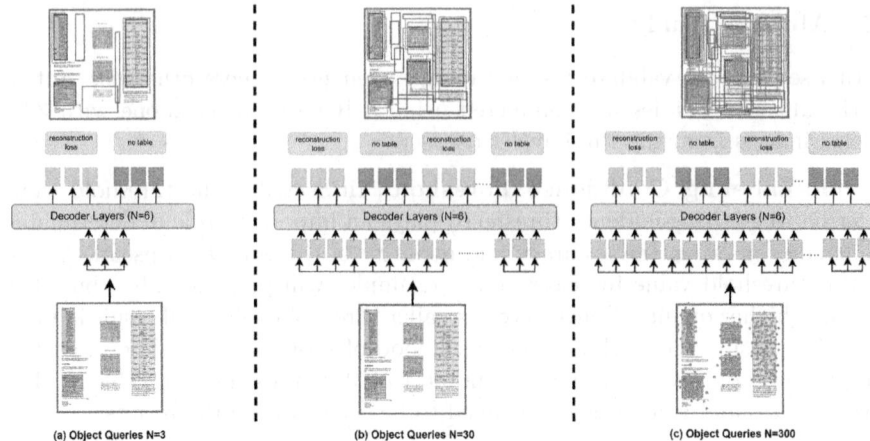

Fig. 7. Comparison of performance by variation of the number of object queries fed as input in the decoder of deformable DETR. Here, (a) takes N = 3 object queries as input, (b) contains N = 30 object queries as input, and (c) has N =300 object queries as input. The optimal performance is achieved by selecting the number of queries N to 30; deviating from this value results in a decrease in performance. Here, blue rectangles denote ground truth (GT), green rectangles indicate object class, and red rectangles show background class. (Color figure online)

from this value results in a decrease in performance. Table 11 presents and analyzes the result for varying object query quantities. Choosing a small value for N could result in the model failing to identify particular objects, negatively impacting its performance. On the other hand, selecting a large value for N may cause the model to perform poorly due to overfitting, as it would incorrectly classify certain regions as objects. Moreover, training complexity $O(Nkc_i^2)$ of this semi-supervised self-attention mechanism in the decoder of student-teacher module depends on the number of object queries and is subsequently improved as complexity is reduced by minimizing the number of object queries.

6   Conclusion

This paper introduces a semi-supervised approach that employs the deformable transformer for table detection in document images. The proposed method mitigates the need of large-scale annotated data and simplifies the process by integrating the pseudo-label generation framework into a streamlined mechanism. The simultaneous generation of pseudo-labels leads to a dynamic process known as the "flywheel effect", where one model continually improves the pseudo-boxes produced by the other model as the training progresses. The pseudo-class labels and pseudo-bounding boxes are improved in this framework using two distinct modules named student and teacher. These modules update each other by the EMA function to provide precise classification and bounding box predictions.

The results indicate that this approach surpasses the performance of supervised models when applied to labeling ratios of 10%, 30%, and 50% on TableBank all splits and the PubLayNet training data. Furthermore, when trained on the 10% labeled data of PubLayNet, the model performed comparably to the current CNN-based semi-supervised baseline. In future, we aim to investigate the impact of the proportion of annotated data on the ultimate performance and develop models that function effectively with a minimal quantity of labeled data. Additionally, we intend to employ the transformer-based semi-supervised learning mechanism for table structure recognition task.

Acknowledgement. The work leading to this publication has been partially funded by the European project AIRISE (https://airise.eu/) under Grant Agreement ID 101092312.

References

1. Zhao, Z., Jiang, M., Guo, S., Wang, Z., Chao, F., Tan, K.C.: Improving deep learning based optical character recognition via neural architecture search. In: IEEE Congress on Evolutionary Computation (CEC 2020), pp. 1–7 (2020)
2. Van Strien, D., Beelen, K., Ardanuy, M.C., Hosseini, K., McGillivray, B., Colavizza, G.: Assessing the impact of OCR quality on downstream NLP tasks (2020)
3. Coüasnon, B., Lemaitre, A.: Recognition of tables and forms. In: Handbook of Document Image Processing and Recognition (2014)
4. Zanibbi, R., Blostein, D., Cordy, J.R.: A survey of table recognition. Doc. Anal. Recogn. **7**(1), 1–16 (2004)
5. Jorge, A.M., Torgo, L., et al.: Design of an end-to-end method to extract information from tables. IJDAR **8**(2), 144–171 (2006)
6. Fang, J., Gao, L., Bai, K., Qiu, R., Tao, X., Tang, Z.: A table detection method for multipage pdf documents via visual seperators and tabular structures. Int. Conf. Doc. Anal. Recogn. **2011**, 779–783 (2011)
7. Chen, J., Lopresti, D.: Table detection in noisy off-line handwritten documents. Int. Conf. Doc. Anal. Recogn. **2011**, 399–403 (2011)
8. Hashmi, K.A., Bymana Ponnappa, R., Bukhari, S.S., Jenckel, M., Dengel, A.: Feedback learning: automating the process of correcting and completing the extracted information. In: 2019 International Conference on Document Analysis and Recognition Workshops (ICDARW), vol. 5, pp. 116–121 (2019)
9. Saha, R., Mondal, A., Jawahar, C.V.: Graphical object detection in document images. CoRR abs/2008.10843 (2020). arXiv:2008.10843
10. Girshick, R.B.: Fast R-CNN CoRR, abs/1504.08083 (2015). arXiv:1504.08083
11. Ren, S., He, K., Girshick, R.B., Sun, J.: Faster R-CNN: towards real-time object detection with region proposal networks. CoRR. abs/1506.01497 (2015). arXiv:1506.01497
12. Redmon, J., Farhadi, A.: YOLO9000: better, faster, stronger. CoRR. abs/1612.08242 (2016). arXiv:1612.08242
13. He, K., Gkioxari, G., Dollár, P., Girshick, R.: Mask R-CNN. In: IEEE International Conference on Computer Vision (ICCV 2017), pp. 2980–2988 (2017)
14. Orosz, T., Vági, R., Csányi, G.M., Nagy, D., Üveges, I., Vadász, J.P., Megyeri, A.: Evaluating human versus machine learning performance in a LegalTech problem. Appl. Sci. **12**(1), 297 (2022). www.mdpi.com/2076-3417/12/1/297

15. Schreiber, S., Agne, S., Wolf, I., Dengel, A., Ahmed, S.: Deepdesrt: deep learning for detection and structure recognition of tables in document images. In: 2017 14th IAPR International Conference on Document Analysis and Recognition (ICDAR), vol. 01, pp. 1162–1167 (2017)

16. Minouei, M., Hashmi, K.A., Soheili, M.R., Afzal, M.Z., Stricker, D.: Continual learning for table detection in document images. Appl. Sci. **12**(18), 8969 (2022). www.mdpi.com/2076-3417/12/18/8969

17. Hashmi, K.A., Stricker, D., Liwicki, M., Afzal, M.N., Afzal, M.Z.: Guided table structure recognition through anchor optimization. CoRR. abs/2104.10538 (2021). arXiv:2104.10538

18. Hashmi, K.A., Pagani, A., Liwicki, M., Stricker, D., Afzal, M.Z.: Cascade network with deformable composite backbone for formula detection in scanned document images. Appl. Sci. **11**(16), 7610 (2021). www.mdpi.com/2076-3417/11/16/7610

19. Sinha, S., Hashmi, K.A., Pagani, A., Liwicki, M., Stricker, D., Afzal, M.Z.: Rethinking learnable proposals for graphical object detection in scanned document images. Appl. Sci. **12**(20), 10578 (2022). www.mdpi.com/2076-3417/12/20/10578

20. Naik, S., Hashmi, K.A., Pagani, A., Liwicki, M., Stricker, D., Afzal, M.Z.: Investigating attention mechanism for page object detection in document images. Appl. Sci. **12**(15), 7486 (2022). www.mdpi.com/2076-3417/12/15/7486

21. Wang, K., Yan, X., Zhang, D., Zhang, L., Lin, L.: Towards human-machine cooperation: self-supervised sample mining for object detection. CoRR. abs/1803.09867 (2018). arXiv:1803.09867

22. Tang, P., Ramaiah, C., Xu, R., Xiong, C.: Proposal learning for semi-supervised object detection. CoRR. abs/2001.05086 (2020). arXiv:2001.05086

23. Rhee, P.K., Erdenee, E., Kyun, S.D., Ahmed, M.U., Jin, S.: Active and semi-supervised learning for object detection with imperfect data. Cogn. Syst. Res. 45, 109–123 (2017). www.sciencedirect.com/science/article/pii/S1389041716301127

24. Xie, Q., Dai, Z., Hovy, E.H., Luong, M., Le, Q.V.: Unsupervised data augmentation. CoRR. abs/1904.12848 (2019). arXiv:1904.12848

25. Zhu, X., Su, W., Lu, L., Li, B., Wang, X., Dai, J.: Deformable DETR: deformable transformers for end-to-end object detection. CoRR. abs/2010.04159 (2020). arXiv:2010.04159

26. Itonori, K.: Table structure recognition based on textblock arrangement and ruled line position. In: Proceedings of 2nd International Conference on Document Analysis and Recognition (ICDAR 1993), pp. 765–768 (1993)

27. Tupaj, S., Shi, Z., Chang, C.H., Alam, H.: Extracting tabular information from text files. EECS Department, Tufts University, Medford, USA, vol. 1 (1996)

28. Chandran, S., Kasturi, R.: Structural recognition of tabulated data. In: Proceedings of 2nd International Conference on Document Analysis and Recognition (ICDAR 1993), pp. 516–519 (1993)

29. Hirayama, Y.: A method for table structure analysis using DP matching. In: Proceedings of 3rd International Conference on Document Analysis and Recognition, vol. 2, pp. 583–586 (1995)

30. Kieninger, T.G.: Table structure recognition based on robust block segmentation. In: Lopresti, D.P., Zhou, J., Document Recognition V (Eds.), vol. 3305, International Society for Optics and Photonics. SPIE, pp. 22–32 (1998). https://doi.org/10.1117/12.304642

31. Siddiqui, S.A., Malik, M.I., Agne, S., Dengel, A., Ahmed, S.: Decnt: deep deformable CNN for table detection. IEEE Access. **6**, 74151–74161 (2018)

32. Hashmi, K.A., Pagani, A., Liwicki, M., Stricker, D., Afzal, M.Z.: Castabdetectors: cascade network for table detection in document images with recursive feature pyramid and switchable atrous convolution. J. Imaging. **7**, 214 (2021)
33. Nazir, D., Hashmi, K.A., Pagani, A., Liwicki, M., Stricker, D., Afzal, M.Z.: HybridTabNet: towards better table detection in scanned document images. Appl. Sci. **11**(18), 8396 (2021). www.mdpi.com/2076-3417/11/18/8396
34. Pyreddy, P., Croft, W.B.: Tintin: a system for retrieval in text tables. In: Digital Library (1997)
35. Pivk, A., Cimiano, P., Sure, Y., Gams, M., Rajkovič, V., Studer, R.: Transforming arbitrary tables into logical form with tartar. Data Knowl. Eng. **60**(3), 567–595 (2007). www.sciencedirect.com/science/article/pii/S0169023X06000620
36. Hu, J., Kashi, R.S., Lopresti, D.P., Wilfong, G.: Medium-independent table detection. In: Lopresti, D.P., Zhou, J., (Eds.) Document Recognition and Retrieval VII, vol. 3967, International Society for Optics and Photonics. SPIE, pp. 291–302 (1999). https://doi.org/10.1117/12.373506
37. Khusro, S., Latif, A., Ullah, I.: On methods and tools of table detection, extraction and annotation in pdf documents. J. Inf. Sci. **41**(1), 41–57 (2015)
38. Embley, D.W., Hurst, M., Lopresti, D., Nagy, G.: Table-processing paradigms: a research survey. IJDAR **8**(2), 66–86 (2006)
39. Cesarini, F., Marinai, S., Sarti, L., Soda, G.: Trainable table location in document images. In: 2002 International Conference on Pattern Recognition, vol. 3, pp. 236–240 (2002)
40. Silva, A.C.: Learning rich hidden Markov models in document analysis: table location. In: 2009 10th International Conference on Document Analysis and Recognition, pp. 843–847 (2009)
41. Silva, A.: Parts that Add up to a Whole: a Framework for the Analysis of Tables. Edinburgh University, UK (2010)
42. Kasar, T., Barlas, P., Adam, S., Chatelain, C., Paquet, T.: Learning to detect tables in scanned document images using line information. In: 2013 12th International Conference on Document Analysis and Recognition. IEEE, pp. 1185–1189 (2013)
43. Yang, X., Yümer, M.E., Asente, P., Kraley, M., Kifer, D., Giles, C.L.: Learning to extract semantic structure from documents using multimodal fully convolutional neural network. CoRR, abs/1706.02337 (2017). arXiv:1706.02337
44. He, D., Cohen, S., Price, B., Kifer, D., Giles, C.L.: Multi-scale multi-task FCN for semantic page segmentation and table detection. In: 2017 14th IAPR International Conference on Document Analysis and Recognition (ICDAR), vol. 01, pp. 254–261 (2017)
45. Kavasidis, I., et al.: A saliency-based convolutional neural network for table and chart detection in digitized documents. CoRR. abs/1804.06236 (2018). arXiv:1804.06236
46. Paliwal, S.V.D., Rahul, R., Sharma, M., Vig, L.: TableNet: deep learning model for end-to-end table detection and tabular data extraction from scanned document images. CoRR, abs/2001.01469 (2020). arXiv:2001.01469
47. Gao, L., et al.: ICDAR 2019 competition on table detection and recognition (CTDAR). In: 2019 International Conference on Document Analysis and Recognition (ICDAR), pp. 1510–1515. IEEE (2019)
48. Zhong, X., Tang, J., Yepes, A.J.: Publaynet: largest dataset ever for document layout analysis. In: 2019 International Conference on Document Analysis and Recognition (ICDAR), pp. 1015–1022. IEEE, September 2019
49. Mondal, A., Lipps, P., Jawahar, C.V.: IIIT-AR-13K: a new dataset for graphical object detection in documents. CoRR, abs/2008.02569 (2020). arXiv:2008.02569

50. Göbel, M.C., Hassan, T., Oro, E., Orsi, G.: ICDAR 2013 table competition. In: 2013 12th International Conference on Document Analysis and Recognition, pp. 1449–1453 (2013)
51. Gao, L., Yi, X., Jiang, Z., Hao, L., Tang, Z.: ICDAR 2017 competition on page object detection. In: 2017 14th IAPR International Conference on Document Analysis and Recognition (ICDAR), vol. 01, pp. 1417–1422 (2017)
52. Li, M., Cui, L., Huang, S., Wei, F., Zhou, M., Li, Z.: TableBank: a benchmark dataset for table detection and recognition (2019)
53. Smock, B., Pesala, R., Abraham, R.: PubTables-1M: towards comprehensive table extraction from unstructured documents. In: Proceedings of the IEEE/CVF Conference on Computer Vision and Pattern Recognition (CVPR), pp. 4634–4642, June 2022
54. Long, J., Shelhamer, E., Darrell, T.: Fully convolutional networks for semantic segmentation. CoRR, abs/1411.4038 (2014). arXiv:1411.4038
55. Li, X.-H., Yin, F., Liu, C.-L.: Page object detection from pdf document images by deep structured prediction and supervised clustering. In: 2018 24th International Conference on Pattern Recognition (ICPR), pp. 3627–3632 (2018)
56. Holecek, M., Hoskovec, A., Baudis, P., Klinger, P.: Line-items and table understanding in structured documents. CoRR. abs/1904.12577 (2019). arXiv:1904.12577
57. Riba, P., Goldmann, L., Terrades, O.R., Rusticus, D., Fornés, A., Lladós, J.: Table detection in business document images by message passing networks. Pattern Recogn. **127**, 108641 (2022). www.sciencedirect.com/science/article/pii/S0031320322001224
58. Hao, L., Gao, L., Yi, X., Tang, Z.: A table detection method for pdf documents based on convolutional neural networks. In: 2016 12th IAPR Workshop on Document Analysis Systems (DAS), pp. 287–292 (2016)
59. Yi, X., Gao, L., Liao, Y., Zhang, X., Liu, R., Jiang, Z.: CNN based page object detection in document images. In: 2017 14th IAPR International Conference on Document Analysis and Recognition (ICDAR), vol. 01, pp. 230–235 (2017)
60. Lin, T., Goyal, P., Girshick, R.B., He, K., Dollár, P.: Focal loss for dense object detection. CoRR. abs/1708.02002 (2017). arXiv:1708.02002
61. Fang, Y., et al.: You only look at one sequence: rethinking transformer in vision through object detection. CoRR, abs/2106.00666 (2021). arXiv:2106.00666
62. He, K., Gkioxari, G., Dollár, P., Girshick, R.B.: Mask R-CNN. CoRR. abs/1703.06870 (2017). arXiv:1703.06870
63. Cai, Z., Vasconcelos, N.: Cascade R-CNN: delving into high quality object detection. CoRR abs/1712.00726 (2017). arXiv:1712.00726
64. Vo, N.D., Nguyen, K., Nguyen, T.V., Nguyen, K.: Ensemble of deep object detectors for page object detection. In: Proceedings of the 12th International Conference on Ubiquitous Information Management and Communication, ser. IMCOM 2018, Association for Computing Machinery. New York, NY, USA (2018). https://doi.org/10.1145/3164541.3164644
65. Gilani, A., Qasim, S.R., Malik, I., Shafait, F.: Table detection using deep learning. In: 2017 14th IAPR International Conference on Document Analysis and Recognition (ICDAR), vol. 01, pp. 771–776 (2017)
66. Huang, Y., et al.: A yolo-based table detection method. In: International Conference on Document Analysis and Recognition (ICDAR 2019), pp. 813–818 (2019)
67. Zheng, X., Burdick, D., Popa, L., Wang, N.X.R.: Global table extractor (GTE): a framework for joint table identification and cell structure recognition using visual context. CoRR. abs/2005.00589 (2020). arXiv:2005.00589

68. Prasad, D., Gadpal, A., Kapadni, K., Visave, M., Sultanpure, K.: Cascadetabnet: an approach for end to end table detection and structure recognition from image-based documents. CoRR. abs/2004.12629 (2020). arXiv:2004.12629

69. Agarwal, M., Mondal, A., Jawahar, C.V.: CDEC-Net: composite deformable cascade network for table detection in document images. In: 2020 25th International Conference on Pattern Recognition (ICPR), pp. 9491–9498 (2021)

70. Arif, S., Shafait, F.: Table detection in document images using foreground and background features. In: Digital Image Computing: Techniques and Applications (DICTA 2018), pp. 1–8 (2018)

71. Dai, J., et al.: Deformable convolutional networks. CoRR. abs/1703.06211 (2017). arXiv:1703.06211

72. Liu, Y., et al.: CBNet: a novel composite backbone network architecture for object detection. CoRR. abs/1909.03625 (2019). arXiv:1909.03625

73. Jeong, J., Lee, S., Kim, J., Kwak, N.: Consistency-based semi-supervised learning for object detection. In: Wallach, H., Larochelle, H., Beygelzimer, A., d' Alché-Buc, F., Fox, E., Garnett, R. (Eds.) Advances in Neural Information Processing Systemsvol, vol. 32. Curran Associates Inc, (2019). www.proceedings.neurips.cc/paper/2019/file/d0f4dae80c3d0277922f8371d5827292-Paper.pdf

74. Radosavovic, I., Dollár, P., Girshick, R.B., Gkioxari, G., He, K.: Data distillation: towards omni-supervised learning. CoRR. abs/1712.04440 (2017). arXiv:1712.04440

75. Zoph, B., et al.: Rethinking pre-training and self-training. In: Larochelle, H., Ranzato, M., Hadsell, R., Balcan, M., Lin, H. (Eds.) Advances in Neural Information Processing Systems, vol. 33. Curran Associates Inc, 2020, pp. 3833–3845. www.proceedings.neurips.cc/paper/2020/file/27e9661e033a73a6ad8cefcde965c54d-Paper.pdf

76. Li, Y., Huang, D., Qin, D., Wang, L., Gong, B.: Improving object detection with selective self-supervised self-training. CoRR. abs/2007.09162 (2020). arXiv:2007.09162

77. Shehzadi, T., Hashmi, K.A., Pagani, A., Liwicki, M., Stricker, D., Afzal, M.Z.: Mask-aware semi-supervised object detection in floor plans. Appl. Sci. **12**(19), 9398 (2022). www.mdpi.com/2076-3417/12/19/9398

78. Kallempudi, G., Hashmi, K.A., Pagani, A., Liwicki, M., Stricker, D., Afzal, M.Z.: Toward semi-supervised graphical object detection in document images. Future Internet. **14**(6), 176 (2022). www.mdpi.com/1999-5903/14/6/176

79. Sohn, K., Zhang, Z., Li, C., Zhang, H., Lee, C., Pfister, T.: A simple semi-supervised learning framework for object detection. CoRR. abs/2005.04757 (2020). arXiv:2005.04757

80. Carion, N., Massa, F., Synnaeve, G., Usunier, N., Kirillov, A., Zagoruyko, S.: End-to-end object detection with transformers. In: Vedaldi, A., Bischof, H., Brox, T., Frahm, J.-M. (eds.) ECCV 2020. LNCS, vol. 12346, pp. 213–229. Springer, Cham (2020). https://doi.org/10.1007/978-3-030-58452-8_13

81. Parmar, N., Vaswani, A., Uszkoreit, J., Kaiser, L., Shazeer, N., Ku, A.: Image transformer. CoRR. abs/1802.05751 (2018). arXiv:1802.05751

82. Bello, I., Zoph, B., Vaswani, A., Shlens, J., Le, Q.V.: Attention augmented convolutional networks. CoRR. abs/1904.09925 (2019). arXiv:1904.09925

83. Vaswani, A., et al.: Attention is all you need. In: Advances in Neural Information Processing Systems, Guyon, I., et al. (Eds.), vol. 30. Curran Associates Inc. (2017). www.proceedings.neurips.cc/paper/2017/file/3f5ee243547dee91fbd053c1c4a845aa-Paper.pdf

84. Zhu, X., Hu, H., Lin, S., Dai, J.: Deformable convnets v2: More deformable, better results. CoRR. abs/1811.11168 (2018). arXiv:1811.11168

85. Lin, T.-Y., Dollár, P., Girshick, R., He, K., Hariharan, B., Belongie, S.: Feature pyramid networks for object detection. In: IEEE Conference on Computer Vision and Pattern Recognition (CVPR 2017), pp. 936–944 (2017)

86. Zhao, Q., et al.: M2det: a single-shot object detector based on multi-level feature pyramid network. CoRR. abs/1811.04533 (2018). arXiv:1811.04533

87. Tarvainen, A., Valpola, H.: Weight-averaged consistency targets improve semi-supervised deep learning results. CoRR. abs/1703.01780 (2017). arXiv:1703.01780

88. Liu, Y., et al.: Unbiased teacher for semi-supervised object detection. CoRR. abs/2102.09480 (2021). arXiv:2102.09480

89. Kuhn, H.W.: The Hungarian method for the assignment problem. Naval Res. Logist. Q. **2**(1–2), 83–97 (1955)

90. Lin, T.-Y., et al.: Microsoft COCO: common objects in context. In: Fleet, D., Pajdla, T., Schiele, B., Tuytelaars, T. (eds.) ECCV 2014. LNCS, vol. 8693, pp. 740–755. Springer, Cham (2014). https://doi.org/10.1007/978-3-319-10602-1_48

91. Li, M., et al.: DocBank: a benchmark dataset for document layout analysis. CoRR. abs/2006.01038, 2020. arXiv:2006.01038

92. Powers, D.M.W.: Evaluation: from precision, recall and f-measure to ROC, informedness, markedness and correlation. CoRR. abs/2010.16061 (2020). arXiv:2010.16061

93. Szegedy, C., Ioffe, S., Vanhoucke, V.: Inception-v4, inception-resnet and the impact of residual connections on learning. CoRR. abs/1602.07261 (2016). arXiv:1602.07261

94. Krizhevsky, A., Sutskever, I., Hinton, G.E.: Imagenet classification with deep convolutional neural networks. In: Pereira, F., Burges, C., Bottou, L., Weinberger, K. (Eds.) Advances in Neural Information Processing Systems, vol. 25. Curran Associates Inc. (2012). www.proceedings.neurips.cc/paper/2012/file/c399862d3b9d6b76c8436e924a68c45b-Paper.pdf

95. Tang, Y., Chen, W., Luo, Y., Zhang, Y.: Humble teachers teach better students for semi-supervised object detection. CoRR. abs/2106.10456 (2021). arXiv:2106.10456

96. Xu, M., et al.: End-to-end semi-supervised object detection with soft teacher. CoRR. abs/2106.09018 (2021). arXiv:2106.09018

97. Zhang, P., et al.: VSR: a unified framework for document layout analysis combining vision, semantics and relations. CoRR. abs/2105.06220 (2021). arXiv:2105.06220

SpaDen: Sparse and Dense Keypoint Estimation for Real-World Chart Understanding

Saleem Ahmed[(✉)] [ID], Pengyu Yan [ID], David Doermann [ID], Srirangaraj Setlur [ID], and Venu Govindaraju [ID]

Department of Computer Science and Engineering, University at Buffalo, SUNY, Buffalo, USA
{sahmed9,pyan4,doermann,setlur,govind}@buffalo.edu

Abstract. We introduce a novel bottom-up approach for the extraction of chart data. Our model utilizes images of charts as inputs and learns to detect keypoints (KP), which are used to reconstruct the components within the plot area. Our novelty lies in detecting a fusion of continuous and discrete KP as predicted heatmaps. A combination of sparse and dense per-pixel objectives coupled with a uni-modal self-attention-based feature-fusion layer is applied to learn KP embeddings. Further leveraging deep metric learning for unsupervised clustering, allows us to segment the chart plot area into various objects. By further matching the chart components to the legend, we are able to obtain the data series names. A post-processing threshold is applied to the KP embeddings to refine the object reconstructions and improve accuracy. Our extensive experiments include an evaluation of different modules for KP estimation and the combination of deep layer aggregation and corner pooling approaches. The results of our experiments provide extensive evaluation for the task of real-world chart data extraction. Our Code is publicly available (https://github.com/cse-ai-lab/SpaDen).

Keywords: Charts · Document Understanding · Reasoning

1 Introduction

Data visualizations are effective constructs to efficiently convey knowledge in documents. Most documents consist of semantically structured textual content and complementary visualizations in the form of figures, generic infographics, technical diagrams, charts, etc. Our work focuses on the problem of reconstructing tabular data used to plot the visualization, given the image and structural information as input. We focus on the family of visualizations called charts, specifically from the scientific literature.

S. Ahmed and P. Yan — Equal contribution

ⓒ The Author(s), under exclusive license to Springer Nature Switzerland AG 2023
G. A. Fink et al. (Eds.): ICDAR 2023, LNCS 14188, pp. 77–93, 2023.
https://doi.org/10.1007/978-3-031-41679-8_5

1.1   Charts

Charts are visual representations of data composed of simple abstract shapes arranged to have a semantic meaning. They are often used to make it easier to understand large amounts of data and to see patterns and trends. Charts can represent many types of data, including numerical, categorical, and time-based data. Standard chart types include bar, line, scatter, and box plots. The elements of a chart are the visual components used to represent data. These can include the title of the chart, plot text labels, the axes, the legend, and the data points or lines.

The chart's title is typically a brief phrase or sentence describing the data being plotted. The plot text labels include text such as the actual values represented on the chart. The axes are the lines along the bottom/top and left/right sides of the chart and show the scale and range of the data being plotted. The axis usually has an axis title, major/minor tick marks, and tick labels. The legend is a key that explains the meaning of different colors, symbols, or other visual elements used in the chart. These are usually in a horizontal or vertical box with patches representing the key and text representing a data series. The data points or lines are the individual elements of the chart that represent the data. For example, in a line chart, the data points would be plotted along the line, while in a bar chart, they would be represented by the individual bars, grouped bars, or stacked bars; box plots show the distribution as a box with three whiskers - first quartile, central tendency, and third quartile, most box plots also have two whiskers for minimum and maximum value whereas, in a scatter plot, they are just the points.

Other chart elements can include gridlines, which are lines that help to divide the chart into smaller sections and make it easier to read, and data labels, which show the exact values for each data point or line. The overall design of the chart, including the colors, fonts, and layout, can also be considered an element of the chart.

1.2   Chart Data Extraction

There has been decent strides made in the space of document understanding [2,8,18,20], with specialised tasks for non-textual understanding such as parsing reason over mathematical expressions [1,14] One such focus has been automation of chart parsing, which originated from public challenges for chart data extraction [4–6]. Multiple iterations of this competition have spurred significant community interest in this highly challenging task. While earlier challenges comprised large-scale (100k images) synthetic charts on which deep models can achieve very high accuracies, the latest iterations feature only real-world charts, which continue to be extremely challenging especially as it pertains to the end-to-end data extraction task. Our focus is on Task-6 of the challenge where the input is a chart image, the text corresponding to the chart, and structural properties such as the role of each text element and legend and axes elements, and the output is a table with the data used to generate the original chart image. Other works outside this competition have also been published on the chart data extraction task but remain severely constrained, as discussed in Sect. 2.4.

2   Background

In this section, We describe our problem, popular model architectures used in the literature for keypoint(KP) and object detection tasks and techniques for improving such models. We also discuss relevant prior works in this domain.

2.1   Chart Infographics: Chart Data Extraction Challenge

This challenge [6] aims to evaluate and promote the development of automated chart data processing systems. This involves the extraction of structured data from chart images. The challenge is divided into six sub-tasks, which mimic the common steps used in manual chart data extraction.

The first task is chart type classification. The second task is the detection and recognition of text regions in the input chart image. In the third task, text elements are classified according to their semantic role in the chart, such as chart title, axis title, or tick label. The fourth task requires associating tick labels with specific pixel coordinates. The fifth task involves pairing the textual labels in the legend with the associated graphical markers in the chart.

The sixth task is data extraction, where the goal is to extract the original data used to create the chart. This is further divided into two parts: plot element detection and classification, and data extraction. The former involves segmenting the chart image into atomic elements such as bars, points, and lines, while the latter involves producing named sequences of (x, y) pairs that represent the data points used to create the chart. Subsequent tasks assume output of previous tasks as available input. For our work, we focus on the sixth task of this challenge.

2.2   Keypoint Estimation Architectures

Keypoint estimation is a common task in computer vision, involving the detection of specific points or landmarks in an image. Popular models for this task include Hourglass Network (HGN), [15], Cascade Pyramid Network (CPN) [3], and Simple Pose Network (SPN) [10]. HGN has a bottleneck shape that compresses and expands data through downsampling and upsampling layers. Its symmetrical shape allows the model to learn spatial information at multiple scales. CPN is a variant of HGN that combines the strengths of bottom-up and top-down approaches using multiple cascaded hourglass modules. SPN, on the other hand, is a simpler and more efficient model than HGN and CPN, using devonvolution learnable layers for upsampling and a bottom-up approach to KP estimation. In the realm of KP localization benchmarks, including tasks such as human pose, facial landmark detection, and document corner detection, it is generally observed that while HGN and CPN are more accurate, they are also more complex and resource-intensive. Conversely, SPN is faster and simpler to train, but may sacrifice some degree of accuracy.

We conduct experiments with all three variants to provide an exhaustive comparison.

2.3 Anchor-Free Object Detection for Keypoints

In the domain of object detection, a promising approach is to treat it as a KP detection problem and bypass the need for predefined anchors or bounding boxes. This anchor-free paradigm has been successfully implemented in the Single Shot Object Detection framework. CornerNet [9] is based on an HGN backbone and uses two heads to detect the top-left and bottom-right KP of an object. The KP are combined using corner pooling layers. CentreNet [7]extends CornerNet by using an additional center KP. Thus three KP are combined using centre pooling layers. Object as Points [21] proposes that the center point is sufficient to detect objects. It uses the expected center of a box as both an object and a KP to determine the coordinates and offsets of the bounding box. They combine DLA-based architectures [19], with deformable convolutions replacing the upsampling layers in SPN.

Inspired by these methods, we propose an extension that combines all four top-down, bottom-up, left and right directional pooling for improved KP features. Furthermore, we modify the downsampling convolution layers in popular anchor-free object detection models such as HGN, CPN, and SPN with a DLA-34 architecture to experiment with alternative backbones. To achieve this, we implement a hierarchical deep-aggregation strategy for each of the downsampling layers in a KP-backbone encoder, which we believe will enhance the overall performance of the model.

2.4 Chart Data Extraction Models

Domain-specific models have been proposed for bottom-up chart data extraction techniques. In [13], the authors propose an ensemble of different popular models for each task of box detection, point detection, and legend matching. For bar plots and box-plots, they use a Feature Pyramid Network with a ResNet backbone, and for points, they use a Fully Convolutional Network for producing heatmaps. Further, they train a separate feature extractor with triplet loss over legend patches for the legend linking task. In the same 2020 challenge, another submission [12] uses a CentreNet model for bar plots and a CentreNet with DLA-34 [19] connections for box, line, and scatter plots having a different number of final layers per chart type. They do legend matching through HOG features.

Authors of [11] propose using a CornerNet model with an added head for chart-type classification. Current literature [6] provides a general framework for bottom-up chart data extraction. They primarily utilize off-the-shelf computer vision models that are either disconnected and trained for separate tasks, have different architectures for different chart types, or solve only half the task of visual element detection and no legend matching. Extracting data from a chart without its contextual legend information is ineffective. Also, since these works were benchmarked on older versions of datasets or datasets with missing chart types, it is hard to compare across methods.

We provide a systematic study that encompasses this family of architectures and techniques and propose the first model, to our knowledge, for complete chart data extraction.

2.5   Contrastive Loss for Visual Element Reconstruction

For plot elements such as lines, the predicted KP need to be clustered or 'reconstructed' in a bottom-up fashion. To train these clustering embeddings we experiment with two types of contrastive losses.

Push-pull loss operates by comparing the distances between the reference points and the actual data points in the embedding space. For data points similar to their reference point (i.e., belonging to the same class), the push-pull loss function will try to minimize the distance between them while maximizing for the rest.

Mathematically, the push-pull loss for KP detection can be defined as:

$$L = (1 - Y) * (max(0, d_p - d_n + m))^2 + Y * (max(0, d_n - d_p + m))^2$$

where L is the loss, Y is a binary label indicating whether the inputs are similar $(Y = 1)$ or dissimilar $(Y = 0)$, d_p is the distance between the KP detected for the positive input, d_n is the distance between the KP detected for the negative input, and m is a margin hyperparameter that determines how far apart the KP should be.

Multi-Similarity Loss (MS) is designed to encourage learning deep features that are discriminative between classes and similar within each class.

The discriminative term of the loss function is given by:

$$L_{dis} = -\sum_{i=1}^{N} \log \frac{e^{f_{y_i}}}{\sum_{j=1}^{C} e^{f_j}}$$

where N is the number of training examples, C is the number of classes, y_i is the class label of the ith training example, and f_j is the predicted class score for the jth class.

The similarity term of the loss function is given by:

$$L_{sim} = \frac{1}{2N} \sum_{i=1}^{N} \sum_{j=1}^{N} [y_i = y_j] \left(1 - \frac{f_{y_i} - f_{y_j}}{\max(0, f_{y_i} - f_{y_j}) + \alpha}\right)$$

where α is a hyperparameter that controls the strength of the similarity regularization.

The sum of the discrimination and similarity terms then gives the overall loss function:

$$L = L_{dis} + L_{sim}$$

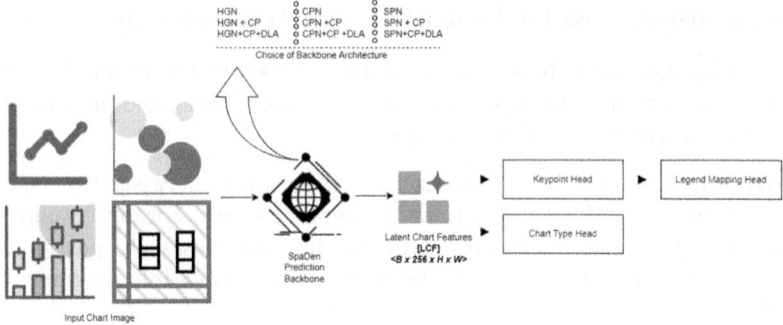

(a) The Unified Data Extraction(UDE) Framework

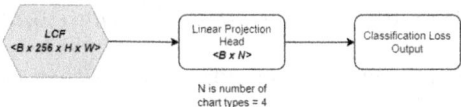

(b) Chart Type Head

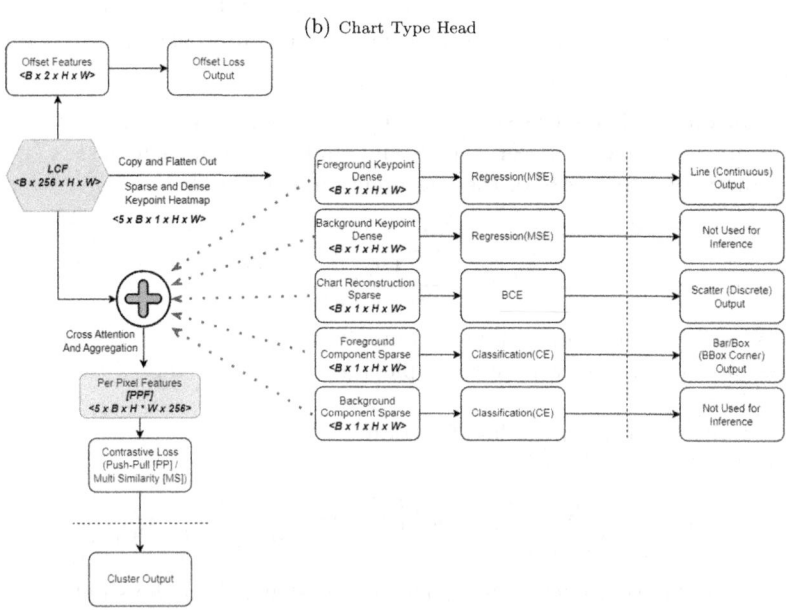

(c) SpaDen Keypoint Head : All 5 heatmaps are trained for each chart type. During inference selective output is used. The dense/sparseness comes from the segmentaion mask used for training.

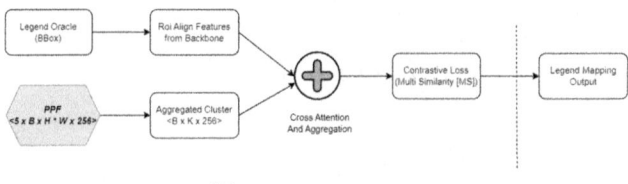

(d) Legend Mapping Head

Fig. 1. Chart Data Extraction using SpaDen Model in the UDE Framework.

3   Unified Data Extraction (UDE) Framework

We propose a generic framework for bottom-up parsing for chart data extraction. Figure 2 shows the general blocks of UDE and Fig. 1 shows a more detailed view of the overall architecture. The UDE framework consists of a generalized backbone chart feature extractor, chart type prediction, KP localization, chart component reconstruction by KP grouping, and a legend mapping block. We discuss these blocks in detail in this section. We also describe a custom segmentation mask formulation, and making the feature extractor invariant to chart text.

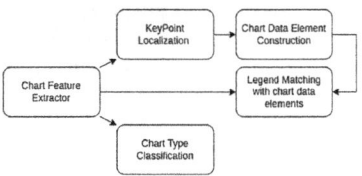

Fig. 2. UDE Framework Blocks

3.1   Backbone Feature Extractor

Inspired by architectures used for KP extraction, our baselines consist of HGN, CPN, SPN, DLA-34, and general-purpose Resnet. We further propose a novel architecture using DLA techniques. We modify the connection layers of HGN, CPN, and SPN by adding iterative deep aggregations. Another way of understanding these is (i) Hourglass blocks where each stage is a DLA encoder, (ii) DLA model with hybrid top-down and bottom-up output layers similar to CPN and (iii) DLA model with deconvolution layers instead of upsampling as in SPN. We conduct ablation studies with each variant and corner pooling layers.

3.2   Chart Type Classification

A linear projection head takes base features and passes them through a series of 1x1Conv-BN layers to get the chart-type output. This is shown in Fig. 1b.

3.3   Keypoint Localization

These networks are trained with an L2 custom loss for KP and an L1 distance loss for offset. We refer readers to [9] for their detailed implementations. Generally, KP estimation literature makes an architectural decision based on the number of output classes for this head. In charts, we have an unconstrained number of KP for types like scatter and line. In our implementation, we predict a single heatmap for all KP. As depicted in Fig. 1c we predict 5 different 'views' of the KP heatmap, trained using different segmentation masks discussed in detail below.

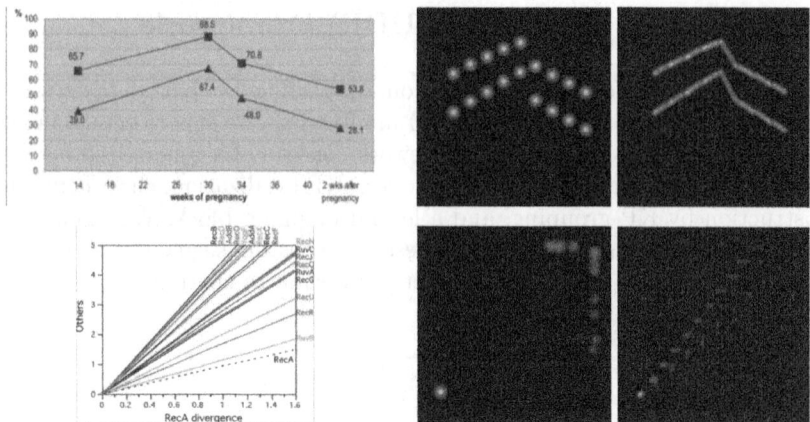

Fig. 3. Custom Groundtruth Masks. First Column: Input Chart, Second Column: Generic Gaussian Mask (Classification), Third Column: Dense Directional Mask (Regression).

Dense Directional Keypoint Masks are developed to learn regression heatmaps for unconstrained KP in charts. This differs from the sparse KP masks typically used in other tasks, such as human pose estimation. Sparse KP estimation involves assigning a probability density centered at each sparse KP location, achieved by applying a Gaussian kernel to the KP. Where each KP has a 'class' and each class has a different output layer. Due to the unconstrained nature of chart KP, we output a single heatmap which makes it harder for the model to segment each point by only using sparse labels.

To address this issue, we create dense directional KP masks by interpolating KP between inflection points. Specifically, we interpolate 10 fixed KP between each ground truth point and use a Gaussian kernel with a spread of 2.0 to generate the dense mask. Figure 3 illustrates an example of the resulting dense mask.

To compute the loss during training, we use the dense masks to focus on informative pixels in the image. We binarize the masks to create classification labels for line plots using a threshold of 0.6. For bar plots, we only use the top-left, center, and bottom-right KP, and for box plots, we only use five KP for each marking. For scatter plots, we use all KP. To generate background masks, we invert the foreground masks.

SpaDen Keypoint-Loss utilizes a combination of direct regression, binary classification, and even multi-class classification. While each loss has its merits, each faces unique challenges on unconstrained KP when trained individually. The lack of data makes regression overfit. Multiple KP without class discriminator makes binary cross entropy hard to converge on a single heatmap, whereas a general lack of informative pixels (99.9% pixels in plot area are background) makes cross entropy an imbalanced classification problem. Even dice loss for

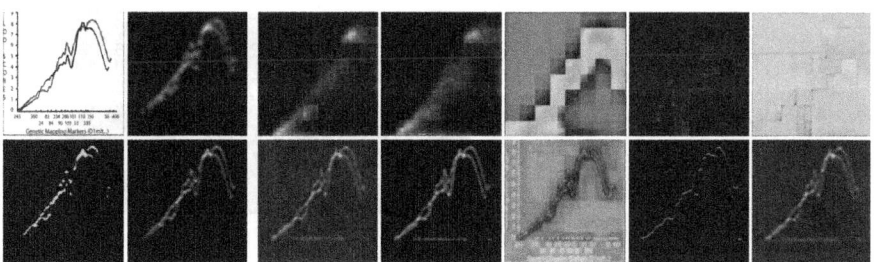

Fig. 4. Affect of Keypoint loss. Top row shows output when trained in isolation: input chart, gaussian mask, predicted BCE heatmap, predicted MSE foreground, predicted MSE background, predicted CE foreground, predicted CE background. Bottom Row shows outputs when trained in combination, first two show the groundtruth Classification Mask and Regression Mask, rest are counterparts of top row outputs.

tackling imbalanced classification severely affects the training process of the model when unconstrained KP from different plot area components are projected on the same single layer instead of the per-class output layer.

Our final model consists of 5 heads for the different 'views' of the chart. Direct binary reconstruction, fore/back-ground regression, and fore/back-ground classifier. Figure 4 shows output predictions for each loss individually as well as when combined. We weigh the background pixels to 0.01 and the foreground to 0.99. We find that using a mixture of sparse masks for classification, and dense directional masks for regression, works best. This architecture choice helps further with chart component reconstruction by providing specific feature outputs to calculate global chart vs. KP attention and also in post-processing for discrete and continuous data in the same model, as described in the next section. The final loss value is an alpha blend of $0.7\times$ Aggregated KP Loss from 5 heads and $0.2\times$ contrastive loss for learning associations and $0.1\times$ chart type classification.

3.4   Keypoint Clustering

In this stage, we first post-process the output heatmaps and then use a per-pixel contrastive loss to cluster the KP.

Foreground Keypoint Heatmaps are critical for our method. The regression output enables us to predict the complete contour of chart elements accurately. The classification output gives us corners/points for discrete elements. Foreground regression output is also used to extract clusters of relevant pixels of individual chart elements. To avoid noise interference in the clustering process, we apply a post-processing procedure shown in Fig. 5. Our approach involves identifying pixel clusters by selecting the top 1,000 pixels with the highest confidence values and reducing each connected component to individual points by thresholding at $0.85\times$ max-intensity.

We also find the median RGB color of the chart and discard all foreground KP within 0.25 min-distance. For each point, we obtain the RGB color distribution

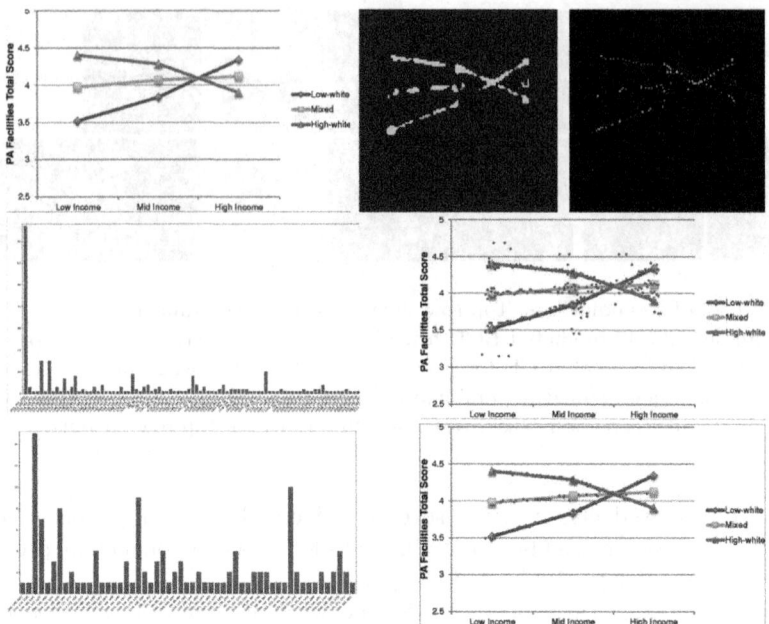

Fig. 5. Post-Processing Regression Prediction: (Top row) Input image, Thresholded islands, Points from connected components, (Middle Row) Colour Histogram over all points, All points from cc before filtering, (Bottom Row) Histogram after discarding pixels close to median color, Final points chosen for contrastive grouping.

and identify the median color value, as well as the peaks in the distribution except for the median. We discard all points with a color distance greater than the minimum distance between the median color and peaks. If the legend patch is available, we use the centroid distance from the median color to the mean legend patch color else mean peak value from cluster. Color distance is computed as the L2 distance between RGB tuples.

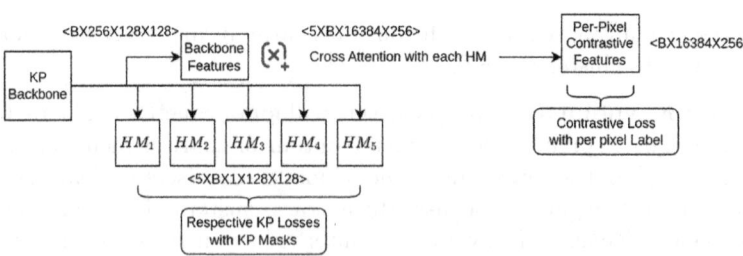

Fig. 6. Global chart vs key-point attention implementation to generate Contrastive Feature Map

Chart Component Reconstruction is done from the pixel clusters learnt through contrastive loss. First, we calculate the cross-attention between the global chart features from the backbone and each of the individual heatmap heads HM_i. The resulting per-pixel embeddings are summed and aggregated to perform clustering, as shown in Fig. 6.

We treat all KP belonging to the same instance of a line, bar, box, or scatter as positive samples, and rest as negative. During inference, we filter the regression heatmap output to obtain K informative pixels and then calculate their exhaustive $K \times K$ similarity. A threshold value of 0.85 is used for cosine similarity in the MS loss or $1e^{-5}$ for Euclidean distance in the push-pull loss.

Each resulting cluster is reconstructed to its original chart component using heuristic rules. Lines are assumed to have a vertical axis as the dependent variable and are joined from left to right. For bars, boxes, and scatter plots, the closest point from each cluster with the highest probability output from the foreground classification heatmap is taken. For bars and boxes, we take the top 2 (corner) and top 5 (whiskers) points, respectively. For scatter plots, we threshold by selecting all points with a confidence interval of 0.25× the maximum value.

To retrieve the data, we use task-6 inputs of a text box, axis tick, and text role to get the data value at the pixel location. We also classify horizontal/vertical bars and boxes as such.

3.5   Legend Mapping

To effectively associate the chart components with their legend patches, we introduce a ROI-align layer that operates on the backbone features. We utilize a legend oracle to obtain bounding box coordinates, then the roi-align layer provides uniform embeddings as illustrated in Fig. 1d. We aggregate the KP clusters using concatenation and 1x1Conv-BN and use the MS-Loss function to measure their similarity with each of the legend patches. During training, the legend oracle provides the association label. At inference time, we match the KP cluster with the most similar legend patch.

3.6   Invariance to Chart Text

KP based backbones exhibit sensitivity to individual pixels, making invariance to text a desirable characteristic. To achieve this, the heatmap features for the plot area must generate KP exclusively for components within the chart area, and not for text within the chart. In order to evaluate the robustness of our approach, we introduced 'easy' and 'hard' samples by selectively adding or removing text boxes from the chart. During each iteration of the training process, there is a 25% chance that all text boxes and content are replaced with the median color from the chart, and a 25% chance of adding skewed and cropped contextual text boxes to random positions within the chart. The model may receive different augmented chart inputs in each epoch, but the same mask labels. The efficacy of this augmentation strategy is demonstrated in Fig. 7, where the text box information is obtained from an oracle.

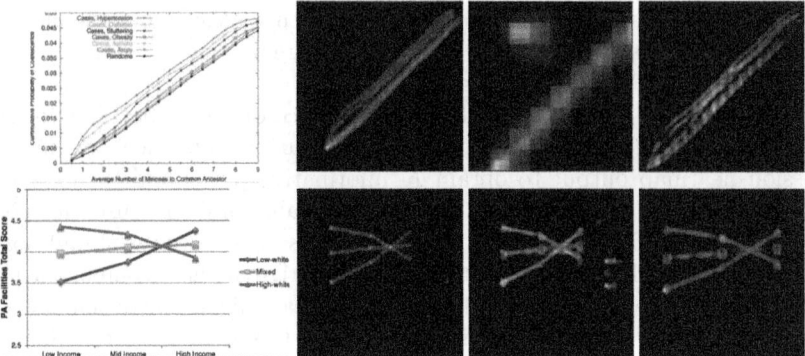

Fig. 7. Text box invariance for chart Keypoint-Estimation. TopRow: Input Chart, Directional Mask, Predicted heatmap using generic mask w/o augmentation, Predicted heatmap using custom mask with augmentation. BottomRow: Input Chart, Directional Mask, Predicted heatmap using custom mask w/o augmentation, Predicted heatmap using custom mask with augmentation.

4   Experiments

We first describe the dataset and evaluation and then provide an exhaustive quantitative evaluation. We summarize our findings and conclusions comparing different architectures, pooling, aggregation techniques, objective functions, and post-processing methods for chart reconstruction.

4.1   Dataset and Evaluation

The Chart-Infographics Challenge Dataset is a rigorously curated collection of over 86,815 real charts sourced from the Open Access (OA) section of PubMed Central (PMC). The dataset comprises both multi-panel and single-panel charts, with the latter forming the focus of downstream tasks. Among these, over 36,000 images are categorized as single-panel charts and are utilized for training and testing. Multi-panel charts are divided to extract underrepresented chart classes. The final training set comprises 22,923 images, and the testing set comprises 13,260 images. In addition, we incorporate 999 charts from Arχiv publications, as provided in [16], along with 100k synthetic charts from the previous Chart-Infographics '19 challenge, to enhance the training data. During each epoch, training is conducted over all real charts and an equivalent number of randomly selected synthetic charts.

The Evaluation is conducted on the publicly released test set [6], with the same splits as the challenge, ensuring a fair comparison in our study. The test set contains a diverse range of chart types, including line plots, scatter plots, horizontal and vertical bar charts, and vertical box-plots. The charts in the dataset have been annotated for hierarchical image classification and further categorized as containing charts or not. For the downstream tasks, the output and ground-truth for each chart are defined as the visual location in terms of (x, y) coordinates for task-6a and a set of (name, data series) pairs for task-6b.

The evaluation metrics for the tasks are designed to capture the nuances of different chart types. For continuous data series, represented by line charts, the metric quantifies the difference between two functions as an integral of their point-wise differences. Bi-variate point set data, such as scatter plots, is evaluated using a capped distance function, where the distance is scaled by the inverse covariance matrix of the ground truth points. Discrete data, represented by bar charts and some line plots, is evaluated using two cases: exact text match and fuzzy text matching. The distance is calculated based on string equality for the exact match case and the normalized edit distance between predicted and ground truth strings for fuzzy text matching. Boxplots are evaluated using exact text matching between the predicted and ground truth strings for real-numbered summary statistics such as minimum, maximum, first quartile, third quartile, and median, ignoring outliers in this representation. The final metric is calculated as 1 minus the distance for each task. We refer the readers to the challenge for exhaustive details[1] and publicly available script [2] used for our evaluation.

4.2 Experiment Result

Qualitative outputs were shown in previous sections alongside the description of our implementation in Figs. 4, 6, 7. These are all conducted with the same simple HGN backbone without added pooling or aggregation. We further discuss quantitative metrics below.

Exhaustive quantitative evaluation is provided in Table 1, grouped by rows.

We report most runs for the 'Line' type chart as they are the hardest to reconstruct and further provide evaluation over all types for the best-performing combination for each flavor of backbone.

We evaluated different reconstruction strategies, ranging from heatmap output to chart objects, using various heuristics such as connected component analysis [CC], [HOG] features extracted from the original image after thresholding and clustering points, and low-level feature matching techniques using correlation, local binary patterns, and cross-correlation. Additionally, we tested push-pull vs. multi-similarity loss objectives for contrastive learning. All hyperparameters for backbone models is kept the same as the original implementations. All thresholding multipliers mentioned in Sect. 3 are calculated using Otsu's hysteresis on randomly sampled heatmaps during validation, minimising val loss and maximising val accuracy.

Rows 1–2 benchmark the public cornernet (HGN+CP) implementation of [11]. Their performance on 6a is comparable out-of-the-box, but for 6b, they

---

[1] https://chartinfo.github.io/metrics/metric.pdf.
[2] https://github.com/chartinfo/chartinfo.github.io/blob/master/metrics/.

Table 1. Data Extraction Results for different models on the chart element detection (6a) and data extraction task (6b). NN - Neural Net, CC - Connected Component, HOG - Histogram of Gradient, FM - Feature Matching, PP - Push Pull Loss, MS - Multi Similarity Loss

#	Model Backend	Element Reconstruction	Chart Type	6a	6b-Data	6b-Name
1	Chart-OCR (pretrained)	NN + Rules	Line	0.71	0.25	-
2	Chart-OCR (pretrained)	NN + Rules	Bar	0.88	0.28	-
3	IIT_CVIT (as reported)	Heuristic	Line	0.773	-	-
4	IIT_CVIT (as reported)	Heuristic	Bar	0.906	-	-
5	IIT_CVIT (as reported)	Heuristic	Box	**0.970**	0.834	**0.921**
6	IIT_CVIT (as reported)	Heuristic	Scatter	0.773	-	-
7	ResNet32	Rule based [CC]	Line	0.29	0.256	0.342
8	FCN	Rule based [CC]	Line	0.38	0.286	0.33
9	HGN	Rule based [CC]	Line	0.49	0.294	0.34
10	HGN	Rule based [HOG]	Line	0.56	0.292	0.34
11	HGN	Rule based [FM]	Line	0.14	0.285	0.34
12	HGN	Learnt [PP]	Line	0.68	0.611	0.741
13	HGN	Learnt [MS]	Line	0.699	0.683	0.756
14	HGN + CP	Rule based [CC]	Line	0.49	0.324	0.34
15	HGN + CP	Learnt [PP]	Line	0.69	0.62	0.756
16	HGN + CP	Learnt [MS]	Line	0.71	0.685	0.77
17	HGN + CP + DLA	Learnt [PP]	Line	0.74	0.66	0.75
18	HGN + CP + DLA	Learnt [MS]	**Line**	**0.83**	**0.69**	**0.77**
19	HGN + CP + DLA	Learnt [MS]	**Bar**	**0.912**	**0.81**	**0.86**
20	HGN + CP + DLA	Learnt [MS]	Box	0.965	**0.882**	0.88
21	HGN + CP + DLA	Learnt [MS]	**Scatter**	**0.782**	**0.62**	**0.55**
22	HGN + CP + DLA	Learnt [MS]	**ALL**	**0.8722**	**0.7505**	**0.765**
23	CPN	Learnt [PP]	Line	0.67	0.581	0.701
23	CPN	Learnt [MS]	Line	0.66	0.592	0.69
24	CPN + CP	Learnt [PP]	Line	0.67	0.585	0.701
25	CPN + CP	Learnt [MS]	Line	0.66	0.592	0.69
26	CPN + CP + DLA	Learnt [PP]	Line	0.675	0.58	0.69
27	CPN + CP + DLA	Learnt [MS]	Line	0.68	0.61	0.56
28	CPN + CP + DLA	Learnt [MS]	Bar	0.89	0.78	0.72
29	CPN + CP + DLA	Learnt [MS]	Box	0.952	0.83	0.772
30	CPN + CP + DLA	Learnt [MS]	Scatter	0.71	0.601	0.661
31	CPN + CP + DLA	Learnt [MS]	ALL	0.808	0.705	0.678
32	SPN	Learnt [PP]	Line	0.59	0.448	0.342
33	SPN	Learnt [MS]	Line	0.62	0.51	0.35
34	SPN + CP	Learnt [PP]	Line	0.587	0.44	0.342
35	SPN + CP	Learnt [MS]	Line	0.613	0.505	0.32
36	SPN + CP + DLA	Learnt [PP]	Line	0.577	0.421	0.338
37	SPN + CP + DLA	Learnt [MS]	Line	0.601	0.499	0.336
38	SPN + CP + DLA	Learnt [MS]	ALL	0.623	0.59	0.482
39	SPN	Learnt [MS]	Bar	0.82	0.782	0.68
40	SPN	Learnt [MS]	Box	0.85	0.812	0.778
41	SPN	Learnt [MS]	Scatter	0.58	0.324	0.27
42	SPN	Learnt [MS]	All	0.7175	0.607	0.519

assume axis position and calculate the range of data values as the max difference between OCR outputs, which does not scale for real charts and has a low 6b-data score. The model also cannot perform legend matching and does not report a 6b-name score. We used an open-source implementation [17] to replace their proprietary text box localization and recognition software while keeping everything else the same.

Rows 3–6 present partial results from the latest challenge participants, referred to as 'IIT_CVIT' in the challenge report [6].

Rows 7–42 present our implementations of the proposed strategies described in this paper:

1. Rows 7–8: Baselines with simple ResNet32 and FCN based backbones.
2. Rows 9–13: We tested the best reconstruction strategy with a simple HGN backbone, and learned embedding through MS Loss provided the best results.
3. Rows 14–22: We added pooling and aggregation to the HGN backbone and compared across objective and chart types.
4. Rows 23–31: We compared CPN backbone-based models with contrastive loss, pooling, aggregation, and different chart types.
5. Rows 32–42: We conducted a similar study for SPN-based models. Since improving encoder architecture through pooling and aggregation did not help, we used a base model for different chart types.

The results show that the best-performing model for line charts is the HGN + CP + DLA model with MS loss for both element detection and data extraction, achieving F1 scores of 0.83 and 0.69, respectively.

The best-performing model for bar charts is also the HGN + CP + DLA model with the MS loss function, achieving an F1 score of 0.912 for element detection and 0.81 for data extraction.

For box charts, the best-performing model is the IIT_CVIT model with heuristic-based element reconstruction, achieving an F1 score of 0.97 for element detection and 0.834 for data extraction.

For scatter charts, the HGN + CP + DLA model with the MS loss function achieves the highest F1 score of 0.782 for element detection and 0.62 for data extraction.

The results also show that rule-based methods for element detection, such as connected components and histogram of gradients, perform poorly compared to learned methods. Furthermore, combining different models, such as the HGN + CP + DLA models, improves the overall performance of the system. It is interesting to note that SPN (upsampling is learned through deconvolution) performs better without added pooling and aggregation. This shows that while these techniques help improve encoders, their contribution might be outperformed by a much more sophisticated decoder, which remains invariant to these features in the deconvolution operation.

4.3   Conclusion

We have described our approaches to chart data extraction through bottom-up KP parsing methods. We present an end-to-end framework for chart visual element detection, data series extraction, and legend matching. We have provided exhaustive experimentation with multiple backbones, pooling, and layering strategies. We find that our approach using HG-Net KP backbone augmented with the proposed pooling and aggregation techniques performs the best.

References

1. Ahmed, S., Davila, K., Setlur, S., Govindaraju, V.: Equation attention relationship network (earn): a geometric deep metric framework for learning similar math expression embedding. In: 2020 25th International Conference on Pattern Recognition (ICPR), pp. 6282–6289 (2021). https://doi.org/10.1109/ICPR48806.2021.9412619
2. Appalaraju, S., Jasani, B., Kota, B.U., Xie, Y., Manmatha, R.: Docformer: end-to-end transformer for document understanding. In: Proceedings of the IEEE/CVF International Conference on Computer Vision, pp. 993–1003 (2021)
3. Chen, Y., Wang, Z., Peng, Y., Zhang, Z., Yu, G., Sun, J.: Cascaded pyramid network for multi-person pose estimation. In: Proceedings of the IEEE Conference on Computer Vision and Pattern Recognition, pp. 7103–7112 (2018)
4. Davila, K., et al.: ICDAR 2019 competition on harvesting raw tables from infographics (chart-infographics). In: 2019 International Conference on Document Analysis and Recognition (ICDAR), pp. 1594–1599 (2019). https://doi.org/10.1109/ICDAR.2019.00203
5. Davila, K., Tensmeyer, C., Shekhar, S., Singh, H., Setlur, S., Govindaraju, V.: ICPR 2020 - competition on harvesting raw tables from infographics. In: Del Bimbo, A., et al. (eds.) ICPR 2021. LNCS, vol. 12668, pp. 361–380. Springer, Cham (2021). https://doi.org/10.1007/978-3-030-68793-9_27
6. Davila, K., Xu, F., Ahmed, S., Mendoza, D., Setlur, S., Govindaraju, V.: ICPR 2022 - challenge on harvesting raw tables from infographics. In: International Conference on Pattern Recognition (2022)
7. Duan, K., Bai, S., Xie, L., Qi, H., Huang, Q., Tian, Q.: Centernet: keypoint triplets for object detection. In: Proceedings of the IEEE/CVF International Conference on Computer Vision, pp. 6569–6578 (2019)
8. Gu, J., Kuen, J., Morariu, V.I., Zhao, H., Jain, R., Barmpalios, N., Nenkova, A., Sun, T.: Unidoc: unified pretraining framework for document understanding. Adv. Neural. Inf. Process. Syst. **34**, 39–50 (2021)
9. Law, H., Deng, J.: Cornernet: detecting objects as paired keypoints. In: Proceedings of the European Conference on Computer Vision (ECCV), pp. 734–750 (2018)
10. Li, J., Su, W., Wang, Z.: Simple pose: rethinking and improving a bottom-up approach for multi-person pose estimation. In: Proceedings of the AAAI Conference on Artificial Intelligence, vol. 34, pp. 11354–11361 (2020)
11. Luo, J., Li, Z., Wang, J., Lin, C.Y.: Chartocr: data extraction from charts images via a deep hybrid framework. In: Proceedings of the IEEE/CVF Winter Conference on Applications of Computer Vision, pp. 1917–1925 (2021)
12. Luo, Z., Zhang, Z., Li, G., Che, L., He, J., Xu, Z.: A benchmark for analyzing chart images. In: Del Bimbo, A., et al. (eds.) ICPR 2021. LNCS, vol. 12668, pp. 390–400. Springer, Cham (2021). https://doi.org/10.1007/978-3-030-68793-9_29
13. Ma, W., et al.: Towards an efficient framework for data extraction from chart images. In: Lladós, J., Lopresti, D., Uchida, S. (eds.) ICDAR 2021. LNCS, vol. 12821, pp. 583–597. Springer, Cham (2021). https://doi.org/10.1007/978-3-030-86549-8_37

14. Mansouri, B., Agarwal, A., Oard, D.W., Zanibbi, R.: Advancing math-aware search: the ARQMath-3 lab at CLEF 2022. In: Hagen, M., et al. (eds.) ECIR 2022. LNCS, vol. 13186, pp. 408–415. Springer, Cham (2022). https://doi.org/10.1007/978-3-030-99739-7_51
15. Newell, A., Yang, K., Deng, J.: Stacked hourglass networks for human pose estimation. In: Leibe, B., Matas, J., Sebe, N., Welling, M. (eds.) ECCV 2016. LNCS, vol. 9912, pp. 483–499. Springer, Cham (2016). https://doi.org/10.1007/978-3-319-46484-8_29
16. Siegel, N., Horvitz, Z., Levin, R., Divvala, S., Farhadi, A.: FigureSeer: parsing result-figures in research papers. In: Leibe, B., Matas, J., Sebe, N., Welling, M. (eds.) ECCV 2016. LNCS, vol. 9911, pp. 664–680. Springer, Cham (2016). https://doi.org/10.1007/978-3-319-46478-7_41
17. Smith, R.: An overview of the tesseract OCR engine. In: Ninth International Conference on Document Analysis and Recognition (ICDAR 2007), vol. 2, pp. 629–633. IEEE (2007)
18. Xu, Y., Li, M., Cui, L., Huang, S., Wei, F., Zhou, M.: Layoutlm: pre-training of text and layout for document image understanding. In: Proceedings of the 26th ACM SIGKDD International Conference on Knowledge Discovery & Data Mining, pp. 1192–1200 (2020)
19. Yu, F., Wang, D., Shelhamer, E., Darrell, T.: Deep layer aggregation. In: Proceedings of the IEEE Conference on Computer Vision and Pattern Recognition, pp. 2403–2412 (2018)
20. Zhong, X., Tang, J., Yepes, A.J.: Publaynet: largest dataset ever for document layout analysis. In: 2019 International Conference on Document Analysis and Recognition (ICDAR), pp. 1015–1022. IEEE (2019)
21. Zhou, X., Wang, D., Krähenbühl, P.: Objects as points. CoRR arXiv:1904.07850 (2019)

Generalization of Fine Granular
Extractions from Charts

Shubham Paliwal[✉], Manasi Patwardhan, and Lovekesh Vig

TCS Research, New Delhi, India
{shubham.p3,manasi.patwardhan,lovekesh.vig}@tcs.com

Abstract. Current approaches for high precision fine-grained visual extractions from charts are highly data intensive requiring thousands of annotated samples. Annotating a dataset and retraining for every new chart type with a shift in the spatial composition of chart elements, text role regions, legend preview styles, chart element shapes and text-role definitions, is a time-consuming and costly affair. Current approaches struggle to generalize to new chart types with a distributional shift across the above dimensions. In this paper, we define a novel attention and dynamic filtering based approach for extracting chart elements and identifying text-role regions, achieving SOTA results for visual extraction on charts for the PlotQA dataset, surpassing existing approaches by an mAP of 2.81% @0.90 IOU. More importantly, the methods proposed are designed to adapt to unseen chart types having the above mentioned shifts in chart element distributions. We demonstrate the generalization capabilities of our models trained on the PlotQA train set by providing chart extraction results on out-of-distribution charts selected from the LeafQA dataset. We achieve an mAP of 90.64% and 92.18% for @0.90 IOU for this well-curated out-of-distribution chart data in zero and few-shot settings, respectively.

Keywords: Chart Extraction · Information extraction

1 Introduction

Charts are compact visualization techniques frequently used for illustrating facts in scientific and financial documents in order to summarize observations and draw conclusions about the underlying data. The fine grained perception capabilities required to interpret the elements of a chart are one of the main bottlenecks towards automated fact extraction from charts. High precision detection of visual elements is critical, as errors in this step would cascade to downstream inference tasks leading to substantial discrepancies in the final conclusions, especially in case of numerical data. These chart elements are distributed across textual elements such as the chart title, X/Y-axis labels, X/Y-ticks and tick labels, legend preview, legend labels, in addition to visual such as bars, lines, and dots. Existing approaches for information extraction from charts fine-tune object detection networks such as Fast-RCNN [6], Faster-RCNN [29], YOLO [28],

G. A. Fink et al. (Eds.): ICDAR 2023, LNCS 14188, pp. 94–110, 2023.
https://doi.org/10.1007/978-3-031-41679-8_6

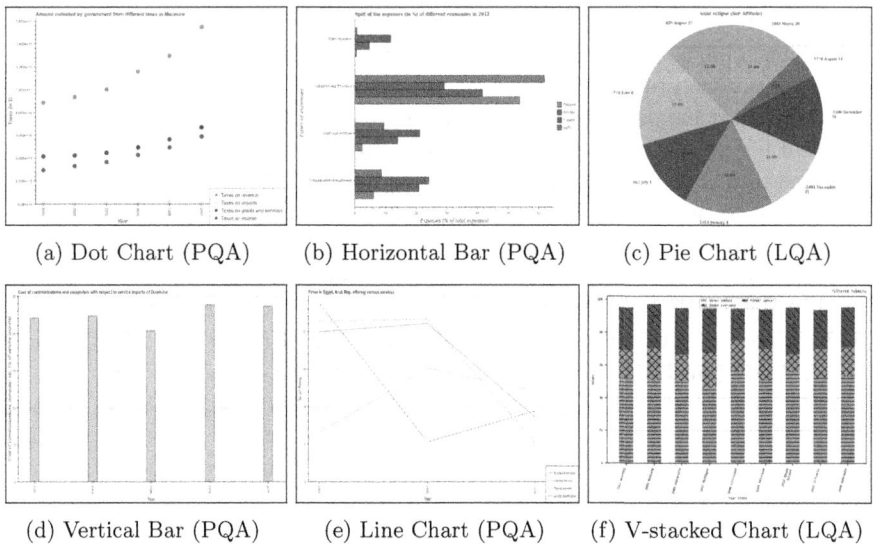

(a) Dot Chart (PQA)   (b) Horizontal Bar (PQA)   (c) Pie Chart (LQA)

(d) Vertical Bar (PQA)   (e) Line Chart (PQA)   (f) V-stacked Chart (LQA)

Fig. 1. Chart Types from PlotQA (PQA) [26] and LeafQA (LQA) [3]

SSD [21], and Mask-RCNN [9]. Fine-tuning is done over a large number of charts (157K for PlotNet [5] and 200K for LeafQA [3]) either with [3,5,22,23] or without [26] modification to the construction of the region proposals and/or the loss function. These approaches achieve a maximum mAP (mean Average Precision) of 93.44% @0.90 IOU (Intersection Over Union) for visual element detection which is not good enough for consistent downstream extraction quality. Thus, these models not only struggle with fine-grained Chart Visual Element Extraction but are also highly data intensive.

Charts have distinct spatial compositions for certain chart elements such as the chart title which is predominantly positioned on the top or bottom of the chart. Similarly, chart legends can be at distinct positions in different charts (Charts (a), (b) and (f) in Fig. 1), X/Y axis labels may change their positions based on whether graphs are horizontal or vertical (Charts (d), (f) versus chart (b) in Fig. 1) and bars in bar graphs might be adjacent or stacked (Charts (b) versus chart (f) in Fig. 1). Also, charts have distinct types of legend preview styles with usage of different colors (Charts (a), (b), (c), (d) in Fig. 1), texture patterns or combination of both (Charts (e) and (f) in Fig. 1). Moreover, charts are of distinct types (Fig. 1) such as vertical/horizontal bar graphs, stacked bar graphs, line charts, dot charts, pie charts, donut charts, box-plots and others. These distinct types of chart have chart elements of distinct shapes such as rectangular bars (Charts (b), (d) and (f) in Fig. 1), pie shaped pie charts (Chart (c) in Fig. 1), circular dots (Chart (a) in Fig. 1), etc. Also, the distinct chart types may have distinct text role definitions, for instance bar (Charts (b), (d) and (f) in Fig. 1), line (Chart (c) in Fig. 1) and dot charts (Chart (a) in Fig. 1) have text roles corresponding to X/Y axis labels, X/Y axis tick labels, whereas for Pie (Chart (c) in Fig. 1) or Donut charts, the only valid text roles are region labels. Thus, charts are subject to variations across the following dimensions: (i) spatial composition of chart elements and text role regions, (ii) legend preview styles, (iii) chart element shapes and (iv) text-role definitions.

In real life scenarios, there can be a shift in the distribution over the above mentioned dimensions. Current architectures mentioned above are not designed to generalize to unseen charts with these distributional shifts. Realistically, it is infeasible to create an annotated dataset with thousands of samples for each new chart dimension/type and re-train the model each time to achieve satisfactory fine-grained extraction. Thus, we need an approach which can be easily adapted to unseen chart types with no (zero-shot) or minimal (few-shot) training data to provide comparable performance for chart visual extractions.

In this paper, we define two novel architectures for (i) Extraction of chart elements such as bars, lines, and dots (ii) Text Region and Role Detection which allows for segmentation of textual components of charts and classification of those segments to distinct text roles.

The main contributions of this work are as follows:

- The novel architectures achieve State-of-the-Art(SOTA) results for precise fine-grained chart visual extractions on the PlotQA dataset [26] with an mAP of 96.25% @0.90 IOU, surpassing existing approaches [5] by an mAP of ∼3% @0.90 IOU.
- Our approach provides results comparable with the SOTA achieving an mAP of 96.17% @0.90 IOU with significantly less data (only 2.50% of training data).
- Our Chart Element Extraction (CEE) model uses an attention mechanism to segment the regions in the chart following the same style as that of the legend preview, which serves as the query patch (Sect. 4.2). This mechanism facilitates the detection of chart elements independent of their spatial compositions and shapes, style definitions of legend previews and text-role definitions, making it generalizable to unseen charts. We demonstrate that the model yields comparable results on a subset of charts from the LeafQA dataset [3] by rapidly adapting in a few shot setting (Sect. 5).
- Our Text Region and Role Detection (TRR) model uses dynamic filtering with a text role patch as the trigger to detect the remaining text patches belonging to the same text role (Sect. 4.1). This mechanism allows the model to exploit the spatial relationship between the regions belonging to the same text role and makes it invariant to the changes in i) the spatial composition of the text regions, and ii) chart element shapes and styles. We demonstrate the generalization capabilities of the model by showcasing comparable results on a subset of charts from the LeafQA dataset [3], which share the same text roles, in both zero and few shot settings (Sect. 5).

2   Related Work

We examine existing techniques for extraction of visual as well as textual elements from charts and their generalization capabilities. We mention the prior work on dynamic filters in the Supplementary section.

Chart Extraction approaches such as [15], which use techniques like the VegaLite interpreter [31], D3 deconstructor [7,8] or ReVision [32] to extract chart elements require charts be constructed from a predefined grammar. [25] use image processing for chart extractions but the simplistic chart images they tackle are synthetically generated from the FigureQA [13] and CQAC[1] datasets. Recent extraction approaches fine-tune object detection networks such as DenseNet [10], Fast-RCNN [29], Fast-RCNN with Feature Pyramind Network [18], Yolo2 [4], SSD [21], Mask-RCNN [34] and cascade RCNN [23], on a large number of charts. [5,22,26] demonstrate that existing object detection networks perform reasonably well at an Intersection Over Union (IOU) of 0.50. However, for higher IOUs of 0.75 and 0.90, the accuracy falls drastically. For the fine grained visual element detection required for charts, having a high margin-of-error (IOU of 0.50) is unacceptable.

To address this issue, (i) [22] adopt a modified version of CornerNet [17] with an Hourglass Net [27] backbone for key point detection. (ii) The latest end-to-end chart extraction approach named PlotNet [5] uses a hybrid network employing a Fast R-CNN (FRCNN) along with FPN and ROI-Align layers. These approaches [5,22] achieve an mAP of up to 93.44% @0.90 IOU. However, these models are extremely data hungry with 157K samples needed for training PlotNet [5], 200K for LeafQA [3] and 400K for ExcelChart400K. Thus, these models are infeasible for real-world applications, where large scale annotated chart images are not available. In this paper, we design a novel pipeline of neural architectures to extract visual elements from charts which yields SOTA results for chart extraction with an mAP of 96.25% @0.90 IOU on the PlotQA dataset. More importantly, our approach provides comparable results (mAP of 96.17% @0.90 IOU) with significantly less data (only 2.50% of PlotQA training data).

We further evaluate the generalization capability of the above approaches [5, 22]. [22] apply a set-of rule-based algorithms specific for a chart-type in order to extract chart data elements from the detected key-points. As these algorithms are specialized, they do not generalize to unseen chart-types. The prediction heads of PlotNet [5] consists of a Classification (CH), Regression (RH) and a Linking head (LH). CH classifies the given region proposal into 10 different classes (bar, dot-line, legend-preview, legend-label, plot-title, x/y-axis label, x/y-axis ticks, background), whose bounding boxes are obtained from the RH. LH predicts whether a given proposal needs to be merged with its immediate 4 (top, left, right and bottom) neighbours. As the CH is constrained with a fixed number of classes, the method cannot be generalized to charts of other types (pie, donut, box-plots) with out-of-distribution (OOD) text-role definitions. Secondly, PlotNet does not use an RPN network. Instead, it proposes a combination of vision methods (edge

---

[1] https://cqaw.github.io/challenge.

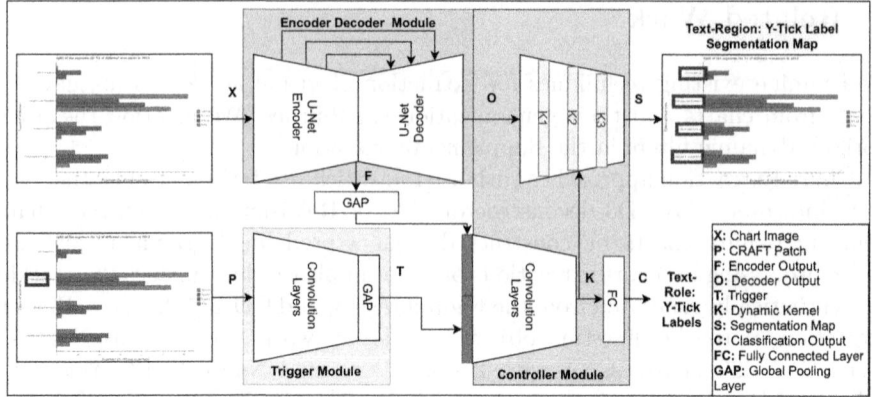

Fig. 2. Text Region and Role Detection Model

detectors and same color/intensity segmentation) to generate region proposals. The authors claim that the method yields fewer and more accurate proposals as compared to RPN. However, while these image processing based methods work for the solid colored PlotQA [26] chart elements, they would struggle to adapt to texture rich chart elements containing patterns (such as crossed lines and checks) or to shapes of unseen chart types. Lastly, PlotNet [5] treats every entity present in the chart as an independent object. Thus, there is no conceptual understanding of the relationships between chart entities. For instance, PlotNet is oblivious to the fact that the legend preview depicts the style of the visual chart elements or that the text regions for the same text roles (X/Y axis tick labels or legend preview labels) are related. In this paper we demonstrate that our chart element extraction approach works well on OOD LeafQA charts with distinct legend preview styles and unseen chart element shapes in both zero and few-shot settings. Moreover, our models exploit the generic relationships between the chart elements aiding generalization over a wide variety of chart types not seen during training.

3   Datasets

To cater to line of research of Q&A over charts, the community has created synthetic datasets such as DVQA [12] or FVQA [13]. However, these simulated datasets are unrealistic with limited variability in the data-values and chart elements. To avoid these biases, Leaf-QA [3] and PlotQA [26] datasets are constructed from real-world sources like government census and financial data, world bank open data, and the global terrorism database. Both these datasets have good variations in their chart-types which include horizontal/vertical group bars, stacked bars, pie charts, donut charts, horizontal/vertical box-plots, dot-graphs, line-graph, and scatter-plots. The LeafQA++ [34] dataset also contains real world charts, however, to the best of our knowledge, is not publicly available. Recent works [14,19,20,24] have used datasets of real charts. However, these approaches are end-to-end and directly extract the data series table out of the plot. Thus, unavailability of the intermediate annotations and results prevent us from benchmarking against these approaches.

In this paper, we train our models using the PlotQA [26] dataset which contains horizontal/vertical bar charts, dot and line charts and test on the PlotQA test split. In PlotQA there are 157,070, 33,650 and 33,657 plot images in train, valid and test splits respectively. To demonstrate the data efficiency of our pipeline we also train our models with only 4K representative images (1K per chart type), which are randomly sampled from the train split of PlotQA. We use a few OOD samples from the LeafQA [3] dataset, having stacked bar charts, pie charts, donut charts and box-plots to demonstrate the generalization capabilities of our models. Two sets of test samples are selectively chosen from the LeafQA [3] dataset for testing our models in a zero-shot setting: (i) 500 charts which have a distributional shift in the spatial composition of visual and textual elements and legend preview styles, yet follow the PlotQA distribution for chart shapes and text role class definitions, (ii) 500 charts of distinct chart types with a shift in the distribution of shapes of chart elements and text role class definitions, yet follow the PlotQA distribution for spatial composition of chart visual and textual elements and legend preview styles. We call these test sets *LeafQA-spatial-style* and *LeafQA-shape-roles*, respectively. We make these test sets publicly available for further research[2]. We test our models on both the test sets to evaluate for generalization across all four dimensions, viz. spatial composition of chart elements, legend preview styles, chart shapes and text role class definitions. Generalization capability of the Text Region Role Detection model is evaluated on text-role classes seen during training. For few shot adaptation results, we fine-tune the models on N (10 in our experiments) samples from the test sets with the worst zero-shot performance and compute the results on the remaining $(500 - N)$ samples.

4   Approach

We extract visual chart information by employing modules for: (i) Chart Element Extraction (CEE) (ii) Text Region and Role Detection (TRR). We describe both the modules here along with the novel architectures defined and their important features which allow for generalization.

4.1   Text Region and Role Detection (TRR)

The text present in the image is detected by employing the CRAFT model [1]. However, CRAFT frequently misses isolated characters and often yields partial detection of text regions. We propose a novel approach depicted in Fig. 2 for TRR, which corrects partially detected text, segments out the corrected text region and identifies text-role labels (such as chart title, legend, X/Y-axis, and

---

[2] https://drive.google.com/drive/folders/100FbK_CliT7fehQEvuS18TyDSmLeWZDx.

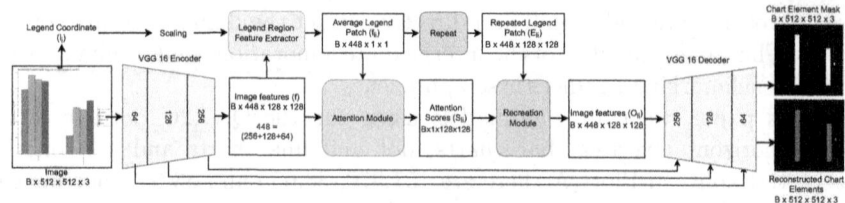

Fig. 3. Architecture of Chart Element Extraction Model

X/Y-tick labels) for each segmented text region. As we have multiple set-of segments (text-regions) getting detected for a single image, we have defined an approach which uses dynamic kernels. The dynamic kernels use a patch based triggering mechanism where given a text role patch, the model detects the remaining patches belonging to the same text role. This mechanism allows the model to be invariant to the spatial composition of text regions belonging to distinct roles, and chart element shapes/styles, thus allowing for generalization to unseen charts sharing the same set of text roles. The architecture consists of two modules described below.

Encoder-Decoder Module. For this module we use the U-net model [30] where the encoder has four down-sampling steps, each consisting of a convolution layer with a 3×3 kernel, followed by a ReLU activation and a batch normalization layer. In order to minimize information loss, we replace the Max-Pooling layer, which employs batch normalization, with a convolution layer with Stride 2. Starting from 32 features in the first step, each down-sampled step contains twice the number of features until the latent representation layer. The latent representation for an image x obtained from the U-net encoder is:

$$r = UNET_E(x) \tag{1}$$

The encoder filters out irrelevant information by squeezing the feature map to a latent space. The decoder is symmetric to the encoder block. At each up-sampling layer, the feature maps are reduced by a factor of 2, and subsequently, the symmetric layer from the encoder block is summed to the feature map, as a skip connection. The output of the decoder for an image x is:

$$o = UNET_D(r) \tag{2}$$

Trigger-Controller Module. The image x is appended with the trigger patch p of a text region detected by CRAFT along the channel dimension, with highlighted patch contours. This patch has the same dimensions as the image so after appending we obtain an updated image x_p. With this input, the trigger module extracts features:

$$t = GAP(NN_T(x_p)) \tag{3}$$

where NN_T is a convolutional feature extractor followed by a Global Average Pooling (GAP) layer. The features of the trigger patch t are concatenated ($\|$)

with the extracted encoder output features r, and are fed to the controller module to generate dynamic kernels:

$$k = NN_C(GAP(r)||t)))\tag{4}$$

The next part of the controller module is the dynamic head which contains three stacked 1×1 convolution layers, whose kernel weights are derived from the dynamic kernel output k to get the final segmentation map s as explained in [11].

$$s = (((o \circ k_1) \circ k_2) \circ k_3)\tag{5}$$

where $\circ$ represents the convolution operation and k_1, k_2, k_3 are the weights of the dynamic kernel k distributed across the three convolutions. The dynamic kernel output k is then fed to a fully connected linear layer to determine the text role of the region. Thus, the trigger-controller module exploits the spatial relationships between text-roles to generate dynamic kernels and obtain text-role specific segmentation maps from the decoded image.

The whole network is trained with cross entropy loss by comparing the predicted and actual text-role class labels c and the predicted and the actual text-role segmentation maps s. During inference, given a trigger patch p of an image x for a detected text-region belonging to an unknown text-role, the model provides the actual text-role classification output c and the segmentation map s of the text-regions for that text-role. Trigger patches overlapping with detected text-role regions are removed before repeating the process for the remaining trigger patches. This process may lead to multiple segmentation maps for each text-role, over which a union operation is performed.

4.2   Chart Element Extraction (CEE)

Prior works have performed the chart extractions in two steps: (i) different chart elements are detected using object detection methodologies and (ii) extracted elements are mapped with corresponding legend previews assigning semantic meaning to extracted chart elements. While some approaches rely on color matching, usually, the second step involves training via a contrastive loss to learn models capable of aligning the legend pattern with chart elements [2,23].

In our work, we propose an end-to-end model, which extracts chart elements and also assigns semantic meaning to them. Our novel method requires only a single patch of the chart element style (colour, patterns, etc.), preferably from the detected legend preview, to segment out all the matching chart element regions from the chart image. The intuition for the proposed approach is derived from the fact that, irrespective of the chart type, legend previews replicate the pattern representations within the chart elements. The architecture uses an attention mechanism to segment the regions in the chart with the same style as that of the legend preview which serves as the query patch. This allows the approach to be invariant to chart types and thus generalize to accommodate the distributional shift in spatial composition of the chart elements, legend preview styles, chart element shapes and text-role definitions. The architecture consists of the following three modules:

Legend Region Feature Extractor Module. Our proposed architecture utilizes the VGG16 architecture [33] as the encoder backbone, to extract a rich image representation. As, the relevant features for our task are more localized, the shallow features extracted from the first three pooling layers of VGG16 are appended and used as the image feature representation f. The detected legend labels extracted via the TRR model (Sect. 4.1) are used to identify the location of legend previews. We provide the detected legend preview bounding box, l_i, to the model after scaling it down to the spatial size of the chart feature map f. For the details of the exact dimensions refer to Fig. 3. We take the mean coordinates of the bounding box of l_i and the features corresponding to these coordinates in f to generate the extracted features f_{l_i} for the legend preview l_i, having its style information. As the field of view (FOV) of the VGG Encoder when using 3×3 pooling filters is restricted to local regions, the legend features in f which are accumulated from shallow layers are not contaminated by other nearby legend entries. This leads to f_{l_i} spanning only the local neighbourhood.

Legend Attention Module. The legend attention module clusters regions which are similar to the legend preview. To achieve this, it utilizes the legend preview features and finds their similarity with the rest of the image features. Traditionally, *cosine similarity* is used for similarity computations. However, in traditional cosine similarity there is no control over the cosine curve. In our use-case, usually the charts have very similar patterns, thus using traditional cosine method might not facilitate definite demarcations, leading to different chart elements having close similarity scores or similar chart elements having distinct similarity scores with the given legend preview. To overcome this issue, we leverage *von Mises-Fisher(vMF)* based measuring function. However, one serious drawback of the vMF based formulations is that it suffers from its light tail. This creates difficulty in learning as back-propagation yields null gradients for the light tailed region of the distribution. Thus, along similar lines as [16] we extend the vMF with the heavy-tailed student-t distribution and scale to obtain similarity values by using Eq. 6.

$$sim(x,y) = \frac{1 + \hat{x}.\hat{y}}{2(1 + k.(1 - \hat{x}.\hat{y})))} \tag{6}$$

where, $\hat{x}$, $\hat{y}$ are unit vectors of x and y respectively, whose similarity has to be computed and k is the parameter to control the compactness of the similarity curve. Using (6), we compute the similarity score S_{l_i} of the image's feature map f with respect to f_{l_i} in the range 0 to 1. This helps to obtain an attention heat map based on similarity value.

$$S_{l_i} = sim(f_{l_i}, f(x,y)) \ \forall x \in [1...h_f] \ \forall y \in [1...w_f] \tag{7}$$

where h_f and w_f are the height and width of feature map f.

Figure 5 shows the comparison of how different attention functions align similarity. Here the X-axis is the dot product of the normalized vectors representing the features of the query preview patch and the features of a chart element and the Y-axis depicts the attention score.

Re-creation Module. As a part of this module we segment the regions matching the legend preview styles from the chart image. Exploiting the key-value relationship inherent in charts between the legend preview styles and the corresponding chart image patches, we use the legend preview as a unit style patch to identify the corresponding legend style distribution in the chart image. This is performed by applying the following two steps.

We first perform a 'REPEAT' operation of the legend preview features f_{l_i} across the image feature map f. The 'REPEAT' function creates an empty template with dimensions of f, filled with f_{l_i}. The resulting output E_{l_i} matches the resolution of the image's feature f.

$$E_{l_i} = REPEAT(f_{l_i}, f(x,y)) \; \forall x \in [1...h_f] \; \forall y \in [1...w_f] \qquad (8)$$

Finally, the obtained feature map E_{l_i} is scaled with respect to a similarity score S_{l_i} obtained from 6, such that

$$O_{l_i} = (E_{l_i} \times S_{l_i}) + \epsilon \qquad (9)$$

where ϵ is a constant adjustment factor. This allows for suppression of the chart regions not matching with the legend preview features and amplification of the regions matching the preview style. Obtained output from Eq. (9) is scaled up along-with skip connections from the encoder to smooth out the boundary regions. The final decoder layer decodes the regions corresponding to the legend preview pattern from the chart region. We compare this with the ground truth chart element mask to compute the Binary Cross Entropy loss. To facilitate better training of the model we also consider an auxiliary reconstruction loss.

Handling Non-legend Cases. As mentioned, the above model assumes the presence of legend preview entries, which serve as an input to the model as a query patch. However, there are also special cases where charts representing a single entity do not have a legend (For example, Chart (d) in Fig. 1). To handle such cases, we create a processing pipeline where we select random patches from a chart. We plot the color histogram of the chart image and observe that the background color has the highest frequency. With this histogram as the reference, depending on the occurrence of the color in a sampled patch, it is labeled as background and foreground. With each foreground patch as the query along with the corresponding chart image as an input to the trained CEE, we extract regions similar to a query patch from the chart image. We remove the other query patches which are covered by the regions of the given query patch. Repeating this process on all the selected patches yields a set-of segments of chart elements which share a common style. There is a possibility of noisy segments getting detected due to the chart element edges and textual regions. To get rid of such noisy segments we use a threshold which defines the minimum number of pixels required within a segment, for it to be non-noisy. This thresholding mechanism is mainly useful for charts like line charts, where the chart element segments contain fewer pixels.

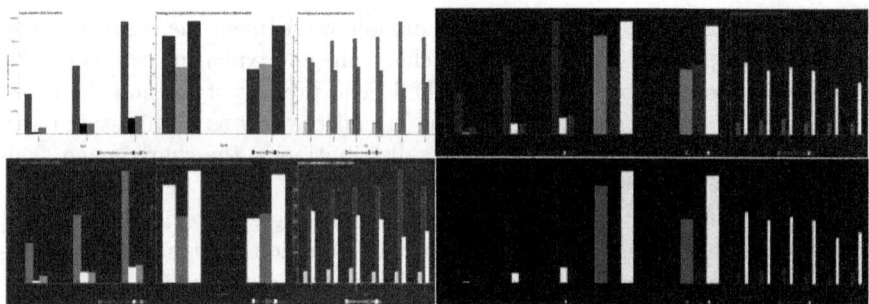

Fig. 4. Attention Similarity Score Visualization (Left Top) Input image, (Left Bottom) Cosine, (Right Top) t-vMF k = 8, (Right Bottom) t-vMF k = 32

5 Results and Discussion

We first discuss hyper-parameter settings used for training with the PlotQA as well as fine-tuning for few-shot adaptation on the LeafQA. The TRR model encoder has the pre-trained VGG19 backbone. We train TRR with a batch size of 8, for 2 epochs over the entire PlotQA training set, using the Adam optimizer with an initial learning rate of 0.0005. While creating the training tuples, we under-sample the axes-label classes to address class imbalance. For the CEE encoder, a pre-trained VGG16 is set as the backbone further conditioned with pre-training in an auto-encoder setting on the PlotQA train set. We train the model with a batch size of 6 and use the *ReduceLROnPlateau* scheduler on the number of iterations, with patience and factor value of 64 and 0.5, respectively. For both the models the spatial resolution of an input image is set to 512×512. We employ an augmentation scheme for both the models to include spatial transforms (rotation in range $-\pi/6$ to $\pi/6$, with translation between -50 to $+50$ pixels along both axes, for the MirrorTransform and ZoomTransform, we use a flipping ratio of 0.5 and a scale factor ranging from 0.5 to 2, respectively), noise transforms (Gaussian noise and Gaussian blur transforms with noise variance from 0 to 0.2 and blur sigma in range of 0.5 to 1, respectively), resample transforms (simulate low resolution input data by scaling down the image and resizing back to the original size, zoom scale used in the transform lies in range of 0.5 to 1) and color transforms (brightness, contrast, gamma and color exchange). For the legend attention module, we used the optimal value of $k = 32$ (Eq. 6) after experimenting with different values of k (Fig. 4).

Table 1 illustrates the SOTA chart visual extraction results on the PlotQA Test Set with 96.25% mAP @0.90 IOU when trained with 157K images, surpassing the baseline PlotNet [5] by 2.81% mAP @0.90 IOU. With Our data efficient architectures, we get comparable results with only 2.5% ($\sim$1K per chart-type, total 4K) of the training samples, illustrated in the last row of Table 1. For Table 1 and 3, the results of the first two columns (Bar, Dot-line) are based on the inference performed with the CEE Model (Sect. 4.2). The results of columns Leg Lbl, Leg PV, Plot title, X/Y axis Lbl and Ticks are based on the inference

Table 1. Results on the PlotQA Test Set with mAP scores (in %). Leg: Legend, Lbl: Labels, PV: Preview. Bar, Dot-line results are based on the inference with the CEE and Leg Lbl, Leg PV, Plot title, X/Y axis Lbl and Ticks results are based on the inference with the TRR. Both trained with PlotQA Train set. **Bold Underlined**: Best Performance, **Bold**: Second-Best

IOU	@0.90									@0.75	@0.50
Existing Models[5]	Bar	Dot-line	Leg Lbl	Leg PV	Plot Title	X-axis Lbl	X-axis Ticks	Y-axis Lbl	Y-axis Ticks	mAP	mAP	mAP
FRCNN (FPN+RA)	87.59	31.62	79.05	66.39	0.22	69.78	88.29	46.63	84.60	61.57	69.82	72.18
FrCNN (RA)	63.86	14.79	70.95	60.61	0.18	83.89	60.76	93.47	50.87	55.49	89.14	96.80
FrRCNN (FPN+RA)	85.54	27.86	93.68	96.30	0.22	99.09	96.04	99.46	96.80	77.22	94.58	97.76
PlotNet	**92.80**	**70.11**	**98.47**	**96.33**	**99.52**	**97.31**	**94.29**	**97.66**	**94.48**	**93.44**	**97.93**	**98.32**
Ours (Train: All)	**96.24**	**74.57**	**99.89**	**98.67**	**99.99**	**99.90**	**99.45**	**99.89**	**97.69**	**96.25**	**98.63**	**99.66**
Ours (Train: 4K)	93.83	73.95	96.31	96.31	99.63	96.35	96.84	99.48	96.58	95.92	98.09	99.18

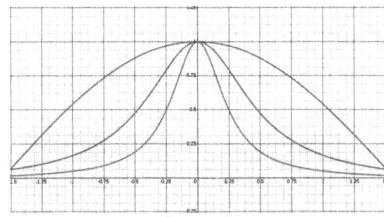

Fig. 5. Attention graphs: cosine similarity (red), t-vMF $k = 8$ (blue), $k = 32$ (green) (Color figure online)

Table 2. CEE Results: ablation with and without legend charts for the PlotQA Test Set. Leg: Legend, w/o: Without Legend, mAP: Cumulative mAP

IOU	Bar charts			Dot/Line charts		
	Leg	w/o	mAP	Leg	w/o	mAP
@0.9	96.53	94.52	96.24	75.07	72.15	74.57
@0.75	98.87	99.01	98.89	89.26	90.41	89.46
@0.5	99.58	99.34	99.54	98.02	97.84	97.99

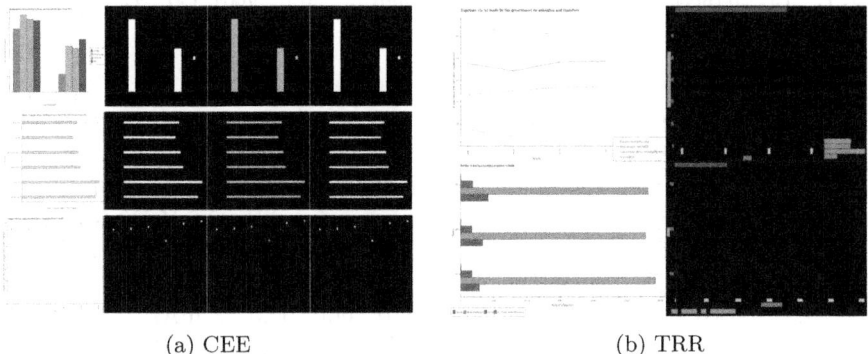

(a) CEE   (b) TRR

Fig. 6. Predictions for PlotQA Test Examples. Columns for 6a: Original image, Ground Truth for a chart element, auxiliary reconstruction, mask prediction

performed with the TRR Model (Sect. 4.1) on the overlapping Text-Role classes. We prefer mAP @0.90 IOU over mAP @0.75 IOU as an evaluation metric as we require precise fine-grained extractions and the acceptable error margin is very small as the resulting data errors can propagate to the downstream reasoning tasks.

Due to unavailability of a trained model and code, we could not use PlotNet [5] as a benchmark for generalizability results. However, the first row of the

Table 3. Results for OOD charts in *LeafQA-spatial-style* (Sect. 3)

IOU						@0.90					@0.75	@0.50
LeafQA-spatial-style [3]	Bar	Dot-line	Leg Lbl	Leg PV	Plot Title	X-axis Lbl	X-axis Ticks	Y-axis Lbl	Y-axis Ticks	mAP	mAP	mAP
0-Shot [26]	9.13	18.56	0.0	0.0	0.0	0.0	2.11	0.0	0.0	3.31	14.85	71.76
0-Shot	79.34	70.44	96.14	95.75	93.60	95.20	95.84	95.80	93.68	90.64	93.73	95.83
10-shot	85.84	72.55	96.62	96.38	95.71	95.71	95.97	96.12	94.71	92.18	94.77	96.67

Table 4. Results for OOD charts in *LeafQA-shape-roles* (Sect. 3) H: Horizontal, V: Vertical, Box: BoxPlot, mAP: Cumulative mAP. TRR performance is computed for the text roles common in PlotQA Train Set and *LeafQA-shape-roles* *Average of the zero-shot results on the available text role labels

IOU	@0.90									@0.75	@0.50
LeafQA-shape-roles [3]	Donut	Pie	H_Box	V_Box	Leg Lbl	Leg PV	Plot Title	X-axis Ticks	Y-axis Ticks	mAP	mAP	mAP
0-Shot [26]	-	-	-	-	0.0	0.0	0.0	0.0	0.0	0.0*	0.06*	0.14*
0-Shot	90.51	92.57	89.77	89.06	95.62	94.92	92.80	96.41	94.52	92.91	93.79	94.77
10-shot	93.44	93.46	90.71	90.82	95.58	95.24	94.28	96.48	95.14	93.91	94.68	95.37

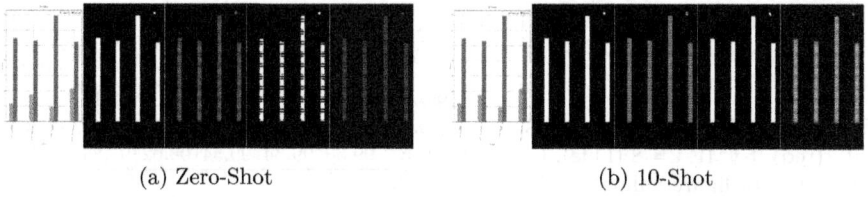

(a) Zero-Shot   (b) 10-Shot

Fig. 7. CEE predictions on *LeafQA-spatial-style* Row: Input chart, GT, reconstruction, prediction mask and attention map (left to right).

Tables 3 and 4 depict the zero-shot result with the FRCNN-FPN object detection model trained with PlotQA training data [26] and tested on *LeafQA-spatial-style* and *LeafQA-shape-roles* Test Sets, respectively. The zero-shot performance for fine-granular extractions is very poor with near zero mAP @0.9 IOU, showcasing bad generalization capabilities of the model. Also, the model can not provide extractions for the classes unseen during training, such as 'Donut', 'Pie', 'H-Box' and 'V-Box'. On the other hand, zero shot (90.64% mAP @0.90 IOU) and 10-shot (92.18% mAP @0.90 IOU) results on *LeafQA-spatial-style* Test Set (Table 3), are comparable with the full scale training results, thus demonstrating the generalization capability of the models with respect to spatial composition of chart elements and legend preview styles. Table 4 showcases the results on *LeafQA-shape-roles*. Some text roles are applicable to only a subset of the charts in *LeafQA-shape-roles*. Thus, the results of Leg Lbl and Leg PV columns are for donut and pie charts and the results of X/Y axis Ticks columns are for Box Plots. 92.91% cumulative mAP @0.90 IOU in zero-shot settings demonstrates our approach can handle OOD charts with distinct chart element shapes such as box-plots, pie and donut charts.

Table 2 illustrates the ablation for charts with and without legends (Sect. 4.2). The performance for charts without legend drops by a small amount (∼2.5 % mAP @0.90 IOU), however is still comparable with the charts having pre-defined

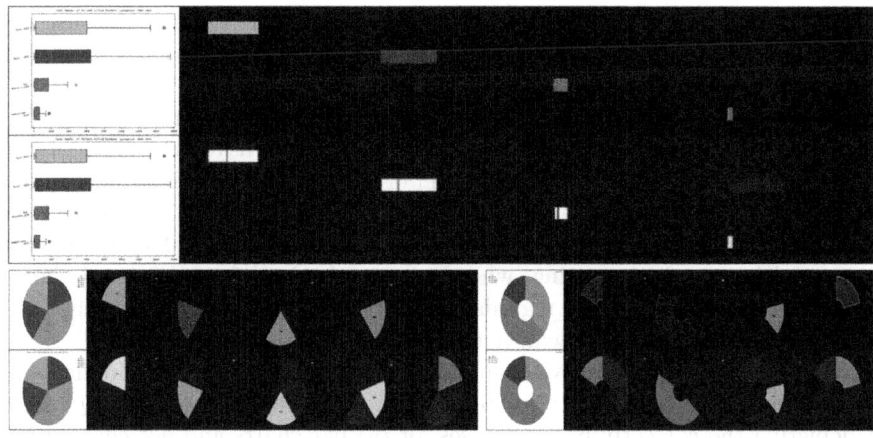

Fig. 8. CEE Zero-Shot Results on *LeafQA-shape-roles* for unseen Box-plots, Pie-charts and Donut-charts. Rows: Reconstruction, Attention Map

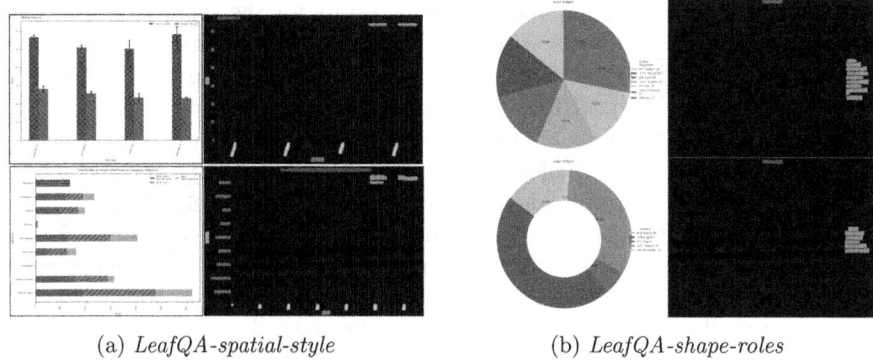

(a) *LeafQA-spatial-style*   (b) *LeafQA-shape-roles*

Fig. 9. Zero-Shot TRR Predictions. For 9b examples of text-role class labels overlapping with the PlotQA train set

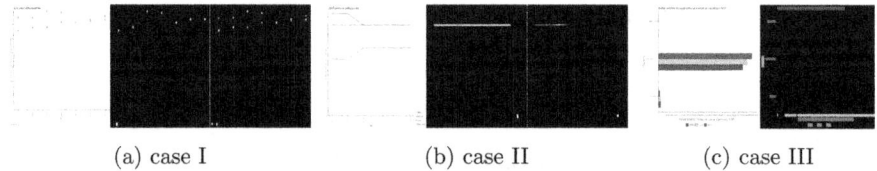

(a) case I   (b) case II   (c) case III

Fig. 10. Failure Cases. For 10a 10b Row: Input chart, GT, prediction mask

legend previews. From Tables 1 and 2 we observe that the extraction of dot/line regions is challenging because of their small size, sparse distribution and very close coloring scheme of legend styles. Moreover, the extractions primarily fail in cases where the dots or lines are eclipsed or intersected by the dot/lines of other series (Fig. 1(c)). The generalizability results in Table 3 demonstrate more drop (from 96.24 to 79.34% mAP @0.90 IOU) in zero-shot settings for Bars as compared to the other chart elements. This drop is mainly because the bar charts in PlotQA train set are all filled with continuous colors (Charts (b) and (d) in Fig. 1), whereas plots in *LeafQA-spatial-style* have bars filled with a combination of color and pattern (chart (f) in Fig. 1). However, as demonstrated in Table 3, the performance improves from 79.34 to 85.84% mAP @0.90 IOU just by adapting the model with fine-tuning on 10 samples (10-shot). This is very useful in a realistic setting, where we can obtain a few annotated charts for a shifted distribution with regards to, for example, new styles and adapt the model on those few-shots to achieve comparable performance.

Following are the qualitative examples for TRR and CEE model predictions. Figure 6a and Fig. 6b illustrate predictions on examples from the PlotQA Test set depicting the correctly handled cases for (a) bar charts and dot charts with and without legends by CEE and (b) Line and Bar charts by TRR, respectively. Figure 7 and Fig. 9a illustrate correct predictions on examples from *LeafQA-spatial-style* (i) stacked bar graphs with patterned style unseen during training by CEE in a zero-shot setting and improved results for the same after few-shot adaptation and (ii) bar charts with distinct composition of legends (top right corner) and X/Y tick labels (slanted) unseen during training by TRR model in zero shot setting, respectively. Figure 8 and Fig. 9b provide predictions on examples from *LeafQA-shape-roles* for (i) Charts with completely unseen shapes of char elements such as box-plots, pie-charts and donut charts by CEE model (ii) overlapping text-roles for charts unseen by the TRR model.

Figure 10 illustrates the most commonly occurring failure cases of CEE and TRR models. We observe that the CEE model typically fails at getting correct extractions for charts (a) having their defined styles (colors in the illustrated example) very close to each other, (b) char elements (lines in the illustrated examples) of one style (color in the illustrated example) occlude or intersect the chart elements of another style, (C) The TRR model typically fails to get correct detections for text regions for text roles (X-axis labels illustrated in the example), where the text-regions are highly cluttered causing merging of segments with neighbouring regions.

6   Conclusion

Models for real life chart extraction must be able to rapidly adapt to unseen chart distributions with only a few labelled examples. We present novel architectures for generalizable detection of Text Regions and their Roles, in addition to Chart Element Extraction and demonstrate that our models are data efficient and seamlessly adaptable to new chart distributions. Further, these models produce SOTA results for Chart Visual and Textual Element Extraction and offer accurate results in zero and few shot settings for unseen charts.

References

1. Baek, Y., Lee, B., Han, D., Yun, S., Lee, H.: Character region awareness for text detection. In: Proceedings of the IEEE Conference on Computer Vision and Pattern Recognition, pp. 9365–9374 (2019)
2. Carderas, A., et al.: Automated data extraction of bar chart raster images. arXiv preprint arXiv:2011.04137 (2020)
3. Chaudhry, R., Shekhar, S., Gupta, U., Maneriker, P., Bansal, P., Joshi, A.: Leaf-QA: locate, encode & attend for figure question answering. In: Proceedings of the IEEE/CVF Winter Conference on Applications of Computer Vision, pp. 3512–3521 (2020)
4. Choi, J., Jung, S., Park, D.G., Choo, J., Elmqvist, N.: Visualizing for the non-visual: enabling the visually impaired to use visualization. In: Computer Graphics Forum, vol. 38, pp. 249–260. Wiley Online Library (2019)
5. Ganguly, P., Methani, N., Khapra, M.M., Kumar, P.: A systematic evaluation of object detection networks for scientific plots. arXiv preprint arXiv:2007.02240 (2020)
6. Girshick, R.: Fast R-CNN. In: Proceedings of the IEEE International Conference on Computer Vision, pp. 1440–1448 (2015)
7. Harper, J., Agrawala, M.: Deconstructing and restyling D3 visualizations. In: Proceedings of the 27th Annual ACM Symposium on User Interface Software and Technology, pp. 253–262 (2014)
8. Harper, J., Agrawala, M.: Converting basic D3 charts into reusable style templates. IEEE Trans. Visual Comput. Graphics **24**(3), 1274–1286 (2017)
9. He, K., Gkioxari, G., Dollár, P., Girshick, R.: Mask R-CNN. In: Proceedings of the IEEE International Conference on Computer Vision, pp. 2961–2969 (2017)
10. Huang, G., Liu, Z., Van Der Maaten, L., Weinberger, K.Q.: Densely connected convolutional networks. In: Proceedings of the IEEE Conference on Computer Vision and Pattern Recognition, pp. 4700–4708 (2017)
11. Jia, X., De Brabandere, B., Tuytelaars, T., Gool, L.V.: Dynamic filter networks. Adv. Neural. Inf. Process. Syst. **29**, 667–675 (2016)
12. Kafle, K., Price, B., Cohen, S., Kanan, C.: DVQA: understanding data visualizations via question answering. In: Proceedings of the IEEE Conference on Computer Vision and Pattern Recognition, pp. 5648–5656 (2018)
13. Kahou, S.E., Michalski, V., Atkinson, A., Kádár, Á., Trischler, A., Bengio, Y.: Figureqa: an annotated figure dataset for visual reasoning. arXiv preprint arXiv:1710.07300 (2017)
14. Kantharaj, S., Do, X.L., Leong, R.T.K., Tan, J.Q., Hoque, E., Joty, S.: Opencqa: open-ended question answering with charts. arXiv preprint arXiv:2210.06628 (2022)
15. Kim, D.H., Hoque, E., Agrawala, M.: Answering questions about charts and generating visual explanations. In: Proceedings of the 2020 CHI Conference on Human Factors in Computing Systems, pp. 1–13 (2020)
16. Kobayashi, T.: T-vMF similarity for regularizing intra-class feature distribution. In: Proceedings of the IEEE/CVF Conference on Computer Vision and Pattern Recognition, pp. 6616–6625 (2021)
17. Law, H., Deng, J.: Cornernet: detecting objects as paired keypoints. In: Proceedings of the European Conference on Computer Vision (ECCV), pp. 734–750 (2018)
18. Lin, T.Y., Dollár, P., Girshick, R., He, K., Hariharan, B., Belongie, S.: Feature pyramid networks for object detection. In: Proceedings of the IEEE Conference on Computer Vision and Pattern Recognition, pp. 2117–2125 (2017)

19. Liu, F., et al.: DePlot: one-shot visual language reasoning by plot-to-table translation. arXiv preprint arXiv:2212.10505 (2022)
20. Liu, F., et al.: Matcha: enhancing visual language pretraining with math reasoning and chart derendering. arXiv preprint arXiv:2212.09662 (2022)
21. Liu, W., et al.: SSD: single shot MultiBox detector. In: Leibe, B., Matas, J., Sebe, N., Welling, M. (eds.) ECCV 2016. LNCS, vol. 9905, pp. 21–37. Springer, Cham (2016). https://doi.org/10.1007/978-3-319-46448-0_2
22. Luo, J., Li, Z., Wang, J., Lin, C.Y.: Chartocr: data extraction from charts images via a deep hybrid framework. In: Proceedings of the IEEE/CVF Winter Conference on Applications of Computer Vision, pp. 1917–1925 (2021)
23. Ma, W., et al.: Towards an efficient framework for data extraction from chart images. In: Lladós, J., Lopresti, D., Uchida, S. (eds.) ICDAR 2021. LNCS, vol. 12821, pp. 583–597. Springer, Cham (2021). https://doi.org/10.1007/978-3-030-86549-8_37
24. Masry, A., Long, D.X., Tan, J.Q., Joty, S., Hoque, E.: Chartqa: a benchmark for question answering about charts with visual and logical reasoning. arXiv preprint arXiv:2203.10244 (2022)
25. Masry, A., Prince, E.H.: Integrating image data extraction and table parsing methods for chart question answering (2021)
26. Methani, N., Ganguly, P., Khapra, M.M., Kumar, P.: PlotQA: reasoning over scientific plots. In: Proceedings of the IEEE/CVF Winter Conference on Applications of Computer Vision, pp. 1527–1536 (2020)
27. Newell, A., Yang, K., Deng, J.: Stacked hourglass networks for human pose estimation. In: Leibe, B., Matas, J., Sebe, N., Welling, M. (eds.) ECCV 2016. LNCS, vol. 9912, pp. 483–499. Springer, Cham (2016). https://doi.org/10.1007/978-3-319-46484-8_29
28. Redmon, J., Farhadi, A.: Yolov3: an incremental improvement. arXiv preprint arXiv:1804.02767 (2018)
29. Ren, S., He, K., Girshick, R., Sun, J.: Faster R-CNN: towards real-time object detection with region proposal networks. In: Advances in Neural Information Processing Systems, vol. 28 (2015)
30. Ronneberger, O., Fischer, P., Brox, T.: U-Net: convolutional networks for biomedical image segmentation. In: Navab, N., Hornegger, J., Wells, W.M., Frangi, A.F. (eds.) MICCAI 2015. LNCS, vol. 9351, pp. 234–241. Springer, Cham (2015). https://doi.org/10.1007/978-3-319-24574-4_28
31. Satyanarayan, A., Moritz, D., Wongsuphasawat, K., Heer, J.: Vega-lite: a grammar of interactive graphics. IEEE Trans. Visual Comput. Graphics 23(1), 341–350 (2016)
32. Savva, M., Kong, N., Chhajta, A., Fei-Fei, L., Agrawala, M., Heer, J.: Revision: automated classification, analysis and redesign of chart images. In: Proceedings of the 24th Annual ACM Symposium on User Interface Software and Technology, pp. 393–402 (2011)
33. Simonyan, K., Zisserman, A.: Very deep convolutional networks for large-scale image recognition. arXiv preprint arXiv:1409.1556 (2014)
34. Singh, H., Shekhar, S.: STL-CQA: structure-based transformers with localization and encoding for chart question answering. In: Proceedings of the 2020 Conference on Empirical Methods in Natural Language Processing (EMNLP), pp. 3275–3284 (2020)

Document Information Extraction

Improving Information Extraction from Semi-structured Documents Using Attention Based Semi-variational Graph Auto-Encoder

Djedjiga Belhadj[(✉)][iD], Abdel Belaïd[iD], and Yolande Belaïd[iD]

Université de Lorraine-LORIA, Campus Scientifique,
54500 Vandoeuvre-Lès-Nancy, France
djedjiga.belhadj@gmail.com, {abdel.belaid,yolande.belaid}@loria.fr

Abstract. In this paper, we propose a semi-supervised system for information extraction from administrative documents, that learns from both labeled and unlabeled data. The document is modeled as a words graph where each node contains the textual, layout and visual features of the word and it is connected to its spatially close neighbors. Semi-supervised variational graph auto-encoders (VGAE) have proven efficient on graph-based tasks, but they usually separate the classifier from the encoder and decoder and don't take full advantage of the VGAE model for the benefit of the classification. To optimize the classification as much as possible, we propose a semi-VGAE with an attention-based classifier that shares its layers with the VGAE encoder. This is further enhanced by proposing a VGAE loss managed by the classification loss. Experiments show that our model helps improve nodes prediction accuracy. We tested the architecture on two artificially generated datasets: Gen-Invoices and Gen-Payslips and one real dataset: receipts issued from the SROIE ICDAR 2019 competition. The latter data set yielded an important F1 score of 97.94%, placing our system among the best systems on this dataset.

Keywords: Semi-supervised · Multi-GAT · VGAE · Labeled and Unlabeled document · Semi-Structured Document

1 Introduction

Information extraction (IE) from administrative documents like invoices and payslips is a crucial task in document management for almost all companies (for mail management, information manipulation, etc.). These administrative documents, called semi-structured documents (SSD), are usually presented in a specific format. Automatic processing of this type of documents remains a challenge due to the diversity of their content and layout.

Various systems of IE from administrative documents are based on supervised deep learning approaches, like [2,6,16,20,21,25,30,31]. These systems model documents in a variety of ways, such as text sequences, graphs and grids. Each

---

Supported by BPI DeepTech.

G. A. Fink et al. (Eds.): ICDAR 2023, LNCS 14188, pp. 113–129, 2023.
https://doi.org/10.1007/978-3-031-41679-8_7

modeling can focus on several text features, including textual, visual, and positional ones. The two main obstacles of this category of systems are the complexity and costliness of the document labeling task and the lack of real training datasets.

To cope with the lack of training datasets, alternative methods have been proposed such as [7,9,19,28] which perform transfer learning by adding a pre-training step on unlabeled data from out-domain (free text / ordinary documents). They pre-train a model using tasks like input reconstruction, next sentence prediction, etc., in an unsupervised mode and then they fine tune the result on labeled documents. However, these methods require huge training datasets for pre-training and form complex models (hundreds of millions parameters).

In this regard, we have developed a less complex semi-supervised system for IE from SSDs, that learns from labeled and unlabeled in-domain documents. Our system is based on a multi-GAT graph nodes classifier and a variational graph auto-encoder (VGAE). We model the SSD as a graph of words that takes into account all of its unique characteristics. To make use of the availability of indicative keywords, we provide an effective nearest neighbors selection strategy. In addition, we incorporate into it multiple multi-modal word features, such as word region information. We choose a GAT based classifier to extract as much information as possible from the graph to aid in predicting the entities classes of the words. To fully maximize the classification optimization process on the labeled and in-domain unlabeled data, we propose a customized Semi-VGAE architecture, with a classifier and an encoder that share GAT layers and an objective function controlled by the classification loss. This most effectively steers the model toward optimum optimization for the classification purposes. Finally, compared to the state-of-the-art systems, our suggested model is significantly less complex with only 41M parameters.

The paper consists of the following parts: Sect. 2 briefly describes IE from administrative documents methods and the various training modes in the literature; Sect. 3 expands on the proposed approach by describing the components of the global architecture in detail; Sect. 4 presents the experiments and results obtained, and Sect. 5 concludes the paper and shows the global contribution of our system.

2   Related Work

Systems that learn from both labeled and unlabeled data to extract information from documents can be classified into two types: pre-training based methods trained on out-of-domain unlabeled data and semi-supervised methods trained on in-domain unlabeled data.

2.1   Pre-training Based Models

Most state-of-the-art information extraction systems such as [7,9,19,27,28] apply a two-step process on token sequences. The first step concerns learning the tokens features and their context from unlabeled dataset using various pre-training tasks. The second step, known as fine-tuning, allows the system to specialize on a downstream task by learning on labeled data. The first step's pre-trained model is fine-tuned by adding an output layer, that allows token classification or tagging, to the pre-trained layers. These systems are based on transformer architecture [24]. The baseline model in this category is BERT [7] which employs a multi-layer bidirectional Transformer encoder architecture based on attention mechanism [24]. BERT is pre-trained on a huge unlabeled dataset (BOOKCORPUS + English WIKIPEDIA) using a masked language model and Next Sentence Prediction tasks. Another system, roBERTa [19] takes the same architecture of BERT and widens the learning dataset used in the pre-training step. Additionally, it eliminates one task from the pre-training step and modifies the model's hyperparameters such as the learning rate, regularization rate, number of epochs, batch size, etc. This update has improved the BERT results on the different tasks after the fine-tuning. LayoutLM [28] and LAMBERT [9] propose adding the 2D position of the token in the document to the pre-training of BERT and RoBERTa. LAMBERT [9] that records an important score of information extraction on SROIE (their best model has reached a F1 score of 0,98), uses the RoBERTa model and it adds in addition to the position information of the token in the sequence (1D), the information of the 2D position in the document and then integrates biases relating to these two 1D and 2D positions in the calculation of the attention mechanism of the transformers. LayoutLM [28] adds the 2D position of the token (in 4 coordinates) to the BERT input, and then it adds the result of a faster RCNN+FC layers (image embedding) at the pre-trained BERT output. LayoutLMV2 [27] integrates the visual information in the pre-training step for a multi-modal training more efficient.

However, the pre-training step in these systems requires a very large mass of training data and the final models are considered as complex models (at least 200 millions parameters). This makes it challenging to deploy and maintain them in practical contexts.

2.2   Semi-supervised VAE Based Methods

Deep generative methods have recently made semi-supervised learning with in-domain unlabeled data appealing. They are less complex than pre-training based methods and have shown promise in text classification applications. [4,8,10,29] employ the VAE (variational auto-encoder), a specific kind of generative model, to perform text classification tasks within token sequences. The authors in [8] introduce a semi-supervised VAE based on LSTM in which the Kullback-Leibler divergence is removed from the loss to simplify the model. In [4], they use a BiGRU based semi VAE for sequence labeling. The authors in [10] pre-train a VAE based on feed-forward networks on word frequencies of unlabeled data. The learned representations were then used concatenated to word vectors in a downstream classifier. Another semi-supervised VAE was suggested in [29], where gated convolutional neural networks (GCNN) served as both the encoder and the decoder. They use a layer called Scalar after Batch Normalization (BN) to scale the BN's output as a solution to the KL-divergence vanishing problem of the VAE. The semi-VAE have proven efficient on other text analysis tasks, such as relation extraction and sentiment analysis. The authors in [32] present a semi-supervised VAE, to extract biomedical relation from biomedical text, based on BLSTM encoder and CNN decoder, with a separate CNN classifier. In [26], one bring a BLSTM based semi-supervised VAE for sentiment analysis, while in [5] one present a sentiment analysis system built with a transformer encoder and decoder and various kinds of classifiers such as LSTM and attention based one. Some of the methods mentioned above suggest improvements to the loss function calculation. [5, 26, 32]'s authors use an hyperparameter that balances the relative weight between generative and discriminative learning and controls the weight of the additional classification loss. In [8], they remove the KL-divergence and the components associated with the latent variable in order to simplify the model and improve the speed of the learning process. In fact, the VAE architecture is not entirely oriented by these methods to improve the classification task.

In our approach, we study the effectiveness of this type of architectures for IE from graph modeled SSDs. We aim to maximize the exploitation of the VAE architecture for the benefit of the classification by proposing an adapted semi-VGAE model and an improved loss function.

3   Proposed Method

This section provides a description of our semi-supervised approach with the use of labeled and unlabeled SSDs by the different parts of the model. As shown in Fig. 1, our system is composed mainly of three parts: graph modeling, multi-GAT classifier and VGAE. We will detail the components of the model, the equivalent inference problem and its derived loss function.

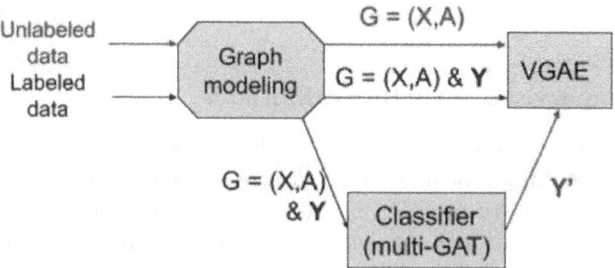

Fig. 1. The flow of labeled and unlabeled data across the three architecture's components (Graph modeling, classifier and VGAE)

3.1   Graph Modeling

As described in [1], a SSD can be modeled as a set of information blocks distributed over the layout of the document, such as company information, shipping and billing addresses in an invoice. Each information block regroups a set of elementary description presented in a pair format (keywords, information). The SSD modeling can be simplified to $SSD = \{P_i\}, P_i = (KW, Info), i \in [1, l]$, where l is the number of pairs in the document and KW and $Info$ are sequences of words. P_i, KW and $Info$ have some common characteristics in all SSDs: each $Info$'s paired KW belongs to a limited lexicon; Some Infos may have a very specific format without being accompanied by KW, such as dates (dd-mm-yyyy; dd/mm/yy, etc.) and company/client identifiers (given by a combination of numbers, letters and special characters); despite the fact that P_i's position varies in the same class of documents (invoices from different providers), it may vary within the same region or limited variable zones. For instance, the total is always at the bottom of an invoice, and the invoice number is always at the top of the page (on either the left or the right side). To predict the class of $Info$ in a SSD, we are primarily interested in the close spatial neighborhood that should contain the KW which introduces the information and helps to recognize its correct class. Thereby, KW can be fully aligned with $Info$ horizontally (placed on the left side of $Info$), vertically (placed above $Info$), but can also be partially aligned vertically or horizontally or completely not aligned, but keep close spacing.

Three-Lines Neighborhood. For an improved neighborhood calculation, that prioritizes the word's spatially close neighbors in the SSD, we suggest a three line based neighbors calculation. We take the "k" nearest neighbors in over three consecutive lines as shown in the neighborhood selection of Fig. 2 (same line, line above and line below if they exist) for each word in the SSD. We select the "k" closest words (using Euclidean distance between bounding boxes) on the left and right of the word on the same line, and the "k" closest words on the lines above and below. The total number of word's neighbors is at most $n = 4 * k$. n is set experimentally in Sect. 4. The selected neighborhood is said to contain the indicative keywords that introduce the information we want to extract from the SSD. Unlike [2, 20] that use a global graph of all the training dataset words, which makes the model very complex and increases the learning time, we propose to limit the graphs to a maximum of n_max=256 nodes (256 is the closest power of 2 to the average words number across all SSD graphs studied here). This is done by either associating several graphs of less than 256 nodes in the same graph, or dividing a graph having more than 256 nodes into several graphs of less than 256 nodes each, as shown in Fig. 2. This is possible thanks to the fact that the neighborhood of the nodes is limited and can be dissociated from each other.

Let $G = (V, E)$ be undirected graph modeling the SSD words, where V is the set of the graph nodes and E the set of connections between those nodes. Although E is limited to indicating the adjacency relations between nodes, V contains a maximum of information on the node, as explained below.

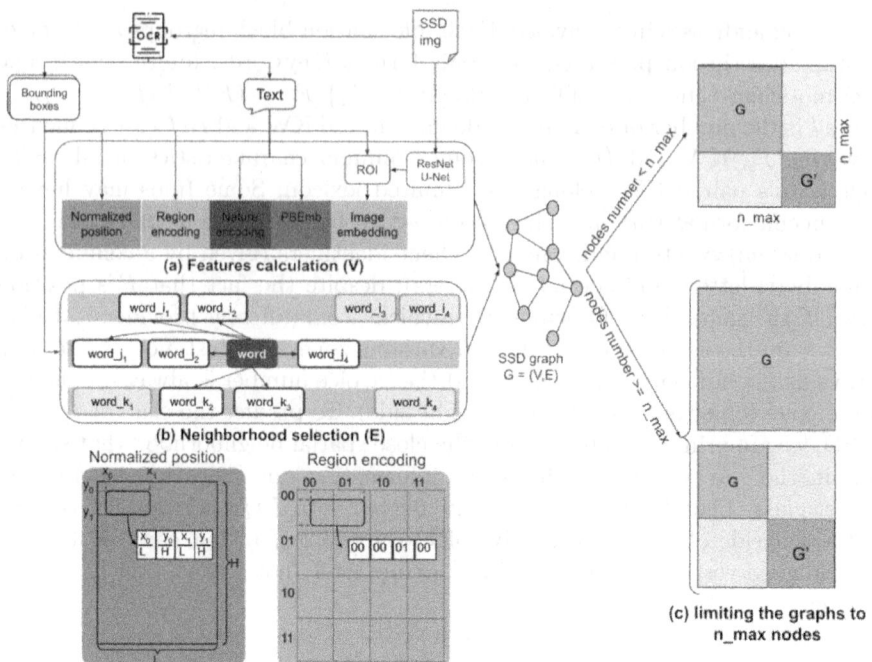

Fig. 2. Graph modeling: (a) Features calculation to generate the multimodal features vectors. (b) Neighbors selection to choose the word's neighbors over three lines. (c) Graph nodes limitation.

Word's Features. V corresponds to a set of multimodal features, comprising:

– Textual features: we choose BPEmb [12], a pre-trained subword unit embeddings based on Byte Pair Encoding [23] to calculate the word embedding vectors. BPEmb allows us to have the same vector size of each word in the graph (300) and is available in 275 different languages among others: English and French. It also performs better than other subword approaches.
A Boolean vector representing the word nature (alphabetic, alphanumeric, etc.) is added to it. We reuse the 8 sized vector proposed in [2].
– Positional features (normalized position): first, the area that surrounds the text is extracted and then the new coordinates of the words are calculated according to these new references. We pick the two points that define each word bounding box (the top left point and the bottom right one) and we calculate their normalized coordinates in the new reference as can be seen in the Fig. 2.
– Region encoding: we noticed that the information have a variable position from a document to another, but floats usually in the same part of the document i.e. the top right of the SSD, etc. To exploit this fact, a new region encoding is added. The area surrounding the text in the document is divided horizontally and vertically by 4. Each part is then represented by two binary

numbers (x and y) encoding the SSD part number. x and y stand in two positions (00,01,10,11). The bounding boxes coordinates are encoded according to the SSD region containing them.

- Visual features: an image embedding vector is computed, for each word in the SSD. The ResNet Unet [11,22] network is used for this purpose. The input image is resized to unify the model input dimension, and then it is passed to a pre-trained ResNet Unet encoder. Using the result features map and the original words bounding boxes coordinates, we extract for each word, its region of interest (ROI). Finally, after flatting the ROI results, an image embedding vector of size 21 is obtained.

The global features vector is a concatenation of all these multi-modal features like in [2, 20], forming a 337 sized vector.

The final graph G is modeled by a features matrix "X" that represents the graph nodes set (V) and an adjacency matrix "A" that refers to the graph edges (E).

3.2   Multi-GAT Classifier

G is fed into a classifier (Multi-GAT) that ensures a graph nodes classification task. The multi-GAT is a multi-layer convolutional network [2] based on multi-head attention. Each layer of the Multi-GAT takes into account a different level of word neighborhood. The first layers apply the multi-head attention with "k" heads and ReLU as a non linearity activation function, while the output layer applies a Softmax function to classify the graph nodes, as shown in Fig. 3. For each dataset, a different "k" is fixed according to the hypothesis proposed in [2] about the correlation with the extracted entities number. The attention mechanism in one layer is used to learn which neighbors are most important and which neighbors are least important for each node in the SSD graph, making prediction of node classes more reliable. The number of GAT layers is fixed in the experiments section for each dataset. It represents the words' neighborhood level to consider in order to capture the useful distant information belonging to the same information block, since generally the information in each block comes in a logical order (like: "sub-total", "total", "without tax", "payment", etc.).

3.3   Variational Graph Auto-Encoder: VGAE

The VGAE is composed of an encoder and two decoders, as can be seen in Fig. 3. The encoder is made up primarily of multi-GAT and dense layers. It shares its first GAT layers with the classifier and outputs Gaussian variables Z. These shared layers allow to associate the task of classification to the data distribution learning. This lead to the classifier becoming more generic as it learns the data characteristics within their distribution. The encoder first learns the mapping of the input data into latent space. The distribution of the latent variables is modeled by a Gaussian distribution with its mean and variance generated by the output layers of the encoder. From these Gaussian variables and the classifier result (y') or the ground truth (y), we reconstruct X and A using two decoders. The first decoder which reconstructs A consists mainly of an Inner-product layer, as proposed in [15], while the second decoder which reconstructs X, consists of several 2D convolution layers.

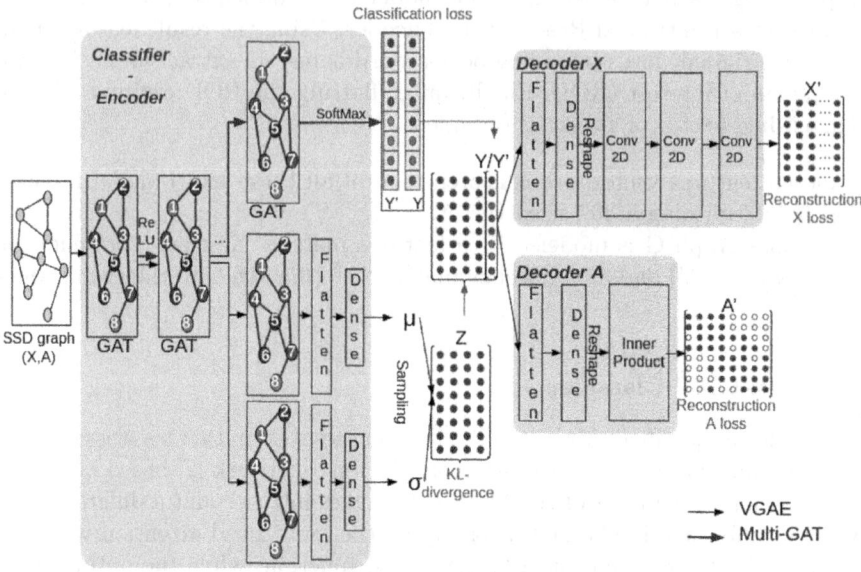

Fig. 3. Overview of our semi-supervised VGAE. The SSD graph goes through shared GAT layers, followed by separate layers to output the nodes predictions and the latent variable Z. These outputs are concatenated and fed into two decoders, which reconstruct A and X.

The input of the decoders is the concatenation of the latent variable z generated by the encoder and the vector of the nodes classes as shown in Fig. 3. The nodes classes vector is either y which represents the "ground truth (GT)" if the input is labeled or y' representing the nodes classes predictions if the input is unlabeled. The decoders try to reconstruct the input from the latent variable z and the vector (y/y'). The classifier predictions directly affect the output of both decoders. When the input graphs are unlabeled data, the classifier weights are updated using the VGAE losses. This means that our system learns useful features from labeled and unlabeled data.

To ensure that the decoder input is as accurate as possible, we first train the classifier on the labeled SSDs, then we continue the training by adding the other VGAE components as well as the unlabeled SSDs.

Optimisation. In the associated inference problem of this architecture, each component is considered as a probability distribution. The encoder refers to the inference network with $p(z/x, a)$. The decoders represent generative networks which reconstruct X $(p(x/z, y))$ and A $(p(a/z, y))$, while the classifier represents the discriminative network with $p(y/x, a)$. The objective of our model is to maximize the log-likelihood $\log p_\theta(x, a, y)$. As it is difficult to directly maximize $\log p_\theta(x, a, y)$, we maximize its variational lower bound (ELBO).

We rely on the bound J on the marginal likelihood for an entire dataset of labeled LD and unlabeled data UD proposed by [14]. We extend their J bound to take into account both our architecture choice of using two decoders rather than one as well as our data modeling, which involves adding "a" to "x" input data and a new parametrization of the different losses parts. So we get:

$$J = \alpha E_{(x,a,y) \sim LD} \left[\log(q_\phi(y \mid x, a)) \right] + E_{(x,a,y) \sim LD} \left[L(x, a, y) \right] + E_{(x,a) \sim UD} \left[U(x, a) \right] \tag{1}$$

For one data point in LD, we have:

$$\log p_\theta(x, a, y) \geq E_{q_\phi(z|x,a,y)} \left[\log p_\theta(x \mid y, z) \right] + E_{q_\phi(z|x,a,y)} \left[\log p_\theta(a \mid y, z) \right]$$
$$+ \log p_\theta(y) - \beta KL(q_\phi(z \mid x, a, y) \parallel p_\theta(z)) = -L(x, a, y) \tag{2}$$

where KL is the Kullabck-Leibler divergence of $p_\theta(z)$ and $q_\phi(z \mid x, a, y)$ which measures the difference between the latent z variables and the Gaussian distribution. For UD, we get:

$$\log p_\theta(x, a) \geq \sum_y q_\phi(y \mid x, a) \left(-L(x, a, y) \right) + H \left[q_\phi(y \mid x, a) \right] = -U(x, a)$$
$$H \left[q_\phi(y \mid x, a) \right] = \sum_y q_\phi(y \mid x, a) \left[\log q_\phi(y \mid x, a) \right] \tag{3}$$

In this paper, we suggest a novel β parameterization based on the classification loss $\beta = f(E_{(x,a,y) \sim LD} \left[\log(q_\phi(y \mid x, a)) \right])$, and $f(x) = 1 + \frac{l}{u} * x$. l refers to the size of LD and u to the size of UD. The main objective of this setting is to penalize the KL loss according to the classification loss and force the semi-VGAE to learn better distribution estimation of z, that improves the classification task. Furthermore, we modify the α introduced in [14] by setting it to $\frac{l}{u}$, in order to prioritize the classification loss.

4   Experiments

In this section, experiments are conducted to test the effectiveness of the implemented method on different SSDs datasets.

4.1   Datasets

We experiment our approach on one real dataset SROIE and two generated other datasets: Gen-Invoices, Gen-Payslips. The layout of the three datasets is variable within different levels. SROIE: is a dataset of real English receipts proposed in [13]. It is labeled with 4 entities: Address, Company, Data and Total.

Table 1. Datasets statistics and characteristics, * refers to real documents

Characteristic	SROIE	Gen-Invoices	Gen-Payslips
Training set size	626*	700	700
Validation set size	-	200	200
Test set size	437*	300	300
Unlabeled set size	300	300	300
Entities number	4	27	25
language	English	English/French	French

Gen-Invoices, Gen-Payslips and Gen-Receipts are artificial datasets generated using a generic SSD generator [1,3]. We make sure to have as much content and layout diversity as possible during the generation process in order to be as realistic of the SSD as possible. For this purpose, various diversity scores are utilized, including the BLEU score for text diversity, blocks alignment and overlapping for layout diversity. The three datasets's scores closely resemble those of the SROIE. Gen-Receipts is only used to run semi-supervised mode experiments on SROIE.

Gen-Invoices contains 27 entities to extract among others: invoice number, company address, company ids like: siren number and siret number, client information, product information: serial number, description, unit price, etc., the totals and the payment mode. As for Gen-payslips, it contains 25 classes to predict, including employee information: name, address, registration number, etc.; company information: name, address, Siret, etc.; salary information: payment period, earned leave, net salary, etc.

The details concerning the element numbers taken in the three datasets are shown in the Table 1.

4.2   Implementation Details

The proposed model has been implemented using Tensorflow and Keras frameworks. The ResNet-50 [11] Unet is adopted as our visual encoder to construct the image embedding. All the experiments are performed using a mini-batch size of 4. The Adam optimizer is used to optimize both multi-GAT and VGAE. The learning rate is set to 0.002 for the Multi-GAT and 1.e-5 for the VGAE. The maximum number of epoch is set to 1000 and the early stopping to 50. The n-heads in the Multi-GAT layers is set to 26 for the Gen-invoices and Gen-Payslips and to 8 for SROIE while in the GAT layers of the encoder, it's equal to 8. The latent dimension used in the encoder is set to 5. To compare the different results, we use the F1 score metric.

4.3   Tests and Results

Different ablation studies are performed to show the impact of features vector components; the number of the neighbors in the graph and the number of

Multi-GAT layers. We show also the results of our semi-supervised model. Finally, we compare our results on the three datasets with those of state of the art systems.

Features Vector Components. To show the effect of each features vector component, we compare the multi-GAT results on the three datasets by varying the feature vector components. The basis vector initially contains the word embedding part and the word nature boolean vector. We test the effect of normalized position, layout encoding and image embedding on the Multi-GAT predictions. The Table 2 shows the F1 score obtained on SROIE, Gen-Invoices and Gen-Payslips.

Table 2. F1 score (%) obtained by varying the vector features component in the three datasets. These experiments are obtained using a 4 layers Multi-GAT. n in the graph modeling is set to 4 for Gen-Payslips and Gen-Invoices and 8 for SROIE. "+" indicates the presence and "−" the absence of the corresponding feature.

Added features to the basis vector			Dataset		
Norm. pos	Region encod	Image embed	SROIE	Gen-Invoices	Gen-Payslips
−	+	+	96.37	95.84	99.46
+	−	+	95.98	95.18	99.36
+	+	−	96.53	96.12	99.44
+	+	+	**97.77**	**96.87**	**99.48**

As can be seen in the Table 2, each feature improves the classification score on the three datasets. The best combination for the three datasets is the vector composed of word's nature +word embedding+ normalized position (Norm. pos.) + region encoding (Region encod.) + image embedding (Image embed.). These results confirm that the multi-modal representation helps to improve the results of the words classification in the SSDs.

Word's Neighborhood in the Graph. We also tested the effect of the first-level word's neighborhood in the three datasets graphs. We use a 4 layers multi-GAT for the three datasets. The purpose of this experiment is to find the number of the first-level neighbors needed to capture the local neighbors needed for a better graph nodes prediction.

The results in Table 3 show that the model performs better with n = 4 (k = 1) for the invoices and payslips dataset and around n = 8 (k = 2) for the SROIE dataset. This can be explained by the fact that the first level of word's neighborhood, which is the most important in the SSD, should contain the keywords that introduce the information. Analysis of the number of keywords introducing entities in the three datasets reveals that most of the entities are introduced using a single keyword in the payslips and invoices datasets.

Table 3. F1 score (%) obtained with the supervised Multi-GAT by varying the number of word's neighbors (n = 4*k) in the graph

Neighbors number	SROIE	Gen-Invoices	Gen-Payslips
4	97.57	**96.87**	**99.48**
8	**97.77**	94.12	98.10
12	96.33	92.08	97.13

The configuration of 8 neighbors (two on the left, on the right, above and below the word) gives the best score for SROIE. Improvements in F1 scores are noted in the total and date entities. The date in SROIE is generally introduced by two keywords "Date:" and also followed by the time, usually composed of 2 words ("Time 00:00:00" or "00:00:00 AM"). The total entity also can be introduced by one or more than one word, so taking a single word will exclude the other cases, while taking 2 words (max) on each side will consider the case of one keyword.

We also study the effect of the variation of the number of layers in the Multi-GAT on the prediction of the classes of the graph nodes in each dataset.

Table 4. F1 score (%) obtained by varying the number of layers in the multi-GAT

Nb layers	SROIE	Gen-Invoices	Gen-Payslips
2	95.99	91.54	98.06
3	97.56	93.88	99.25
4	97.77	**96.87**	**99.48**
5	97.85	96.62	99.20
6	**97.91**	96.59	99.05
7	97.45	95.70	98.72

As can be seen in the Table 4, the model achieves the best results with 4 layers for payslips and invoices and 6 for SROIE. Here, each layer takes into account a higher neighborhood level than the previous one. With each new layer, we take into consideration more distant neighbors located in the same line and in more distant lines. With each new layer, we add information from another distant line. We need at least four neighborhood levels to get the best results. Indeed, the information represented in a SSDs generally follows a logical order, and it is found in information blocks, such as company information (name, address, identifiers, etc.), order identifiers (invoice or receipt), payment information that follow the total or the detail of the total which precedes it (VAT, discount, etc.). They thus form a local context which is important for the entities prediction. On the other hand, taking a very wide neighborhood (≥ 5 for Gen-Invoices and Gen-Payslips and ≥ 7 for SROIE) leads to the loss of the local block context and can cause a divergence.

Learning Modes. In this part, the contribution of the semi-supervised mode to the performance of the model is evaluated. We compare the F1 scores obtained on the two generated datasets (Gen-Invoices and Gen-Paysplis) using our supervised model Multi-GAT (MG), two semi-supervised VGAE baslines and our final semi-supervised model (MG+VGAE). Baseline1 is the MG+VGAE without the shared GAT layers between encoder and classifier and the α/β parameterization. Baseline2 is our MG+VGAE without the regularization hyperparameter β.

Table 5. F1 score % obtained by varying the learning mode, the multi-GAT contains 4 layers for both Gen-Invoices and Gen-Payslips (in the hyperparameters α and β, l refers to the size of the labeled dataset, u to the unlabeled dataset and L_s is the classification loss)

Model	Shared GATs	α	β	Gen-Invoices	Gen-Payslips
Multi-GAT	-	-	-	96.87	99.48
Baseline1	No	1	1	97.14	99.56
Baseline2	Yes	l/u	1	97.27	99.56
MG+VGAE	Yes	l/u	$1 + (l/u) * L_s$	**97.34**	**99.58**

As can be seen in Table 5, the MG+VGAE outperforms the supervised multi-GAT and the two semi-VGAE baselines. This demonstrates that our architecture with the shared GAT layers is advantageous and our proposed loss function optimizes better the classification results. By using the Gen-Receipts dataset, we also improve the SROIE results to an F1 score of 97.94%. Considering that the layouts of the generated receipts differ slightly from the SROIE ones, the improvement is less than it is for the other two datasets.

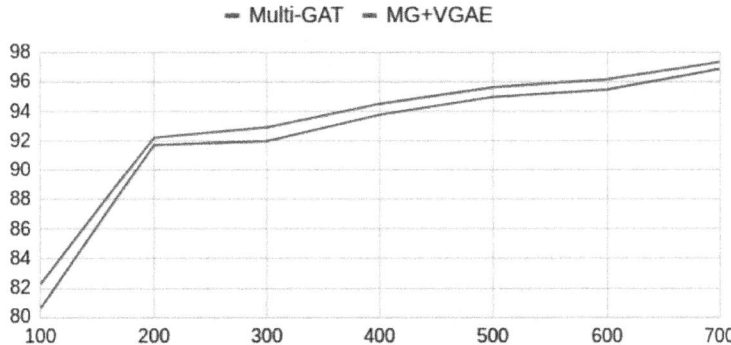

Fig. 4. F1 score (%) obtained on Multi-GAT and MG+VGAE by varying the labeled Gen-Invoices training size

Figure 4 shows that our MG+VGAE improves the F1 score obtained by the supervised Multi-GAT on Gen-Invoices for all the labeled dataset's sizes.

Despite the limited labeled data, the F1 score increases from 80.58 to 82.26 using MG+VGAE for the smallest dataset of 100 invoices. This demonstrates that our MG+VGAE can improve prediction score even with limited labeled data.

Overall Results. To evaluate our model, we compared our results on three datasets to the results of the two graph nodes prediction systems proposed in [2] and [20].

Table 6. F1 score (%) obtained with the graph nodes prediction systems

Model	SROIE	Gen-Invoices	Gen-Payslips
[20]	96.71	92.87	97.98
[2]	97.62	93.53	99.28
Ours	**97.94**	**97.34**	**99.58**

As can be seen in the Table 6, our system outperforms the other two proposed systems in [2] and [20]. Our system is richer in word's features and it provides more meaningful graph nodes neighborhood. In addition, it integrates unlabeled SSDs in a semi-supervised scheme. We notice that the scores recorded for Gen-Payslips are the highest, it results from the fact that this dataset has less variation in content and layout than the two other ones.

Table 7. F1 score comparison between our system and the other systems (M refers to million)

System	Params	F1 score
LAMBERT [9]	125M	**98.17**
LayoutLMV2(L) [27]	426M	97.81
LayoutLMV2(B) [27]	200M	96.25
StrucText [17]	107M	96.88
ViBERTGrid [18]	157M	96.40
TRIE [31]	–	96.18
PICK [30]	–	96.12
LayoutLM(L) [28]	343M	95.24
LayoutLM(B) [28]	113M	94.38
BERT(B) [7]	340M	92
Ours	41M	**97.94**

We also compared our results (F1 score) on SROIE with several state-of-the-art systems as well as the complexity of the models (number of parameters in the model) in Table 7. Our system achieves competitive results, it comes in the second position after the state-of-the-art model [9]. We want to point that the best model registered on the public leaderboard for the Key Information Extraction from SROIE belongs to [17] with 98.70%. Our system otherwise is much less complex than all these systems, the classifier contains almost 41M parameters (around 40.6M for the ResNet-UNet and BPEmb pre-trained models and 277K for the multi-GAT). This last point makes our system much more easier and faster than the others.

5   Conclusion

In this paper, we presented a semi-supervised entity extraction system from semi-structured documents based on a multi-GAT graph nodes classifier and a variational graph auto-encoder. We proposed a graph modeling of the SSD based on its characteristics, with an efficient three-line-based neighborhood calculation and the inclusion of the information clock region in the SSD layout. By using the multi-head attention mechanism, our Multi-GAT classifier maximizes the exploitation of the efficient nearest neighborhood and includes more useful distant information. In addition, we have integrated a VGAE that optimizes our classifier with both labeled and unlabeled in-domain data. Our VGAE is oriented as much as possible toward the classification task optimization by proposing an encoder and a classifier that share their first GAT layers and parameterizing the VGAE loss based on classification loss. Our proposed semi-VGAE outperforms the supervised version of multi-GAT, two additional graph-based extraction systems, and other semi-GVAE baselines on both Gen-Invoices and Gen-Payslips datasets. It extracts 28 entities from Gen-Invoices with an interesting global F1 score of 97.34%, 25 entities from Gen-Payslips with a score of 99.58%, and 4 entities from the SROIE dataset with a score of 97.94%. In the future, we intend to test our system on additional real data from other SSD classes and further reduce its complexity to make it faster and easier to deploy.

Acknowledgements. This work was conducted within the BPI DeepTech project, in collaboration between the University of Lorraine (Ref. UL: GECO/2020/00331), the CNRS, the INRIA Lorraine and the company FAIR&SMART. The authors want to thank all of their collaborators for their successful cooperation.

References

1. Belhadj, D., Belaïd, Y., Belaïd, A.: Automatic generation of semi-structured documents. In: Barney Smith, E.H., Pal, U. (eds.) ICDAR 2021. LNCS, vol. 12917, pp. 191–205. Springer, Cham (2021). https://doi.org/10.1007/978-3-030-86159-9_13
2. Belhadj, D., Belaïd, Y., Belaïd, A.: Consideration of the word's neighborhood in GATs for information extraction in semi-structured documents. In: Lladós, J., Lopresti, D., Uchida, S. (eds.) ICDAR 2021. LNCS, vol. 12822, pp. 854–869. Springer, Cham (2021). https://doi.org/10.1007/978-3-030-86331-9_55

3. Blanchard, J., Belaïd, Y., Belaïd, A.: Automatic generation of a custom corpora for invoice analysis and recognition. In: 2019 International Conference on Document Analysis and Recognition Workshops (ICDARW), vol. 7, p. 1. IEEE (2019)
4. Chen, M., Tang, Q., Livescu, K., Gimpel, K.: Variational sequential labelers for semi-supervised learning. In: Proceedings of the 2018 Conference on Empirical Methods in Natural Language Processing, pp. 215–226 (2019)
5. Cheng, X., Xu, W., Wang, T., Chu, W.: Variational semi-supervised aspect-term sentiment analysis via transformer. In: Proceedings of the 23rd Conference on Computational Natural Language Learning, pp. 961–969 (2019)
6. Chiu, J.P., Nichols, E.: Named entity recognition with bidirectional LSTM-CNNs. Trans. Assoc. Comput. Linguist. 4, 357–370 (2016)
7. Devlin, J., Chang, M.W., Lee, K., Toutanova, K.: Bert: pre-training of deep bidirectional transformers for language understanding. In: Proceedings of North American Chapter of the Association for Computational Linguistics: Human Language Technologies (2019)
8. Felhi, G., Roux, J.L., Seddah, D.: Challenging the semi-supervised VAE framework for text classification. arXiv preprint arXiv:2109.12969 (2021)
9. Garncarek, Ł, et al.: LAMBERT: layout-aware language modeling for information extraction. In: Lladós, J., Lopresti, D., Uchida, S. (eds.) ICDAR 2021. LNCS, vol. 12821, pp. 532–547. Springer, Cham (2021). https://doi.org/10.1007/978-3-030-86549-8_34
10. Gururangan, S., Dang, T., Card, D., Smith, N.A.: Variational pretraining for semi-supervised text classification. In: Proceedings of the 57th Annual Meeting of the Association for Computational Linguistics, pp. 5880–5894 (2019)
11. He, K., Zhang, X., Ren, S., Sun, J.: Deep residual learning for image recognition. In: Proceedings of the IEEE Conference on Computer Vision and Pattern Recognition, pp. 770–778 (2016)
12. Heinzerling, B., Strube, M.: BPEmb: tokenization-free pre-trained subword embeddings in 275 languages. arXiv preprint arXiv:1710.02187 (2017)
13. Huang, Z., et al.: ICDAR 2019 competition on scanned receipt OCR and information extraction. In: 2019 International Conference on Document Analysis and Recognition (ICDAR), pp. 1516–1520. IEEE (2019)
14. Kingma, D.P., Mohamed, S., Jimenez Rezende, D., Welling, M.: Semi-supervised learning with deep generative models. In: Advances in Neural Information Processing Systems, vol. 27 (2014)
15. Kipf, T.N., Welling, M.: Variational graph auto-encoders. arXiv preprint arXiv:1611.07308 (2016)
16. Lample, G., Ballesteros, M., Subramanian, S., Kawakami, K., Dyer, C.: Neural architectures for named entity recognition. arXiv preprint arXiv:1603.01360 (2016)
17. Li, Y., et al.: Structext: structured text understanding with multi-modal transformers. In: Proceedings of the 29th ACM International Conference on Multimedia, pp. 1912–1920 (2021)
18. Lin, W., et al.: ViBERTgrid: a jointly trained multi-modal 2D document representation for key information extraction from documents. In: Lladós, J., Lopresti, D., Uchida, S. (eds.) ICDAR 2021. LNCS, vol. 12821, pp. 548–563. Springer, Cham (2021). https://doi.org/10.1007/978-3-030-86549-8_35
19. Liu, Y., et al.: Roberta: a robustly optimized BERT pretraining approach. arXiv preprint arXiv:1907.11692 (2019)
20. Lohani, D., Belaïd, A., Belaïd, Y.: An invoice reading system using a graph convolutional network. In: Carneiro, G., You, S. (eds.) ACCV 2018. LNCS, vol. 11367, pp. 144–158. Springer, Cham (2019). https://doi.org/10.1007/978-3-030-21074-8_12

21. Ma, X., Hovy, E.: End-to-end sequence labeling via Bi-directional LSTM-CNNS-CRF. arXiv preprint arXiv:1603.01354 (2016)
22. Ronneberger, O., Fischer, P., Brox, T.: U-Net: convolutional networks for biomedical image segmentation. In: Navab, N., Hornegger, J., Wells, W.M., Frangi, A.F. (eds.) MICCAI 2015. LNCS, vol. 9351, pp. 234–241. Springer, Cham (2015). https://doi.org/10.1007/978-3-319-24574-4_28
23. Sennrich, R., Haddow, B., Birch, A.: Neural machine translation of rare words with subword units. arXiv preprint arXiv:1508.07909 (2015)
24. Vaswani, A., et al.: Attention is all you need. In: Advances in Neural Information Processing Systems, vol. 30 (2017)
25. Wang, J., et al.: Towards robust visual information extraction in real world: new dataset and novel solution. In: Proceedings of the AAAI Conference on Artificial Intelligence, vol. 35, pp. 2738–2745 (2021)
26. Wu, C., Wu, F., Wu, S., Yuan, Z., Liu, J., Huang, Y.: Semi-supervised dimensional sentiment analysis with variational autoencoder. Knowl.-Based Syst. **165**, 30–39 (2019)
27. Xu, Y., et al.: Layoutlmv2: multi-modal pre-training for visually-rich document understanding. arXiv preprint arXiv:2012.14740 (2020)
28. Xu, Y., Li, M., Cui, L., Huang, S., Wei, F., Zhou, M.: Layoutlm: pre-training of text and layout for document image understanding. In: Proceedings of the 26th ACM SIGKDD International Conference on Knowledge Discovery & Data Mining, pp. 1192–1200 (2020)
29. Ye, H., Zhang, W., Nie, M.: An improved semi-supervised variational autoencoder with gate mechanism for text classification. Int. J. Pattern Recognit. Artif. Intell. **36**(10), 2253006 (2022)
30. Yu, W., Lu, N., Qi, X., Gong, P., Xiao, R.: Pick: processing key information extraction from documents using improved graph learning-convolutional networks. In: 2020 25th International Conference on Pattern Recognition (ICPR), pp. 4363–4370. IEEE (2021)
31. Zhang, P., et al.: TRIE: end-to-end text reading and information extraction for document understanding. In: Proceedings of the 28th ACM International Conference on Multimedia, pp. 1413–1422 (2020)
32. Zhang, Y., Lu, Z.: Exploring semi-supervised variational autoencoders for biomedical relation extraction. Methods **166**, 112–119 (2019)

Language Independent Neuro-Symbolic Semantic Parsing for Form Understanding

Bhanu Prakash Voutharoja[ID], Lizhen Qu[✉][ID], and Fatemeh Shiri[ID]

Monash University, Clayton, VIC 3800, Australia
{lizhen.qu,fatemeh.shiri}@monash.edu

Abstract. Recent works on form understanding mostly employ multimodal transformers or large-scale pre-trained language models. These models need ample data for pre-training. In contrast, humans can usually identify key-value pairings from a form only by looking at layouts, even if they don't comprehend the language used. No prior research has been conducted to investigate how helpful layout information alone is for form understanding. Hence, we propose a unique entity-relation graph parsing method for scanned forms called LAGNN, a language-independent Graph Neural Network model. Our model parses a form into a word-relation graph in order to identify entities and relations jointly and reduce the time complexity of inference. This graph is then transformed by deterministic rules into a fully connected entity-relation graph. Our model simply takes into account relative spacing between bounding boxes from layout information to facilitate easy transfer across languages. To further improve the performance of LAGNN, and achieve isomorphism between entity-relation graphs and word-relation graphs, we use integer linear programming (ILP) based inference. Code is publicly available at https://github.com/Bhanu068/LAGNN.

Keywords: Document Layout Analysis · Graph Neural Network · Language Independent · Deep Learning

1 Introduction

Despite the growing popularity of e-forms, paper forms are still widely used to collect data by various types of organizations, from government agencies to private companies. A large body of collected data, especially historical data, is still available only in paper forms or scanned document images. To digitize such data, we introduce the task of *entity-relation graph parsing for form understanding*, which maps the document image of a form to a structured entity-relation graph. As a result, users can explore and analyze semantic information in such entity-relation graphs without any need to read the original document images.

Entity relation graphs are introduced for forms in FUNSD [12], which are annotated on a small sample of scanned forms. Herein, an entity is a group of words representing a semantic and spatial standpoint, such as question and answer, and a relation is a directed edge between two entities, as illustrated in Fig. 1. Such graphs are layout-agnostic. However, such graphs are not always fully connected via those relations due

G. A. Fink et al. (Eds.): ICDAR 2023, LNCS 14188, pp. 130–146, 2023.
https://doi.org/10.1007/978-3-031-41679-8_8

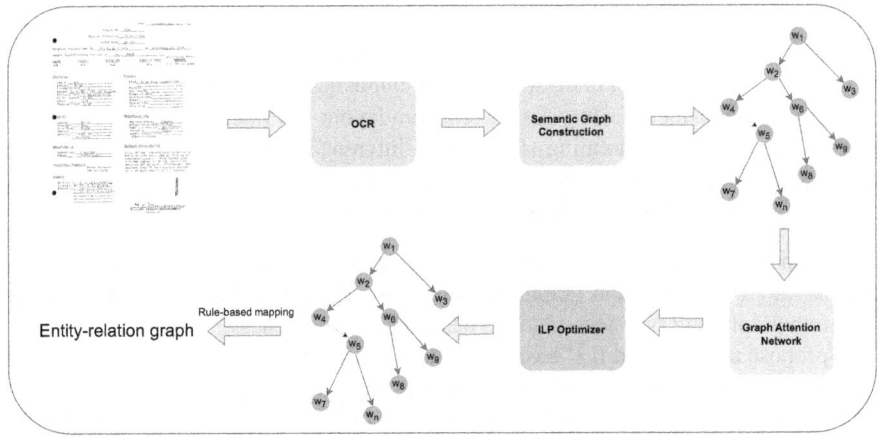

Fig. 1. Illustration of the pipeline of our proposed method.

to neglecting the layout information between entities. In contrast, form designers often put semantically relevant entities close to each other in a form, hence spatial proximateness is informative for semantic relevance between entities or relations.

Prior models on form understanding tackle at least two subtasks in the sequel, which are entity recognition and relation extraction [6]. The former identifies a group of words belonging to the same semantic entity and the label of the entity, while the latter predicts the relation between any two entities. Such a pipeline may easily lead to error propagation because the models for relation extraction are not able to fix errors of entity recognition, not to mention the exploitation of proximateness between relations to build a fully connected entity-relation graph. However, entity-relation graph parsing by tacking both subtasks jointly is challenging because the time complexity of inference is quadratic to the number of tokens in a form, as shown in Sect. 3.

The use of large-scale pre-trained language models or multimodal transformers dominates in the recent studies on form understanding [6,11,26]. However, such models require large-scale data for pre-training, especially for multilingual document understanding. For example, LANGUAGEXLM needs 30 million documents in 53 languages for pre-training [6]. Adding any new languages or new document collections often requires re-training of the models. In contrast, humans are often capable of recognizing key-value pairs from a form by only using layout information, even though they may not understand the language in the form. However, no prior studies explore to what degree layout information is useful for language-agnostic form understanding.

In this work, we propose a novel language-agnostic Graph Neural Networks (GNN) model for entity-relation graph parsing of scanned forms, coined LAGNN. To facilitate navigation in an entity-relation graph and retain proximateness of entities and relations after parsing, we add two types of relations for proximateness to entity-relation graphs: i) vertically proximate, coined *proximate_(V)*, and ii) horizontally proximate, coined *proximate_(H)*. To mitigate error propagation in pipeline approaches, we introduce a word-relation graph representation so that our model parses a document image into such a graph, which is subsequently converted to a fully connected entity-relation graph by deterministic rules. This simplification enables linear inference time. To support language-agnostic form understanding without using pre-trained language models, our model only considers features extracted from layout information. Furthermore, we apply integer linear programming (ILP) based inference to ensure that the generated graphs satisfy the properties of entity-relation graphs.

The main contributions of this paper are summarised as follows:

- We propose a language-agnostic GNN model, coined LAGNN, for parsing scanned forms into a word-relation graph, which is isomorphic to a fully connected entity-relation graph with two new relations based on proximateness.
- We propose a designated ILP-based inference method to ensure that i) the generated graphs are fully connected, and ii) relations in a graph are logically coherent.
- The extensive experimental results show that our model significantly outperforms the competitive baselines in terms of all metrics in the monolingual settings and the averaged metrics in the zero-shot multilingual settings.

2   Related Work

Deep learning techniques have dominated document interpretation tasks [1,21,28] in the past decade. Grid-based techniques [7,13,17] were suggested for representing 2D documents. In these techniques, first character-level or word-level embeddings are used to represent text, and later CNNs are used to categorize them into different field types.

Self-supervised pre-training has had a lot of success lately. Recent work on structured document pre-training [6,15,23,25,26] has pushed the boundaries, drawing inspiration from the success of pre-trained language models on multiple downstream NLP tasks. The BERT architecture was altered by LayoutLM [26] by including 2D spatial coordinate embeddings. By considering the visual aspects as independent tokens, LayoutLMv2 [25] outperformed LayoutLM. To optimize the use of unlabeled document data, extra pre-training activities were investigated. In contrast to StructuralLM [15], who suggested cell-level 2D position embeddings and the accompanying pre-training target, SelfDoc [16] developed the contextualization across a block of text. To unify many issues surrounding natural language, TILT [19] suggests a pre-trained layout-aware multimodal encoder-decoder Transformer. The useful coarse-grained information like natural units and salient visual regions are ignored by the current layout-aware multimodal Transformers. In an effort to include coarse-grained information into pre-trained layout-aware multimodal Transformers, [24] argues that both fine-grained and coarse-grained multimodal information is useful for document understanding and proposes a multi-grained and multimodal transformer, ERNIE-mmLayout.

However, the preceding Structured Document Understanding (SDU) methods mostly rely on a single language, which is usually English, making them rather constrained in terms of multilingual application scenarios. LayoutXLM [6] was the first to incorporate a multilingual text model InfoXLM [3] initialization to LayoutLMv2 framework for multilingual pre-training with structured documents. However, a laborious procedure of multilingual data collecting, cleansing, and pre-training was necessary. To address this issue, LiLT [23], a straightforward yet powerful language-independent layout Transformer for monolingual/multilingual structured document interpretation was proposed. LiLT employs BiACM to achieve language-independent cross-modality interaction and an efficient asynchronous optimization technique for both textual and non-textual flows in pre-training using two pre-training objectives.

Current state-of-the-art approaches to these document understanding challenges have made use of the power of large pre-trained language models, focusing on language more than the visual and geometrical information in a text, and end up using hundreds of millions of parameters in the process [9]. Additionally, the majority of these models are trained using a massive transformer pipeline, which necessitates the pre-training of enormous amounts of data. In this sense, models that are independent of language were proposed [5,20]. [5] concentrated on identifying entity relationships in forms using a straightforward CNN as a text line detector, and then they find key-value relationship pairs using heuristics based on the model's scores for each connection candidate. Later, [20] reformulated the issue as a semantic segmentation (pixel labelling) task with a focus on extracting the form structure. They employed a U-Net based architecture pipeline, which was quite effective at concurrently predicting all levels of the document hierarchy. For form understanding, [2] employed GCNs to solve the entity grouping, labelling, and entity linking tasks. They did not utilize any visual features and instead used word embeddings and bounding box information as the main node features, and k-nearest neighbours to obtain edge features. The FUDGE [4] framework was then created as an extension of [5] to help with form understanding. It proposes relationship pairings using the same detection CNN as in [5], considerably improving the state-of-the-art on both the semantic entity labeling and entity linking tasks. Then, because predicting key-value connection pairs and the semantic labels for text entities are two tasks that are closely associated, a GCN was implemented using plugged visual features from the CNN. Inspire by FUDGE [4], a task-agnostic GNN-based framework called Doc2Graph [9] that adopts a similar joint prediction of both the tasks, semantic entity labeling and entity linking utilizing a node classification and edge classification module, respectively, without relying on heuristics to establish associations between words or entities was developed. To take advantage of the relative location of document objects via polar coordinates, a novel GNN architecture pipeline with node and edge aggregation functions is implemented.

3   Methodology

Entity-relation graph parsing for form understanding is concerned with mapping a document image to an entity-relation graph. To reduce inference time complexity, we propose to map a scanned form image to a novel word-relation graph, which is isomorphic to the corresponding entity-relation graph. Then entity-relation graphs can be directly constructed from word-relation graphs using deterministic rules.

Formally, an entity-relation graph is denoted by $\mathcal{G}^e = \{\mathcal{V}^e, \mathcal{E}^e\}$, composed of a set of entities $\mathcal{V}^e$ and a set of relations between entities $\mathcal{E}^e$. Each entity is a word sequence

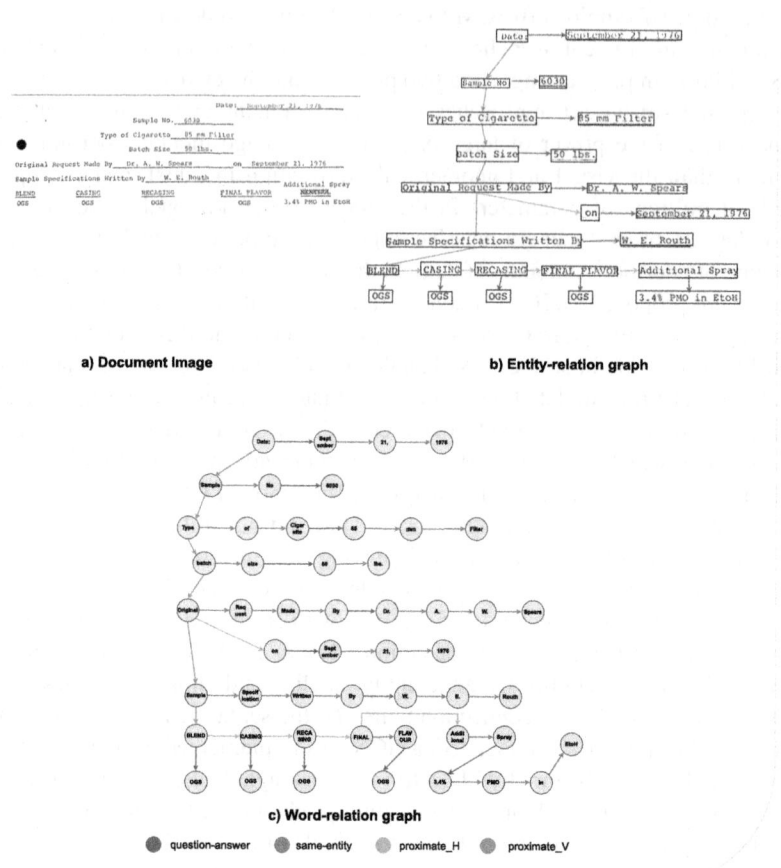

a) Document Image

b) Entity-relation graph

c) Word-relation graph

● question-answer ● same-entity ● proximate_H ● proximate_V

Fig. 2. Illustration of our proposed entity-relation graph and word-relation graph.

$[w_0, ..., w_m]$ labeled by a type $z^e \in \mathcal{Z}^e$, and each word is associated with a bounding box $(x_{left}, y_{top}, x_{right}, y_{bottom})$ [12]. The set $\mathcal{Z}^e$ includes *question*, *answer*, and *section*. A relation $(v_a, y, v_b) \in \mathcal{E}^e$ is denoted by a directed edge from an entity v_a to another entity v_b labeled by $z \in \mathcal{Z}^r$, where $\mathcal{Z}^r$ includes the following relations.

- *Question-answer*, denoted by a directed edge from a question to the corresponding answer.
- *Header-question*, denoted by a directed edge from a section and a question belonging to that section.
- *Proximate_(V)*, denoted by an undirected edge between an entity A and the first entity B in the first line below it, which is not an answer and there is no relation between A and B or the parent of B.
- *Proximate_(H)*, denoted by an undirected edge between an entity A and the closest entity labeled as question or header to its right in the same line.

In contrast, a word-relation graph $\mathcal{G}^w = \{\mathcal{V}^w, \mathcal{E}^w\}$ depicted in Fig. 2 comprises a set of nodes $\mathcal{V}^w$ consisting of only the words in a document, and a set of relations $\mathcal{E}^w$ between those words. A relation in $\mathcal{E}^w$ is a tuple (w_a, z, w_b) from a word w_a to another word w_b labeled by $z \in \mathcal{Z}^w$, where the edge label set $\mathcal{Z}^w$ augments $\mathcal{Z}^r$ by adding word-level relations. More details are provided in Sect. 3.1

As illustrated in Fig. 1, to parse document images into word-relation graphs, we first apply an off-the-shelf Optical Character Recogniser docTR [18] to map a document image into a set of cells, whereby each cell is a sequence of bounding boxes and each bounding box corresponds to a word. Then we apply our model LAGNN to estimate the probabilities of relations between each possible pair of words largely based on relative distances between bounding boxes. To construct a word-relation graph, we apply a designated ILP inference algorithm to assign relation labels to word pairs *jointly*.

3.1 Word-Relation Graph

A naive solution that directly parses a document image to an entity-relation graph results in a time complexity quadratic to the number of words in a document in the worst case. First, a model needs to identify if a word belongs to the same group of its adjacent words, followed by predicting relations between any pairs of word groups. In the worst case, if each entity comprises only one word, the time complexity is $O(n^2)$ because the number of edge predictions is $n \times n - 1$, where n is the number of words in a document. In fact, the average word lengths of entities in FUNSD [12] is 3.4, the actual inference time complexity is not far from the worst case. The cost is estimated without considering the one for estimating word groups for each entity.

We map entity-relation graphs to word-relation graphs so that entity-relation graph parsing becomes the task of predicting relations between words in a word-relation graph. To eliminate the task of entity recognition, we introduce an undirected relation *same-entity* to link two adjacent words in the same word sequence of an entity. For a relation (v_a, z, v_b) between two entities in an entity-relation graph, we adapt the relation labels in entity-relation graphs to word-level relations:

- *Question-answer*, the last word in a question is linked to the first word in the corresponding answer.
- *Header-question*, the first word in a section is linked to the first word of the corresponding question.
- *Proximate_(V)*, the first word in v_a is linked to the first word in v_b.
- *Proximate_(H)*, the last word in v_a is linked to the first word in v_b.
- *Same-entity*, the adjacent words within an entity are connected with this relation.

The conversion process is deterministic because i) for a relation between two entities, either the first or the last word of an entity is linked, and ii) *same-entity* links only adjacent words in an entity. Thus it is straightforward to revert the process to map a word-relation graph and an entity-relation graph by using rules. As a result, an word-relation graph is isomorphic to an entity-relation graph.

3.2 LAGNN Model

In this section, we present LAGNN that parses a document image to a word-relation graph, which is reformulated as predicting relations between words in a document. In contrast to prior studies [6,26], our model relies on relative distances between the bounding boxes of words, which are robust across languages, rather than linguistic features. Moreover, we adopt Graph Attention Network (GAT) [22] to capture the similarities of relations between the neighbours of a word. However, we slightly modify GAT model to include edge features in addition to node features in message passing. We do this by concatenating our edge features with node features before performing the message passing. The coherence of predicted relations and graph properties of a word-relation graph are ensured by ILP at the inference stage.

Similar to prior works [6,26], we apply the off-the-shelf OCR model [18] to map document images to bags of words. The OCR outputs are reorganized by sorting lines from top to bottom and arranging words from left to right in their original order by using their bounding-box coordinates. For each word w, we define its neighbourhood $\mathcal{N}_w$ as the set of words, to which it can potentially have relations. The set $\mathcal{N}_w$ consists of at most K nearest words, including the word to the right of w and the words below it. This definition of neighbourhood is motivated by the fact that the edge score between two words is independent of the orientation of the edge. Hence, during inference, we only compute edge scores between a word and each word in its neighbourhood $\mathcal{N}_w$ to avoid repeated computations.

We observe that a significant proportion of the neighbours of a word have the same relations. For example, all words in an answer composed of multiple words are associated with the relation *same-entity*. By viewing a form as a word-relation graph, we take the GAT model as our backbone because it supports both node features and edge features. Herein, each node w is represented by a normalized bounding box coefficient vector $\boldsymbol{h}_w = (x_{left}/W, y_{top}/H, x_{right}/W, y_{bottom}/H)$, where W and H denotes the width and the height of a document image respectively. To support cross-lingual parsing, we represent an edge (v_a, v_b) only by a relative spacing feature vector $\boldsymbol{d}_{ij}$. Such a feature vector $\boldsymbol{d}_{ij}$ is computed based on both horizontal and vertical relative distances between their normalized bounding boxes. Specifically, given two words w_i and w_j with the bounding boxes $(x_{i1}, y_{i1}, x_{i2}, y_{i2})$, and $(x_{j1}, y_{j1}, x_{j2}, y_{j2})$ respectively, the spacing between w_i and w_j is calculated along x and y-axis by

$$d_{x1}, d_{x2}, d_{x3} = (x_{i1} - x_{j1}), (x_{i2} - x_{j1}), (x_{i2} - x_{j12})$$
$$d_{y1}, d_{y2}, d_{y3} = (y_{i1} - y_{j1}), (y_{i2} - y_{j1}), (y_{i2} - y_{j12})$$

As a result, the spacing feature vector $\boldsymbol{d}_{ij} = [d_{x1}, d_{x2}, d_{x3}, d_{y1}, d_{y2}, d_{y3}]$.

In comparison with the prior works [6,11,25,26], which use text, image, and layout features altogether, our model is more parameter efficient by adopting only language-independent and layout-agnostic features. Due to those simple features, our model contains only 7.3K parameters, significantly less than those transformer-based models, which have approximately 200-400M parameters.

3.3 Inference

We formulate the inference problem as an integer linear program that aims to identify the most likely word-relation graph satisfying the graph properties outlined in Sect. 3.1.

Given an edge embedding e_{ij}, we compute a score s_{ijz} for each possible label $z \in \mathcal{Z}^w$ by using a linear layer. As a word-relation graph is sparse, we extend $\mathcal{Z}^w$ to the set $\mathcal{Z}^+$ by adding a relation *no-relation*, which indicates there is no relation between two words. Let $u_{jiz} \in \{0, 1\}$ be a binary variable indicating if there is an edge with a label z between v_i and v_j, the solution to the following integer linear program is the most likely word-relation graph.

$$\min \ s^{\mathsf{T}} u \tag{1}$$
$$\text{s.t.} \ \ Au \leq b$$

where A denotes the constraint matrix and b is the corresponding constant vector.

We formulate five constraints specific to documents for efficient joint inference using integer linear programming:

- **Connectivity constraint (C1):** The generated word-relation graphs are fully connected such that there is a path between any pair of nodes in a graph.

$$\sum_{v_j \in \mathcal{N}_i} \sum_{z \in \mathcal{Z}^r} u_{ijz} \geq 1, \forall v_i \in \mathcal{V}^w$$

- **QA constraint (C2):** If a relation between two words u_{ijz^q} is *question-answer*, where z^q denotes *question-answer*, then the next immediate relation must be either *proximate_(H)* (z^h) or *same-entity* (z^w). Let the next relation is denoted by $u_{i(j+1)z}$, this constraint is defined as:

$$u_{i(j+1)z^h} + u_{i(j+1)z^w} \geq u_{ijz^q}$$

- **Single label constraint (C3):** There is only one relation in $\mathcal{Z}^w$ between any pair of nodes in a word-relation graph.

$$\sum_{z \in \mathcal{Z}^+} u_{jiz} = 1, \forall (v_i, v_j) \in \mathcal{V}^w \times \mathcal{V}^w$$

- **At least one semantic relations (C4):** A word is part of an entity. Therefore, it is linked to another word via *same-entity* if the entity contains multiple words. If the word is at the beginning or the end of an entity, it points to another entity via either *question-answer* or *header-question*. In other words, a word is involved in at least one relation in $\mathcal{Z}^s = \{question\text{-}answer, same\text{-}entity, header\text{-}question\}$. If an entity contains a single word, it should be associated with either *question-answer* or *header-question*. Otherwise, if a word is associated only with $proximate_V$ or $proximate_H$, we cannot determine which entity it belongs to and the type of the corresponding entity.

$$\sum_{j \in \mathcal{N}} \sum_{z \in \mathcal{Z}^s} u_{jiz} \geq 1, \forall v_i$$

– **At least one semantic relation in the neighbourhood (C5):** The previous constraint can still fail to exclude the cases that a word is only involved in *same-entity* in $\mathcal{Z}^s$. In such a case, the word is expected to be at either end of an entity. If we need to infer the type of entity, this word should be linked to another word via either *question-answer* or *header-question*. Therefore, if a word is linked with a *same-entity* relation, it should be linked to a different word with a relation in $\mathcal{Z}^s$.

$$u_{i(i+1)z^h} + u_{i(i+1)z^w} + u_{i(i+1)z^q} \geq u_{(i-1)iz^w}$$

After parsing a document image into a word-relation graph, we apply the deterministic rules introduced in Sect. 3.1 to convert the graph into an entity-relation graph.

3.4 Model Training

Inspired by [8], we apply the cross-entropy loss to each pair of nodes during training. The construction of word-relation graphs is realized by the ILP-based inference method detailed above.

$$\mathcal{L} = -\sum_{ij} \sum_{z \in \mathcal{Z}^+} z_{ij} \log \hat{z}_{ij} \tag{2}$$

where z_{ij} and $\hat{z}_{ij}$ denote the ground-truth labels and predicted labels respectively.

4   Experiments

We compare our model with the state-of-the-art methods on both monolingual and multilingual form datasets. The results show that our models significantly outperform the baselines in terms of relation prediction on word-relation graphs in both settings, which leads further to superior performance on entity-relation graphs. The extensive ablation studies demonstrate the effectiveness of incorporating structural information using GAT and the constraints during ILP-based inference.

4.1   Datasets

FUNSD [12]. It is a form understanding dataset consisting of 199 noisy documents which are fully annotated. It has a total of 9,707 semantic entities over 31,485 words. In the official data split, the 199 samples are divided into 149 training samples and 50 testing samples. However, we further split the 149 samples into 139 training and 10 validation samples. Our test split is the same as the official split. Each entity is labelled with one of the four semantic entity labels - "question", "answer", "header", and "other". This dataset is widely employed for semantic entity labelling and relation extraction tasks.

XFUND [27]. This is a multilingual benchmark dataset comprising 199 forms that are labeled by humans in 7 languages, which are Chinese (ZH), Japanese (JA), Spanish (ES), French (FR), Italian (IT), German (DE), Portuguese (PT). The dataset is divided into 149 forms for training and 50 for testing. In the ground-truth annotations, each entity is labeled with either "question", "answer", "header", or "other". Following previous works [6, 23], we use this dataset to compare our model with existing state-of-the-art models in the zero-shot settings.

To obtain the relation annotations in entity-relation graphs on both datasets, we apply rules to generate the relation annotations based on entity annotations, followed by manually checking all document images for correctness. Due to the isomorphism between word-relation graphs and entity-relation graphs, we map the resulting entity-relation graphs to word-relation graphs by using the deterministic rules.

Table 1. Relation extraction (RE) results on the word-relation graphs of FUNSD dataset, where T, L, I denotes if models use text, layout, and image features respectively. Ours denotes LAGNN without applying ILP-based inference. M stands for million and K refers to thousand. T, L, and I refer to text, layout, and image modalities respectively.

Model	#Parameters	Modality	Precision ($\uparrow$)	Recall ($\uparrow$)	F1 ($\uparrow$)
XLM-RoBERTa$_{BASE}$	-	T	0.563	0.561	0.561
InfoXLM$_{BASE}$	-	T	0.593	0.603	0.598
LayoutLM	11M	T+L+I	0.664	0.666	0.665
LayoutXLM$_{BASE}$	30M	T+L+I	0.709	0.715	0.712
LayoutLMv2	11M	T+L+I	0.722	0.712	0.717
StructuralLM	11M	T+L	0.741	0.749	0.745
LiLT[InfoXLM]$_{BASE}$	11M	T+L	0.792	0.786	0.789
LayoutLMv3	11M	T+L+I	0.801	0.809	0.805
Ours	8.1K	L	0.837	0.854	0.845
Ours + Constraints	8.1K	L	**0.848**	**0.861**	**0.854**

Table 2. Evaluation of entity recognition on the entity-relation graphs of the FUNSD dataset.

Model	Precision ($\uparrow$)	Recall ($\uparrow$)	F1 ($\uparrow$)
LayoutLM	0.753	0.757	0.755
LayoutXLM$_{BASE}$	0.791	0.796	0.793
LayoutLMv2	0.847	0.852	0.849
LiLT[InfoXLM]$_{BASE}$	0.864	0.881	0.872
LayoutLMv3	0.898	0.903	0.900
LAGNN	**0.921**	**0.936**	**0.928**

4.2 Implementation Details

The entity-relation graphs are constructed using Deep Graph Library (DGL). We use a single-layer GAT with 3 heads and a hidden dimension size of 64 for both node and edge features. We train our model using Adam optimizer for 500 iterations with a learning rate of $1 \times e^{-3}$. If the performance on the validation data does not improve after 100 iterations, training stops early. During training, we save the model checkpoint based on its performance on validation data. For inference on the test set, the checkpoint with the best performance across all training epochs is loaded. We train our model on 1 GTX 1080Ti 12 GB GPU.

4.3 Monolingual Results on FUNSD

We first evaluate our model on the word-relation graphs on FUNSD by considering the task as relation extraction between words. More specifically, we run the state-of-the-art models LayoutLM [26], LayoutLMv2 [25], LayoutLMv3 [11], StructuralLM [15], LayoutXLM [6], InfoXLM$_{BASE}$ [3], and LiLT[InfoXLM]$_{BASE}$ [23] to predict entities, followed by relation extraction.

For baselines, we perform relation extraction by following the approach in [6]. First, we create all possible entity pairs as relation candidates. Each candidate is represented by the concatenation of the corresponding entity representations. Furthermore, the representation of an entity is first constructed by concatenating the embedding of the first token of each entity and the entity type embedding, followed by feeding them through two position-wise feed-forward networks (FFN) modules. The resulting relation candidate representations are fed into a bi-affine classifier for relation classification. The conversion from entity-relation graphs to word-relation graphs is performed by the same set of deterministic rules introduced in Sect. 3.1.

Table 1 reports the relation extraction results on word-relation graphs in terms of *Precision*, *Recall*, and *F1*. For each metric, we take the micro-average among the relations in $\mathcal{Z}^w$. The two variations of our model achieve superior performance over the baselines by only using the relative spacing features. In contrast, all baselines use textual features extracted from large language models that require pre-training on large-scale datasets. Some of the baselines, such as variations of LayoutLM, require even vision features. The number of parameters of our models is also significantly smaller than their competitors. Although pre-trained language models are widely used in a number of AI applications, our work raises the basic question for future research "Are language models necessary for form recognition?".

Apart from relation extraction, we also evaluate the models in terms of entity recognition on converted entity-relation graphs. Table 2 summarizes the results based on an exact match in terms of Precision, Recall, and F1.

4.4 Zero-Shot Multilingual Results

We further evaluate model performance by applying the models trained on FUNSD directly to the scanned forms in other languages on XFUND. Herein, we compare our models with the state-of-the-art methods: XLM-RoBERTa$_{BASE}$, InfoXLM$_{BASE}$, LayoutXLM$_{BASE}$ [6], and LiLT[InfoXLM]$_{BASE}$ [23]. The latter two models are pre-trained on large-scale document datasets of size 30M and 11M respectively.

Table 3 reports the corresponding zero-shot multilingual relation extraction results on the word-relation graphs of XFUND dataset. Overall, our best model outperforms the baselines in 7 out of 8 languages in terms of *F1*. The geometric mean of the F1

Table 3. Multilingual relation extraction (RE) results on word-relation graphs of XFUND dataset.

Model	Pretraining		FUNSD	XFUND								Avg
	Language	Size	EN	JA	ZH	DE	FR	PT	ES	IT	
XLM-RoBERTa$_{BASE}$	-	-	0.587	0.116	0.132	0.335	0.400	0.354	0.281	0.286	0.311
InfoXLM$_{BASE}$	-	-	0.601	0.132	0.159	0.357	0.398	0.352	0.295	0.300	0.324
LayoutXLM$_{BASE}$	Multilingual	30M	0.701	0.258	0.240	0.443	0.568	0.549	0.461	0.499	0.464
LiLT[InfoXLM]$_{BASE}$	English	11M	0.771	0.303	0.349	0.562	**0.691**	0.613	0.554	0.586	0.553
Ours	×	-	0.845	0.611	0.625	0.636	0.679	0.641	**0.651**	0.625	0.664
Ours + Constraints	×	-	**0.853**	**0.625**	**0.626**	**0.647**	0.673	**0.641**	0.644	**0.638**	**0.669**

Table 4. Ablation studies. C1, C2, C4 and C5 are the constraints defined in Sect. 3.3

Model	Precision	Recall	F1
GraphSAGE	0.756	0.768	0.761
GCNs	0.814	0.823	0.818
LaGNN- Edge$_{feats}$	0.705	0.717	0.710
LaGNN + Edge$_{feats}$	0.837	0.854	0.845
LaGNN + Edge$_{feats}$ + all constrs	**0.848**	**0.861**	**0.854**
LaGNN + Edge$_{feats}$ + all constrs - C4+C5	0.846	0.860	0.852
LaGNN + Edge$_{feats}$ + all constrs - C1	0.845	0.858	0.851
LaGNN + Edge$_{feats}$ + all constrs - C2	0.840	0.851	0.845

is more than 10% better than the strongest baseline. It is noteworthy that performance improvement is achieved without any time-consuming pre-training. We only use the 139 forms from the FUNSD dataset to train our models. Our models based on relative spacing features are more transferrable than those multilingual language models on this task. A further inspection shows that our models benefit from the fact that the space between two words within an entity, as well as the distance between a question word and an answer word, are similar across languages. In this zero-shot multilingual setting, incorporating the constraints into inference does not always help, though it leads to improvements in 6 out of 8 languages (Fig. 3).

4.5   Ablation Study

We evaluate the effectiveness of using GAT, edge features, constraints in ILP, and the size of neighbourhood in our model. To show the usefulness of GAT, we compare it with GraphSAGE [10] and graph convolutional networks (GCNs) [14] by using the same features. For edge features, we run ablation studies by removing them with or without applying ILP. To understand the usefulness of the constraints during inference, we remove the connectivity constraint (C1), QA constraint (C2), and the last two semantic constraints (C4+C5) respectively from the full model.

As shown in Table 4, it is expected that our full model performs the best among all variations. Removing relative spacing features leads to the largest drop in terms of

Table 5. This table illustrates the number of predictions by LAGNN that violated a constraint. These are corrected by applying ILP inference described in Sect. 3.3.

Constraints	Number of violations
C1	71
C2	169
C3	0
C4	51
C5	10

Table 6. Effectiveness of edge features.

Model	Precision	Recall	F1
LAGNN + Edge$_{feats}$ in lower-level and at classifier	**0.837**	**0.854**	**0.845**
LAGNN + Edge$_{feats}$ in lower-level and not at classifier	0.713	0.725	0.718
LAGNN + Edge$_{feats}$ not in lower-level but at classifier	0.831	0.849	0.840
LAGNN- Edge$_{feats}$	0.705	0.717	0.710

all metrics. GAT demonstrates its strengths over the two alternative neural structures. Removing any of the constraints, the model performance drops slightly. Among them, C2 is clearly the most useful one among them in terms of improving performance. As such, C2 is the qa constraint that is designed to correct the wrongly predicted *question-answer* relation labels, which are one of the most frequent model prediction errors based on our evaluation. Before applying ILP, we observe that there were 169 out of 51775 word pairs in the test set that violate this constraint. Table 5 illustrates the number of violations in the predictions of LAGNN before applying the constraint-based inference.

The constraints C4 and C5 are designed to ensure isomorphism between entity-relation graphs and word-relation graphs. This constraint assists in preventing any isolated word pairs that are not connected to an entity (question, or header). For instance, if a question has more than one word in the answer, the words within the answer are linked together using the *same-entity* relation. The last word of the question and the first word of the answer are linked by the *question-answer* relation. When transforming a word-relation graph to an entity-relation graph, we take into account the rule that any word pairs with a *same-entity* connection following a *question-answer* relation are mapped to the entity relation "answer" in the entity-relation graph. However, if LAGNN predicts a different relation for a word pair within the answer than *same-entity*, this will result in an inaccurate mapping of word-relation graph to entity-relation graph. Additionally, a word pair can only have a *same-entity* relation if it is part of a chain of words whose head is linked to either a "question" or a "header". It is impossible to have an isolated word pair with a *same-entity* relation that isn't related to either the "question" or "header" in our proposed way of word-relation graph construction. However, a few of the LAGNN model's relation predictions could result in isolated word pairs. In order to prevent this, our uniquely designed semantic constraints correct these mistakes and aid

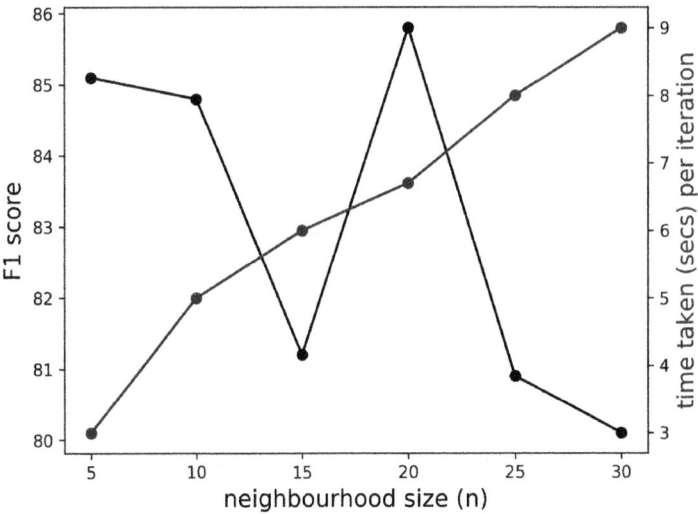

Fig. 3. Trade off between neighbourhood size and computational time.

in the accurate mapping of the word-relation graph to the entity-relation graph. There are 134 total isolated word pairings before semantic constraints are applied. However, with the application of the constraints, there were no violations, and the performance benefit from using semantic constraints-which are made particularly to achieve isomorphism between word-relation graph and entity-relation graph—is very little.

Where to Put Edge Features. To demonstrate the significance of using spacing between pairs of nodes as its edge feature, we conduct four experiments with and without edge features. The results for these experiments are available in Table 6. Our key objective is to determine whether it is more effective to concatenate edge features with node features when computing the node attention rather than doing so only at the model's final classifier layer. It is evident that employing edge features solely while computing node attention (row 2) results in much lower performance than using the model's final (classifier) layer (row 3). Nevertheless, it is still marginally preferable to not use edge features at all (row 4). The optimal performance can be obtained by employing edge features while computing node attention as well as at the last layer of the model (row 1).

Effect of Neighbourhood Size. On both datasets, the "linking" key found in the JSON annotation files of the ground-truth documents is used to construct the ground-truth graphs. There are no ground-truth annotation files accessible during the inference. Therefore, the model must decide which nodes should be connected to which nodes. The "no-link" relation label enables the model to discover which nodes ought to be connected by an edge. Each word or node in the graph is linked to the n following words or nodes in the document using an edge labelled *no-relation*. The edges predicted with a "no-link" relation label are removed prior to applying the ILP inference. We experiment with various n to examine the impact of neighbourhood size on the model's performance. As n increases, the computational time increases. We determine an ideal value for n by finding an optimal trade-off between computation time and performance.

5   Conclusion

In this work, we propose a novel entity-relation graph parsing model called LAGNN that is language independent. Our model parses a document image into a word-relation graph in order to minimize error propagation of pipeline approaches and reduce the time complexity of inference. This graph is then transformed by deterministic rules into a fully linked entity-relation graph. Due to this simplification, inference time is sharply reduced. Our model simply takes into account relative spacing features extracted from layout information in order to allow language-independent form understanding without the use of pre-trained language models. To ensure that the generated graphs match the specifications of entity-relation graphs, we use ILP-based inference by incorporating designated constraints for this task. Our experimental results on the multilingual XFUND and FUSND datasets demonstrate that our proposed approach produces superior results over competitive baselines. In particular, on the zero-short multilingual form understanding task, our model surpasses recent strong baselines by a large margin. Additionally, we conduct extensive ablation studies that demonstrate the effectiveness of each new design choice we proposed.

References

1. Borges Oliveira, D.A., Viana, M.P.: Fast CNN-based document layout analysis. In: 2017 IEEE International Conference on Computer Vision Workshops (ICCVW), pp. 1173–1180 (2017). https://doi.org/10.1109/ICCVW.2017.142
2. Carbonell, M., Riba, P., Villegas, M., Fornés, A., Lladós, J.: Named entity recognition and relation extraction with graph neural networks in semi structured documents. In: 2020 25th International Conference on Pattern Recognition (ICPR), pp. 9622–9627 (2021). https://doi.org/10.1109/ICPR48806.2021.9412669
3. Chi, Z., et al.: InfoXLM: an information-theoretic framework for cross-lingual language model pre-training. In: Proceedings of the 2021 Conference of the North American Chapter of the Association for Computational Linguistics: Human Language Technologies, pp. 3576–3588. Association for Computational Linguistics, Online (2021). https://doi.org/10.18653/v1/2021.naacl-main.280. http://aclanthology.org/2021.naacl-main.280
4. Davis, B., Morse, B., Price, B., Tensmeyer, C., Wiginton, C.: Visual FUDGE: form understanding via dynamic graph editing. In: Lladós, J., Lopresti, D., Uchida, S. (eds.) ICDAR 2021. LNCS, vol. 12821, pp. 416–431. Springer, Cham (2021). https://doi.org/10.1007/978-3-030-86549-8_27
5. Davis, B.L., Morse, B., Cohen, S.D., Price, B.L., Tensmeyer, C.: Deep visual template-free form parsing. In: 2019 International Conference on Document Analysis and Recognition (ICDAR), pp. 134–141 (2019)
6. Déjean, H., Clinchant, S., Meunier, J.: Layoutxlm vs. GNN: an empirical evaluation of relation extraction for documents. CoRR abs/2206.10304 (2022)

7. Denk, T.I., Reisswig, C.: BERTgrid: contextualized embedding for 2D document representation and understanding. In: Workshop on Document Intelligence at NeurIPS 2019 (2019).
   http://openreview.net/forum?id=H1gsGaq9US
8. Domke, J.: Structured learning via logistic regression. In: Advances in Neural Information
   Processing Systems, vol. 26 (2013)
9. Gemelli, A., Biswas, S., Civitelli, E., Lladós, J., Marinai, S.: Doc2graph: a task agnostic
   document understanding framework based on graph neural networks (2022). https://doi.org/
   10.48550/ARXIV.2208.11168. http://arxiv.org/abs/2208.11168
10. Hamilton, W., Ying, Z., Leskovec, J.: Inductive representation learning on large graphs. In:
    Advances in Neural Information Processing Systems, vol. 30 (2017)
11. Huang, Y., Lv, T., Cui, L., Lu, Y., Wei, F.: Layoutlmv3: pre-training for document AI with
    unified text and image masking. In: Magalhães, J., et al. (eds.) MM 2022: The 30th ACM
    International Conference on Multimedia, Lisboa, Portugal, 10–14 October 2022, pp. 4083–
    4091. ACM (2022)
12. Jaume, G., Ekenel, H.K., Thiran, J.: FUNSD: a dataset for form understanding in noisy
    scanned documents. In: 2nd International Workshop on Open Services and Tools for Doc-
    ument Analysis, OST@ICDAR 2019, Sydney, Australia, 22–25 September 2019, pp. 1–6.
    IEEE (2019)
13. Katti, A.R., et al.: Chargrid: towards understanding 2D documents. In: Proceedings of the
    2018 Conference on Empirical Methods in Natural Language Processing, Brussels, Belgium,
    pp. 4459–4469. Association for Computational Linguistics (2018). https://doi.org/10.18653/
    v1/D18-1476. http://aclanthology.org/D18-1476
14. Kipf, T.N., Welling, M.: Semi-supervised classification with graph convolutional networks.
    In: International Conference on Learning Representations (2017). http://openreview.net/
    forum?id=SJU4ayYgl
15. Li, C., et al.: StructuralLM: structural pre-training for form understanding. In: Proceedings
    of the 59th Annual Meeting of the Association for Computational Linguistics and the 11th
    International Joint Conference on Natural Language Processing (Volume 1: Long Papers),
    pp. 6309–6318. Association for Computational Linguistics, Online (2021). https://doi.org/
    10.18653/v1/2021.acl-long.493. http://aclanthology.org/2021.acl-long.493
16. Li, P., et al.: Selfdoc: self-supervised document representation learning. In: 2021 IEEE/CVF
    Conference on Computer Vision and Pattern Recognition (CVPR), pp. 5648–5656 (2021)
17. Lin, W., et al.: ViBERTgrid: a jointly trained multi-modal 2D document representation for
    key information extraction from documents. In: Lladós, J., Lopresti, D., Uchida, S. (eds.)
    ICDAR 2021. LNCS, vol. 12821, pp. 548–563. Springer, Cham (2021). https://doi.org/10.
    1007/978-3-030-86549-8_35
18. Mindee: docTR: document text recognition (2021). http://github.com/mindee/doctr
19. Powalski, R., Borchmann, Ł., Jurkiewicz, D., Dwojak, T., Pietruszka, M., Pałka, G.: Going
    full-TILT boogie on document understanding with text-image-layout transformer. In: Lladós,
    J., Lopresti, D., Uchida, S. (eds.) ICDAR 2021. LNCS, vol. 12822, pp. 732–747. Springer,
    Cham (2021). https://doi.org/10.1007/978-3-030-86331-9_47
20. Sarkar, M., Aggarwal, M., Jain, A., Gupta, H., Krishnamurthy, B.: Document structure
    extraction using prior based high resolution hierarchical semantic segmentation. In: Vedaldi,
    A., Bischof, H., Brox, T., Frahm, J.-M. (eds.) ECCV 2020. LNCS, vol. 12373, pp. 649–666.
    Springer, Cham (2020). https://doi.org/10.1007/978-3-030-58604-1_39
21. Siegel, N., Lourie, N., Power, R., Ammar, W.: Extracting scientific figures with distantly
    supervised neural networks. In: Proceedings of the 18th ACM/IEEE on Joint Conference on
    Digital Libraries (2018)
22. Veličković, P., Cucurull, G., Casanova, A., Romero, A., Liò, P., Bengio, Y.: Graph atten-
    tion networks. In: International Conference on Learning Representations (2018). http://
    openreview.net/forum?id=rJXMpikCZ

23. Wang, J., Jin, L., Ding, K.: LiLT: a simple yet effective language-independent layout trans-former for structured document understanding. In: Proceedings of the 60th Annual Meeting of the Association for Computational Linguistics (Volume 1: Long Papers), Dublin, Ireland, pp. 7747–7757. Association for Computational Linguistics (2022). https://doi.org/10.18653/v1/2022.acl-long.534. http://aclanthology.org/2022.acl-long.534
24. Wang, W., et al.: Mmlayout: multi-grained multimodal transformer for document under-standing. In: Proceedings of the 30th ACM International Conference on Multimedia, MM 2022, pp. 4877–4886. Association for Computing Machinery, New York (2022). https://doi.org/10.1145/3503161.3548406
25. Xu, Y., et al.: LayoutLMv2: multi-modal pre-training for visually-rich document understand-ing. In: Zong, C., Xia, F., Li, W., Navigli, R. (eds.) Proceedings of the 59th Annual Meeting of the Association for Computational Linguistics and the 11th International Joint Confer-ence on Natural Language Processing, ACL/IJCNLP 2021, (Volume 1: Long Papers), Virtual Event, 1–6 August 2021, pp. 2579–2591. Association for Computational Linguistics (2021)
26. Xu, Y., Li, M., Cui, L., Huang, S., Wei, F., Zhou, M.: Layoutlm: pre-training of text and layout for document image understanding. CoRR abs/1912.13318 (2019)
27. Xu, Y., et al.: XFUND: a benchmark dataset for multilingual visually rich form understand-ing. In: Findings of the Association for Computational Linguistics: ACL 2022, Dublin, Ire-land, pp. 3214–3224. Association for Computational Linguistics (2022). https://doi.org/10.18653/v1/2022.findings-acl.253. http://aclanthology.org/2022.findings-acl.253
28. Yang, X., Yumer, E., Asente, P., Kraley, M., Kifer, D., Giles, C.L.: Learning to extract seman-tic structure from documents using multimodal fully convolutional neural networks. 2017 IEEE Conference on Computer Vision and Pattern Recognition (CVPR), pp. 4342–4351 (2017)

DocILE Benchmark for Document Information Localization and Extraction

Štěpán Šimsa[1]([✉])[iD], Milan Šulc[1][iD], Michal Uřičář[1][iD], Yash Patel[2][iD],
Ahmed Hamdi[3][iD], Matěj Kocián[1][iD], Matyáš Skalický[1][iD], Jiří Matas[2][iD],
Antoine Doucet[3][iD], Mickaël Coustaty[3][iD], and Dimosthenis Karatzas[4][iD]

[1] Rossum, Prague, Czech Republic
stepan.simsa@rossum.ai
[2] Visual Recognition Group, Czech Technical University in Prague, Prague, Czechia
[3] University of La Rochelle, La Rochelle, France
[4] Computer Vision Center, Universitat Autónoma de Barcelona, Barcelona, Spain

Abstract. This paper introduces the DocILE benchmark with the largest dataset of business documents for the tasks of *Key Information Localization and Extraction* and *Line Item Recognition*. It contains 6.7k annotated business documents, 100k synthetically generated documents, and nearly 1M unlabeled documents for unsupervised pretraining. The dataset has been built with knowledge of domain- and task-specific aspects, resulting in the following key features: (i) annotations in 55 classes, which surpasses the granularity of previously published key information extraction datasets by a large margin; (ii) Line Item Recognition represents a highly practical information extraction task, where key information has to be assigned to items in a table; (iii) documents come from numerous layouts and the test set includes zero- and few-shot cases as well as layouts commonly seen in the training set. The benchmark comes with several baselines, including RoBERTa, LayoutLMv3 and DETR-based Table Transformer; applied to both tasks of the DocILE benchmark, with results shared in this paper, offering a quick starting point for future work. The dataset, baselines and supplementary material are available at https://github.com/rossumai/docile.

Keywords: Document AI · Information Extraction · Line Item Recognition · Business Documents · Intelligent Document Processing

1 Introduction

Automating information extraction from business documents has the potential to streamline repetitive human labour and allow data entry workers to focus on more strategic tasks. Despite the recent shift towards business digitalization, the majority of Business-to-Business (B2B) communication still happens through the interchange of semi-structured[1] business documents such as invoices,

---

[1] We use the term *semi-structured documents* as [62,69]; visual structure is strongly related to the document semantics, but the layout is variable.

© The Author(s), under exclusive license to Springer Nature Switzerland AG 2023
G. A. Fink et al. (Eds.): ICDAR 2023, LNCS 14188, pp. 147–166, 2023.
https://doi.org/10.1007/978-3-031-41679-8_9

tax forms, orders, etc. The layouts of these documents were designed for human readability, yet the downstream applications (i.e. accounting software) depend on data in a structured, computer-readable format. Traditionally, this has been solved by manual data entry, requiring substantial time to process each document. The automated process of data extraction from such documents goes far beyond Optical Character Recognition (OCR) as it requires understanding of semantics, layout and context of the information within the document. The machine learning field dealing with this is called *Document Information Extraction* (IE), a sub-category of *Document Understanding* (DU).

Information Extraction from business documents lacks practical large-scale benchmarks, as noted in [11,35,56,69,75]. While there are several public datasets for document understanding, as reviewed in Sect. 2.2, only a few of them focus on information extraction from business documents. They are typically small-scale [49,74,79], focusing solely on receipts [58,74], or limit the task, e.g., to *Named Entity Recognition* (NER), missing location annotation [5,29,72,73]. Many results in the field are therefore published on private datasets [9,21,26,33,56], limiting the reproducibility and hindering further research. Digital semi-structured documents often contain sensitive information, such as names and addresses, which hampers the creation of sufficiently-large public datasets and benchmarks.

The standard problem of *Key Information Extraction* (KIE) should be distinguished [69] from *Key Information Localization and Extraction* (KILE) as the former lacks the positional information, required for effective human-in-the loop verification of the extracted data. Business documents often come with a list of items, e.g. a table of invoiced goods and services, where each item is represented by a set of key information, such as name, quantity and price. Extraction of such items is the target of the *Line Item Recognition* (LIR) [69], which was not explicitly targeted by existing benchmarks.

In this work, we present the DocILE (**Doc**ument **I**nformation **L**ocalization and **E**xtraction) dataset and benchmark with the following contributions:
(i) the largest dataset for KILE and LIR from semi-structured business documents both in terms of the number of labeled documents and categories; (ii) rich set of document layouts, including layout cluster annotations for all labeled documents; (iii) the synthetic subset being the first large synthetic dataset with KILE and LIR labels; (iv) detailed information about the document selection, processing and annotations, which took around 2, 500 h of annotation time, (v) baseline evaluations of popular architectures for language modelling, visually-rich document understanding and computer vision. (vi) is used both for a research competition, as well as a long-term benchmark of key information extraction and localization, and line item recognition systems; (vii) can serve other areas of research thanks to the rich annotations (table structure, layout clusters, metadata, and the HTML sources for synthetic documents).

The paper is structured as follows: Sect. 2 reviews the related work. The DocILE dataset is introduced and its characteristics and its collection are described in Sect. 3. Section 4 follows with the tasks and evaluation metrics.

Baseline methods are described and experimented in Sect. 5. Finally, conclusions are drawn in Sect. 6.

2 Related Work

To address the related work, we first introduce general approaches to document understanding, before specifically focusing on information extraction tasks and existing datasets.

2.1 Methods for Document Understanding

Approaches to document understanding have used various combinations of input modalities (text, spatial layout, image) to extract information from structurally rich documents. Such approaches have been successfully applied to understanding of forms [8,22,90], receipts [27,29], tables [24,64,89], or invoices [45,46,62].

Convolutional neural networks based approaches such as [33,42] use character or word vector-based representations to make a grid-style prediction similar to semantic segmentation. The pixels are classified into the field types for invoice documents. LayoutLM [84] modifies the BERT [10] language model to incorporate document layout information and visual features. The layout information is passed in the form of 2D spatial coordinate embeddings, and the visual features for each word token are obtained via Faster-RCNN [61]. LayoutLMv2 [83] treats visual tokens separately, instead of adding them to the text tokens, and incorporates additional pre-training tasks. LayoutLMv3 [28] introduces more pre-training tasks such as masked image modeling, or word-patch alignment. BROS [27] also uses a BERT-based text encoder equipped with SPADE [30] based graph classifier to predict the entity relations between the text tokens. Document understanding has also been approached from a question-answering perspective [47,48]. Layout-T5 [76] uses the layout information with the generative T5 [60] language model, and TILT [59] uses convolutional features with the T5 model. In UDOP [77], several document understanding tasks are formulated as sequence-to-sequence modelling in a unified framework. Recently, GraphDoc [86], a model based on graph attention networks pre-trained only on 320k documents, has been introduced for document understanding tasks, showing satisfactory results.

Transformer-based approaches typically rely on large-scale pre-training on unlabeled documents while the fine-tuning of a specific downstream task is sufficient with much smaller annotated datasets. Noticeable amount of papers have focused on the pre-training aspect of document understanding [1,15,17,18,27, 36,38,40,59]. In this paper we use the popular methods [7,28,44] to provide the baselines for KILE and LIR on the proposed DocILE dataset.

2.2 Information Extraction Tasks and Datasets

Extraction of information from documents includes many tasks and problems from basic OCR [13,20,31,50,54,70] up to visual question answering

Table 1. Datasets with KILE and LIR annotations for semi-structured business documents.

name	document type	# docs labeled	classes	source	multi page	lang.	task
DocILE *(ours)*	invoice-like	**106680**	**55**	digital, scan	yes	en	KILE, LIR
CORD [58]	receipts	11000	$30-42^a$	photo	no	id	≈KILE, ≈LIR^b
WildReceipt [74]	receipts	1740	25	photo	no	en	KILE
EPHOIE [79]	chinese forms	1494	10	scan	no	zh	KILE
Ghega [49]	patents, datasheets	246	11/8	scan	yes	en	KILE

^{a}54 classes mentioned in [58], but the repository https://github.com/clovaai/cord only considers 30 out of 42 listed classes, as of January 2023.
^{b}COORD annotations contain classification of word tokens (as in NER) but with the additional information which tokens are grouped together into fields or menu items, effectively upgrading the annotations to KILE/LIR field annotations.

(VQA) [47, 48]. The landscape of IE problems and datasets was recently reviewed by Borchmann et al. [5], building the DUE Benchmark for a wide range of document understanding tasks, and by Skalický et al. [69], who argue that the crucial problems for automating B2B document communication are Key Information Localization and Extraction and Line Item Recognition.

Key Information Extraction (KIE) [15, 29, 72] aims to extract pre-defined key information (categories of "fields" – name, email, the amount due, etc.) from a document. A number of datasets for KIE are publicly available [29, 49, 72–74, 74, 79]. However, as noted by [69], most of them are relatively small and contain only a few annotated field categories.

Key Information Localization and Extraction (KILE) [69] additionally requires precise localization of the extracted information in the input image or PDF, which is crucial for human-in-the-loop interactions, auditing, and other processing of the documents. However, many of the existing KIE datasets miss the localization annotations [5, 29, 72]. Publicly available KILE datasets on business documents [49, 58, 74, 79] and their sizes are listed in Table 1. Due to the lack of large-scale datasets for KILE from business documents, noted by several authors [11, 35, 56, 69, 75], many research publications use private datasets [9, 26, 33, 43, 55, 56, 65, 87].

Line Item Recognition (LIR) [69] is a part of table extraction [3, 9, 25, 46, 56] that aims at finding Line Items (LI), localizing and extracting key information for each item. The task is related to Table Structure Recognition [64, 71, 78], which typically aims at detecting table rows, columns and cells. However, sole table structure recognition is not sufficient for LIR: an enumerated item may span several rows in a table; and columns are often not sufficient to distinguish all semantic information. There are several datasets [14, 52, 66, 71, 88, 89] for Table Detection and/or Structure Recognition, PubTables-1M [71] being the largest with a million tables from scientific articles. The domain of scientific articles is prevailing among the datasets [14, 66, 71, 89], due to easily obtainable annotations from the LaTeXsource codes. However, there is a non-trivial domain

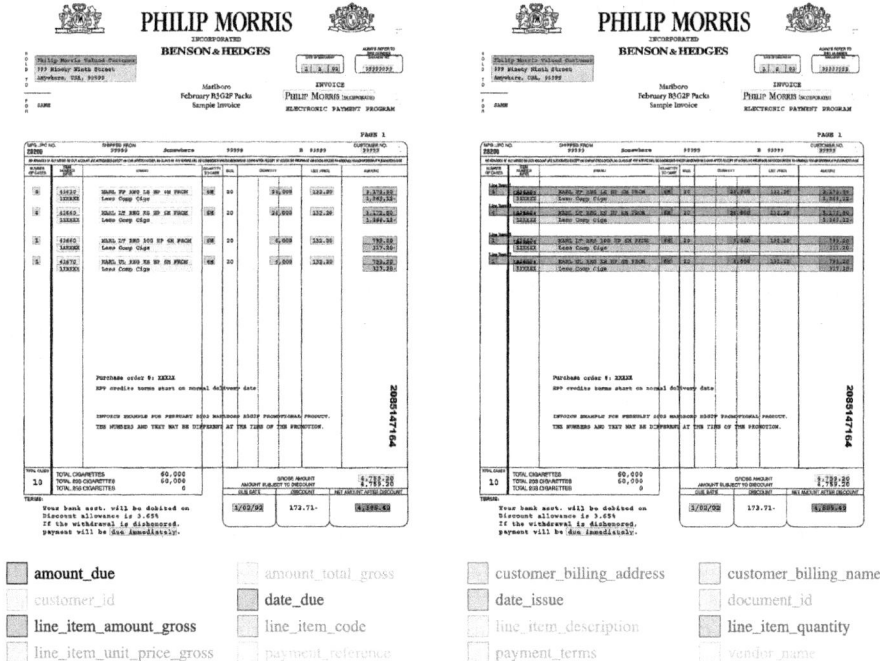

Fig. 1. DocILE: a document with KILE and LIR annotations (left) and the Line Item areas emphasized (right) by alternating blue ▇ and green ▢ for odd and even items, respectively. Bottom: color legend for the KILE and LIR classes. (Color figure online)

shift introduced by the difference in the Tables from scientific papers and business documents. FinTabNet [88] and SynthTabNet [52] are closer to our domain, covering table structure recognition of complex financial tables. These datasets, however, only contain annotations of the table grid/cells. From the available datasets, CORD [58] is the closest to the task of Line Item Recognition with its annotation of sub-menu items. The documents in CORD are all receipts, which generally have simpler structure than other typical business documents, which makes the task too simple as previously mentioned in [5].

Named Entity Recognition (NER) [39] is the task of assigning one of the pre-defined categories to entities (usually words or word-pieces in the document) which makes it strongly related to KILE and LIR, especially when these entities have a known location. Note that the task of NER is less general as it only operates on word/token level, and using it to solve KILE is not straightforward, as the classified tokens have to be correctly aggregated into fields and fields do not necessarily have to contain whole word-tokens.

3   The DocILE Dataset

In this section, we describe the DocILE dataset content and creation.

3.1   The Annotated, the Unlabeled, and the Synthetic

The DocILE dataset and benchmark is composed of three subsets:

1. an *annotated set* of 6,680 real business documents from publicly available sources which were annotated as described in Sect. 3.3.
2. an *unlabeled set* of 932k real business documents from publicly available sources, which can be used for unsupervised (pre-)training.
3. a *synthetic set* of 100k documents with full task labels generated with a proprietary document generator using layouts inspired by 100 fully annotated real business documents from the *annotated set*.

The *labeled* (i.e., *annotated* and *synthetic*) subsets contain annotations for the tasks of *Key Information Localization and Extraction* and *Line Item Recognition*, described below in Sects. 4.1 and 4.2, respectively. An example document with such annotations is shown in Fig. 1. Table 2 shows the size of the dataset and Fig. 2 shows the distribution of document lengths in the dataset.

3.2   Data Sources

Documents in the DocILE dataset come from two public data sources: UCSF Industry Documents Library [80] and Public Inspection Files (PIF) [82]. The UCSF Industry Documents Library contains documents from industries that influence public health, such as tobacco companies. This source has been used to create the following document datasets: RVL-CDIP [23], IIT-CDIP [37], FUNSD [32], DocVQA [48] and OCR-IDL [4]. PIF contains a variety of information about American broadcast stations. We specifically use the "political files" with documents (invoices, orders, "contracts") from TV and radio stations for political campaign ads, previously used to create the Deepform [73]. Documents from both sources were retrieved in the PDF format.

Documents for DocILE were selected from the two sources as follows. For UCSF IDL, we used the public API [81] to retrieve only publicly available documents of type `invoice`. For documents from PIF, we retrieved all "political

Table 2. DocILE dataset — the three subsets.

	annotated	synthetic	unlabeled
documents	6 680	100 000	932 467
pages	8 715	100 000	3.4M
layout clusters	1 152	100	*Unknown*
pages per doc.	1-3	1	1-884

files" from tv, fm and am broadcasts. We discarded documents with broken PDFs, duplicates[2], and documents not classified as *invoice-like*[3]. Other types of documents, such as budgets or financial reports, were discarded as they typically contain different key information. We refer to the selected documents from the two sources as *PIF* and *UCSF* documents.

3.3   Document Selection and Annotation

To capture a rich distribution of documents and make the dataset easy to work with, expensive manual annotations were only done for documents which are:

1. short (1-3 pages), to annotate many different documents rather than a few long ones;
2. written in English, for consistency and because the language distribution in the selected data sources is insufficient to consider multilingual analysis;
3. dated[4] 1999 or later in *UCSF*, as older documents differ from the more recent ones (typewritten, etc.);
4. representing a rich distribution of layout clusters, as shown in Fig. 3.

We clustered the document layouts[5] based on the location of fields detected by a proprietary model for KILE. The clustering was manually corrected for the annotated set. More details about the clustering can be found in the Supplementary Material.

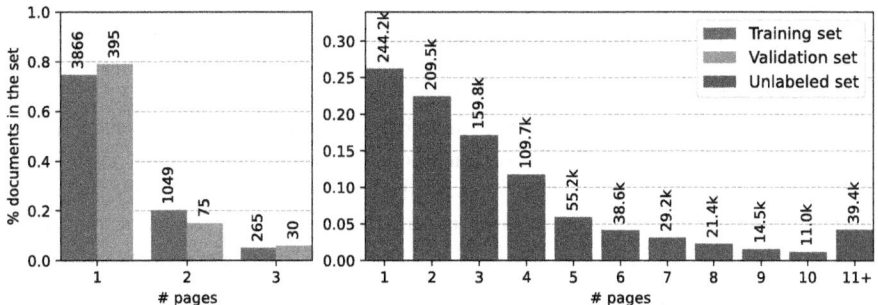

Fig. 2. Distribution of the number of document pages in the training, validation and unlabeled sets. The numbers of documents are displayed above the bars.

---

[2] Using hash of page images to capture duplicates differing only in PDF metadata.

[3] Invoice-like documents are tax invoice, order, purchase order, receipt, sales order, proforma invoice, credit note, utility bill and debit note. We used a proprietary document-type classifier provided by Rossum.ai.

[4] The document date was retrieved from the UCSF IDL metadata. Note that the majority of the documents in this source are from the 20th century.

[5] We loosely define layout as the positioning of fields of each type in a document. We allow, e.g., different length of values, missing values, and resulting translations of whole sections.

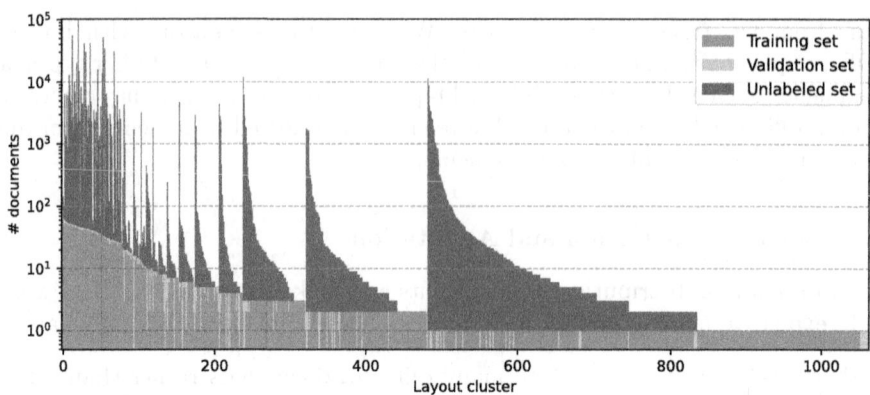

Fig. 3. The number of documents of each layout cluster in the training, validation and unlabeled sets, on a logarithmic scale. While some clusters have up to 100k documents, the largest cluster in *train. + val.* contains only 90 documents.

In the annotation process, documents were skipped if they were not *invoice-like*, if they contained handwritten or redacted key information or if they listed more than one set of line items (e.g. several tables listing unrelated types of items). Additionally, PDF files composed of several documents (e.g., a different invoice on each page) were split and annotated separately.

For KILE and LIR, *fields* are annotated as a triplet of *location* (bounding box and page), *field type* (class) and *text*. LIR fields additionally contain the *line item ID*, assigning the line item they belong to. If the same content is listed in several tables with different granularity (but summing to the same total amount), the less detailed set of line items is annotated.

Notice that the fields can overlap, sometimes completely. A field can be multi-line or contain only parts of words. There can be multiple fields with the same field type on the same page, either having the same value in multiple locations or even having different values as well as multiple fields with the same field type in the same line item. The full list of field types and their description are in the Supplementary Material.

Additional annotations, not necessary for the benchmark evaluation, are available and can be used in the training or for other research purposes. Table structure annotations include: 1) line item headers, representing the headers of columns corresponding to one field type in the table, and 2) the table grid, containing information about rows, their position and classification into header, data, gap, etc., and columns, their position and field type when the values in the column correspond to this field type. Additionally, metadata contain: document type, currency, layout cluster ID, source and original filename (linking the document to the source), page count and page image sizes.

Annotating the 6,680 documents took approx. 2,500 h of annotators' time including the verification. Of the annotated documents, 53.7% originate from

PIF and the remaining 46.3% from *UCSF* IDL. The annotated documents underwent the image pre-processing described in the Supplementary Material.

All remaining documents from *PIF* and *UCSF* form the *unlabeled* set.

3.4 Dataset Splits

The *annotated* documents in the DocILE dataset are split into *training* $(5, 180)$, *validation* (500), and *test* $(1, 000)$ sets. The *synthetic* set with 100k documents and *unlabeled* set with 932k documents are provided as an optional extension to the training set, as unsupervised pre-training [85] and synthetic training data [6,12,19,52,53] have been demonstrated to improve results of machine learning models in different domains.

The training, validation and test splitting was done so that the validation and test sets contain 25% of zero-shot samples (from layouts unseen during training[6]), 25% of few-shot samples (from layouts with ≤ 3 documents seen during training) and 50% of many-shot samples (from layouts with more examples seen during training). This allows to measure both the generalization of the evaluated methods and the advantage of observing documents of known layouts.

The test set annotations are not public and the test set predictions will be evaluated through the RRC website[7], where the benchmark and competition is hosted. The validation set can be used when access to annotations and metadata is needed for experiments in different tasks.

As inputs to the document synthesis described in Sect. 3.5, 100 one-page documents were chosen from the training set, each from a different layout cluster. In the test, resp. validation sets, roughly half of the few-shot samples are from layouts for which synthetic documents were generated. There are no synthetic documents generated for zero-shot samples. For many-shot samples, $35 - 40\%$ of documents are from layouts with synthetic documents.

3.5 Synthetic Documents Generation

To generate synthetic documents with realistic appearance and content, we used the following procedure: First, a set of template documents from different layout clusters was selected, as described in Sect. 3.4. All elements in the selected documents, including all present keys and values, notes, sections, borders, etc., were annotated with layout (bounding box), semantic (category) and text (where applicable) annotations. Such full annotations were the input to a rule-based document synthesizer, which uses a rich set of content generators[8] to fill seman-

---

[6] For the test set, documents in both training and validation sets are considered as seen during training. Note that some test set layouts may be present in the validation set, not the training set.

[7] https://rrc.cvc.uab.es/.

[8] Such as generators of names, emails, addresses, bank account numbers, etc. Some utilize the Mimesis library [16]. Some content, such as keys, is copied from the annotated document.

tically relevant information in the annotated areas. Additionally, a style generator controls and enriches the look of the resulting documents (via font family and size, border styles, shifts of the document contents, etc.). The documents are first generated as HTML files and then rendered to PDF. The HTML source code of all generated documents is shared with the dataset and can be used for future work, e.g., for generative methods for conversion of document images into a markup language.

3.6   Format

The dataset is shared in the form of pre-processed[9] document PDFs with task annotations in JSON. Additionally, each document comes with DocTR [51] OCR predictions with word-level text and location[10].

A python library *docile*[11] is provided to ease the work with the dataset.

4   Benchmark Tasks and Evaluation Metrics

Sections 4.1 and 4.2 describe the two benchmark tasks as introduced in the teaser [67] along with the challenge evaluation metrics used for the leaderboard ranking. Additional evaluation metrics are described in the Supplementary Material.

4.1   Track 1: Key Information Localization and Extraction

The goal of the first track is to localize key information of pre-defined categories (field types) in the document. It is derived from the task of *Key Information Localization and Extraction*, as defined in [69] and motivated in Sect. 2.2.

We focus the challenge on detecting semantically important values corresponding to tens of different field types rather than fine-tuning the underlying text recognition. Towards this focus, we provide word-level text detections for each document, we choose an evaluation metric (below) that does not pay attention to the text recognition part, and we simplify the task in the challenge by only requiring correct localization of the values in the documents in the primary metric. Text extractions are checked, besides the locations and field types, in a separate evaluation — the leaderboard ranking does not depend on it. Any post-processing of values — deduplication, converting dates to a standardized format etc., despite being needed in practice, is not performed. With the simplifications, the main task can also be viewed as a detection problem. Note that when several instances of the same field type are present, all of them should be detected.

---

[9] Pre-processing consists of correcting page orientation, de-skewing scanned documents and normalizing them to 150 DPI.

[10] Axis-aligned bounding boxes, optionally with additional snapping to reduce white space around word predictions, described in the Supplementary Material.

[11] https://github.com/rossumai/docile.

Fig. 4. Each word is split uniformly into pseudo-character boxes based on the number of characters. Pseudo-Character Centers are the centers of these boxes.

(a) Correct extraction examples.   (b) Incorrect extraction examples.

Fig. 5. Visualization of correct and incorrect bounding box predictions to capture the phone number. Bounding box must include exactly the Pseudo-Character Centers that lie within the ground truth annotation. Note: in 5a, only one of the predictions would be considered correct if all three boxes were predicted.

The Challenge Evaluation Metric. Since the task is framed as a detection problem, the standard *Average Precision* metric is used as the main evaluation metric. Unlike the common practice in object detection, where true positives are determined by thresholding the Intersection-over-Union, we use a different criterion tailored to evaluate the usefulness of detections for text read-out. Inspired by the CLEval metric [2] used in text detection, we measure whether the predicted area contains nothing but the related character centers. Since character-level annotations are hard to obtain, we use CLEval definition of Pseudo-Character Center (PCC), visualized in Fig. 4. Examples of correct and incorrect detections are depicted in Fig. 5.

4.2 Track 2: Line Item Recognition

The goal of the second track is to localize key information of pre-defined categories (field types) and group it into line items [3,9,25,46,56]. A *Line Item* (LI) is a tuple of fields (e.g., *description*, *quantity*, and *price*) describing a single object instance to be extracted, e.g., a row in a table, as visualized in Fig. 1 and explained in Sect. 2.2.

The Challenge Evaluation Metric. The main evaluation metric is the micro F1 score over all line item fields. A predicted line item field is correct if it fulfills the requirements from Track 1 (on field type and location) and if it is assigned to the correct line item. Since the matching of ground truth (GT) and predicted line items may not be straightforward due to errors in the prediction, our evaluation metric chooses the best matching in two steps:

1. for each pair of predicted and GT line items, the predicted fields are evaluated as in Track 1,

2. the maximum matching is found between predicted and GT line items, maximizing the overall recall.

4.3 Benchmark Dataset Rules

The use of external document datasets (and models pre-trained on such datasets) is prohibited in the benchmark in order to focus on clear comparative evaluation of methods that use the provided collection of labeled and unlabeled documents. Usage of datasets and pre-trained models from other domains, such as images from ImageNet [63] or texts from BooksCorpus [91], is allowed.

5 Baseline Methods

We provide as baselines several popular state-of-the-art transformer architectures, covering text-only (RoBERTa), image-only (DETR) and multi-modal (LayoutLMv3) document representations. The code and model checkpoints for all baseline methods are distributed with the dataset.

5.1 Multi-label NER Formulation for KILE and LIR Tasks

The baselines described in Sects. 5.2 and 5.3 use a joint multi-label NER formulation for both KILE and LIR tasks. The LIR task requires not only to correctly classify tokens into one of the LIR classes, but also the assignment of tokens to individual Line Items. For this purpose, we add classes <B-LI>, <I-LI>, <O-LI> and <E-LI>, representing the beginning of the line item, inside and outside tokens and end token of the line item. We found it is crucial to re-order the OCR tokens in top-down, left-to-right order for each predicted text line of the document. We provide a detailed description of the OCR tokens re-ordering in the Supplementary Material. For the LIR and KILE classification, we use the standard BIO tagging scheme. We use the binary cross entropy loss to train the model.

The final KILE and LIR predictions are formed by the merging strategy as follows. We group the predicted tokens based on the membership to the predicted line item (note that we can assign the tokens which do not belong to any line item to a special group $\emptyset$), then we use the predicted OCR text lines to perform the horizontal merging of tokens assigned to the same class. Next, we construct a graph from the horizontally merged text blocks, based on the thresholded x and y distances of the text block pairs (as a threshold we use the height of the text block with a 25% margin). The final predictions are given by merging the graph components. By merging, we mean taking the union of the individual bounding boxes of tokens/text blocks and for the text value, if the horizontal merging is applied, we join the text values with space, and with the new line character when vertical merging is applied.

Note that this merging strategy is rather simplistic and its proper redefinition might be of interest for the participants who cannot afford training of big models as we publish also the baselines model checkpoints.

5.2 RoBERTa

RoBERTa [44] is a modification of the BERT [10] model which uses improved training scheme and minor tweaks of the architecture (different tokenizer). It can be used for NER task simply by adding a classification head after the RoBERTa embedding layer. Our first baseline is purely text based and uses RoBERTa$_{BASE}$ as the backbone of the joint multi-label NER model described in Sect. 5.1.

5.3 LayoutLMv3

While the RoBERTa-based baseline only operates on the text input, LayoutLMv3 [28] is multi-modal transformer architecture that incorporates image, text, and layout information jointly. The images are encoded by splitting into non-overlapping patches and feeding the patches to a linear projection layer, after which they are combined with positional embeddings. The text tokens are combined with one-dimensional and two-dimensional positional embeddings, where the former accounts for the position in the sequence of tokens, and the latter specifies the spatial location of the token in the document. The two-dimensional positional embedding incorporates the layout information. All these tokens are then fed to the transformer model. We use the LayoutLMv3$_{BASE}$ architecture as our second baseline, also using the multi-label NER formulation from Sect. 5.1. Since LayoutLMv3$_{BASE}$ was pre-trained on an external document dataset, prohibited in the benchmark, we pre-train a checkpoint from scratch in Sect. 5.4.

5.4 Pre-training for RoBERTa and LayoutLMv3

We use the standard masked language modeling [10] as the unsupervised pre-training objective to pre-train RoBERTa$_{OURS}$ and LayoutLMv3$_{OURS}$[12] models. The pre-training is performed from scratch using the 932k unlabeled samples introduced in Sect. 3. Note that the pre-training uses the OCR predictions provided with the dataset (with reading order re-ordering).

Additionally, RoBERTa$_{BASE/OURS+SYNTH}$ and LayoutLMv3$_{OURS+SYNTH}$ baselines use supervised pre-training on the DocILE synthetic data.

---

[12] Note that LayoutLMv3$_{BASE}$ [28] used two additional pre-training objectives, namely masked image modelling and word-patch alignment. Since pre-training code is not publicly available and some of the implementation details are missing, LayoutLMv3$_{OURS}$ used only masked language modelling.

Table 3. Baseline results for KILE & LIR. LayoutLMv3$_{BASE}$, achieving the best results, was pre-trained on another document dataset – IIT-CDIP [37], which is prohibited in the official benchmark. The best results among permitted models are underlined. The primary metric for each task is shown in **bold**.

Model	KILE				LIR			
	F1	**AP**	Prec.	Recall	**F1**	AP	Prec.	Recall
RoBERTa$_{BASE}$	0.664	0.534	0.658	0.671	0.686	0.576	0.695	0.678
RoBERTa$_{OURS}$	0.645	0.515	0.634	0.656	0.686	0.570	0.693	0.678
LayoutLMv3$_{BASE}$ (prohibited)	0.698	0.553	0.701	0.694	0.721	0.586	0.746	0.699
LayoutLMv3$_{OURS}$	0.639	0.507	0.636	0.641	0.661	0.531	0.682	0.641
RoBERTa$_{BASE+SYNTH}$	0.664	0.539	0.659	0.669	0.698	0.583	0.710	0.687
RoBERTa$_{OURS+SYNTH}$	0.652	0.527	0.648	0.656	0.675	0.559	0.696	0.655
LayoutLMv3$_{OURS+SYNTH}$	0.655	0.512	0.662	0.648	0.691	0.582	0.709	0.673
NER upper bound	0.946	0.897	1.000	0.897	0.961	0.926	1.000	0.926
DETR^{table} + RoBERTa$_{BASE}$	-	-	-	-	0.682	0.560	0.706	0.660
DETR^{table} + DETR^{LI} + RoBERTa$_{BASE}$	-	-	-	-	0.594	0.407	0.632	0.560

5.5 Line Item Detection via DETR

As an alternative approach to detecting Line Items, we use the DETR [7] object detector, as proposed for table structure recognition on the PubTables-1M dataset [71]. Since pretraining on other document datasets is prohibited in the DocILE benchmark, we initialize DETR from a checkpoint[13] pretrained on COCO [41], not from [71].

Two types of detectors are fine-tuned independently. DETR^{table} for table detection and DETR^{LI} for line item detection given a table crop — which in our preliminary experiments lead to better results than one-stage detection of line items from the full page.

5.6 Upper Bound for NER-Based Solutions

All our baselines use NER models with the provided OCR on input. This comes with limitations as a field does not have to correspond to a set of word tokens — a field can contain just a part of some word and some words covering the field might be missing in the text detections. A theoretical upper bound for NER-based methods that classify the provided OCR words is included in Table 3. The upper bound constructs a prediction for each ground truth field by finding all words whose PCCs are covered by the field and replacing its bounding box with a union of bounding boxes of these words. Predicted fields that do not match their originating ground truth fields are discarded.

---

[13] https://huggingface.co/facebook/detr-resnet-50.

5.7 Results

The baselines described above were evaluated on the DocILE test set, the results are in Table 3. Interestingly, from our pre-trained models (marked $_{OURS}$), the RoBERTa baseline outperforms the LayoutLMv3 baseline utilizing the same RoBERTa model in its backbone. We attribute this mainly to differences in the LayoutLMv3 pre-training: 1) our pre-training used only the masked language modelling loss, as explained in Sect. 5.4, 2) we did not perform a full hyperparameter search, and 3) our pre-training performs image augmentations not used in the original LayoutLMv3 pre-training, these are described in the Supplementary Material.

Models pre-trained on the synthetic training data are marked with $_{SYNTH}$. Synthetic pre-training improved the results for both KILE and LIR in all cases except for LIR with RoBERTa$_{OURS+SYNTH}$, validating the usefulness of the synthetic subset.

The best results among the models permitted in the benchmark – i.e. not utilizing additional document datasets – were achieved by RoBERTa$_{BASE+SYNTH}$.

6 Conclusions

The DocILE benchmark includes the largest research dataset of business documents labeled with fine-grained targets for the tasks of Key Information Localization and Extraction and Line Item Recognition. The motivation is to provide a practical benchmark for evaluation of information extraction methods in a domain where future advancements can considerably save time that people and businesses spend on document processing. The baselines described and evaluated in Sect. 5, based on state-of-the-art transformer architectures, demonstrate that the benchmark presents very challenging tasks. The code and model checkpoints for the baselines are provided to the research community allowing quick start for the future work.

The benchmark is used for a research competition hosted at ICDAR 2023 and CLEF 2023 and will stay open for post-competition submission for long-term evaluation. We are looking forward to contributions from different machine learning communities to compare solutions inspired by document layout modelling, language modelling and question answering, computer vision, information retrieval, and other approaches.

Areas for future contributions to the benchmark include different training objective statements — such as different variants of NER, object detection, or sequence-to-sequence modelling [77], or graph reasoning [74]; different model architectures, unsupervised pre-training [28,77], utilization of table structure — e.g., explicitly modelling regularity in table columns to improve in LIR; addressing dataset shifts [57,68]; or zero-shot learning [34].

Acknowledgements. We acknowledge the funding and support from Rossum and the intensive work of its annotation team, particularly Petra Hrdličková and Kateřina Večerková. YP and JM were supported by Research Center for Informatics (project

CZ.02.1.01/0.0/0.0/16_019/0000765 funded by OP VVV), by the Grant Agency of the Czech Technical University in Prague, grant No. SGS20/171/OHK3 /3T/13, by Project StratDL in the realm of COMET K1 center Software Competence Center Hagenberg, and Amazon Research Award. DK was supported by grant PID2020-116298GB-I00 funded by MCIN/AE/NextGenerationEU and ELSA (GA 101070617) funded by EU.

References

1. Appalaraju, S., Jasani, B., Kota, B.U., Xie, Y., Manmatha, R.: Docformer: end-to-end transformer for document understanding. In: ICCV (2021)
2. Baek, Y., et al.: Cleval: character-level evaluation for text detection and recognition tasks. In: CVPR workshops (2020)
3. Bensch, O., Popa, M., Spille, C.: Key information extraction from documents: evaluation and generator. In: Abbès, S.B., et al. (eds.) Proceedings of DeepOntoNLP and X-SENTIMENT (2021)
4. Biten, A.F., Tito, R., Gomez, L., Valveny, E., Karatzas, D.: OCR-IDL: OCR annotations for industry document library dataset. In: Karlinsky, L., Michaeli, T., Nishino, K. (eds.) Computer Vision – ECCV 2022 Workshops. ECCV 2022. LNCS, vol. 13804, pp. 241–252. Springer, Cham (2022). https://doi.org/10.1007/978-3-031-25069-9_16
5. Borchmann, Ł., et al.: DUE: end-to-end document understanding benchmark. In: NeurIPS (2021)
6. Bušta, M., Patel, Y., Matas, J.: E2E-MLT - an unconstrained end-to-end method for multi-language scene text. In: ACCV workshops (2019)
7. Carion, N., Massa, F., Synnaeve, G., Usunier, N., Kirillov, A., Zagoruyko, S.: End-to-end object detection with transformers. In: Vedaldi, A., Bischof, H., Brox, T., Frahm, J.-M. (eds.) ECCV 2020. LNCS, vol. 12346, pp. 213–229. Springer, Cham (2020). https://doi.org/10.1007/978-3-030-58452-8_13
8. Davis, B., Morse, B., Cohen, S., Price, B., Tensmeyer, C.: Deep visual template-free form parsing. In: ICDAR (2019)
9. Denk, T.I., Reisswig, C.: BERTgrid: contextualized embedding for 2d document representation and understanding. arXiv (2019)
10. Devlin, J., Chang, M.W., Lee, K., Toutanova, K.: BERT: pre-training of deep bidirectional transformers for language understanding. arXiv (2018)
11. Dhakal, P., Munikar, M., Dahal, B.: One-shot template matching for automatic document data capture. In: Artificial Intelligence for Transforming Business and Society (AITB) (2019)
12. Dosovitskiy, A., et al.: Flownet: learning optical flow with convolutional networks. In: ICCV (2015)
13. Du, Y., et al.: PP-OCR: a practical ultra lightweight OCR system. arXiv (2020)
14. Fang, J., Tao, X., Tang, Z., Qiu, R., Liu, Y.: Dataset, ground-truth and performance metrics for table detection evaluation. In: Blumenstein, M., Pal, U., Uchida, S. (eds.) DAS (2012)
15. Garncarek, Ł., et al.: Lambert: layout-aware language modeling for information extraction. In: ICDAR (2021)
16. Geimfari, L.: Mimesis: the fake data generator (2022). http://github.com/lk-geimfari/mimesis
17. Gu, J., et al.: Unidoc: Unified pretraining framework for document understanding. In: NeurIPS (2021)

18. Gu, Z., et al.: XYLayoutLM: towards layout-aware multimodal networks for visually-rich document understanding. In: CVPR (2022)
19. Gupta, A., Vedaldi, A., Zisserman, A.: Synthetic data for text localisation in natural images. In: CVPR (2016)
20. Hamad, K.A., Mehmet, K.: A detailed analysis of optical character recognition technology. Int. J. Appl. Math. Electron. Comput. **2016**, 244–249 (2016)
21. Hamdi, A., Carel, E., Joseph, A., Coustaty, M., Doucet, A.: Information extraction from invoices. In: ICDAR (2021)
22. Hammami, M., Héroux, P., Adam, S., d'Andecy, V.P.: One-shot field spotting on colored forms using subgraph isomorphism. In: ICDAR (2015)
23. Harley, A.W., Ufkes, A., Derpanis, K.G.: Evaluation of deep convolutional nets for document image classification and retrieval. In: ICDAR (2015)
24. Herzig, J., Nowak, P.K., Müller, T., Piccinno, F., Eisenschlos, J.M.: Tapas: weakly supervised table parsing via pre-training. arXiv (2020)
25. Holeček, M., Hoskovec, A., Baudiš, P., Klinger, P.: Table understanding in structured documents. In: ICDAR Workshops (2019)
26. Holt, X., Chisholm, A.: Extracting structured data from invoices. In: Proceedings of the Australasian Language Technology Association Workshop 2018, pp. 53–59 (2018)
27. Hong, T., Kim, D., Ji, M., Hwang, W., Nam, D., Park, S.: Bros: a pre-trained language model focusing on text and layout for better key information extraction from documents. In: AAAI (2022)
28. Huang, Y., Lv, T., Cui, L., Lu, Y., Wei, F.: Layoutlmv3: pre-training for document AI with unified text and image masking. In: ACM-MM (2022)
29. Huang, Z., et al.: ICDAR2019 competition on scanned receipt OCR and information extraction. In: ICDAR (2019)
30. Hwang, W., Yim, J., Park, S., Yang, S., Seo, M.: Spatial dependency parsing for semi-structured document information extraction. arXiv (2020)
31. Islam, N., Islam, Z., Noor, N.: A survey on optical character recognition system. arXiv (2017)
32. Jaume, G., Ekenel, H.K., Thiran, J.P.: FUNSD: a dataset for form understanding in noisy scanned documents. In: ICDAR (2019)
33. Katti, A.R., et al.: Chargrid: towards understanding 2d documents. In: EMNLP (2018)
34. Kil, J., Chao, W.L.: Revisiting document representations for large-scale zero-shot learning. arXiv (2021)
35. Krieger, F., Drews, P., Funk, B., Wobbe, T.: Information extraction from invoices: a graph neural network approach for datasets with high layout variety. In: Innovation Through Information Systems: Volume II: A Collection of Latest Research on Technology Issues (2021)
36. Lee, C.Y., et al.: FormNet: structural encoding beyond sequential modeling in form document information extraction. In: ACL (2022)
37. Lewis, D., Agam, G., Argamon, S., Frieder, O., Grossman, D., Heard, J.: Building a test collection for complex document information processing. In: SIGIR (2006)
38. Li, C., et al.: StructuralLM: structural pre-training for form understanding. In: ACL (2021)
39. Li, J., Sun, A., Han, J., Li, C.: A survey on deep learning for named entity recognition. IEEE Trans. Knowl. Data Eng. **34**, 50–70 (2020)
40. Li, Y., et al.: Structext: structured text understanding with multi-modal transformers. In: ACM-MM (2021)

41. Lin, T.-Y., et al.: Microsoft COCO: common objects in context. In: Fleet, D., Pajdla, T., Schiele, B., Tuytelaars, T. (eds.) ECCV 2014. LNCS, vol. 8693, pp. 740–755. Springer, Cham (2014). https://doi.org/10.1007/978-3-319-10602-1_48

42. Lin, W., et al.: VibertGrid: a jointly trained multi-modal 2d document representation for key information extraction from documents. In: ICDAR (2021)

43. Liu, W., Zhang, Y., Wan, B.: Unstructured document recognition on business invoice. Technical report (2016)

44. Liu, Y., et al.: RoBERTa: A Robustly Optimized BERT Pretraining Approach. arXiv (2019)

45. Lohani, D., Belaïd, A., Belaïd, Y.: An invoice reading system using a graph convolutional network. In: ACCV workshops (2018)

46. Majumder, B.P., Potti, N., Tata, S., Wendt, J.B., Zhao, Q., Najork, M.: Representation learning for information extraction from form-like documents. In: ACL (2020)

47. Mathew, M., Bagal, V., Tito, R., Karatzas, D., Valveny, E., Jawahar, C.: InfographicVQA. In: WACV (2022)

48. Mathew, M., Karatzas, D., Jawahar, C.: DocVQA: a dataset for VQA on document images. In: WACV (2021)

49. Medvet, E., Bartoli, A., Davanzo, G.: A probabilistic approach to printed document understanding. In: ICDAR (2011)

50. Memon, J., Sami, M., Khan, R.A., Uddin, M.: Handwritten optical character recognition (OCR): a comprehensive systematic literature review (SLR). IEEE Access. **8**, 142642–142668 (2020)

51. Mindee: docTR: Document text recognition (2021). http://github.com/mindee/doctr

52. Nassar, A., Livathinos, N., Lysak, M., Staar, P.W.J.: TableFormer: table structure understanding with transformers. arXiv (2022)

53. Nayef, N., et al.: ICDAR 2019 robust reading challenge on multi-lingual scene text detection and recognition-RRC-MLT-2019. In: ICDAR (2019)

54. Olejniczak, K., Šulc, M.: Text detection forgot about document OCR. In: CVWW (2023)

55. Palm, R.B., Laws, F., Winther, O.: Attend, copy, parse end-to-end information extraction from documents. In: ICDAR (2019)

56. Palm, R.B., Winther, O., Laws, F.: CloudScan - a configuration-free invoice analysis system using recurrent neural networks. In: ICDAR (2017)

57. Pampari, A., Ermon, S.: Unsupervised calibration under covariate shift. arXiv (2020)

58. Park, S., et al.: Cord: a consolidated receipt dataset for post-OCR parsing. In: NeurIPS Workshops (2019)

59. Powalski, R., Borchmann, L., Jurkiewicz, D., Dwojak, T., Pietruszka, M., Pałka, G.: Going full-tilt boogie on document understanding with text-image-layout transformer. In: ICDAR (2021)

60. Raffel, C., et al.: Exploring the limits of transfer learning with a unified text-to-text transformer. JMLR. **21**, 5485–5551 (2020)

61. Ren, S., He, K., Girshick, R., Sun, J.: Faster R-CNN: towards real-time object detection with region proposal networks. In: NeurIPS (2015)

62. Riba, P., Dutta, A., Goldmann, L., Fornés, A., Ramos, O., Lladós, J.: Table detection in invoice documents by graph neural networks. In: ICDAR (2019)

63. Russakovsky, O., et al.: Imagenet large scale visual recognition challenge. IJCV. **115**, 211–252 (2015)

64. Schreiber, S., Agne, S., Wolf, I., Dengel, A., Ahmed, S.: DeepDeSRT: deep learning for detection and structure recognition of tables in document images. In: ICDAR (2017)
65. Schuster, D., et al.: Intellix-end-user trained information extraction for document archiving. In: ICDAR (2013)
66. Siegel, N., Lourie, N., Power, R., Ammar, W.: Extracting scientific figures with distantly supervised neural networks. In: Chen, J., Gonçalves, M.A., Allen, J.M., Fox, E.A., Kan, M., Petras, V. (eds.) Proceedings of the 18th ACM/IEEE on Joint Conference on Digital Libraries, JCDL (2018)
67. Šimsa, Š., Šulc, M., Skalický, M., Patel, Y., Hamdi, A.: Docile 2023 teaser: document information localization and extraction. In: ECIR (2023)
68. Šipka, T., Šulc, M., Matas, J.: The hitchhiker's guide to prior-shift adaptation. In: WACV (2022)
69. Skalický, M., Šimsa, Š., Uřičář, M., Šulc, M.: Business document information extraction: Towards practical benchmarks. In: CLEF (2022)
70. Smith, R.: An overview of the tesseract OCR engine. In: ICDAR (2007)
71. Smock, B., Pesala, R., Abraham, R.: PubTables-1M: towards comprehensive table extraction from unstructured documents. In: CVPR (2022)
72. Stanisławek, T., et al.: Kleister: key information extraction datasets involving long documents with complex layouts. In: ICDAR (2021)
73. Stray, J., Svetlichnaya, S.: DeepForm: extract information from documents (2020). http://wandb.ai/deepform/political-ad-extraction, benchmark
74. Sun, H., Kuang, Z., Yue, X., Lin, C., Zhang, W.: Spatial dual-modality graph reasoning for key information extraction. arXiv (2021)
75. Sunder, V., Srinivasan, A., Vig, L., Shroff, G., Rahul, R.: One-shot information extraction from document images using neuro-deductive program synthesis. arXiv (2019)
76. Tanaka, R., Nishida, K., Yoshida, S.: VisualMRC: machine reading comprehension on document images. In: AAAI (2021)
77. Tang, Z., et al.: Unifying vision, text, and layout for universal document processing. arXiv (2022)
78. Tensmeyer, C., Morariu, V.I., Price, B., Cohen, S., Martinez, T.: Deep splitting and merging for table structure decomposition. In: ICDAR (2019)
79. Wang, J., et al.: Towards robust visual information extraction in real world: new dataset and novel solution. In: AAAI (2021)
80. Web: Industry Documents Library. www.industrydocuments.ucsf.edu/. Accessed 20 Oct 2022
81. Web: Industry Documents Library API. www.industrydocuments.ucsf.edu/research-tools/api/. Accessed 20 Oct 2022
82. Web: Public Inspection Files. http://publicfiles.fcc.gov/. Accessed 20 Oct 2022
83. Xu, Y., et al.: Layoutlmv2: Multi-modal pre-training for visually-rich document understanding. In: ACL (2021)
84. Xu, Y., Li, M., Cui, L., Huang, S., Wei, F., Zhou, M.: LayoutLM: pre-training of text and layout for document image understanding. In: KDD (2020)
85. Xu, Y., et al.: LayoutXLM: multimodal pre-training for multilingual visually-rich document understanding. arXiv (2021)
86. Zhang, Z., Ma, J., Du, J., Wang, L., Zhang, J.: Multimodal pre-training based on graph attention network for document understanding. IEEE Trans. Multimed. (2022)
87. Zhao, X., Wu, Z., Wang, X.: CUTIE: learning to understand documents with convolutional universal text information extractor. arXiv (2019)

88. Zheng, X., Burdick, D., Popa, L., Zhong, X., Wang, N.X.R.: Global table extractor (GTE): a framework for joint table identification and cell structure recognition using visual context. In: WACV (2021)
89. Zhong, X., Tang, J., Jimeno-Yepes, A.: PublayNet: largest dataset ever for document layout analysis. In: ICDAR (2019)
90. Zhou, J., Yu, H., Xie, C., Cai, H., Jiang, L.: IRMP: from printed forms to relational data model. In: HPCC (2016)
91. Zhu, Y., et al.: Aligning books and movies: Towards story-like visual explanations by watching movies and reading books. In: ICCV (2015)

Robustness Evaluation of Transformer-Based Form Field Extractors via Form Attacks

Le Xue[✉], Mingfei Gao, Zeyuan Chen, Caiming Xiong, and Ran Xu

Salesforce AI, Palo Alto, USA
lxue@salesforce.com

Abstract. We propose a novel framework to evaluate the robustness of transformer-based form field extraction methods via form attacks. We introduce 14 novel form transformations to evaluate the vulnerability of the state-of-the-art field extractors against form attacks from both OCR level and form level, including OCR location/order rearrangement, form background manipulation and form field-value augmentation. We conduct robustness evaluation using real invoices and receipts, and perform comprehensive research analysis. Experimental results suggest that the evaluated models are very susceptible to form perturbations such as the variation of field-values ($\sim 15\%$ drop in F1 score), the disarrangement of input text order($\sim 15\%$ drop in F1 score) and the disruption of the neighboring words of field-values($\sim 10\%$ drop in F1 score). Guided by the analysis, we make recommendations to improve the design of field extractors and the process of data collection. Code will be available at here.

Keywords: Document Understanding · Robustness Evaluation

1 Introduction

Forms such as invoices and receipts are essential in business workflows. Extracting target values for fields of interest from forms (see an example in Fig. 1) is among the most important tasks in document understanding. There are large amounts of forms processed every day, but most current systems still rely on human labor to manually capture field-values from massively irrelevant information. Developing a method that automatically extracts field-values based on understanding the forms is crucial to reduce human labor, thus improve business efficiency.

Existing works [1,2,6,10,13,15,19,24,26] focus on improving the modeling of field extractors and have made great progress. However, their evaluation paradigms are limited. First, most of the methods are evaluated using internal datasets. Internal datasets usually have very limited variations and are often biased towards certain data distributions due to the constraints of the data collection process. For example, the forms might be collected from just a few

G. A. Fink et al. (Eds.): ICDAR 2023, LNCS 14188, pp. 167–184, 2023.
https://doi.org/10.1007/978-3-031-41679-8_10

vendors in a relatively short time which leads to similar semantics and layouts across the forms. Second, public datasets lack for diversity in terms of both textual expression and form layouts. Take the most frequently used dataset, SROIE [7], as an example. The fields, *company* and *address*, are always on the very top in all receipts. Although the existing models achieve decent performance on these datasets, it is difficult to know whether they can generalize well. This issue can be solved by collecting large-scale diverse forms for evaluation, but it is very challenging since real forms usually contain customers' private information, thus are not publicly accessible.

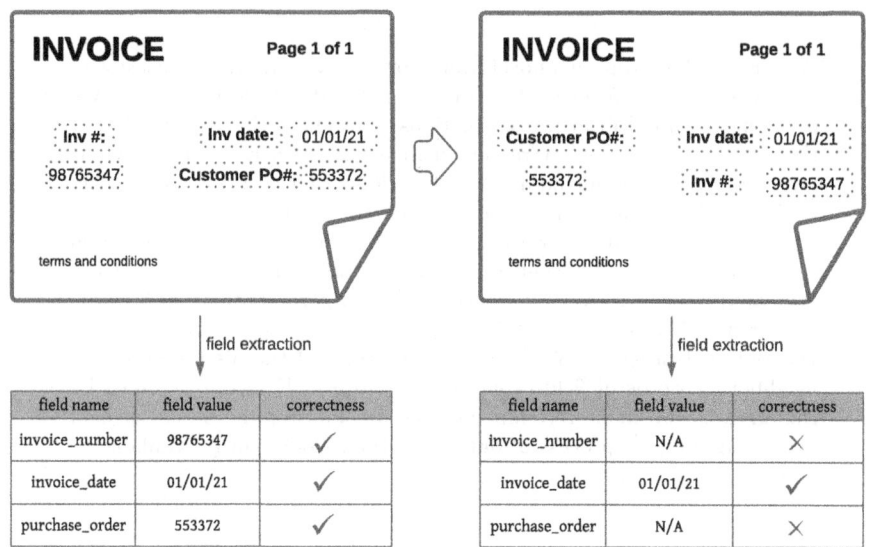

Fig. 1. A form field extraction system may fail due to a slight modification to the form. Keys (concrete text-expressions of fields) are marked in blue boxes. Values are marked in red boxes. (Color figure online)

To tackle this dilemma, we propose a novel framework to evaluate the robustness of form field extractors by attacking the models using form transformations. We consider form perturbations from both OCR level and form level, including OCR text location/order rearrangement, form background manipulation, and form field-value augmentation. Fourteen form transformations are proposed to impose these attacks. Using the proposed framework, we conduct robustness analysis on two commonly used form types, i.e., invoices and receipts. Experimental results demonstrate that the state-of-the-art (SOTA) methods are particularly vulnerable to form perturbations, including the variation of field-values, the disarrangement of input text order, and the disruption of the neighboring words of field-values. Recommendations for model design and data collection/augmentation are made accordingly.

Our contributions are summarized as follows. First, we introduce a framework to measure the robustness of form field extractors by attacking the models

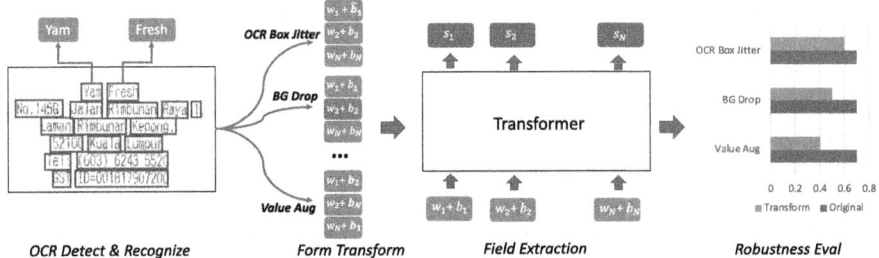

Fig. 2. An illustration of our evaluation pipeline. An OCR engine is first used to extract texts and locations. 14 transformations are applied to the texts and their locations to generate diverse form variants. Then, each transformed set will input to a transformer-based field extractor. Robustness evaluation results are finally generated.

using the proposed form transformations. To the best of our knowledge, this is the first work studying form attacks to field extraction methods. Second, we identify the susceptibilities of the SOTA methods by comprehensive robustness analysis on two form types using the proposed framework and make insightful recommendations.

2   Related Work

Information Extraction from Forms is a widely researched area. [3,8] encode each page of a form as a two-dimensional grid and extract header and line items from it using fully convolutional networks. DocStruct [22] conducts document structure inference by encoding the form structure as a graph-like hierarchy of text fragments. Research works specifically focusing on form field extraction are more related to our work. Earlier methods [2,18] relied on pre-registered templates in the system for information extraction. [15] extract field-values of invoices via an Attend, Copy, Parse architecture. Recent methods formulate the field extraction problem as field-value pairing [6,13] and field tagging [24] tasks, where transformer [21] based structures are used to extract informative form representation via modeling interactions among text tokens. More recently, Donut, an OCR-free method has also been developed [10] and shows promising results. We focus on evaluating transformer-based field extraction methods given their great predictive capability for the task.

Robustness Evaluation of models has received considerable attention. Errudite [23] introduces model and task agnostic principles for informative error analysis of NLP models. [12] propose NLPAug, which contains simple textual augmentations to improve model robustness. Some works aim at robustness of text attacks [9,14,25]. A recent work, Robustness Gym [5], presents a simple and extensible evaluation toolkit that unifies standard evaluation paradigms. There are also recent methods studying robustness of visual models [16,17,20]. To the best of our knowledge, this work is the first one focusing on robustness evaluation of form field extraction systems.

3   Preliminary: Transformer-Based Form Field Extractor

We are focusing on the robustness evaluation of transformer-based form field extractors due to their undisputedly outstanding performance. Before discussing the robustness evaluation, we first illustrate the field extraction pipeline.

In a standard field extraction system, an OCR engine is used to extract a set of words, $\{w_1, w_2, ..., w_N\}$ and their bounding box locations, $\{b_1, b_2, ..., b_N\}$, where N indicates the total number of words. Then, a transformer-based feature backbone is used to model the interactions between text tokens and generate informative token representations, f_i. Since both semantics and layouts are essential for field-value inference, we use LayoutLM [24] as the feature backbone. We also experiment with two more transformers, i.e., BERT [4] and RoBERTa [11] which only take text as input in Sect. 6. Finally, a fully connected (FC) layer is used to project the token features to field space and generate, $s_i = FC(f_i)$, where $s_i \in R^{1 \times (M+1)}$ indicates the predicted field score and M denotes the total number of positive fields. During training, cross entropy loss between s_i and field label is utilized for model optimization. During inference, a post-processing method is applied to the predicted field scores to get the value for each field. We follow the simple criteria to generate field-values: (1) we find the predicted field label for each word by $\hat{f}d = \underset{c}{argmax}(s_{ic})$, where c corresponds to fields (2) by default, for each field, we only keep the word as the value if its prediction score is the highest among all the words and larger than a threshold ($\theta = 0.1$). For fields that often include multi-word values, e.g., address and company, we keep all the words exceeding the threshold and group nearby ones with the same predicted field.

Evaluation Metric. End-to-end F1 score averaging over fields is used to evaluate models. We use exact string matching between the predicted values and the ground-truth to count true positives, false positives, and false negatives. Precision, recall and F1 scores are obtained accordingly for each field.

4   Robustness Evaluation via Form Attacks

We propose OCR level and form level transformations to attack field extractors. As shown in Fig. 2, our transformations are performed after OCR extraction, and the transformed data is input to a transformer-based field extractor. An analysis is conducted via performance comparison between the original set and transformed sets. Each transformation and the principles behind it are introduced as follows.

4.1   OCR Location and Order Rearrangement

We transform the original data to meaningful variants by slightly altering the OCR locations and the text order arrangement. These transformations simulate scenarios where we may obtain different OCR results before inputting to field extractors due to various reasons, e.g., the quality of OCR engines.

Center Shift and Box Stretch. To evaluate model robustness to OCR text location jittering, we propose two word-level location transformations. In *Center Shift*, we keep the box size and randomly shift the center of a box. The shifting is in proportional to the width (horizontally) and height (vertically) of the box, and the ratio is a random number drawn from a normal distribution, $\mathcal{N}(0, \delta_{cent})$. *Box Stretch* randomly changes the four coordinates of a box in a similar way using $\mathcal{N}(0, \delta_{xy})$.

Margin Padding. Scanning forms may introduce white margins, which globally changes text locations. We use *Margin Padding* to manipulate the locations of all the words in a form. We pad white margins in the left, right, up, and down sides of a form where the margin length is a generated random number between 1 and r_{mp} of the page size.

Global Shuffle. We observe that organizing a transformer's inputs in reading order is particularly beneficial to understanding the form structure. However, the reading order is not always guaranteed by OCR engines. Hence, it is interesting to investigate model robustness to poor reading order quality. We use *Global Shuffle* to shuffle the order of words before inputting to transformers. Note that the words and their locations are not changed at all and the only difference is the order of the word sequence input to the transformer.

Neighbor Shuffle and Non-neighbor Shuffle. Intuitively, local neighbors of a value make more contributions to its prediction. So, we propose *Neighbor Shuffle* which shuffles the order of each value's neighbors and keeps the order of the rest. Oppositely, we also have *Non-neighbor Shuffle*. A word, w_i, is defined as a value's neighbor if the IoU between b_i and the neighbor zone of the value is larger than 0.5. The neighbor zone is a box that shares the same center as the value box with expanded width and height (expand rate denoted as r_{nb}). We also include n_{nb} nearby words from the original reading order as the neighbors.

4.2 Form Background Manipulation

Form background generally affects the model performance in two ways: (1) some background words are strong indicators to improve field-values' recall and (2) accurate prediction of background reduces false positives, thus increase model precision. We propose the following transformations to evaluate model robustness to background perturbations.

BG Drop. Background (BG) Drop mimics the scenario that some words are completely missed by OCR detection. This transformation removes background words together with the corresponding boxes at a probability of p_{bgd}.

Neighbor BG Drop is similar to *Neighbor Shuffle*, which drops all background words if they are neighbors of a field-value.

Key Drop. Keys are concrete text-representations of fields in a form. For example, the field *invoice_number* may be represented as "INV #", "Invoice No." etc. in a form. A key is a very important feature for value localization since the value is often located near the key. We propose *Key Drop* to see the model performance change if keys are accidentally missed by OCR detection.

BG Typo. OCR recognition usually makes errors. *BG Typo* simulates word-level string typos. We select each background word at a probability of p_{typo}. For each selected word, we apply one of the error types, including swapping, deleting, adding, and replacing a random or a specific character.[1]

BG Synonyms. Similar semantics may be represented using different word synonyms. *BG Synonyms* randomly replaces each background word at a probability of p_{bgs} with their synonyms.[2]

BG Adversarial. Some forms contain only one word in the same data type as field-values. For example, there might be only one word with the date type of *date*. It is less challenging for a model to recognize it as the *invoice_date*. However, this type of easy case is not always guaranteed in real-world applications. *BG Adversarial* is used to increase the difficulty level by adding distraction. Concretely, we select background words at a probability of p_{bga} and use adversarial words for replacement. For each replacement, we randomly choose a data type and then generate a random value of the corresponding data type. We focus on three data types, i.e., *date*, *number* and *money*. We generate random dates using Faker.[3] For numbers, we first generate the number length randomly and then a random number of the length accordingly. For *money*, we obtain a random amount within lower and upper bounds. Then, we make the amount in money format, where we add a decimal point at the second to the left digit, place a comma at every third digit to the left of the decimal point and randomly insert $ at the beginning. To protect strong indicators for values, neighbor words are not replaced.

4.3   Form Field-Value Augmentations

Modifying field-values is a more direct way to increase the diversity of the evaluation set. We augment field-values in both text and locations.

Value Text Augment. Field-values of forms may be biased due to the limitation of the data collection process. For example, the *invoice_date* may be restricted to the year the form is collected, and the *invoice_number* may be biased towards the vendor's numbering system. *Value Text Augment* transformation targets at augmenting the field-values based on their data types. For each field-value, we randomly generate a substitute with the same data type following the same value generation procedure as we do in *BG Adversarial*.

---

[1] We utilize the implementation of the string typos provided in https://pypi.org/project/typo/.

[2] We generate word synonyms using WordNet Interface (https://www.nltk.org/howto/wordnet.html).

[3] https://faker.readthedocs.io/en/master/.

Value Location Augment. Form layouts can be very diverse in real-world scenarios. Intuitively, we should be able to infer a field-value as long as a key is represented properly no matter where we place the key-value pair in the document. We introduce *Value Location Augment* to increase layout diversity. To maintain the form format to the most, we keep the background as it is and shuffle the key-value pair's locations in the form. For example, for the field *invoice_number* (key: Invoice No., value: 1234) and *invoice_date* (key: Invoice date, value: 01/01/2021), we swap the box locations of "Invoice No." and "Invoice date", and also the locations of "1234" and "01/01/21".

5   Experiments

We evaluate the robustness of transformation-based field extractors using our framework on two commonly used form types, i.e., invoices and receipts.

5.1   Datasets

Our evaluation models are trained using a labeled train set, and the best performing model is picked based on a validation set. We prepare a separate test set to perform the robustness evaluation. To perform the proposed transformations, we annotate both the key and value of each field of interest with their bounding box locations in a form.

Invoice. The train, valid, and test sets contain 158, 348 and 338 real invoices. They are collected from 111, 222, and 222 vendors, respectively. We sample at most 5 forms from the same vendor and the vendors of train, valid and test sets do not overlap. We consider 7 frequently used fields including *invoice_number*, *purchase_order*, *invoice_date*, *due_date*, *amount_due*, *total_amount* and *total_tax*.

Receipt. We use the publicly available receipt dataset, SROIE. The annotations of their original test set are not publicly available, so we split the original train set to train, valid, and test sets based on their company names and sample at most 5 forms per company following [13]. Finally, we get 237 receipts for training, 76 for validation, and 74 for testing. The fields of interest are *company*, *address*, *date* and *total*. We add value boxes and annotate keys according to the text-level annotations provided by the original dataset.

5.2   Implementation Details

Our evaluation framework is implemented using Pytorch and the experiments are conducted on a single Tesla V100 GPU. The strength of a transformation is controlled by parameters. We set parameters to make moderate perturbations. In *BG Typo*, *BG Drop*, *BG Synonyms* and *BG Adversarial*, transformations are only applied to some selected words. We fix the pre-defined probabilities, $p_{typo}, p_{bgs}, p_{bga}$, to 0.1. θ_{center} is set to 0.5 and θ_{xy} is 0.1. r_{mp} in Margin Padding is set to 0.3. When determining the value's neighbor zone, we set the expand rate as $r_{nb} = 0.02$ and $n_{nb} = 2$.

We generate random values based on data types in *BG Adversarial* and *Value Text Augment*. For dates, we randomly pick a date from the year 2001 to 2021 in one of the formats, including mm/dd/yy, yy-mm-dd, dd/month/yy, and dd/mon/yy. For numbers, the number length is randomly generated from 3 to 12. The amount of money is randomly selected from 1 to 10,000,000.

We use a commercial OCR engine[4] for OCR extraction and utilize Tesseract[5] to rank the words in reading order. Our default transformer is LayoutLM [24] with text and boxes as inputs. We also evaluate using BERT [4] and RoBERTa [11] that take only text tokens as the inputs in Sect. 6. All models are finetuned from the corresponding base models. During training, we set the batch size to 8 and use the Adam optimizer with a learning rate of $5e^{-5}$.

5.3 Robustness Evaluation of Invoices

OCR Location Modification. Robustness evaluation of the LayoutLM model to OCR text location jittering is shown in Table 1. We obtain comparable results to the original performance when applying *Center Shift* and *Box Stretch* which indicates that slight box jittering is tolerable in our case. *Margin Pad* shifts all text locations by adding random margins around a form. This transformation also just slightly decrease the model performance.

Table 1. Evaluation of robustness to OCR location modifications on invoices. LayoutLM is used as the transformer model.

Transforms	Precision	Recall	F1
Original	70.9	69.3	70.0
Center Shift	70.8	69.9	70.3
Box Stretch	70.0	68.5	69.1
Margin Pad	69.6	68.3	68.9

Table 2. Evaluation of robustness to OCR text order on invoices. LayoutLM is used as the transformer model.

Transforms	Precision	Recall	F1
Original	70.9	69.3	70.0
Global Shfl	58.5	53.9	55.9
Neighbor Shfl	66.9	61.9	64.1
Non-neighbor Shfl	67.0	65.3	65.9

---

[4] https://api.einstein.ai/signup.
[5] https://github.com/tesseract-ocr/tesseract.

OCR Text Order is essential to transformer-based field extractors since the order can serve as an important feature to improve model performance. We show the model performance when applying order shuffle to different places in Table 2. The results show that if we shuffle all text orders, the performance drops dramatically by 14.1% in F1 score. When only shuffle value neighbors, we obtain ~6% lower F1 score. We get ~4% lower F1 score if we shuffle non-neighbor words, even though we keep the text order of all values and their neighbors the same. The results demonstrate the importance of the text order. If the model is trained using texts with a good reading order, we may also want to ensure a good reading order during inference.

A natural question is, what if we break the reading orders during training. Will this help us save the effort of ensuring test reading order during inference? We re-train field extractors using texts with random orders. We obtain 58.7% in F1 score, which is 11.3% lower than our baseline. The comparison result suggests that the text reading order is a very important feature. How to use it without overfitting to it is an interesting research topic.

Table 3. Evaluation of robustness to BG manipulations on invoices. LayoutLM is used as the transformer model.

Transforms	Precision	Recall	F1
Original	70.9	69.3	70.0
BG Drop	69.8	66.2	67.7
Neighbor BG Drop	67.6	57.1	61.3
Key Drop	62.8	56.4	59.1
BG Typo	69.6	66.9	68.1
BG Synonyms	70.4	68.8	69.5
BG Adversarial	66.2	67.7	66.9

Background Drop related transformations simulate the scenarios where an OCR detector accidentally misses some background words. *BG Drop* removes words randomly selected in the background. As shown in Table 3, global *BG Drop* leads to slight performance decrease. When we apply *Neighbor BG Drop*, the performance largely drops by ~9% in F1 score. Comparing *BG Drop* and *Neighbor BG Drop*, we find that neighbor words are indeed more important for value extraction. For a fair comparison, we have adjusted the dropping rate of *BG Drop* to 0.13, such that the total number of dropped words is roughly equal to the number of neighbor words. Further, *Key Drop* results in a similar performance decrease as *Neighbor BG Drop*, although the total number of Keys is less than half of that for neighbor words.

Other Background Manipulations. Some background words, e.g., keys and other useful indicators, are important features to localize values. Adding typos

Table 4. Evaluation of robustness to background and field-value augmentations on invoices. LayoutLM is used as the transformer model.

Transforms	Precision	Recall	F1
Original	70.9	69.3	70.0
Value Text Aug	56.3	53.5	54.5
Value Location Aug	61.4	56.8	58.8

to the background words may be harmful to these good features, thus *BG Typo* leads to a 2.4% drop in recall rate.

Forms from different vendors may use different words even when represent similar semantics. We attack the models using *BG Synonyms*. As shown in Table 3, our field extractors are quite robust to this transformation with only a negligible drop in F1 score.

BG Adversarial is used to add background words (serve as distractions) with similar data types as field-values. As shown in Table 3, *BG Adversarial* leads to 4.7% drop in model precision.

Value Augmentations. Field-values of forms may be limited in diversity. *Value Text Augment* transformation augments field-values by replacing them with randomly generated values in the same data type. We augment the values of all the fields, except for *total_amount* and *amount_due*, since these two fields may involve complicated mathematical computations. For *total_tax*, we randomly select a number between 0 and 15% of the *total_amount*. The comparison results in Table 4 show that the model performance drop significantly by 15.5% F1 score.

Value Location Augment changes the spatial arrangements of key-value pairs. In practice, we only shuffle the key-value pairs if they have the same number of key words and the same number of value words, resulting in more than 75% key-value pairs relocated. The results in Table 4 demonstrate that *Value Location Augment* significantly reduces F1 scores by 11.2%.

Multiple Transformations. The proposed transformations can be combined together to generate more diverse sets. We conduct exhaustive combinations of every two and three transformations which result in 91 2-transformation combinations and 364 3-transformation combinations.[6] The top-10 most impactful combinations are shown in Fig. 3. The comparison results suggest the following conclusions.

First, generally if an individual transformation drops more performance, it also contributes more drop when combined with other transformations. The most impactful combination is (*Value text Augment, Global Shuffle, Value Location Augment*) with a F1 score of 25.7. They are the top-3 impactful transformations suggested in Fig. 4.

---

[6] We observe that different orders of transformations in a combination result in ignorable differences.

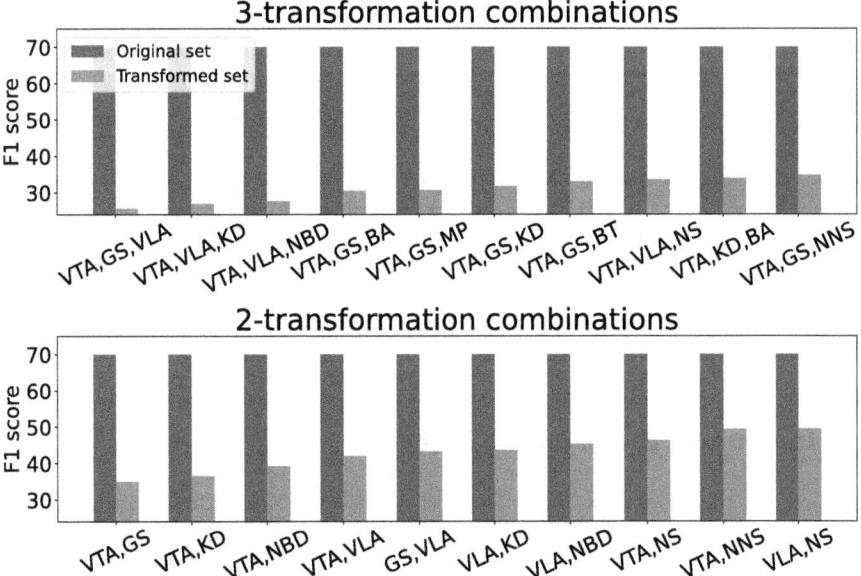

Fig. 3. Top-10 most impactful 2-transformation and 3-transformation combinations. VTA: Value Text Augment, GS: Global Shift, VLA: Value Location Augment, KD: Key Drop, NBD: Neighbor Background Drop, BA: Background Adversarial, MP: Margin Pad, BT: Background Typo, NS: Neighbor Shuffle, NNS: Non-Neighbor Shuffle.

Second, some individual transformations are less impactful, but they affect more when combined with some specific transformations. For example, individual *Margin Pad* ranks low in Fig. 4. However, it leads to more performance drop when combined with *Value Text Augment* and *Global Shuffle*. Their performance drop ranks 5 out of 364 combinations (F1 is 30.8). This may due to that *Margin Pad* (changes all words' locations), *Value Text Augment* (changes values' texts) and *Global Shuffle* (changes text input order) are three complementary transformations. When we do *Margin Pad* alone, the model resorts to the information of value texts and text orders. However, when we do these three transformations together, the model becomes inevitably confused.

Third, if the transformations have overlapping effects, their combination has a lower impact. For example, *Key Drop*, *Neighbor BG Drop* and *Neighbor Shuffle* all manipulate neighbor words. The performance drop on their combination ranks 246 out of 364 combinations although their individual transformation is impactful (see Fig. 4). The F1 score is 58.1 which is very close to an individual *Key Drop* transformation (59.1).

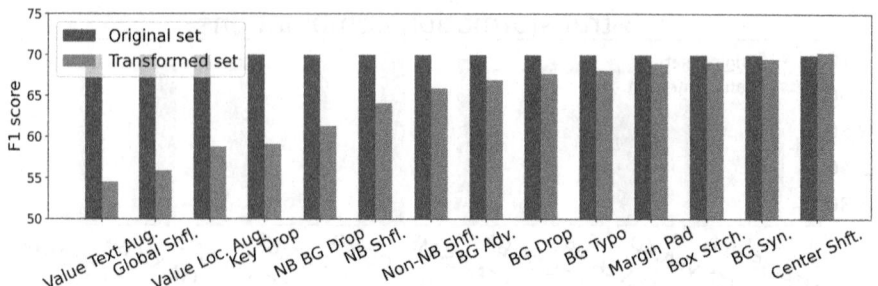

Fig. 4. An overview of LayoutLM-based model performance drop due on different transformed dataset. The results are sorted by the performance gap.

5.4 Robustness Evaluation of Receipts

There are two interesting features of the SROIE dataset. First, a significant amount of field-values have no keys, for example, all values of *company* and *address*, and some values of *date* and *total*. Consequently, changing the context of values has a minor effect on model performance. Second, the layouts of different receipts are very fixed. For example, *company* and *address* are always on the very top of every receipt. So, models could easily overfit to field-value locations and the text order. As shown in Table 5, *Global Shuffle* leads to significant performance drop by 38.7% F1 score. Specifically, the fields of *address* and *company* become 0% F1 score when the text's order is completely shuffled before inputting to the transformer. The results demonstrate that the model is overfitting to the input text order, especially for *address* and *company*.

Table 5. Robustness evaluation on SROIE dataset. LayoutLM is used as the transformer model.

Transforms	Precision	Recall	F1
Original	81.8	80.1	80.9
Global Shfl	42.5	41.9	42.2
Value Location Aug.*	54.3	48.0	49.7
Value Text Aug	75.2	73.6	74.4

Most of the fields in SROIE have no keys. To augment receipt layouts, we design a dedicated method that locally moves field-value locations. Specifically, for *company* and *address* in SROIE receipts, we move the values to the bottom of the form and shift the rest above to fill the gap as shown in Fig. 5. We refer to this transformation as *Value Location Augment**. This transformation changes the location of the values without breaking the text order within each value. We obtain 21.6% and 1.7% F1 score for *company* and *address*, respectively, which are around 60% and 68% lower than the original numbers.

Fig. 5. An illustration of *Value Location Augment** transformation on a receipt in SROIE dateset.

Besides, we also evaluate models on test set transformed by *Value Text Augment*. We replace values of *company, address* and *date* using substitute randomly generated by Faker. Same as what we do for *total_amount* and *amount_due* for invoices, we keep the values of *total* as they are. To maintain the layout structure, we only replace *company* and *address* if we are able to get a randomly generated sample with the same number of words as the original sample. This results in about 69% *company* values and 31% *address* values changed, respectively. As shown in Table 5, the *Value Text Augment* largely decreases the model performance by 5.5% in F1 score.

5.5   Observations and Suggestions

An overview comparison of all the transformations of invoices is summarized in Fig. 4. As we can see, the top-3 substantial transformations are *Value Text*

Augment, *Global Shuffle* and *Value Location Augment*. Experiments on receipts also show the effectiveness of these three transformations.

We make the following recommendations based on the analysis. For data collection/augmentation, forms with more diverse values are preferable. For example, we may want *dates* covering a wide range of time periods with more types of formats and *numbers* being more extensive. Varying forms' layouts is also beneficial. Especially, we may want to focus on varying the arrangement of field-values instead of altering individual word locations locally.

For the design of field extractors, we suggest making better utilization of the text order. As shown in our experiments, the text order is a very useful feature. How to utilize the text reading order without overfitting to it is an interesting topic. Besides, *Key Drop* and *Neighbor BG Drop* result in significant performance decreases as shown in Fig. 4. This suggests that value's neighbors, especially the keys, are essential for value extractions. Current state-of-the-art models use transformers to model interactions between all words. We believe paying attention to keys and neighbors in the model design has the potential to improve the existing field extraction systems.

6   Experiments on More Transformers

6.1   Robustness Evaluation of Invoices

The results of BERT and RoBERTa on invoices are summarized in Table 6 and Table 7.

Table 6. Robustness evaluation on invoice dataset. BERT is used as the transformer model.

Transforms	Precision	Recall	F1
Original	58.4	58.1	57.8
Global Shfl	40.5	38.2	38.6
Neighbor Shfl	51.5	50.7	50.6
Non-neighbor Shfl	57.9	56.4	56.6
BG Drop	57.7	57.2	56.9
Neighbor BG Drop	49.9	44.7	46.7
Key Drop	50.2	46.5	47.6
BG Typo	56.0	54.1	54.7
BG Synonyms	58.8	58.3	58.0
BG Adversarial	48.7	51.6	49.7
Value Text Aug	36.2	33.8	34.3
Value Location Aug	55.5	54.4	54.4

Table 7. Robustness evaluation on invoice dataset. RoBERTa is used as the transformer model.

Transforms	Precision	Recall	F1
Original	63.4	59.1	61.1
Global Shfl	40.5	33.3	36.4
Neighbor Shfl	58.9	53.2	55.9
Non-neighbor Shfl	60.6	57.0	58.7
BG Drop	63.6	58.8	61.0
Neighbor BG Drop	58.8	49.5	53.5
Key Drop	57.7	49.8	53.2
BG Typo	62.1	57.7	59.8
BG Synonyms	62.7	58.5	60.4
BG Adversarial	56.2	55.2	55.5
Value Text Aug	44.9	39.2	41.5
Value Location Aug	57.9	53.9	55.7

OCR Text Order. BERT and RoBERTa do not have text locations as inputs, so they rely more on text order than LayoutLM does. When deploying *Global Shuffle*, we observe more performance drop, i.e., by 19.2% F1 score for BERT and by 24.7% F1 score for RoBERTa.

Background Drop. We observe similar results on BERT and RoBERTa as compared to that on LayoutLM. *Key Drop* and *Neighbor BG Drop* have more impact than global *BG Drop*.

Other Background Manipulations. We observe 3.1% and 1.3% drop in F1 score on *BG Typo*, when using BERT and RoBERTa. Similar to LayoutLM, BERT and RoBERTa are robust to *BG Synonyms*. *BG Adversarial* leads to 9.7% and 7.2% drop in model precision for BERT and RoBERTa based methods, respectively.

Value Augmentations. *Value Text Augment* results in 23.5% (BERT) and 19.6% (RoBERTa) drop in F1 score. Since BERT and RoBERTa do not rely on location input, the performance drop on *Value Location Augment* of these two models is much less than that of the LayoutLM (drop 11.2%).

6.2   More Robustness Evaluation of Receipts

The results of BERT and RoBERTa on receipts are summarized in Table 8 and Table 9. The experimental results suggest that these three transformations have significant impact to the performance of BERT and RoBERTa.

Table 8. Robustness evaluation on SROIE dataset. BERT is used as the transformer model.

Transforms	Precision	Recall	F1
Original	75.4	73.3	74.3
Global Shfl	37.9	36.1	37.0
Value Location Aug.*	51.9	44.9	46.0
Value Text Aug	71.2	69.2	70.2

Table 9. Robustness evaluation on SROIE dataset. RoBERTa is used as the transformer model.

Transforms	Precision	Recall	F1
Original	77.3	74.7	75.9
Global Shfl	40.6	37.1	38.8
Value Location Aug.*	48.6	41.9	43.4
Value Text Aug	72.5	70.6	71.5

7   Conclusion and Future Work

We proposed a novel framework to evaluate the robustness of transformer-based form field extractors via form attacks. We introduced 14 transformations that transform forms in different aspects, including OCR-level location and order, background contexts, and field-value text and layouts. We conducted studies on real invoices and receipts with three types of transformer-based models using our proposed framework. Research recommendations were made based on the robustness analysis.

Improving field extraction from forms using the research analysis generated by the robustness evaluation is a very meaningful research area. The proposed transformations are potentially useful for increasing the diversity of training samples, thus improving model robustness. We will consider this in the future work.

8   Broader Impact

This work targets at robustness evaluation of form information extraction systems, so it has positive impacts such as identifying bias of existing information extractors and improving the fairness of model comparison. On the opposite side, our method may have unintended negative consequences in that we have proposed transformations that evaluate various aspects of model robustness, but the metrics we have selected may not be comprehensive. As a result, there is likely some degree of model bias present that has been missed by the proposed framework. However, this negative impact is not specific to our work and should be considered in general in the field of robustness AI.

The invoice dataset is for internal use only and does not contain any personally identifiable data. The SROIE dataset is a public dataset under MIT license. All forms were annotated by the authors. Consequently, we are confident that the datasets do not have ethical issues.

References

1. Appalaraju, S., Jasani, B., Kota, B.U., Xie, Y., Manmatha, R.: DocFormer: end-to-end transformer for document understanding. In: Proceedings of the IEEE/CVF International Conference on Computer Vision, pp. 993–1003 (2021)
2. Chiticariu, L., Li, Y., Reiss, F.R.: Rule-based information extraction is dead! Long live rule-based information extraction systems! In: EMNLP (2013)
3. Denk, T.I., Reisswig, C.: BertGrid: contextualized embedding for 2d document representation and understanding. In: NeurIPS Workshop (2019)
4. Devlin, J., Chang, M.W., Lee, K., Toutanova, K.: BERT: pre-training of deep bidirectional transformers for language understanding. In: NAACL (2019)
5. Goel, K., et al.: Robustness gym: Unifying the NLP evaluation landscape. arXiv preprint arXiv:2101.04840 (2021)
6. Hong, T., Kim, D., Ji, M., Hwang, W., Nam, D., Park, S.: Bros: a pre-trained language model focusing on text and layout for better key information extraction from documents. In: Proceedings of the AAAI Conference on Artificial Intelligence, vol. 36, pp. 10767–10775 (2022)
7. Huang, Z., et al.: ICDAR 2019 competition on scanned receipt OCR and information extraction. In: ICDAR (2019)
8. Katti, A.R., et al.: CharGrid: towards understanding 2d documents. In: EMNLP (2018)
9. Kiela, D., et al.: Dynabench: rethinking benchmarking in NLP. In: NAACL (2021)
10. Kim, G., et al.: OCR-free document understanding transformer. In: European Conference on Computer Vision. Brostow, G., Cissé, M., Farinella, G.M., Hassner, T. (eds.) Computer Vision – ECCV 2022. ECCV 2022. LNCS, vol. 13688, pp. 498–517. Springer, Cham (2022). https://doi.org/10.1007/978-3-031-19815-1_29
11. Liu, Y., et al.: Roberta: a robustly optimized BERT pretraining approach. arXiv preprint arXiv:1907.11692 (2019)
12. Ma, E.: NLP augmentation (2019). http://github.com/makcedward/nlpaug
13. Majumder, B.P., Potti, N., Tata, S., Wendt, J.B., Zhao, Q., Najork, M.: Representation learning for information extraction from form-like documents. In: ACL (2020)
14. Morris, J.X., Lifland, E., Yoo, J.Y., Qi, Y.: TextAttack: a framework for adversarial attacks in natural language processing. arXiv preprint arXiv:2005.05909 (2020)
15. Palm, R.B., Laws, F., Winther, O.: Attend, copy, parse end-to-end information extraction from documents. In: ICDAR (2019)
16. Salman, H., Ilyas, A., Engstrom, L., Kapoor, A., Madry, A.: Do adversarially robust imagenet models transfer better? In: NeurIPS (2020)
17. Santurkar, S., Tsipras, D., Madry, A.: Breeds: Benchmarks for subpopulation shift. arXiv preprint arXiv:2008.04859 (2020)
18. Schuster, D., et al.: Intellix - end-user trained information extraction for document archiving. In: ICDAR (2013)
19. Schuster, D., et al.: Intellix-end-user trained information extraction for document archiving. In: 2013 12th International Conference on Document Analysis and Recognition, pp. 101–105. IEEE (2013)

20. Taori, R., Dave, A., Shankar, V., Carlini, N., Recht, B., Schmidt, L.: Measuring robustness to natural distribution shifts in image classification. In: NeurIPS (2020)
21. Vaswani, A., et al.: Attention is all you need. In: NeurIPS (2017)
22. Wang, Z., Zhan, M., Liu, X., Liang, D.: DocStruct: a multimodal method to extract hierarchy structure in document for general form understanding. In: EMNLP (2020)
23. Wu, T., Ribeiro, M.T., Heer, J., Weld, D.S.: Errudite: Scalable, reproducible, and testable error analysis. In: ACL (2019)
24. Xu, Y., Li, M., Cui, L., Huang, S., Wei, F., Zhou, M.: LayoutLM: pre-training of text and layout for document image understanding. In: KDD (2020)
25. Zeng, G., et al.: OpenAttack: an open-source textual adversarial attack toolkit. arXiv preprint arXiv:2009.09191 (2020)
26. Zhang, Z., et al.: Layout-aware information extraction for document-grounded dialogue: dataset, method and demonstration. In: Proceedings of the 30th ACM International Conference on Multimedia, pp. 7252–7260 (2022)

Key-Value Information Extraction from Full Handwritten Pages

Solène Tarride[1(✉)] [ID], Mélodie Boillet[1,2] [ID], and Christopher Kermorvant[1,2] [ID]

[1] TEKLIA, Paris, France
starride@teklia.com
[2] LITIS, Normandy University, Rouen, France

Abstract. We propose a Transformer-based approach for information extraction from digitized handwritten documents. Our approach combines, in a single model, the different steps that were so far performed by separate models: feature extraction, handwriting recognition and named entity recognition. We compare this integrated approach with traditional two-stage methods that perform handwriting recognition before named entity recognition, and present results at different levels: line, paragraph, and page. Our experiments show that attention-based models are especially interesting when applied on full pages, as they do not require any prior segmentation step. Finally, we show that they are able to learn from key-value annotations: a list of important words with their corresponding named entities. We compare our models to state-of-the-art methods on three public databases (IAM, ESPOSALLES, and POPP) and outperform previous performances on all three datasets.

Keywords: Key-value extraction · Named-Entity Recognition · Handwritten Document · Segmentation-free Approach

1 Introduction

Although machine learning and deep learning techniques are nowadays commonly used in the field of automatic processing of historical documents [12], scientific work still often focuses on some specific processing steps in isolation. It is common to develop models either for page analysis or line detection, for handwriting recognition or for information extraction. Processing chains are still often developed as a sequence of these steps independently. However, these processing chains suffer from several drawbacks. Firstly, errors accumulate along the chain: if the line detection step is bad, write recognition will be highly impacted and information extraction impossible. On the other hand, the implementation of these chains and their maintenance is complex: each step requires specific skills and annotated data for each model and any update of a part of the chain has an impact on all downstream processes. Finally, the different modules are developed independently and there is no global optimization of the processing chain. For all these reasons, the development of models allowing the extraction of information directly from the image, by an end-to-end approach, with a single model, would be very beneficial.

© The Author(s), under exclusive license to Springer Nature Switzerland AG 2023
G. A. Fink et al. (Eds.): ICDAR 2023, LNCS 14188, pp. 185–204, 2023.
https://doi.org/10.1007/978-3-031-41679-8_11

As far as automatic recognition is concerned, three main types of projects are currently being carried out on collections of historical documents, depending on the intended use. The first type of project aims to carry out a complete transcription of the documents to allow full-text searches [16,27]. The processing chain then focuses on the page analysis stage to extract a maximum number of lines of text and the handwriting recognition stage to best recognize the text. The result of the processing is then exploited thanks to a search engine that allows queries to be made and documents to be identified according to their content. The second type of processing aims to produce electronic editions of documents [11]. In this case, the emphasis is obviously on the quality of the recognition, but also the fidelity to the text of the document and the reading order. The result of the automatic processing is in this case always submitted to the correction of an expert before publication. The last type of project aims at extracting information from documents in order to populate a database with the information they contain [24]. In addition to the document analysis and handwriting recognition stages, these projects also incorporate an information extraction stage, often in the form of named entity extraction. It is this third type of project, the most complex in its implementation, that we are interested in this work.

Information extraction chains for historical handwritten documents are usually composed of the following steps: line detection or document layout analysis (DLA), handwriting recognition (HTR) and named entity extraction (NER). In this paper, we first reconsider the possibility of combining the HTR and NER models into a single model. Then we study whether it is possible to extend this model to the processing of a complete page without going through a line detection step. Finally, we show that it is possible to go even further and train a single model for the extraction of target information, of the key-value type, without going through an explicit transcription.

The rest of this paper is organized as follows. In Sect. 2, we review the state-of-the-art for information extraction in handwritten documents. We describe our methodology and experiments in Sect. 3. The experimental results are presented and analyzed in Sect. 4. Finally, in Sect. 5 we discuss the conclusions and outline future works.

2   Related Work

Recent advances in computer vision and natural language processing have led to major breakthroughs in the field of automatic document understanding. Deep learning-based systems are now capable of automatically extracting relevant information from historical documents. Interest in this field has been encouraged by the emergence of competitions, such as the Information Extraction on Historical Handwritten Records competition [8] on the ESPOSALLES database [20], as well as the publication of named entity recognition annotations for other databases, such as IAM-NER [26] and POPP [4].

Two main approaches exist to address automatic information extraction from handwritten documents:

- **Sequential approaches** consist in dividing the problem into two successive tasks: handwritten text recognition, and named entity recognition;
- **Integrated approaches** consist in combining text and named entity recognition in a single-step.

Each of these approaches can work at several levels: either on words, lines, paragraphs, or directly on full pages. Segmentation-based systems work on pre-segmented text zones (words, lines, or paragraphs), while segmentation-free systems work directly on full pages. Performing handwriting recognition on smaller zones is usually easier to achieve, but requires a prior segmentation step. As opposed, handwriting recognition on full pages is more challenging (memory management, reading order), but does not require any prior segmentation.

2.1 Sequential Approaches

In sequential approaches, HTR is performed first, then, NER is applied on recognized text. Note that HTR and NER can be applied at different levels: HTR is usually performed at line-level, and NER at paragraph or page-level.

Segmentation-Based Systems. Five systems were introduced during the ICDAR2017 Competition on Information Extraction in Historical Handwritten Records [8] on ESPOSALLES. Most participants used CRNN trained with CTC to recognize handwritten text. Named entity recognition was then performed using logical rules based on regular expressions or CRF tagging. Other methods were proposed after the competition.

Prasad et al. [17] propose a two-stage system combining a CRNN-CTC neural network for HTR on text line images, followed by a BLSTM layer over the feature layer for NER.

Tuselmann et al. [26] also introduce a two-stage system for information extraction that combines a Transformer model [10] for HTR on word images, and a LSTM-CRF model with word embeddings obtained using a pre-trained RoBERTa for NER. They highlight the advantages of two-stage methods for information extraction, as these methods yield state-of-the art results and are easy to improve using post-processing techniques, close dictionary, or pre-trained embeddings.

Monroc et al. [15] compare different off-the-shelf NER libraries on handwritten historical documents: SpaCy [9], FLAIR [1], and Stanza [19]. They perform experiments on three datasets in an end-to-end setting, and study the impact of text line detection and text line recognition on NER performances. Their results highlight that line detection errors have a greater impact than handwriting recognition errors. This conclusion suggests that working directly on pages could prevent segmentation errors from impacting the final entity recognition.

Segmentation-Free Systems. To the best of our knowledge, no system performing sequential HTR and NER at page-level has been proposed so far. However, many segmentation-free HTR models working directly at page-level [5,6,28] have been introduced recently. Any of these models could easily be combined with off-the-shelf NER libraries for segmentation-free information extraction.

2.2   Integrated Approaches

Integrated approaches combine HTR and NER in a single step by modeling named entities with special tokens. This can be achieved with or without prior segmentation.

Segmentation-Based Systems. Toledo et al. [25] and Rowtula et al. [22] introduce models that work at word-level. Their systems are able to recognize and classify word images into semantic categories.

Both Carbonell et al. [3] and Tarride et al. [23] propose neural networks that predict characters and semantic tags from line images, respectively, using a CRNN model trained with CTC and an attention-based network. Both of these studies suggest that working on records would allow the model to capture more contextual information.

In [4], the authors use the same approach on French census images from the POPP dataset, as they predict text characters and special tokens for empty cells and column separators. Although, this dataset does not directly include named entities, each word is linked to a specific column and can be seen as a named entity (name, surname, date of birth, place of birth...).

Finally, Rouhou et al. [21] are the first to introduce a Transformer model for combined HTR and NER at record-level on the ESPOSALLES database. They highlight the interest of performing this task on records to benefit from more contextual information. As each page contains several records, this model still requires record segmentation. Moreover, they use a special token for line breaks, as they observe this improves performance.

Segmentation-Free Systems. Carbonell et al. [2] are the first to propose a model that works directly at page-level on ESPOSALLES. Their system is able to jointly learn word bounding boxes, word transcription and word semantic category on ESPOSALLES. However, a major limitation of this method is that it requires word bounding boxes during training.

The Transformer proposed by Rouhou et al. [21] could be applied to full pages in its current stage, although this task has not been tackled by the authors.

Finally, the Document Attention Network (DAN) [5] is able to recognize text on full pages with reading order. It is based on the Transformer architecture and jointly learns characters and special tokens that represent layout information. It is likely that this method is also able to recognize named entities, or in other words, tokens that are not spatially localized but have a semantic meaning. However, the authors did not perform any experiments on named entity recognition.

2.3   Discussion

The literature review opens up three main questions that are discussed in the following.

What Is the Best Approach for Information Extraction? Although this question has been well studied in the past, no consensus has been reached. On the one hand, researchers have shown the interest of sequential methods which can be optimized at every stage (with a language model, a dictionary, or pre-trained embeddings) [15, 26]. On the other hand, the advantages of integrated methods have also been demonstrated [23], notably because they benefit from shared contextual features and avoid cascade errors.

Can We Extract Relevant Information from Full Pages? Different methods were designed to work at different levels, some of them requiring prior segmentation of text lines or paragraphs. However, in real-world scenarios, text areas are not known and must therefore be detected automatically, which can introduce segmentation errors. It has been established that segmentation errors have a greater impact on information extraction than handwriting recognition errors [15]. Recently, Transformers have proved their ability to learn from paragraphs and pages [5, 21], enabling segmentation-free information extraction. Learning directly from pages increases the task difficulty, but avoids the need for prior segmentation. Moreover, working directly on pages makes it possible to benefit from a larger context [21].

Are Integrated Models Able to Learn from Key-Value Annotations? As sequential approaches rely on HTR, they require the entire transcription before retrieving named entities. However, integrated methods could potentially learn from key-value annotations, which corresponds to a list of words with their corresponding named entities. In this scenario, ground-truth is also easier and faster to produce, as annotators would only have to annotate important words as well as their semantic category. This approach could also be applied in a lot of practical applications where full transcriptions are not available, such as genealogical crowdsourced information (civil status, or personal records). This question has not been studied yet in the context of information extraction.

In the next section, we describe the experiments designed to address these three questions.

3   Methodology and Experiments

In this section, we introduce the datasets used during our experiments, present our methodology, and describe the different experiments conducted in this study.

3.1   Datasets

During our experiments, we worked on three public datasets of different kind.

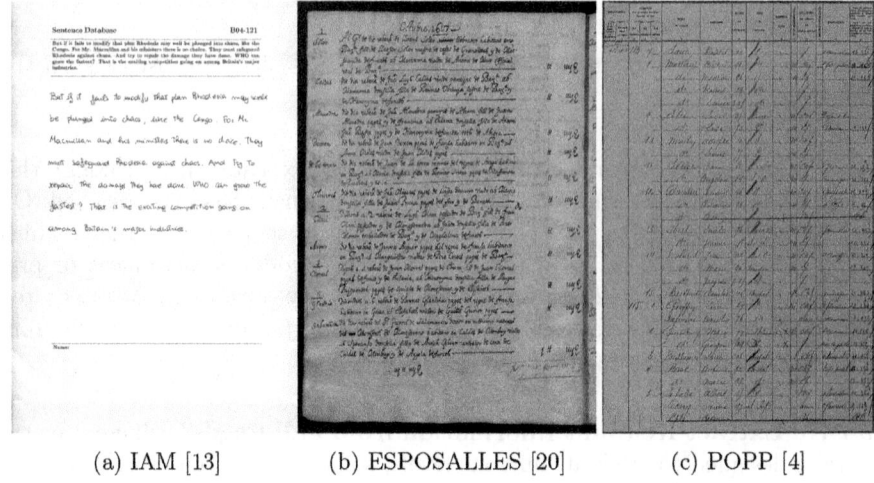

(a) IAM [13]   (b) ESPOSALLES [20]   (c) POPP [4]

Fig. 1. Examples of pages from the three datasets used in this work

IAM. The IAM dataset [13] is composed of modern documents written in English by 500 writers. It includes 747 training pages with corresponding transcriptions. NER annotations have been made available by Tüselmann et al. [26]. A page from IAM is presented in Fig. 1a.

For our experiments, we use the RWTH split with 18 entities: Cardinal, Date, Event, FAC, GPE, Language, Law, Location, Money, NORP, Ordinal, Organization, Person, Percent, Product, Quantity, Time and Work of art. The details of this split are provided in the appendix. Less than 10% of words are associated to an entity. Due to the large number of classes, some entities have very few examples in the training set. We perform experiments at two levels: text line and page. When working on pages, we remove the header so that the model does not see the printed transcription instruction.

ESPOSALLES. The ESPOSALLES dataset [8] is a collection of historical marriage records from the archives of the Cathedral of Barcelona. The corpus is composed of 125 pages. Each document is written in old Catalan by a single writer. It includes 125 pages with word, line and record segmentations. The details of this split are provided in the appendix. A page from ESPOSALLES is presented in Fig. 1b.

Each word is transcribed and labeled with a semantic category (name, surname, occupation, location, state, other) and a person (husband, wife, husband's father, husband's mother, wife's father, wife's mother, other person, none). More than 50% of words are associated to an entity. As there is no validation set, we keep 25% of training pages for validation. We perform experiments at three levels: text line, record, and page.

POPP. The POPP dataset contains tabular documents from the 1926 Paris census whose statistics are detailed in Table 1. It contains 160 pages written in French, each page contains 30 lines. A page from POPP is presented in Fig. 1c.

Each row is divided in 10 columns: surname, name, birthdate, birthplace, nationality, civil status, link, education level, occupation, employer. In our experiments, we use the column name as a named entity. As a consequence, 100% words are associated to an entity. We perform experiments at two levels: text line and page.

Table 1. Statistics of the POPP dataset

(a) Pages, lines, words, and entities by split

	Train	Validation	Test
Pages	128	16	16
Lines	3,837	480	479
Words	29,581	3,681	3,569
Entities	29,581	3,681	3,569

(b) Entities by split

	Train	Validation	Test
Surname	3,100	392	375
First name	3,853	476	478
Birthdate	3,824	469	466
Location	4,789	600	584
Nationality	283	17	30
Civil status	2,277	292	225
Link	3,667	449	412
Education level	25	4	12
Occupation	4,488	529	535
Employer	3,275	453	452

3.2 Methods

Three methods are introduced and compare in this work.

Two-Stage Workflow. The first method is a traditional two-stage workflow for information extraction that combines two steps. First, an HTR system is applied for text recognition on line-level images, then, SpaCy[1] [9] is used for named entity recognition. We compare two systems for the HTR task: PyLaia [18] and DAN [5].

– PyLaia[2] is an open source model for handwritten text recognition. It combines 4 convolutional layers and 3 recurrent layers, and is trained with the CTC loss function. The last layer is a linear layer with a softmax activation function that computes probabilities associated with each character of

---

[1] https://spacy.io.
[2] https://github.com/jpuigcerver/PyLaia.

the vocabulary. We use early stopping to avoid overfitting: the training is stopped after 50 epochs without improvement. PyLaia is trained on text line images.

– DAN[3] is an open source attention-based Transformer model for handwritten text recognition that can work directly on images of paragraph or page. It is trained with the cross-entropy loss function. The last layer is a linear layer with a softmax activation function that computes probabilities associated with each character of the vocabulary. For each dataset, we train DAN on zones with the strongest semantic consistency: on records for ESPOSALLES, on pages for IAM, and lines for POPP.

For NER, we use SpaCy, a production-oriented NLP library that includes transformer-based pipelines with support of English (for IAM), Catalan (for ESPOSALLES), and French (for POPP). Like DAN, SpaCy is trained on records for ESPOSALLES, pages for IAM, and lines for POPP. For ESPOSALLES, we train two SpaCy models: one for the *category* label and one for the *person* label. Comparing two HTR systems with the same SpaCy model allows us to study the impact of handwriting recognition errors on the overall performance.

Integrated Workflow. The second method consists in training a model to recognize directly characters and NER tokens.
We train DAN models for this task, later referred to as *HTR+NER*. The model is trained at different levels to evaluate the impact of context: on lines and pages for IAM, on lines, records and pages for ESPOSALLES, on lines and pages for POPP. NER tokens are considered like characters by the network and are localized before relevant words, as illustrated in Table 2. For ESPOS-ALLES, we use a unique tag combining the *category* and *person* information (ex: `<name_wife>Maria`), as we found out that using two separate tags led to poorer performance. This observation is consistent with the findings of Carbonell et al. [3] and Rouhou et al. [21]. Finally, we also trained DAN with curriculum learning, e.g. trained for *HTR* and fine-tuned for *HTR+NER* and found out that the network reach similar performance. For clarity, we only provide results without curriculum learning.

Integrated Workflow with Key-Value Annotations. Our last experiment consists in training DAN on key-value annotations, so as to only predict relevant information with the relevant text and the corresponding named-entity. This task is referred to as *Key-value HTR+NER* in the rest of the article. To achieve this, words that are not linked to any entities are removed from transcriptions, as illustrated in Table 2. As a result, the model must learn to directly extract important words with their named entities, and ignore any other word. In this

---

[3] https://github.com/FactoDeepLearning/DAN.

scenario, the two-stage approach cannot be used, as the full transcription is not available. This task is very challenging on IAM, as 90% of words are not linked to any entities, and more than 5% of pages do not have any entities. As a result, the training data is very sparse. The task is easier for ESPOSALLES, as 50% of words are linked to an entity. Finally, in POPP, every word is related to a named entity, so the *HTR+NER* and *Key-value HTR+NER* tasks are the same.

Table 2. Example of different transcriptions of the same record from the Esposalles database. Each transcription is used for a different task. *HTR*: the model predicts characters, *HTR+NER*: the model predicts characters and NER tokens, *Key-value HTR+NER*: the model predicts characters and NER tokens only for relevant words, ignoring words that are not associated with NER tokens.

Task	Transcription
HTR	dit dia rebere de Jua Oliveres pages de Llissa demunt viudo ab Maria donsella filla de Juan Pruna pages del far y de Beneta
HTR+NER	dit dia rebere de \<N-H>Jua \<SN-H>Oliveres \<O-H>pages de \<L-H>Llissa demunt \<S-H>viudo ab \<N-W>Maria \<S-W>donsella filla de \<N-WF>Juan \<SN-WF>Pruna \<O-WF>pages del \<L-WF>far y de \<N-WM>Beneta
Key-value HTR+NER	\<N-H>Jua \<SN-H>Oliveres \<O-H>pages \<L-H>Llissa \<S-H>viudo \<N-W>Maria \<S-W>donsella \<N-WF>Juan \<SN-WF>Pruna \<O-WF>pages \<L-WF>far \<N-WM>Beneta

4 Experimental Results

In this section, we introduce the evaluation metrics and present the results obtained on each dataset. We also compare our work with state-of-the-art methods and discuss the results.

4.1 Metrics

For all three datasets, performances are evaluated by the same standard character recognition and entity recognition metrics, as detailed in the following paragraphs. An additional metric is used to evaluate the experiments on ESPOSALLES.

4.2   HTR Metrics

The quality of handwriting recognition is evaluated using the character error rate (CER) and word error rate (WER). The full text is evaluated, and named entity tokens are ignored in integrated methods at this step of the evaluation.

4.3   NER Metrics

We use the Nerval[4] evaluation toolkit to evaluate named entity recognition results. In Nerval [14], the automatic transcription is aligned with the ground truth at character level. Predicted and ground truth words are considered a match if their edit distance is less than 30%. From this alignment, precision, recall and F1-score are computed.

4.4   IEHHR Metrics

Finally, for the ESPOSALLES dataset, we also compute the IEHHR metric that was introduced in the ICDAR 2017 Competition on Information Extraction in Historical Handwritten [8]. This metric jointly evaluates HTR and NER. Only words associated with named entities are taken into account in this evaluation. The "basic" score is equal to 100-CER if the *category* tag is correct, 0 otherwise. The "complete" score is equal to 100-CER if both the *category* and *person* tags are correct, 0 otherwise.

4.5   Evaluation Results

We present handwritten text recognition results in Table 3 and named entity recognition results in Table 4. For ESPOSALLES, we also provide the results for information extraction in Table 5 and obtain state-of-the-art results on the public IEHHR benchmark[5].

What is the Best Model for HTR? Results in Table 3 show that DAN is better than PyLaia for HTR on all three datasets. The DAN model trained only for *HTR* is generally better than the model directly trained for *HTR+NER*. The results show that DAN is always better than PyLaia for handwriting recognition: CER and WER are always lower with DAN. The WER reaches 1.37% on ESPOSALLES, 13.66% on IAM, and 18.09% on POPP. Finally, we note that DAN can be more performant on larger text zones. Indeed, DAN performs better on pages on IAM, and on records on ESPOSALLES. On the other hand, on POPP, the best performances are obtained on text lines. This observation can be explained by the fact that POPP documents are tables in which the lines are independent.

---

[4] https://gitlab.com/teklia/ner/nerval.
[5] https://rrc.cvc.uab.es/?ch=10&com=evaluation&task=1.

Table 3. Evaluation results for handwritten text recognition on IAM, ESPOSALLES, and POPP. Results are given for test sets. NER tokens are not taken into account for this evaluation.

(a) IAM (RWTH split)

Model	Task	CER (%)	WER (%)	Input
VAN [6]	*HTR*	4.45	14.55	Line
PyLaia	*HTR*	7.79	24.73	Line
DAN	*HTR*	**4.30**	**13.66**	Page
DAN	*HTR+NER*	5.12	16.17	Line
DAN	*HTR+NER*	4.82	14.61	Page

(b) ESPOSALLES

Method	Task	CER (%)	WER (%)	Input
Seq2seq [23]	*HTR*	2.82	8.33	Line
Seq2seq [23]	*HTR+NER*	1.81	6.10	Line
PyLaia	*HTR*	0.76	2.62	Line
DAN	*HTR*	0.46	**1.37**	Record
DAN	*HTR+NER*	0.48	1.75	Line
DAN	*HTR+NER*	**0.39**	1.51	Record
DAN	*HTR+NER*	3.61	4.23	Page

(c) POPP

Model	Task	CER (%)	WER (%)	Input
VAN [4]	*HTR*	**7.08**	19.05	Line
PyLaia	*HTR*	17.19	37.43	Line
DAN	*HTR*	8.18	**18.09**	Line
DAN	*HTR+NER*	7.83	24.57	Line
DAN	*HTR+NER*	11.74	30.78	Page

Table 4. Evaluation results for named entity recognition on IAM, ESPOSALLES, and POPP. Results are given for test sets. Evaluation results are computed using Nerval, which computes an alignment between ground truth and predicted entities.

(a) IAM (RWTH split)

Method	P (%)	R (%)	F1 (%)	Input Type
Tülselmann et al.* [26]	60.4	50.9	54.2	Word/Record
Rowtula et al.* [22]	33.8	30.9	32.3	Word/Record
Todelo et al.* [25]	26.4	10.8	14.9	Word/Record
Dessurt [7]	-	-	40.4	Page
Ground-truth + SpaCy	74.9	76.2	75.5	-/Page
PyLaia + SpaCy	56.5	49.0	52.5	Line/Page
DAN + SpaCy	**61.8**	**57.9**	**59.8**	Page/Page
DAN	37.1	30.8	33.7	Line
DAN	37.2	27.0	31.3	Page
DAN	0	0	0	Page (key-value)

* Different computation method due to pre-existing word alignment.

(b) ESPOSALLES

Method	Person			Category			Input Type
	P (%)	R (%)	F1 (%)	P (%)	R (%)	F1 (%)	
Tülselmann et al.* [26]	**99.3**	**99.2**	**99.3**	**98.5**	**98.2**	**98.3**	Word/Record
Rowtula et al.* [22]	97.0	96.2	96.6	97.1	97.0	97.0	Word/Record
Todelo et al.* [25]	98.5	97.8	98.1	98.5	97.8	98.1	Word/Record
Ground-truth + SpaCy	98.6	98.4	98.5	98.3	98.7	98.5	-/Record
PyLaia + SpaCy	95.9	94.0	94.9	95.6	94.3	95.0	Line/Record
DAN + SpaCy	**97.9**	97.9	97.9	**97.6**	**98.1**	**97.8**	Record/Record
DAN	96.0	96.1	96.1	96.9	97.0	96.9	Line
DAN	**97.9**	98.2	**98.1**	97.4	97.8	97.6	Record
DAN	95.0	**98.4**	96.6	94.2	97.6	95.9	Page
DAN	97.0	97.4	97.2	96.7	97.1	96.9	Record (key-value)

* Different computation method due to pre-existing word alignment.

(c) POPP

Method	P (%)	R (%)	F1 (%)	Input type
Ground-truth + SpaCy	95.6	97.3	96.4	-/Line
PyLaia + SpaCy	75.6	77.0	76.3	Line/Line
DAN + SpaCy	82.8	85.3	84.0	Line/Line
DAN	**85.6**	86.2	**85.9**	Line
DAN	83.8	**86.9**	85.3	Page

Table 5. IEHHR scores given for the test set of ESPOSALLES dataset.

Method	Basic (%)	Complete (%)	Input Type
Baseline HMM [8]	80.28	63.11	Line/Line
CITlab ARGUS-1 [8]	89.54	89.17	Line/Line
CITlab ARGUS-2 [8]	91.63	91.19	Line/Line
CITlab ARGUS-3 [8]	91.94	91.58	Line/Line
CVC [25]	90.59	89.40	Line/Line
Naver Labs [17]	95.46	95.03	Line/Line
IRISA [23]	94.7	94.0	Line
IRISA multi-task [23]	95.2	94.4	Line
InstaDeep GNN/Transformer[8]	96.22	96.24	Record
InstaDeep Transformer [21]	96.25	95.54	Record
TEKLIA Kaldi + Flair [15]	96.96	-	Line/Record
Ground-truth + SpaCy	97.51	97.57	-/Record
PyLaia + SpaCy	96.58	96.58	Line/Record
DAN + SpaCy	**97.13**	**97.11**	Record/Record
DAN	96.26	94.47	Line
DAN	97.03	96.93	Record
DAN	95.45	95.04	Page
DAN (key-value)	96.48	96.31	Record (key-value)

What is the Impact of HTR Errors on NER? Results in Table 4 help us understand the impact of handwriting recognition errors on NER performance. The second block of each subtable compares the results using ground transcription or predicted transcriptions (PyLaia or DAN). On ESPOSALLES, both HTR systems are very performant with CER below 1%. As a result, NER performance remains very good. However, on IAM and POPP, PyLaia and DAN yield a higher CER. As a consequence, the F1 score drops by 15 points for a 5% CER on IAM, and by 10 points for a 10% CER on POPP.

What is the Best Approach for Information Extraction? The best performance on IAM is achieved with a two-stage method, combining DAN (HTR) and SpaCy (NER). These results support the observations of Tüselmann et al. [26], and can be explained because there are few entities in the dataset. As a result, DAN struggles to learn semantic information, while SpaCy benefits from pretrained embeddings for the English language. However, on POPP, DAN trained for *HTR+NER* outperforms the two-stage approach combining DAN and SpaCy, although SpaCy does benefit from pre-trained French embeddings. There are two possible explanations for this result. First, POPP documents contain mostly names and surnames, which may not be included in the embeddings. Second, since these are tabular documents, word localization determines the semantic category, as each column corresponds to a specific named entity. Unlike DAN, SpaCy does not have any information regarding the word localization. Finally, on ESPOSALLES, both approaches yield similar results: SpaCy recognizes the *category* labels better while DAN recognizes the *person* labels better.

What is the Performance of Segmentation-Free Methods? It is interesting to note that DAN often performs better on pages (IAM) or records (ESPOS-ALLES) than on text lines. And yet, the text recognition task is traditionally done on text lines, which requires prior automatic or manual segmentation. But manual segmentation is time-consuming, and automatic segmentation can introduce many errors that affect the performance of handwriting or named entity recognition [15]. Therefore, results presented on pages cannot be directly compared to the results on text lines or records, as the task is much harder. In order to compare these results fairly, segmentation-based workflows should be evaluated on automatically segmented text lines or records. It is likely that segmentation-free workflows will outperform segmentation-based workflows in an end-to-end evaluation setting.

Is DAN Able to Learn from Key-Value Annotations? Finally, we evaluate the ability of DAN to learn from key-value annotations. On ESPOSALLES, where 50% of words are linked to an entity, DAN manages to learn from key-value annotations. It learns to recognize relevant words and to ignore the others. Although its performances are slightly lower than when trained with full transcripts, they remain very competitive. In contrast, DAN fails to learn on IAM, in which only 10% of words are linked to an entity. The model can be trained for a few epochs before overfitting. As a result, it does not predict anything on the test set. Finally, on POPP, all words are linked to an entity, so this experiment is similar to the one with full annotations, as there are no words to ignore during training.

5   Conclusion

In this paper, we focus on information extraction in digitized handwritten documents. We compare an integrated approach trained for joint HTR and NER with a traditional two-stage approach that performs HTR before NER. We present results at different levels: pages, paragraphs and lines and reach state-of-the-art performance on three datasets.

Our experiments show that integrated approaches trained jointly for HTR and NER can outperform two-stage approaches when word localization has an impact on the NER label (POPP). As opposed, two-stage approaches are better when applied on datasets with few entities (IAM) as they can benefit from pre-trained embeddings. In other cases (ESPOSALLES), two-stage and integrated approaches reach similar performance 97.11% and 96.93% respectively, for the complete IEHHR score on records on ESPOSALLES. We also demonstrate that applying these models directly on pages leads to very acceptable performances, either better than when applied on lines (ESPOSALLES, IAM), or with a minor performance loss (POPP). The interest of this method is enhanced by the lack of need for prior automatic segmentation, which is known to impact handwriting recognition performances [15]. Finally, we show that, under certain conditions, integrated methods are able to learn from key-value annotations, e.g. from a list of relevant words with their corresponding named entities. On ESPOSALLES, the model trained on key-value annotations reaches a complete recognition score of 96.31%. This observation is encouraging as it would allow training models from incomplete information manually, which considerably reduces the effort needed for manual transcription.

In future works, we are interested is measuring the impact of segmentation errors when evaluating end-to-end systems for information extraction. We also would like to identify the conditions needed to train a model on key-value annotations. Finally, we want to improve DAN for the task of information extraction. For example, the training loss could also be adapted to differentiate NER tokens from characters. Performance could be improved by using pre-trained embeddings like in SpaCy. Since DAN and SpaCy rely on character- and word-embeddings respectively, it would be interesting to find a common representation at sub-word level.

Appendix

Detailed splits for IAM and Esposalles

We provide the detailed splits used for IAM in Table 6 and ESPOSALLES in Table 7. For IAM, we use the RWTH split. For ESPOSALLES, we use the official split, with 25% of training data used for validation.

Impact of curriculum learning

We evaluate the impact of curriculum learning for the task of *HTR+NER* in Table 8 and 9. The DAN model trained with curriculum learning is pre-trained on the *HTR* task, then fine-tuned on the *HTR+NER* task. The results show that curriculum learning does not always have a positive impact on final performances.

Table 6. Statistics of the IAM dataset (RWTH split)

(b) Entities by split

	Train	Validation	Test
Person	1,399	252	603
GPE	731	38	129
Organization	825	39	100
NORG	282	19	79
Date	1,000	57	178
Cardinal	409	75	130
Work of Art	294	41	110
Time	167	24	114
FAC	126	37	71
Quantité	107	17	66
Location	124	16	41
Ordinal	104	19	38
Product	78	6	24
Percent	91	6	4
Event	61	2	15
Law	43	6	0
Language	15	0	5
Money	12	0	6

(a) Pages, lines, words, and entities by split

	Train	Validation	Test
Pages	747	116	336
Lines	6,482	976	2,915
Words	55,111	8,900	25,931
Entities	5,868	654	1,713

Table 7. Statistics of the ESPOSALLES dataset

(b) Entities by split

	Train	Validation	Test
Name	3,774	1,223	1,312
Surname	2,033	634	694
Location	3,440	1,069	1,087
Occupation	2,273	737	797
State	868	274	319
Wife	2,093	678	768
Wife's father	2,745	847	908
Wife's mother	566	188	189
Husband	4,334	1,493	1,563
Husband's father	1,838	476	518
Husband's mother	462	1401	156
Other person	350	115	136

(a) Pages, records, lines, words, and entities by split

	Train	Validation	Test
Pages	75	25	25
Records	731	267	253
Lines	2,328	742	757
Words	23,893	7,608	8,026
Entities	12,388	3,937	4,238

For POPP, we also trained the model for key-value *HTR+NER* in a random order, e.g. with named entities in a random order. Results show that DAN is also able to learn with a random reading order, although the error rates are a bit higher than when the model is trained with the correct reading order.

Table 8. Impact of curriculum learning on handwritten text recognition on IAM, ESPOSALLES, and POPP. Results are given for test sets. NER tokens are not taken into account for this evaluation.

(a) IAM (RWTH split)

Model	Task	CER (%)	WER (%)	Input
DAN	*HTR*	4.86	15.78	Line
DAN	*HTR*	**4.30**	**13.66**	Page
DAN	*HTR+NER*	5.12	**16.17**	Line
DAN curriculum	*HTR+NER*	**5.01**	16.32	Line
DAN	*HTR+NER*	4.82	14.61	Page
DAN curriculum	*HTR+NER*	**4.30**	**13.65**	Page

(b) ESPOSALLES

Method	Task	CER (%)	WER (%)	Input
DAN	*HTR*	0.54	2.13	Line
DAN	*HTR*	0.46	**1.37**	Record
DAN	*HTR*	2.77	3.58	Page
DAN	*HTR+NER*	**0.48**	**1.75**	Line
DAN curriculum	*HTR+NER*	0.64	2.02	Line
DAN	*HTR+NER*	**0.39**	**1.51**	Record
DAN curriculum	*HTR+NER*	0.89	1.97	Record
DAN	*HTR+NER*	3.61	4.23	Page
DAN curriculum	*HTR+NER*	**2.23**	**3.15**	Page

(c) POPP

Model	Task	CER (%)	WER (%)	Input
DAN	*HTR*	8.18	**18.09**	Line
DAN	*HTR+NER*	*7.83*	*24.57*	Line
DAN curriculum	*HTR+NER*	8.06	24.85	Line
DAN curriculum + random order	*HTR+NER*	9.53	27.01	Line

Table 9. Impact of curriculum learning for named entity recognition on IAM, ESPOS-ALLES, and POPP. Results are given for test sets. Evaluation results are computed using Nerval, which computes an alignment between ground truth and predicted entities.

(a) IAM (RWTH split)

Method	P (%)	R (%)	F1 (%)	Input Type
DAN	37.1	30.8	33.7	Line
DAN curriculum	33.0	23.3	27.3	Line
DAN	37.2	27.0	31.3	Page
DAN curriculum	38.2	29.1	33.1	Page

(b) ESPOSALLES

Method	Person			Category			Input Type
	P (%)	R (%)	F1 (%)	P (%)	R (%)	F1 (%)	
DAN	96.0	96.1	96.1	96.9	97.0	96.9	Line
DAN curriculum	95.6	94.0	94.8	96.3	95.5	95.9	Line
DAN	**97.9**	98.2	**98.1**	97.4	97.8	97.6	Record
DAN curriculum	97.3	97.5	97.4	96.5	97.3	96.9	Record
DAN	95.0	**98.4**	96.6	94.2	97.6	95.9	Page
DAN curriculum	96.4	97.3	96.9	95.4	97.2	96.3	Page
DAN	97.0	97.4	97.2	96.7	97.1	96.9	Record (key-value)
DAN curriculum	96.7	96.3	96.5	96.0	96.1	96.0	Record (key-value)

(c) POPP

Method	P (%)	R (%)	F1 (%)	Input type
DAN	**85.6**	**86.2**	**85.9**	Line
DAN curriculum	85.4	**86.2**	85.8	Line
DAN curriculum + random order	84.6	84.8	84.7	Line

References

1. Akbik, A., Bergmann, T., Blythe, D., Rasul, K., Schweter, S., Vollgraf, R.: FLAIR: an easy-to-use framework for state-of-the-art NLP. In: 2019 Annual Conference of the North American Chapter of the Association for Computational Linguistics (NAACL)(Demonstrations), pp. 54–59 (2019)
2. Carbonell, M., Fornés, A., Villegas, M., Lladós, J.: A neural model for text localization, transcription and named entity recognition in full pages. Pattern Recogn. Lett. **136**, 219–227 (2020). https://doi.org/10.1016/j.patrec.2020.05.001
3. Carbonell, M., Villegas, M., Fornés, A., Lladós, J.: Joint recognition of handwritten text and named entities with a neural end-to-end model. In: 2018 13th IAPR International Workshop on Document Analysis Systems (DAS), pp. 399–404. IEEE Computer Society, Los Alamitos, CA, USA, April 2018. https://doi.org/10.1109/DAS.2018.52
4. Constum, T., et al.: Recognition and information extraction in historical handwritten tables: toward understanding early 20th century Paris census. In: 15th International Workshop on Document Analysis Systems (DAS), pp. 143–157, May 2022. https://doi.org/10.1007/978-3-031-06555-2_10

5. Coquenet, D., Chatelain, C., Paquet, T.: DAN: a segmentation-free document attention network for handwritten document recognition. IEEE Trans. Pattern Anal. Mach. Intell. **45**, 1–17 (2023). https://doi.org/10.1109/TPAMI.2023.3235826

6. Coquenet, D., Chatelain, C., Paquet, T.: End-to-end handwritten paragraph text recognition using a vertical attention network. IEEE Trans. Pattern Anal. Mach. Intell. **45**, 508–524 (2023). https://doi.org/10.1109/TPAMI.2022.3144899

7. Davis, B., Morse, B., Price, B., Tensmeyer, C., Wigington, C., Morariu, V.: End-to-end Document Recognition and Understanding with Dessurt (2022). https://doi.org/10.48550/ARXIV.2203.16618

8. Fornés, A., Romero, V., Baro, A., Toledo, J., Sánchez, J.A., Vidal, E., Lladós, J.: ICDAR2017 competition on information extraction in historical handwritten records. In: 2017 14th IAPR International Conference on Document Analysis and Recognition (ICDAR), pp. 1389–1394, November 2017. https://doi.org/10.1109/ICDAR.2017.227

9. Honnibal, M., Montani, I., Van Landeghem, S., Boyd, A.: spaCy: Industrial-strength Natural Language Processing in Python (2020). https://doi.org/10.5281/zenodo.1212303

10. Kang, L., Toledo, J.I., Riba, P., Villegas, M., Fornés, A., Rusiñol, M.: Convolve, attend and spell: an attention-based sequence-to-sequence model for handwritten word recognition. In: German Conference on Pattern Recognition, pp. 459–472 (2019)

11. Kiessling, B., Tissot, R., Stokes, P., Stökl Ben Ezra, D.: eScriptorium: an open source platform for historical document analysis. In: 2019 International Conference on Document Analysis and Recognition Workshops (ICDARW) (2019). https://doi.org/10.1109/ICDARW.2019.10032

12. Lombardi, F., Marinai, S.: Deep learning for historical document analysis and recognition-a survey. J. Imaging **6**, 110 (2020)

13. Marti, U.V., Bunke, H.: The IAM-database: an English sentence database for offline handwriting recognition. Int. J. Doc. Anal. Recogn. **5**, 39–46 (2002). https://doi.org/10.1007/s100320200071

14. Miret, B., Kermorvant, C.: Nerval: a python library for named-entity recognition evaluation on noisy texts (2021). http://gitlab.com/teklia/ner/nerval

15. Monroc, C.B., Miret, B., Bonhomme, M.L., Kermorvant, C.: A comprehensive study of open-source libraries for named entity recognition on handwritten historical documents. In: Document Analysis Systems, pp. 429–444 (2022). https://doi.org/10.1007/978-3-031-06555-2_29

16. Muehlberger, G., et al.: Transforming scholarship in the archives through handwritten text recognition: Transkribus as a case study. J. Doc. **75**, 954–976 (2019). https://doi.org/10.1108/JD-07-2018-0114

17. Prasad, A., Déjean, H., Meunier, J., Weidemann, M., Michael, J., Leifert, G.: Bench-marking information extraction in semi-structured historical handwritten records. In: CoRR (2018). http://arxiv.org/abs/1807.06270

18. Puigcerver, J.: Are multidimensional recurrent layers really necessary for handwritten text recognition? In: 2017 14th IAPR International Conference on Document Analysis and Recognition (ICDAR), vol. 01, pp. 67–72 (2017). https://doi.org/10.1109/ICDAR.2017.20

19. Qi, P., Zhang, Y., Zhang, Y., Bolton, J., Manning, C.D.: Stanza: a python natural language processing toolkit for many human languages. In: 58th Annual Meeting of the Association for Computational Linguistics: System Demonstrations, pp. 101–108, January 2020. https://doi.org/10.18653/v1/2020.acl-demos.14

20. Romero, V., et al.: The ESPOSALLES database: an ancient marriage license corpus for off-line handwriting recognition. Pattern Recogn. **46**, 1658–1669 (2013). https://doi.org/10.1016/j.patcog.2012.11.024

21. Rouhou, A.C., Dhiaf, M., Kessentini, Y., Salem, S.B.: Transformer-based approach for joint handwriting and named entity recognition in historical document. Pattern Recogn. Lett. **155**, 128–134 (2022). https://doi.org/10.1016/j.patrec.2021.11.010

22. Rowtula, V., Krishnan, P., Jawahar, C.V.: POS tagging and named entity recognition on handwritten documents. In: Proceedings of the 15th International Conference on Natural Language Processing (2018)

23. Tarride, S., Lemaitre, A., Coüasnon, B., Tardivel, S.: A comparative study of information extraction strategies using an attention-based neural network. In: Document Analysis Systems, pp. 644–658 (2022). https://doi.org/10.1007/978-3-031-06555-2_43

24. Tarridea, S., et al.: Large-scale genealogical information extraction from handwritten Quebec parish records. Int. J. Document Anal. Recogn. (2023)

25. Toledo, J.I., Carbonell, M., Fornés, A., Lladós, J.: Information extraction from historical handwritten document images with a context-aware neural model. Pattern Recogn. **86**, 27–36 (2019). https://doi.org/10.1016/j.patcog.2018.08.020

26. Tüselmann, O., Wolf, F., Fink, G.A.: Are end-to-end systems really necessary for ner on handwritten document images? In: Document Analysis and Recognition - ICDAR 2021, pp. 808–822 (2021). https://doi.org/10.1007/978-3-030-86331-9_52

27. Vidal, E., et al.: The Carabela project and manuscript collection: large-scale probabilistic indexing and content-based classification. In: In proceedings of the 17th International Conference on Frontiers in Handwriting Recognition (ICFHR 2020) (2020)

28. Yousef, M., Bishop, T.: OrigamiNet: weakly-supervised, segmentation-free, one-step, full page text recognition by learning to unfold. In: IEEE Conference on Computer Vision and Pattern Recognition (CVPR), pp. 14698–14707, June 2020

Information Extraction from Documents: Question Answering Vs Token Classification in Real-World Setups

Laurent Lam[(✉)], Pirashanth Ratnamogan, Joël Tang, William Vanhuffel, and Fabien Caspani

BNP Paribas, Paris, France
{laurent.lam,pirashanth.ratnamogan,joel.tang,william.vanhuffel,
fabien.caspani}@bnpparibas.com

Abstract. Research in Document Intelligence and especially in Document Key Information Extraction (DocKIE) has been mainly solved as Token Classification problem. Recent breakthroughs in both natural language processing (NLP) and computer vision helped building document-focused pre-training methods, leveraging a multimodal understanding of the document text, layout and image modalities.

However, these breakthroughs also led to the emergence of a new DocKIE subtask of extractive document Question Answering (DocQA), as part of the Machine Reading Comprehension (MRC) research field.

In this work, we compare the Question Answering approach with the classical token classification approach for document key information extraction. We designed experiments to benchmark five different experimental setups : raw performances, robustness to noisy environment, capacity to extract long entities, fine-tuning speed on Few-Shot Learning and finally Zero-Shot Learning.

Our research showed that when dealing with clean and relatively short entities, it is still best to use token classification-based approach, while the QA approach could be a good alternative for noisy environment or long entities use-cases.

Keywords: Document Key-Information Extraction · Machine Reading Comprehension · Named Entity Recognition · Token Classification · Document Question Answering

1 Introduction

Document understanding is a key research area with a growing industrial interest. Many businesses manually process thousands of documents for recurrent tasks. As part of document understanding, information extraction is a complex task, targeting to extract key structured information from unstructured documents. It is an important but complex process as documents can take several shapes (contracts, invoices, reports ...) with their various inherent challenges (long documents, complex layouts, tables ...)

G. A. Fink et al. (Eds.): ICDAR 2023, LNCS 14188, pp. 205–220, 2023.
https://doi.org/10.1007/978-3-031-41679-8_12

Recently, multi-modal approaches combining natural language processing, computer vision and layout understanding proved to be effective in this configuration.

In the literature two standard approaches are used. The classical approach formulates the information extraction task as a token classification approach (i.e. we classify each token to belong to a specific entity or not). The second formulates the information extraction task as a span extraction task (i.e. we search for the beginning and end of a given entity), often represented as a question - answer with fixed questions.

Our study is the first one undertaking an empirical study comparing the two approaches in complex scenarios using multiple datasets. It relies on LayoutLM [23], a standard backbone commonly used in the document understanding tasks.

Our contributions are as follows:

- We propose multiple scenarios in order to emulate real-world complexity on open source datasets,
- We pursue the first study comparing token classification and question answering approaches in the information extraction task,
- We state in which setting to use one approach or the other one.

2   Problem Definition

2.1   Information Extraction from Documents

Information Extraction is a sub-task of document understanding that aims at extracting structured information from unstructured data.

Traditionally, extracting information from documents consisted in classifying each token of the text as belonging to a certain class (one per attribute/entity). The IOB (Inside-Outside-Beginning) tagging [14] introduced a B(eggining) token class declaring the start of a new entity.

Several benchmark datasets are widely used, such as FUNSD, CORD or SROIE [5,6,11]. We note that existing benchmarks are heavily biased towards short entities. For instance, clauses to be extracted from long documents are not covered by these datasets.

In our study, we will explore two different approaches to perform such Information Extraction:

- via classical Token Classification,
- via a Question Answering approach.

2.2   Machine Reading Comprehension

Machine reading comprehension is an active research field, belonging to the Natural Language Processing (NLP) field [26] and overlaps with the Question Answering task.

The QA task is a NLP task and consists in processing a given question and outputting the answer. Both the question and the answer are in natural language. More specifically, the question is free-text (prompt) and the answer is extracted as one or multiple spans from a given text (the *context*) since we consider the **Extractive QA** setup. In that case, the concatenation of the question and the context forms the input of the model and the output is one or multiple contiguous spans of the text (denoted by start and end token indexes).

3   Related Work

The Information Extraction problem was first turned into a sequence labeling task which is a conceptually simple approach and led to good results but it does not allow models to separate two consecutive entities of the same class.

When the IOB tagging [14] was introduced, it led to more accurate tagging and many variants of tagging schemes exist with variable performances over the models and the dataset typology [24].

Recent breakthroughs in both natural language processing and Computer Vision led to notable improvements in the key-information extraction (KIE) task with the emergence of various efficient document pre-training methods such as the Transformer-based LayoutLM family of models [4,22,23]. These models introduced a multimodal pre-training approach, standing out not only due to their difference in architecture but mostly from their incorporation of text, layout and image modalities into their pre-training for document image understanding and information extraction tasks. It leverages both text and layout features and incorporates them into a single framework which is why this family of models or variants are used by many research works. [2,10,15]

However the data quality remains essential and noisy datasets, missing tags or errors in labels are common in various real-life use-cases. Research work have been performed to detect or learn from such noisy settings [21,27] in terms of Token Classification.

Another approach could also be considered in order to perform Information Extraction.

The machine reading comprehension field has also known many breakthroughs [26], in particular for the question answering task. At the intersection of information extraction and question answering, a few research work focused on reframing the classical token classification problem as a MRC one [8].

Some propose to convert each attribute into a question, and to identify the answer span corresponding to the attribute value in the context [9,20]. This idea was followed by several incremental improvements, such as asking multiple questions in a single pass [16].

We note the scarcity of question answering datasets designed for data and questions typically framed as long document information extraction problems. For instance, SQuAD [13], SQuADv2 [12] and Natural Questions [7] all propose questions related to text comprehension, but a few include classes to be directly extracted from the text. The CUAD dataset [3] a comprehensive QA dataset on legal documents which are long by their nature: some questions ask to retrieve contract dates, parties names or non-competing clauses. Such entities to be extracted can be short (person names, dates) or very long (clauses to be retrieved in a long contract).

Li et al. [8] studied this particular topic by focusing on the adaptation of the specific NER into a MRC task.

Therefore, our research work introduces a new review of both Token Classification and machine reading comprehension approaches, beyond the NER adaptation previously presented. In such field of Document Information Retrieval, we experiment with various settings from standard benchmarks, noisy environment, long entities, to few-shot or zero-shot learning setups, an extensive comparison that has not been done so far in the literature.

All our experiments will be using a LayoutLM model backbone [23] since it represents a standard baseline for Information Retrieval. It is also easy to set up towards a token classification task or a question answering task.

4   Experimental Setups

Open-source datasets usually contain clean labels and data which facilitates the benchmark of various models. However, in the industry, the data quality does not necessarily reach such standards and can present various difficulties for the model to actually learn to extract the correct entities.

In our work, we will focus on 4 different experimental settings.

4.1   Noisy Environments

Annotating documents and creating training data is an expensive and essential part for building a supervised model.

As part of the information extraction task, annotating without any human errors is extremely costly. For example, the annotation of 510 contracts in the legal CUAD dataset is valued at $2 million [3].

Hence, it is common to assume a certain level of noise in the dataset. For example, the original version of FUNSD contains too many annotation errors [18].

In the information extraction task, the noise may be as follows:

- some annotations may be missing,
- some texts may be incorrectly annotated (i.e. "Barack Obam" instead of "Barack Obama"),
- an entity can be partially annotated (i.e. "Obama" instead of "Barack Obama").

Therefore, it is essential to be able to create robust and performing models even if noise is present in the dataset.

4.2 Long Entities and Long Documents

Extracting long entities represents a common challenge for information extraction and QA tasks. Often, they are found on long documents like contracts and other legal-binding documents (often composed of multiple dozens of pages). In this case, one entity frequently corresponds to multiple sentences.

The first issue is when the number of input tokens in the document exceeds the maximum capacity of the model. Transformer architectures, which are the state-of-the-art for the defined tasks, cannot exceed a fixed size (pretrained positional embeddings size is often set at 512 for common architectures such as BERT [1]). This is mainly because of their quadratic complexity with respect to the number of input tokens ($O(n^2)$).

The common fallback is to divide the text into chunks such that they can properly fit in the model. However, this is an unsatisfying approach when answers are expected to be long entities, as the probability that the answer is overlapping on at least two chunks is significant.

Another common approach is to alleviate the memory consumption issue, often by truncating the attention matrix format, by using diverse inductive biases. BigBird [25] proposes a sampling methodology to choose the tokens used for self-attention at any position. Finally, general approaches to reduce memory consumption of models can be used, such as gradient checkpointing, at the expense of increased time complexity.

For Token Classification, long entities are also a challenge when defining the task. Models are typically trained to classify the tokens given the IOB tagging scheme [14]. As entities become longer, this model output becomes very sensitive to errors. For example, one common issue is when the model predicts a sequence like "BIIIIOIIIII".

While the token classification task requires the entity classification for each token, the question answering task is only trying to classify a token as the start or the end of an entity. Long entities that overlap over different chunks could therefore be treated more independently with less dependency to contiguous context in QA than token classification.

4.3 Few-Shot Learning

Few-Shot Learning consists in feeding a machine learning model with very few training data to guide and focus its future predictions, as opposed to a classical fine-tuning which require a large amount of training labeled data samples for the pre-trained model to adapt to the desired task with accuracy.

It represents a major challenge in the industry since large dataset labelling is extremely costly. Therefore, achieving great performances with only a few labeled data samples is an essential goal to reach.

4.4 Zero-Shot Learning

Zero-Shot Learning is a problem setup which consists in a model learning how to perform a task it never did before, here: classify unseen classes.

5   Datasets

In order to evaluate the different approaches and the various learning settings, multiple datasets are available. Some of them allow to benchmark different Document Information Extraction tasks, and in our work we will use FUNSD [6], SROIE [5], Kleister-NDA [17] and CUAD [3] that are key information extraction datasets respectively from forms, receipts and contracts. We will also use an internal dataset of Trade Confirmations.

In this work, we focus on the key information extraction task as a common benchmark between visually-rich document understanding and machine reading comprehension.

Table 1. Dataset statistics comparison between QA and Token Classification approaches for SROIE, FUNSD, Kleister NDA, CUAD and Trade Confirmations

Datasets	Split	Token Classification			QA
		Documents	Entities	Categories	QA Samples
SROIE	train	626	2655	4	2498
	test	346	1462		1384
FUNSD	train	149	6536	3	440
	test	50	1983		117
Kleister	train	254	2861	4	744
NDA	validation	83	928		254
CUAD	custom train	408	1849	10	787
	custom test	102	421		168
Trade	custom train	170	1970	12	1857
Confirmations	custom test	42	544		465

5.1   SROIE

The Scanned Receipts OCR and key Information Extraction (SROIE) [5] dataset was introduced at the ICDAR 2019 conference for competition. Three tasks were set up for the competition: Scanned receipt text localisation, scanned receipt optical character recognition and key information extraction from scanned receipts. In this work, SROIE will be mainly used for the different benchmarks, with entities to extract among *company, address, total* or *date*.

5.2   FUNSD

The Form Understanding in Noisy Scanned Documents (FUNSD) [6] dataset was introduced in 2019 and has been a classical benchmark in recent Document KIE research works. The different entities are labeled as *header, question* or *answer*. In this work, we will be using a revised version of this dataset [19] with cleaned annotations.

5.3   Kleister NDA

The Kleister NDA [17] dataset is composed of long formal born-digital documents of US Nondisclosure Agreements, also known as Confidentiality Agreements, with labels such as *Party, Jurisdiction, Effective Date* or *Term*.

5.4   CUAD

Contract Understanding Atticus Dataset (CUAD) [3] is a dataset introduced in 2021, with classes representing information of interest to lawyers and other legal workers when analyzing such legal documents. For instance, short labels (*contract date, parties names*, etc.) exist, just as long labels (*outsourcing agreement, outsourcing agreement*, etc.) where clauses are labeled.

5.5   Trade Confirmations

Trade confirmations are financial documents reporting the details of a completed trade. It is well structured as the document comes from a limited number of counterparties. It is composed of one-page PDFs detailing derivatives products and 13 different entities to extract (*price, volume, trade date* ...).

6   Method and Experiments

6.1   Question Answering

Extractive QA datasets are based on tuples of contexts, queries and answers (c, q, a). The answers can be found in the contexts and are associated with their respective start indices and lengths in each context.

In order to adapt the QA task as an Information Extraction task into an entity retrieval task, we turned the usual query into a generic question in natural language about the given label: **What is the $<$LABEL$>$?**.

For each label to be found in a document, we create a QA data sample composed of the document's context, the generated query and the start indices and lengths of the answers. One document could therefore appear multiple times in the QA dataset since it would be associated with different queries and answers.

This procedure allows us to convert token classification-based datasets into question answering datasets. As shown in Table 1, the number of samples for each dataset is significantly larger when converted to QA task due to the dependence on the number of entities and labels.

6.2   Common Setup

In this work, as mentioned, we will base our experiment on a unique LayoutLM [23] backbone and especially the style-based embeddings variant introduced by Oussaid et al. [10] due to its efficiency and effectiveness.

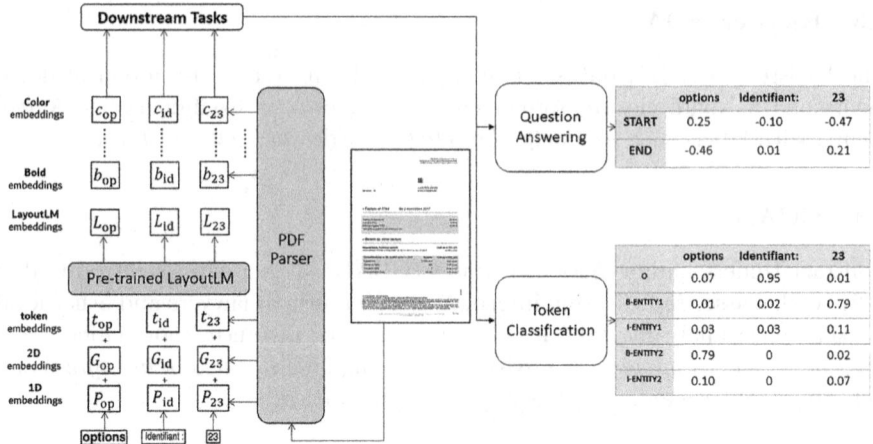

Fig. 1. Processes of LayoutLM + style-based embeddings with QA and Token Classification tasks

Based on that backbone as shown on Fig. 1, the difference lies on the last classification layer when trying to classify the different entities for each dataset in token classification whereas the last layer in QA classifies start and end.

For the token classification approach, we use a batch size of 2 and an Adam optimizer with an initial learning rate of $2 * 10^{-5}$. Then throughout the training, if there is no increase in the validation F1-score after 10 epochs, the learning rate is divided by 2. We stop fine-tuning when the learning rate goes below 10^{-7}.

Regarding the experimental setup for fine-tuning, for the QA approach, we use a batch size of 4 and an Adam optimizer with an initial learning rate of $2 * 10^{-5}$. The experiments also ran with gradient accumulation steps of 2. We run the fine-tuning with early stopping on F1-score as well.

6.3   Vanilla Setting

In order to compare both approaches, we started by benchmarking the performances of LayoutLM as token classification and LayoutLM as QA on the datasets.

As represented in Table 2, the question answering approach's performances are not consistent across datasets since it performed very well on SROIE whereas it achieved only poor results on Kleister NDA and did not succeed in capturing some of the labels on FUNSD dataset. On the other hand, the token classification approach achieved from acceptable to very good results on each dataset.

The poor results from QA approach on FUNSD could be explained by the lack of specific semantics from its labels (*header, question* or *answer*). Indeed this approach will first process the question query via its semantics but the *question* or *answer* is not necessarily a refined or accurate query for the model, especially when taken independently.

Table 2. Performance of Token Classification and QA LayoutLM information extraction models on SROIE, FUNSD, Kleister NDA and Trade Confirmations

Model	LayoutLM$_{base}$ TC			LayoutLM$_{base}$ QA		
	F1	Precision	Recall	F1	Precision	Recall
SROIE	**95.79**	95.36	96.24	93.78	93.78	93.78
FUNSD	**86.57**	87.23	86.03	6.84	55.90	4.03
Kleister NDA	**76.81**	77.60	76.40	32.58	53.71	24.03
Trade Confirmations	**97.14**	97.24	97.06	87.88	94.45	83.09

Another drawback of the QA approach that may lead to poorer performances on datasets with multiple tags per label is its multi-responses handling for a given label. Since the SQuAD v2 [12] dataset with the introduction of unanswerable questions, the model must also determine when no viable answer can be extracted from the context and should abstain from answering. But when there are multiple (k) expected answers, taking the k-top answer outputs while filtering out the considered non-viable answers from the model does not necessarily lead to the wanted tags. As in SQuAD v2, the sum of the logits of the *start* and *end* tokens must then be positive in order to be considered.

On the internal Trade Confirmations dataset, the question answering dataset achieved correct performances with 87.88 weighted average F1-score whereas the token classification model achieved almost perfect predictions with 97.14 weighted average F1-score.

6.4 Noisy Environments

In order to generate a noisy dataset from the SROIE dataset, we randomly subsampled from the tags of each document using different sub-sampling ratios. That setting allows us to recreate an environment where datasets are not fully but only partially tagged.

We decided to take as sub-sampling tag ratios *10%, 30%, 50%, 70%, 90%*. This sub-sampling procedure is applied on the training and validation sets but the test set remains the same fully annotated split. It can also happen that all tags from a document are discarded using this procedure, the document is then also discarded from the training or validation set.

We also performed the sub-sampling and training using 5 different random seeds in order to assess the stability and statistical significance of the results.

As shown in Fig. 2, we notice that the Token Classification approach is seriously affected by this noisy environment with more than 50% of average decrease in weighted F1-score when using only 10% of the tags (40.66%) whereas for the QA approach, the F1-score only decreased on average of 9% achieving on average 83.09% of weighted F1-score.

We notice that the results from the token classification approach have a greater variance than the results from QA.

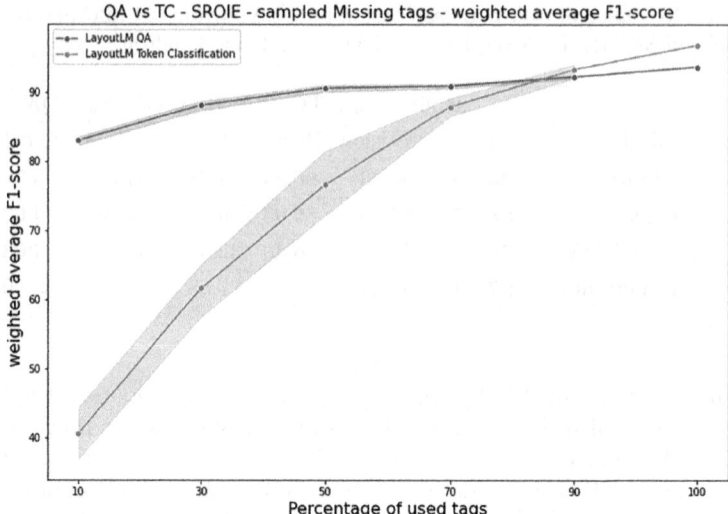

Fig. 2. SROIE: QA vs Token Classification performances using partial random missing tags

This is due to the fact that the QA approach is only fed with positive training samples compared to the token classification approach that is provided all token classifications, even where there are no tags. It therefore learns not to predict certain tokens or values.

The performances of the QA approach is therefore more consistent using different ratios of tag sub-sampling than the token classification approach.

However the token classification approach reaches slightly higher performances than the QA approach with the fully annotated dataset as stated in the Vanilla setting (Table 3).

6.5 Long Entities and Long Documents

In order to assess the performances of both approaches on Long Entities Information Extraction, we based our experiment on the CUAD dataset using only its top 10 labels with the longest entities on average.

This dataset represents a particular business use-case when dealing with contracts or more generally Legal documents, in which entities to extract can be several sentences or even entire paragraphs.

As shown in Table 4, the token classification approach has a lot of difficulties at capturing long entities whereas the QA approach can predict some of them correctly, even though the F1-score is still low (weighted average F1-score at 37.51).

This could be explained by the fact that in the regular token classification approach, for long documents, since the **O** tag is predominant among all token

Table 3. CUAD: Chosen labels & tag character lengths

Label	Count	Average #characters	Median #characters
Affiliated License Licensor	96	576	485
Source Code Escrow	59	500	257
Affiliate License Licensee	96	559	475
Post Termination Services	378	461	370
Non Transferable License	237	412	344
Uncapped Liability	131	456	406
Irrevocable Or Perpetual License	128	594	510
Most Favored Nation	30	455	370
License Grant	639	431	355
Competitive Restriction Exception	96	433	361

Table 4. Performance of Token Classification and QA LayoutLM information extraction models on CUAD dataset

CUAD	Token Classification			QA		
Model	F1	Recall	Precision	F1	Recall	Precision
Affiliated License Licensor	0.00	0.00	0.00	0.00	0.00	0.00
Source Code Escrow	0.00	0.00	0.00	**20.00**	14.29	33.33
Affiliate License Licensee	0.00	0.00	0.00	**38.46**	26.32	71.42
Post Termination Services	0.00	0.00	0.00	**39.64**	30.56	56.41
Non Transferable License	0.00	0.00	0.00	**42.55**	32.79	60.61
Uncapped Liability	0.00	0.00	0.00	**39.22**	27.78	66.67
Irrevocable Or Perpetual License	0.00	0.00	0.00	**43.63**	32.43	66.67
Most Favored Nation	0.00	0.00	0.00	**40.00**	25.00	99.99
License Grant	0.00	0.00	0.00	**36.76**	24.64	72.34
Competitive Restriction Exception	0.00	0.00	0.00	**34.04**	25.81	50.00
Weighted average	0.00	0.00	0.00	**37.51**	27.08	63.06

classification entities so the model may be focused on reducing that loss overall and therefore predicting no tags.

6.6   Few-Shot Learning

In order to assess Few-Shot Learning capabilities, we used the SROIE dataset and we randomly sub-sampled from the documents using different sub-sampling ratios. That setting allows us to recreate an environment where the number of labeled documents is largely reduced.

We decided to take as sub-sampling document ratios *10%, 30%, 50%, 70%, 90%*. This sub-sampling procedure is applied on the training and validation sets but the test set remains the same.

Similarly to the noisy environment setting, we also performed the sub-sampling and training using 5 different random seeds in order to assess the stability and statistical significance of the results.

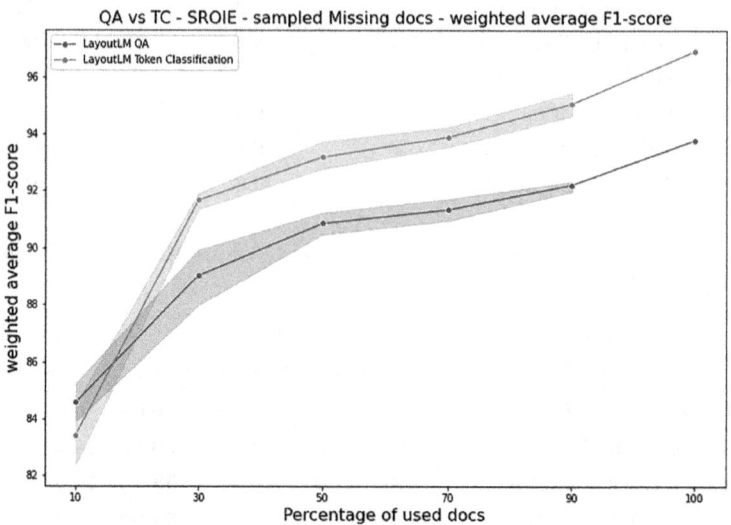

Fig. 3. SROIE: QA vs Token Classification performances using partial random missing documents

As shown in the performance graph using different sub-sampling ratios in Fig. 3, both approaches are impacted when trained with fewer documents, but still achieving correct results. It is particularly true with the token classification approach which decreased on average from 95.07% to 83.43% of weighted F1-score when using only 10% of documents. On the other hand, the performance drop for the QA approach was only from 92.76% to 84.58%.

The QA approach seems then slightly more robust to the lack of documents compared to the token classification approach, even though the token classification approach outperforms the QA approach when provided with sufficient documents.

Once again, we notice that the results from the token classification approach have a greater variance than the results from QA.

6.7   Zero-Shot Learning

The token classification approach cannot classify unseen classes. We could only assess the Zero-Shot capabilities of the QA approach.

Indeed as mentioned above, we assessed the QA approach using the prepared QA datasets, with potentially generated questions with unseen labels. For example, the model did not necessarily learn entities such as *Header, Party* or *Market* during pre-training.

Using the base pre-trained LayoutLM or after fine-tuning on other datasets, we tested the MRC performances on the different datasets (Tables 5, 6 and 7).

Table 5. Zero-Shot performances of QA LayoutLM information extraction models pre-trained or fine-tuned on different datasets and evaluated on SROIE

SROIE	Pre-trained			FUNSD			Kleister NDA		
Model	F1	Precision	Recall	F1	Precision	Recall	F1	Precision	Recall
Company	0.00	0.00	0.00	**5.33**	3.08	20.00	0.00	0.00	0.00
Address	0.00	0.00	0.00	0.00	0.00	0.00	0.00	0.00	0.00
Total	0.00	0.00	0.00	**1.54**	0.86	6.82	0.00	0.00	0.00
Date	54.63	86.39	39.95	**79.67**	70.70	91.25	22.32	12.59	98.11
weighted average	15.43	24.40	11.28	**24.17**	20.93	32.27	6.30	3.56	27.72

Table 6. Zero-Shot performances of QA LayoutLM information extraction models pre-trained or fine-tuned on different datasets and evaluated on FUNSD

FUNSD	Pre-trained			SROIE			Kleister NDA		
Model	F1	Precision	Recall	F1	Precision	Recall	F1	Precision	Recall
Header	0.00	0.00	0.00	**0.38**	3.00	50.00	0.00	0.00	0.00
Question	0.00	0.00	0.00	0.00	0.00	0.00	0.00	0.00	0.00
Answer	0.00	0.00	0.00	0.00	0.00	0.00	0.00	0.00	0.00
weighted average	0.00	0.00	0.00	**0.19**	0.10	2.52	0.00	0.00	0.00

Table 7. Zero-Shot performances of QA LayoutLM information extraction models pre-trained or fine-tuned on different datasets and evaluated on Kleister NDA

Kleister NDA	Pre-trained			SROIE			FUNSD		
Model	F1	Precision	Recall	F1	Precision	Recall	F1	Precision	Recall
Party	0.00	0.00	0.00	0.00	0.00	0.00	**1.62**	0.82	57.14
Jurisdiction	0.00	0.00	0.00	0.00	0.00	0.00	**0.68**	0.36	5.00
Effective Date	24.71	43.75	17.21	49.26	40.98	61.73	**64.96**	62.30	67.73
Term	0.00	0.00	0.00	0.00	0.00	0.00	0.00	0.00	0.00
weighted average	3.25	5.75	2.26	6.48	5.39	8.11	**9.59**	8.73	40.32

As shown in the different benchmarks above, the QA approach has only little to none Zero-Shot capabilities depending on the labels it looks for. Especially *Date* labels seems to be easier to detect since it might have seen similar Date labels during fine-tuning. Also it is a relatively standard label since its format has few variants. Otherwise, predicting totally new labels is not very effective, without any prompt tuning or customization.

7 Conclusion

In this paper, we presented an extensive benchmark between the token classification approach and the question answering approach in the Document Information Retrieval context. We showed that the token classification approach is best suited with clean and relatively short entities in terms of both effectiveness and efficiency, whereas the question answering approach can actually be used for this task and represent an credible and robust alternative especially in the cases of noisy datasets or with long entities to extract.

This represents a first study in the literature opposing those approaches on this Key Information Extraction task, that can be completed with further works such as question prompt-tuning or conditional search for the question answering approach, additional experimental settings with nested entities, or even more model architectures that could include image features as shown in the document visual question answering task.

References

1. Devlin, J., Chang, M.W., Lee, K., Toutanova, K.: BERT: pre-training of deep bidirectional transformers for language understanding. In: Proceedings of the 2019 Conference of the North American Chapter of the Association for Computational Linguistics: Human Language Technologies, Volume 1 (Long and Short Papers) (2019). http://aclanthology.org/N19-1423.pdf
2. Douzon, T., Duffner, S., Garcia, C., Espinas, J.: Improving information extraction on business documents with specific pre-training tasks. In: Document Analysis Systems 15th IAPR International Workshop, DAS 2022. LNCS, vol. 13237, pp. 111–125. Springer International Publishing, La Rochelle, France (2022). https://doi.org/10.1007/978-3-031-06555-2_8, http://hal.archives-ouvertes.fr/hal-03676134
3. Hendrycks, D., Burns, C., Chen, A., Ball, S.: CUAD: an expert-annotated NLP dataset for legal contract review. arXiv preprint arXiv:2103.06268 (2021)
4. Huang, Y., Lv, T., Cui, L., Lu, Y., Wei, F.: Layoutlmv3: pre-training for document AI with unified text and image masking. In: Proceedings of the 30th ACM International Conference on Multimedia, pp. 4083–4091. MM 2022, Association for Computing Machinery, New York, NY, USA (2022). https://doi.org/10.1145/3503161.3548112
5. Huang, Z., et al.: ICDAR 2019 competition on scanned receipt OCR and information extraction, pp. 1516–1520 (2019). http://arxiv.org/pdf/2103.10213.pdf
6. Jaume, G., Kemal Ekenel, H., Thiran, J.P.: FUNSD: a dataset for form understanding in noisy scanned documents. In: 2019 International Conference on Document Analysis and Recognition Workshops (ICDARW), vol. 2, pp. 1–6 (2019). https://doi.org/10.1109/ICDARW.2019.10029
7. Kwiatkowski, T., et al.: Natural questions: a benchmark for question answering research. Trans. Assoc. Comput. Linguist. 7, 452–466 (2019). https://doi.org/10.1162/tacl_a_00276, http://aclanthology.org/Q19-1026
8. Li, X., Feng, J., Meng, Y., Han, Q., Wu, F., Li, J.: A unified MRC framework for named entity recognition (2019). https://doi.org/10.48550/ARXIV.1910.11476, http://arxiv.org/abs/1910.11476

9. Mengge, X., Yu, B., Zhang, Z., Liu, T., Zhang, Y., Wang, B.: Coarse-to-fine pre-training for named entity recognition. In: Proceedings of the 2020 Conference on Empirical Methods in Natural Language Processing (EMNLP), pp. 6345–6354. Association for Computational Linguistics, November 2020. https://doi.org/10.18653/v1/2020.emnlp-main.514, http://aclanthology.org/2020.emnlp-main.514

10. Oussaid, I., Vanhuffel, W., Ratnamogan, P., Hajaiej, M., Mathey, A., Gilles, T.: Information extraction from visually rich documents with font style embeddings. CoRR abs/2111.04045 (2021). http://arxiv.org/abs/2111.04045

11. Park, S., et al.: Cord: a consolidated receipt dataset for post-OCR parsing (2019)

12. Rajpurkar, P., Jia, R., Liang, P.: Know what you don't know: Unanswerable questions for SQuAD. In: Proceedings of the 56th Annual Meeting of the Association for Computational Linguistics (Volume 2: Short Papers), pp. 784–789. Association for Computational Linguistics, Melbourne, Australia, July 2018. https://doi.org/10.18653/v1/P18-2124, http://aclanthology.org/P18-2124

13. Rajpurkar, P., Zhang, J., Lopyrev, K., Liang, P.: SQuAD: 100,000+ questions for machine comprehension of text. In: Proceedings of the 2016 Conference on Empirical Methods in Natural Language Processing, pp. 2383–2392. Association for Computational Linguistics, Austin, Texas, November 2016. https://doi.org/10.18653/v1/D16-1264, http://aclanthology.org/D16-1264

14. Ramshaw, L.A., Marcus, M.P.: Text chunking using transformation-based learning (1995). https://doi.org/10.48550/ARXIV.CMP-LG/9505040, http://arxiv.org/abs/cmp-lg/9505040

15. Saha, A., Finegan-Dollak, C., Verma, A.: Position masking for improved layout-aware document understanding. CoRR abs/2109.00442 (2021). http://arxiv.org/abs/2109.00442

16. Shrimal, A., Jain, A., Mehta, K., Yenigalla, P.: NER-MQMRC: formulating named entity recognition as multi question machine reading comprehension. In: Proceedings of the 2022 Conference of the North American Chapter of the Association for Computational Linguistics: Human Language Technologies: Industry Track, pp. 230–238. Association for Computational Linguistics, Hybrid: Seattle, Washington + Online, July 2022. https://doi.org/10.18653/v1/2022.naacl-industry.26

17. Stanislawek, T., et al.: Kleister: key information extraction datasets involving long documents with complex layouts. CoRR abs/2105.05796 (2021). http://arxiv.org/abs/2105.05796

18. Vu, H.M., Nguyen, D.T.N.: Revising FUNSD dataset for key-value detection in document images. arXiv preprint arXiv:2010.05322 (2020)

19. Vu, H.M., Nguyen, D.T.: Revising FUNSD dataset for key-value detection in document images. CoRR abs/2010.05322 (2020). http://arxiv.org/abs/2010.05322

20. Wang, Q., et al.: Learning to extract attribute value from product via question answering: a multi-task approach. In: Proceedings of the 26th ACM SIGKDD International Conference on Knowledge Discovery and Data Mining, pp. 47–55. KDD 2020, Association for Computing Machinery, New York, NY, USA (2020). https://doi.org/10.1145/3394486.3403047

21. Wang, W.C., Mueller, J.: Detecting label errors in token classification data (2022). https://doi.org/10.48550/ARXIV.2210.03920, http://arxiv.org/abs/2210.03920

22. Xu, Y., et al.: LayoutLMv2: multi-modal pre-training for visually-rich document understanding. In: Proceedings of the 59th Annual Meeting of the Association for Computational Linguistics and the 11th International Joint Conference on Natural Language Processing (Volume 1: Long Papers), pp. 2579–2591. Association for Computational Linguistics, August 2021. https://doi.org/10.18653/v1/2021.acl-long.201

23. Xu, Y., Li, M., Cui, L., Huang, S., Wei, F., Zhou, M.: LayoutLM: pre-training of text and layout for document image understanding. In: Proceedings of the 26th ACM SIGKDD International Conference on Knowledge Discovery and Data Mining, pp. 1192–1200. KDD 2020, Association for Computing Machinery, New York, NY, USA (2020). https://doi.org/10.1145/3394486.3403172

24. Zaheer, M., et al.: Big bird: transformers for longer sequences. In: Larochelle, H., Ranzato, M., Hadsell, R., Balcan, M., Lin, H. (eds.) Advances in Neural Information Processing Systems. vol. 33, pp. 17283–17297. Curran Associates, Inc. (2020). www.proceedings.neurips.cc/paper/2020/file/c8512d142a2d849725f31a9a7a361ab9-Paper.pdf

25. Zaheer, M., et al.: Big bird: transformers for longer sequences. In: Larochelle, H., Ranzato, M., Hadsell, R., Balcan, M., Lin, H. (eds.) Advances in Neural Information Processing Systems. vol. 33, pp. 17283–17297. Curran Associates, Inc. (2020). www.proceedings.neurips.cc/paper/2020/file/c8512d142a2d849725f31a9a7a361ab9-Paper.pdf

26. Zeng, C., Li, S., Li, Q., Hu, J., Hu, J.: A survey on machine reading comprehension: tasks, evaluation metrics and benchmark datasets (2020). https://doi.org/10.48550/ARXIV.2006.11880, http://arxiv.org/abs/2006.11880

27. Zhou, W., Chen, M.: Learning from noisy labels for entity-centric information extraction. In: Proceedings of the 2021 Conference on Empirical Methods in Natural Language Processing, pp. 5381–5392. Association for Computational Linguistics, Online and Punta Cana, Dominican Republic, November 2021. https://doi.org/10.18653/v1/2021.emnlp-main.437

Document Analysis and Recognition 2: Camera + Scene Text

ViSA: Visual and Semantic Alignment for Robust Scene Text Recognition

Zhenru Pan[1(✉)], Zhilong Ji[1], Xiao Liu[2], Jinfeng Bai[1], and Cheng-Lin Liu[3,4]

[1] Tomorrow Advancing Life, Beijing, China
panzhenru0797@gmail.com, jizhilong@tal.com
[2] Tencent, Beijing, China
[3] State Key Laboratory of Multimodal Artificial Intelligence Systems, Institute of Automation of Chinese Academy of Sciences, Beijing 100190, China
liucl@nlpr.ia.ac.cn
[4] School of Artificial Intelligence, University of Chinese Academy of Sciences, Beijing 100049, China

Abstract. Unsupervised domain adaptation is studied to meet the challenge of scene text recognition in diverse scenarios. Existing methods try to align source and target domain at the image or character level. However, these approaches are somewhat coarse-grained as they involve irrelevant information or ignore category attributes. To address the above issues, we propose a novel **Vi**sual and **S**emantic **A**lignment (ViSA) method to reduce the domain shifts in the high-frequency domain and category space. Specifically, the high-frequency domain alignment extracts the high-frequency components of global visual features, which allows the domain classifier to focus on the text-relevant features. Furthermore, the category space alignment is introduced to align character features at the category level. In the category space alignment, cross-domain contrastive learning and prototype-consistency matching are adopted to minimize the distance between domains. ViSA is flexible to be plugged into various existing recognizers. In addition, ViSA can be conducted in training and dropped in evaluation, which means no impact on inference speed. Adequate experiments verify the superiority of each module of ViSA, and our method achieves state-of-the-art results on several benchmarks, such as SVT, IC13 and IC15.

Keywords: Scene Text Recognition · Unsupervised Domain Adaptation · High-Frequency Domain · Contrastive Learning · Prototype Matching

1 Introduction

Understanding texts in scene images has wide applications in human life, and poses the problem of scene text recognition (STR). A practical challenge in STR is that the recognizer is expected to possess the strong generalization ability to diverse scenarios. As yet, the majority of existing methods [6,27,32,35,36,42,45,51] attempted to improve the model structures and

© The Author(s), under exclusive license to Springer Nature Switzerland AG 2023
G. A. Fink et al. (Eds.): ICDAR 2023, LNCS 14188, pp. 223–242, 2023.
https://doi.org/10.1007/978-3-031-41679-8_13

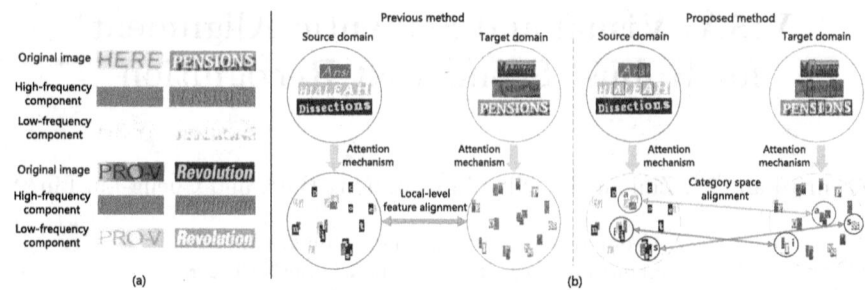

Fig. 1. (a) Text images, and the high-frequency and low-frequency components of them by wavelet transformation. (b) The intuition behind previous local-level feature alignment through category-agnostic marginal distribution and our category space alignment through domain-category joint distribution. For a better view, only the alignments of categories "a", "i", "s" are indicated.

trained them with massive synthetic data to acquire robust recognizers. However, the domain shift caused by the discrepancy between synthetic and real data inevitably leads to the performance degradation of recognizers in real scenarios. A reasonable solution to the above problem is unsupervised domain adaptation. In unsupervised domain adaptation [8,15,24,28,34], the labeled synthetic data and unlabeled real data are respectively regarded as the source and target domain, and the domain shift can be reduced employing feature alignment [46,61]. Recently, some efforts on unsupervised domain adaptation [54,57,58] have been made in STR. Nevertheless, these methods have two obvious drawbacks: 1) Text images carry mixed information such as text content, writing style and background style, some of which are irrelevant or even detrimental to domain adaptation. 2) The utilization of category information is inadequate, because the local-level feature alignment only judges whether a character belongs to the source or target domain but neglects the category attribute, as shown in Fig. 1(b). As a consequence, character features of different categories are prone to mismatches across domains, which can be summarized as a negative transfer phenomenon.

In this paper, to address the aforementioned issues, we propose a novel Visual and Semantic Alignment (ViSA) method which takes a closer insight into the domain shift problems in STR from visual and semantic perspectives. On the one hand, we introduce the high-frequency domain alignment to alleviate the domain shift at the visual level. On the other hand, we employ the category space alignment to reduce the character-level discrepancy between domains at the semantic level.

At the visual level, the low-frequency information in text images generally reflects the overall appearance of images, while the high-frequency information tends to describe the texture details of texts, as shown in Fig. 1(a). Based on this phenomenon, we argue that focusing on high-frequency information in domain adaptation can facilitate scene text recognition. Hence, we introduce a High-

Frequency Feature Extraction Module (HFFEM) which decomposes the visual features of overall images into multi-frequency components by the wavelet transformation and extracts the high-frequency components by the high-pass filter. Then, the high-frequency domain alignment is executed by a domain classifier.

At the semantic level, the attention-based LSTM decomposes the global visual features to character-level semantic features. To reduce the character-level domain shift more fine-grained, we align the domain-category joint distributions via two novel alignment strategies: cross-domain contrastive learning and prototype-consistency matching. To maintain categorial awareness of character features during domain adaptation, cross-domain contrastive learning is adopted to minimize the distance between character features of the same category and maximize the distance between character features of different categories. Beyond that, the prototype-consistency matching closes the paired prototypes of source and target domain together to further reduce the discrepancy of feature distributions of the same category. With the assistance of the above two strategies, the category-level character feature alignment can be accomplished.

In the end, under the action of high-frequency domain alignment and category space alignment, the strong generalization ability of the recognizer can be guaranteed. Moreover, since ViSA is decoupled from the recognition model, it can be flexibly plugged into various STR models to improve their performance.

The main contributions of this paper are summarized as follows.

- A Visual and Semantic Alignment method (ViSA) is proposed to alleviate the domain shift problems in STR from the perspectives of the high-frequency domain and category space in domain adaptation.
- A high-frequency domain alignment module is introduced, employing a HFFEM to extract high-frequency features from text images to make the domain classifier focus on the textual features.
- A category space alignment module is introduced, employing cross-domain contrastive learning and prototype-consistency matching to reduce domain shift more fine-grained via domain-category joint distribution.
- Experimental results on six public benchmarks, including IIIT5K, SVT, IC13, IC15, SVTP and CUTE, demonstrate the effectiveness and superiority of ViSA, and our method achieves state-of-the-art performance on several benchmarks.

2   Related Work

2.1   Scene Text Recognition

As a widely applied visual task, STR has developed rapidly driven by deep learning technology in recent years. STR methods can be generally divided into two categories: semantic context-free methods and semantic context-aware methods.

In semantic context-free methods [22,44,49,50,56,59], Connectionist-Temporal Classification (CTC)-based methods [11,14,35] are a kind of mainstream method. As a classical CTC-based model, CRNN [35] combined both

CNN and RNN to acquire visual features with sequential properties, then obtained the classification results of character sequence through a CTC decoder.

Except for normal visual modules, semantic context-aware methods [21, 25, 30, 31, 41, 42, 45, 47, 52, 53, 55] invariably introduce additional semantic modules to improve the capacity of recognizers. In attention-based methods, an encoder-decoder structure with attention mechanisms was regarded as an implicit language model, through which the dependencies between characters can be extracted. Aster [36], a typical attention-based method, added a rectification module to handle text with distortions and irregular layouts. Based on Aster, Mou et al. [27] introduced a pluggable super-resolution branch to alleviate the degradation of recognition performance caused by low-quality text images.

With Transformer [38] in full swing for natural language processing (NLP) tasks, Transformer blocks are gradually incorporated into STR methods [2, 7, 43, 51]. To overcome the inefficiency caused by the time-dependent RNN methods, Yu et al. [51] adopted a multiplex parallel transmission way to capture global semantic information. ABINet [7] exploited a Transformer-based language model to capture extra linguistic information to facilitate the process of STR. Xie et al. [43] used the corner point to assist Transformer for artistic text recognition.

2.2 Unsupervised Domain Adaptation in Scene Text Recognition

In recent years, the study of unsupervised domain adaptation [15, 23, 24, 28, 34] has been gradually deepened in various high-level visual tasks, especially in digit classification. Compared to digit classification, the STR task is more complicated and challenging, in which the sequential character information is contained in images under diverse appearances and backgrounds. Zhan et al. [54] proposed a geometry-aware domain adaptation network to reduce domain shifts in spatial and appearance spaces. Focusing on the sequential attribute of text images, Zhang et al. [58] exploited a sequence-to-sequence domain adaptation method to align distributions between source and target domain from a character-level perspective. On the basis of [58], a more remarkable domain adaptation performance was achieved by adversarial learning [57]. However, the aforementioned works merely applied domain adaptation measures coarsely in the process of STR. In this paper, we take a closer insight into domain shift problems in STR from the perspective of the high-frequency domain and character-category space.

3   Proposed Method

The proposed method is aimed to better utilize the visual and semantic information of real scene text in unsupervised domain adaptation. To achieve this goal, we probe into the process of domain adaptation in-depth from the high-frequency domain and category space.

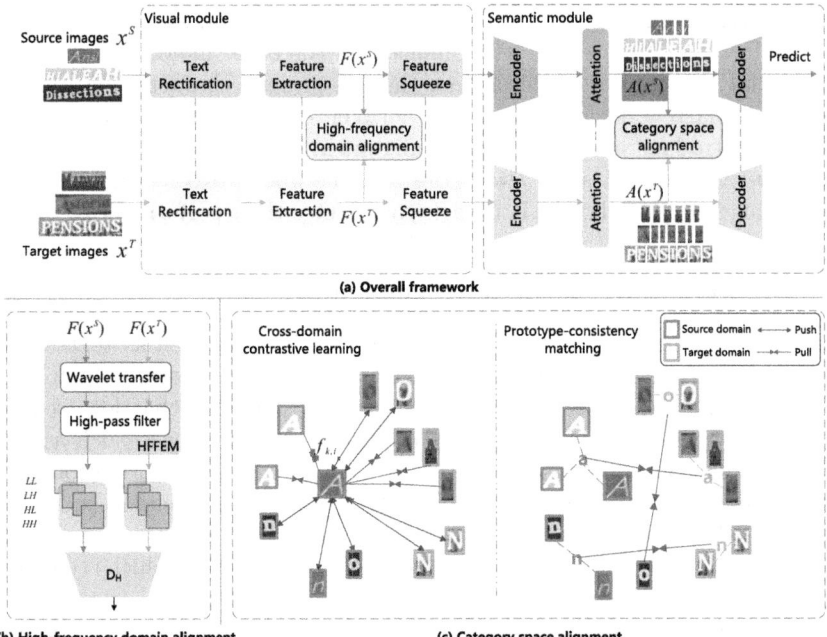

Fig. 2. (a) The overall framework of ViSA, where the model parameters of the source and target domains are shared; (b) The high-frequency domain alignment which applies an HFFEM to extract high-frequency features and aligns the distributions between domains through a domain classifier; (c) The category space alignment, including cross-domain contrastive learning and prototype-consistency matching, aligns character-level features at the category level.

3.1 Overall Framework

We adopt a CNN-LSTM pipeline as the base recognizer which can be divided into two parts: the visual and semantic modules, as shown in Fig. 2(a). Specifically, the visual module is composed of text rectification, feature extraction, and feature squeeze. The text rectification, which is derived from Aster [36], is applied to rectify irregular text images. The feature extraction shared a similar ResNet structure as Aster does, except that the down-sampling convolutional layers in the last three blocks of ResNet are removed. We employ the feature squeeze module [27] to maintain more spatial information. In this module, a 1×1 convolutional layer is used for channel reduction, and a reshape layer is adopted to generate the one-dimension vectors for subsequent recognition. The semantic module is an encoder-decoder structure, in which the visual features are transferred to sequential semantic features. The encoder is a two-layer Bidirectional LSTM (BiLSTM), and the decoder is a two-layer attention-based LSTM. The detailed information of the base recognizer is described in Sect. 4.1.

On the one hand, as mentioned in [57], the visual feature alignment generally reduces the domain gap from the aspect of image appearance. Considering the property of text image, we conduct a novel high-frequency domain alignment to further promote the performance. On the other hand, as a text image representing a sequence of characters, the fine-grained character feature alignment could narrow down the distance between domains at the semantic level. Thus, we introduce a category space alignment to enhance the robustness of character-level domain adaptation. Besides, the above two feature alignments can be plugged into other STR models to enhance their performance.

3.2   High-Frequency Domain Alignment

There contains multifaceted information in text images, such as text content, writing style, background style, etc. If we analyze this information from the frequency domain perspective, a pattern can be concluded that low-frequency information generally depicts the overall appearance of images (such as image style, illumination and image quality) and high-frequency information invariably highlights the texture details of texts (such as text content and writing style). Based on this finding, we claim that the high-frequency information has a stronger impact on the domain adaptation of text recognition, which is confirmed in subsequent experiments in Sect. 4.4.

To extract the high-frequency information, we introduce a High-Frequency Feature Extraction Module (HFFEM) which includes two operations: the wavelet transformation and high-pass filtration, as shown in Fig. 2(b). Specifically, a classical wavelet transformation method, the Haar wavelet transform [9], is employed to decompose the visual features, output from the feature extraction module, into four components: LL, LH, HL and HH. Among these multi-frequency components, LL indicates the low-frequency information, and LH, HL, HH indicate the high-frequency information. In the high-pass filtration, LL is abandoned, while LH, HL, HH are gathered as the high-frequency features. To encourage the domain invariant of high-frequency features, we constrain the high-frequency features of the source and target domain with a domain classifier D_H.

Formally, we assume data from the source domain as $D^S = \{(x^S, y^S)\}$ and data from the target domain as $D^T = \{x^T\}$, where x^S, x^T and y^S respectively denote the text images of the source and target domain and the text label corresponding to x^S. Here, we represent the 2D visual features of the source and target domain as $F(x^S)$ and $F(x^T)$, respectively. To narrow down the distance of high-frequency feature distributions between domains, the adversarial learning [10] is adopted. The high-frequency adversarial loss can be written as:

$$
\begin{aligned}
L_{hf} = &- E_{x^S \sim D^S}\left[\log(1 - D_H(H_{wavelet}(F(x^S))))\right] \\
&- E_{x^T \sim D^T}\left[\log(D_H(H_{wavelet}(F(x^T))))\right]
\end{aligned}
\tag{1}
$$

where $H_{wavelet}$ denotes the wavelet transformation and high-pass filtration operations in HFFEM. In this manner, the recognition capacity on target data can be immediately promoted.

3.3   Category Space Alignment

Apart from the high-frequency domain alignment at the visual level, we focus on the character feature alignment at the semantic level in this part. It is aware that the previous local-level feature alignment in Fig. 1(b) only considers whether a character feature belongs to the source or target domain but neglects the category attribute of the character. As a result, this category-agnostic alignment strategy tends to the false match between character features of different categories cross domains, which leads to the negative transfer problem. Aiming at the above problem, we introduce the cross-domain contrastive learning and prototype-consistency matching strategies to align the domain-category joint distributions between domains, instead of the category-agnostic marginal distributions.

Specifically, in the semantic module, each character region in the text image can be derived through the attention mechanism, where the character features are obtained. Formally, the character representations of the source domain can be denoted as $A(x^S) = \{f_i^S\}_{i=1}^{N_S}$, where N_S represents the number of characters in image x^S and f_i^S denotes the ith character feature. Likewise, the character representations of the target domain can be represented as $A(x^T) = \{f_i^T\}_{i=1}^{N_T}$. In the implementation, we gather the character features in a batch as a collection. For simplicity of expression, we still use $A(x^S)$ and $A(x^T)$ to represent the character feature collections of the source and target domain, as shown in Fig. 2(a). With these character features, the character feature alignment can be achieved.

Cross-Domain Contrastive Learning. Our goal is to maintain categorical awareness of character features in domain adaptation. A spontaneous thought is to increase the inter-class gap and reduce the intra-class gap of character features. In consideration of the advantage of contrastive learning [1,17,40], which minimizes the feature distance of the same category and maximizes the feature distance of different categories by a simple InfoNCE loss [5], we introduce a cross-domain contrastive learning framework to achieve the category-aware character feature alignment.

Technically, the category-aware character feature alignment needs both domain and category information. However, the target data is unlabeled. Hence, we assign the pseudo labels to character features of target data according to the maximum confidence category in predicted results. With the basic materials of domain and category information, the cross-domain contrastive learning can be implemented. Specifically, a character feature of the kth category, denoted as $f_{k,i}$, is considered as a query sample, as shown in Fig. 2(c). Corresponding to $f_{k,i}$, all character features belonging to the kth category in both source and target domain are selected as positive samples. In contrast, the remaining character features are regarded as negative samples. Note that, as the number of negative samples in a batch is adequate, the selection of negative samples is only carried out in the current batch. In the process of cross-domain contrastive learning, the character features belonging to the same category are pulled together, and the character features belonging to different categories are pushed away. The cross-domain contrastive loss can be defined as:

$$L_{con} = \sum_{i \in I} \frac{1}{N_P} \sum_{p \in P} -\log \frac{\exp(f_{k,i} \cdot f_{k,p}/\tau)}{\sum_{p \in P} \exp(f_{k,i} \cdot f_{k,p}/\tau) + \sum_{n \in N, k \neq k'} \exp(f_{k,i} \cdot f_{k',n}/\tau)}$$

(2)

where I is the set of overall character features in both source and target domain, P and N respectively denote the sets of positive and negative samples, N_P is the number of positive samples, $f_{k,p}$ represents a positive sample corresponding to $f_{k,i}$, and $f_{k',n}$ represents a negative sample of the k'th category. Besides, a temperature hyper-parameter τ is utilized in the softmax operation. With the assistance of cross-domain contrastive learning, the mismatch problem can be alleviated.

Prototype-Consistency Matching. We observed a phenomenon in our experiments that the character features of the same category from the same domain (i.e., source or target domain) are aligned well, whereas the character features of the same category across domains (i.e., source and target domains) are aligned slightly worse. To further reduce the discrepancy of feature distributions of the same category across domains, we introduce a prototype-consistency matching strategy which is based on the concept of representing category-level features by prototypes [28,37]. In general, prototype-consistency matching implicitly makes the feature distributions of each category in the source and target domain consistent by pairing all category prototypes of two domains separately, as shown in Fig. 2(c). To be specific, the feature distribution of each category can be represented by a prototype that is the average of all character features in the same category:

$$m_k = \frac{1}{N_k} \sum_{i=1}^{N_k} f_{k,i}$$

(3)

where N_k is the number of character features in the kth category. In [28], the calculation of a prototype is on account of the current batch statistics. Considering the limited number of character features in a batch, simply representing the category-level feature distribution with those samples is inadequate, which can lead to a relatively large estimation error. Hence, we adopt a momentum-like strategy to adaptively update prototypes during iterations:

$$\hat{m}_k \leftarrow \beta \hat{m}_k + (1 - \beta)\dot{m}_k$$

(4)

where $\dot{m}_k$ indicates the prototype of the kth category calculated by the current batch, $\hat{m}_k$ denotes the global prototype of the kth category, and $\beta \in [0,1)$ is a momentum coefficient. In Eq. 4, the global prototype is obtained by the moving average calculation between the historical and the current-batch prototype.

In reference to [4], we utilize MSE loss to proceed one-to-one matching of prototypes of all categories between the source and target domain:

$$L_{proto} = \sum_{k=1}^{C} ||\hat{m}_k^S - \hat{m}_k^T||_2 \tag{5}$$

where C is the number of categories, $\hat{m}_k^S$ and $\hat{m}_k^T$ indicate the global prototypes of the kth category in the source and target domain, respectively. Under the dynamic alignment of prototypes during training, the prototype-consistency matching can be implemented. Ultimately, the category-level character feature alignment is achieved with the support of both cross-domain contrastive learning and prototype-consistency matching.

3.4 Training and Inference

The training of the entire network is implemented under the combined action of supervised text recognition and unsupervised domain adaptation. For the supervised text recognition, a decoding loss is applied to optimize the recognizer with the labeled source data:

$$L_{rec} = E_{(x^S, y^S) \sim D^S} \big[-\log p(y^S | x^S) \big] \tag{6}$$

As for the domain adaptation, there include the high-frequency adversarial loss, cross-domain contrastive loss and prototype-consistency loss. The overall loss of the network is defined as follow:

$$L = L_{rec} + \lambda_{hf} L_{hf} + \lambda_{con} L_{con} + \lambda_{proto} L_{proto} \tag{7}$$

where λ_{hf}, λ_{con} and λ_{proto} represent the weighting factors. In the inference stage, after removing the high-frequency domain classifier, the adaptive model can be directly applied to target data.

4 Experiments

4.1 Experimental Setup

All experiments in this paper involve two synthetic datasets and six real datasets. The synthetic datasets are Synth90k [16] and SynthText [12], and the real datasets include IIIT5K-words (IIIT5K) [26], Street View Text (SVT) [39], ICDAR-2013 (IC13) [19], ICDAR-2015 (IC15) [18], SVT Perspective (SVTP) [29] and CUTE80 (CUTE) [33]. On account of the various versions of several datasets, we follow the protocol used in [45].

During training, the labeled Synth90k and SynthText are used as the source data. Meanwhile, the training sets of the real datasets are adopted as the unlabeled target data. Since there are no training sets in SVTP and CUTE, the unlabeled target data only contains the training sets of IIIT5K, SVT, IC13 and

IC15. After training, we evaluate the model on the test sets of all six real datasets. These test sets are strictly consistent with those of other SOTA methods. Furthermore, the target and test data do not share the same images, but belong to the same domain. In summary, we guarantee a fair comparison in the dataset as well as all other aspects.

Following the configuration of ASTER [36], the output image size of the text rectification module is 32×100. Because the down-sampled convolutional layers are removed from the last three blocks of ResNet, the output features of the feature extraction module have the dimensions as $\frac{W}{4} \times \frac{H}{4} \times C$, where W, H and C denote the width, height and channel of the input image and $W = 100, H = 32$. The feature squeeze module reduces the channel dimension of input features from 512 to 128 by a 1×1 convolutional layer and then reshapes the features to the dimensions of 25×1024. More details of the feature squeeze module can be found in [27]. As for the semantic module, the BiLSTM of the encoder has 512 hidden units and the attention-based LSTM of the decoder has 1024 hidden units. Moreover, the domain classifier is composed of three fully connected layers with the dimension of 1024. The momentum coefficient β is set to 0.9. For joint optimization, we set λ_{con} and λ_{proto} to 0.005 and 0.005, respectively. The setting of λ_{hf} follows the dynamic alignment strategy in [57], decreasing from 0.5 to 0.

The proposed ViSA is implemented with Pytorch. We primitively pre-train our recognizer on labeled source data (i.e., Synth90k and SynthText), and then train the overall model end-to-end on both labeled source data and unlabeled target data (i.e., real scene data) by the manner of unsupervised domain adaptation. Without loss of generality, no data augmentation is used during training. The training batch size is 512. The overall model is optimized with ADADELTA for the minimization of the objective function. We initialize the learning rate as 2.0, and respectively decay it to 0.2 and 0.02 after 9 and 11 epochs.

4.2   Comparisons with State-of-the-Art Methods

We compare our proposed ViSA with other outstanding methods on six benchmarks, and the results are shown in Table 1. To illustrate the contribution of our approach to domain adaptation for STR, we set a baseline by the "source-only". The "source-only" is trained on synthetic labeled source data only, where no domain adaptation is performed. It can be observed that ViSA outperforms the baseline by 0.6%, 1.4%, 1.0%, 1.8%, 2.7% and 4.9%, respectively, on these benchmarks. The above enhancement for domain adaptation is attributed to the high-frequency domain alignment and category space alignment in ViSA. It is noteworthy that there contains no training data of SVTP and CUTE in unlabeled target data (i.e., dataset 'U'), yet the accuracies of these two datasets are also significantly improved. This phenomenon indicates that ViSA has strong robustness not only for the target data but also for the unknown open data.

Furthermore, ViSA can be flexibly plugged in other recognizers to improve their performance. To verify the adaptability of ViSA, we plug ViSA into the typical language-free recognizer–Aster, as well as two typical language-based

Table 1. Comparisons with previous STR methods on benchmarks. "Avg" means the average accuracy on six benchmarks; "90K", "ST" and "R" represent Synth90k, SynthText and the labeled real data, "U" is the training sets of benchmarks as unlabeled target data ; † indicates the methods equipped with language models; * indicates the reproduced results; the best results are marked in **bold**. The second group of table denotes the results of plugging ViSA into different models.

	Method	Year	Training data	IIIT5K	SVT	IC13	IC15	SVTP	CUTE	Avg
SOTA methods	CRNN [35]	2016	90K	81.2	82.7	89.6	-	-	-	-
	ASTER [36]	2018	90K+ST	93.4	89.5	91.8	76.1	78.5	79.5	86.7
	ACE [44]	2019	90K	82.3	82.6	89.7	68.9	70.1	82.6	78.8
	SSDAN [58]	2019	90K+U	83.8	84.5	91.8	-	-	-	-
	SAR [20]	2019	90K+ST+R	95.0	91.2	94.0	78.8	86.4	89.6	89.5
	DAN [41]	2020	90K+ST	94.3	89.2	93.9	74.5	80.0	84.4	87.2
	RobustScanner [52]	2020	90K+ST	95.3	88.1	94.8	77.1	79.5	90.3	88.4
	PlugNet [27]	2020	90K+ST	94.4	92.3	95.0	82.2	84.3	85.0	90.0
	SRN [51]†	2020	90K+ST	94.8	91.5	95.5	82.7	85.1	87.8	90.4
	GA-SPIN [55]	2020	90K+ST	95.2	90.9	94.8	82.8	83.2	87.5	90.3
	ASSDA [57]	2021	90K+ST+U	88.3	88.6	93.7	78.7	-	83.3	-
	JVSR [3]	2021	90K+ST	95.2	92.2	95.5	84.0	85.7	89.7	91.1
	VisionLAN [42]	2021	90K+ST	95.8	91.7	95.7	83.7	86.0	88.5	91.2
	PREN2D [45]	2021	90K+ST	95.6	94.0	96.4	83.0	87.6	91.7	91.5
	S-GTR [13]†	2021	90K+ST	95.8	94.1	96.8	84.6	87.9	**92.3**	92.1
	ABINet [7]†	2021	90K+ST	96.2	93.5	97.4	86.0	89.3	89.2	92.7
	SGBANet [60]	2022	90K+ST	95.4	89.1	95.1	78.4	83.1	88.2	89.2
	CornerTransformer [43]	2022	90K+ST	95.9	94.6	96.4	86.3	**91.5**	92.0	92.9
	PARSeq$_A$ [2]	2022	90K+ST	**97.0**	93.6	97.0	86.5	88.9	92.2	93.2
	DiG-ViT [48]	2022	90K+ST	96.7	94.6	96.9	**87.1**	91.0	91.3	93.3
Ours	Baseline (Source-only)	-	90K+ST	94.5	92.7	96.3	81.9	86.0	83.3	90.2
	ViSA	-	90K+ST+U	95.1	94.1	97.3	83.7	88.7	88.2	91.6
	Aster*	-	90K+ST	92.7	90.1	94.7	77.0	81.9	81.2	87.4
	Aster+**ViSA**	-	90K+ST+U	93.5	92.0	95.9	80.0	86.5	84.0	89.3
	ABINet-SV†*	-	90K+ST	95.4	94.4	96.4	84.0	87.6	85.8	91.5
	ABINet-SV† +**ViSA**	-	90K+ST+U	95.9	95.4	96.4	84.9	88.8	87.5	92.2
	ABINet†*	-	90K+ST	96.4	94.6	96.6	85.8	89.3	87.2	92.6
	ABINet†+**ViSA**	-	90K+ST+U	96.6	**94.7**	**97.5**	**87.1**	91.2	89.6	**93.4**

recognizers–ABINet-SV and ABINet. As shown in the second group of Table 1, the performance of these recognizers is significantly improved. The average accuracies are increased by 1.9%, 0.7% and 0.8% for Aster, ABINet-SV and ABINet, respectively. Throughout Table 1, ABINet plugged with ViSA has state-of-the-art performance on SVT, IC13 and IC15 with the accuracies of 94.7%, 97.5% and 87.1% respectively, and it achieves the best average performance of 93.4%.

In addition, as mentioned in Sect. 3.4, ViSA is only adopted during training, and does not exist in the inference phase. Hence, ViSA does not increase any labor to the model, no matter latency or FLOPs.

4.3 Comparisons with Other Domain Adaptation Methods

The proposed ViSA is designed to alleviate the degradation of recognition performance in real scenarios, which is caused by the discrepancy between source

Table 2. Comparisons with other domain adaptation methods in STR. * denotes that the domain adaptation method is applied to the base recognizer.

Method	IIIT5K	SVT	IC13	IC15	SVTP	CUTE
SSDAN [58]	83.8	84.5	91.8	-	-	-
ASSDA [57]	88.3	88.6	93.7	78.7	-	83.3
Source-only	94.5	92.7	96.3	81.9	86.0	83.3
RevGrad [8]*	94.4	93.0	96.4	81.8	87.1	85.1
SSDAN*	94.5	93.0	96.5	82.4	87.8	86.1
ASSDA*	94.7	93.1	96.8	82.5	87.4	86.8
Ours	**95.1**	**94.1**	**97.3**	**83.7**	**88.7**	**88.2**

and target domain. To demonstrate the domain adaptation capacity of ViSA in STR, we compare it with other domain adaptation methods. In Table 2, the results of SSDAN and ASSDA are directly drawn from [57,58]. In both SSDAN and ASSDA, the local-level feature alignment in Fig. 1(b) was employed for character feature alignment. Except that, ASSDA added a global-level feature alignment for multi-granularity domain adaptation. The remaining experiments all adopt the base recognizer, described in Sect. 3.1, as the recognition model for a fair comparison.

We analyze Table 2 from the following three folds: First, the recognition performance of ViSA is boosted enormously over the original SSDAN and ASSDA; Second, the results which are obtained utilizing the same base recognizer but different domain adaptation methods of RevGrad, SSDAN and ASSDA, are inferior to ours; Third, the variant SSDAN and ASSDA by our implementation, outperform the original SSDAN and ASSDA. The above analysis states that our method alleviates the domain shift problem to a greater extent than the other three domain adaptation methods. Furthermore, ViSA is equipped with a more advanced base recognizer and has superior performance on STR.

4.4   Ablation Study

Effect of High-Frequency Domain Alignment. To illustrate the effectiveness of high-frequency domain alignment, we conduct experiments to compare our high-frequency domain alignment with low-frequency domain alignment and global-level alignment. Besides, in the experiments of high-frequency domain alignment, two feature fusion methods: concatenation and sum, are utilized for comparison. As demonstrated in Table 3, the low-frequency domain alignment shows the poorest results, which indicates that it is difficult to capture valuable information from low-frequency features only. In contrast, the results of high-frequency domain alignment are higher than that of global-level alignment. Beyond that, the sum-based feature fusion shows a more prominent effect than the concatenation-based feature fusion. Compared to the global-level feature alignment, the high-frequency domain alignment using sum-based feature fusion

Table 3. Ablation study of high-frequency domain alignment. "GLFEM" means the global-level feature alignment without HFFEM; "LFFEM" means the low-frequency domain alignment; "HFFEM-C" and "HFFEM-S" mean the high-frequency domain alignment by HH, HL, LH concatenated or summed, respectively.

Module	IIT5K	SVT	IC13	IC15	SVTP	CUTE
GLFEM	94.4	93.0	96.4	81.8	87.1	85.1
LFFEM	94.1	92.3	96.0	81.4	86.2	84.4
HFFEM-C	94.7	93.5	**96.5**	82.3	87.1	**85.4**
HFFEM-S	**94.9**	**93.8**	96.4	**82.8**	**87.6**	**85.4**

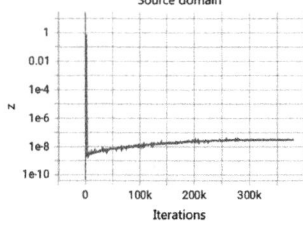 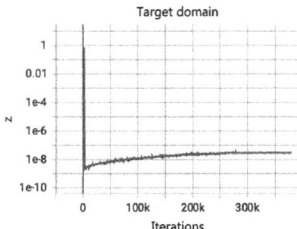

Fig. 3. The ratio of low-frequency features in the source (left) and target domain (right) during iterations. Note that, the y-axis is log scaled.

achieves gains of 0.5%, 0.8%, 1.0%, 0.5% and 0.3% in IIT5K, SVT, IC15, SVTP and CUTE, respectively.

To further explore the influence of high-frequency and low-frequency features on the process of domain adaptation, another experiment is conducted, in which the ratio of high-frequency and low-frequency features is dynamically adjusted by a gated unit during training: $z = \sigma(W_z \cdot [f_L, f_H]), f = z * f_L + (1-z) * f_H$, where f_L and f_H are the low-frequency and high-frequency features, and W_z is trainable weight. It can be observed in Fig. 3, the low-frequency features are barely used in the course of feature alignment. On account of the above experiment results, we conclude that the high-frequency features of text images are more conducive to domain adaptation.

Effect of Category Space Alignment. In this section, we perform a series of ablation experiments to verify the superiority of category space alignment. Analyzing Table 4, we observe the following phenomena: (1) The cross-domain contrastive learning outperforms the "source-only" in all datasets, and significantly improves the accuracies by 1.3%, 2.2% and 3.5% on IC15, SVTP and CUTE; (2) The prototype-consistency matching also gets improvements over the "source-only" in the majority of datasets, while displaying inferior performance to the cross-domain contrastive learning; (3) The category space alignment scheme, the combination of the above two methods, achieves the best performance.

The above experimental results demonstrate that the cross-domain contrastive learning and prototype-consistency matching in the category space

Table 4. Ablation study of category space alignment.

Method	L_{con}	L_{proto}	IIIT5K	SVT	IC13	IC15	SVTP	CUTE
Source-only	✗	✗	94.5	92.7	96.3	81.9	86.0	83.3
Contrast	✓	✗	94.7	93.5	97.0	83.2	88.2	86.8
Proto	✗	✓	94.6	93.0	96.7	82.9	87.8	86.8
Category	✓	✓	**94.9**	**93.8**	**97.2**	**83.4**	**88.5**	**87.5**

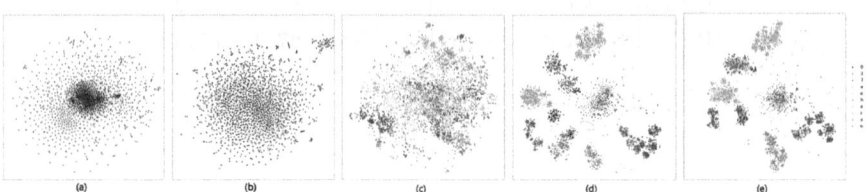

(a)   (b)   (c)   (d)   (e)

Fig. 4. The visualizations of global visual feature distributions (a) for source-only and (b) after high-frequency domain alignment; and character feature distributions (c) for source-only, (b) after cross-domain contrastive learning, and (d) after category-consistency alignment. In visual feature distributions of (a) and (b), the blue and red dots denote the source and target domain. In character feature distributions of (c), (d) and (e), the dots and triangles denote the source and target domain. The character features in the middle are hard samples. (Color figure online)

alignment jointly promote the character-level domain adaptation. In addition, we infer the reason why the prototype-consistency matching is inferior to the cross-domain contrastive learning as follow: without the "push" action in Fig. 2(c), the prototype-consistency matching leads to blurred category boundaries, which reduces the discriminability of character features. Besides, the effect of prototype-consistency matching on the cross-domain contrastive learning is further explained in Sect. 4.5.

4.5   Visualization and Analysis

Visualization on the Feature Distribution. To intuitively illustrate the domain adaptation effect of ViSA, we employ t-SNE to visualize the visual and semantic feature distributions of source and target domain, and the results are shown in Fig. 4. As a note here, for visualization, we randomly select about 8,000 samples from Synth90k and SynthText as source data and utilize all real data as target data.

It can be observed that the visual feature distributions of two domains are much closer after high-frequency domain alignment, which verifies the effectiveness of high-frequency domain alignment. Besides, to facilitate observation, we choose the character features of 10 categories from 38 categories (i.e., digits, letters, "UNKNOWN" and "EOS") to visualize the character feature distributions. In Fig. 4, the character feature distributions of the "source-only" are confusing, which implies the serious problem of misalignment between the character features of different categories. After employing cross-domain contrastive learning, the boundaries of feature distributions become clear, which indicates that the discriminability of the categories of character features is enhanced. However, there still exist domains gaps for some categories. With the assistance of prototype-consistency matching, the feature distributions of two domains converge in the majority of categories. This proves that prototype-consistency matching can further reduce the distribution discrepancy between domains.

Visualization on the Attention Results. In category space alignment, attention mechanism is utilized to extract character regions in the text image. We visualize the attention maps of text images at each time step, some attention results are shown in Fig. 5. We can perceive that the character regions are relatively accurate, which provides a reliable guarantee for character feature alignment.

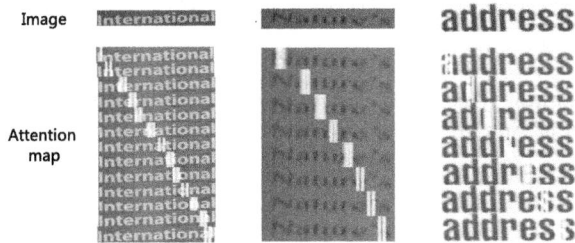

Fig. 5. The attention results of text images.

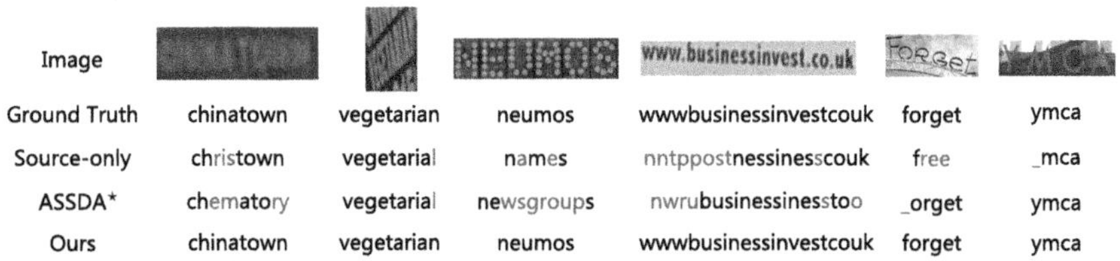

Image						
Ground Truth	chinatown	vegetarian	neumos	wwwbusinessinvestcouk	forget	ymca
Source-only	christown	vegetarial	names	nntppostnessinesscouk	free	_mca
ASSDA*	chematory	vegetarial	newsgroups	nwrubusinessinesstoo	_orget	ymca
Ours	chinatown	vegetarian	neumos	wwwbusinessinvestcouk	forget	ymca

Fig. 6. The prediction results of different methods.

Qualitative Comparison of Different Domain Adaptation Methods. We present the prediction results of different domain adaptation methods for qualitative comparison, as shown in Fig. 6. The "source-only" shows a poor recognition performance on the displayed text images. The robustness of the adversarial domain adaptation of ASSDA is not strong enough to adapt to various real scenarios. On the contrary, ViSA is capable to recognize texts that suffered from degradation and distortion, even long texts and artistic words. For example, the word "chinatown" can be recognized, even though it is difficult for humans.

5 Conclusions

In this paper, we propose a novel Visual and Semantic Alignment method for domain adaptation in scene text recognition. The proposed ViSA reduces the domain gap at fine granularities in the high-frequency domain and category space. In high-frequency domain alignment, a HFFEM is used to extract high-frequency features from global features to make the domain classifier focus on text-relevant information. In category space alignment, cross-domain contrastive learning and prototype-consistency matching are adopted to align the character feature distributions at category level. Extensive experimental results and analysis demonstrate the effectiveness of ViSA: plugging ViSA into three representative STR models, our method achieves state-of-the-art performance in several benchmarks.

Acknowledgements. This work was supported by National Key R&D Program of China, under Grant No. 2020AAA0104500.

References

1. Aberdam, A., et al.: Sequence-to-sequence contrastive learning for text recognition. In: Proceedings of the IEEE/CVF Conference on Computer Vision and Pattern Recognition, pp. 15302–15312 (2021)
2. Bautista, D., Atienza, R.: Scene text recognition with permuted autoregressive sequence models. In: Avidan, S., Brostow, G., Cissé, M., Farinella, G.M., Hassner, T. (eds.) Computer Vision – ECCV 2022. LNCS, vol. 13688, pp. 178–196. Springer, Cham (2022). https://doi.org/10.1007/978-3-031-19815-1_11
3. Bhunia, A.K., Sain, A., Kumar, A., Ghose, S., Chowdhury, P.N., Song, Y.Z.: Joint visual semantic reasoning: Multi-stage decoder for text recognition. In: Proceedings of the IEEE/CVF International Conference on Computer Vision, pp. 14940–14949 (2021)
4. Chen, C., et al.: Progressive feature alignment for unsupervised domain adaptation. In: Proceedings of the IEEE/CVF Conference on Computer Vision and Pattern Recognition, pp. 627–636 (2019)
5. Chen, X., Fan, H., Girshick, R., He, K.: Improved baselines with momentum contrastive learning. arXiv preprint arXiv:2003.04297 (2020)
6. Chen, Y., Rohrbach, M., Yan, Z., Shuicheng, Y., Feng, J., Kalantidis, Y.: Graph-based global reasoning networks. In: Proceedings of the IEEE/CVF Conference on Computer Vision and Pattern Recognition, pp. 433–442 (2019)
7. Fang, S., Xie, H., Wang, Y., Mao, Z., Zhang, Y.: Read like humans: autonomous, bidirectional and iterative language modeling for scene text recognition. In: Proceedings of the IEEE/CVF Conference on Computer Vision and Pattern Recognition, pp. 7098–7107 (2021)

8. Ganin, Y., Lempitsky, V.: Unsupervised domain adaptation by backpropagation. In: International Conference on Machine Learning, pp. 1180–1189. PMLR (2015)
9. Gao, Y., Wei, F., Bao, J., Gu, S., Chen, D., Wen, F., Lian, Z.: High-fidelity and arbitrary face editing. In: Proceedings of the IEEE/CVF Conference on Computer Vision and Pattern Recognition, pp. 16115–16124 (2021)
10. Goodfellow, I., et al.: Generative adversarial nets. In: Advances in Neural Information Processing Systems, vol. 27 (2014)
11. Graves, A., Fernández, S., Gomez, F., Schmidhuber, J.: Connectionist temporal classification: labelling unsegmented sequence data with recurrent neural networks. In: Proceedings of the 23rd International Conference on Machine Learning, pp. 369–376 (2006)
12. Gupta, A., Vedaldi, A., Zisserman, A.: Synthetic data for text localisation in natural images. In: Proceedings of the IEEE Conference on Computer Vision and Pattern Recognition, pp. 2315–2324 (2016)
13. He, Y., et al.: Visual semantics allow for textual reasoning better in scene text recognition. arXiv preprint arXiv:2112.12916 (2021)
14. Hu, W., Cai, X., Hou, J., Yi, S., Lin, Z.: GTC: Guided training of CTC towards efficient and accurate scene text recognition. In: Proceedings of the AAAI Conference on Artificial Intelligence, vol. 34, pp. 11005–11012 (2020)
15. Isobe, T., et al.: Multi-target domain adaptation with collaborative consistency learning. In: Proceedings of the IEEE/CVF Conference on Computer Vision and Pattern Recognition, pp. 8187–8196 (2021)
16. Jaderberg, M., Simonyan, K., Vedaldi, A., Zisserman, A.: Synthetic data and artificial neural networks for natural scene text recognition. arXiv preprint arXiv:1406.2227 (2014)
17. Jeon, S., Hong, K., Lee, P., Lee, J., Byun, H.: Feature stylization and domain-aware contrastive learning for domain generalization. In: Proceedings of the 29th ACM International Conference on Multimedia, pp. 22–31 (2021)
18. Karatzas, D., et al.: ICDAR 2015 competition on robust reading. In: 2015 13th International Conference on Document Analysis and Recognition (ICDAR), pp. 1156–1160. IEEE (2015)
19. Karatzas, D., et al.: ICDAR 2013 robust reading competition. In: 2013 12th International Conference on Document Analysis and Recognition, pp. 1484–1493. IEEE (2013)
20. Li, H., Wang, P., Shen, C., Zhang, G.: Show, attend and read: a simple and strong baseline for irregular text recognition. In: Proceedings of the AAAI Conference on Artificial Intelligence, vol. 33, pp. 8610–8617 (2019)
21. Litman, R., Anschel, O., Tsiper, S., Litman, R., Mazor, S., Manmatha, R.: Scatter: selective context attentional scene text recognizer. In: Proceedings of the IEEE/CVF Conference on Computer Vision and Pattern Recognition, pp. 11962–11972 (2020)
22. Liu, W., Chen, C., Wong, K.Y.K., Su, Z., Han, J.: Star-net: a spatial attention residue network for scene text recognition. In: BMVC, vol. 2, p. 7 (2016)
23. Liu, Z., et al.: Open compound domain adaptation. In: Proceedings of the IEEE/CVF Conference on Computer Vision and Pattern Recognition, pp. 12406–12415 (2020)
24. Luo, Y., Zheng, L., Guan, T., Yu, J., Yang, Y.: Taking a closer look at domain shift: category-level adversaries for semantics consistent domain adaptation. In: Proceedings of the IEEE/CVF Conference on Computer Vision and Pattern Recognition, pp. 2507–2516 (2019)

25. Lyu, P., Liao, M., Yao, C., Wu, W., Bai, X.: Mask TextSpotter: an end-to-end trainable neural network for spotting text with arbitrary shapes. In: Ferrari, V., Hebert, M., Sminchisescu, C., Weiss, Y. (eds.) Computer Vision – ECCV 2018. LNCS, vol. 11218, pp. 71–88. Springer, Cham (2018). https://doi.org/10.1007/978-3-030-01264-9_5
26. Mishra, A., Alahari, K., Jawahar, C.: Scene text recognition using higher order language priors. In: BMVC-British Machine Vision Conference. BMVA (2012)
27. Mou, Y., Tan, L., Yang, H., Chen, J., Liu, L., Yan, R., Huang, Y.: PlugNet: degradation aware scene text recognition supervised by a pluggable super-resolution unit. In: Vedaldi, A., Bischof, H., Brox, T., Frahm, J.-M. (eds.) ECCV 2020. LNCS, vol. 12360, pp. 158–174. Springer, Cham (2020). https://doi.org/10.1007/978-3-030-58555-6_10
28. Pan, Y., et al.: Transferrable prototypical networks for unsupervised domain adaptation. In: Proceedings of the IEEE/CVF Conference on Computer Vision and Pattern Recognition, pp. 2239–2247 (2019)
29. Phan, T.Q., Shivakumara, P., Tian, S., Tan, C.L.: Recognizing text with perspective distortion in natural scenes. In: Proceedings of the IEEE International Conference on Computer Vision, pp. 569–576 (2013)
30. Qiao, Z., Qin, X., Zhou, Y., Yang, F., Wang, W.: Gaussian constrained attention network for scene text recognition. In: 2020 25th International Conference on Pattern Recognition (ICPR), pp. 3328–3335. IEEE (2021)
31. Qiao, Z., et al.: PimNet: a parallel, iterative and mimicking network for scene text recognition. In: Proceedings of the 29th ACM International Conference on Multimedia, pp. 2046–2055 (2021)
32. Qiao, Z., Zhou, Y., Yang, D., Zhou, Y., Wang, W.: Seed: semantics enhanced encoder-decoder framework for scene text recognition. In: Proceedings of the IEEE/CVF Conference on Computer Vision and Pattern Recognition, pp. 13528–13537 (2020)
33. Risnumawan, A., Shivakumara, P., Chan, C.S., Tan, C.L.: A robust arbitrary text detection system for natural scene images. Expert Syst. Appl. 41(18), 8027–8048 (2014)
34. Saito, K., Watanabe, K., Ushiku, Y., Harada, T.: Maximum classifier discrepancy for unsupervised domain adaptation. In: Proceedings of the IEEE Conference on Computer Vision and Pattern Recognition, pp. 3723–3732 (2018)
35. Shi, B., Bai, X., Yao, C.: An end-to-end trainable neural network for image-based sequence recognition and its application to scene text recognition. IEEE Trans. Pattern Anal. Mach. Intell. 39(11), 2298–2304 (2016)
36. Shi, B., Yang, M., Wang, X., Lyu, P., Yao, C., Bai, X.: Aster: an attentional scene text recognizer with flexible rectification. IEEE Trans. Pattern Anal. Mach. Intell. 41(9), 2035–2048 (2018)
37. Snell, J., Swersky, K., Zemel, R.: Prototypical networks for few-shot learning. In: Advances in Neural Information Processing Systems, vol. 30 (2017)
38. Vaswani, A., et al.: Attention is all you need. In: Advances in Neural Information Processing Systems, vol. 30 (2017)
39. Wang, K., Babenko, B., Belongie, S.: End-to-end scene text recognition. In: 2011 International Conference on Computer Vision, pp. 1457–1464. IEEE (2011)
40. Wang, L., et al.: Unsupervised degradation representation learning for blind super-resolution. In: Proceedings of the IEEE/CVF Conference on Computer Vision and Pattern Recognition, pp. 10581–10590 (2021)
41. Wang, T., et al.: Decoupled attention network for text recognition. In: Proceedings of the AAAI Conference on Artificial Intelligence, vol. 34, pp. 12216–12224 (2020)

42. Wang, Y., Xie, H., Fang, S., Wang, J., Zhu, S., Zhang, Y.: From two to one: a new scene text recognizer with visual language modeling network. In: Proceedings of the IEEE/CVF International Conference on Computer Vision, pp. 14194–14203 (2021)
43. Xie, X., Fu, L., Zhang, Z., Wang, Z., Bai, X.: Toward understanding WordArt: corner-guided transformer for scene text recognition. In: Avidan, S., Brostow, G., Cissé, M., Farinella, G.M., Hassner, T. (eds.) Computer Vision – ECCV 2022. LNCS, vol. 13688, pp. 303–321. Springer, Cham (2022). https://doi.org/10.1007/978-3-031-19815-1_18
44. Xie, Z., Huang, Y., Zhu, Y., Jin, L., Liu, Y., Xie, L.: Aggregation cross-entropy for sequence recognition. In: Proceedings of the IEEE/CVF Conference on Computer Vision and Pattern Recognition, pp. 6538–6547 (2019)
45. Yan, R., Peng, L., Xiao, S., Yao, G.: Primitive representation learning for scene text recognition. In: Proceedings of the IEEE/CVF Conference on Computer Vision and Pattern Recognition, pp. 284–293 (2021)
46. Yang, B., Ma, A.J., Yuen, P.C.: Domain-shared group-sparse dictionary learning for unsupervised domain adaptation. In: Thirty-Second AAAI Conference on Artificial Intelligence (2018)
47. Yang, M., et al.: Symmetry-constrained rectification network for scene text recognition. In: Proceedings of the IEEE/CVF International Conference on Computer Vision, pp. 9147–9156 (2019)
48. Yang, M., Liao, M., Lu, P., Wang, J., Zhu, S., Luo, H., Tian, Q., Bai, X.: Reading and writing: Discriminative and generative modeling for self-supervised text recognition. In: Proceedings of the 30th ACM International Conference on Multimedia, pp. 4214–4223 (2022)
49. Yao, C., Bai, X., Liu, W.: A unified framework for multioriented text detection and recognition. IEEE Trans. Image Process. **23**(11), 4737–4749 (2014)
50. Yao, C., Bai, X., Shi, B., Liu, W.: Strokelets: a learned multi-scale representation for scene text recognition. In: Proceedings of the IEEE Conference on Computer Vision and Pattern Recognition, pp. 4042–4049 (2014)
51. Yu, D., Li, X., Zhang, C., Liu, T., Han, J., Liu, J., Ding, E.: Towards accurate scene text recognition with semantic reasoning networks. In: Proceedings of the IEEE/CVF Conference on Computer Vision and Pattern Recognition, pp. 12113–12122 (2020)
52. Yue, X., Kuang, Z., Lin, C., Sun, H., Zhang, W.: RobustScanner: dynamically enhancing positional clues for robust text recognition. In: Vedaldi, A., Bischof, H., Brox, T., Frahm, J.-M. (eds.) ECCV 2020. LNCS, vol. 12364, pp. 135–151. Springer, Cham (2020). https://doi.org/10.1007/978-3-030-58529-7_9
53. Zhan, F., Lu, S.: ESIR: end-to-end scene text recognition via iterative image rectification. In: Proceedings of the IEEE/CVF Conference on Computer Vision and Pattern Recognition, pp. 2059–2068 (2019)
54. Zhan, F., Xue, C., Lu, S.: GA-DAN: geometry-aware domain adaptation network for scene text detection and recognition. In: Proceedings of the IEEE/CVF International Conference on Computer Vision, pp. 9105–9115 (2019)
55. Zhang, C., at al.: Spin: structure-preserving inner offset network for scene text recognition. arXiv preprint arXiv:2005.13117 (2020)
56. Zhang, C., Gupta, A., Zisserman, A.: Adaptive text recognition through visual matching. In: Vedaldi, A., Bischof, H., Brox, T., Frahm, J.-M. (eds.) ECCV 2020. LNCS, vol. 12361, pp. 51–67. Springer, Cham (2020). https://doi.org/10.1007/978-3-030-58517-4_4

57. Zhang, Y., Nie, S., Liang, S., Liu, W.: Robust text image recognition via adversarial sequence-to-sequence domain adaptation. IEEE Trans. Image Process. **30**, 3922–3933 (2021)
58. Zhang, Y., Nie, S., Liu, W., Xu, X., Zhang, D., Shen, H.T.: Sequence-to-sequence domain adaptation network for robust text image recognition. In: Proceedings of the IEEE/CVF Conference on Computer Vision and Pattern Recognition, pp. 2740–2749 (2019)
59. Zhang, Z., Zhang, C., Shen, W., Yao, C., Liu, W., Bai, X.: Multi-oriented text detection with fully convolutional networks. In: Proceedings of the IEEE Conference on Computer Vision and Pattern Recognition, pp. 4159–4167 (2016)
60. Zhong, D., et al.: Sgbanet: Semantic gan and balanced attention network for arbitrarily oriented scene text recognition. In: Avidan, S., Brostow, G., Cissé, M., Farinella, G.M., Hassner, T. (eds.) Computer Vision – ECCV 2022. LNCS, vol. 13688, pp. 464–480. Springer, Cham (2022). https://doi.org/10.1007/978-3-031-19815-1_27
61. Zhuo, J., Wang, S., Zhang, W., Huang, Q.: Deep unsupervised convolutional domain adaptation. In: Proceedings of the 25th ACM international conference on Multimedia, pp. 261–269 (2017)

DQ-DETR: Dynamic Queries Enhanced Detection Transformer for Arbitrary Shape Text Detection

Chixiang Ma[1(✉)], Lei Sun[1], Jiawei Wang[1,2], and Qiang Huo[1]

[1] Microsoft Research Asia, Beijing, China
chixiangma@gmail.com, wangjiawei@mail.ustc.edu.cn, qianghuo@microsoft.com
[2] Department of EEIS, University of Science and Technology of China, Hefei, China

Abstract. We propose a new Transformer-based text detection model, named Dynamic Queries enhanced DEtection TRansformer (DQ-DETR), to detect arbitrary shape text instances from images with high localization accuracy. Unlike previous Transformer-based methods which take all control points on the boundaries/center-lines of all text instances as the queries of each Transformer decoder layer, we extend the query set for each decoder layer gradually, allowing the DQ-DETR to achieve higher localization accuracy by detecting control points for each text instance progressively. Specifically, after refining the positions of existing control points from the preceding decoder layer, each decoder layer further appends a new point on each side of each center-line segment, which are input to the next decoder layer as additional queries for detecting new control points. As offsets from the new control points to the added reference points are small, their positions can be predicted more precisely, leading to higher center-line detection accuracy. Consequently, our DQ-DETR achieves state-of-the-art performance on five public text detection benchmarks, including MLT2017, Total-Text, CTW1500, ArT and DAST1500.

Keywords: Arbitrary shape text detection · Dynamic query · DETR

1 Introduction

Automatic text reading and understanding from images are playing an increasingly more important role in various visual intelligence applications, such as Robotic Process Automation (RPA), autonomous driving and OCR translation. Robust text detection as a prerequisite has attracted increasing attention in recent years. Advances in deep learning and the availability of massive data have greatly improved the accuracy and capability of existing text detection methods (e.g., [1–33]). However, arbitrary shape text detection remains challenging due to the high variations of text font, shape, scale, orientation, language,

---

This work was done when Jiawei Wang was an intern in MMI Group, Microsoft Research Asia, Beijing, China.

G. A. Fink et al. (Eds.): ICDAR 2023, LNCS 14188, pp. 243–260, 2023.
https://doi.org/10.1007/978-3-031-41679-8_14

and extremely complex backgrounds, as well as various distortions and artifacts caused by image capturing such as non-uniform illumination, low contrast, blur, and occlusion.

Recently, DEtection TRansformer (DETR) [34] has made a profound impact to object detection as latest DETR family object detection models (e.g., [35–40]) have achieved better performance than previously popular object detection frameworks, like Faster R-CNN [41], without relying on many hand-crafted components like anchor generation, rule-based training target assignment and non-maximum suppression (NMS). These models have also been introduced to solve the text detection problem and achieved remarkable results [42–46]. For example, Raisi et al. [42] first adapted DETR [34] to detect multi-oriented texts by introducing a rotated version of the GIoU loss [47]. To detect the boundaries of arbitrary shape text instances more precisely, TESTR [44] followed the two-stage Deformable DETR [35] framework to use Transformer encoder to detect an axis-aligned bounding-box for each text instance first. Then, it generated a set of control points for each bounding-box and took them as the queries of Transformer decoder to enhance their representation ability. Finally, each enhanced query embedding was input to a text/non-text classifier to predict its textness score and a regressor to refine the position of its corresponding control point, respectively. Inspired by TESTR, DPText-DETR [45] and DeepSolo [46] proposed to update the point queries on the boundaries/center-lines of all text instances iteratively between decoder layers to achieve higher accuracy. Despite the superior performance achieved on many public benchmark datasets, we observe that the quality of the proposals predicted by the Transformer encoder is not high enough for long text-lines, which affects their text/non-text classification accuracy. Moreover, we also find that the localization accuracy is still unsatisfactory. Some examples are shown in Fig. 3(a).

To address these issues, we propose a new Transformer-based text detection approach, named DQ-DETR, by introducing the concept of dynamic queries into the DETR framework. Unlike previous methods that directly initialize all queries before inputting them into the Transformer decoder, the query set in DQ-DETR is extended gradually for each decoder layer. Specifically, each decoder layer first refines the positions of already detected control points on text center-lines from the preceding decoder layer, and then appends a new point on each side of each center-line segment, which are input to the next decoder layer as additional queries for detecting new control points. As offsets from the new control points to the added reference points are small, their positions can be predicted more precisely. Consequently, our DQ-DETR achieves state-of-the-art performance on five public text detection benchmarks, including MLT2017, Total-Text, CTW1500, ArT and DAST1500.

2   Related Works

2.1   Scene Text Detection

Existing deep learning based text detection methods can be categorized into three groups: bottom-up methods, segmentation-based methods and regression-based methods.

Bottom-Up Methods usually use object detection models to detect text components (e.g., characters or text segments) first and then group these components into text instances. The major difficulty of these methods lies in how to robustly group detected components into words/text-lines. Earlier works, like CTPN [26] and WordSup [27], used rule-based methods to group text segments into horizontal or multi-oriented text instances, which are not robust to curved texts. Some follow-up works (e.g., [28–31]) first preformed pixel-wise text/non-text classification, text segment box regression and optionally inter-pixel linkage relationship prediction on each feature map simultaneously, then grouped text segments into arbitrary shape text instances by using the local pixel connectivity information on textness score maps and optionally inter-pixel linkage relationships. Later, DRRG [32] and ReLaText [33] leveraged GCNs to improve linkage relationship prediction accuracy with wider contextual information further. These methods tend to use only local information to perform segment-wise text/non-text classification, so they cannot reject text-like objects robustly and generate more false alarms.

Segmentation-Based Methods can be further classified into two categories: two-stage methods and one-stage methods. Two-stage methods (e.g., [14–18]) usually borrow two-stage instance segmentation frameworks like Mask R-CNN [48] to detect text region proposals first, then predict a segmentation mask and optionally extra geometric attributes for the corresponding text instance in each positive region proposal. These methods are not robust to nearby long curved text instances as in the DAST1500 dataset [29], because the detected rectangular proposals in this scenario are highly overlapped, which will cause some of them to be wrongly suppressed by the NMS algorithm and hurt the recall rate. One-stage methods (e.g., [19–25]) usually leverage FCN-based semantic segmentation frameworks to predict a pixel-level textness score map from the input image first, then use different methods to group text pixels into words/text-lines. In order to avoid merging nearby words/text-lines together or over-segmenting words/text-lines into pieces, these approaches tried to leverage other auxiliary information, e.g., link prediction [21], progressive scale expansion [22], text border prediction [23], direction field prediction [24], to enhance pixel merging performance. Despite these efforts, these methods still tend to over segment text instances with large inter-character spacing into pieces [24,25]. Segmentation-based methods require manually designed post-processing algorithms to calculate text boundaries from segmentation mask(s).

Regression-Based Methods (e.g., [1–11]) adopt box-regression based object detection frameworks to detect text instances. Earlier works, like EAST [2] and TextBoxes++ [3], only consider straight texts, which are represented by quadrilaterals. Later, many methods explored better representations for arbitrary shape texts, from polygon vertices coordinates [7–9] to parameterized contours by Bezier curve fitting [10] or Fourier tranformation [11]. To improve localization accuracy, PCR [12] proposed to progressively evolve the initial axis-aligned text bounding boxes to arbitrarily shaped text contours in a top-down manner. TextBPN [49] and TextBPN++ [13] also detected arbitrary shape texts in a coarse-to-fine manner. They first segmented text center regions as proposals, and then gradually refined the positions of points on these proposals via iterative boundary deformation. Recently, the DETR framework and its variants [34,35] have been introduced into the text detection field and significantly improved the performance of regression-based methods. Raisi et al. [42] first adapted DETR [34] to detect multi-oriented texts by introducing a rotated version of the GIoU loss [47]. Tang et al. [43] leveraged a Transformer encoder to model the relationships of a few sampled representative features, then used these enhanced feature vectors to predict the control points of Bezier curves to detect arbitrary shape texts. Based on deformable DETR [35], TESTR [44] proposed a single-encoder dual-decoder model for jointly performing curved text instance detection and character recognition. Their bounding-box guided polygon detection procedure allows the effective detection of arbitrarily shaped texts. DPText-DETR [45] and DeepSolo [46] further improved the detection accuracy of TESTR by directly using the coordinates of control points on text contours [45] or center-lines [46] as queries and dynamically updating them between decoder layers.

2.2 DETR and Its Variants

Carion et al. [34] proposed a new Transformer-based object detector, named DETR (DEtection TRansformer), which introduced the concept of object queries and set prediction loss to object detection to eliminate many manually designed components in previous object detectors like anchor generation and NMS. However, DETR has three issues: 1) Slow training convergence; 2) Unclear physical meaning of object queries; 3) Hard to leverage high-resolution feature maps due to high computational complexity. Deformable DETR [35] proposed effective techniques to address these issues: 1) Formulating queries as 2D anchor points; 2) Designing a deformable attention module that only attends to certain sampling points around a reference point to efficiently leverage multi-scale feature maps; 3) Proposing a two-stage DETR framework and an iterative bounding box refinement algorithm to further improve accuracy. Inspired by the concept of reference point, some follow-up works attempted to address the slow convergence issue by giving spatial priors to object queries. For instance, Conditional DETR [36] proposed a conditional spatial query to make each cross-attention head in each decoder layer focus on a different part of an object. Anchor DETR [37] generated object queries from 2D anchor points directly. DAB-DETR [38] proposed to directly use 4D anchor box coordinates as queries, which will be updated dynamically in each decoder layer. DN-DETR [39] found that the instability of bipartite graph matching used in set prediction loss is another reason for slow convergence and proposed a novel denoising training method to solve this problem. DINO [40] improved DN-DETR in performance and efficiency further by introducing a contrastive way for denoising training, a mixed query selection method for anchor initialization, and a look-forward-twice scheme for box prediction.

3   Methodology

3.1   Overview

As depicted in Fig. 1, our DQ-DETR based text detector consists of a CNN backbone network, a deformable Transformer encoder, a dynamic queries enhanced Transformer decoder, and a set of prediction heads for text/non-text classification and control point coordinate regression. Given an image, we follow Deformable DETR [35] to extract four multi-scale feature maps with a ResNet-50 backbone and feed them into the deformable Transformer encoder to obtain an enhanced embedding for each pixel on each feature map, based on which a prediction head is used to detect a set of text center-line proposals first, each containing 3 control points. The control points of these center-line proposals are taken as the initial queries of the DQ-DETR decoder, based on which the decoder predicts the positions of other control points on each center-line progressively. Specifically, given a set of queries output by the preceding decoder layer, each decoder layer (except the last one) enhances their embeddings first and then uses a detection head to refine the positions of their corresponding control points. Based on the refined control points, we append one more point on both sides of each detected text center-line segment. These new points are taken as additional queries of the next decoder layer for detecting new control points for each center-line. Finally, the query embeddings output by the last decoder layer are fed into a detection head to reject non-text proposals and predict the relative positions of control points on the boundary of each text instance with respect to their corresponding reference points on the detected center-line.

3.2   Text Center-Line Proposal Generation

A number of stacked encoder layers are first used to enhance the image features from the ResNet-50 backbone network. Each encoder layer contains a multi-scale and multi-head deformable self-attention module [35] and a feed-forward network (FFN). To retain positional and scale information, we add a 2D position embedding and a learnable level embedding to the feature vector of each pixel on each feature map before feeding them into the Transformer encoder. The 2D position embedding is calculated by using the sinusoidal positional encoding function [50] which takes the normalized coordinates of each pixel on each feature map as input. Feature vectors from a same feature map share a same level

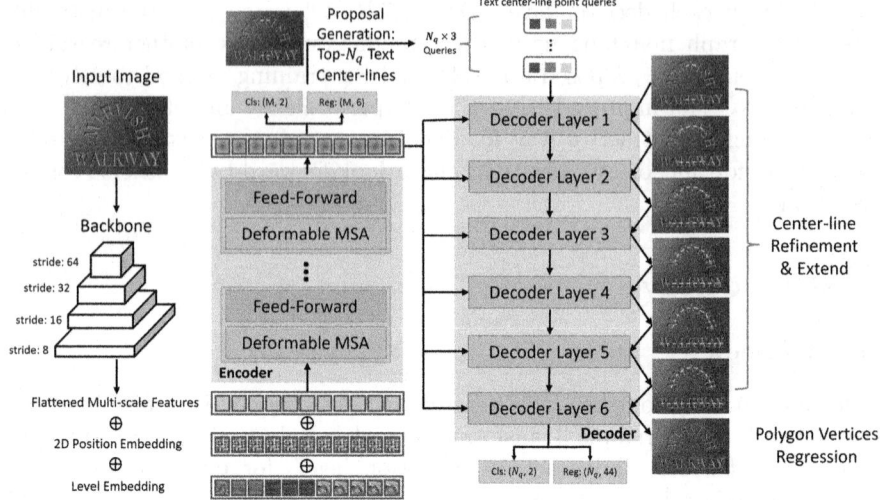

Fig. 1. An overview of the proposed DQ-DETR based text detector.

embedding. Then, each enhanced feature vector is fed into a prediction head to perform text/non-text classification and text center-line regression, respectively. The text/non-text classifier is implemented with a fully-connected (fc) layer followed by a sigmoid activation function to determine whether a pixel $\mathbf{v}_i$ on a feature map corresponds to the midpoint of a text center-line or not. Here, we use $K = 11$ ordered control points to represent a text center-line. Let $\mathbf{c}_{j,k}$ denote the k-th control point on the j-th text center-line $\mathbf{l}_j$ in an image, then $\mathbf{l}_j$ can be denoted as follows:

$$\mathbf{l}_j = \{\mathbf{c}_{j,k} | k = 1, 2, ..., K\}. \tag{1}$$

If a pixel $\mathbf{v}_i$ is classified as corresponding to a text center-line midpoint $\mathbf{c}_{j,mid}$ ($mid = \frac{K+1}{2}$), we will use a regressor to predict the offsets from $\mathbf{v}_i$ to $\mathbf{c}_{j,mid}$ and its two neighbors $\mathbf{c}_{j,mid-1}$ and $\mathbf{c}_{j,mid+1}$ simultaneously, so each text center-line proposal contains 3 control points initially. The regressor is implemented with a 3-layer MLP whose output channel dimension is 6. Finally, the control points of N_q top scored center-line proposals are used to construct the initial query set for the following DQ-DETR decoder.

3.3   DQ-DETR Decoder for Text Center-Line Extension

Query Initialization. The first decoder layer in DQ-DETR takes the control points on all selected center-line proposals as queries. Let $\mathbf{q}_{j,k}^0$ denote the initial embedding of the k-th control point on the j-th selected center-line proposal. $\mathbf{q}_{j,k}^0$ is initialized as follows:

$$\mathbf{q}_{j,k}^0 = \mathbf{ce}_{j,k} + \mathbf{pe}_{j,k}, \tag{2}$$

where $\mathbf{ce}_{j,k}$ is a learnable content embedding and $\mathbf{pe}_{j,k}$ is a positional embedding, which is calculated by using the sinusoidal positional encoding function with the normalized coordinates of the control point as input. In this way, we can initialize $3N_q$ queries from the N_q center-line proposals.

Query Embedding Enhancement. The queries input to a decoder layer form a tensor $\mathbf{Q}$ with the shape of $N_q \times N_p \times D$, where N_p is the number of already detected control points on each text center-line proposal before the current decoder layer and D is the dimension of each query embedding. Here, each decoder layer is composed of a factorized self-attention (SA) module [51], a deformable cross-attention module [35] (CA) and an FFN. In the factorized self-attention module, instead of conducting self-attention among all queries, it first conducts intra-line self-attention and then inter-line self-attention. Denote the j-th center-line proposal input to the l-th decoder layer as an ordered point set $\{\mathbf{p}_{j,k}^{l-1} | k = start, ..., mid, ..., end\}$, where $start = mid - \frac{N_p-1}{2}$, $mid = \frac{K+1}{2}$, $end = mid + \frac{N_p-1}{2}$, and denote the input embedding of the k-th control point on the j-th center-line proposal, i.e., $\mathbf{p}_{j,k}^{l-1}$, as $\mathbf{q}_{j,k}^{l-1}$. Then, the intra-line self-attention operator f_{intra} and inter-line self-attention operator f_{inter} are formulated as follows:

$$f_{intra}(\mathbf{q}_{j,*}^{l-1}) = \left[\bar{\mathbf{q}}_{j,start}^{l-1}, ..., \bar{\mathbf{q}}_{j,mid}^{l-1}, ..., \bar{\mathbf{q}}_{j,end}^{l-1}\right]$$
$$= SA(\mathbf{q}_{j,start}^{l-1}, ..., \mathbf{q}_{j,mid}^{l-1}, ..., \mathbf{q}_{j,end}^{l-1}),$$
$$f_{inter}(\bar{\mathbf{q}}_{*,k}^{l-1}) = \left[\hat{\mathbf{q}}_{1,k}^{l-1}, \hat{\mathbf{q}}_{2,k}^{l-1}, ..., \hat{\mathbf{q}}_{N_q,k}^{l-1}\right]$$
$$= SA(\bar{\mathbf{q}}_{1,k}^{l-1}, \bar{\mathbf{q}}_{2,k}^{l-1}, ..., \bar{\mathbf{q}}_{N_q,k}^{l-1}),$$
$$j \in \{1, 2, ..., N_q\}, \quad k \in \{start, ..., mid, ..., end\}, \tag{3}$$

where $\hat{\mathbf{q}}$ is the enhanced embedding of each query output by the factorized self-attention module. Compared with vanilla self-attention, factorized self-attention can reduce the computational complexity from $O(N_q^2 N_p^2 D)$ to $O(N_q N_p^2 D + N_q^2 N_p D)$. The updated queries are further sent into the deformable cross-attention module [35] to aggregate multi-scale image features from the Transformer encoder to enhance their embeddings.

Dynamic Query Generation. The updated query embeddings from each decoder layer (except the last one) will first be fed into a regressor to refine the locations of current reference points. Given a query embedding $\mathbf{q}_{j,k}^l$ output from the l-th decoder layer, we denote its current position in the input image as $\mathbf{p}_{j,k}^{l-1}$ and its refined position as $\mathbf{p}_{j,k}^l$, respectively. $\mathbf{p}_{j,k}^l$ is calculated as follows:

$$\mathbf{p}_{j,k}^l = \left(\sigma\left(\sigma^{-1}(px_{j,k}^{l-1}) + \Delta px_{j,k}^l\right), \sigma\left(\sigma^{-1}(py_{j,k}^{l-1}) + \Delta py_{j,k}^l\right)\right), \tag{4}$$

where $(px_{j,k}^{l-1}, py_{j,k}^{l-1})$ is the normalized coordinates of $\mathbf{p}_{j,k}^{l-1}$, $\sigma(\cdot)$ is the sigmoid function and $(\Delta px_{j,k}^l, \Delta py_{j,k}^l)$ is the predicted offset by the regressor, which is implemented by a 3-layer MLP. After control point position refinement, we add one more point on each end of each text center-line proposal. Specifically, denote a refined center-line proposal as an ordered point set $\{\mathbf{p}_{j,k}^l | k = start, ..., mid, ...end\}$, we will insert a new point $\mathbf{p}_{j,start-1}^l$ before the first point and append a new point $\mathbf{p}_{j,end+1}^l$ after the last point. The locations of the newly added points will be simply initialized as $\mathbf{p}_{j,start-1}^l = \mathbf{p}_{j,start}^l$ and $\mathbf{p}_{j,end+1}^l = \mathbf{p}_{j,end}^l$. With these extended points, we can dynamically generate $2N_q$ new queries following Eq. 2 and concatenate them with the existing query tensor. In this way, the center-lines can be extended progressively and refined iteratively by the following decoder layers to achieve higher localization accuracy. As depicted in Fig. 1, we use the first 5 decoder layers to generate a complete center-line with $K = 11$ points. The above-mentioned regressor is shared by these 5 decoder layers.

3.4   Final Prediction Head

The output query embeddings $\mathbf{Q} \in \mathbb{R}^{N_q \times K \times D}$ from the last decoder layer will be fed into a prediction head with two parallel FFNs for text/non-text classification and text-line polygon vertices regression, respectively. Specifically, the text/non-text classifier is implemented by an fc layer followed by a sigmoid activation function. It takes the mean of query embeddings from a same text center-line as input and predict whether there is a text-line. If so, a regressor will take each query embedding as input and predict the offsets from the reference point to the corresponding control points on the top boundary and bottom boundary of a text-line, respectively. Here, the regressor is implemented by a 3-layer MLP with an output channel dimension of 4.

4   Optimization

4.1   Bipartite Matching

We adopt the Hungarian algorithm [52] to solve the bipartite matching problem to find the optimal matching between the predictions $\{P\}$ and the ground-truths $\{G\}$. Specifically, we need to find an injective function $\sigma : \{G\} \mapsto \{P\}$ that minimizes the matching cost as follows:

$$\underset{\sigma}{\arg\min} \sum_{i=1}^{|G|} \mathcal{C}(G^{(i)}, P^{(\sigma(i))}), \tag{5}$$

where $|\cdot|$ denotes the number of elements in a set. Furthermore, $\mathcal{C}(G^{(i)}, P^{(\sigma(i))})$ is defined as follows:

$$\mathcal{C}(G^{(i)}, P^{(\sigma(i))}) = \lambda_{cls} FL'(s^{(\sigma(i))}) + \lambda_{reg} \mathcal{L}_{reg}(\hat{\mathbf{b}}^{(i)}, \mathbf{b}^{(\sigma(i))}), \tag{6}$$

where λ_{cls} and λ_{reg} are hyper-parameters to balance different tasks, and $s^{(\sigma(i))}$ is the classification score for the $\sigma(i)$-th predicted box/center-line. FL' is derived from the focal loss [53], and is defined as follows:

$$FL'(s) = -\alpha(1-s)^\gamma log(s) + (1-\alpha)s^\gamma log(1-s). \tag{7}$$

$\mathcal{L}_{reg}$ is the cost for coordinates regression. To deal with the order ambiguity problem in predicting the sequence of control points, inspired by [49], we introduce an order-insensitive point matching method when calculating the regression cost. Specifically, we first enumerate all the potential point orders for each ground-truth control point set, and calculate the L1 loss between these GT point sequences with the predicted one. Then, we take the minimum one as the final cost. Denote the ground-truth control point set as $\hat{\mathbf{b}} = \{\hat{\mathbf{p}}_i | i = 1, 2, ..., N\}$ and the predicted control point set as $\mathbf{b} = \{\mathbf{p}_i | i = 1, 2, ..., N\}$, then the regression cost between $\hat{\mathbf{b}}$ and $\mathbf{b}$ is defined as follows:

$$\mathcal{L}_{reg}(\hat{\mathbf{b}}, \mathbf{b}) = \min_\pi L_1(\{\hat{\mathbf{p}}_\pi\}, \{\mathbf{p}_i\}), \tag{8}$$

where $L_1(\cdot, \cdot)$ is the L1 loss between two point sequences, and π is the permutation of the point sequence. We use the same bipartite matching scheme to match the predictions from both encoder and decoder with the ground-truths.

4.2   Training Losses

Classification Loss. We adopt the focal loss for the classification task. Specifically, for the i-th predicted box/center-line, whose classification score is s_i, the classification loss can be calculated as follows:

$$\mathcal{L}_{cls}^{(i)} = -\mathbb{1}_{\{i \in \mathrm{Im}(\sigma)\}} \alpha(1-s_i)^\gamma log(s_i) - \mathbb{1}_{\{i \notin \mathrm{Im}(\sigma)\}}(1-\alpha)(s_i)^\gamma log(1-s_i), \tag{9}$$

where $\mathbb{1}$ is the indicator function and $\mathrm{Im}(\sigma)$ is the image of the mapping σ.

Regression Loss. The regression loss is calculated between the matched ground-truth box/center-line and the predicted box/center-line. For the i-th predicted box/center-line, the corresponding regression loss can be defined as follows:

$$\mathcal{L}_{reg}^{(i)} = \mathbb{1}_{\{i \in \mathrm{Im}(\sigma)\}} \mathcal{L}_{reg}(\hat{\mathbf{b}}^{\sigma^{-1}(i)}, \mathbf{b}^{(i)}), \tag{10}$$

where $\mathcal{L}_{reg}(\cdot, \cdot)$ is defined as in Eq. 8.

Total Loss. The model is trained in an end-to-end manner, and we also add deep supervision in each intermediate decoder layer. Therefore, the total loss is the summation of the losses from the encoder $\mathcal{L}_{enc}$ and each decoder layer $\mathcal{L}_{dec}^l$:

$$\mathcal{L} = \mathcal{L}_{enc} + \sum_l \mathcal{L}_{dec}^l, \tag{11}$$

where $\mathcal{L}_{enc}$ and $\mathcal{L}_{dec}^l$ share the same formulation, which is defined as follows:

$$\lambda_{cls} \sum_i \mathcal{L}_{cls}^{(i)} + \lambda_{reg} \sum_i \mathcal{L}_{reg}^{(i)}. \tag{12}$$

5   Experiments

5.1   Datasets

We conduct experiments on five representative scene text detection benchmarks to evaluate our DQ-DETR based text detector, including MLT2017 [54], Total-Text [55], CTW1500 [7], ArT [56] and DAST1500 [29].

MLT2017 [54] is proposed for multilingual and multi-oriented text detection. It contains $7,200$ training images, $1,800$ validation images and $9,000$ testing images. Text instances are labeled with quadrilaterals.

Total-Text [55] contains $1,255$ training images and 300 testing images. Text instances are labeled in word-level with polygons.

CTW1500 [7] contains $1,000$ training images and 500 testing images. Text instances are annotated in text-line level with 14-point polygons.

ArT [56] is a large scale arbitrary shape scene text detection dataset, which contains $5,603$ training images and $4,563$ testing images. Text instances in this dataset are labeled in word level with polygons.

DAST1500 [29] is proposed for dense and long curved text detection. It contains $1,038$ training images and 500 testing images. Text instances are labeled in text-line level with polygons.

SynthText150K [10] is a synthetic dataset and consists of $94,723$ images with multi-oriented texts and $54,327$ images with curved texts. We use the word-level annotations to pre-train our text detection models.

5.2   Implementation Details

We use ResNet-50 as the backbone network, which is pre-trained on the ImageNet-1K dataset [57]. By default, the number of Transformer encoder and decoder layers are both set to 6. The number of attention head and sampling points in each deformable attention module are set to 8 and 4, respectively. The number of selected center-line proposals N_q is set to 300 in all experiments. We implement our model based on Detectron2[1] and conduct experiments on a workstation with 8 Nvidia V100 GPUs. During training, the batch size is set to 16 in all experiments. AdamW [58] is used as the optimizer and the weight decay is $1e^{-4}$. The loss weights λ_{cls} and λ_{reg} are set to 1.0 and 5.0, respectively. The hyper-parameters α and γ in the focal loss are set to 0.25 and 2.0. We first pre-train our model on a mixture of SynthText150K, MLT2017 and Total-Text for 350K iterations. The base learning rate is $1e^{-4}$ and is divided by 10 at 280K iterations. Then, we finetune it on each target dataset with a base learning rate $5e^{-5}$ and the learning rate is divided by 10 at 80% of the total number of iterations. The number of finetuning iterations is set according to the size of different datasets. Specifically, we finetune 100K, 50K and 20K iterations on MLT2017, ArT and other small datasets (Total-Text, CTW1500 and DAST1500), respectively. Data augmentations are used for better performance, including random

---

[1] https://github.com/facebookresearch/detectron2.

Table 1. Performance comparison on MLT2017.

Method	P(%)	R(%)	F(%)
Lyu et al. [20]	83.8	55.6	66.8
CRAFT [31]	80.6	68.2	73.9
PSENet [22]	73.8	68.2	70.7
DB [60]	83.1	67.9	74.7
DRRG [32]	75.0	61.0	67.3
Xiao et al. [61]	84.2	72.8	78.1
MOST [62]	82.0	72.0	76.7
Raisi et al. [42]	84.8	63.2	72.4
FSGNet [43]	**87.3**	73.2	79.6
DQ-DETR	85.9	**77.6**	**81.5**

geometrical distortion [59] with a probability of 0.1, randomly rotating the image by 90° with a probability of 0.2, randomly resizing both sides of an image to sizes ranging from 128 to 2,560 pixels without keeping aspect ratios, instance-aware random cropping and color jittering. In the testing phase, the longer side of each testing image in all datasets except CTW1500 is resized to be 1,600 pixels while keeping the aspect ratio. As the resolutions of raw images in CTW1500 are relatively lower, the longer side of each image in this dataset is resized to be 800 pixels.

5.3   Comparison with State-of-the-Art Methods

We compare our DQ-DETR based text detector with previous scene text detection methods on MLT2017, Total-Text, CTW1500, ArT and DAST1500. For the sake of fair comparisons, all the reported results are based on single-model and single-scale testing.

Multilingual and Multi-oriented Text Detection. We first evaluate our approach for multilingual and multi-oriented text detection on MLT2017. The quantitative results are listed in Table 1. Our approach achieves the best F1-score of 81.5%, surpassing the recent best performing method, FSGNet [43], by 1.9%. The superior performance achieved on this challenging dataset demonstrates the effectiveness of our approach.

Curved Text Detection. To validate the effectiveness of our approach on curved text detection, we further conduct experiments on three representative datasets, i.e., Total-Text, CTW1500 and ArT. The experimental results are listed in Table 2. On Total-Text and CTW1500, our approach achieves the highest F1-scores of 89.2% and 89.6%, respectively. On the larger scale dataset, ArT, our approach outperforms previous methods by a large margin, achieving the best performance of 86.1%, 78.4% and 82.1% in terms of precision, recall and F1-score, respectively.

Table 2. Performance comparison on Total-Text, CTW1500 and ArT. * indicates the results on ArT are collected from the official website.

Method	Total-Text			CTW1500			ArT		
	P(%)	R(%)	F(%)	P(%)	R(%)	F(%)	P(%)	R(%)	F(%)
TextSnake [30]	82.7	74.5	78.4	67.9	85.3	75.6	–	–	–
PAN [63]	89.3	81.0	85.0	86.4	81.2	83.7	–	–	–
CRAFT* [31]	87.6	79.9	83.6	86.0	81.1	83.5	77.2	68.9	72.9
TextFuseNet* [64]	87.5	83.2	85.3	85.8	85.0	85.4	82.6	69.4	75.4
DB [60]	87.1	82.5	84.7	86.9	80.2	83.4	–	–	–
PCR [12]	88.5	82.0	85.2	87.2	82.3	84.7	84.0	66.1	74.0
ABCNet-v2 [65]	90.2	84.1	87.0	85.6	83.8	84.7	–	–	–
I3CL [66]	89.2	83.7	86.3	87.4	84.5	85.9	82.7	71.3	76.6
TextBPN++ [13]	91.8	85.3	88.5	87.3	83.8	85.5	81.1	71.1	75.8
FSGNet [43]	90.7	85.7	88.1	88.1	82.4	85.2	–	–	–
TESTR-polygon [44]	93.4	81.4	86.9	92.0	82.6	87.1	–	–	–
SwinTextSpotter [67]	–	–	88.0	–	–	88.0	–	–	–
DPText-DETR [45]	91.8	86.4	89.0	91.7	86.2	88.8	83.0	73.7	78.1
DQ-DETR	93.5	85.2	**89.2**	90.4	88.8	**89.6**	86.1	78.4	**82.1**

Table 3. Performance comparison on DAST1500. *indicates the results are from [29].

Method	P(%)	R(%)	F(%)
SegLink* [28]	66.0	64.7	65.3
CTD+TLOC* [7]	73.8	60.8	66.6
PixelLink* [21]	74.5	75.0	74.7
SegLink++ [29]	79.6	79.2	79.4
ReLaText [33]	89.0	82.9	85.8
MAYOR [18]	87.8	85.5	86.6
DQ-DETR	**91.1**	**88.2**	**89.6**

Dense and Long Curved Text Detection. We further evaluate our approach on DAST1500, which contains a large number of dense and long curved text-lines. As the aspect ratios of text-lines in this dataset are much larger than that of text instances in other datasets, we use more points (i.e., $K = 15$) to represent each text center-line, which also requires two more decoder layers to detect all control points on each center-line. The quantitative results are shown in Table 3. Our approach achieves the new state-of-the-art performance of 91.1%, 88.2% and 89.6% in terms of precision, recall and F1-score, respectively, outperforming the most competitive method, MAYOR [18], by 3.0% in terms of F1-score. The superior performance achieved on this dataset also demonstrates the advantages of the proposed dynamic query design.

Qualitative Results. Some qualitative results of our approach on these datasets are presented in Fig. 2, from which we can observe that our approach can work robustly under various challenging conditions such as perspective distortion, occlusion, non-uniform illumination, low contrast, low resolution, extremely large aspect ratio, dense distribution and arbitrary shapes.

Fig. 2. Qualitative detection results of DQ-DETR. (a-b) are from MLT2017, (c-d) are from Total-Text, (e-f) are from CTW1500, (g-h) are from ArT, and (i-j) are from DAST1500.

5.4 Ablation Study

Effectiveness of Dynamic Queries. In this section, we investigate the influence of the proposed dynamic query design on text detection accuracy. To this end, we have implemented a baseline model without dynamic queries. In detail,

Table 4. Ablation study for the effectiveness of dynamic queries on MLT2017 and ArT.

Method	MLT2017			ArT		
	P(%)	R(%)	F(%)	P(%)	R(%)	F(%)
DQ-DETR	85.9	77.6	**81.5**	86.1	78.4	**82.1**
− dynamic query	84.3	76.3	80.1	84.6	77.7	81.0

Table 5. Ablation study for the effectiveness of dynamic queries on DAST1500.

Method	IoU@0.5			IoU@0.8			FPS
	P(%)	R(%)	F(%)	P(%)	R(%)	F(%)	
DQ-DETR	91.1	88.2	**89.6**	80.8	78.2	**79.4**	**6.1**
− dynamic query	89.7	87.7	88.7	78.0	76.3	77.1	5.7

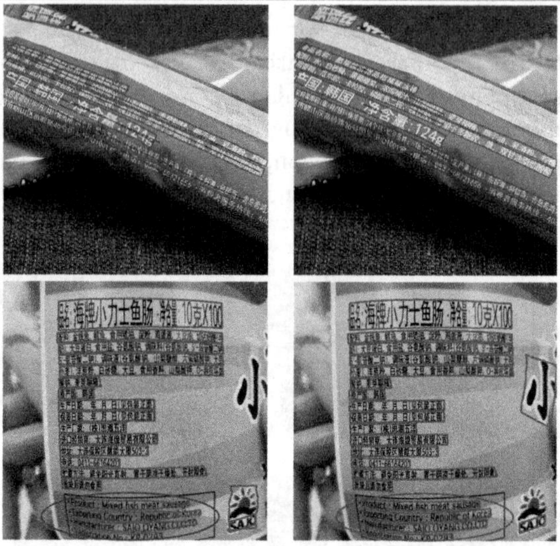

(a)   (b)

Fig. 3. Some comparison examples between (a) baseli vne model (without dynamic queries) and (b) the proposed DQ-DETR. The line segments in blue are center-line proposals from Transformer encoder and the polygons in red are the final detection results from Transformer decoder. (Color figure online)

we let the encoder of the baseline model detect all control points on each text center-line proposal directly. Then, we take all these control points as queries and feed them into the Transformer decoder to refine their positions iteratively. Therefore, the number of queries is constant in all decoder layers. We keep all the other settings unchanged to make a fair comparison. The experimental results on MLT2017 and ArT are listed in Table 4 and the results on DAST1500 are shown in Table 5. We can observe that dynamic queries can consistently improve the performance in terms of precision, recall and F1-score on different datasets. Specifically, with dynamic queries, the F1-scores on MLT2017 and ArT can be improved by 1.4% and 1.1%, respectively. On DAST1500, dynamic queries can improve the F1-score by 0.9% when evaluated at the IoU threshold of 0.5, and the performance gap will be much larger, i.e., 2.3%, when compared at a higher IoU threshold of 0.8. These results demonstrate that the proposed dynamic query design can improve the localization accuracy effectively, especially for dense and long curved text-lines. Some comparison examples can be found in Fig. 3.

6 Limitations

The proposed approach has shown promising results in previous experiments, exhibiting superior capabilities in most challenging scenarios. However, there are still some limitations to be addressed. One of the challenges is caused by text-lines with ambiguous layouts, where the directions of text-lines cannot be determined by appearance features only. Additionally, the current model cannot detect text-lines overlaid by watermarks or stamps robustly. Some failure examples are illustrated in Fig. 4. Finding effective solutions to these problems will be our future work.

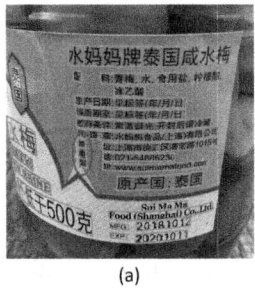 

(a)   (b)

Fig. 4. Some typical failure cases. (a) text-lines with ambiguous layout, which should be detected as three isolated characters in the red circle; (b) overlaid text-lines. (Color figure online)

7 Conclusion and Future Work

In this paper, we introduce a new Transformer-based text detection model, named DQ-DETR, by introducing the concept of dynamic queries into the DETR framework. With the help of dynamic queries, our DQ-DETR based text detector can achieve higher localization accuracy than the previous Transformer-based text detection models by detecting control points for each text instance progressively. Consequently, our DQ-DETR based text detector has achieved state-of-the-art performance on five public text detection benchmarks, namely MLT2017, Total-Text, CTW1500, ArT and DAST1500.

For future work, we will explore how to incorporate textual information into our DQ-DETR decoder to detect text-lines with ambiguous layouts. Moreover, we will explore to use different number of control points to represent text instances with different aspect ratios, which may bring further improvement for both accuracy and efficiency.

References

1. Zhong, Z., Jin, L., Huang, S.: DeepText: a new approach for text proposal generation and text detection in natural images. In: ICASSP, pp. 1208–1212 (2017)

2. Zhou, X., et al.: EAST: an efficient and accurate scene text detector. In: CVPR, pp. 5551–5560 (2017)
3. Liao, M., Shi, B., Bai, X.: TextBoxes++: a single-shot oriented scene text detector. IEEE Trans. Image Process. **27**(8), 3676–3690 (2018)
4. Gupta, A., Vedaldi, A., Zisserman, A.: Synthetic data for text localisation in natural images. In: CVPR, pp. 2315–2324 (2016)
5. Ma, J., et al.: Arbitrary-oriented scene text detection via rotation proposals. IEEE Trans. Multimedia **20**(11), 3111–3122 (2018)
6. Liu, Y., Jin, L.: Deep matching prior network: toward tighter multi-oriented text detection. In: CVPR, pp. 1962–1969 (2017)
7. Liu, Y., Jin, L., Zhang, S., Luo, C., Zhang, S.: Curved scene text detection via transverse and longitudinal sequence connection. Pattern Recogn. **90**, 337–345 (2019)
8. Wang, X., Jiang, Y., Luo, Z., Liu, C., Choi, H., Kim, S.: Arbitrary shape scene text detection with adaptive text region representation. In: CVPR, pp. 6449–6458 (2019)
9. Wang, F., Chen, Y., Wu, F., Li, X.: TextRay: contour-based geometric modeling for arbitrary-shaped scene text detection. In: ACM MM, pp. 111–119 (2020)
10. Liu, Y., Chen, H., Shen, C., He, T., Jin, L., Wang, L.: ABCNet: real-time scene text spotting with adaptive bezier-curve network. In: CVPR, pp. 9809–9818 (2020)
11. Zhu, Y., Chen, J., Liang, L., Kuang, Z., Jin, L., Zhang, W.: Fourier contour embedding for arbitrary-shaped text detection. In: CVPR, pp. 3123–3131 (2021)
12. Dai, P., Zhang, S., Zhang, H., Cao, X.: Progressive contour regression for arbitrary-shape scene text detection. In: CVPR, pp. 7393–7402 (2021)
13. Zhang, S.X., Zhu, X., Yang, C., Yin, X.C.: Arbitrary shape text detection via boundary transformer. arXiv preprint arXiv:2205.05320 (2022)
14. Lyu, P., Liao, M., Yao, C., Wu, W., Bai, X.: Mask TextSpotter: an end-to-end trainable neural network for spotting text with arbitrary shapes. In: Ferrari, V., Hebert, M., Sminchisescu, C., Weiss, Y. (eds.) Computer Vision – ECCV 2018. LNCS, vol. 11218, pp. 71–88. Springer, Cham (2018). https://doi.org/10.1007/978-3-030-01264-9_5
15. Xie, E., Zang, Y., Shao, S., Yu, G., Yao, C., Li, G.: Scene text detection with supervised pyramid context network. In: AAAI, pp. 9038–9045 (2019)
16. Zhang, C., et al.: Look more than once: an accurate detector for text of arbitrary shapes. In: CVPR, pp. 10552–10561 (2019)
17. Wang, Y., Xie, H., Zha, Z.J., Xing, M., Fu, Z., Zhang, Y.: ContourNet: taking a further step toward accurate arbitrary-shaped scene text detection. In: CVPR, pp. 11753–11762 (2020)
18. Qin, X., et al.: Mask is all you need: rethinking mask R-CNN for dense and arbitrary-shaped scene text detection. In: ACM MM, pp. 414–423 (2021)
19. Zhang, Z., Zhang, C., Shen, W., Yao, C., Liu, W., Bai, X.: Multi-oriented text detection with fully convolutional networks. In: CVPR, pp. 4159–4167 (2016)
20. Lyu, P., Yao, C., Wu, W., Yan, S., Bai, X.: Multi-oriented scene text detection via corner localization and region segmentation. In: CVPR, pp. 7553–7563 (2018)
21. Deng, D., Liu, H., Li, X., Cai, D.: PixelLink: detecting scene text via instance segmentation. In: AAAI, pp. 6773–6780 (2018)
22. Wang, W., et al.: Shape robust text detection with progressive scale expansion network. In: CVPR, pp. 9336–9345 (2019)
23. Wu, Y., Natarajan, P.: Self-organized text detection with minimal post-processing via border learning. In: ICCV, pp. 5000–5009 (2017)

24. Xu, Y., Wang, Y., Zhou, W., Wang, Y., Yang, Z., Bai, X.: TextField: learning a deep direction field for irregular scene text detection. IEEE Trans. Image Process. **28**(11), 5566–5579 (2019)
25. Xue, C., Lu, S., Zhang, W.: MSR: multi-scale shape regression for scene text detection. In: IJCAI, pp. 20–36 (2019)
26. Tian, Z., Huang, W., He, T., He, P., Qiao, Yu.: Detecting text in natural image with connectionist text proposal network. In: Leibe, B., Matas, J., Sebe, N., Welling, M. (eds.) ECCV 2016. LNCS, vol. 9912, pp. 56–72. Springer, Cham (2016). https://doi.org/10.1007/978-3-319-46484-8_4
27. Hu, H., Zhang, C., Luo, Y., Wang, Y., Han, J., Ding, E.: WordSup: exploiting word annotations for character based text detection. In: ICCV, pp. 4940–4949 (2017)
28. Shi, B., Bai, X., Belongie, S.: Detecting oriented text in natural images by linking segments. In: CVPR, pp. 2550–2558 (2017)
29. Tang, J., Yang, Z., Wang, Y., Zheng, Q., Xu, Y., Bai, X.: SegLink++: detecting dense and arbitrary-shaped scene text by instance-aware component grouping. Pattern Recogn. **96**, 106954 (2019)
30. Long, S., Ruan, J., Zhang, W., He, X., Wu, W., Yao, C.: TextSnake: a flexible representation for detecting text of arbitrary shapes. In: Ferrari, V., Hebert, M., Sminchisescu, C., Weiss, Y. (eds.) ECCV 2018. LNCS, vol. 11206, pp. 19–35. Springer, Cham (2018). https://doi.org/10.1007/978-3-030-01216-8_2
31. Baek, Y., Lee, B., Han, D., Yun, S., Lee, H.: Character region awareness for text detection. In: CVPR, pp. 9365–9374 (2019)
32. Zhang, S.X., et al.: Deep relational reasoning graph network for arbitrary shape text detection. In: CVPR, pp. 9699–9708 (2020)
33. Ma, C., Sun, L., Zhong, Z., Huo, Q.: ReLaText: exploiting visual relationships for arbitrary-shaped scene text detection with graph convolutional networks. Pattern Recogn. **111**, 107684 (2021)
34. Carion, N., Massa, F., Synnaeve, G., Usunier, N., Kirillov, A., Zagoruyko, S.: End-to-end object detection with transformers. In: Vedaldi, A., Bischof, H., Brox, T., Frahm, J.-M. (eds.) ECCV 2020. LNCS, vol. 12346, pp. 213–229. Springer, Cham (2020). https://doi.org/10.1007/978-3-030-58452-8_13
35. Zhu, X., Su, W., Lu, L., Li, B., Wang, X., Dai, J.: Deformable DETR: deformable transformers for end-to-end object detection. In: ICLR (2021)
36. Meng, D., et al.: Conditional DETR for fast training convergence. In: ICCV, pp. 3651–3660 (2021)
37. Wang, Y., Zhang, X., Yang, T., Sun, J.: Anchor DETR: query design for transformer-based detector. In: AAAI, pp. 2567–2575 (2022)
38. Liu, S., et al.: DAB-DETR: dynamic anchor boxes are better queries for DETR. In: ICLR (2022)
39. Li, F., Zhang, H., Liu, S., Guo, J., Ni, L.M., Zhang, L.: DN-DETR: accelerate DETR training by introducing query denoising. In: CVPR, pp. 13619–13627 (2022)
40. Zhang, H., et al.: DINO: DETR with improved denoising anchor boxes for end-to-end object detection. arXiv preprint arXiv:2203.03605 (2022)
41. Ren, S., He, K., Girshick, R., Sun, J.: Faster R-CNN: towards real-time object detection with region proposal networks. In: NeurIPS, pp. 91–99 (2015)
42. Raisi, Z., Naiel, M.A., Younes, G., Wardell, S., Zelek, J.S.: Transformer-based text detection in the wild. In: CVPR, pp. 3162–3171 (2021)
43. Tang, J., et al.: Few could be better than all: feature sampling and grouping for scene text detection. In: CVPR, pp. 4563–4572 (2022)
44. Zhang, X., Su, Y., Tripathi, S., Tu, Z.: Text spotting transformers. In: CVPR, pp. 9519–9528 (2022)

45. Ye, M., Zhang, J., Zhao, S., Liu, J., Du, B., Tao, D.: DPText-DETR: towards better scene text detection with dynamic points in transformer. arXiv preprint arXiv:2207.04491 (2022)
46. Ye, M., et al.: DeepSolo: let transformer decoder with explicit points solo for text spotting. arXiv preprint arXiv:2211.10772 (2022)
47. Rezatofighi, H., Tsoi, N., Gwak, J., Sadeghian, A., Reid, I., Savarese, S.: Generalized intersection over union: a metric and a loss for bounding box regression. In: CVPR, pp. 658–666 (2019)
48. He, K., Gkioxari, G., Dollár, P., Girshick, R.: Mask R-CNN. In: ICCV, pp. 2961–2969 (2017)
49. Zhang, S.X., Zhu, X., Yang, C., Wang, H., Yin, X.C.: Adaptive boundary proposal network for arbitrary shape text detection. In: ICCV, pp. 1305–1314 (2021)
50. Vaswani, A., et al.: Attention is all you need. In: NeurIPS (2017)
51. Dong, Q., Tu, Z., Liao, H., Zhang, Y., Mahadevan, V., Soatto, S.: Visual relationship detection using part-and-sum transformers with composite queries. In: ICCV, pp. 3550–3559 (2021)
52. Kuhn, H.W.: The Hungarian method for the assignment problem. Nav. Res. Logist. Q. **2**(1–2), 83–97 (1955)
53. Lin, T.Y., Goyal, P., Girshick, R., He, K., Dollár, P.: Focal loss for dense object detection. In: ICCV, pp. 2980–2988 (2017)
54. Nayef, N., et al.: ICDAR2017 robust reading challenge on multi-lingual scene text detection and script identification-RRC-MLT. In: ICDAR, pp. 1454–1459 (2017)
55. Ch'ng, C.K., Chan, C.S.: Total-text: a comprehensive dataset for scene text detection and recognition. In: ICDAR, pp. 935–942 (2017)
56. Chng, C.K., et al.: ICDAR2019 robust reading challenge on arbitrary-shaped text-RRC-art. In: ICDAR, pp. 1571–1576 (2019)
57. He, K., Zhang, X., Ren, S., Sun, J.: Deep residual learning for image recognition. In: CVPR, pp. 770–778 (2016)
58. Loshchilov, I., Hutter, F.: Decoupled weight decay regularization. In: ICLR (2019)
59. Liu, H., Li, X., Liu, B., Jiang, D., Liu, Y., Ren, B.: Neural collaborative graph machines for table structure recognition. In: CVPR, pp. 4533–4542 (2022)
60. Liao, M., Wan, Z., Yao, C., Chen, K., Bai, X.: Real-time scene text detection with differentiable binarization. In: AAAI, pp. 11474–11481 (2020)
61. Xiao, S., Peng, L., Yan, R., An, K., Yao, G., Min, J.: Sequential deformation for accurate scene text detection. In: Vedaldi, A., Bischof, H., Brox, T., Frahm, J.-M. (eds.) ECCV 2020. LNCS, vol. 12374, pp. 108–124. Springer, Cham (2020). https://doi.org/10.1007/978-3-030-58526-6_7
62. He, M., et al.: MOST: a multi-oriented scene text detector with localization refinement. In: CVPR, pp. 8813–8822 (2021)
63. Wang, W., et al.: Efficient and accurate arbitrary-shaped text detection with pixel aggregation network. In: ICCV, pp. 8440–8449 (2019)
64. Ye, J., Chen, Z., Liu, J., Du, B.: TextFuseNet: scene text detection with richer fused features. In: IJCAI, pp. 516–522 (2020)
65. Liu, Y., et al.: ABCNet v2: adaptive bezier-curve network for real-time end-to-end text spotting. IEEE Trans. Pattern Anal. Mach. Intell. **44**(11), 8048–8064 (2021)
66. Du, B., Ye, J., Zhang, J., Liu, J., Tao, D.: I3CL: intra- and inter-instance collaborative learning for arbitrary-shaped scene text detection. Int. J. Comput. Vision **130**(8), 1961–1977 (2022)
67. Huang, M., et al.: SwinTextSpotter: scene text spotting via better synergy between text detection and text recognition. In: CVPR, pp. 4593–4603 (2022)

Decoupling Visual-Semantic Features Learning with Dual Masked Autoencoder for Self-Supervised Scene Text Recognition

Zhi Qiao⬛, Zhilong Ji$^{(\boxtimes)}$⬛, Ye Yuan⬛, and Jinfeng Bai⬛

Tomorrow Advancing Life, Beijing, China
{qiaozhi1,jizhilong,yuanye8,baijinfeng1}@tal.com

Abstract. Self-supervised text recognition has attracted more and more attention since it provides an effective way to utilize unlabeled real text images. Nowadays, Masked Image Modeling (MIM) shows superiority in visual representation learning, and several works introduce it into text recognition. In this paper, we take a further step and design a method for text-recognition-friendly self-supervised feature learning. Specifically, we propose to decouple visual and semantic feature learning with different masking strategies. For the visual features, intra-window random masking is proposed where the reconstruction is applied on a local image region with random masking, which prevents the model from the help of much context information. In the meanwhile, semantic feature learning is based on a window random masking, which removes more visual clues and boosts the sequence modeling of the model. Based on this idea, we first propose a siamese network that aligns dual features with each other, then we explore the dual distillation with a co-teacher framework. Our proposed method shows the effectiveness of self-supervised scene text recognition with state-of-the-art performances on most benchmarks.

Keywords: Scene Text Recognition · Self-Supervised Learning · Masked Image Modeling

1 Introduction

Nowadays, scene text recognition has been a hot topic due to the various practical applications, which aims to transcribe the text in the image to computer-editable text format. Existing text recognition methods [17,46,48,61] have achieved significant improvements with the help of deep learning. However, deep learning based text recognition methods also highly depend on large-scale training data. Fortunately, the synthetic text images can be a solution for data-hungry training, but recent works [2,4] also indicate the superiority of the real training images. Considering the expensive annotations of the text images, it is valuable to explore the usage of unlabeled real data.

G. A. Fink et al. (Eds.): ICDAR 2023, LNCS 14188, pp. 261–279, 2023.
https://doi.org/10.1007/978-3-031-41679-8_15

(a) Random Masking   (b) Block-wise Masking   (c) Intra-Window Random Masking   (d) Window Random Masking

Fig. 1. Illustrations of different masking strategies. The first of two are existing masking strategies: (a) random masking and (b) block-wise masking, the former reserves more visual details and the latter fully masks some characters. The last of two are our proposed strategies: (c) intra-window random masking and (d) window random masking. The red rectangles in the figure represent the windows. These two strategies first split the image into several windows, and apply the masking based on the window. Intra-window random masking uniformly masks the patches within a window, and the window random masking masks all patches in some windows. (Color figure online)

Thanks to Self-Supervised Learning, the model can achieve a strong feature extractor with unlabeled data, the contrastive learning [7,21] and Masked Image Modeling (MIM) [20,57] are two common techniques. The effectiveness of self-supervised learning for text recognition is also verified, where SeqCLR [1] proposes sequence-level contrastive learning with some corresponding data augments. PerSec [33] performs contrastive learning on different stages of the backbone to achieve multi-level feature learning. DiG [59] integrates contrastive learning and MIM into a unified framework, where the techniques of MIM follow the SimMIM [57]. We explain that MIM is more suitable for text recognition for three reasons: (1) The instance discrimination used in contrastive learning is not very flexible for text recognition, since the characters are the individual elements, the number of which varies in the text images; (2) MIM helps the model build the intra-characters and inter-character dependencies; (3) Masked Language Modeling (MLM) [13] has been widely adopted in Natural Language Processing (NLP), and the text recognition is a cross-modal task so that MIM and MLM could be integrated to achieve cross-modal pre-training.

In this paper, we propose to improve the existing MIM methods for scene text recognition. The motivation is that visual and semantic information are two key parts for the identification of each character, where a character can be recognized by the corresponding visual clues and context information from other characters. Based on this observation, we aim to decouple and improve the learning of these two features during self-supervised pre-training with specifically designed masks. As shown in Fig. 1(a) random masking and (b) block-wise masking are two common masking strategies used in MIM. The random masking follows a uniform distribution and it can reserve some visual details for characters. The block-wise masking removes a large continuous region, where some characters are fully masked. We explain that random masking can benefit the model with the learning of character visual details, and block-wise masking helps the modeling of context information among characters. In this paper, we unify these two masking strategies in a single model with some improvements and pay attention to both these two kinds of features during pre-training.

Furthermore, we modify the two masking strategies to achieve a better adaption for text recognition. As shown in Fig. 1(c), different from traditional random masking, we limit the visible regions in a local window. In particular, the encoder only performs attention operations among the patches in the same window. This intra-window random masking can prevent the model from much context information of the whole text image but reserve some visual details of characters. Specifically, we first split the image into several non-overlapped windows and the random masking is applied to each window independently with the same masking ratio. Compared with traditional random masking, the window-independent masking can decrease the possibility that characters are completely masked. For the window random masking, since most text images are single-line with horizontal direction, we simplify the block-wise masking with vertical rectangle mask regions as shown in Fig. 1(d). Note that the window random masking removes all visual clues for some characters, and lets the model pay attention to the context information. Based on the integration of two masking strategies, we propose Dual Masked Autoencoder (Dual-MAE for short) to improve the learning of the visual and semantic features. We start from a siamese network including dual branches to reconstruct the pixel values of two proposed masking strategies. Additionally, we align the features from two branches with bi-directional supervision. In this way, the features of visual and semantic information can benefit from each other. Then we try to replace the low-level pixel supervision with high-level feature supervision using a dual distillation framework. Specifically, we pre-trained two teacher networks adopting intra-window random masking and window random masking respectively, which provide feature targets focusing on the visual and semantic information. In summary, we first propose to decouple the learning of visual and semantic features for self-supervised text recognition with a dual masking strategy, and our contributions are as follows:

1. We aim to decouple visual and semantic features learning for self-supervised scene text recognition. A dual masking strategy of intra-window random masking and window random masking is proposed for two kinds of features;

2. We design a siamese network integrating the dual masking strategies into a single framework, and align the features belonging to each other;

3. A dual distillation framework with co-teacher is then proposed to adopt the high-level feature as targets. The two teacher networks are pre-trained with different masking strategies to provide two kinds of targets for distillation.

4. We conduct extensive experiments to verify the effectiveness of our Dual-MAE. State-of-the-art (SOTA) performance is achieved compared with other self-supervised text recognition methods.

2   Related Works

2.1   Scene Text Recognition

Scene text recognition is a hot topic and existing methods can be divided into traditional and deep learning based methods. Traditional methods [60] usually adopt a bottom-to-up framework, which detects the characters first and then groups them with heuristic rules. Deep learning based methods attract more attention nowadays, which can be distinguished from a different view of decoding strategies. Connectionist Temporal Classification (CTC for short) based methods [6,22,25,46,50] first extract the visual and context features, then transcribe the target text with CTC. With the development of attention mechanism, many works [10,30,31,40,47,48,56] introduce attention mechanism into text recognition. Due to the flexibility of the attention mechanism, recent works propose various improvements for text recognition. Such as irregular text recognition [31,37,48,63], the usage of semantic or linguistic information [4,17,42,54,61], novel decoding strategies [11,12,41,61], and so on. Additionally, Segmentation based methods [23,32,51] treat text recognition as a task of semantic segmentation. Note that most of the above work adopts synthetic text images as the training data for the data-hungry challenge.

2.2 Self-Supervised Learning

Self-supervised learning is one of the most concerned research topics recently. Pretext tasks [18,65] and contrastive learning [7,21] based methods have been dominant with significant performance. With the development of Vision Transformer (ViT) [15] and MLM in NLP, MIM has attracted more and more attention. Existing methods can be roughly classified based on the reconstruction targets. For example, MAE [20] and SimMIM [57] adopt the pixel as the target. To achieve high-level supervision for the mask tokens, BEiT [3] and CAE [8] adopt visual discrete tokenizer to provide a corresponding label, and MVP [55] uses CLIP [44] as the tokenizer. Without the pre-trained tokenizer, some methods [9,14] construct the target from the momentum updated teacher online.

For self-supervised text recognition, SeqCLR [1] extends the SimCLR [7] with sequence-to-sequence contrastive learning, where the feature sequence is divided into several instances with different mapping strategies. PerSec [33] applies contrastive learning on two feature levels in a hierarchical manner. To combine the advantages of contrastive learning and MIM, DiG [59] tries to integrate these two techniques, and achieves a significant improvement. Apart from contrastive learning, SimAN [36] first attempts to use generative learning for self-supervised text recognition, which lets the model recover the augmented images. In summary, different from PerSec, we deal with the visual and semantic feature representation from a MIM view with novel dual masking strategies instead of hierarchical contrastive learning. Compared with DiG, we take a further step for the MIM-based text recognition, where a siamese network and a dual distillation framework are proposed.

3   Method

In this section, we will introduce the details of our Dual-MAE. We first illustrate the adopted ViT-based architecture in Sect. 3.1. The proposed intra-window random masking and window random masking are described in Sect. 3.2. Based on

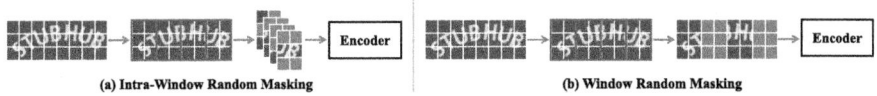

Fig. 2. The pipeline of the proposed intra-window random masking and window random masking. Besides the differences of the mask patches sampling, the intra-window random masking treats each window as an independent instance for the encoder. We indicate that the gray patches are masked, which will not be input into the encoder. (Color figure online)

the two kinds of masking strategies, the proposed siamese network and dual distillation framework are described in Sect. 3.3 and Sect. 3.4 respectively. Finally, we introduce the details of fine-tuning in the last sub-section.

3.1   Architecture

For self-supervised pre-training, we follow most settings from MAE. An asymmetric encoder-decoder architecture is adopted, the encoder and decoder are both a fully Transformer based structure. The input image is first split into several non-overlapped patches, and the patch size is set to 4×4. Same as MAE, we remove the masked patches and only input the visible patches to the encoder. The unmasked patches are then embedded by a linear projection layer and added with position embedding. Just as the standard ViT, several Transformer blocks are adopted to extract extensive features with the input of embedded patches.

The inputs of the decoder are the outputs of the encoder combined with mask tokens, and the decoder aims to predict the corresponding values for the mask tokens. More details of the decoder will be described in Sect. 3.3 and Sect. 3.4. Note that the decoder only exists during pre-training, when fine-tuning, the decoder is replaced with other text recognition decoders.

3.2   Masking Strategy

As mentioned above, we propose two kinds of masking strategies, intra-window random masking, and window random masking. To implement these two masking strategies, we first split the input image into windows and the masking is applied according to the windows. Denote the size of an image as $H \times W$, and the patch size is 4×4, so the number of patch tokens is $H' \times W'$ where the $H' = H/4, W' = W/4$. We adopt horizontal windows to split the patch tokens, a single window contains $H' \times 4$ patches, thus the number of windows is $W'/4$.

As shown in Fig. 2(a), the intra-window random masking treats each window independently, and the mask ratio is set to 75% as MAE, which is widely adopted by most MIM methods. The major concern of our work is to design the corresponding masking strategy for feature learning, so we fix the masking ratio without tuning. The encoder performs self-attention between each patch within the same window, which prevents the encoder from much context information.

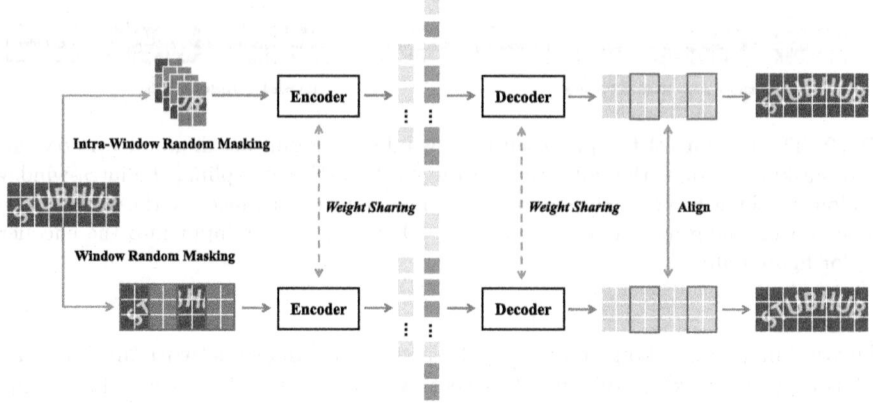

Fig. 3. The pipeline of our siamese network for Dual-MAE. There are two branches for different masked inputs and the parameters are shared during training. Both two branches reconstruct the pixel values of the image, and the features belonging to each other are aligned. The red rectangles on the features represent the part to be aligned. (Color figure online)

Since the masking is applied to each window independently, there are always some patches in a window visible, which will reserve some visual clues for a local region contributing to reconstruction. In this way, intra-window random masking lets the model focus on the visual features in the corresponding window.

Window random masking aims to improve the modeling ability of the global context information. Different from other masking strategies, the masking elements are the windows instead of patches, where a masked window represents all patches in it that are not visible. In this way, window random masking removes more visual clues and the context information plays an important role to reconstruct the target. The mask ratio of window random masking is also set to 75%, and the encoder applies attention operations on all visible patches without the limitation of windows, as shown in Fig. 2(b).

3.3   Siamese Network Based Dual-MAE

The motivation of siamese network based architecture is to align the predictions from two kinds of masking strategies, which lets visual and semantic features benefit from each other. As shown in Fig. 3, there are two branches in the siamese network, which share the parameters. The upper branch adopts intra-window random masking and the bottom adopts window random masking. Then the upper branch encodes the visible patches within a window and the bottom branch encodes all visible patches. With the outputs of the encoder, the decoder first inserts the mask tokens into corresponding positions, and additional Transformer blocks are applied. We denote the outputs of the decoder as F_u and F_b, which represent the outputs from the upper and bottom branches respectively. A linear

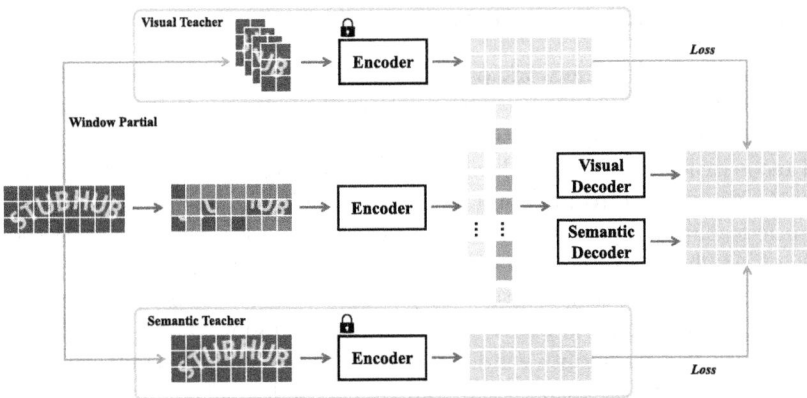

Fig. 4. The pipeline of our proposed dual distillation based Dual-MAE. The visual and semantic teachers are pre-trained with intra-window random masking and window random masking respectively. Instead of low-level pixel values, the target network aims to reconstruct the high-level features from two teachers with two decoders. The parameters of the dual teachers are frozen without updating. The red rectangles represent the windows, where the encoder treats them independently. (Color figure online)

layer is adopted to predict the pixel values, which are denoted as P_u and P_b. The objective function for the pixel regression is as follow:

$$L_{pixel} = \sum_{k \in \mathcal{M}_u} ||P_u^k - \overline{x}^k||_2^2 + \sum_{k \in \mathcal{M}_b} ||P_b^k - \overline{x}^k||_2^2 \qquad (1)$$

where the loss contains two parts for upper and bottom branches, $\mathcal{M}_u$ and $\mathcal{M}_b$ represent the intra-window and window random masked tokens, $\overline{x}$ indicates the normalized pixel values. Besides the loss for pixel reconstruction, we align the outputs of the decoder with a mean squared error:

$$L_{align} = \sum_{k \in \mathcal{M}_b} ||F_u^k - F_b^k||_2^2. \qquad (2)$$

where $\mathcal{M}_b$ indicates that we only align the window random masked patches, which are predicted with the context information. In other words, the siamese network tries to align the predictions based on the visual and semantic information, and let them benefit from each other. The final objective function of the siamese network consists of two parts:

$$L_{siam} = L_{pixel} + L_{align}. \qquad (3)$$

Moreover, several kinds of weights for two losses are compared, where treating them equally achieves better results.

3.4 Dual Distillation Based Dual-MAE

The proposed siamese network provides an effective way to model visual and semantic information, but the low-level pixel reconstruction task may limit the performance of self-supervised pre-training. Another idea is to replace the pixel reconstruction with high-level features. Inspired by the knowledge distillation [24], we propose to pre-train two teacher networks focusing on visual and semantic features respectively, then the target network is supervised by the two teachers. As shown in Fig. 4, the two teachers adopt different masking strategies and are pre-trained with the task of pixel reconstruction in advance. Specifically, the visual teacher is pre-trained with intra-window random masking, and the semantic teacher adopts window random masking on the contrary.

The target network utilizes the traditional random masking same as MAE, and a dual decoder is applied to predict the corresponding features provided by the encoders of pre-trained teachers. During distillation, the inputs to the visual teacher are the window partial patches according to the intra-window random masking, so the encoder will perform the attention operations within a window. The semantic teacher applies the attention operations on all patches. We denote the outputs of two teachers as T_v and T_s, the prediction of the target network are S_v and S_s, the objective function for the distillation is:

$$L_{distill} = \sum_{k \in \mathcal{M}} ||T_v^k - S_v^k||_2^2 + \sum_{k \in \mathcal{M}} ||T_s^k - S_s^k||_2^2 \qquad (4)$$

the $\mathcal{M}$ is random masked patches of the target network. With the dual distillation, the target network can obtain visual and semantic knowledge from a high-level view with the help of two teachers.

3.5 Text Recognition

After self-supervised pre-training, we apply the pre-trained encoder to the downstream task of text recognition. To verify the generalization of our Dual-MAE, we adopt three kinds of decoders for text recognition: CTC-based, attention-based, and Transformer-based. The CTC-based decoder contains a linear layer to predict a distribution for each character from the features extracted by the encoder. Considering CTC can only be applied on 1D features sequence, we convert the 2D feature map from ViT encoder to 1D with a vertical average-pooling. The attention-based and Transformer-based decoders use the attention mechanism to align the features for each corresponding character, which can be directly performed on the 2D feature map. Specifically, the attention based decoder contains an LSTM to transcribe the text autoregressively with the attention weights calculated between the hidden states of the LSTM and the feature map. The Transformer based decoder adopts the Transformer blocks as the main components, the self-attention and cross-attention are included for decoding. The settings of hyper-parameters are followed with DiG for a fair comparison.

4 Experiments

We conduct extensive experiments to analyze the effectiveness of our proposed Dual-MAE. In this section, we first introduce the datasets we used for pre-training and fine-tuning, then the implementation details are described in Sect. 4.1. The ablation studies are shown in Sect. 4.3 to analyze the effectiveness of our proposed modules. Finally, we compare our Dual-MAE with other SOTA methods.

4.1 Datasets

We follow the training and evaluation datasets with DiG including three parts:

Synthetic Text Data (STD). Synth90K [26] and SynthText [19] are two common synthetic datasets for text recognition, which contain 9 million and 8 million images respectively. We adopt these two datasets for self-supervised pre-training and text recognition fine-tuning.

Annotated Real Data (ARD). We use the same annotated real data as DiG, which includes 2.78 million images from TextOCR [49] and Open Images Dataset v5 [29]. They are also adopted for pre-training and fine-tuning.

Evaluation Benchmarks. We evaluate our proposed method on several public benchmarks, including IIIT5K-Words (IIIT5K) [38], Street View Text (SVT) [52], SVT-Perspective (SVTP) [43], ICDAR2013 (IC13) [28], ICDAR2015 (IC15) [27] and CUTE80 (CUTE) [45]. Specifically, the IIIT5K, IC13, and SVT mainly contain regular text, and IC15, SVTP, and CUTE focus on the irregular text.

In summary, we pre-train our Dual-MAE on STD and ARD, then fine-tune the text recognition methods on STD or ARD. Note that, compared with PerSec and DiG, we use much fewer images for pre-training (about 100 million and 15 million fewer.), which also verifies the effectiveness of our method.

4.2 Implementation Details

Model Settings. The feature dimension of Dual-MAE is 256 with a single channel image as the input. The encoder contains 12 Transformer blocks, 8 blocks are used for the decoder of the siamese network. The dual distillation based Dual-MAE adopts 2 blocks for the decoder to achieve more efficient training with less memory. The input images are resized to 32×128 without keeping ratio.

Self-Supervised Pre-Training. We follow the most training settings from MAE, where the optimizer is AdamW [34], the learning rate is set to $1.5e^{-4}$ with a cosine decay schedule [35]. The batch size is 512, and the model is optimized for 50 epochs with 4 warm-up epochs. No augments are adopted for pre-training.

Text Recognition Fine-Tuning. 36 symbols are covered for recognition, including digits, and lower-case characters. For attention-based and Transformer-based decoders, the additional ⟨EOS⟩ token is appended, and ⟨Blank⟩ token is

Table 1. Fine-tuning evaluation on the public benchmarks. All models are trained or fine-tuned with synthetic text images (STD). "Scratch" represents without self-supervised pre-training, "Dec." is short for "Decoder" and "Avg." represents the average accuracy. "Attn" and "Trans" are the short for "Attention" and "Transformer".

Methods	Dec.	IIIT5K	SVT	IC13		IC15		SVTP	CUTE	Avg.
		3000	647	857	1015	1811	2077	645	288	
Scratch	CTC	95.1	89.6	95.1	93.7	83.0	79.2	81.4	84.7	88.2
MAE		95.3	89.8	95.3	93.4	84.8	80.9	84.8	86.4	89.1
Dual-MAE (Siam)		**96.0**	90.7	95.0	93.7	84.8	**81.1**	84.8	87.8	89.5
Dual-MAE (Distill)		95.8	**91.6**	**95.7**	**94.2**	**84.9**	81.0	**85.3**	**88.9**	**89.7**
Scratch	Attn	96.2	91.0	96.0	95.3	85.5	81.8	86.7	89.2	90.2
MAE		96.3	91.9	96.0	94.8	86.1	82.6	87.7	89.2	90.6
Dual-MAE (Siam)		96.2	90.7	96.5	**95.7**	85.6	81.8	87.9	90.6	90.4
Dual-MAE (Distill)		**96.6**	**92.7**	**96.7**	95.3	**86.5**	**82.9**	**89.5**	**92.4**	**91.1**
Scratch	Trans	96.0	93.3	97.0	95.6	85.2	81.2	89.1	89.6	90.4
MAE		96.8	93.0	97.0	95.2	86.6	82.7	90.4	91.3	91.2
Dual-MAE (Siam)		96.7	**94.6**	97.2	95.8	**87.1**	83.2	89.6	91.0	**91.5**
Dual-MAE (Distill)		**97.0**	93.5	**97.7**	**96.4**	86.9	**83.3**	90.9	91.6	**91.5**

introduced to CTC-based decoding. The optimizer is also AdamW, with the learning rate $1e^{-4}$. The total epochs are 10 with 1 warm-up epoch. We adopt the same data augments as ABINet [17], including rotation, distortion, blur, etc.

4.3   Ablation Studies

About Dual-MAE. To analyze the effectiveness of our proposed Dual-MAE, we evaluate the models with different pre-training initialization. As shown in Table 1, when fine-tuned with synthetic text images, our Dual-MAE improves the performance on public benchmarks significantly with different decoders. Specifically, our distillation based Dual-MAE works best on average when combined with all three kinds of decoders. For the CTC-based decoder, 1.5% and 0.6% improvements are achieved compared with training from scratch and traditional MAE. The performance increases are also consistent with the other two decoders. We observe that Dual-MAE works better on some challenging benchmarks, such as IC15 and SVTP, which contains more low-quality images. To deal with low-quality images, the coordination of visual and semantic information plays an important role. Dual-MAE decouples the visual and semantic information learning, which improves the modeling capability of the features.

The superiority of our Dual-MAE is more significant when fine-tuned with real text images. As shown in Table 2, the siamese network based Dual-MAE achieves the best performance on nearly all benchmarks. Compared with the model without self-supervised pre-training, our proposed method improves the performance with 3.8%, 2.2%, and 3.1% on average with different decoders.

Table 2. Fine-tuning evaluation on the public benchmarks. All models are trained or fine-tuned with real text images (ARD).

Methods	Dec.	IIIT5K	SVT	IC13		IC15		SVTP	CUTE	Avg.
		3000	647	857	1015	1811	2077	645	288	
Scratch	CTC	94.6	91.3	93.7	93.6	84.0	81.3	82.6	88.5	88.8
MAE		96.3	93.3	95.2	95.2	85.6	83.7	85.3	92.4	92.4
Dual-MAE (Siam)		96.9	94.7	96.7	96.4	**88.1**	**86.8**	**89.1**	**94.1**	**92.6**
Dual-MAE (Distill)		**97.2**	**94.9**	**97.0**	**96.7**	87.5	86.1	89.0	93.8	92.5
Scratch	Attn	96.5	93.7	94.7	94.9	85.7	84.3	88.1	92.7	91.0
MAE		97.1	95.0	96.2	96.2	87.7	86.5	89.4	96.2	92.5
Dual-MAE (Siam)		97.4	**95.4**	**97.5**	**97.3**	**88.4**	**87.4**	**90.1**	**96.5**	**93.2**
Dual-MAE (Distill)		**97.5**	**95.4**	96.6	96.6	88.1	86.6	89.1	95.5	92.8
Scratch	Trans	96.2	93.3	94.5	94.8	86.6	84.8	88.7	90.6	91.1
MAE		97.7	96.0	97.1	97.0	88.8	87.7	**93.2**	97.2	93.6
Dual-MAE (Siam)		**97.8**	**96.6**	**97.2**	**97.2**	**90.2**	**89.0**	93.0	96.5	**94.2**
Dual-MAE (Distill)		**97.8**	96.3	97.1	**97.2**	89.7	88.7	93.0	**97.9**	94.0

Table 3. Comparison with different masking strategies. "Random" represents the traditional random masking, and "Dual" and "Dual w/o alignment" indicate integrating intra-window random and window random masking with or without feature alignment.

Mask	IIIT5K	SVT	IC13		IC15		SVTP	CUTE	Avg.
	3000	647	857	1015	1811	2077	645	288	
Random	96.3	93.3	95.2	95.2	85.6	83.7	85.3	92.4	92.4
Intra-Window Random	94.5	89.6	93.9	93.9	82.6	80.5	81.4	89.6	88.2
Window Random	94.5	91.2	93.3	93.4	83.5	81.2	82.2	87.8	88.5
Dual w/o align	96.5	92.6	94.3	94.4	85.3	83.6	85.4	93.8	90.6
Dual	**96.9**	**94.7**	**96.7**	**96.4**	**88.1**	**86.8**	**89.1**	**94.1**	**92.6**

Furthermore, our Dual-MAE outperforms traditional MAE with 0.2%, 0.7%, and 0.6% performance gains. An observation is that the siamese network based Dual-MAE works slightly better than distillation based Dual-MAE when fine-tuned with real text images. We explain that the pre-training dataset includes large-scale synthetic images, and the distillation based method may suffer from the domain gap due to the teachers pre-trained on the synthetic images.

About Dual Masking Strategies. We conduct experiments to analyze the proposed intra-window random masking and window random masking based on the siamese network. As shown in Table 3, only adopting intra-window random masking or window random masking harm the performance of fine-tuning, since the visual and semantic information are not both considered. Integrating two window masking strategies improves the performance with 2.1% on average, and the additional visual and semantic feature alignment further improves the accuracy of 2%. Traditional random masking also works better than window-based masking, because the two kinds of features can be learned implicitly. However, the worse performance of traditional MAE compared with our Dual-MAE also verifies the effectiveness of the features decoupling learning, where our Dual-MAE aims to learn the visual and semantic information explicitly.

About Distillation. We compare our distillation based Dual-MAE with different teachers and decoder structures. Table 4 illustrates the performance with different teachers. Compared with pixel reconstruction, using high-level features as the targets achieve better performance. Our dual distillation based Dual-MAE outperforms other teachers with 0.5% improvements on average, where the teachers are pre-trained with different masking strategies listed in Table 4. For example, "MAE" represents the teacher pre-trained with traditional MAE. Another choice is adopting the siamese network based Dual-MAE as the teacher. As shown in Table 4, the second-best performance is achieved since it also contains redundant visual and semantic information

As shown in Fig. 4, we adopt a dual decoder for better learning from two teachers respectively, and the parameters of the two decoders are not shared during training. Specifically, the decoder contains several Transformer blocks and a prediction head. As shown in Fig. 5, the architecture of the decoder can be modified according to blocks and heads. Figure 5(a) shows a "Dual Block and Dual Head" structure as our Dual-MAE adopted. Sub-figure (b) and (c) illustrate the other two designs with the parameters sharing of the blocks and head. As shown in Table 5, the blocks and heads without weight sharing achieve the best performance on all benchmarks, which indicates that the model without parameters sharing adapts to feature decoupling learning better.

Table 4. The comparison of fine-tuning with different distillation teachers. "Siamese" represents adopting the siamese network based Dual-MAE as the teacher, and "Dual" denotes the Dual distillation with co-teacher.

Teacher	IIIT5K	SVT	IC13		IC15		SVTP	CUTE	Avg.
	3000	647	857	1015	1811	2077	645	288	
MAE	96.9	94.0	96.6	96.3	87.1	85.7	87.9	93.7	92.0
Intra-Window Random MAE	97.1	94.3	96.7	96.5	86.7	85.2	88.2	**94.4**	92.0
Window Random MAE	96.2	92.9	95.2	95.3	86.2	84.8	87.3	92.7	91.1
Siamese	97.1	**95.1**	96.7	**96.7**	87.0	85.5	**90.1**	**94.4**	92.3
Dual	**97.2**	94.9	**97.0**	**96.7**	**87.5**	**86.1**	89.0	93.8	**92.5**

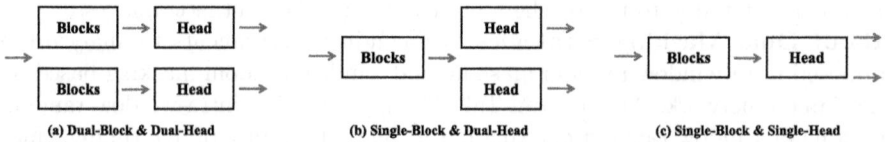

(a) Dual-Block & Dual-Head　　(b) Single-Block & Dual-Head　　(c) Single-Block & Single-Head

Fig. 5. Different decoders for dual distillation. The difference is whether the parameters of the blocks and heads are shared. For example, the single-block and dual-head based decoder adopts parameter-shared blocks and separated heads.

4.4 Comparisons with Existing Self-Supervised Methods

In this section, we compare our proposed Dual-MAE with other self-supervised text recognition methods. As shown in Table 6, our Dual-MAE outperforms Seq-CLR and PerSec with significant improvements, even though we use fewer real text images for self-supervised pre-training compared with PerSec. Compared with DiG, our Dual-MAE uses less pre-training data (13M real text images fewer.), but our proposed method achieves better or comparable fine-tuning performance on both three decoders. For example, our Dual-MAE works better on SVTP and CUTE with 1.4% and 2.4% improvements respectively when combined with a CTC-based decoder.

Table 5. The comparison of fine-tuning with the different decoder architecture of our dual distillation based Dual-MAE.

Dual Head	Dual Blocks	IIIT5K	SVT	IC13		IC15		SVTP	CUTE	Avg.
		3000	647	857	1015	1811	2077	645	288	
		96.6	91.6	95.4	95.1	86.0	84.1	86.5	93.1	90.9
✓		96.4	92.4	95.7	95.6	85.9	84.0	85.7	92.7	90.9
✓	✓	**97.2**	**94.9**	**97.0**	**96.7**	**87.5**	**86.1**	**89.0**	**93.8**	**92.5**

Table 6. Comparison with other self-supervised text recognition methods. "UTI-100M" is a large-scale unlabeled real dataset adopted by PerSec, "*-Real" represents the number of the real text images used for pre-training. "DiG*" indicates the "DiG-Small", the parameters of which are similar with our Dual-MAE for a fair comparison.

Method	Training Data	Dec.	IIIT5K	SVT	IC13	IC15	SVTP	CUTE
SeqCLR [1]	STD	CTC	80.9	–	86.3	–	–	–
PerSec [33]	STD + UTI-100M		85.4	86.1	92.8	70.3	73.9	69.2
DiG* [59]	STD + 16M-Real		95.5	**91.8**	95.0	84.1	83.9	86.5
Dual-MAE (Siam)	STD + 3M-Real		**96.0**	90.7	95.0	84.8	84.8	87.8
Dual-MAE (Distill)	STD + 3M-Real		95.8	91.6	**95.7**	**84.9**	**85.3**	**88.9**
SeqCLR [1]	STD	Attn	82.9	–	87.9	–	–	–
PerSec [33]	STD + UTI-100M		88.1	86.8	94.2	73.6	77.7	72.7
SimAN [36]	STD + Real-300K		87.5	–	89.9	–	–	–
DiG* [59]	STD + 16M-Real		96.4	**94.6**	96.6	86.0	89.3	88.9
Dual-MAE (Siam)	STD + 3M-Real		96.2	90.7	96.5	85.6	87.9	90.6
Dual-MAE (Distill)	STD + 3M-Real		**96.6**	92.7	**96.7**	**86.5**	**89.5**	**92.4**
DiG* [59]	STD + 16M-Real	Trans	96.7	93.4	97.1	**87.1**	90.1	88.5
Dual-MAE (Siam)	STD + 3M-Real		96.7	**94.6**	97.2	**87.1**	89.6	91.0
Dual-MAE (Distill)	STD + 3M-Real		**97.0**	93.5	**97.7**	86.9	**90.9**	**91.6**

4.5   Comparisons with Existing Scene Text Recognition Methods

We compare our Dual-MAE with recent scene text recognition methods. Without additional language model [17], permutation language modeling training [33] or specific decoding designs [11,41], our proposed method just adopts the plain ViT-based encoder and Transformer-based decoder. As shown in Table 7, our Dual-MAE achieves the best performance on most benchmarks with a 0.3% improvement on average. Especially, our Dual-MAE works best on some low-quality benchmarks such as IC15, which verifies the effectiveness of the visual and semantic features learning.

Table 7. Comparison with state-of-the-art scene text recognition methods. **Bold** represents the best performance. <u>Underline</u> represents the second best performance.

Method	IIT5K	SVT	IC13		IC15		SVTP	CUTE	Avg.
	3000	647	857	1015	1811	2077	645	288	
DAN [53]	94.3	89.2	–	93.9	–	74.5	80.0	84.4	86.9
RobustScanner [62]	95.3	88.1	–	94.8	–	77.1	79.5	90.3	88.2
SRN [61]	94.8	91.5	95.5	–	82.7	–	85.1	87.8	90.4
PIMNet [41]	95.2	91.2	95.2	93.4	83.5	81.0	84.3	84.4	90.5
ABINet [17]	95.4	93.2	96.8	–	84.0	–	87.0	88.9	91.5
GA-SPIN [64]	95.2	90.9	–	94.8	82.8	79.5	83.2	87.5	90.4
PREN2D [58]	95.6	94.0	96.4	–	83.0	–	87.6	91.7	91.5
VisionLAN [54]	95.8	91.7	95.7	–	83.7	–	86.0	88.5	91.2
JVSR [5]	95.2	92.2	–	95.5	84.0	–	85.7	89.7	91.2
S-GTR [23]	95.8	94.1	96.8	–	84.6	–	87.9	92.3	92.1
SVTR [16]	96.3	91.7	<u>97.2</u>	–	86.6	–	88.4	**95.1**	92.8
CornerTransformer [56]	95.9	**94.6**	96.4	–	86.3	–	**91.5**	92.0	92.6
LevOCR [11]	96.6	92.9	96.9	–	86.4	–	88.1	91.7	92.8
ParSeq [4]	97.0	<u>93.6</u>	97.0	<u>96.2</u>	86.5	82.9	88.9	92.2	93.2
Dual-MAE (Siam)	<u>96.7</u>	**94.6**	<u>97.2</u>	95.8	**87.1**	<u>83.2</u>	89.6	91.0	93.4
Dual-MAE (Distill)	**97.0**	93.5	**97.7**	**96.4**	<u>86.9</u>	**83.3**	90.9	91.6	**93.5**

4.6   Qualitative Analysis

In this section, we visualize the pixel reconstruction and recognition results to illustrate the effectiveness of our Dual-MAE qualitatively. As shown in Fig. 6, we visualize the pixel reconstruction results of our proposed window masking. Even though the window random masking removes all visual details of some characters, the reconstruction can also be achieved, which verifies the contributions of the semantic information. However, the reconstructed images from the window random masking are not as well as the intra-window random masking, which indicates that the former may not adapt to learn visual clues of characters. We

Fig. 6. Pixel reconstruction results of the two proposed masking strategies. The upper image is the result of intra-window random masking, the bottom is the result of window random masking. For better visualization, we concatenate the windows of the image.

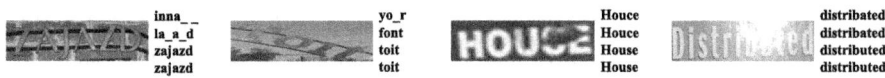

Fig. 7. The visualization of the text recognition results. For each image, the strings from top to bottom are the results from the model without pre-training, with MAE, with siamese net based Dual-MAE, and with dual distillation based Dual-MAE.

explain that the visual details are the concern of intra-window random masking, which is exactly the motivation of our Dual-MAE.

Figure 7 shows some text recognition results from different pre-training methods. Due to the better learning of the visual and semantic features, our proposed Dual-MAE can help the text recognition model with some challenges of recognition, such as occluded characters, curved text, and complex backgrounds.

5   Limitations

Compared with contrastive learning based methods, our Dual-MAE may suffer from instance discrimination because no negative samples are used during pre-training. Therefore, Dual-MAE may not be sensitive to the different characters. We will improve the MIM with the idea of instance discrimination to achieve better self-supervised text recognition in the future.

Another issue is that the MIM may not adapt to a CNN-based encoder. However, it is not a severe problem, since the ViT-based encoder has been mainstream for scene text recognition [4,12,16,39] recently. We will also regard the Dual-MAE for CNN-based encoder as our future work.

6   Conclusion

In this paper, we focus on visual and semantic feature representation learning for scene text recognition. We try to decouple the learning of these two features during the self-supervised pre-training. Based on this motivation, we first propose two novel masking strategies, intra-window random masking, and window random masking, which aim at the learning of visual features and semantic features respectively. Furthermore, we propose the Dual Masked Autoencoder (Dual-MAE) with a siamese network based and a dual distillation based architecture. Specifically, the siamese network based Dual-MAE aligns the features from two different maskings, and the dual distillation based Dual-MAE introduces the high-level visual and semantic information with two corresponding teachers. We first compare our Dual-MAE with traditional MAE, the improvements for different decoders demonstrate the effectiveness of our decoupling features learning. State-of-the-art or comparable performance is also reached compared with other self-supervised text recognition methods. Combined with our proposed self-supervised techniques, our Dual-MAE achieves satisfactory performance, compared with other scene text recognition methods, even if only the plain ViT-based encoder and Transformer-based decoder are adopted. In addition to dealing with the limitations mentioned above, we will increase the scale of model size and training data for large text recognition models.

Acknowledgments. This work was supported by National Key R&D Program of China, under Grant No. 2020AAA0104500.

References

1. Aberdam, A., et al.: Sequence-to-sequence contrastive learning for text recognition. In: CVPR, pp. 15302–15312 (2021)
2. Baek, J., Matsui, Y., Aizawa, K.: What if we only use real datasets for scene text recognition? Toward scene text recognition with fewer labels. In: CVPR, pp. 3113–3122 (2021)
3. Bao, H., Dong, L., Piao, S., Wei, F.: BEiT: BERT pre-training of image transformers. In: ICLR (2022)
4. Bautista, D., Atienza, R.: Scene text recognition with permuted autoregressive sequence models. In: Avidan, S., Brostow, G., Cissé, M., Farinella, G.M., Hassner, T. (eds.) Computer Vision – ECCV 2022. LNCS, vol. 13688, pp. 178–196. Springer, Cham (2022). https://doi.org/10.1007/978-3-031-19815-1_11
5. Bhunia, A.K., Sain, A., Kumar, A., Ghose, S., Chowdhury, P.N., Song, Y.Z.: Joint visual semantic reasoning: multi-stage decoder for text recognition. In: ICCV, pp. 14940–14949 (2021)
6. Chao, L., Chen, J., Chu, W.: Variational connectionist temporal classification. In: Vedaldi, A., Bischof, H., Brox, T., Frahm, J.-M. (eds.) ECCV 2020. LNCS, vol. 12373, pp. 460–476. Springer, Cham (2020). https://doi.org/10.1007/978-3-030-58604-1_28
7. Chen, T., Kornblith, S., Norouzi, M., Hinton, G.E.: A simple framework for contrastive learning of visual representations. In: ICML, pp. 1597–1607. ACM (2020)
8. Chen, X., et al.: Context autoencoder for self-supervised representation learning. arXiv preprint arXiv:2202.03026 (2022)
9. Chen, Y., et al.: SdAE: self-distillated masked autoencoder. In: Avidan, S., Brostow, G., Cissé, M., Farinella, G.M., Hassner, T. (eds.) Computer Vision – ECCV 2022. LNCS, vol. 13690, pp. 108–124. Springer, Cham (2022). https://doi.org/10.1007/978-3-031-20056-4_7
10. Cheng, Z., Bai, F., Xu, Y., Zheng, G., Pu, S., Zhou, S.: Focusing attention: towards accurate text recognition in natural images. In: ICCV, pp. 5076–5084. IEEE (2017)

11. Da, C., Wang, P., Yao, C.: Levenshtein OCR. In: Avidan, S., Brostow, G., Cissé, M., Farinella, G.M., Hassner, T. (eds.) Computer Vision – ECCV 2022. LNCS, vol. 13688, pp. 322–338. Springer, Cham (2022). https://doi.org/10.1007/978-3-031-19815-1_19

12. Da, C., Wang, P., Yao, C.: Multi-granularity prediction for scene text recognition. In: Avidan, S., Brostow, G., Cissé, M., Farinella, G.M., Hassner, T. (eds.) Computer Vision – ECCV 2022. LNCS, vol. 13688, pp. 339–355. Springer, Cham (2022). https://doi.org/10.1007/978-3-031-19815-1_20

13. Devlin, J., Chang, M.W., Lee, K., Toutanova, K.: BERT: pre-training of deep bidirectional transformers for language understanding. In: NAACL, pp. 4171–4186 (2019)

14. Dong, X., et al.: Bootstrapped masked autoencoders for vision BERT pretraining. In: Avidan, S., Brostow, G., Cissé, M., Farinella, G.M., Hassner, T. (eds.) Computer Vision – ECCV 2022. LNCS, vol. 13690, pp. 247–264. Springer, Cham (2022). https://doi.org/10.1007/978-3-031-20056-4_15

15. Dosovitskiy, A., et al.: An image is worth 16x16 words: transformers for image recognition at scale. In: ICLR (2021)

16. Du, Y., et al.: SVTR: scene text recognition with a single visual model. In: IJCAI, pp. 884–890 (2022)

17. Fang, S., Xie, H., Wang, Y., Mao, Z., Zhang, Y.: Read like humans: autonomous, bidirectional and iterative language modeling for scene text recognition. In: CVPR, pp. 7098–7107 (2021)

18. Gidaris, S., Singh, P., Komodakis, N.: Unsupervised representation learning by predicting image rotations. In: ICLR (2021)

19. Gupta, A., Vedaldi, A., Zisserman, A.: Synthetic data for text localisation in natural images. In: CVPR, pp. 2315–2324. IEEE (2016)

20. He, K., Chen, X., Xie, S., Li, Y., Dollár, P., Girshick, R.B.: Masked autoencoders are scalable vision learners. In: CVPR, pp. 16000–16009 (2022)

21. He, K., Fan, H., Wu, Y., Xie, S., Girshick, R.: Momentum contrast for unsupervised visual representation learning. In: CVPR, pp. 9726–9735 (2020)

22. He, P., Huang, W., Qiao, Y., Chen, C.L., Tang, X.: Reading scene text in deep convolutional sequences. In: AAAI, pp. 3501–3508. AAAI (2016)

23. He, Y., et al.: Visual semantics allow for textual reasoning better in scene text recognition. In: AAAI. AAAI (2021)

24. Hinton, G.E., Vinyals, O., Dean, J.: Distilling the knowledge in a neural network. arXiv preprint arXiv:1503.02531 (2015)

25. Hu, W., Cai, X., Hou, J., Yi, S., Lin, Z.: GTC: guided training of CTC towards efficient and accurate scene text recognition. In: AAAI, pp. 11005–11012 (2020)

26. Jaderberg, M., Simonyan, K., Vedaldi, A., Zisserman, A.: Reading text in the wild with convolutional neural networks. IJCV 116(1), 1–20 (2016)

27. Karatzas, D., et al.: ICDAR 2015 competition on robust reading. In: ICDAR, pp. 1156–1160. IEEE (2015)

28. Karatzas, D., et al.: ICDAR 2013 robust reading competition. In: ICDAR, pp. 1484–1493. IEEE (2013)

29. Krylov, I., Nosov, S., Sovrasov, V.: Open images V5 text annotation and yet another mask text spotter. In: ACML, vol. 157, pp. 379–389. PMLR (2021)

30. Lee, C.Y., Osindero, S.: Recursive recurrent nets with attention modeling for OCR in the wild. In: CVPR, pp. 2231–2239. IEEE (2016)

31. Li, H., Wang, P., Shen, C., Zhang, G.: Show, attend and read: a simple and strong baseline for irregular text recognition. In: AAAI, pp. 8610–8617. AAAI (2019)

32. Liao, M., et al.: Scene text recognition from two-dimensional perspective. In: AAAI, pp. 8714–8721 (2019)
33. Liu, H., et al.: Perceiving stroke-semantic context: hierarchical contrastive learning for robust scene text recognition. In: AAAI, pp. 1702–1710. AAAI (2021)
34. Loshchilov, I., Hutter, F.: Fixing weight decay regularization in adam. arXiv preprint arXiv:1711.05101 (2017)
35. Loshchilov, I., Hutter, F.: SGDR: stochastic gradient descent with warm restarts. In: ICLR (2016)
36. Luo, C., Jin, L., Chen, J.: SimAN: exploring self-supervised representation learning of scene text via similarity-aware normalization. In: CVPR, pp. 1039–1048 (2022)
37. Luo, C., Jin, L., Sun, Z.: MORAN: a multi-object rectified attention network for scene text recognition. PR **90**, 109–118 (2019)
38. Mishra, A., Alahari, K., Jawahar, C.: Scene text recognition using higher order language priors. In: BMVC. BMVA (2012)
39. Qiao, Z., Ji, Z., Yuan, Y., Bai, J.: A vision transformer based scene text recognizer with multi-grained encoding and decoding. In: Porwal, U., Fornés, A., Shafait, F. (eds.) ICFHR 2022. LNCS, pp. 198–212. Springer, Cham (2022). https://doi.org/10.1007/978-3-031-21648-0_14
40. Qiao, Z., Qin, X., Zhou, Y., Yang, F., Wang, W.: Gaussian constrained attention network for scene text recognition. In: ICPR, pp. 3328–3335 (2020)
41. Qiao, Z., et al.: PIMNet: a parallel, iterative and mimicking network for scene text recognition. In: MM, pp. 2046–2055. ACM (2021)
42. Qiao, Z., Zhou, Y., Yang, D., Zhou, Y., Wang, W.: SEED: semantics enhanced encoder-decoder framework for scene text recognition. In: CVPR, pp. 13525–13534. IEEE (2020)
43. Quy Phan, T., Shivakumara, P., Tian, S., Lim Tan, C.: Recognizing text with perspective distortion in natural scenes. In: ICCV, pp. 569–576. IEEE (2013)
44. Radford, A., et al.: Learning transferable visual models from natural language supervision. In: ICML, vol. 139, pp. 8748–8763. PMLR (2021)
45. Risnumawan, A., Shivakumara, P., Chan, C.S., Tan, C.L.: A robust arbitrary text detection system for natural scene images. ESA **41**(18), 8027–8048 (2014)
46. Shi, B., Bai, X., Yao, C.: An end-to-end trainable neural network for image-based sequence recognition and its application to scene text recognition. TPAMI **39**(11), 2298–2304 (2016)
47. Shi, B., Wang, X., Lyu, P., Yao, C., Bai, X.: Robust scene text recognition with automatic rectification. In: CVPR, pp. 4168–4176. IEEE (2016)
48. Shi, B., Yang, M., Wang, X., Lyu, P., Yao, C., Bai, X.: ASTER: an attentional scene text recognizer with flexible rectification. TPAMI **41**(9), 2035–2048 (2018)
49. Singh, A., Pang, G., Toh, M., Huang, J., Galuba, W., Hassner, T.: TextOCR: towards large-scale end-to-end reasoning for arbitrary-shaped scene text. In: CVPR, pp. 8802–8812 (2021)
50. Su, B., Lu, S.: Accurate recognition of words in scenes without character segmentation using recurrent neural network. In: PR, pp. 397–405 (2017)
51. Wan, Z., He, M., Chen, H., Bai, X., Yao, C.: TextScanner: reading characters in order for robust scene text recognition. In: AAAI, pp. 12120–12127. AAAI (2020)
52. Wang, K., Babenko, B., Belongie, S.: End-to-end scene text recognition. In: ICCV, pp. 1457–1464. IEEE (2011)
53. Wang, T., et al.: Decoupled attention network for text recognition. In: AAAI, pp. 12216–12224 (2020)

54. Wang, Y., Xie, H., Fang, S., Wang, J., Zhu, S., Zhang, Y.: From two to one: a new scene text recognizer with visual language modeling network. In: ICCV, pp. 14194–14203 (2021)
55. Wei, L., Xie, L., Zhou, W., Li, H., Tian, Q.: MVP: multimodality-guided visual pre-training. In: Avidan, S., Brostow, G., Cissé, M., Farinella, G.M., Hassner, T. (eds.) Computer Vision – ECCV 2022. LNCS, vol. 13690, pp. 337–353. Springer, Cham (2022). https://doi.org/10.1007/978-3-031-20056-4_20
56. Xie, X., Fu, L., Zhang, Z., Wang, Z., Bai, X.: Toward understanding wordArt: corner-guided transformer for scene text recognition. In: Avidan, S., Brostow, G., Cissé, M., Farinella, G.M., Hassner, T. (eds.) Computer Vision – ECCV 2022. LNCS, vol. 13688, pp. 303–321. Springer, Cham (2022). https://doi.org/10.1007/978-3-031-19815-1_18
57. Xie, Z., et al.: SimMIM: a simple framework for masked image modeling. In: CVPR, pp. 9653–9663 (2022)
58. Yan, R., Peng, L., Xiao, S., Yao, G.: Primitive representation learning for scene text recognition. In: CVPR, pp. 284–293 (2021)
59. Yang, M., et al.: Reading and Writing: discriminative and generative modeling for self-supervised text recognition. In: MM, pp. 4214–4223. ACM (2022)
60. Ye, Q., Doermann, D.: Text detection and recognition in imagery: a survey. TPAMI 37(7), 1480–1500 (2014)
61. Yu, D., et al.: Towards accurate scene text recognition with semantic reasoning networks. In: CVPR, pp. 12110–12119. IEEE (2020)
62. Yue, X., Kuang, Z., Lin, C., Sun, H., Zhang, W.: RobustScanner: dynamically enhancing positional clues for robust text recognition. In: Vedaldi, A., Bischof, H., Brox, T., Frahm, J.-M. (eds.) ECCV 2020. LNCS, vol. 12364, pp. 135–151. Springer, Cham (2020). https://doi.org/10.1007/978-3-030-58529-7_9
63. Zhan, F., Lu, S.: ESIR: end-to-end scene text recognition via iterative image rectification. In: CVPR, pp. 2059–2068. IEEE (2019)
64. Zhang, C., et al.: SPIN: structure-preserving inner offset network for scene text recognition. In: AAAI, pp. 3305–3314 (2021)
65. Zhang, R., Isola, P., Efros, A.A.: Colorful image colorization. In: Leibe, B., Matas, J., Sebe, N., Welling, M. (eds.) ECCV 2016. LNCS, vol. 9907, pp. 649–666. Springer, Cham (2016). https://doi.org/10.1007/978-3-319-46487-9_40

Re-Thinking Text Clustering for Images with Text

Shwet Kamal Mishra$^{(\boxtimes)}$, Soham Joshi , and Viswanath Gopalakrishnan

International Institute of Information Technology Bangalore, Bengaluru, India
shwet.mishra@iitb.ac.in

Abstract. Text-VQA refers to the set of problems that reason about the text present in an image to answer specific questions regarding the image content. Previous works in text-VQA have largely followed the common strategy of feeding various input modalities (OCR, Objects, Question) to an attention-based learning framework. Such approaches treat the OCR tokens as independent entities and ignore the fact that these tokens often come correlated in an image representing a larger 'meaningful' entity. The 'meaningful' entity potentially represented by a group of OCR tokens could be primarily discerned by the layout of the text in the image along with the broader context it appears. In the proposed work, we aim to cluster the OCR tokens using a novel spatially-aware and knowledge-enabled clustering technique that uses an external knowledge graph to improve the answer prediction accuracy of the text-VQA problem. Our proposed algorithm is generic enough to be applied to any multi-modal transformer architecture used for text-VQA training. We showcase the objective and subjective effectiveness of the proposed approach by improving the performance of the M4C model on the Text-VQA datasets.

Keywords: Text VQA · Scene Text Clustering · Knowledge Graph

1 Introduction

Text-VQA plays an integral role in the automatic understanding of images that come along with rich contextual text data. Specific questions regarding the content of text in an image can only be answered with the contextual understanding of the various objects in the image along with the detected text. The success of text-VQA approaches not only relies on proper reasoning regarding the inter-dependency between visual and textual content but also on the correlation between different words present in the textual content.

Previous works in text-VQA have focused on establishing inter-relationships between multiple modalities [1,12,18,22] involving objects in the image, OCR-detected text [10], and questions asked about the textual content. The different modalities are fed as inputs to a multi-modal attention framework involving transformers and learned in an end-to-end fashion with the answer as the ground truth data. While it makes sense to learn the cross-correlation between the question, image content (objects) and the detected OCR tokens using the guidance from

G. A. Fink et al. (Eds.): ICDAR 2023, LNCS 14188, pp. 280–294, 2023.
https://doi.org/10.1007/978-3-031-41679-8_16

answer ground truths, the correlation between various OCR tokens in an image cannot be learned in a similar way. Though the OCR tokens are detected separately, in many cases they form a group or cluster with a larger context involved. The understanding of this broader group of OCR tokens is imperative to rightly answer many questions involved in a text-VQA task. In this work, we focus on understanding the broader context in which the OCR tokens can be grouped and subsequently feed the grouping information to a transformer-based attentional framework with the aim to improve the accuracy of the text-VQA task.

Fig. 1. Our approach clusters the OCR tokens based on their spatial layout and an external knowledge graph. The clustering information functions as an additional input to a multi-modal transformer framework to improve the accuracy of the text-VQA problem.

The grouping of OCR tokens under the broader context can be better understood by considering the example shown in Fig. 1. In Fig. 1, the OCR tokens detected are 'Making', 'Visible', 'Mira', 'Schendel', and so on. Though individual OCR tokens 'Mira' and 'Schendel' can occur with independent meanings, in Fig. 1, they represent the name of the person 'Mira Schendel'. To correctly answer the question 'Whose name is written on the white booklet?', the training network will need knowledge of the aforementioned grouping. Our work proposes a novel method for OCR token grouping using a joint approach that exploits the spatial layout of the tokens as well as the information available from external knowledge bases. We can easily verify from Fig. 1 that the spatial layout of the OCR token indeed holds a strong clue regarding their grouping while the presence of such a meaningful entity could be further established with the help of an external knowledge graph.

The proposed work can be summarized in the following key points:

i) We improve the scene text clustering technique proposed in [26] by considering the impact of spatial layout in grouping the tokens
ii) We modify the clustering algorithm parameters based on the inputs from external knowledge bases and propose a novel method to ingrain the external knowledge into the clustering technique.

iii) We devise a novel method to incorporate the OCR clustering information into the multi-modal transformer architecture and train it in an end-to-end fashion to showcase improved objective and subjective results in text-VQA problems.

2   Background

2.1   Text Visual Question Answering (Text-VQA)

Text-VQA involves answering questions that require models to explicitly reason about the text present in the image. Multiple datasets [1,7,9,10,18] have been proposed for the text-VQA task. Subsequently, several methods have also been proposed for this task. Broadly, the popular methods can be categorized into vanilla attention-based models [1], graph-based models [21], and transformer-based models [12,22]. An example of attention-based models is LoRRA [1] which extends the Pythia framework [19] by including an additional attention mode for OCR tokens to reason over a combined list of vocabulary and detected OCR tokens. Multiple approaches have also incorporated the OCR tokens as a modality in their models [10,18,20]. [21] builds a three-layer multi-modal graph comprising numeric, semantic and visual features and then trains a graph neural network for the Text-VQA task. Recently, transformer-based methods [12,22] have been widely re-used for text VQA tasks by adapting the framework to accept multiple modalities for OCR tokens, Visual features and Question tokens. We elaborate on [12] (which is our baseline) in the following subsection.

There are other recent works that leverage large pre-trained encoder-decoder language models and Vision Transformers (ViTs) which are topping the leaderboard of the text-VQA tasks [23–25]. Also, these models use OCR systems (Google-OCR, Azure-OCR, etc.) that are more accurate than the Rosetta-OCR results provided with the Text-VQA dataset [1]. We chose M4C as a baseline model and used Rosetta-OCR results [12] to best demonstrate the advantage of the proposed clustering strategy. However, our proposed method is modular and scalable and thus can be plug-and-played with any transformer-based architecture.

2.2   Multimodal Multi-Copy Mesh (M4C)

M4C [12] builds on top of the [13] to create a multimodal transformer module. This multimodal transformer has input modalities namely, question tokens, object embeddings, and OCR tokens. The feature extraction procedure for these three modalities is as follows:

1. Question Words: The question words are encoded using a pre-trained BERT model [14]. The question embedding is fed into the transformer through the question modality.
2. Detected Objects: The objects in the image are detected by passing the image through a Faster R-CNN network [15] to detect the proposals. The object embedding is generated by adding positional information (about the bounding box) about the normalized coordinates of the object, thus making the embedding richer.

3. OCR tokens: The OCR features are extracted through FastText [16], Faster R-CNN detector [15] and Pyramidal Histogram of Characters (PHOC) [17]. These features form parts of the OCR embedding. Additionally, positional information (about the bounding box) is also added to enrich the OCR embedding.

The M4C model projects the feature representations from these three modalities as vectors in a learned common embedding space. The model learns to predict the answer through iterative decoding accompanied by a dynamic pointer network. The work M4C was the first breakthrough in text-VQA which demonstrated the use of multimodal transformer architecture. M4C was benchmarked on text-VQA datasets like Text-VQA [1] and ST-VQA [18] and achieved SOTA results on the same.

2.3 Scene Text Clustering

Scene Text clustering is relatively a new idea in the text-VQA domain. There has been only one such attempt in the past where Lu et al. [26] clustered the tokens based on the bounding box coordinates and passed on that information to a multimodal transformer through positional embeddings. However, this approach does not cluster the tokens at a local level, due to which a larger set of tokens are grouped together. It is also prone to grouping unrelated tokens with significant differences in font size together just because they might be in close vicinity. Apart from this, it uses positional embeddings for token numbers and line numbers as well. In contrast to the clustering approach in [26], we leverage the spatial alignment of the OCR tokens and the related external knowledge to make the clusters more localized and meaningful. Furthermore, we explored the idea of passing the clustering information through a simple mechanism instead of positional encodings.

3 Spatially-Aware and Knowledge-Enabled Clustering

In this paper, we propose an approach that uses features of OCR bounding boxes to cluster tokens together. Our contribution is novel in the following ways:

1. Implemented a localized clustering that works at an entity level unlike the approach proposed in [26].
2. Utilized the spatial layout of OCR tokens by introducing a height penalty parameter in the clustering method.
3. Clustered the OCR tokens to meaningful entities by combining the spatial layout information of tokens with the external knowledge of WikiData.

Thus, our approach is spatially-aware by clustering the tokens based on spatial features and knowledge-enabled by identifying the group of tokens based on their actual meaning and presence in the knowledge graph.

Clustering is done in the pre-processing stage and the outputs of clustering are then fed to a multimodal transformer.

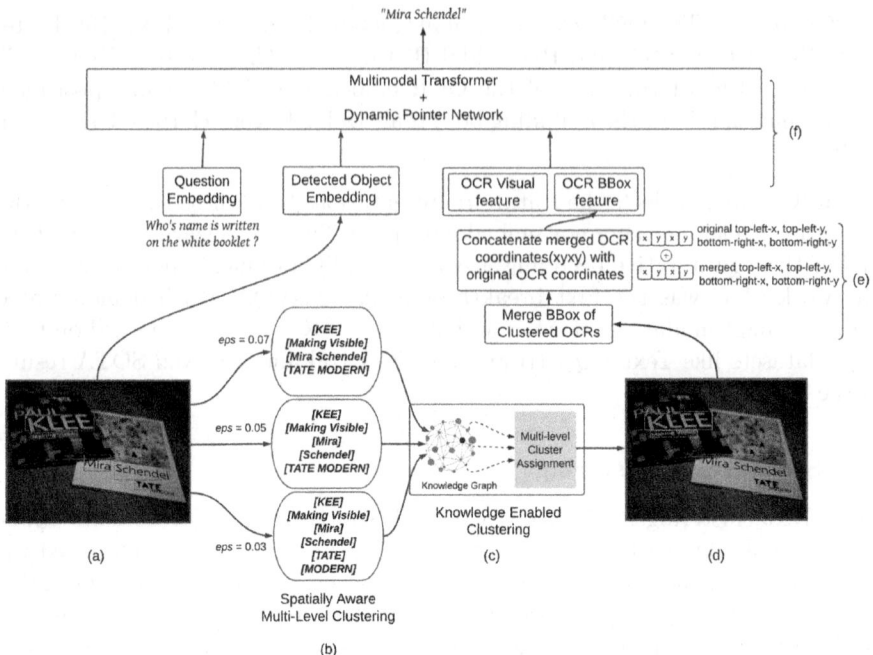

Fig. 2. An overview of our spatially-aware and knowledge-enabled approach. (a) Image with OCR Bounding boxes. (b) The OCR tokens are clustered at different levels based on the *eps* parameter that produces clusters with larger group sizes to smaller group sizes. (c) WikiData Knowledge graph is used to identify real-world entities from the clustered tokens. (d) The clusters are identified based on KG modifications and, (e) bounding boxes of grouped OCR tokens are concatenated with the original bounding boxes of tokens. (f) The concatenated bounding box vector is passed on to a multimodal transformer that eventually uses a Dynamic Pointer Network to produce the output.

3.1 Spatially Aware Clustering

We use the DBSCAN [29] algorithm to cluster bounding boxes in each image during the preprocessing stage. Each bounding box is represented by 17 features, 16 features coming from x, and y coordinates of the top left, top right, bottom left, bottom right, top midpoint, bottom midpoint, left midpoint, and right midpoint points of the bounding box and 17th feature is the height of the box. These features are passed to the DBSCAN algorithm for all the images and the clustering is tuned by the epsilon(*eps*) parameter that specifies how close boxes should be to each other to be considered a part of a cluster. The difference from traditional DBSCAN is that we use custom distance computation for clustering.

To compute the distance we first choose the two nearest points between the boxes using euclidean distance, and then the distance is penalized by the height difference between the two boxes. The idea here is to increase the distance between two boxes for clustering that has significant differences in their height even after being in close vicinity.

$$distance = d + \lambda \times \Delta H \qquad (1)$$

Here, d is the Euclidean distance between the two nearest points of bounding boxes, λ is a penalty parameter and ΔH is the height difference between the boxes.

The clustering is tuned in such a way that it focuses on grouping together tokens at a localized level (Fig. 3).

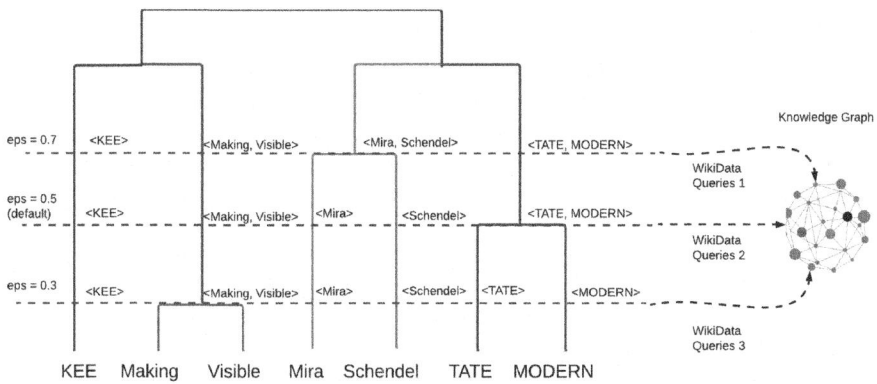

Fig. 3. Multi-level clustering happens at three *eps* values 0.07, 0.05, and, 0.03. Starting with *eps* = 0.07, the clustered tokens are concatenated and queried on WikiData. The same is done for other *eps* values and the largest string that finds a match in the WikiData query is retained in the final clustering. Refer to Sect. 3.2 for more details.

3.2   Knowledge Enabled Clustering

As demonstrated in Fig. 2, oftentimes, the text present in the image contains a real-world entity like the author's name, organization name, brand name, etc. Such real-world entities are stored in publicly available knowledge graphs. Our idea is to leverage a knowledge graph - WikiData to cluster real-world entities that are present in the image.

To cluster larger entities and to reduce the number of queries hit on WikiData APIs we take a sequential clustering approach where multi-level spatially aware clustering is done for each image. Initial clustering is done with a larger *eps* value, creating clusters with more tokens. In subsequent clustering, the *eps* value is decreased and clusters get smaller. This is done for *eps* values 0.07, 0.05, and 0.03. The sequence of steps in our proposed clustering is described below:

1. A global clustering variable(*global_cluster*) is initialized with −1 for all the tokens.

2. The clustering result with the same order is picked, and tokens from each cluster are joined into a single string S.
3. S is then queried on WikiData and if S exists as an entity then the tokens present in S are assigned the same cluster in the *global_cluster* and this cluster assignment cannot be changed by any further smaller cluster.
4. The above (3) process continues until the last cluster.
5. Finally, the tokens that still have no cluster assigned are passed through a cluster reassignment method, which simply assigns new cluster ids based on the *eps* 0.05 results.

Refer to Algorithm 1 for the pseudo-code of the approach.

3.3 Feature Engineering of Clustering Annotations

Scene Text Clustering is done during the preprocessing stage where every token is assigned an additional key *cluster_bbox* that stores the bounding box coordinate of the cluster that the token belongs to. During the training, this additional information is concatenated with every token's original bounding box coordinates vector. Thus, making the final OCR bounding vector 8-dimensional

$$
\begin{aligned}
x_n^{b'} = [&x_{top\_left}, y_{top\_left}, x_{bottom\_right}, y_{bottom\_right}, \\
&x'_{top\_left}, y'_{top\_left}, x'_{bottom\_right}, y'_{bottom\_right}]
\end{aligned} \tag{2}
$$

Algorithm 1. Algorithm to Assign Cluster Ids to each token in a single Image

---

$n \leftarrow$ number of OCR tokens in the image
token_list $\leftarrow$ list of OCR Tokens
global_cluster $\leftarrow [-1, -1, ..., -1]_{1 \times n}$
for eps_value in $[0.07, 0.05_{default}, 0.03]$ **do**   ▷ Decreasing order of *eps* value to ensure multi-level clustering
 cluster_ids = DBSCAN(eps=*eps_value*)
 for S in stringify(token_list, cluster_ids) **do**   ▷ Stringify joins tokens in the same cluster and creates a list of Strings
  if S in WikiData **then**
   for *token* in S.tokens() **do**
    cluster_id = *cluster_ids*[*token*]
    if *global_cluster*[*token*] == -1 **then**
     global_cluster[*token*] $\leftarrow$ *cluster_id*
    end if
   end for
  end if
 end for
end for
for token in token_list **do**
 if *global_cluster*$_{token}$ == -1 **then**
  global_cluster[*token*] $\leftarrow$ *token_cluster_id*$_{default}$ ▷ Cluster ids assigned here does not overlap with any existing id in the global cluster
 end if
end for

This information is further linearly projected (into the same hidden-size space as described in [12]) before adding it to the embedding formed from the FastText, Faster-RCNN and PHOC features. The OCR feature embedding is created as follows (similar to [12]):

$$x_n^{ocr} = LN(W_3 x_n^{ft} + W_4 x_n^{fr} + W_5 x_n^{p}) + LN(W_6 x_n^{b'}) \qquad (3)$$

where W_3, W_4, W_5 and W_6 are learned projection matrices and $LN()$ is a layer normalization; x_n^{ft}, x_n^{fr}, x_n^{p} and $x_n^{b'}$ are FastText vector, Faster-RCNN (appearance) feature, Pyramidal Histogram of Character (PHOC) feature and concatenation of location feature and cluster aggregation result respectively.

This embedding is now richer in terms of the positioning of the related "meaningful" entity to which the OCR token belongs.

Another variation experimented with the OCR cluster embedding was feeding it through a linear layer instead of concatenating it with the OCR token embedding. In this approach, we project the OCR cluster Bounding box $(x_n^{b''})$ and the OCR token Bounding Box (x_n^{b}) differently before adding them as described in the equation below.

$$x_n^{ocr} = LN(W_3 x_n^{ft} + W_4 x_n^{fr} + W_5 x_n^{p}) + LN(W_6 x_n^{b}) + LN(W_7 x_n^{b''}) \qquad (4)$$

where W_7 is a learned projection matrices and $LN()$ is a layer normalization; $x_n^{b''}$ is cluster aggregation result.

4 Experiments

4.1 Datasets

Text-VQA. One of the datasets extensively used for Text Visual Question Answering experiments is Text-VQA [1]. Text-VQA dataset contains images from the Open Images dataset [2] from categories containing text like "billboard", "traffic sign" and "whiteboard". The dataset contains 28,408 images and 45,336 questions asked by (sighted) humans over them. Each question-image pair has 10 ground truth annotations (given by humans). The training set contains 34,602 questions based on 21,953 images whereas the validation set contains 5,000 questions based on 3,166 images.

ST-VQA. The ST-VQA dataset [18] contains images from a combination of public datasets used for scene text understanding and general computer vision tasks. The ST-VQA comprises images from six datasets namely: ICDAR 2013 [3] and ICDAR 2015 [4], ImageNet [5], VizWiz [6], IIIT Scene Text Retrieval [7], Visual Genome [8], and COCO-Text [9]. ST-VQA dataset contains a total of 31,791 questions over 23,038 images. The training set contains 26,308 questions based on 19,027 images. We only use the training set for our experiments.

OCR-VQA-200K. The dataset OCR-VQA [10] is derived from [11]. This dataset contains cover images of the books including meta-data containing author names, titles and genres. The OCR-VQA dataset comprises 207,572 images and 1,002,146 question-answer pairs. The training set contains approximately 800,000 question-answer pairs whereas the validation set contains 100,000 pairs.

4.2   Implementation Details

Concatenation of OCR Bounding Boxes (Concat-Boxes). The image consists of multiple OCR tokens which are clustered by the Multi-level clustering module (refer to Fig. 2) according to the information from the Knowledge Graph. This process is described in greater detail in Sect. 3. The clustering algorithm identifies the labels of the OCR tokens and their corresponding cluster. Thus, every OCR token forms a part of a larger group of OCR tokens. In this experiment, the OCR bounding box feature is concatenated with the information of the cluster (minimum and maximum boundaries of the box in both dimensions). The aim is to find the tightest bounding box which can cover all the OCR tokens in the cluster. We utilise the annotations of the Rosetta OCR system [28] for our experiments. The resultant bounding box OCR embedding $(x_n^{b'})$ is thus 8-dimensional (4 (token) + 4 (concat box)). This information is further linearly projected (into the same hidden-size space as described in [12]) before adding it to the embedding formed from the FastText, Faster-RCNN and PHOC features. The OCR feature embedding is created as discussed in Subsect. 3.3 (similar to [12]).

Following this, we apply a similar training strategy as described by [12] for the transformer module. We conduct two runs with this model, (i) Only Text-VQA training data, and (ii) Text-VQA + ST-VQA training data.

The experiment results are presented in Table 2. The validation set questions were divided into three subsets to evaluate the impact of the clustering strategy: (i) Single-word answers – QA pairs with the answer as a single word, (ii) Multi-word answers – QA pairs with the answer as two or more words, and (iii) Limited OCR tokens – QA pairs where the number of OCR tokens is within the 75th percentile of the overall number of OCR tokens distribution. The third subset was selected to further evaluate the effectiveness of clustering, as clustering is most effective when the number of tokens in the scene text is limited.

This strategy (Concat-Boxes) helps the model increase the single-word accuracy (in the first run with only Text-VQA training data) as compared to the baseline by **0.03%**. Additionally, it also boosted the **overall model accuracy** by nearly **0.4%** in the second run. Moreover, it also pushed the **multi-word** accuracy by **0.72%**, and **0.61%** in **limited OCR tokens** set. The results of the experiments on the dataset OCR-VQA-200K [10] are presented in Table 1.

Linear Projection of the Concatenation of OCR Bounding Boxes (Concat-Boxes Linear Projection). The first experiment involved directly concatenating the OCR cluster Bounding Box with the OCR token Bounding Box. In this approach, we project the OCR cluster Bounding box $(x_n^{b''})$ and the

OCR token Bounding Box (x_n^b) differently before adding them as described in Subsect. 3.3.

This experiment was designed to investigate whether different projection matrices for OCR tokens and cluster Bounding Box would help the transformer get better information. Similar to the previous experiment, we conduct two runs with this model, (i) Only Text-VQA training data and (ii) Text-VQA + ST-VQA training data.

The results of the experiment are tabulated in Table 2. The results demonstrate that we only improve the single-word accuracy at a marginal cost of overall accuracy, in the first run. In the second run, there is a marginal increment in the single-word and multi-word data subsets.

Thus, we can conclude that the addition of the linear projection of the cluster bounding box coordinates makes the model less attentive to the OCR clusters (Figs. 4 and 5).

Table 1. Accuracy scores for experiments 1 and 2. Results in the second row represent the validation accuracy of LOGOS [26]. The improvements over the baseline are shown in bold. The best configuration is Concat-Boxes with training data from TextVQA and ST-VQA. There is a considerable improvement of nearly 0.4% in the overall accuracy, 0.72% in the Multi-word subset, and 0.61% in the limited OCR (as compared to the baseline). As the code for [26] is not public, we created our own implementation of the algorithm.

Model	Dataset for training	Validation Accuracy			
		TextVQA (entire)	Single-word	Multi-word	Limited OCR
Baseline	TextVQA	39.65	47.13	34.28	41.52
	TextVQA + STVQA	40.24	**48.36**	34.42	41.82
LOGOS[a] [26]	TextVQA	38.55	46.57	32.81	40.11
	TextVQA + STVQA	39.51	47.6	33.7	41.2
(1) Concat-Boxes	TextVQA	39.32	47.16	33.7	40.92
	TextVQA + STVQA	**40.64**	48.31	**35.14**	**42.43**
(2) Concat-Boxes Linear Projection	TextVQA	39.28	47.94	33.07	40.72
	TextVQA + STVQA	40.19	47.88	34.68	41.32

[a] Our implementation of LOGOS [26]

Table 2. Accuracy numbers for Concat-Boxes model on OCR-VQA-200K citeocrvqa dataset. There is an improvement of 0.13% in the overall accuracy. Additionally, the multi-word analysis also shows an improvement of 0.18%.

Model	Dataset for training	Validation Accuracy			
		OCR-VQA-200K (entire)	Single-word	Multi-word	Limited OCR
Baseline	OCR-VQA-200K	63.48	86.80	44.80	63.47
(1) Concat-Boxes	OCR-VQA-200K	**63.61**	**86.87**	**44.98**	**63.53**

Q: What band is featured on all three items?
Baseline: beatles beates
Ours: the beatles

Q: This establishment is called coach and what?
Baseline: coach youngers
Ours: coach & horses

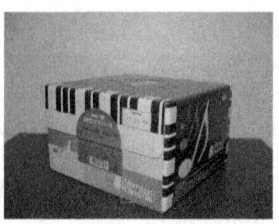

Q: What does the box say at the bottom?
Baseline: christmes
Ours: merry christmas

Q: Who is the author of this book?
Baseline: brian aldiss
Ours: brian w. aldiss

Q: Where is this mug featuring?
Baseline: daytona beach fla
Ours: daytona beach

Q: What's the book title?
Baseline: american tales
Ours: american gothic tales

Q: What is this publication called?
Baseline: global book
Ours: the open book

Q: What does the light sign read on the farthest right window?
Baseline: all light
Ours: bud light

Q: What is on the bottom left?
Baseline: ouble bonus
Ours: double bonus

Fig. 4. Qualitative Examples: These examples show the QA pairs (from Text-VQA dataset [1]) along with the bounding box highlighting the captured meaningful entity based on the question. The baseline considered was M4C [12] and our model is as shown in Table 1(1).

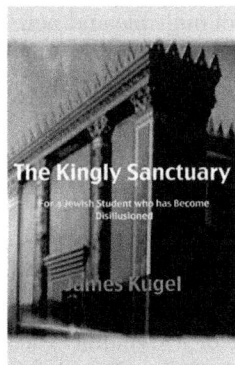

Q: What is the title of this book?
Baseline: the kingly sanctuary for a jew-
ish student has become disillusioned
Ours: the kingly sanctuary

Q: Who is the author of this book?
Baseline: wayne w. dyer
Ours: dr. wayne w. dyer

Q: Who wrote this book?
Baseline: ursula k. guin
Ours: ursula k. le guin

Q: What is the title of this book?
Baseline: the aesthetics of aquinas
Ours: the aesthetics of thomas aquinas

Fig. 5. More Qualitative Examples: These examples demonstrate the subjective improvement on OCR-VQA-200K [10] dataset. The baseline considered was M4C [12] and our model is as shown in Table 2(1).

5   Discussion

1. Text-KVQA dataset [27] contains questions that need to be answered using external knowledge whereas Text-VQA [1] doesn't need the same. The QA pairs in Text-VQA dataset mostly refer to a "meaningful entity" present in the scene text data. As opposed to the knowledge-graph-based works on the Text-KVQA datasets, this work focuses on finding a meaningful entity in the scene text data using external knowledge. To the best of our understanding, our knowledge-graph-based clustering approach is the first such attempt to answer questions on Text-VQA dataset. The clustering algorithm is transferable to a

different domain which can expect the integration of multiple and new knowledge graphs. This emphasizes the scalability and modularity of the algorithm.

2. Although, clustering the meaningful entities in the scene text improves the
accuracy for many QA pairs, for some examples as shown in Fig. 6, the answers
need to be extracted partially from that entity. In those scenarios, clustering information can force the multimodal transformer to output the entire
clustered entity as the answer.

3. Aside from the height discrepancy penalty, which aims to address variations
in size among OCR tokens, features based on font style can also enhance clustering and open up the potential for further investigation. Another potential
avenue for model improvement is the integration of a dynamic vocabulary
of OCR clusters, similar to the approach used with OCR tokens in the M4C
decoder [12]. This can lead to improved predictions and provide an alternative
option to simply choosing a cluster as the answer.

Q: How many sides of this in-
tersection have a stop sign?
Baseline: 4
Ours: 4-way

Q: What city is listed on the
bottle?
Baseline: milan
Ours: milan mich.

Q: what does the sign say is
"back"?
Baseline: driving
Ours: driving is back.

Fig. 6. Failure cases of the clustering approach.

6   Conclusion

The paper presents a new approach to scene text clustering, based on external
knowledge, to address the text-VQA problem. By clustering OCR tokens and
prioritizing spatial alignment of scene text, our approach generates more informative queries for the knowledge graph. We also show that the information about
grouped tokens can be efficiently transmitted to a multimodal transformer-based
framework through box concatenation embeddings. Our objective and subjective evaluations on the Text-VQA dataset demonstrate the significance of the
proposed method, particularly for multi-word answers.

References

1. Singh, A., et al.: Towards VQA models that can read. arXiv 2019, arxiv.org/abs/1904.08920
2. Krasin, I., et al.: OpenImages: a public dataset for large-scale multi-label and multi-class image classification (2016)
3. Karatzas, D., et al.: ICDAR 2013 robust reading competition. In: 2013 12th International Conference on Document Analysis and Recognition, pp. 1484–1493 (2013)
4. Karatzas, D., et al.: ICDAR 2015 competition on Robust Reading (2015)
5. Deng, J., Dong, W., Socher, R., Li, L., Li, K., Fei-Fei, L.: ImageNet: a large-scale hierarchical image database. In: 2009 IEEE Conference on Computer Vision and Pattern Recognition, pp. 248–255 (2009)
6. Gurari, D., et al.: VizWiz grand challenge: answering visual questions from blind people. arXiv 2018, arxiv.org/abs/1802.08218
7. Mishra, A., Alahari, K., Jawahar, C.: Image retrieval using textual cues. In: 2013 IEEE International Conference On Computer Vision, pp. 3040–3047 (2013)
8. Krishna, R., et al.: Visual genome: connecting language and vision using crowd-sourced dense image annotations. arXiv 2016, arxiv.org/abs/1602.07332
9. Veit, A., Matera, T., Neumann, L., Matas, J., Belongie, S.: COCO-text: dataset and benchmark for text detection and recognition in natural images. arXiv 2016, arxiv.org/abs/1601.07140
10. Mishra, A., Shekhar, S., Singh, A., Chakraborty, A.: OCR-VQA: visual question answering by reading text in images. In: ICDAR (2019)
11. Iwana, B., Rizvi, S., Ahmed, S., Dengel, A., Uchida, S.: Judging a book by its cover. arXiv 2016, arxiv.org/abs/1610.09204
12. Hu, R., Singh, A., Darrell, T., Rohrbach, M.: Iterative answer prediction with pointer-augmented multimodal transformers for TextVQA. arXiv 2019, arxiv.org/abs/1911.06258
13. Vaswani, A., et al.: Attention is all you need. arXiv 2017, arxiv.org/abs/1706.03762
14. Devlin, J., Chang, M., Lee, K., Toutanova, K.: BERT: pre-training of deep bidirectional transformers for language understanding. arXiv 2018, arxiv.org/abs/1810.04805
15. Ren, S., He, K., Girshick, R., Sun, J.: Faster R-CNN: towards real-time object detection with region proposal networks. arXiv 2015, arxiv.org/abs/1506.01497
16. Bojanowski, P., Grave, E., Joulin, A., Mikolov, T.: Enriching word vectors with subword information. arXiv 2016, arxiv.org/abs/1607.04606
17. Almazán, J., Gordo, A., Fornés, A., Valveny, E.: Word spotting and recognition with embedded attributes. IEEE Trans. Pattern Anal. Mach. Intell. **36**, 2552–2566 (2014)
18. Biten, A., et al.: Scene text visual question answering. arXiv 2019, arxiv.org/abs/1905.13648
19. Jiang, Y., Natarajan, V., Chen, X., Rohrbach, M., Batra, D., Parikh, D.: Pythia v0.1: the winning entry to the VQA challenge 2018. arXiv 2018, arxiv.org/abs/1807.09956
20. Biten, A., et al.: ICDAR 2019 competition on scene text visual question answering (2019)
21. Gao, D., Li, K., Wang, R., Shan, S., Chen, X.: Multi-modal graph neural network for joint reasoning on vision and scene text. arXiv 2020, arxiv.org/abs/2003.13962
22. Kant, Y., et al.: Spatially aware multimodal transformers for TextVQA. arXiv 2020, arxiv.org/abs/2007.12146

23. Chen, X., et al.: PaLI: a jointly-scaled multilingual language-image model. arXiv 2022, arxiv.org/abs/2209.06794
24. Wang, J., et al.: GIT: a generative image-to-text transformer for vision and language. arXiv 2022, arxiv.org/abs/2205.14100
25. Kil, J., et al.: PreSTU: pre-training for scene-text understanding. arXiv 2022, arxiv.org/abs/2209.05534
26. Lu, X., Fan, Z., Wang, Y., Oh, J., Rose, C.: Localize, group, and select: boosting text-VQA by scene text modeling (2021). arxiv.org/abs/2108.08965
27. Singh, A., Mishra, A., Shekhar, S., Chakraborty, A.: From strings to things: knowledge-enabled VQA Model that can Read and Reason. In: ICCV (2019)
28. Borisyuk, F., Gordo, A., Sivakumar, V.: Rosetta: large scale system for text detection and recognition in images. CoRR abs/1910.05085 (2019). arxiv.org/abs/1910.05085
29. Ester, M., Kriegel, H., Sander, J., Xu, X.: A density-based algorithm for discovering clusters in large spatial databases with noise. In: Knowledge Discovery and Data Mining (1996)

Scene Table Structure Recognition with Segmentation and Key Point Collaboration

Zhuoming Li[1], Fan Peng[1], Yang Xue[1]([✉]), Ni Hao[2], and Lianwen Jin[1]

[1] School of Electronic and Information Engineering, South China University of Technology, Guangzhou, China
{202121014036,202020112442,yxue,eelwjin}@mail.scut.edu.cn
[2] Department of Mathematics, University College London, London, England
h.ni@ucl.ac.uk

Abstract. This paper proposes a Segmentation and Key point Collaboration Network (SKCN) for structure recognition of complex tables with geometric deformations. First, we combine the cell regions of the segmentation branch and the corner locations of the key point regression branch in the SKCN to obtain more reliable detection bounding box candidates. Then, we propose a Centroid Filtering-based Non-Maximum Suppression algorithm (CF-NMS) to deal with the problem of overlapping detected bounding boxes. After obtaining the bounding boxes of all cells, we propose a post-processing method to predict the logical relationships of cells to finally recover the structure of the table. In addition, we design a module for online generation of tabular data by applying color, shading and geometric transformation to enrich the sample diversity of the existing natural scene table datasets. Experimental results show that our method achieves state-of-the-art performance on two public benchmarks, TAL_OCR_TABLE and WTW.

Keywords: table structure recognition · segmentation and key point collaboration · centroid filtering NMS · online generation of tabular data

1 Introduction

Table is widely used as an effective representation of structured data in various types of documents in daily life. With the rise digitalization, table recognition has become an important research topic in the field of document understanding. How to correctly recognize the structure of a table is an important step in table recognition, whose main task is to identify the internal structure of a table. It aims to locate all the physical position of cells in the table and obtain information about the rows and columns in order to better understand the table as a whole.

Z. Li and F. Peng—Authors contributed equally as first author.

G. A. Fink et al. (Eds.): ICDAR 2023, LNCS 14188, pp. 295–310, 2023.
https://doi.org/10.1007/978-3-031-41679-8_17

However, it's a challenging task for natural scene tables which can be complex in structure, vary in style and content, and may cause geometric distortions or even bending during the image acquisition process. With the explosive growth in the number of documents, applying table detection and table recognition techniques to reconstruct tables from document images has become one of the important techniques in current document understanding systems that can facilitate many downstream tasks and has significant research value.

Early table recognition studies mainly focused on hand-crafted features and heuristic rules [7–9]. Most of them were applied to simple table structures or specific data formats, such as PDF. Recently, research scholars have proposed more general models for structure recognition, such as LGPMA [22] and Flag-Net [14]. The advantage is only one model is needed for all types of wired and wireless tables in document and natural scenes. However, these models are generally complex and the feature that can be utilized is the intersection of features extracted from different types of tables, thus ignoring the unique feature of each type of table. In real life, table recognition tasks are usually applied to fixed scenes, which require more targeted table recognition models. For example, for distorted wired table recognition in natural scenes, in order to obtain accurate cell boundaries, it is necessary to take full advantage of the most obvious visual features of the cells, i.e., the box lines and four corner points of each cell; whereas generic models, in order to be applicable to both wired and wireless table recognition, often do not take full advantage of these most salient visual features. Therefore, it also makes sense to design a specialized table recognition model to take full advantage of the salient features of each type of table.

Cycle-CenterNet [17] proposed a detection-based table structure recognition method that works well for wired table recognition in seven sub-scenarios. It first locates the four corner points of each cell and further infers the overall logical structure of the table from the coordinates of the cell. However, it only utilizes the corner point features of the wired tables and ignores the box line features of the tables. For table recognition of complex natural scenes with challenges such as geometric distortion, overlay, occlusion and blurring, it is inadequate to completely describe the overall position information of a cell by only four corner points. Better results can be achieved if a scheme can be proposed to extract both corner point features and box line features of wired tables.

Based on this, we propose a Segmentation and Key point Collaboration Network (SKCN) that combines the cell region of the segmentation branch and the corner locations of the key point regression branch to obtain a more reliable detection bounding box for better recognition performance. On the one hand, these two branches can assist each other during training. On the other hand, their respective results can interact and fuse to obtain refined detection results. In order to effectively filter redundant detected bounding boxes, we propose a centroid filtering algorithm based on the standard NMS algorithm, which achieves accurate cell detection results. Base on the refined cell boxes, we design a post-processing scheme to predict the logical relationship of the cells to recover the structure of the table.

The main contributions of this paper are as follows:

1. We combine the cell region of the segmentation branch and the corner locations of the key point regression branch to obtain a refined detected bounding box.

2. We propose a Centroid Filtering-based Non-Maximum Suppression algorithm (CF-NMS). To address the challenge of overlapping bounding boxes in table recognition tasks, we use CF-NMS to filter out prediction results with high IOU values that overlap with the target cell, thus improving the model detection performance.

3. We propose a module to generate tabular data online by applying color, shading and geometric transformation to enrich the sample diversity of existing natural scene table datasets.

2 Related Work

Early methods for table structure recognition [7–9, 24, 26] were mainly based on well-designed handcrafted features and heuristic rules. Most of these methods were applied to specific data formats, such as PDF files. However, in these traditional methods, there are strong assumptions about the layout of the tables, which limits their generality. With the rapid development of deep neural networks, image-based table structure recognition methods have shown great potential and outperform traditional methods by a large margin. We roughly divide these methods into four categories: image-to-token generation method, graph-based method, segmentation-based method, and object detection-based method.

2.1 Image-to-Token Generation Method

This method treats table structure recognition as an image-to-token generation problem, typically using an encoder-decoder structure that directly converts the source table image into target token to adequately describe tabular data structure and its cell content. Existing approaches have tried several attempts to convert table images into symbols or HTML sequences [3, 11, 30, 33]. However, these methods usually rely on a large amount of data to train for convergence. In some cases, especially with large and complex tables, this approach may lead to performance degradation. Due to the limited length of the sequences, these methods usually adopt certain trade-off strategies for large tables and have difficulty in tuning parameter and network design with their weep explanatory.

2.2 Graph-Based Method

The graph-based approach [21, 23] treats the bounding boxes of cell regions or text regions as nodes in a graph and uses graph neural networks to predict the logical relationship of each sampled node pair. GraphTSR [1] introduces the attention module to predict whether the sampled node pair belong to same row or same column. FLAG-Net [14] combines Transformer with graph-based context aggregator in an adaptive way to exploit the advantages of both. NCGM [13] leverages graphs and modality interaction to enhance the multi-modal representation of text embeddings. However, these methods rely on bounding boxes of cell regions or text regions used as additional input, which are not available directly from the table images, thus bringing extra network cost.

2.3 Segmentation-Based Method

The segmentation-based approach first obtains the segmentation results from the table image and then parses the segmentation results to reconstruct the table structure. There are two broad types of this approach. One is to first obtain the segmentation of the rows and columns, and then use the segmentation results to grid out the cell boundaries. DeepdeSRT [25] and TableNet [19] semantically segment rows and columns, and intersect the segmentation results of rows and columns to obtain cell segmentation. To deal with spanning cells, SPLERGE [28] uses the split model to segment cell boundaries and then uses the merge model to further merge adjacent cells to obtain spanning cell boxes. SEM [31] follows the idea of multimodality and introduces textual feature to fuse with visual feature for each cell. The other is to recover cell boundaries to obtain cell boxes directly. CascadeTabNet [20] classifies tables into bordered and borderless tables, then predicts cell segmentation for borderless tables and extracts cells from bordered tables using traditional algorithms. LGPMA [22] combines local and global feature to accurately reconstruct cell boundaries by using soft pyramidal masks. However, these methods cannot handle distorted tables because they rely on table-axis alignment.

2.4 Object Detection-Based Method

The method based on object detection first obtains the basic cells of a table from a table image by directly detecting the bounding box of a cell or text. Heuristic rules are then used to predict the logical relationships between detected cells to further reconstruct the logical structure of the table. [23,27,32] propose to detect the bounding boxes of table cells directly. After obtaining the bounding boxes of cells, [23,32] designed some rules for clustering cells into rows and columns. However, the methods mentioned above assume that the table is well aligned and the target bounding boxes are rectangular, which are not suitable for natural scene tables. Cycle-CenterNet [17] introduces a cyclic pairing module to predict quadrilateral bounding boxes. Our method also uses quadrilateral bounding boxes for detection, which are more adaptable to the complexity of natural scene tables and achieve better performance in experiments. However, quadrilateral bounding boxes are still difficult to accurately describe curved cells and also may bring the potential of degrading detection performance. Sequential-free box discretization (SBD) [16] parameterizes bounding boxes as key edges and predicts the coordinates of four key points of the box from which the box is subsequently recovered. It can output more qualitative and accurate results in natural scene table recognition. Therefore, we use SBD to predict four corner points in our method. The model is built based on the box discretization network [16], which use SBD as an additional branch to Mask-RCNN [6].

3   Methodology

Our approach consists of two main components: the Segmentation and Key point Collaboration Network (SKCN) and the Centroid Filtering-based Non-Maximum Suppression module (CF-NMS). The former is to obtain cell regions from both the segmentation branch and the key point regression branch to generate refined bounding boxes. The latter deals with the problem of overlapping cell boxes under the natural scene table. After obtaining bounding boxes, we use a post-processing algorithm to cluster cells into rows and columns and then parse the table structure. The details of our approach are described separately in the following sections.

3.1   SKCN

As shown in Fig. 1, the input image is first transformed into the output of four branches, i.e., box classification, box regression, box segmentation and point regression. The box regression branch outputs the minimum area bounding rectangle about the cell. The box classification branch predicts the category of the cell, such as images, text, formulas and other categories. Among them, the box segmentation branch and the point regression branch play an important role. The box segmentation branch focuses on the box-line characteristics of the wired table to get the segmentation result of the cell region wrapped by boundaries, which better adapts to the arbitrary deformation of the cell. The point regression branch mainly locates the four key points by using the key-point characteristics of the cell. The advantage of our model is to fully capture the feature of table elements to achieve a more accurate detection.

Since box segmentation and point regression serve for the same task of cell detection, previous studies usually selected only one of the two in this case. However, we believe that each of these two branches has its own characteristics. The box segmentation branch outputs pixel-level instance segmentation of cells, so the predicted box will be closer to ground truth. However, when it comes to complex tables with geometrical distortions or incomplete linear characteristic, it is difficult to separate out closely adjacent cell instances, which can easily lead to missed detection. The point regression branch only needs to return the four key points of the target. We first predict the eight boundary key edges of the cell in the process of locating the key points, and then combine them into four key points, which makes it easier to learn. The SBD branch tends to predict cells more completely, but the drawback is also obvious. If a predicted error occurs at one of the four key points, the detected bounding box becomes imprecise. Based on this, we propose to fuse results by a proper process to achieve better performance. We give these two branches different priorities in different

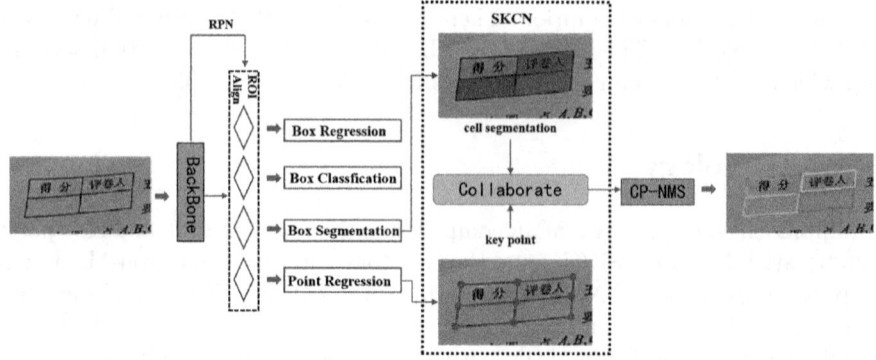

Fig. 1. The architecture of SKCN

confidence ranges. Firstly, we add a small constant of 0.03 to the segmentation results with confidence higher than 0.9. In this way, the high confidence segmentation results are preferred by the Non-Maximum Suppression. Then we select the results of SBD with confidence higher than 0.2 and mix the results of both branches together into the CF-NMS module to obtain the final results.

3.2 Key Point Prediction

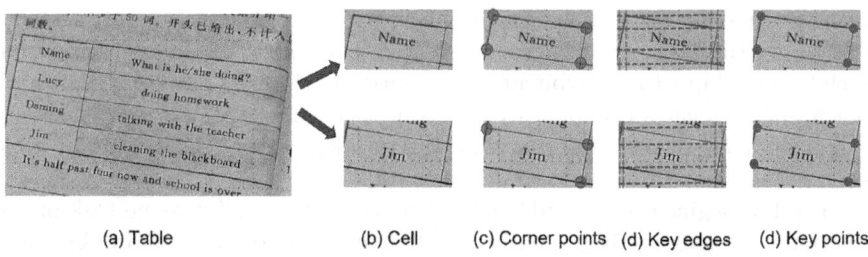

(a) Table   (b) Cell   (c) Corner points (d) Key edges   (d) Key points

Fig. 2. We visualize the results of the corner point detection method and our key point detection method. For the cell "Name", both methods predict correctly. However, for the cell "Jim", the corner point detection method predicts incorrectly because the cell misses a lower-left corner point. However, our key point detection method avoids this error by obtaining the critical point from the critical edge with the help of SBD's edge detection.

The most intuitive way of key point regression is to directly predict the corner points of the target to localize it like CornerNet [10], and then Liu et al. [16] proposed a method called SBD to solve the LC (Learning Confusion) problem [15], which first predicts the eight boundary values of the target and then combines them to obtain the four key points of the target. We refer to it to design our key point regression branch. Compared to direct corner point prediction, our method has certain advantages. As shown in Fig. 2, the corner characteristics of the cells may be incomplete for defective tables and wireless tables. In this case, it is difficult for the direct corner point prediction method to accurately predict the four corner points, which brings a large error. But for SBD, the eight bounding key edges of the cell are obtained first. Key points can be predicted with the help of border information, text information, not just relying on the four corners. Our key point prediction method locates four key points of a cell with the help of the location of eight boundaries, which is more adaptive in natural scene table recognition.

3.3 Centroid Filtering Non Maximum Suppression (CF-NMS)

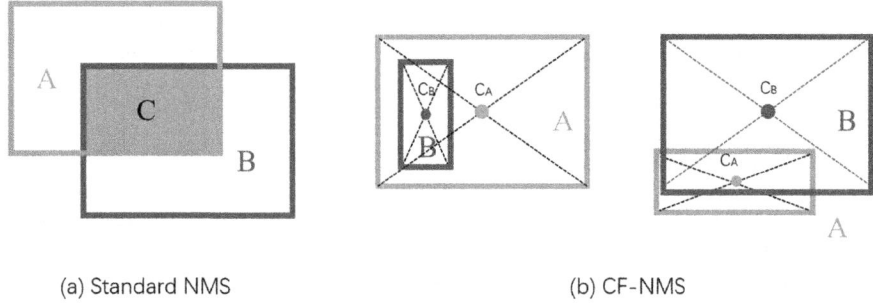

(a) Standard NMS   (b) CF-NMS

Fig. 3. Illustration of NMS and our CF-NMS. (a) is the standard NMS and (b) is our proposed CF-NMS.

In the process of fusion of two branches of SKCN, there's bound to be many overlapped and redundant bounding boxes. How to effectively filter the wrong detection boxes is the key to ensure the model performance. The process of the standard NMS algorithm (Fig. 3(a)) consists of: (1) set the confidence threshold for the target box, (2) arrange the list of candidate boxes in descending order of their confidence, (3) select the box A with the highest confidence and add it to the output list, while removing it from the list of candidate boxes, (4) calculate the IOU value of all boxes in the list of candidate boxes with A, and remove the candidate boxes with IOU values greater than the threshold value, (5) repeat the above process until the list of candidate boxes is empty, and return the output list. The effectiveness of the standard NMS algorithm depends on the setting of the IoU threshold. A relatively high threshold can result in a large number of false positives, while a lower threshold can result in missing highly overlapped correct results.

The standard NMS algorithm is not fully applicable to the filtering of candidate boxes in natural scene tables where many complexities exist. On the one hand, the threshold of NMS cannot be set too low, because the bounding boxes of adjacent cells in the distorted skew table often have a large part of overlapping areas. In order to ensure the integrity of cell detection, we have to set the NMS threshold higher, which results in many redundant detection boxes not being filtered out. On the other hand, in some large tables, the GT boxes of cells are relatively dense and the area of each box is small. Therefore, it can also be regarded as a dense detection task of small targets. In this case, it is easy to predict some bounding boxes that are wrapped internally or around the perimeter of the correct box. Such errors are unavoidable due to the relatively high NMS threshold. Therefore, in order to solve the aforementioned problems of standard NMS in cell detection tasks, a new NMS algorithm is urgently needed to solve the existing problems of redundant bounding box detection.

Therefore, we propose a new non-maximum suppression algorithm based on centroid filtering (CF-NMS) to avoid threshold setting. CF-NMS filters overlapping bounding boxes by centroids, as shown in Fig. 3(b). Assuming that box A is the correct bounding box to be picked out, if the center of box B is inside box A, or conversely the center of box A is inside box B, then box B is judged to be redundant. This can effectively eliminate the case of nested detection boxes and make the detection results more accurate.

3.4 Tabular Data Augmentation

To address the lack of tabular datasets, we propose a tabular data enhancement module (TabSynth) to expand the number and diversity of tables and improve the performance of the model online. We propose three types of enhancement methods. The first is color variation, whose change is achieved by changing the HSV value of the table image. The second is shading transformation, which changes the lighting conditions of the table by combining the collected shadow photos with the table image to get a new table image with shading. And the third is geometric transformation that changes the degree of tilt and distortion of the table. The steps to achieve it are similar to the document image composition process described in DocUNet [18]. Our enhancement is not a random enhancement, but a targeted solution to two problems. Firstly, the distribution of various types of tables in the existing datasets is not uniform. Some types of tables with larger percentages are better trained and therefore more likely to yield better results than others, while some types of tables with smaller percentages are not sufficiently trained, which often leads to poorer performance. Therefore, we address the problem of data distribution by using TabSynth to enrich the sample diversity of the existing natural scene table datasets to enable each type of table to be adequately trained. Secondly, the existing datasets also have some tables with extreme aspect ratios, distortions and skews, and the structure recognition of these tables is also very challenging with existing methods. Therefore, our augmentation module can increase the number of these difficult samples in a targeted manner, so that the structure recognition model can fully learn the characteristics of the difficult samples and achieve better performance Fig. 4.

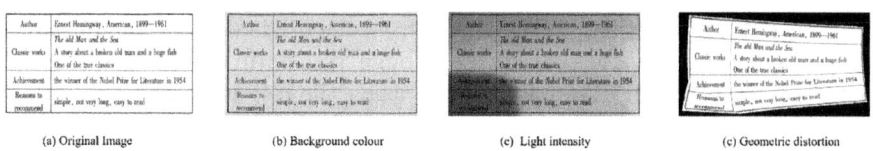

| (a) Original Image | (b) Background colour | (c) Light intensity | (c) Geometric distortion |

Fig. 4. Example of some augmented tabular data.

3.5 Table Structure Recovery

After obtaining refined detection cells, we further designed an adaptive adjacency matching algorithm to reconstruct the table structure. First, the four corner points of all cells in the table are arranged in the order of top-left, top-right, bottom-right and bottom-left. Then, we propose a center line matching strategy to perform row/column matching on these cells. For example, in row matching we first use the center point of the right boundary to match the rights, and adjust the coordinates of the right boundary used for matching according to the size of the matched cells, and then pass them to the right side one by one. Left row matching is similarly. Since cells in natural scenes are not usually aligned, and the idea of dealing with cross-row and cross-column cases is to use small cells to match large cells, we use an adaptive boundary matching strategy, which means that the current cell boundary used for matching will be adjusted according to the matched cells. For paired cell boxes in right matching, $\{(x_1, y_1), (x_2, y_2), (x_3, y_3), (x_4, y_4)\}$ and $\{(x'_1, y'_1), (x'_2, y'_2), (x'_3, y'_3), (x'_4, y'_4)\}$, if $\frac{y_2 + y_3}{2} >= y'_1$ and $\frac{y_2 + y_3}{2} <= y'_4$, the paired boxes are predicted to belong to the same row. Then the coordinates of the right border used for matching are adjusted by $y_2 = min(y_2, y'_2)$ and $y_3 = max(y_3, y'_3)$.

4 Experiments

In this section, we conducted experiments on two publicly available natural scene table datasets to evaluate the performance of our proposed table structure recognition method. To verify the effectiveness of the SKCN and the CF-NMS for the table structure recognition task, we conducted ablation experiments. The following are the relevant details of the experiments.

4.1 Datasets and Evaluation Metrics

Datasets. We evaluate our method on two publicly available natural scene table datasets, WTW and TAL_OCR_TABLE.

WTW [17] is a challenging and complex dataset for table structure recognition in the wild with 10970 training images and 3611 testing images, a sum of 14581 images. WTW divides the data into 7 cases by their own characteristics and unique challenges: simple, inclined, extreme aspect ratio, occluded and blurred, overlaid, multi-color and gird, and curved. The dataset annotation contains table ids, table coordinates, cell coordinates and row/column information about cells. We cropped out table regions from the original images and used the tilt angle of the table regions obtained by Hough Transform to rotation correction for training and testing. We followed [17] using the cell adjacency relationship (IoU = 0.6) [4] as the evaluation metric for this dataset. There are two versions of the evaluation metric for cell adjacent relationship, ICDAR2013 [5] and ICDAR2019 [4]. Because some tabular datasets do not have textual annotations, such as WTW, the previous version cannot be used in this case. We used the more general version of ICDAR2019 without exact text-matching.

TAL_OCR_TABLE (TAL) [2] is a natural scene table dataset provided in the PRCV2021 TAL table recognition competition, which focuses on wired tables for educational scenarios. The dataset contains 18,000 images, 16,000 of which have provided annotations for training and 2,000 for testing. The annotation of the dataset includes the physical location of the cells and the HTML code of the table. The physical location of the table is annotated by the four vertices of the quadrilateral. We also cropped out the table regions from the original images for training and testing.

Evaluation Metrics. There are two common evaluation metrics used in table recognition tasks, TEDS and cell adjacent relationship.

Tree Edit Distance based Similarity (TEDS) [33] represents the logical structure of a table with a tree structure and examines the table structure recognition results at the global tree-structure level. It uses the tree edit distance to evaluate the accuracy of table structure recognition, with higher values being better. The TEDS results contain the extra results of text recognition, and taking OCR errors into account may lead to unfair comparisons, since previous work used different OCR models. Therefore, the TEDS metrics in this paper only calculate the results for the logical structure of the table, without considering the OCR recognition results.

Cell adjacent relationship [4] is used to evaluate the effectiveness of structure recognition by the accuracy of the physical location and the row/column coordinates of each sampled adjacent cell pair. The adjacency relationship of each cell is generated with its horizontal and vertical adjacent cells. Then precision, recall and F1 scores are calculated to compare the predicted relationships with the ground truth.

4.2   Implementation Details

All experiments were implemented in PyTorch with 4x2080Ti GPUs. In Table 1, we compared the experimental results of different backbones. Since the difference between them is not very significant, we regard ResNet-50 as the backbone of network by default in the subsequent experiments. From the comparison of the results of different cell detection strategies, the accuracy of the box segmentation branch is higher than that of the key point regression branch, while the recall is lower. We further find that the SKCN after the collaborative processing of these two strategies can improve the prediction results synthetically. After adding

Table 1. Results with different backbones and cell detection strategies on TAL dataset. Here, * denotes using our online tabular data generation module TabSynth. SEG refers to the box segmentation. KEY refers to the key point regression.

Training data	Backbone	Strategy	Prec. (%)	Rec. (%)	F1. (%)
TAL	ResNet-50	SEG	99.5	98.89	99.2
		KEY	99.12	99.41	99.27
		SKCN	99.54	99.4	99.47
TAL	ResNet-101	SEG	99.6	98.89	99.24
		KEY	99.18	99.43	99.31
		SKCN	99.6	99.39	99.49
TAL*	ResNet-50	SEG	99.87	99.36	99.61
		KEY	99.78	99.84	99.81
		SKCN	**99.88**	**99.82**	**99.86**

our TabSynth module for training, the detection performance is also further improved, and the effect of this improvement is greater than replacing ResNet-50 with ResNet-101. This module allows us to adequately explore the potential of the model and achieve better results with a smaller cost for the model.

4.3 Comparisons with Prior Arts

We have compared our proposed method with several state-of-the-art methods on the public datasets TAL and WTW. our method achieves a state-of-the-art performance of 99.35% in terms of TEDS, as shown in Table 2. The experiments for SPLERGE [28] and CascadeTabNet [20] were reproduced based on the authors' original design. Since they are designed for scanned tables, they could not perform well in natural scenarios. To validate the effectiveness of our method on boundary warping or bending tables in natural scenarios, we conducted experiments on the WTW dataset. The results in Table 3 show that our method outperforms existing methods in terms of F1 scores for cell adjacent relationship, improving by 1.2% over Cycle-CenterNet, designed specifically for natural scenes, and by 0.2% over TSRFormer [12], which is able to robustly identify the structure of distorted tables with and without borders.

To better verify the robustness of our approach to complex situations, we analyzed the F1 scores of different types of tables on WTW, as shown in Table 3. Although our performance is slightly lower than Cycle-CenterNet on three ordinary table subsets, our method shows significant improvements on complex scenarios. In particular, for the subset "overlaid", we achieve a 14% improvement with the mainly contribution of CF-NMS. The experiments on these subsets fully demonstrate the superiority of our method and the ability to deal with complex scenarios in table structure recognition.

Table 2. Comparison of TEDS on TAL dataset

Method	TEDS (%)
SPLERGE [28]	53.14
CascadeTabNet [20]	66.71
Table-Master [30]	94.30
SCAN [29]	98.45
TAL_First_Place [2]	99.20
Ours	**99.35**

Table 3. Comparison of cell adjacent relationship on WTW dataset

Method	Curved	Overlaid	Simple	Occluded and blurred	Extreme aspect ratio	Inclined	Multi color and grid	All		
				F1. (%)				Prec. (%)	Rec. (%)	F1. (%)
Cycle-CenterNet [17]	76.1	84.1	**99.3**	77.4	91.9	**97.7**	**93.7**	93.3	91.5	92.4
FLAG-Net [14]	–	–	–	–	–	–	–	91.6	89.5	90.5
TSRFormer [12]	–	–	–	–	–	–	–	93.7	**93.2**	93.4
Ours	83.4	98.1	98.9	**82.6**	**96.3**	97.2	92.7	**94.2**	93.1	**93.6**

Table 4. Ablation experiments on TAL dataset

Training data	TabSynth	SEG	KEY	CF-NMS	TEDS (%)
TAL		✓			93.2
	✓	✓			97.7
	✓		✓		98.5
	✓	✓	✓		98.7
	✓	✓	✓	✓	**99.4**

4.4 Ablation Studies

We conducted a series of experiments on the TAL and WTW datasets to verify the effectiveness of the proposed modules, and the experimental results are shown in Table 4. For TAL dataset, after adding our tabular data generation module TabSynth for training, the training data of the model can simulate complex scenarios with random distortions, random light and multi-color background to overcome the difficulty of the lack of natural scene dataset, thus making the model more robust and achieving a 4.5% improvement. We thus use TabSynth to assist in training by default. The results show that the TEDS metric of the KEY branch is higher than that of the SEG branch, and the interactive results of the two branches outperform the results of the two branches individually, providing support for our method and demonstrating that our method is more suitable for challenging table structure recognition tasks in natural scenes. What's more, our CF-NMS module, designed for the dense detection in table scenes, contributes a 0.7% improvement over the standard NMS, whose threshold is set to 0.5 in our experiment (Table 5).

For the WTW dataset, our approach also achieves considerable improvements in F1 scores. Replacing the standard NMS with our CF-NMS can effectively handle the challenges of dense cell detection scenarios and can yield improvements in precision and recall, improving the F1 score by nearly 1.8%. This also shows

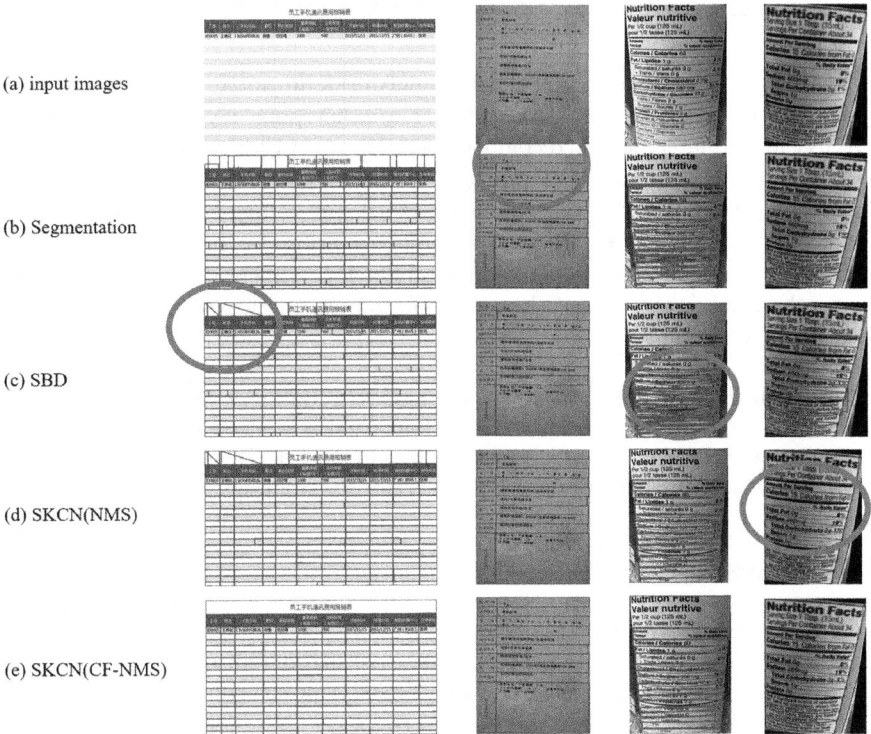

(a) input images

(b) Segmentation

(c) SBD

(d) SKCN(NMS)

(e) SKCN(CF-NMS)

Fig. 5. Qualitative results of our approach.

Table 5. Ablation experiments on WTW dataset for cell detection (IOU = 0.6)

Training data	SEG	KEY	CF-NMS	Prec. (%)	Rec. (%)	F1. (%)
WTW	✓			87.0	91.67	89.27
		✓		89.44	94.01	91.67
	✓	✓		93.4	93.59	93.5
	✓	✓	✓	**96.87**	**93.84**	**95.28**

that the CF-NMS is designed to be very friendly for cell detection tasks. Figure 5 gives a demonstration of the qualitative results of our method, and it can be seen that for large dense table detection tasks, the collaboration of the SEG and KEY branches outperforms both in terms of refinement results. Our proposed SKCN and CF-NMS modules can even be applied to other dense target detection tasks in the future.

5 Conclusion

In this paper, we consider that existing networks do not fully exploit the features of tables and propose a segmentation and key point collaboration network (SKCN) for table structure recognition in the wild. Unlike previous detection methods, the two branches of our model can complement each other during training process and the results of each branch can be fused to obtain a refined result. To better cope with complex table scenarios, we further propose CF-NMS and a tabular data generation module. Experimental results show that our method achieves state-of-the-art performance on two public benchmarks, including TAL and WTW.

Acknowledgments. This research is supported in part by GD-NSF (No. 2021A1515011870), NSFC (Grant no. 61771199), Zhuhai Industry Core and Key Technology Research Project (No. 2220004002350), and the Science and Technology Foundation of Guangzhou Huangpu Development District (Grant 2020GH17).

References

1. Chi, Z., Huang, H., Xu, H.D., Yu, H., Yin, W., Mao, X.L.: Complicated table structure recognition. arXiv preprint arXiv:1908.04729 (2019)
2. TAL Contributors: TAL_OCR_TABLE: a scene table structure recognition benchmark (2021). https://ai.100tal.com/dataset
3. Deng, Y., Rosenberg, D., Mann, G.: Challenges in end-to-end neural scientific table recognition. In: 2019 International Conference on Document Analysis and Recognition (ICDAR), pp. 894–901. IEEE (2019)
4. Gao, L., et al.: ICDAR 2019 competition on table detection and recognition (CTDAR). In: 2019 International Conference on Document Analysis and Recognition (ICDAR), pp. 1510–1515. IEEE (2019)
5. Göbel, M., Hassan, T., Oro, E., Orsi, G.: A methodology for evaluating algorithms for table understanding in pdf documents. In: Proceedings of the 2012 ACM Symposium on Document Engineering, pp. 45–48 (2012)
6. He, K., Gkioxari, G., Dollár, P., Girshick, R.: Mask R-CNN. In: Proceedings of the IEEE International Conference on Computer Vision, pp. 2961–2969 (2017)
7. Itonori, K.: Table structure recognition based on textblock arrangement and ruled line position. In: Proceedings of 2nd International Conference on Document Analysis and Recognition (ICDAR 1993), pp. 765–768. IEEE (1993)
8. Kieninger, T., Dengel, A.: The T-Recs table recognition and analysis system. In: Lee, S.-W., Nakano, Y. (eds.) DAS 1998. LNCS, vol. 1655, pp. 255–270. Springer, Heidelberg (1999). https://doi.org/10.1007/3-540-48172-9_21
9. Laurentini, A., Viada, P.: Identifying and understanding tabular material in compound documents. In: International Conference on Pattern Recognition, p. 405. IEEE Computer Society Press (1992)
10. Law, H., Deng, J.: CornerNet: detecting objects as paired keypoints. In: Ferrari, V., Hebert, M., Sminchisescu, C., Weiss, Y. (eds.) Computer Vision – ECCV 2018. LNCS, vol. 11218, pp. 765–781. Springer, Cham (2018). https://doi.org/10.1007/978-3-030-01264-9_45

11. Li, M., Cui, L., Huang, S., Wei, F., Zhou, M., Li, Z.: TableBank: table benchmark for image-based table detection and recognition. In: Proceedings of The 12th Language Resources and Evaluation Conference, pp. 1918–1925 (2020)
12. Lin, W., et al.: TSRFormer: table structure recognition with transformers. In: Proceedings of the 30th ACM International Conference on Multimedia, pp. 6473–6482 (2022)
13. Liu, H., Li, X., Liu, B., Jiang, D., Liu, Y., Ren, B.: Neural collaborative graph machines for table structure recognition. In: Proceedings of the IEEE/CVF Conference on Computer Vision and Pattern Recognition, pp. 4533–4542 (2022)
14. Liu, H., et al.: Show, read and reason: table structure recognition with flexible context aggregator. In: Proceedings of the 29th ACM International Conference on Multimedia, pp. 1084–1092 (2021)
15. Liu, Y., Jin, L.: Deep matching prior network: toward tighter multi-oriented text detection. In: Proceedings of the IEEE Conference on Computer Vision and Pattern Recognition, pp. 1962–1969 (2017)
16. Liu, Y., Zhang, S., Jin, L., Xie, L., Wu, Y., Wang, Z.: Omnidirectional scene text detection with sequential-free box discretization. arXiv preprint arXiv:1906.02371 (2019)
17. Long, R., et al.: Parsing table structures in the wild. In: Proceedings of the IEEE/CVF International Conference on Computer Vision, pp. 944–952 (2021)
18. Ma, K., Shu, Z., Bai, X., Wang, J., Samaras, D.: DocUNet: document image unwarping via a stacked U-Net. In: Proceedings of the IEEE Conference on Computer Vision and Pattern Recognition, pp. 4700–4709 (2018)
19. Paliwal, S.S., Vishwanath, D., Rahul, R., Sharma, M., Vig, L.: TableNet: deep learning model for end-to-end table detection and tabular data extraction from scanned document images. In: 2019 International Conference on Document Analysis and Recognition (ICDAR), pp. 128–133. IEEE (2019)
20. Prasad, D., Gadpal, A., Kapadni, K., Visave, M., Sultanpure, K.: CascadeTabNet: an approach for end to end table detection and structure recognition from image-based documents. In: Proceedings of the IEEE/CVF Conference on Computer Vision and Pattern Recognition Workshops, pp. 572–573 (2020)
21. Qasim, S.R., Mahmood, H., Shafait, F.: Rethinking table recognition using graph neural networks. In: 2019 International Conference on Document Analysis and Recognition (ICDAR), pp. 142–147. IEEE (2019)
22. Qiao, L., et al.: LGPMA: complicated table structure recognition with local and global pyramid mask alignment. In: Lladós, J., Lopresti, D., Uchida, S. (eds.) ICDAR 2021. LNCS, vol. 12821, pp. 99–114. Springer, Cham (2021). https://doi.org/10.1007/978-3-030-86549-8_7
23. Raja, S., Mondal, A., Jawahar, C.V.: Table structure recognition using top-down and bottom-up cues. In: Vedaldi, A., Bischof, H., Brox, T., Frahm, J.-M. (eds.) ECCV 2020. LNCS, vol. 12373, pp. 70–86. Springer, Cham (2020). https://doi.org/10.1007/978-3-030-58604-1_5
24. Rastan, R., Paik, H.Y., Shepherd, J.: Texus: a unified framework for extracting and understanding tables in pdf documents. Inf. Process. Manage. **56**(3), 895–918 (2019)
25. Schreiber, S., Agne, S., Wolf, I., Dengel, A., Ahmed, S.: DeepDeSRT: deep learning for detection and structure recognition of tables in document images. In: 2017 14th IAPR International Conference on Document Analysis and Recognition (ICDAR), vol. 1, pp. 1162–1167. IEEE (2017)

26. Shigarov, A., Mikhailov, A., Altaev, A.: Configurable table structure recognition in untagged pdf documents. In: Proceedings of the 2016 ACM Symposium on Document Engineering, pp. 119–122 (2016)
27. Siddiqui, S.A., Fateh, I.A., Rizvi, S.T.R., Dengel, A., Ahmed, S.: DeepTabStR: deep learning based table structure recognition. In: 2019 International Conference on Document Analysis and Recognition (ICDAR), pp. 1403–1409. IEEE (2019)
28. Tensmeyer, C., Morariu, V.I., Price, B., Cohen, S., Martinez, T.: Deep splitting and merging for table structure decomposition. In: 2019 International Conference on Document Analysis and Recognition (ICDAR), pp. 114–121. IEEE (2019)
29. Wang, H., Xue, Y., Zhang, J., Jin, L.: Scene table structure recognition with segmentation collaboration and alignment. Pattern Recogn. Lett. **165**, 146–153 (2022)
30. Ye, J., et al.: PingAn-VCGroup's solution for ICDAR 2021 competition on scientific literature parsing task B: table recognition to HTML. arXiv preprint arXiv:2105.01848 (2021)
31. Zhang, Z., Zhang, J., Du, J., Wang, F.: Split, embed and merge: an accurate table structure recognizer. Pattern Recogn. **126**, 108565 (2022)
32. Zheng, X., Burdick, D., Popa, L., Zhong, X., Wang, N.X.R.: Global table extractor (GTE): a framework for joint table identification and cell structure recognition using visual context. In: Proceedings of the IEEE/CVF Winter Conference on Applications of Computer Vision, pp. 697–706 (2021)
33. Zhong, X., ShafieiBavani, E., Jimeno Yepes, A.: Image-based table recognition: data, model, and evaluation. In: Vedaldi, A., Bischof, H., Brox, T., Frahm, J.-M. (eds.) ECCV 2020. LNCS, vol. 12366, pp. 564–580. Springer, Cham (2020). https://doi.org/10.1007/978-3-030-58589-1_34

Frontiers in Handwriting Recognition 3 (Synthesis)

Handwritten Text Generation with Character-Specific Encoding for Style Imitation

Jan Zdenek[✉] and Hideki Nakayama

The University of Tokyo, Tokyo, Japan
jan@nlab.ci.i.u-tokyo.ac.jp, nakayama@ci.i.u-tokyo.ac.jp

Abstract. In this paper, we propose a novel method for handwritten text generation that uses a style encoder based on a vision transformer network that encodes handwriting style from reference images and allows the generator to imitate it. The encoder learns to disentangle style information from the content by learning to recognize who wrote the text, and the self-attention mechanism in the encoder allows us to produce character-specific encodings by using characters in the target sequence as queries. Our method can also generate handwritten text images in random styles by sampling random latent vectors instead of encoding style vectors from reference images.

We demonstrate through experiments that our proposed method outperforms existing methods for handwritten text generation in terms of the quality of generated images and their fidelity with respect to the distribution of real images. Furthermore, it achieves significantly better performance at imitating handwriting styles defined by reference images. Our model generalizes well to unseen data and can generate handwritten images of words and character sequences as well as imitate handwriting styles not included in the training data.

Keywords: Handwritten text generation · Handwriting imitation · Handwritten text recognition

1 Introduction

Significant progress has been achieved in image generation in recent years, particularly thanks to the emergence of new approaches such as generative adversarial networks (GANs) [12], variational auto-encoders (VAEs) [26], and more recently also diffusion models [17]. Generative models can now produce very accurate and detailed images that are difficult to discern from real ones [24,32,39]. The original GAN architecture could only generate images from randomly sampled latent vectors, which did not provide a way to control what was generated. However, further research proposed various methods to manipulate the generation process by conditioning on class labels [30], text embeddings [40]+, segmentation maps [36], reference images [32], etc. This extended the possibilities of image generation beyond a simple generation of random objects.

© The Author(s), under exclusive license to Springer Nature Switzerland AG 2023
G. A. Fink et al. (Eds.): ICDAR 2023, LNCS 14188, pp. 313–329, 2023.
https://doi.org/10.1007/978-3-031-41679-8_18

One of the domains that have adopted recent methods for image generation is handwritten text. Multiple fields and applications can benefit from the ability to generate images of handwritten text automatically. In handwritten text recognition (HTR), being able to generate a large number of diverse handwriting samples for training can improve the accuracy and robustness of recognition models. Handwritten text generation can also be used to create handwritten text assets for games and virtual reality applications, and it can also help in product design when handwritten text is required or desirable.

In recent years, several works have proposed methods for generation of handwritten text images. Some of them can only generate handwritten text with random styles of writing [2,9,38], while some can imitate existing styles from reference images [4,10,22]. The earlier proposed approaches suffer from poor visual quality due to limitations such as being able to generate only fixed-size images [2]. However, recent methods can generate images that are difficult for humans to distinguish from real ones [11,38]. JokerGAN [38] achieves outstanding performance, but it can only generate handwritten text in random styles. In this work, we leverage the performance of JokerGAN by using it as a base for our model and modify it to enable style imitation by generating handwritten text with guidance by reference images. To encode style features from reference images, we train a style encoder together with the rest of the model by making it learn to recognize who wrote the text, inspired by [10]. As vision transformers (ViT) [8] have shown to excel at various vision-based tasks [7,34,39], we employ an encoder based on a ViT network. We further significantly improve the performance of our model by using the target character sequence as a query for self-attention in the transformer, which allows us to generate specific encodings for different characters.

As our proposed model extends the original JokerGAN to support imitation of style from reference images, we call it JokerGAN++.

In summary, the main contributions of our work are as follows:

- We propose a new method to imitate handwriting styles by using a ViT-based encoder that uses target character sequences as queries to produce character-specific style encodings. The experiments show that it can imitate handwriting styles more accurately than existing methods and also generate more authentic images with respect to the distribution of real images.
- We conduct experiments on data augmentation for HTR with generative models and show that HTR models trained on data augmented by images generated by our model outperform models trained on data augmented by images generated by existing methods.
- We demonstrate that our method for generation of handwritten text can also be used to erase handwritten text from images.

2 Related Work

Generation of realistic images of handwritten text is a challenging task. Conventional methods in the past required a lot of manual manipulation of the source images, which involved clipping of individual characters. Those were then combined by various rendering techniques to produce new images of handwritten text [15,35].

The past decade has witnessed the success of deep neural networks, leading to their utilization in various domains and applications, including generation of handwritten text. The first attempts at applying deep neural networks to handwritten text generation focused on online handwritten trajectories. Recurrent neural networks were used to learn and generate the temporal data [13], and further improvement was achieved by adding a discriminator network and employing adversarial training [20]. Manipulating the style of generated images was accomplished by disentangling the style and content of handwritten text [1]. Deep neural networks require a lot of data for successful training, but collecting a large amount of online handwritten data is a demanding task that involves special equipment to record handwriting trajectories. However, collecting offline handwritten data is much easier as it only requires obtaining images of handwritten text. It has been demonstrated that generative adversarial networks (GAN) [12] and variational auto-encoders [26] are capable of generating images of handwritten digits, and it is possible to control which digits are generated with conditional GANs [30]. Recent progress in raster image generation showed the potential of generative models to create very realistic and detailed images [6,23,31], and it helped drive the research on offline handwritten text generation.

There are two approaches to offline handwritten text generation: 1) generation of handwritten text in random styles given by randomly sampled latent vectors, and 2) generation of handwritten text in a specific style defined by a reference image. Most existing methods support generation either only in random styles [2,9,38] or only in specific styles given by reference images [4,21,22]. Recently, methods that support both types of generation have also emerged [10,11].

Application of GANs to offline handwritten text generation was first proposed by Alonso et al. [2] and their model consisted of BigGAN [6] for image generation, an LSTM [18] to encode the target word into a fixed-length vector used as a conditional input for the generator, and a text recognition network to ensure that the generator produces legible images of the target word. Due to its design, the model was restricted to generating text images of a fixed size regardless of the length of the generated word, which causes distortions. ScrabbleGAN [9] resolved this problem by replacing the LSTM-based encoder with a bank of base filters for each letter in the alphabet, which allows generation of images in variable sizes. The generator in ScrabbleGAN uses k base filters corresponding to k letters in the target word. The size of the model grows almost linearly with respect to the size of the character set, which makes it unfeasible for languages with large character sets, such as Chinese or Japanese. To alleviate this issue, JokerGAN [38] replaces the bank of base filters with a single base filter for all characters and uses multi-class conditional batch normalization to generate different characters depending on the conditional input, which is obtained by embedding individual characters in the target word. JokerGAN also introduced a new type of conditional input that makes the generator aware of the position of all characters in the target word with respect to the baseline and mean line, which reduces distortions in generated images.

Kang et al. [22] proposed a method that generates handwritten words from style features extracted from reference images in a few-shot setting and textual features of a predefined text length. In a later work [21], they extended their previous method to support generation of long character sequences and text lines. The recent surge of transformers in computer vision tasks inspired [4] to use transformers to generate styled handwritten text. A transformer is used to model the target word and style features extracted from a reference image by a CNN, and the output is passed to a CNN that upsamples and generates a text image in the desired resolution. The method works in a few-shot setting and requires multiple words as a reference to extract and reproduce the style accurately. HiGAN [10] extended ScrabbleGAN to support generation in specific styles by adding a CNN to encode style from reference images, and a later work [11] improved the performance by modifying the network architecture and using contextual loss in training.

Besides methods that generate handwritten text from latent or encoding vectors, there are methods that synthesize handwritten text in a given style from skeleton images of handwriting [14] and from machine-printed text [28] using an image-to-image translation approach.

Table 1. Comparison of functionality of existing methods and our proposed method. *Latent* means that the method can generate images from random latent vectors. *Reference* means that generation process can be guided by a reference image.

Method	Latent	Reference	Few-shot	One-shot
Alonso et al. [2]	✓			
ScrabbleGAN [9]	✓			
GANWriting [22]		✓	✓	
HWT [4]		✓	✓	
JokerGAN [38]	✓			
HiGAN [10]	✓	✓		✓
HiGAN+ [11]	✓	✓		✓
Ours	✓	✓		✓

Table 1 compares the functionality of our method and existing methods for handwritten text generation. While most methods can generate text either only in random styles using random latent vectors or only in specific styles, guided by reference images, our method supports both types and can generate handwritten text in both random styles and styles defined by a reference image. Our method also requires only a single handwritten word as a reference to imitate the style, unlike some of the other methods that work in a few-shot setting and require multiple words as a reference.

3   Proposed Method

Our proposed method, JokerGAN++, is based on JokerGAN [38] model for handwritten text generation. JokerGAN boasts a high quality of generated handwritten text images, but it cannot imitate specific handwriting styles. Therefore, we modify the original architecture and add a style encoder network to control the style of generated handwriting by reference images. We also revise the discriminator to exploit semantic information about text in images to learn whether the image is real or not. In Sect. 3.1, we describe individual parts of our proposed model, and in Sects. 3.2 and 3.3, we introduce our key contributions to the architecture in detail.

3.1   Model Architecture

Generator. Our generator $\mathcal{G}$ is based on [38]. It supports generation of character sequences of arbitrary length by concatenating k identical base filters that are passed into $\mathcal{G}$, corresponding to the length k of the target character sequence. Generation of different characters is achieved by multi-class conditional batch normalization (MCCBN) that is conditioned on the target sequence of characters. MCCBN is also used to generate handwriting in different styles by concatenating the character sequence embeddings with style codes. In [38], style codes are obtained randomly by sampling from a normal distribution. In our work, we use a style encoder to imitate existing handwriting styles given by reference images; however, random styles can also be generated by using randomly sampled style codes. We also empirically find that using identical encoding vectors in each block of the generator yields better results in our task than using hierarchical input [6] for conditional batch normalization, and injecting noise for additional diversity also hurts the performance.

Discriminator. The discriminator $\mathcal{D}$ learns to predict whether an image is real or generated by $\mathcal{G}$. The discriminator used in [9,38] learns to solve this binary classification problem from real and generated image samples without any explicit information about the character sequences in the images. We add a new component into the discriminator that implicitly learns to recognize individual characters in the image from output feature maps of discriminator layers and uses this semantic information to modulate the feature maps. More details follow in Sect. 3.3.

Style Encoder. We use a style encoder network $\mathcal{E}$ to produce encodings of handwriting styles that can be used as conditional input for $\mathcal{G}$. Detailed description of $\mathcal{E}$ can be found in Sect. 3.2.

Text Recognizer. The objective of text recognizer $\mathcal{R}$ is to promote generation of legible text that matches the target character sequence. Following [9,38], we use a simple network that predicts local patterns without global context to focus on legibility of individual characters. The text recognizer is trained only on real labeled images, and text recognition loss calculated on generated images is used to provide guidance and optimize $\mathcal{G}$.

Figure 1 shows a diagram of the whole model. To simplify the diagram and reduce visual clutter, we do not include all loss functions.

The training process is similar to [10, 38], so we simplify the explanation. We alternate between two optimization passes. In the first pass, we optimize $\mathcal{D}$ by adversarial loss, $\mathcal{R}$ by CTC loss for text recognition, and the writer identification module by cross-entropy loss. In the second pass, we optimize $\mathcal{G}$ and the style encoding module in $\mathcal{E}$ by 1) adversarial loss, 2) CTC text recognition loss on generated samples to ensure that $\mathcal{G}$ generates legible images, 3) cross-entropy loss for writer identification using the writer identification module on generated samples, 4) L1 reconstruction loss of style codes calculated between random latent vectors $\mathbf{z}$ and style encoding vectors obtained by $\mathcal{E}$ from images generated by $\mathcal{G}$ and conditioned on $\mathbf{z}$, and 5) KL-Divergence loss to regularize the style latent space so that it matches normal distribution.

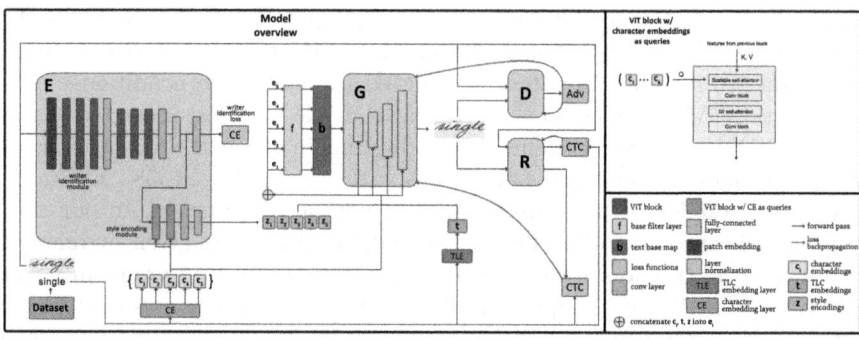

Fig. 1. Overview of the proposed model and the ViT block with character embeddings as queries.

3.2 ViT-Based Style Encoder

To generate images of handwriting that imitate an existing style, we use an encoder network $\mathcal{E}$ to extract the style information from reference images. Inspired by [10], we use $\mathcal{E}$ that is jointly trained by learning to identify the writer of a handwriting sample and learning to encode style encodings for the generator. Instead of using a simple convolutional network as [10] does, our $\mathcal{E}$ is based on ViT. There are two reasons why we employ a ViT-based network for the style encoder. First, ViTs have shown to be exceptionally powerful at modeling local and global information in images, which makes them a great choice to encode handwriting style from an image because a style is defined by both global features (e.g., slant) and local features (e.g., shape of characters, stroke width). Second, ViT allows us to incorporate information about the target character sequence in the encoding mechanism, so we are able to produce style encodings that are not only dependent on the handwriting style, but also on the individual characters appearing in the sequence. As a result, we can create character-specific style encodings, which distinguishes our method from [10] that extracts identical style encoding for the whole character sequence without considering the differences needed to capture to correctly encode the style for different characters.

The identifier module of $\mathcal{E}$ takes handwriting images as input and is composed of a convolutional patch embedding layer and transformer layers from [37] and a single fully-connected layer. It is trained to predict writers of handwriting images, which makes it learn and model differences in handwriting styles.

The encoding module of $\mathcal{E}$ takes features extracted by the transformer layers of the identifier module and disentangles the handwriting styles to produce style encoding vectors. Similarly to the identifier module, it consists of transformer blocks and fully-connected layers to yield fixed-sized length encoding for each character in the target sequence. Each transformer block in the encoding module consists of multi-head scalable self-attention and interactive windowed self-attention introduced in [37] along with fully-connected layers. ScalableViT [37] is a variant of ViT that shows excellent performance across many vision tasks and we empirically found it to yield the best results among several popular state-of-the-art ViT architectures. The feature inputs from the identifier module are averaged across the spatial dimensions before being passed to the encoding module to reduce differences in features based on the characters in the text from the reference image as we want to produce encodings specific to the characters in the target sequence.

To produce character-specific style encodings, we use characters in the target sequence as queries for self-attention in the encoding module. Therefore, unlike regular self-attention where query $Q()$, key $K()$, and value $V()$ all have the same input, the $Q()$ input here is an embedded character c in the target character sequence while $K()$ and $V()$ inputs are features $\mathbf{X}$ from the previous transformer block. The self-attention in our module is thus calculated as

$$Attn\,(\mathbf{X}, c) = softmax(\frac{Q(c)K(\mathbf{X})}{\sqrt{d}})V(\mathbf{X}). \tag{1}$$

We produce a fixed-size style encoding vector for each character in the target sequence. Style encodings of all characters are then employed as conditional input for MCCBN in the generator.

3.3   Character Modulation

The original discriminator in [9,38] does not use any semantic information about characters in the image to predict whether an image is real or generated. We strive to improve the performance of the discriminator by modeling the semantic information. Similarly to techniques such as [19], we add a branch with softmax activation after each block in the discriminator. The feature maps $\mathbf{X}$ where $\mathbf{X} \in \mathbb{R}^{H \times W \times C}$ are passed to a convolutional layer f with the kernel size of 1×1 and K output channels where K corresponds to the number of characters in the character set, such as alphanumeric characters. This aggregates channel-wise features and reduces the number of channels to match the number of characters. In order to utilize the information aggregated in the reduced feature space, we need a gating mechanism that promotes emphasis of a single channel as we ideally want each channel to correspond to one character. To achieve this, we employ a softmax activation function across the channel dimension. The process can be denoted as

$$\hat{\mathbf{X}} = softmax\left(f\left(\mathbf{X}, \mathbf{W}\right)\right), \qquad (2)$$

where $\mathbf{W} \in \mathbb{R}^{C \times K}$. The extracted features $\hat{\mathbf{X}}$ are aggregated with the original features $\mathbf{X}$ by concatenating them across the channel dimension to produce $\tilde{\mathbf{X}} \in \mathbb{R}^{H \times W \times (C+K)}$, and finally the concatenated features are passed to another 1×1 convolutional layer to reduce the number of channel dimensions back to C. Figure 2 illustrates the whole process.

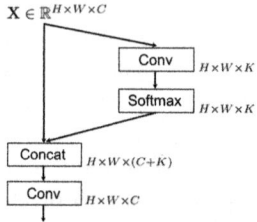

Fig. 2. Diagram of the character modulation component used in the discriminator.

4   Experiments

4.1   Implementation Details

Our model is based on JokerGAN [38], whose core consists of BigGAN [6] layers that are modified for generation of images with a fixed height and variable width. Besides our stated modifications, the architecture of $\mathcal{G}$, $\mathcal{D}$ and $\mathcal{R}$ is identical as in [38]. The identification module of $\mathcal{E}$ consists of a patch embedding layer (filter size 8, stride 4) and two ViT stacks, each comprised of 3 blocks of ViT layers followed by a convolutional layer. The encoding module consists of two blocks of ViT [37] with character embeddings as queries for self-attention, followed by a convolutional layer for downsampling and a fully-connected layer to obtain fixed-size style encoding vectors. The architecture of ViT layers is from [37]. We use the Adam optimizer [25] with a learning rate of 0.0002 for training. Our model is implemented in PyTorch.

4.2   Datasets

We use the following two datasets in our experiments.

- **IAM**. The IAM dataset [29] contains approximately 80k grayscale images of handwritten words in English, divided into training, test and validation sets. The training set consists of about 40k and test set of about 10k images. The words are written by 657 different people and all words written by one person only appear in one of the sets to achieve mutual exclusivity of authors in training, validation, and test sets. The data was created and preprocessed for training of HTR models.
- **GNHK**. The GNHK dataset [27] is composed of images of unconstrained handwritten text in the wild captured by mobile phone cameras. The training set we use in our experiments contains about 28k images of individual handwritten words cropped from the original text images. Due to the nature of the data, the GNHK dataset consists of images with more variety in style and more noise, which makes it more challenging than the IAM dataset.

4.3   Handwritten Text Image Generation

We evaluate the quality of handwritten text image generation and handwriting style imitation using several metrics to measure different performance aspects.

- **Visual Quality**. Our primary metrics of visual quality are Frechet Inception Distance (FID) [16] and Kernel Inception Distance (KID) [5], which are widely used to evaluate GANs. FID and KID compare the distributions of generated images and real samples. We also use the structural similarity index (SSIM) that measures structural similarity between real and generated images.
- **Style Imitation**. We use the writer identification error rate (WIER) [11] to evaluate how well a model can imitate styles. A writer identification network is trained on the test set of images and used to predict writers for images that are generated with test set images as reference. Misclassified samples are deemed as failures of the generative model to accurately imitate the handwriting style. We measure WIER when generating the identical word as in the reference image (WIER-I) as well as when generating a random word (WIER-R). SSIM also indicates how similar the styles in generated and reference images are.
- **Diversity and Fidelity**. To measure how diverse and accurate the generated images are with respect to the distribution of real images, we use GAN-train and GAN-test evaluation [33]. Originally, GAN-train is measured by training an image classifier on generated data and testing on real data and approximates image generation recall, and GAN-test is measured by training on real data and testing on generated data and approximates image generation precision. GAN-tt is an average of GAN-train and GAN-test to consider their trade-off and combine them into one score. In our case, we use a HTR model in place of an image classifier and word recognition accuracy as the underlying metric to calculate GAN-train and GAN-test scores.
- **Readability**. GAN-test also indicates the readability of generated samples as it measures if a HTR model trained on real data can read them.

Table 2. Comparison of performance of handwritten text generation methods on the IAM dataset in terms of metrics for image generation quality based on the distance between real and generated image distributions (FID, KID), structural similarity of images (SSIM), style imitation accuracy measured by WIER, and diversity and quality of image generation measured by training and testing with HTR models (GAN-test, GAN-train, GAN-tt). Evaluation on real data is provided for reference.

Method	FID↓	KID↓	SSIM↑	WIER$_I$↓	WIER$_R$↓	GAN-test↑	GAN-train↑	GAN-tt↑
ScrabbleGAN [9]	19.98	1.359	–	–	–	93.11	27.67	60.39
HWT [4]	18.76	1.214	0.182	0.829	0.855	64.06	22.15	43.11
JokerGAN [38]	4.63	0.186	–	–	–	80.86	54.42	67.64
HiGAN-L [10]	17.69	1.107	–	–	–	**96.15**	31.93	64.38
HiGAN-R [10]	12.43	0.669	0.196	0.628	0.674	96.61	36.03	66.62
HiGAN+ [11]	5.94	0.368	0.332	0.526	0.575	95.65	30.72	63.19
Ours (latent)	3.00	0.098	–	–	–	94.27	55.67	**74.97**
Ours (reference)	**2.14**	**0.078**	**0.429**	**0.327**	**0.499**	81.11	**61.90**	71.51
Real data	0.02	0.002	–	0.043	–	83.49	–	–

As shown in Table 2, our method outperforms existing methods in virtually all metrics. GAN-test is the only metric in which it slightly falls behind. This can be attributed to the fact that our model generates more diverse samples that might be harder for a HTR model to read correctly. The diversity is measured by GAN-train in which our model surpasses other methods with a high GAN-test score by a large margin. In addition, the GAN-test score of our method when using reference images for generation is similar to word recognition accuracy of a HTR model trained and tested on real data, which also suggests that the distribution of images generated with our method is closer to the distribution of real data. Our method also achieves the best WIER scores, indicating that it can imitate styles from reference images better than other methods. We state results of our method when generating both in random styles from randomly sampled latent vectors (latent) and styles from reference images (reference).

SSIM and WIER are only applicable for methods that imitate style from reference images; therefore, we do not use them for evaluation of methods that generate images in random styles. We also include real data results for reference where WIER represents the error rate of a writer identifier trained and tested on real data, and the value in the GAN-test column represents word recognition accuracy of a HTR model trained and tested on real data.

Table 3 shows results of experiments on GNHK. The trends are similar to those in Table 2, further attesting our method outperforms the competition. GNHK dataset is more complex than IAM, so the performance of all models is worse compared to IAM. In particular, the diversity is limited and HTR models trained only on generated images perform poorly as shown by low GAN-train. Due to the nature of the dataset, we use top-5 error rate for WIER for GNHK.

Table 3. Comparison of performance of handwritten text generation methods on the GNHK dataset. The same metrics as in Table 2 are used.

Method	FID↓	KID↓	SSIM↑	WIER$_I$ (top-5)↓	GAN-test↑	GAN-train↑	GAN-tt↑
ScrabbleGAN [9]	20.92	1.603	–	–	92.24	8.98	50.61
JokerGAN [38]	10.35	0.515	–	–	76.97	10.85	43.91
HiGAN-L [10]	14.27	0.840	–	–	**93.89**	8.12	51.00
HiGAN-R [10]	8.12	0.366	**0.327**	0.695	92.09	9.58	50.84
HiGAN+ [11]	8.04	0.459	0.235	0.293	82.76	6.10	44.43
Ours (latent)	8.69	0.395	–	–	89.88	**12.30**	**51.09**
Ours (reference)	**5.99**	**0.194**	0.269	**0.129**	82.39	12.15	47.27
Real data	0.03	0.005	–	0.081	63.32	–	–

4.4 Ablation Study

We perform ablation studies to validate the effectiveness of individual proposed components and modifications. We use the IAM dataset and evaluate the performance in two settings, 1) handwritten text generation guided by a reference image, and 2) handwritten text generation from randomly sampled latent vectors. Baseline refers to [38] with the style encoder from [10].

As can be seen in Table 4 and 5, all of our new components and modifications improve the performance of generation either from reference images or random latent vectors. WIER improves particularly thanks to our newly proposed ViT-based style encoder, which can encode the handwriting styles better than a conventional CNN-based encoder. The performance further improves when we input the target character sequence into the encoder and encode specific style vectors for individual characters in the sequence that we want to generate. The ViT-based style encoder also enhances the overall quality of generated images and their fidelity to the distribution of the original real data as measured by FID and KID, and also similarity to reference images in terms of the structure as measured by SSIM. Since the style encoder is not used for generation with random styles, replacing a CNN-based style encoder with our ViT-based one does not significantly affect performance when generating from random latent vectors instead of reference images as can be seen in Table 5. Using strict style conditioning without introducing additional randomness by appending a random latent vector to the style encoding and using identical style conditions for all blocks in the generator instead of hierarchical input improves the performance of generation from latent vectors. Finally, our newly proposed character modulation in the discriminator improves performance of generation in both settings.

Table 4. Ablation study of individual changes to the baseline model evaluated on the IAM dataset. Style of generated images is defined by reference images.

Method	FID↓	KID↓	SSIM↑	WIER_I↓	GAN-test↑	GAN-train↑
baseline	3.698	0.171	0.266	0.498	69.82	50.22
+ strict style conditioning	3.634	0.154	0.263	0.503	64.99	47.88
+ non-hierarchical conditioning	3.719	0.178	0.271	0.499	75.82	48.53
+ character modulation in $\mathcal{D}$	3.252	0.116	0.284	0.459	85.76	51.35
+ ViT-based style encoder	2.560	0.075	0.327	0.407	88.58	55.86
+ character specific style encoding	2.136	0.078	0.429	0.327	81.11	61.90

Table 5. Ablation study of individual changes to the baseline model evaluated on the IAM dataset. Style of generated images is random, defined by random latent vectors.

Method	FID↓	KID↓	GAN-test↑	GAN-train↑
baseline	7.330	0.575	82.74	48.19
+ strict style conditioning	4.003	0.202	80.92	48.46
+ non-hierarchical conditioning	3.261	0.122	82.05	54.19
+ character modulation in $\mathcal{D}$	3.047	0.104	87.83	50.74
+ ViT-based style encoder	3.042	0.076	95.08	50.57
+ character specific style encoding	3.002	0.098	94.27	55.67

4.5   Data Augmentation for HTR

Creating annotation for training of machine learning models is a demanding and expensive process, so in real life, we may encounter situations where we only have unlabeled or partially labeled data. In this experiment, we follow [38] and simulate the situation that we have partially labeled data by using only 5k images with text annotations from the IAM dataset and the rest without text annotations. We train generative models for handwritten text on both unlabeled and labeled data since only the text recognizer requires text annotation for training, but the rest of the model can be optimized on unlabeled data.

For data augmentation evaluation, we use a HTR model [3] trained only on 5k labeled images from IAM as a baseline (IAM-5k). Each of our trained generative models is then used to augment the training dataset by generating additional 100k images for training of HTR models. Word error rate (WER) and normalized edit distance (NED) are used as metrics for evaluation. Table 6 illustrates that using additional training data generated by handwritten text generation models improves the HTR performance, and in particular, our proposed model achieves the biggest performance boost out of all tested models. We also include the results of a HTR model trained on the complete IAM dataset of 40k labeled images for reference.

Table 6. Comparison of different models when used to generate additional data for training of a handwritten text recognition model.

Data	WER↓	NED↓
IAM-40k	16.52	4.95
IAM-5k	34.17	11.90
IAM-5k + ScrabbleGAN 100k [9]	30.65	10.00
IAM-5k + HWT 100k [4]	33.58	11.59
IAM-5k + JokerGAN 100k [38]	28.50	9.26
IAM-5k + HiGAN-L 100k [10]	30.42	10.01
IAM-5k + HiGAN-R 100k [10]	30.96	10.44
IAM-5k + HiGAN+ 100k [11]	27.76	8.94
IAM-5k + Ours (latent) 100k	27.17	8.73
IAM-5k + Ours (reference) 100k	**25.00**	**8.08**

4.6 Qualitative Evaluation

Figure 3 shows results of handwritten text generation and style imitation with different methods. The models are trained on the training set of the IAM dataset and the reference images used in Fig. 3 are from the test set of IAM. As can be seen, our model can accurately imitate the styles in the reference images, which shows that it generalizes well and it can imitate styles that it did not see during training. Since JokerGAN and ScrabbleGAN cannot imitate styles, the text in Fig. 3 is generated in random styles for these two methods.

Figure 4 shows images generated by models trained on the GNHK dataset. Style imitation is less accurate than in the case of IAM because the GNHK dataset contains a larger variety of images of unrestricted handwritten text in RGB colorspace. However, images produced by our model still exhibit a significant style similarity to reference images.

4.7 Text Erasing

We demonstrate that while our proposed method is primarily intended for handwritten text generation, it can be also used to erase text from documents, as shown in Fig. 5. When we include whitespace in the character set that the model learns to generate, our model can erase text from a reference image by generating whitespace characters with style guided by the reference image. Note that it can erase text while preserving the original background and text lines.

Whitespace characters are not included in the original training data. As a solution, we randomly add a whitespace character to the beginning or the end of a word by padding the image with the left or right edge pixels. The size of the padding corresponds to the set approximate width of one character.

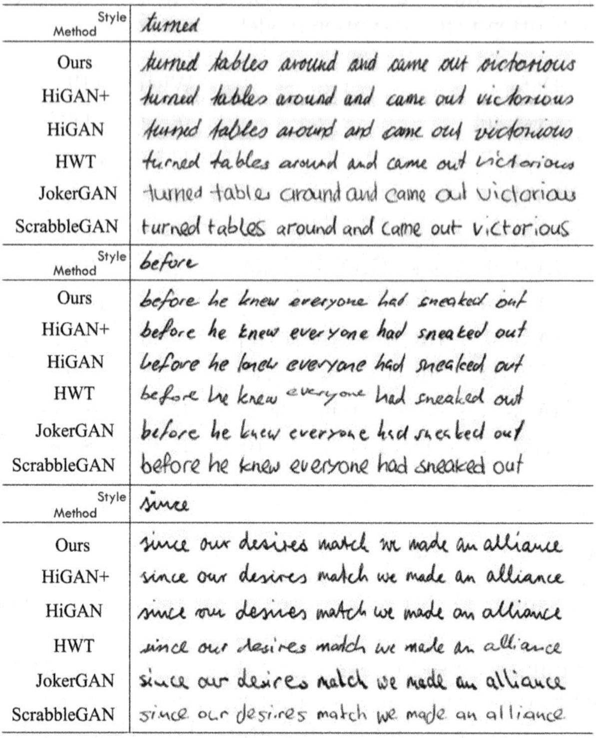

Method	Style: *turned*
Ours	*turned tables around and came out victorious*
HiGAN+	*turned tables around and came out victorious*
HiGAN	*turned tables around and came out victorious*
HWT	*turned tables around and came out victorious*
JokerGAN	*turned table around and came out victorious*
ScrabbleGAN	*turned tables around and came out victorious*

Method	Style: *before*
Ours	*before he knew everyone had sneaked out*
HiGAN+	*before he knew everyone had sneaked out*
HiGAN	*before he knew everyone had sneaked out*
HWT	*before he knew everyone had sneaked out*
JokerGAN	*before he knew everyone had sneaked out*
ScrabbleGAN	*before he knew everyone had sneaked out*

Method	Style: *since*
Ours	*since our desires match we made an alliance*
HiGAN+	*since our desires match we made an alliance*
HiGAN	*since our desires match we made an alliance*
HWT	*since our desires match we made an alliance*
JokerGAN	*since our desires match we made an alliance*
ScrabbleGAN	*since our desires match we made an alliance*

Fig. 3. Results of handwriting style imitation with different methods. The reference style images are from the test set of the IAM dataset and the models did not see those handwriting styles during training.

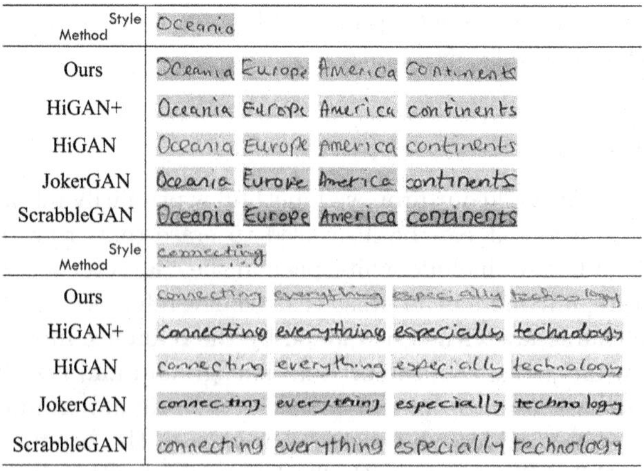

Method	Style: *Oceania*
Ours	*Oceania Europe America Continents*
HiGAN+	*Oceania Europe America continents*
HiGAN	*Oceania Europe america continents*
JokerGAN	*Oceania Europe America continents*
ScrabbleGAN	*Oceania Europe America continents*

Method	Style: *connecting*
Ours	*connecting everything especially technology*
HiGAN+	*connecting everything especially technology*
HiGAN	*connecting everything especially technology*
JokerGAN	*connecting everything especially technology*
ScrabbleGAN	*connecting everything especially technology*

Fig. 4. Results of handwriting style imitation with different methods. The reference style images are from the test set of the GNHK dataset.

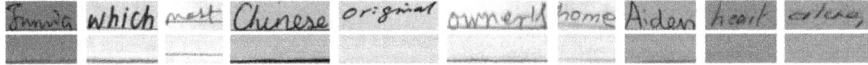

Fig. 5. Results of text erasing with our proposed method.

Table 7. Comparison of existing models and our proposed model in terms of size in megabytes. Only the modules necessary to store to perform generation, that is generator (Gen) and encoder (Enc), are considered.

Method	Size (MB)		
	Gen	Enc	Total
ScrabbleGAN [9]	81.8	N/A	81.8
JokerGAN [38]	11.0	N/A	11.0
GANWriting [22]	95.6	76.5	172.1
HWT [4]	80.7	50.6	131.3
HiGAN [10]	38.6	20.5	59.1
HiGAN+ [11]	15.0	6.7	21.7
Ours	11.6	12.6	24.2

4.8   Model Size

Table 7 shows a comparison of the size of our model and existing models for handwritten text generation. We only consider the size of the modules that are needed at inference time, which is the generator and encoder. In the case of models that only generate handwritten text in random styles, there is no encoder. We denote the size of the models in megabytes. Our model not only achieves better performance in terms of generation quality, but as can be seen, it is also one of the most lightweight models.

5   Conclusion

We have proposed a new method for generation of handwritten text images. Our method can not only generate handwriting in a random style, but it can also imitate a specific handwriting style passed to the model as a reference in the form of a raster image. Experiments show that our method outperforms existing methods in terms of the quality of generation and similarity to the style of handwriting in reference images. The performance has particularly improved thanks to our newly proposed ViT-based style encoder that takes the target character sequence that we want to generate as an additional input to produce character-specific style encodings. We also show that images generated by our model can be used for data augmentation for training of OCR models for handwritten text.

Acknowledgment. This work was supported by JSPS KAKENHI Grant Number JP22H00540.

References

1. Aksan, E., Pece, F., Hilliges, O.: DeepWriting: making digital ink editable via deep generative modeling. In: CHI (2018)
2. Alonso, E., Moysset, B., Messina, R.: Adversarial generation of handwritten text images conditioned on sequences. In: ICDAR (2019)
3. Baek, J., et al.: What is wrong with scene text recognition model comparisons? Dataset and model analysis. In: ICCV (2019)
4. Bhunia, A.K., Khan, S., Cholakkal, H., Anwer, R.M., Khan, F.S., Shah, M.: Handwriting transformers. In: ICCV (2021)
5. Bińkowski, M., Sutherland, D.J., Arbel, M., Gretton, A.: Demystifying MMD GANs. In: ICLR (2018)
6. Brock, A., Donahue, J., Simonyan, K.: Large scale GAN training for high fidelity natural image synthesis. In: ICLR (2018)
7. Carion, N., Massa, F., Synnaeve, G., Usunier, N., Kirillov, A., Zagoruyko, S.: End-to-end object detection with transformers. In: Vedaldi, A., Bischof, H., Brox, T., Frahm, J.-M. (eds.) ECCV 2020. LNCS, vol. 12346, pp. 213–229. Springer, Cham (2020). https://doi.org/10.1007/978-3-030-58452-8_13
8. Dosovitskiy, A., et al.: An image is worth 16x16 words: transformers for image recognition at scale. In: ICLR (2021)
9. Fogel, S., Averbuch-Elor, H., Cohen, S., Mazor, S., Litman, R.: ScrabbleGAN: semi-supervised varying length handwritten text generation. In: CVPR (2020)
10. Gan, J., Wang, W.: HiGAN: handwriting imitation conditioned on arbitrary-length texts and disentangled styles. In: AAAI (2021)
11. Gan, J., Wang, W., Leng, J., Gao, X.: HiGAN+: handwriting imitation GAN with disentangled representations. ACM Trans. Graph. **42**(1), 1–17 (2022)
12. Goodfellow, I.J., et al.: Generative adversarial networks. In: NIPS (2014)
13. Graves, A.: Generating sequences with recurrent neural networks. arXiv preprint arXiv:1308.0850 (2013)
14. Guan, M., Ding, H., Chen, K., Huo, Q.: Improving handwritten OCR with augmented text line images synthesized from online handwriting samples by style-conditioned GAN. In: ICFHR (2020)
15. Haines, T.S.F., Mac Aodha, O., Brostow, G.J.: My text in your handwriting. ACM Trans. Graph. **35**(3), 1–18 (2016)
16. Heusel, M., Ramsauer, H., Unterthiner, T., Nessler, B., Hochreiter, S.: GANs trained by a two time-scale update rule converge to a local Nash equilibrium. In: NIPS (2017)
17. Ho, J., Jain, A., Abbeel, P.: Denoising diffusion probabilistic models. In: NeurIPS (2020)
18. Hochreiter, S., Schmidhuber, J.: Long short-term memory. Neural Comput. **9**(8), 1735–1780 (1997)
19. Hu, J., Shen, L., Sun, G.: Squeeze-and-excitation networks. In: CVPR (2018)
20. Ji, B., Chen, T.: Generative adversarial network for handwritten text. arXiv preprint arXiv:1907.11845 (2019)
21. Kang, L., Riba, P., Rusiñol, M., Fornés, A., Villegas, M.: Content and style aware generation of text-line images for handwriting recognition. TPAMI **44**(12), 8846–8860 (2022)
22. Kang, L., Riba, P., Wang, Y., Rusiñol, M., Fornés, A., Villegas, M.: GANwriting: content-conditioned generation of styled handwritten word images. In: Vedaldi, A., Bischof, H., Brox, T., Frahm, J.-M. (eds.) ECCV 2020. LNCS, vol. 12368, pp. 273–289. Springer, Cham (2020). https://doi.org/10.1007/978-3-030-58592-1_17

23. Karras, T., Aittala, M., Hellsten, J., Laine, S., Lehtinen, J., Aila, T.: Training generative adversarial networks with limited data. In: NeurIPS (2020)
24. Karras, T., et al.: Alias-free generative adversarial networks. In: NeurIPS (2021)
25. Kingma, D.P., Ba, J.: Adam: a method for stochastic optimization. In: ICLR (2015)
26. Kingma, D.P., Welling, M.: Auto-encoding variational bayes. In: ICLR (2014)
27. Lee, A.W.C., Chung, J., Lee, M.: GNHK: a dataset for English handwriting in the wild. In: ICDAR (2021)
28. Luo, C., Zhu, Y., Jin, L., Li, Z., Peng, D.: SLOGAN: handwriting style synthesis for arbitrary-length and out-of-vocabulary text. IEEE Trans. Neural Netw. Learn. Syst. (2022)
29. Marti, U.V., Bunke, H.: The IAM-database: an English sentence database for offline handwriting recognition. IJDAR **5**(1), 39–46 (2002)
30. Mirza, M., Osindero, S.: Conditional generative adversarial nets. arXiv preprint arXiv:1411.1784 (2014)
31. Miyato, T., Koyama, M.: cGANs with projection discriminator. In: ICLR (2018)
32. Park, T., Liu, M.Y., Wang, T.C., Zhu, J.Y.: Semantic image synthesis with spatially-adaptive normalization. In: CVPR (2019)
33. Shmelkov, K., Schmid, C., Alahari, K.: How good is my GAN? In: ECCV (2018)
34. Strudel, R., Garcia, R., Laptev, I., Schmid, C.: Segmenter: transformer for semantic segmentation. In: ICCV (2021)
35. Wang, J., Wu, C., Xu, Y.Q., Shum, H.Y.: Combining shape and physical models for online cursive handwriting synthesis. IJDAR **7**(4), 219–227 (2005)
36. Wang, T.C., Liu, M.Y., Zhu, J.Y., Tao, A., Kautz, J., Catanzaro, B.: High-resolution image synthesis and semantic manipulation with conditional GANs. In: CVPR (2018)
37. Yang, R., et al.: ScalableViT: rethinking the context-oriented generalization of vision transformer. In: Avidan, S., Brostow, G., Cissé, M., Farinella, G.M., Hassner, T. (eds.) Computer Vision – ECCV 2022. LNCS, vol. 13684, pp. 480–496. Springer, Cham (2022). https://doi.org/10.1007/978-3-031-20053-3_28
38. Zdenek, J., Nakayama, H.: JokerGAN: memory-efficient model for handwritten text generation with text line awareness. In: ACM Multimedia (2021)
39. Zhang, B., et al.: StyleSwin: transformer-based GAN for high-resolution image generation. In: CVPR (2022)
40. Zhang, H., et al.: StackGAN: text to photo-realistic image synthesis with stacked generative adversarial networks. In: ICCV (2017)

How to Choose Pretrained Handwriting Recognition Models for Single Writer Fine-Tuning

Vittorio Pippi[1]([✉]) [iD], Silvia Cascianelli[1] [iD], Christopher Kermorvant[2,3] [iD], and Rita Cucchiara[1] [iD]

[1] University of Modena and Reggio Emilia, Modena, Italy
{vittorio.pippi,silvia.cascianelli,rita.cucchiara}@unimore.it
[2] TEKLIA, Paris, France
kermorvant@teklia.com
[3] LITIS, Université de Rouen - Normandie, Sotteville-lés-Rouen, France

Abstract. Recent advancements in Deep Learning-based Handwritten Text Recognition (HTR) have led to models with remarkable performance on both modern and historical manuscripts in large benchmark datasets. Nonetheless, those models struggle to obtain the same performance when applied to manuscripts with peculiar characteristics, such as language, paper support, ink, and author handwriting. This issue is very relevant for valuable but small collections of documents preserved in historical archives, for which obtaining sufficient annotated training data is costly or, in some cases, unfeasible. To overcome this challenge, a possible solution is to pretrain HTR models on large datasets and then fine-tune them on small single-author collections. In this paper, we take into account large, real benchmark datasets and synthetic ones obtained with a styled Handwritten Text Generation model. Through extensive experimental analysis, also considering the amount of fine-tuning lines, we give a quantitative indication of the most relevant characteristics of such data for obtaining an HTR model able to effectively transcribe manuscripts in small collections with as little as five real fine-tuning lines.

Keywords: Document synthesis · Historical document analysis · Handwriting recognition · Synthetic data

1 Introduction

Digitization is becoming a crucial step for the efficient management, preservation, and valorization of documents, both in the cultural and industrial domains. For this reason, Document Analysis (DA) techniques, especially those intended to tackle challenging scenarios of handwritten text, are receiving significant interest from the research community. State-of-the-art Handwritten Text Recognition (HTR) models, trained on large publicly available datasets, can achieve impressive results when applied to documents with characteristics similar to those used during training. However, their performance is unsatisfactory when the data of the domain of interest are too different from the training ones. In this respect,

G. A. Fink et al. (Eds.): ICDAR 2023, LNCS 14188, pp. 330–347, 2023.
https://doi.org/10.1007/978-3-031-41679-8_19

the small but valuable collections of historical manuscripts preserved in many archives pose a challenge for modern HTR models. In fact, such archives often contain few sample pages written by a specific but relevant author, with peculiar characteristics, both visual and linguistic. Thus, a strategy to obtain high HTR performance also for those documents is key to enabling the efficient digitization of such documents. A popular approach to deal with this scenario consists in pretraining the HTR model on large datasets, either real or synthetic, and then fine-tuning on a limited number of real data from the target domain. This strategy has also the potential to enable high-quality on-demand transcriptions of entire single-author collections, which might be a service of interest to the users of digital libraries and archives. In particular, the libraries can store pretrained HTR models, and users can request the transcription of the collection they are interested in by simply providing the annotation for a few lines (*e.g.*, 5–15). At that point, the most suitable pretrained model can be chosen based on the collection characteristics (*e.g.*, language, period, authorship, style) possibly available in the form of metadata and fine-tuned on the user-provided annotations in a limited amount of time. Afterward, the fine-tuned model can transcribe the entire collection with low error. This kind of interaction with the collection can also benefit the overall experience of digital archives users. Note that, so far, interactive transcription enhancement has been explored in terms of language model refinement [36]. With this work, we aim to explore a more holistic approach taking into account both language and appearance.

Note that, in literature, attempts have been made toward the use of synthetic data for pretraining HTR models [2,25,30,32,52]. These strategies are as effective as more similar the synthetic data are to the real ones [12]. In this line, Handwritten Text Generation (HTG) techniques are emerging [5,7,23,31], especially styled HTG ones, which might allow generating training data with the characteristics needed for HTR on specific domains. In fact, models for styled HTG can produce images with arbitrary text in the desired handwriting starting from a few style example images. These models often comprise an encoder to obtain writer-specific style features and a generator, which is fed with the style features and content tokens representing the characters to produce text images conditioned on the desired style and content. In light of this, in this paper, we consider pretraining plus fine-tuning on an automatically generated author-specific synthetic dataset, which is obtained by exploiting a State-of-the-Art styled HTG network. Moreover, we evaluate pretraining on existing benchmark datasets of various languages, with a varied number of authors and of various periods. This way, we investigate whether it is feasible to obtain an effective pipeline for interactive, on-demand HTR of single-author collections. In particular, we provide a set of quantitative guidelines taking into account both visual and linguistic aspects, for designing the most effective pipeline for an HTR model able to transcribe specific manuscript collections with low error after fine-tuning on as little as 5 lines from the target manuscript. Potentially, the defined guidelines for choosing the most suitable pretrained model can either be exploited by the archive management or presented to the user who would be more involved in the transcription process.

2   Related Work

Strategies for HTR. Due to its practical interest in both industrial and cultural domain applications, HTR is a widely-investigated research topic. Despite that, it remains a challenging task. HTR can be performed on single characters, which is a popular choice in the case of idiomatic languages [15], single words [6,48], or entire lines [40,45], paragraphs, and pages [8,9,16,38,55]. The line-level variant is one of the most popular for non-idiomatic language, both standalone and as part of a page-level system [9,38,56] The most used learning-based solutions for HTR rely on Multi-Dimensional Long Short-Term Memory networks (MD-LSTMs) [26] or on the combination of convolutional and one-dimensional LSTMs [10,39,40,45,50] to represent the text image and on the Connectionist Temporal Classifier (CTC) decoding strategy to output the transcription [8,26]. Alternatively to approaches exploiting recurrent models, fully-convolutional networks have been proposed for HTR [18,56], as well as solutions [30,32,53] based on Transformer encoder-decoder architectures [49]. Finally, it is worth noting that explicit language models or lexicons can be exploited to refine the transcription. However, this strategy is all the more effective the more the language of the transcribed images is regular (*i.e.*, it contains no errors, uncommon words, and proper nouns) and well-represented. For this reason, employing language models is not always feasible, especially when dealing with historical manuscripts.

Strategies for HTG. HTG is an increasingly popular research area aimed at producing realistic images of handwritten text. In the styled variant of the task, which we consider in this work, the goal is to generate writer-specific handwritten text images from just a few example images of the writer's style to mimic [5,23,31]. The early approaches to HTG, either styled or not, were able to obtain impressive results, but at the cost of heavy human intervention and feature handcrafting [27,51]. Recently-proposed learning-based solutions, instead, are fully automatic. Usually, these strategies entail using generative adversarial networks (GANs) [24]. In the case of non-styled HTG, these can be unconditioned [1,23]. For styled HTG, instead, the employed GANs are conditioned on style features extracted by an encoder from the handwriting style sample images. Note that the style examples can be line images [20], a few images of words [5,31], or a single image [37]. It is also worth mentioning a more recently-proposed approach based on an encoder-decoder generative Transformer [5].

Synthetic Data for HTR. Lack of training data is a major challenge in HTR, especially in the case of single-author documents or ancient manuscripts that exhibit peculiar characteristics. A possible strategy to tackle this issue is to perform data augmentation either in terms of generic color modifications and geometric distortions [40,50,54] or image modifications carefully designed to match the characteristics of the target data [14]. Another popular strategy entails pretraining the HTR model on large datasets and then fine-tuning it on the target data [25,28,47], which has been proven to be more beneficial than data augmentation for historical manuscripts [2]. The pretraining dataset can be real (*e.g.*, a publicly available benchmark dataset) or synthetic, generally obtained by altering images of text rendered in calligraphic fonts [30,44]. In an attempt to generate more realistic-looking text images, some recent works exploited HTG models, either styled or not, to generate synthetic data for training HTR models and boost their performance on real data. For example, in [46], the authors exploited a compositional approach based on Bayesian Program Learning to generate the symbols in a ciphered corpus and then combined them into realistic-looking text lines for training an HTR model to transcribe historical ciphered manuscripts. The benefits of training HTR models on generated text lines in various styles have been investigated also in [29], where the authors applied a styled HTG model to obtain a pretraining dataset. Finally, as for the single-author scenario, in [12], the authors showed the benefits of pretraining the HTR model on synthetic data that faithfully resemble the real ones over pretraining on generic various-styles images. Their approach, however, heavily relies on human effort to obtain such high-quality synthetic data. In sight of this, in this work, we focus on synthetic data obtained from a fully-automatic HTG model.

3   Proposed Approach

In this work, we explore a pipeline for obtaining good-quality line-level transcriptions of manuscript collections with peculiar characteristics (in terms of handwriting, language, and paper support) by exploiting pretraining on large datasets and fine-tuning on a small amount of samples from the target collection. In particular, we consider pretraining on real datasets and on synthetic ones obtained via a few-shot styled HTG model to better reflect the characteristics of the target data. To build the synthetic datasets, we need a few images of words (15, in this work, as in [5,31]) that can be easily obtained from digitized manuscripts in the small collection of interest. Moreover, we need to specify the text to be rendered in the desired style. In this work, we consider two typical scenarios in digital libraries and archives: one in which only the language of the target manuscript is known and one in which also the information about the author is available. If the author of the collection of interest is known and there exist some other transcribed texts by the same author, we propose to make the HTG model generate these texts. In case only the language is known (or if no other texts by the same author are available), we make the HTG model generate texts in the same language as the target collection. Note that, in both cases, the HTG model outputs images of handwritten words, which we then combine into lines of varying length. In the following, we describe the HTR network used for the transcription and the HTG network used to generate the synthetic pretraining data. An overview of our complete pipeline is depicted in Fig. 1.

3.1   HTR Model

Combining convolutional neural networks and recurrent neural networks for HTR has been the standard choice for years, and many currently available transcription services feature this kind of models for their efficiency. In this work, we

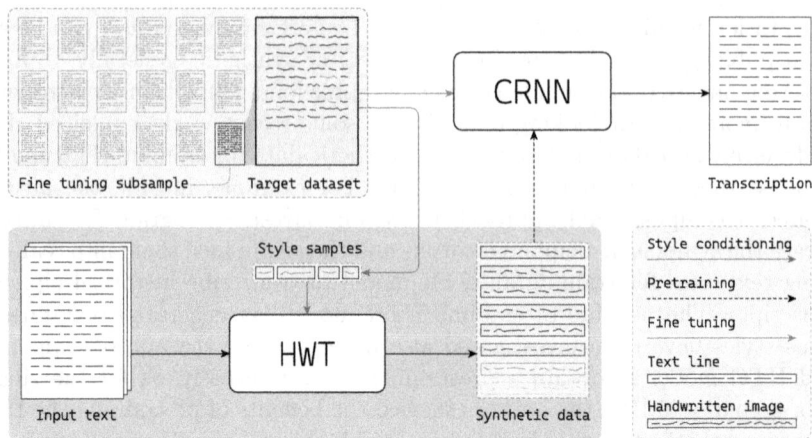

Fig. 1. Overview of our pipeline for synthetic data generation from collection-specific handwritten lines. The generation process renders handwritten line images from a given text conditioned by a few style samples from the target dataset. Then, the synthetic dataset is used to pretrain the CRNN.

consider a model featuring one-dimensional LSTMs, since these have been proven to be comparable or superior to MD-LSTMs [26] and are faster to train [40].

In particular, in this work, we consider a variant of the approach proposed in [45] (referred to as **CRNN** in the following). The convolutional part of the architecture features seven convolutional blocks. For the first six blocks, we adopt the same architecture as in the VGG-11 network, with the difference of applying rectangular pooling in the last two max-pooling layers to better reflect the aspect ratio of text lines images. The seventh convolutional block has a 2×2 kernel. In the adopted variant, the convolutional component features Deformable Convolutions [19], as proposed in [11,12,17], which enhances the performance. The feature map of the last convolutional layer is a $2 \times W \times 512$ tensor, where W depends on the width of the input text line image. This tensor is collapsed along the channel dimension to obtain a sequence of W feature vectors of 1024 elements, which is fed to the recurrent part of the architecture. This consists of two Bidirectional LSTM layers with 512 hidden units each, separated by a dropout layer with probability 0.5. The recurrent part outputs the probability of each feature vector in the sequence to contain each of the characters in a charset.

As customary in HTR, the model is trained to optimize the CTC loss, and thus, a special *blank* character is included in the model charset. Note that we do not use any language model in combination with the HTR network to achieve cross-language adaptability.

3.2   HTG Approach

Styled HTG models allow to efficiently obtain a large number of synthetic text images in the handwriting of the desired author, which can then be used to train an HTR model tailored for the author of interest. In this work, we build upon the transformer-based few-shot styled HTG model recently proposed in [5], namely Handwriting Transformer (**HWT**).

The HTG approach applied is a Convolutional-Transformer encoder-decoder architecture. The handwriting examples are first fed to a convolutional feature extractor in the encoder (namely, a ResNet18), whose outputs are passed through a multi-layer, multi-headed Transformer encoder (with 3 layers and 8 attention heads) and thus enriched with long-range dependencies thanks to the self-attention mechanism. The resulting style vectors are used as keys and values in a multi-layer, multi-headed Transformer decoder (with 3 layers and 8 attention heads as the encoder) that performs cross-attention with vectors representing the characters in the words to be rendered. Normal gaussian noise is added to the resulting vectors to obtain some variability, and those are then fed into a convolutional decoder made of four residual blocks and a tanh activation, which outputs the final styled word images.

The HWT model is trained alongside additional blocks by optimizing a multiple-term loss function. In particular, we follow the adversarial paradigm with the hinge adversarial loss [33] and train HWT together with a convolutional discriminator. Moreover, to enforce the generation of readable word images, we include a CTC loss term obtained by making an HTR model predict the textual content in the generated images. Also this HTR model is inspired by the architecture proposed in [45]. Finally, to force HWT to faithfully render the desired style, we use two additional loss terms. One is the cross-entropy loss of a convolutional classifier aimed at classifying the generated images based on the writers in the HWT training set. The other is a cycle consistency loss term given by the l_1-norm of the difference between the encodings of the real and the generated images obtained by the encoder part of HWT.

4 Experiments

In this section, we describe our experimental analysis. First, we give further implementation details on the adopted HTR and HTG models. Then, we describe the considered small, single-author target datasets, the real benchmark datasets (whose details are reported in Table 1, and some samples in Fig. 2), and the details on the procedure to build the synthetic pretraining datasets. Finally, we describe the evaluation protocol applied and discuss the results obtained.

4.1 Implementation Details

The experiments for this work, both the HTG and HTR part, have been performed on a single NVIDIA RTX 2080 Ti GPU.

CRNN. To train the CRNN model, we rescale all images to a height of 60 pixels, maintaining the original aspect ratio, and then normalize them between -1 and 1. Additionally, when pretraining, we apply the following augmentations.

Table 1. Characteristics of the considered line-level datasets.

	Training Lines	Charset	Period	Language	Authors
Washington [22]	526	68	1755	English	One[1]
Saint Gall [21]	468	49	ca 890–900	Latin	One
Leopardi [12]	1303	76	1818–1832	Italian	One
IAM [35]	6482	79	Modern	English	Many
ICFHR16 [42]	8367	88	1470–1805	German	Many
Rodrigo [43]	9000	105	1545	Spanish	One
ICFHR14 [41]	9198	93	ca 1760–1832	English	One
RIMES [3]	10188	95	Modern	French	Many
NorHand [34]	19653	111	1820–1940	Norwegian	Many
LAM [13]	19830	89	1691–1750	Italian	One
Synthetic for Washington	23121	78	–	English	One
Synthetic for Saint Gall	70494	66	–	Latin	One
Synthetic for Leopardi	89068	113	–	Italian	One

1 - A small number of lines are by another writer.

We modify the brightness of the image with a factor randomly chosen between 0.5 and 5, the contrast with a factor randomly selected from between 0.1 and 10, the saturation with a factor randomly chosen between 0 and 5, and the hue with a factor randomly selected from between −0.1 and 0.1. Moreover, we apply Gaussian blur whose kernel size is set to 5, and the standard deviation is randomly chosen between 0.1 and 2. Finally, we apply a geometric distortion chosen among the following: random rotation (between −1° and 1°), affine transformation (with random rotation between −1° and 1° and random shear between −50° and 30°), and random homography. We use a batch size of 16 when pretraining and a batch size of 8 when fine-tuning and training from scratch. All experiments use a learning rate of 10^{-4}. We train the proposed model with Adam as optimizer, with $\beta_1 = 0.9$ and $\beta_2 = 0.999$, and a scheduler to reduce the learning rate by 10% if the model reaches a plateau for the CER on the validation set. We train the models with a patience of 20 epochs for the CER on the validation set. Note that when fine-tuning, we usually obtain the best CER within the second epoch, which takes roughly less than an hour.

HWT. We train the HWT model on the IAM dataset in the word-level setting (See 4.3). All image samples are grayscale images resized to a height equal to 32 pixels, maintaining the same aspect ratio and normalized between −1 and 1. We train the model with the same settings as those used in the original paper [5] for 7000 epochs. The HWT model generates single-word images whose characters occupy, on average, 16 pixels each. For this reason, we concatenate different images to make a line with a spacing of 16 pixels between the words.

4.2   Target Datasets

Leopardi. [12] The Leopardi dataset consists of a small collection of early 19th Century letters written in Italian by the Romanticism philologist, writer, and poet Giacomo Leopardi. It contains 1303 training lines, 596 validation lines, and 587 test lines. All samples are RGB scans of documents written with ink on ancient paper.

Washington. [22] The George Washington dataset contains 20 English letters written by American President George Washington and one of his collaborators in 1755. The dataset is divided into 526 training lines, 65 validation lines, and 65 test lines. All samples are binary images.

Saint Gall. [21] The Saint Gall dataset contains 60 pages of a handwritten historical manuscript written in Latin by a single author at the end of the 9th Century. The dataset is divided into 468 training lines, 235 validation lines, and 707 test lines. All samples are binary images.

4.3 Pretraining Real Datasets

Rodrigo. [43] The Rodrigo dataset contains 853 pages written in Spanish by a single author. The pages come from a manuscript entitled "Historia de España del arçobispo Don Rodrigo," written in 1545. All samples are grayscale scans of documents written with ink on ancient paper.

NorHand. [34] The NorHand dataset consists of 4144 pages, mainly of diaries and letters written by 15 Norwegian authors from approximately 1820 to 1940. The training set includes 19653 lines. All samples are grayscale scans of documents written with ink on yellowed paper.

LAM. [13] The LAM dataset contains 1171 letters by the historian Ludovico Antonio Muratori in Italian from 1691 to 1750, and thus, exhibits a certain degree of variability due to this wide time-span. The training set includes 19830 lines. All samples are RGB scans of documents written with ink on ancient paper.

ICFHR14. [41] The ICFHR14 dataset contains a collection of 433 pages on law and moral philosophy written by the English philosopher Jeremy Bentham from 1760 to 1832. The training set includes 9198 lines. All samples are grayscale scans of documents written with ink on yellowed paper.

ICFHR16. [42] The ICFHR16 dataset consists of a subset of 400 pages from the Ratsprotokolle collection written from 1470 to 1805 in Early Modern German. The number of writers is unknown. The training set includes 8367 lines. All samples are grayscale scans of documents written with ink on yellowed paper.

RIMES. [3] The RIMES database (Reconnaissance et Indexation de données Manuscrites et de fac similÉS/Recognition and Indexing of handwritten documents and faxes) consists of 12723 pages of scanned letters written in French by 1300 different authors. The training set includes 10188 lines. All samples are binary images.

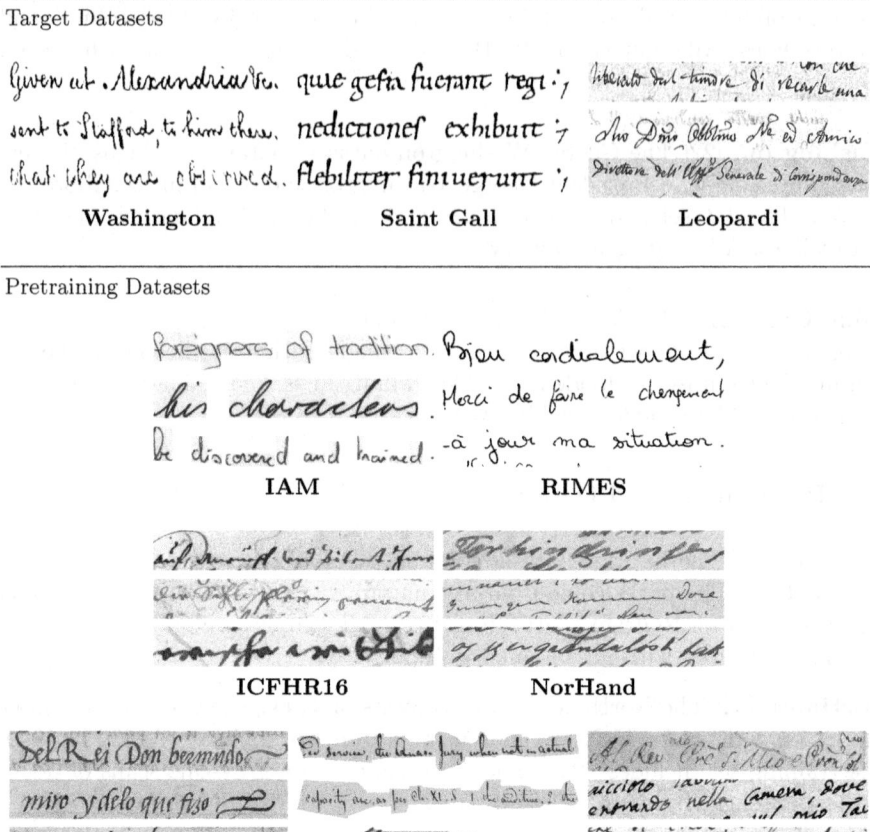

Fig. 2. Exemplar line images from the considered real datasets. These datasets are in different languages, from different periods, and with varied number of writers.

IAM. [35] The IAM Handwriting Database 3.0 is a modern collection of 1539 scanned pages written in English by 657 different authors. The IAM dataset comes in different settings: we use the word-level setting to train the HTG network and the line-level setting for pretraining the HTR network. In particular, we train the HTG network with all the available words from 339 different authors in the word-level training set. For the line-level setting, there are 6482 lines for training, 976 for validation, and 2915 for test. All samples are grayscale scans of documents written with ink on white paper and cleaned digitally.

4.4 Synthetic Data

We use the HWT model, trained on the word-level IAM dataset, to generate synthetic data specific to each of the target datasets. In particular, for each dataset, we isolate a number of word images and repeatedly choose randomly 15 of them to serve as style examples. As for the textual content, we use some of Giacomo Leopardi's proses for the Leopardi dataset, some of George Washington's diaries for the Washington dataset, and a Bible in medieval Latin for the Saint Gall dataset. In this way, the language of the synthetic datasets more closely resembles that of the target datasets. In the following, we refer to these synthetic datasets as **HWT-Generated**. Note that HTW outputs words with characters whose average width is 16 pixels. In our experiments, we also consider a variant of the synthetic datasets in which the generated images are resized in width to match the average character width of the target dataset. The resulting datasets are referred to as **HWT-Generated+WA** (Width Adjustment - WA), in the following. As a final variant, we generate synthetic datasets with the same textual content as the HWT-Generated versions but with style images from the IAM dataset. This variant is referred to as **HWT-Generated+WA+VS** (Varied Styles - VS), in the following. Additionally, for the Leopardi dataset, we exploit the synthetic text lines released alongside the dataset. These have been obtained by rendering the same text that we use for the HWT-Generated Leopardi with a manually built randomized font mimicking the author's handwriting. In the following, we refer to this synthetic dataset as **Human-Synthesized**.

4.5   Evaluation Protocol

To evaluate the effect of the pretraining and fine-tuning strategy in the scenario in which only a few lines in the target dataset are annotated, we perform fine-tuning on a progressively smaller number of training lines, accounting for the 100% (taken as a reference), 50%, 5%, 2.5%, and 1.25%, respectively. As an additional comparison, we train from scratch on the same amount of training lines. Moreover, we consider direct transfer of the pretrained models on the target datasets. The transcription performance is reported in terms of the commonly-used Character Error Rate (CER) and Word Error Rate (WER) scores.

To give further insights into the characteristics of the training data from a linguistic point of view, we calculate the Kullback Leiber Divergence between the distributions of character unigrams, bigrams, and trigrams in each target and pretraining dataset. Additionally, we report the Lexical Similarity [4] between the languages of the considered datasets. This quantifies the lexical similarity between pairs of languages based on words in both languages having a common origin and similar pronunciation and meaning.

4.6   Results

As a baseline experiment, we perform direct transfer of the CRNN model pretrained on the real and synthetic data. The results are reported in Table 2. It can be noticed that, on average, the models pretrained on the synthetic data perform worse than those trained on the real ones, except in the case of the Human-Synthesized dataset for Leopardi. In Tables 3, 4 and 5, we report the results of fine-tuning on a few lines from the real datasets. These experiments

Table 2. Performance of the considered model when pretrained on real datasets or on differently-obtained synthetic datasets (WA stands for 'width adjustment' and VS for 'varied styles') and directly applied to the considered target datasets test set.

	Leopardi		Washington		Saint Gall	
	CER	WER	CER	WER	CER	WER
IAM	57.0	95.6	50.5	**88.7**	**49.9**	**97.7**
RIMES	68.0	97.2	**48.2**	96.2	51.2	98.5
Rodrigo	88.6	99.5	84.3	104.2	55.2	111.4
ICFHR14	75.0	105.6	83.6	128.8	83.7	106.9
ICFHR16	80.5	104.5	87.8	127.4	80.1	106.3
NorHand	49.1	95.4	64.9	101.4	76.0	114.4
LAM	**23.4**	**57.3**	78.8	103.8	78.0	99.5
Human-Synthesized	56.9	95.3	–	–	–	–
HWT-Generated	93.2	99.4	90.9	100.0	83.7	106.9
HWT-Generated+WA	87.5	99.2	87.5	100.0	76.7	100.2
HWT-Generated+WA+VS	87.9	99.3	83.0	98.6	75.3	99.9

aim to reflect the on-demand transcription application. As a reference, we also report the results of CRNN models trained from scratch on the same amount of lines. Note that when less than 230 lines are used for training, the model did not converge, thus enforcing the need for pertaining.

From the results, especially when fine-tuning on 1.25% and 2.5% of the training lines, where the most noticeable differences in performance appear, it emerges that for each target dataset, the best pretraining dataset can be identified, suggesting that the accurate selection of the pretraining data is key in boosting the recognition performance. In particular, the LAM dataset is overall the most suitable when working on the Leopardi dataset (see Table 3), the ICFHR14 is the most helpful when working on the Washington dataset (see Table 4), and Rodrigo when working on Saint Gall (see Table 5).

Note that pretraining a network on a dataset different from the target one induces a bias that depends on some characteristics of the dataset used. In particular, the more the pertaining and target datasets are similar from the linguistic and visual point of view, the more useful the bias will be in terms of the resulting performance. In the following, we explore the main causes of this performance with the aim of tracing some guidelines for the selection of the most suitable pretraining dataset.

Appearance. When choosing the dataset to be utilized during network pretraining, one aspect to consider is the overall visual appearance (*e.g.*, paper support, ink color and thickness, and character width). To analyze the similarities, we refer to Fig. 2, showing some examples of the various datasets. It can be noticed that the samples in the LAM and the Leopardi datasets look similar, and this is reflected in the performance that CRNN reaches when pretrained on LAM and then fine-tuned on Leopardi. Conversely, there is a considerable difference between the samples in Leopardi and Rodrigo. As a result, pretraining

Table 3. Performance of the considered model when pretrained on real datasets or on differently-obtained synthetic ones (WA stands for 'width adjustment' and VS for 'varied styles') and fine-tuned on different portions of the training set of Leopardi.

	Fine-tuning/Training on									
	1.25% (15 l.)		2.5%(32 l.)		5%(65 l.)		50%(652 l.)		100%(1303 l.)	
	CER	WER	CER	WER	CER	WER	CER	WER	CER	WER
Leopardi	–	–	–	–	–	–	4.9	18.4	2.8	10.8
IAM	21.7	63.2	17.1	54.1	12.0	41.0	4.9	19.4	3.3	13.5
RIMES	25.3	68.7	18.2	54.6	13.1	42.3	4.3	15.7	2.7	10.5
Rodrigo	34.9	81.1	23.0	63.7	15.8	48.8	5.2	18.6	2.9	11.0
ICFHR14	23.6	67.7	16.6	53.6	11.7	40.7	4.0	15.9	2.7	10.8
ICFHR16	38.7	85.8	24.4	68.1	16.7	53.4	6.3	24.2	4.3	17.8
NorHand	21.0	63.1	15.2	50.3	11.5	40.2	**3.4**	13.2	**2.3**	8.7
LAM	**12.7**	**42.0**	**12.1**	**39.6**	**8.2**	**28.8**	3.5	**12.7**	**2.3**	**8.6**
Human-Synthesized	25.3	70.1	17.5	53.8	11.7	38.5	4.3	16.1	2.5	9.6
HWT-Generated	66.3	97.8	43.8	84.8	22.5	59.6	5.0	18.3	2.8	10.9
HWT-Generated+WA	43.4	89.2	27.9	70.2	17.4	51.4	5.1	18.6	2.7	10.3
HWT-Generated+WA+VS	35.6	80.9	23.6	63.8	14.8	45.6	4.5	16.9	2.6	10.2

on this latter dataset for HTR on Leopardi leads to poor performance. A similar case can be made for the other two datasets, Saint Gall and Washington. For example, the images in Saint Gall have regular handwriting, similar to those in Rodrigo, which is one of the datasets that leads to the best performance. Moreover, the images in the Washington dataset are visually similar to those in RIMES or IAM. These two datasets are, in fact, those on which performing pretraining leads to the best performance on Washington. From these observations, we can conclude that some visual similarity facilitates transfer learning from the pretraining dataset to the target dataset. However, this is one of many aspects to consider, as we will see below. Another visual aspect to consider is the average character width. As mentioned in Sect. 4.1, HWT generates images that are 32 pixels high and have an average character width of 16 pixels. This aspect ratio is very different from that of the images in the target datasets, which, on average, have a character width of around 8 pixels. For this reason, by using the few examples available, we estimated a form factor to shrink the width of the HWT-generated images to match those of the target datasets. The results of the CRNN models pretrained on the width-adjusted (WA) variant of the synthetics datasets, reported in Tables 2, 3, 4 and 5, highlight that a smaller average character width, which is similar to the character width of the target datasets, leads to better performance compared to the original HWT-generated version.

Handwriting. By observing the results in Tables 2, 3, 4 and 5 alongside the datasets information in Table 1 (especially the number of authors and the time-span), we can highlight a correlation between the performance and the different calligraphies in the pertaining dataset. This latter, in particular, often ensures a high variability of the handwriting style of the images in the dataset. If the

Table 4. Performance of the considered model when pretrained on real datasets or on differently-obtained synthetic ones (WA stands for 'width adjustment' and VS for 'varied styles') and fine-tuned on different portions of the training set of Washington.

	Fine-tuning/Training on									
	1.25%(6 l.)		2.5%(13 l.)		5%(26 l.)		50%(263 l.)		100%(526 l.)	
	CER	WER	CER	WER	CER	WER	CER	WER	CER	WER
Washington	–	–	–	–	–	–	5.3	24.3	3.4	15.9
IAM	**18.8**	**52.7**	14.4	45.9	12.5	40.0	4.9	20.7	3.9	16.1
RIMES	27.1	78.9	20.8	65.2	16.8	55.7	4.6	20.7	3.7	17.1
Rodrigo	48.9	92.6	39.6	84.3	27.3	69.0	6.4	25.8	4.5	19.9
ICFHR14	26.1	64.6	**14.2**	**43.9**	**11.1**	**36.2**	**3.9**	**17.7**	**2.8**	**12.9**
ICFHR16	58.0	97.8	49.1	90.1	30.3	74.6	7.1	28.0	5.2	21.3
NorHand	31.4	76.7	21.7	61.2	16.1	52.9	5.8	23.7	3.7	15.3
LAM	37.2	81.7	27.6	72.6	20.7	60.7	5.9	23.1	4.8	20.3
HWT-Generated	54.9	93.8	43.4	84.7	28.8	70.8	5.3	20.9	3.7	16.5
HWT-Generated+WA	47.4	90.9	34.0	79.7	27.9	70.8	5.5	22.1	3.6	17.3
HWT-Generated+WA+VS	31.3	77.7	23.9	63.4	19.1	56.9	4.8	20.9	3.3	16.1

Table 5. Performance of the considered model when pretrained on real datasets or on differently-obtained synthetic ones (WA stands for 'width adjustment' and VS for 'varied styles') and fine-tuned on different portions of the training set of Saint Gall.

	Fine-tuning/Training on									
	1.25%(5 l.)		2.5%(11 l.)		5%(23 l.)		50%(234 l.)		100%(468 l.)	
	CER	WER	CER	WER	CER	WER	CER	WER	CER	WER
Saint Gall	–	–	–	–	–	–	5.8	38.6	4.5	32.5
IAM	16.5	68.3	13.3	61.0	10.6	54.2	5.4	36.0	4.6	31.4
RIMES	28.2	94.2	19.7	79.7	14.3	66.8	6.5	39.9	5.8	36.9
Rodrigo	**14.4**	**66.4**	**11.3**	**58.7**	**8.8**	**50.8**	**5.3**	35.8	4.6	31.9
ICFHR14	20.4	77.8	16.6	70.5	12.1	54.2	**5.3**	35.4	**4.5**	**30.9**
ICFHR16	32.0	94.2	22.8	83.5	16.1	71.1	6.8	41.9	5.7	36.0
NorHand	27.0	87.9	19.5	75.0	12.8	61.3	**5.3**	**35.1**	4.6	31.5
LAM	20.8	77.8	15.8	68.1	12.2	60.0	**5.3**	35.5	4.6	31.4
HWT-Generated	20.4	77.8	16.6	70.5	12.1	58.8	5.5	37.5	4.8	33.3
HWT-Generated+WA	19.7	80.1	14.3	66.6	11.4	58.6	5.4	36.8	**4.5**	31.2
HWT-Generated+WA+VS	18.8	76.1	13.3	61.2	11.0	55.8	5.4	35.9	**4.5**	31.6

pretraining dataset contains texts written by multiple authors, the network is exposed to high variance and will achieve the ability to handle different styles and calligraphies. On the other hand, pretraining the network on images with only one author's handwriting reduces the variance and makes the network focus on that single author. For example, the LAM dataset has low variance since it is single-author. Nonetheless, it is similar to the Leopardi dataset due to language similarities and the historical period. Thus, pretraining on LAM induces a bias that allows the HTR model to effectively generalize to Leopardi. An example

Table 6. Language comparison between the Leopardi dataset and the considered pertaining datasets, ordered by average Kullback Leiber Divergence of n-grams.

	Lexical Similarity	Kullback Leiber Divergence			FT on 1.25%	
		Unigram	Bigram	Trigram	CER	WER
LAM	–	0.02	0.09	0.23	12.7	42.0
Synthetic for Leopardi	-	0.05	0.19	0.54	35.6	80.9
RIMES	9.54	0.11	0.84	1.89	25.3	68.7
IAM	6.76	0.17	0.85	1.62	21.7	63.2
ICFHR14	6.76	0.17	0.91	1.83	23.6	67.7
Rodrigo	10.45	0.20	0.71	1.48	34.9	81.1
NorHand	3.99	0.34	1.15	2.06	21.0	63.1
ICFHR16	4.19	0.40	1.50	2.50	39.0	86.0

Table 7. Language comparison between the Saint Gall dataset and the considered pertaining datasets, ordered by average Kullback Leiber Divergence of n-grams.

	Lexical Similarity	Kullback Leiber Divergence			FT on 1.25%	
		Unigram	Bigram	Trigram	CER	WER
LAM	5.81	0.17	0.87	1.59	20.8	77.8
Rodrigo	6.08	0.18	0.89	1.74	14.4	66.4
RIMES	5.39	0.19	0.87	1.74	28.2	94.2
ICFHR14	3.50	0.20	0.79	1.43	20.4	77.8
IAM	3.50	0.21	0.74	1.31	16.5	68.3
Synthetic for Saint Gall	-	0.23	0.60	1.08	18.8	76.1
NorHand	2.39	0.42	1.22	1.78	27.0	87.9
ICFHR16	2.73	0.58	1.60	2.10	32.0	94.2

of the opposite case can be observed when pretraining on the ICFHR16 dataset and fine-tuning on Saint Gall (Table 7). ICFHR16 is a German dataset with a significant difference compared to Saint Gall, which is in Latin. Moreover, since ICFHR16 contains texts written by multiple authors, the dataset has a higher variance than Saint Gall, which is single-author. Therefore, during fine-tuning, the network needs to apply more corrections to adjust for the bias and reduce the variance to focus on the Saint Gall texts, and therefore, more samples are needed to achieve good results. Overall, the results in Tables 2, 3, 5, and 4 show that, on average, all single-author datasets bring to better performance on Leopardi and Saint Gall, while in the Washington dataset, which contains the texts written by two authors, the multi-author datasets (*e.g.*, the IAM dataset) with many different styles with a high variance allow obtaining better results.

Language. Tables 6, 7, and 8 compare the language similarity and the Kullback Leiber Divergence (KL) between the pretraining and target datasets. To

Table 8. Language comparison between the Washington dataset and the considered pertaining datasets, ordered by average Kullback Leiber Divergence of n-grams.

	Lexical Similarity	Kullback Leiber Divergence			FT on 1.25%	
		Unigram	Bigram	Trigram	CER	WER
ICFHR14	–	0.05	0.30	0.66	26.1	64.6
Synthetic for Washington	-	0.07	0.31	0.69	23.9	63.4
IAM	–	0.08	0.30	0.59	18.8	52.7
NorHand	4.30	0.29	1.03	1.64	31.4	76.7
RIMES	9.67	0.31	1.36	2.22	27.1	78.9
Rodrigo	7.91	0.35	1.38	2.32	48.9	92.6
ICFHR16	4.72	0.36	1.24	1.83	58.0	97.8
LAM	6.76	0.36	1.52	2.31	37.2	81.7

highlight the correlation between the textual information and the network performance, we sort the tables by KL divergence and include the results obtained after fine-tuning with the 1.25% of the target dataset. In this way, we emphasize a trend where datasets with a slight language variation are more suitable to be used in pretraining than datasets with a significant difference. In particular, ICFHR16, in German, is, on average, the farthest dataset to all the target datasets we compare with in terms of KL. As a result, pretraining on this dataset leads to the highest recognition errors in all three target datasets. On the other hand, pretraining on IAM and ICFHR14, which are in English, allows obtaining good performance thanks to the language similarity to all the target datasets. Notably, from Table 6, we observe that from a lexical point of view, the Leopardi dataset is closer to the LAM dataset than the synthetic one containing proses by the author. The text in all three datasets is in Italian and was written in the same period with a time difference of fewer than 70 years. Arguably, the reason why LAM is closer to the Leopardi dataset is that both datasets are a collection of letters, which share many structural similarities (*e.g.*, dates, openings, salutations). Combining this aspect, the language, and the period makes CRNN pretrained on the LAM dataset to obtain impressive performance on the Leopardi test set, particularly in a direct transfer setting (see Table 2).

5   Conclusion

In this paper, we have explored line-level HTR on historical manuscripts when limited training data are available. To this end, we have proposed to pretrain a dedicated HTR model on existing benchmark datasets or on a large quantity of synthetic data that reflect the characteristics of the handwriting of the target author of the manuscripts, which we built with a fully-automatic procedure, and fine-tune on a portion of real data in the collection of interest. In particular, we have conducted an extensive quantitative analysis of the main characteristics that the pretraining dataset should have in order to obtain a strong HTR model with as little as five lines from the target collection.

The obtained experimental results show that when choosing the real dataset or generating the synthetic one for pretraining, both the overall appearance (given by the paper support, writing tool, and average character width) and the language should be taken into account. Moreover, it has emerged that an HTR model trained on images of text with high variability in handwriting style is more robust and easily adaptable than one trained on a single handwriting style. Nonetheless, when the synthetic data faithfully resemble the real ones in terms of handwriting, satisfactory performance is achievable.

In the sight of these conclusions, this work can help guide the selection of the most suitable pretraining dataset to boost the performance of HTR models on small domain-specific documents and give some insights into the maturity of the HTG field and its potential benefit for HTR. Finally, this work has shed some light on the feasibility of interactive, on-demand HTR on single-author collections, which is a task worthy of further investigation for its application to digital archives use and enhancement.

Acknowledgement. This work was supported by the "AI for Digital Humanities" project (Pratica Sime n.2018.0390), funded by "Fondazione di Modena" and the PNRR project Italian Strengthening of ESFRI RI Resilience (ITSERR) funded by the European Union - NextGenerationEU (CUP: B53C22001770006).

References

1. Alonso, E., Moysset, B., Messina, R.: Adversarial generation of handwritten text images conditioned on sequences. In: ICDAR (2019)
2. Aradillas, J.C., Murillo-Fuentes, J.J., Olmos, P.M.: Boosting offline handwritten text recognition in historical documents with few labeled lines. IEEE Access **9**, 76674–76688 (2021)
3. Augustin, E., Carré, M., Grosicki, E., Brodin, J.M., Geoffrois, E., Prêteux, F.: RIMES evaluation campaign for handwritten mail processing. In: IWFHR (2006)
4. Bella, G., Batsuren, K., Giunchiglia, F.: A database and visualization of the similarity of contemporary lexicons (2021)
5. Bhunia, A.K., Khan, S., Cholakkal, H., Anwer, R.M., Khan, F.S., Shah, M.: Handwriting Transformers. In: ICCV (2021)
6. Bhunia, A.K., Das, A., Bhunia, A.K., Kishore, P.S.R., Roy, P.P.: Handwriting recognition in low-resource scripts using adversarial learning. In: CVPR (2019)
7. Bhunia, A.K., Ghose, S., Kumar, A., Chowdhury, P.N., Sain, A., Song, Y.Z.: MetaHTR: Towards Writer-Adaptive Handwritten Text Recognition. In: CVPR (2021)
8. Bluche, T.: Joint line segmentation and transcription for end-to-end handwritten paragraph recognition. In: NeurIPS (2016)
9. Bluche, T., Louradour, J., Messina, R.: Scan, attend and read: end-to-end handwritten paragraph recognition with MDLSTM attention. In: ICDAR (2017)
10. Bluche, T., Messina, R.: Gated convolutional recurrent neural networks for multilingual handwriting recognition. In: ICDAR (2017)
11. Cascianelli, S., Cornia, M., Baraldi, L., Cucchiara, R.: Boosting modern and historical handwritten text recognition with deformable convolutions. In: IJDAR, pp. 1–11 (2022)

12. Cascianelli, S., Cornia, M., Baraldi, L., Piazzi, M.L., Schiuma, R., Cucchiara, R.: Learning to read L'Infinito: handwritten text recognition with synthetic training data. In: ICPR (2021)
13. Cascianelli, S., et al.: The lam dataset: a novel benchmark for line-level handwritten text recognition. In: ICPR (2022)
14. Chammas, E., Mokbel, C., Likforman-Sulem, L.: Handwriting recognition of historical documents with few labeled data. In: DAS (2018)
15. Cilia, N.D., De Stefano, C., Fontanella, F., di Freca, A.S.: A ranking-based feature selection approach for handwritten character recognition. Pattern Recogn. Lett. **121**, 77–86 (2019)
16. Clanuwat, T., Lamb, A., Kitamoto, A.: KuroNet: pre-modern Japanese Kuzushiji character recognition with deep learning. In: ICDAR (2019)
17. Cojocaru, I., Cascianelli, S., Baraldi, L., Corsini, M., Cucchiara, R.: Watch your strokes: improving handwritten text recognition with deformable convolutions. In: ICPR (2020)
18. Coquenet, D., Chatelain, C., Paquet, T.: Recurrence-free unconstrained handwritten text recognition using gated fully convolutional network. In: ICFHR (2020)
19. Dai, J., Qi, H., Xiong, Y., Li, Y., Zhang, G., Hu, H., Wei, Y.: Deformable convolutional networks. In: CVPR (2017)
20. Davis, B., Tensmeyer, C., Price, B., Wigington, C., Morse, B., Jain, R.: Text and style conditioned GAN for generation of offline handwriting lines. In: BMVC (2020)
21. Fischer, A., Frinken, V., Fornés, A., Bunke, H.: Transcription alignment of Latin manuscripts using hidden Markov models. In: HIP (2011)
22. Fischer, A., Keller, A., Frinken, V., Bunke, H.: Lexicon-free handwritten word spotting using character HMMs. Pattern Recogn. Lett. **33**(7), 934–942 (2012)
23. Fogel, S., Averbuch-Elor, H., Cohen, S., Mazor, S., Litman, R.: ScrabbleGAN: semi-supervised varying length handwritten text generation. In: CVPR (2020)
24. Goodfellow, I.J., et al.: Generative adversarial nets. In: NeurIPS (2014)
25. Granet, A., Morin, E., Mouchère, H., Quiniou, S., Viard-Gaudin, C.: Transfer learning for handwriting recognition on historical documents. In: ICPRAM (2018)
26. Graves, A., Schmidhuber, J.: Offline handwriting recognition with multidimensional recurrent neural networks. In: NeurIPS (2009)
27. Haines, T., Mac Aodha, O., Brostow, G.: My text in your handwriting. ACM Trans. Graphics **35**(3), 1–18 (2016)
28. Jaramillo, J.C.A., Murillo-Fuentes, J.J., Olmos, P.M.: Boosting handwriting text recognition in small databases with transfer learning. In: ICFHR (2018)
29. Kang, L., Riba, P., Rusinol, M., Fornes, A., Villegas, M.: Content and style aware generation of text-line images for handwriting recognition. IEEE Trans. PAMI 1 (2021)
30. Kang, L., Riba, P., Rusiñol, M., Fornés, A., Villegas, M.: Pay attention to what you read: non-recurrent handwritten text-line recognition. Pattern Recogn. **129**, 108766 (2022)
31. Kang, L., Riba, P., Wang, Y., Rusiñol, M., Fornés, A., Villegas, M.: GANwriting: content-conditioned generation of styled handwritten word images. In: ECCV (2020)
32. Li, M., et al.: TrOCR: transformer-based optical character recognition with pre-trained models. arXiv preprint arXiv:2109.10282 (2021)
33. Lim, J.H., Ye, J.C.: Geometric GAN. arXiv preprint arXiv:1705.02894 (2017)
34. Maarand, M., Beyer, Y., Kåsen, A., Fosseide, K.T., Kermorvant, C.: A comprehensive comparison of open-source libraries for handwritten text recognition in norwegian. In: DAS (2022)

35. Marti, U.V., Bunke, H.: The IAM-database: an English sentence database for offline handwriting recognition. IJDAR **5**(1), 39–46 (2002)
36. Martín-Albo Simón, D., Romero Gómez, V., Toselli, A.H., Vidal Ruiz, E.: Multi-modal computer-assisted transcription of text images at character-level interaction. Int. J. Pattern Recognit. Artif. Intell. **26**(05), 1263003 (2012)
37. Mattick, A., Mayr, M., Seuret, M., Maier, A., Christlein, V.: SmartPatch: improving handwritten word imitation with patch discriminators. In: ICDAR (2021)
38. Moysset, B., Kermorvant, C., Wolf, C.: Full-page text recognition: learning where to start and when to stop. In: ICDAR (2017)
39. Pham, V., Bluche, T., Kermorvant, C., Louradour, J.: Dropout improves recurrent neural networks for handwriting recognition. In: ICFHR (2014)
40. Puigcerver, J.: Are multidimensional recurrent layers really necessary for handwritten text recognition? In: ICDAR (2017)
41. Sánchez, J.A., Romero, V., Toselli, A.H., Vidal, E.: ICFHR2014 competition on handwritten text recognition on transcriptorium datasets (HTRtS). In: ICFHR (2014)
42. Sanchez, J.A., Romero, V., Toselli, A.H., Vidal, E.: ICFHR2016 competition on handwritten text recognition on the READ dataset. In: ICFHR (2016)
43. Serrano, N., Castro, F., Juan, A.: The RODRIGO database. In: LREC (2010)
44. Shen, X., Messina, R.: A method of synthesizing handwritten Chinese images for data augmentation. In: ICFHR (2016)
45. Shi, B., Bai, X., Yao, C.: An end-to-end trainable neural network for image-based sequence recognition and its application to scene text recognition. IEEE Trans. PAMI **39**(11), 2298–2304 (2016)
46. Souibgui, M.A., et al.: One-shot compositional data generation for low resource handwritten text recognition. In: WACV (2022)
47. Soullard, Y., Swaileh, W., Tranouez, P., Paquet, T., Chatelain, C.: Improving text recognition using optical and language model writer adaptation. In: ICDAR (2019)
48. Such, F.P., Peri, D., Brockler, F., Paul, H., Ptucha, R.: Fully convolutional networks for handwriting recognition. In: ICFHR (2018)
49. Vaswani, A., et al.: Attention is all you need. In: NeurIPS (2017)
50. Voigtlaender, P., Doetsch, P., Ney, H.: Handwriting recognition with large multidimensional long short-term memory recurrent neural networks. In: ICFHR (2016)
51. Wang, J., Wu, C., Xu, Y.Q., Shum, H.Y.: Combining shape and physical models for on-line cursive handwriting synthesis. IJDAR **7**(4), 219–227 (2005)
52. Wick, C., Zöllner, J., Grüning, T.: Rescoring sequence-to-sequence models for text line recognition with CTC-prefixes. arXiv preprint arXiv:2110.05909 (2021)
53. Wick, C., Zöllner, J., Grüning, T.: Transformer for handwritten text recognition using bidirectional post-decoding. In: ICDAR (2021)
54. Wigington, C., Stewart, S., Davis, B., Barrett, B., Price, B., Cohen, S.: Data augmentation for recognition of handwritten words and lines using a CNN-LSTM network. In: ICDAR (2017)
55. Wigington, C., Tensmeyer, C., Davis, B., Barrett, W., Price, B., Cohen, S.: Start, follow, read: end-to-end full-page handwriting recognition. In: ECCV (2018)
56. Yousef, M., Bishop, T.E.: OrigamiNet: weakly-supervised, segmentation-free, one-step, full page text recognition by learning to unfold. In: CVPR (2020)

Zero-shot Generation of Training Data with Denoising Diffusion Probabilistic Model for Handwritten Chinese Character Recognition

Dongnan Gui[1,2], Kai Chen[1(✉)], Haisong Ding[1], and Qiang Huo[1]

[1] Microsoft Research Asia, Beijing, China
{gdn2001,dinghs11}@mail.ustc.edu.cn, chenkai.cn@hotmail.com,
qianghuo@microsoft.com
[2] University of Science and Technology of China, Hefei, China

Abstract. There are more than 80,000 character categories in Chinese while most of them are rarely used. To build a high performance handwritten Chinese character recognition (HCCR) system supporting the full character set with a traditional approach, many training samples need be collected for each character category, which is both time-consuming and expensive. In this paper, we propose a novel approach to transforming Chinese character glyph images generated from font libraries to handwritten ones with a denoising diffusion probabilistic model (DDPM). Training from handwritten samples of a small character set, the DDPM is capable of mapping printed strokes to handwritten ones, which makes it possible to generate photo-realistic and diverse style handwritten samples of unseen character categories. Combining DDPM-synthesized samples of unseen categories with real samples of other categories, we can build an HCCR system to support the full character set. Experimental results on CASIA-HWDB dataset with 3,755 character categories show that the HCCR systems trained with synthetic samples perform similarly with the one trained with real samples in terms of recognition accuracy. The proposed method has the potential to address HCCR with a larger vocabulary.

Keywords: Denoising Diffusion Probabilistic Model · Handwritten Chinese Character Recognition · Zero-shot Generation

1 Introduction

In the latest National Standards of the People's Republic of China about Chinese coded character set (GB18030-2022), 87,887 Chinese character categories are included. To create a high-performance handwritten Chinese character recognition (HCCR) system that supports the full character set using traditional

---

K. Chen—This work was done when Dongnan Gui was an intern in MMI Group, Microsoft Research Asia, Beijing, China.

approaches, a large number of training samples with various writing styles would be collected for each character category. However, only about 4,000 categories are commonly used in daily life. It is therefore both time-consuming and expensive to collect representative handwritten samples for the remaining 95% rarely-used ones. These categories are often of complicated structures, existing in personal names, addresses, ancient books, historic documents and scientific publications. An HCCR system supporting the full-set of these categories with high accuracy will be beneficial to improve user experience, protect cultural heritages and promote academic exchanges.

Lots of research efforts have been made to build an HCCR system with only real training samples from commonly used characters. A Chinese character consists of radicals/strokes with specific spatial relationships, which are shared across all characters. Rather than encoding each character category as a single one-hot vector, [4,10,44,45] encode it as a sequence of radicals/strokes and spatial relationships to achieve zero-shot recognition goal. In [1,19,21,22], font-rendered glyph images are leveraged to provide reference representations for unseen character categories. There are also some efforts to synthesize handwritten samples for unseen categories. For example, [48] synthesizes unseen character samples with a radical composition network and combines them with real samples to train an HCCR system. However, its recognition accuracy is relatively poor.

We propose to solve this problem by synthesizing diverse and high-quality training samples for unseen character categories with denoising diffusion probabilistic models (DDPMs) [15,38]. Diffusion models have been shown to outperform other generation techniques in terms of diversity and quality [9,29,40–42], due to their powerful modeling capacity of high-dimensional distributions. This also offers a zero-shot generation capability. For example, in diffusion-based text-to-image generation [28,33,36], with all object types and spatial relationships existed in training samples, diffusion models are capable of generating photo-realistic images of in-existence object combinations and layouts. As mentioned above, Chinese characters can be treated as combinations of different radicals/strokes with specific layouts. We can leverage DDPM to achieve the goal of zero-shot handwritten Chinese character image generation.

In this paper, we design a glyph conditional DDPM (GC-DDPM), which concatenates a font-rendered character glyph image with the original input of U-Net used in [9], to guide the model in constructing mappings between font-rendered and handwritten strokes/radicals. To the best of our knowledge, we are the first to apply DDPMs to zero-shot handwritten Chinese character generation. Unlike other image-to-image diffusion model frameworks (e.g., [30,35,43]), which aim at synthesizing images in the target domain while faithfully preserving the content representations, our goal is to learn mappings from rendered printed radicals/strokes to the handwritten ones.

Experimental results on CASIA-HWDB [23] dataset with 3,755 character categories show that the HCCR systems trained with DDPM-synthesized samples outperform other synthetic data based solutions and perform similarly with the one trained with real samples in terms of recognition accuracy. We also visualize the generation effect of both in and out of 3,755 character categories, which indicates that our method has the potential to be extended to a larger vocabulary.

The remainder of the paper is organized as follows. In Sect. 2, we briefly review related works. In Sect. 3, we describe our GC-DDPM design along with sampling methods. Our approach is evaluated and compared with prior arts in Sect. 4. We discuss limitations of our approach and future work in Sect. 5, and conclude the paper in Sect. 6.

2   Related Work

Zero-shot HCCR. Conventional HCCR systems [6,7,20,50,52,53], although achieving superior recognition accuracy, can only recognize character categories that are observed in the training set. Zero-shot HCCR aims to recognize handwritten characters that are never observed. Most of the previous zero-shot HCCR systems can be divided into two categories: structure-based and structure-free methods. In structure-based methods, a Chinese character is represented as a sequence of composing radicals [4,10,44,45] or strokes [5]. Although the character is never observed, the composing radicals, strokes and their spatial relationships have been observed in the training set. Therefore, structure-based methods are able to predict the radical or stroke sequences of unseen Chinese characters and achieve zero-shot recognition. However, in these methods, the radical or stroke sequence representations of Chinese characters require lots of language-specific domain knowledge. In structure-free method, [1,17,21,22] leverage information from the corresponding Chinese character glyph images. Zero-shot HCCR is achieved by choosing the Chinese character whose glyph features are closest to that of the handwritten ones in terms of visual representations. In [19], the radical information is also used to extract the visual representations of glyph images.

Zero-shot Data Synthesis for HCCR. Besides designing zero-shot recognition systems, there are some studies to directly synthesize handwritten training samples for unseen categories. [48] investigates a radical composition network to generate unseen Chinese characters by integrating radicals and their spatial relationships. Although the generated handwritten Chinese characters can increase the recognition rate of unseen handwritten characters, the overall recognition performance is relatively poor. In this work, we propose to use a more powerful diffusion model to generate unseen handwritten Chinese characters given corresponding glyph images.

Zero-shot Chinese Font Generation. Zero-shot Chinese font generation aims to generate font glyph for unseen Chinese characters based on some seen character/font glyph pairs. In [11,25,47,51,54], the image-to-image translation framework is used to achieve this goal. Works in [18,24,31] also leverage the information of composing components, radicals, strokes for better generalization. In this paper, we focus on zero-shot handwritten Chinese character generation with DDPM and we can easily adapt this method to zero-shot Chinese font generation task.

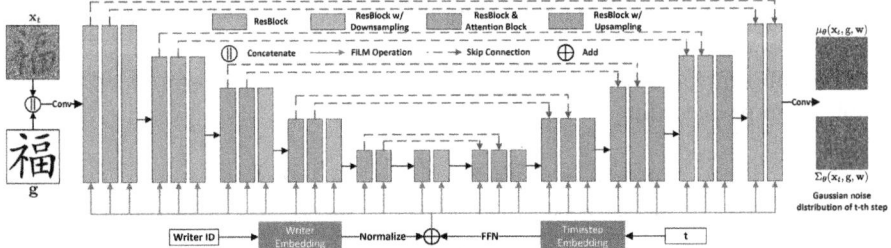

Fig. 1. Architecture of glyph conditional U-Net, which is adapted from the model used in [9]. We concatenate font "kai" rendered character image with original input to provide glyph guidance during generation.

Diffusion Model. DDPM [15,38] has become extremely popular in computer vision and achieves superior performance in image generation tasks. DDPM uses two parameterized Markov chains and variational inference method to reconstruct the data distribution. DDPMs have demonstrated their powerful capabilities to generate high-quality and high-diversity images [9,15,42]. It is shown in [33] that DDPM can perform a great effect on combination of concepts, which can integrate multiple elements. Diffusion models are also applied to other tasks [8,49], including high-resolution generation [34], image inpainting [43], natural language processing [2] and so on. Besides, [27] introduces DDPM to solve the problem of online English handwriting generation. In this work, we propose to leverage DDPM for zero-shot handwritten Chinese character generation and to synthesize training data for unseen Chinese characters to build HCCR systems.

3   Our Approach

3.1   Preliminary

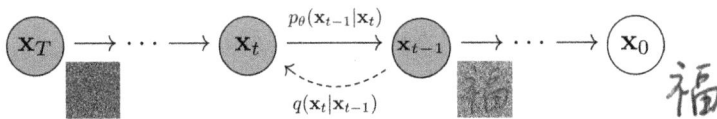

Fig. 2. The Markov chain of forward (reverse) diffusion process of generating a handwritten Chinese character sample by slowly adding (removing) noise. Adapted from [15].

Diffusion model is a new paradigm of data generation. It defines a Markov chain of diffusion steps to slowly add random noise to data and then learn to reverse the diffusion process to construct desired data samples from the noise [46]. As shown in Fig. 2, in our handwritten Chinese character generation scenario, we first sample a character image from the real distribution $\mathbf{x}_0 \sim q(\mathbf{x})$. Then, in forward diffusion process, small amounts of Gaussian noise are added to the sample in steps according to Eq. (1),

$$q(\mathbf{x}_t|\mathbf{x}_{t-1}) = \mathcal{N}(\mathbf{x}_t; \sqrt{1 - \beta_t}\mathbf{x}_{t-1}, \beta_t\mathbf{I}) \tag{1}$$
$$\mathbf{x}_t = \sqrt{\alpha_t}\mathbf{x}_{t-1} + \sqrt{1 - \alpha_t}\epsilon_t$$

where $\alpha_t = 1 - \beta_t$ and $\epsilon_t \sim \mathcal{N}(\mathbf{0}, \mathbf{I})$, producing a sequence of noisy samples. The step sizes are controlled by a variance schedule $\{\beta_t \in (0,1)\}_{t=1}^T$. As t becomes larger, the image gradually loses its distinguishable features. When $t \to \infty$, $\mathbf{x}_t$ becomes a sample of an isotropic Gaussian distribution.

If we can reverse the above process and sample from $q(\mathbf{x}_{t-1}|\mathbf{x}_t)$, we will be able to recreate the true sample from a Gaussian noise $\mathbf{x}_T \sim \mathcal{N}(\mathbf{0}, \mathbf{I})$. If β_t is small enough, $q(\mathbf{x}_{t-1}|\mathbf{x}_t)$ will also be a Gaussian. So we can approximate it with a parameterized model, as shown in Eq. (2)

$$p_\theta(\mathbf{x}_{t-1}|\mathbf{x}_t) = \mathcal{N}(\mathbf{x}_{t-1}; \boldsymbol{\mu}_\theta(\mathbf{x}_t, t), \boldsymbol{\Sigma}_\theta(\mathbf{x}_t, t)). \tag{2}$$

Since $q(\mathbf{x}_{t-1}|\mathbf{x}_t, \mathbf{x}_0)$ is tractable,

$$q(\mathbf{x}_{t-1}|\mathbf{x}_t, \mathbf{x}_0) = \mathcal{N}(\mathbf{x}_{t-1}; \tilde{\boldsymbol{\mu}}(\mathbf{x}_t, \mathbf{x}_0), \tilde{\beta}_t\mathbf{I}) \tag{3}$$

where $\bar{\alpha}_t = \prod_{s=1}^t \alpha_s$, and

$$\tilde{\boldsymbol{\mu}}(\mathbf{x}_t, \mathbf{x}_0) = \frac{1}{\sqrt{\alpha_t}}(\mathbf{x}_t - \frac{1 - \alpha_t}{\sqrt{1 - \bar{\alpha}_t}}\epsilon_t) \tag{4}$$

$$\tilde{\beta}_t = \frac{1 - \bar{\alpha}_{t-1}}{1 - \bar{\alpha}_t} \cdot \beta_t . \tag{5}$$

So we can train a neural network to approximate ϵ_t and the predicted value is denoted as $\epsilon_\theta(\mathbf{x}_t)$. It has been verified that instead of directly setting $\boldsymbol{\Sigma}_\theta(\mathbf{x}_t, t)$ as $\tilde{\beta}_t$, setting it as a learnable interpolation between $\tilde{\beta}_t$, β_t in log domain will yield better log-likelihood [29]:

$$\boldsymbol{\Sigma}_\theta(\mathbf{x}_t, t) = \exp(\boldsymbol{\nu}_\theta(\mathbf{x}_t) \log \beta_t + (1 - \boldsymbol{\nu}_\theta(\mathbf{x}_t)) \log \tilde{\beta}_t) . \tag{6}$$

In this paper, we will train a U-Net to predict $\epsilon_\theta(\mathbf{x}_t)$ and $\boldsymbol{\nu}_\theta(\mathbf{x}_t)$ with the same hybrid loss as in [29].

3.2   Glyph Conditional U-Net Architecture

As shown in Fig. 1, the U-Net architecture we used is borrowed from [9]. With 128×128 image input, there are 5 resolution stages in encoder and decoder respectively, and each stage consists of 2 BigGAN residual blocks (ResBlock) [3]. In addition, BigGAN ResBlocks are also used for downsampling and upsampling activations. We also follow [9] to use multi-head attention at 32×32, 16×16

and 8×8 resolutions. Timestep t will first be mapped to sinusoidal embedding and then processed by a 2-layer feed-forward network (FFN). This processed embedding will then be fed to each convolution layer in U-Net through a feature-wise linear modulation (FiLM) operator [32].

To control the style and content of generated character images, writer information [12] and character category information are also fed to the model. Given a writer $\mathbf{w}$, which is actually the class index of all writer IDs, it will be mapped to a learnable embedding, followed by L2-normalization (denoted as $\mathbf{z}$), which is injected to U-Net together with the timestep embedding [29] as shown in Fig. 1.

If we inject character category information in the same way as writer, the model will not be able to generate samples for unseen categories because their embeddings are not optimized at all. In this paper, we propose to leverage printed images rendered by font "kai" to provide character category information. We denote this glyph image as $\mathbf{g}$. There are several ways to inject $\mathbf{g}$ to the model. For example, it can be encoded as a feature vector by a CNN/ViT and fed to U-Net in FiLM way, or encoded as feature sequences and fed to attention layers of U-Net serving as external keys and values [28]. In this paper, we simply inject $\mathbf{g}$ as model's input by concatenating it with $\mathbf{x}_t$ and leave other ways as future work. We call our approach as **G**lyph **C**onditional DDPM (**GC-DDPM**).

By conditioning model output on glyph image, we expect the model can learn the implicit mapping rules between printed stroke combinations and their handwritten counterparts. Then we can input font-rendered glyph images of unseen characters to the well-trained GC-DDPM and get their handwritten samples of high quality and diversity.

3.3   Multi-conditional Classifier-free Diffusion Guidance

Classifier-free guidance [16] has been proven effective for improving generation quality on different tasks. In this paper, we are also curious about its effects on HCCR system trained with synthetic samples.

There are 2 conditions, glyph $\mathbf{g}$ and writer $\mathbf{w}$, in our model. We assume that given $\mathbf{x}_t$, $\mathbf{g}$ and $\mathbf{w}$ are independent. So we have

$$p_\theta(\mathbf{x}_{t-1}|\mathbf{x}_t, \mathbf{g}, \mathbf{w}) \propto p_\theta(\mathbf{x}_{t-1}|\mathbf{x}_t) p_\theta(\mathbf{g}|\mathbf{x}_t) p_\theta(\mathbf{w}|\mathbf{x}_t) . \tag{7}$$

Following the previous practice in [16], we assume that there is an implicit classifier (ic),

$$p_{ic}(\mathbf{g}, \mathbf{w}|\mathbf{x}_t) \propto \left[\frac{p(\mathbf{x}_t|\mathbf{g})}{p(\mathbf{x}_t)}\right]^\gamma \cdot \left[\frac{p(\mathbf{x}_t|\mathbf{w})}{p(\mathbf{x}_t)}\right]^\eta . \tag{8}$$

Then we have

$$\nabla_{\mathbf{x}_t} \log p_{ic}(\mathbf{g}, \mathbf{w}|\mathbf{x}_t) \propto \gamma \epsilon(\mathbf{x}_t, \mathbf{g}) + \eta \epsilon(\mathbf{x}_t, \mathbf{w}) - (\gamma + \eta)\epsilon(\mathbf{x}_t) . \tag{9}$$

So we can perform sampling with the score formulation

$$\begin{aligned}
\tilde{\epsilon}_\theta(\mathbf{x}_t, \mathbf{g}, \mathbf{w}) = {}& \epsilon_\theta(\mathbf{x}_t, \mathbf{g}, \mathbf{w}) + \gamma \epsilon_\theta(\mathbf{x}_t, \mathbf{g}, \emptyset) \\
& + \eta \epsilon_\theta(\mathbf{x}_t, \emptyset, \mathbf{w}) - (\gamma + \eta)\epsilon_\theta(\mathbf{x}_t, \emptyset, \emptyset) .
\end{aligned} \tag{10}$$

We call γ, η as content and writer guidance scales respectively. When $\mathbf{g} = \emptyset$, an empty glyph image will be fed to U-Net and when $\mathbf{w} = \emptyset$, a special embedding will be used. During training, we set $\mathbf{g}$ and $\mathbf{w}$ to $\emptyset$ with probability 10% independently to get partial/unconditional models.

3.4 Writer Interpolation

Besides generating unseen characters, our model is also able to generate unseen styles by injecting interpolation between different writer embeddings as new writer embedding. Given two normalized writer embeddings $\mathbf{z_i}$ and $\mathbf{z_j}$, we use spherical interpolation [33] to get a new embedding $\mathbf{z}$ with L2-norm being 1, as in Eq. 11:

$$\mathbf{z} = \mathbf{z}_i \cos \frac{\lambda \pi}{2} + \mathbf{z}_j \sin \frac{\lambda \pi}{2}, \quad \lambda \in [0, 1] . \tag{11}$$

4 Experiments

We conduct our experiments on CASIA-HWDB [23] dataset. The detailed experimental setup is comprehensively explained in Sect. 4.1. Experiments on Writer Independent (WI) and Writer Dependent (WD) GC-DDPMs are conducted in Sect. 4.2 and Sect. 4.3, respectively. We further use synthesized samples to augment the training set of HCCR in Sect. 4.4. Finally, we compare our approach with prior arts in Sect. 4.5.

4.1 Experimental Setup

Dataset: The CASIA-HWDB dataset is a large-scale offline Chinese handwritten character database including HWDB1.0, 1.1 and 1.2. We use the HWDB1.0 and 1.1 in experiments, where the former contains 3,866 Chinese character categories written by 420 writers, and the latter contains 3,755 categories written by another 300 writers. We follow the official partition of training and testing sets as in [23], where the training set is written by 576 writers.

Vocabulary Partition: We use the 3,755 categories that cover the standard GB2312-80 level-1 Chinese set in experiments. We denote the set of 3,755 categories as $\mathcal{S}_{3,755}$. Following the setup in [1,45], we select the first 2,000 categories in GB2312-80 set as seen categories (denoted as $\mathcal{S}_{2,000}$), and the remaining 1,755 categories as unseen categories (denoted as $\mathcal{S}_{1,755}$). The diffusion models are trained on training samples of $\mathcal{S}_{2,000}$ and used to generate handwritten Chinese character samples of $\mathcal{S}_{1,755}$ to evaluate the performance of zero-shot training data generation for HCCR.

DDPM Settings: Our DDPM implementation is based on [9]. We use the "kai" as our font library to render printed character images. We conduct experiments on both WI and WD GC-DDPMs. In WI GC-DDPM training, we disable writer embeddings and randomly set content condition $\mathbf{g}$ as $\emptyset$ with probability 10%.

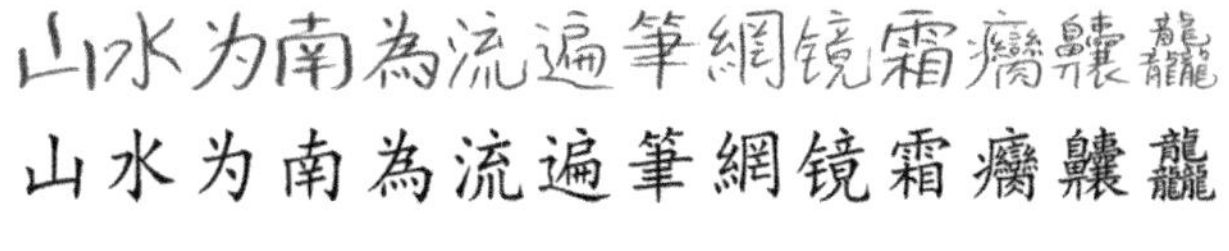

山水为南為流遍筆網鏡霜癗齇龘
山水为南為流遍筆網鏡霜癗齇龘

Fig. 3. Synthetic handwritten Chinese character samples and corresponding glyphs, with stroke numbers increasing from left to right.

And in WD GC-DDPM, writer condition $\mathbf{w}$ is also randomly set to $\emptyset$ with probability 10%. Flip and mirror augmentations are used during training. We set batch size as 256, image size as 128×128, and we use AdamW optimizer [26] with learning rate 1.0e-4. Diffusion step number is set to 1,000 with a linear noise schedule. GC-DDPMs are trained for about 200K steps using a machine with 8 Nvidia V100 GPUs, which takes about 5 d. During sampling, we use the denoising diffusion implicit model (DDIM) [39] sampling method with 50 steps. It takes 62 h to sample 3,755 characters written by 576 writers, which are about 2.2M samples, with the same 8 Nvidia V100 GPUs.

Evaluation Metrics: We evaluate the quality of synthetic samples in three aspects. First, Inception score (IS) [37] and Frechet Inception Distance (FID) [14] are used to evaluate the diversity and distribution similarity of synthetic samples compared with real ones. Second, since samples are synthesized by conditioning on glyph image, the synthetic samples should be consistent with the category of conditioned glyph. Therefore, we introduce a new metric called correctness score (CS). For each synthetic sample, the category of conditioned glyph is used as ground truth, and CS is calculated as the recognition accuracy of synthetic samples using an HCCR model trained with real data, which achieves 97.3% recognition accuracy in real data testing set. Finally, as the purpose of diffusion model here is to generate training data for unseen categories, we also train HCCR models with synthetic samples and evaluate recognition accuracy on the real testing set of unseen categories. Our HCCR model adopts ResNet-18 [13] architecture and is trained with standard SGD optimizer. No data augmentation is applied during HCCR model training. It is noted that starting from different random noise, it is almost impossible to generate exact same handwritten samples even for same conditional character glyphs. So it is not appropriate to adopt pixel-level metrics to evaluate generative effect as [11,18,24,25,31,47,51,54] do (Fig. 3).

4.2   WI GC-DDPM Results

We first conduct experiments on WI GC-DDPM. It is shown in [16] that the classifier guidance scale is able to attain a trade-off between quality and diversity. In order to evaluate the behavior of different content guidance scale γ's, we choose different γ's and generate samples to compute FID, ID and CS. Here we synthesize 50K samples of $\mathcal{S}_{2,000}$, and the HCCR model used to measure CS is trained using real samples of $\mathcal{S}_{3,755}$. $\gamma \in \{0.0, 1.0, 2.0, 3.0, 4.0\}$ are used and the comparison results are summarized in Table 1. We can find that, as γ

Table 1. Comparisons of generation quality using different content guidance scale γ's in terms of IS, FID, and CS.

γ	IS	FID	CS (%)
0.0	2.62	8.07	94.7
1.0	2.51	10.97	99.8
2.0	2.46	18.03	99.9
3.0	2.44	24.34	99.9
4.0	2.39	28.69	99.9

Table 2. Comparisons of generation quality using different content guidance scale γ's in terms of recognition accuracy on testing set of classes in $S_{1,755}$ using generated samples as training set.

γ	0.0	1.0	2.0	3.0	4.0
$Acc_{1,755}$ (%)	93.0	88.6	91.7	63.7	33.2

(a) Failure samples that do not look like any Chinese characters.

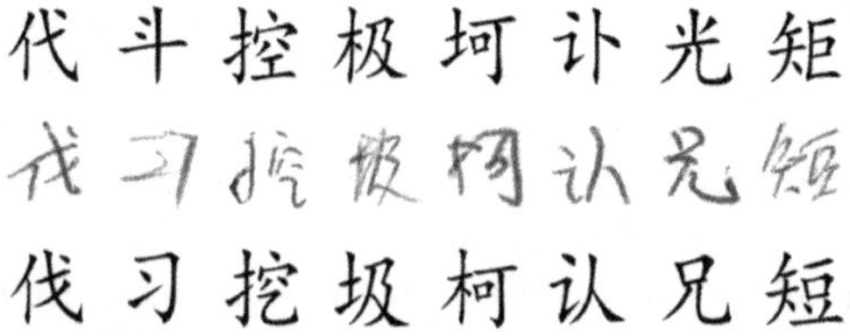

(b) (top) Glyph condition images; (middle) Synthetic samples; (bottom) Most similar characters.

Fig. 4. Synthetic samples that are wrongly recognized by real data trained HCCR model when $\gamma = 0$.

increases, the IS decreases, the FID increases and the CS achieves close to 100% accuracy. This indicates that with a larger γ, the diversity of synthetic samples is decreasing. This behavior is also observed in Fig. 5a where we visualize multiple sampled results of the character class in $S_{2,000}$ using different γ's. The generated samples are less diverse, less cursive and easier to recognize when conditioned on stronger content guidance. According to FID and examples in Fig. 5, the distribution of synthetic samples with $\gamma = 0$ is closer to that of real samples.

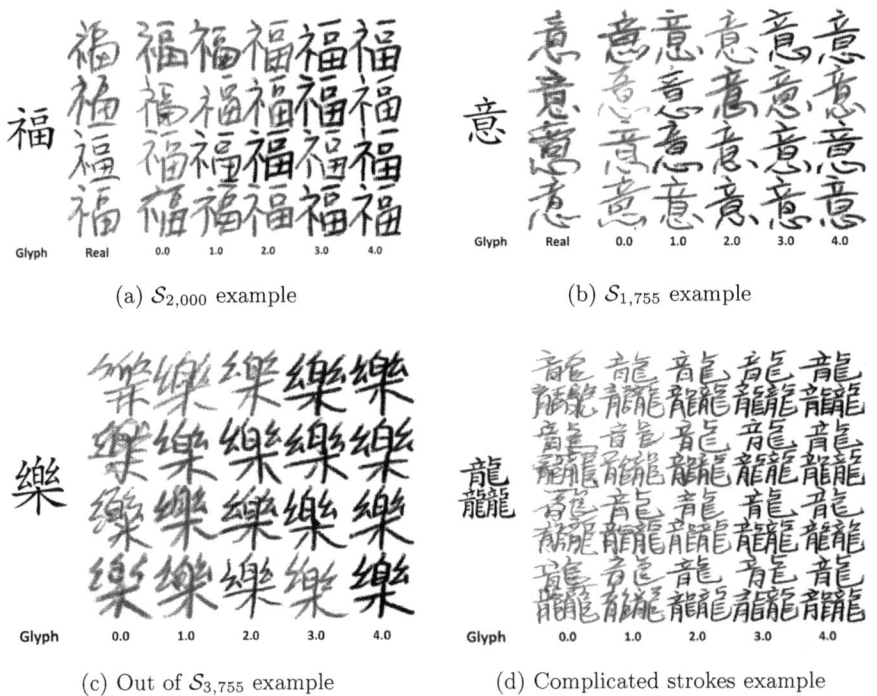

(a) $\mathcal{S}_{2,000}$ example

(b) $\mathcal{S}_{1,755}$ example

(c) Out of $\mathcal{S}_{3,755}$ example

(d) Complicated strokes example

Fig. 5. Multiple synthetic handwritten Chinese character samples with different content guidance scale, where (a), (b) and (c) are characters from classes of $\mathcal{S}_{2,000}$, $\mathcal{S}_{1,755}$, and out of $\mathcal{S}_{3,755}$ Chinese character sets. Samples in each line use the same random seed and initial noise. Samples across lines use different random seeds to visualize diversity.

When $\gamma = 0$, CS achieves 94.7%. In Fig. 4, we show synthetic cases that the trained HCCR model fails to recognize. Failure cases include (a) samples that are unreadable, and (b) samples that are closer to another easily confused Chinese character. They are caused by alignment failures between printed and synthetic strokes, and can be eliminated by improving glyph conditioning method. We leave it as future work.

Then, we evaluate the quality of WI GC-DDPM for zero-shot generation of HCCR training data. We use the trained WI GC-DDPM to synthesize 576 samples for each category in $\mathcal{S}_{1,755}$. Then, the synthetic samples are used along with real samples of categories in $\mathcal{S}_{2,000}$ to train an HCCR model that supports 3,755 categories. We calculate its recognition accuracy on the testing set of category $\mathcal{S}_{1,755}$, which is denoted as $\text{Acc}_{1,755}$. Different γ's are tried, and the results are shown in Table 2. In Fig. 5b, we visualize synthetic samples of one category in $\mathcal{S}_{1,755}$. The best $\text{Acc}_{1,755}$ is achieved when $\gamma = 0$. Although synthetic samples with higher γ are less cursive, they achieve much lower $\text{Acc}_{1,755}$. This is because the lack of diversity makes it difficult to cover the wide distribution of handwritten Chinese character image space.

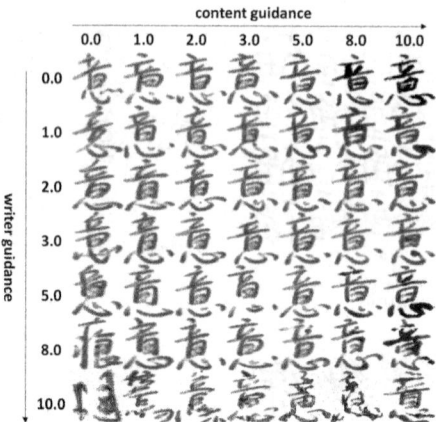

Fig. 6. Generated handwritten Chinese character samples with different content and writer guidance scales, where the character is from the class of $S_{1,755}$. Samples are generated with the same random seed and initial noise.

Table 3. Comparisons of generation quality between WI and WD DDPMs in terms of IS, FID, CS (%) and the recognition accuracy (%) on the testing set of class $S_{1,755}$ using generated samples as training set.

Model	IS	FID	CS	$Acc_{1,755}$
WI	2.62	8.07	94.7	93.0
WD	2.49	6.34	94.8	93.7
WD w/ interpolation	2.53	6.26	95.0	94.7

Clearly, by learning the mapping of radicals and spatial relationship between Chinese printed and handwritten strokes, the diffusion model is capable of zero-shot generation of unseen Chinese character categories. Moreover, a high accuracy of 93.0% is achieved on $S_{1,755}$ by only leveraging the synthetic samples. In Figs. 5c and 5d, we further show the synthetic samples of a Chinese character category that does not belong to $S_{3,755}$. The excellent generation effect implies that our method has the potential to be extended to a larger vocabulary.

4.3   WD GC-DDPM Results

Although WI GC-DDPM can generate desired handwritten characters, we cannot control their writing styles. In this part, we conduct experiments on WD GC-DDPM, which introduces writer information as an additional condition.

Figure 6 shows the visualization results of sampling with different content guidance scale γ's and writer guidance scale η's. It shows that with larger γ, the synthetic samples become less cursive and more similar to the corresponding printed image. This behavior is consistent with that of the WI GC-DDPM in

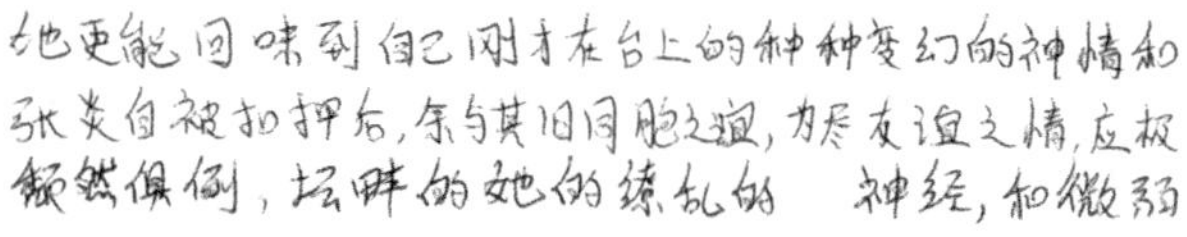

(a) Real text line from [23].

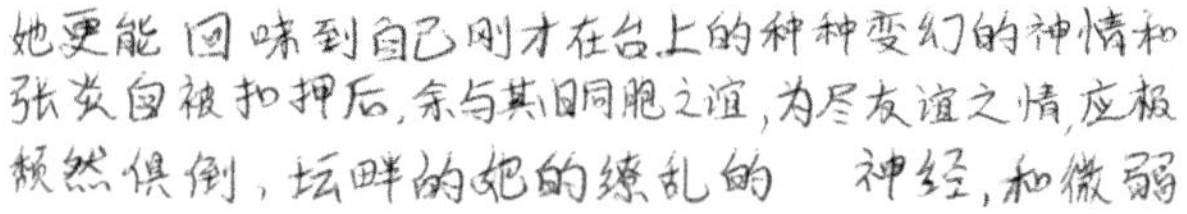

(b) Synthetic samples arranged as a text line.

Fig. 7. Comparisons of real text line images in HWDB2.1 and generated samples arranged in a text line, where we replace the characters from real data with the generated characters. Samples in different lines of (a) and (b) are selected and generated conditioning on the same writer 1001.

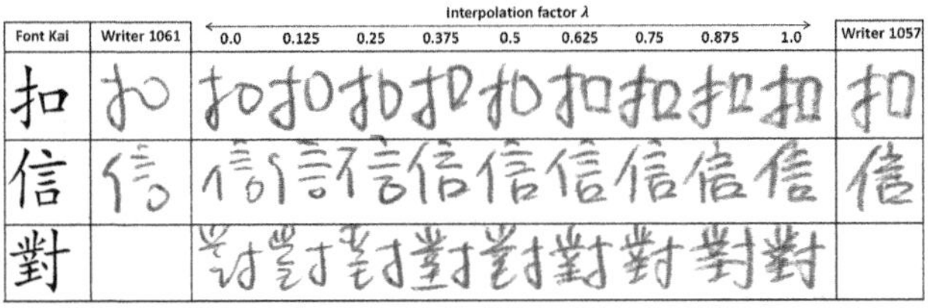

Fig. 8. Interpolation of handwritten Chinese character samples, where the top, middle, bottom lines are characters from classes of $S_{2,000}$, $S_{1,755}$, and out of $S_{3,755}$ Chinese character sets. We choose writer 1061 (left) and writer 1057 (right) for interpolation and interpolation factors are shown at the top of images. Standard glyph images of font "kai" are shown on the left. Samples in each line use the same random seed and initial noise.

Fig. 5. We also find that with large η, the generated sample becomes inconsistent with the conditioned printed image. Since writer information is injected to GC-DDPM in FiLM way, a large guidance scale will cause the mean and variance shift of $\tilde{\mu}_\theta(\mathbf{x}_t, \mathbf{g}, \mathbf{w})$ and $\tilde{\Sigma}_\theta(\mathbf{x}_t, \mathbf{g}, \mathbf{w})$ which hinders the subsequent denoising, leading to over-saturated images with over-smoothed textures [43].

In Fig. 7b, we show several synthetic text line images conditioned on a fixed writer embedding with our WD GC-DDPM. Writing styles of these samples are consistent and quite similar to real samples written by the same writer as shown in Fig. 7a. These results verify the writing style controllability of our model.

Then, we compare the quality of synthetic samples when used as training data for HCCR. For a fair comparison, we also generate 576 samples for each category in $S_{1,755}$, one image for each writer. Recognition performances are shown in Table 3. To improve sampling efficiency and ensure training data diversity, the writer guidance scale of 0 is applied. Compared with using samples synthesized with WI GC-DDPM as HCCR training set, the accuracy on the testing set

Table 4. Comparisons of recognition accuracy (%) on test sets of $S_{2,000}$ and $S_{1,755}$ using real and/or synthetic samples as HCCR training set.

Training set		Accuracy on testing set	
Real	Synthetic	$Acc_{2,000}$	$Acc_{1,755}$
✓	/	97.3	97.2
/	WI	96.3	96.0
/	WD	96.4	96.1
/	WD w/ interpolation	96.5	96.1
✓	WI	97.3	97.3
✓	WD	97.4	97.3
✓	WD w/ interpolation	97.4	97.3

of $S_{1,755}$ is improved from 93.0% to 93.7%. When GC-DDPM is trained without conditioning on writer embedding, it may generate similar samples from different initial noise. Whereas in WD GC-DDPM, by conditioning on different writer embeddings, the model will generate samples with different writing styles. Therefore, the diversity of synthetic samples will be improved. To verify this, we compare the quality of synthetic samples in terms of IS and FID. As shown in Table 3, the FID improves from 8.07 to 6.34. The results demonstrate the superiority of WD GC-DDPM in zero-shot training data generation of unseen Chinese character categories.

Another capability of WD GC-DDPM is that it can interpolate between different writer embeddings and generate samples of new styles. We choose 2 writers and try different interpolation factor λ's and visualize the synthetic samples in Fig. 8. We find that as λ increases from 0 to 1, the style of synthetic samples gradually shifts from one writing style to another. We also observe that with the same λ, the synthetic samples of different Chinese characters share similar writing style as expected. Finally, we use writer style interpolation to generate the training data of $S_{1,755}$ for HCCR, and again 576 samples are generated for each category. For each image, we randomly select 2 writers for interpolation. We simply use an interpolation factor of 0.5. Results are summarized in Table 3. We observe a slight improvement in FID score and a 1% absolute recognition accuracy improvement on $S_{1,755}$, which further verifies the superiority of our WD GC-DDPM.

4.4   Data-Augmented HCCR Results

We also use GC-DDPMs trained on $S_{2,000}$, to synthesize samples for all categories in $S_{3,755}$, and combine them with real samples to build HCCR systems. 3 settings are tried: WI, WD and WD w/ interpolation. And 576 samples for each category are synthesized in each setting. Table 4 summarizes the results. Best accuracies are achieved with samples synthesized by WD w/ interpolation,

Table 5. Comparisons of unseen character categories' recognition accuracy (%) between our method and prior zero-shot HCCR systems. Works with * also use samples from HWDB1.2 for training, while † means online trajectory information is also used.

Method	Accuracy
CM† [1]	86.7
DenseRan [45]	19.5
FewRan* [44]	70.6
HCCR* [4]	73.4
OSOCR* [21]	84.3
OSCCD* [22]	95.6
WI GC-DDPM	96.4
WD GC-DDPM	96.8
WD GC-DDPM w/interpolation	**96.9**

Table 6. Comparisons of unseen character categories' recognition accuracy (%) on CASIA1.2 testing set.

Methods	Accuracy
RCN [48]	46.1
WI GC-DDPM	98.6
WD GC-DDPM	98.6
ResNet-18 trained with real data	97.9

which is consistent with Table 3. The HCCR models trained with only synthetic samples perform slightly worse than the one trained with only real samples. Combining synthetic and real training samples only performs 0.0%~0.1% better than real samples. These results demonstrate the distribution modeling capacity of GC-DDPMs.

4.5 Comparison with Prior Arts

Finally, we compare our method with prior arts. We first compare our method with prior zero-shot HCCR systems. To be consistent with prior works in [4,21,22], we randomly choose 1,000 classes in $S_{1,755}$ as unseen classes and use ICDAR2013 [50] benchmark dataset for testing. Results are shown in Table 5. Here we only list the results from prior arts using 2,000 seen character classes. It is noted that the 2,000/1,000 seen/unseen character class split for training and testing is not exactly the same. So the results are not directly comparable. The results in Table 5 show that our methods achieve the same level recognition accuracy compared with previous state-of-the-art zero-shot HCCR systems. Moreover, our approach directly uses a standard CNN to predict supported categories, which is much simpler compared with the systems in [21,22].

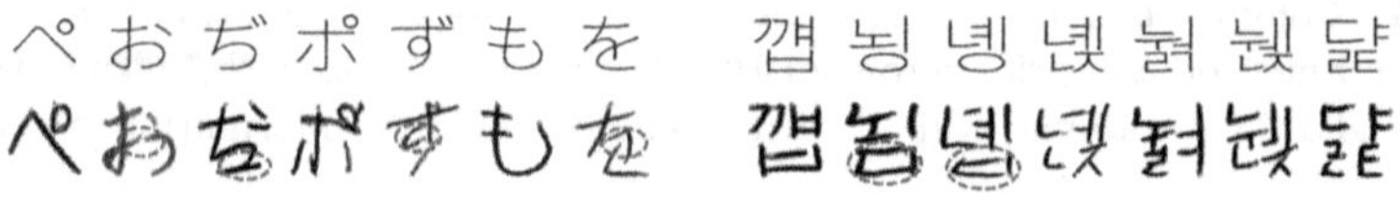

(a) Japanese   (b) Korean

Fig. 9. Synthetic samples of Japanese and Korean characters and standard glyph images in font "SourceHans".

We also compare our approach with [48], which also leverages a generation model to synthesize training samples for unseen classes. We follow the same experimental setups in [48] and use HWDB1.0 and 1.1 as training set, which contains 3,755 categories, to train GC-DDPMs. Unseen 3,319 categories in HWDB1.2 testing set are used as testing set. Results are shown in Table 6. [48] achieves a 46.1% accuracy by adding more than 9.6M generated samples. Our approach achieves a 98.6% accuracy by only adding about 1.9M synthetic samples (576 samples for each unseen category). We also train a classifier using all real samples in HWDB1.2 training set (240 samples for each category). The classifier achieves a 97.9% accuracy, which is slightly worse than ours due to less diverse training samples.

These results verify the zero-shot generation capability of our methods again. It is easy to extend to larger vocabularies, which makes it possible to build a high-quality HCCR system for 87,887 categories.

5 Limitations and Future Work

Although GC-DDPM-synthesized images are quite helpful for building a high-quality HCCR system, there are still some failure cases. The blur and dislocation phenomena in these samples reveal that there exist better ways to inject glyph information. It is also possible to encode radical/stroke sequences with spatial relationships as the condition of DDPM. We will investigate these methods and report the results elsewhere.

Another limitation of our approach is the long training time of DDPMs. We will try to reduce the number of character categories and sample numbers per category to find a better trade-off between synthesis quality and training cost.

Japanese and Korean characters share most strokes with Chinese, so we also try to synthesize handwritten Japanese and Korean samples with our Chinese-trained DDPM. As Fig. 9 shows, except for some circle and curve strokes, the results are quite reasonable. As future work, we will combine handwritten samples of CJK languages to build a new DDPM, which is expected to synthesize samples for each language with higher diversity and quality.

6 Conclusion

We propose WI and WD GC-DDPM solutions to achieve zero-shot training data generation for HCCR. Experimental results have verified their effectiveness in terms of generation quality, diversity and HCCR accuracies of unseen categories. WD performs slightly better than WI due to its better distribution modeling capability and writing style controllability. These solutions can be easily extended to larger vocabularies and other languages, and provide a feasible way to build an HCCR system supporting 87,887 categories with high recognition accuracy.

References

1. Ao, X., Zhang, X.Y., Yang, H.M., Yin, F., Liu, C.L.: Cross-modal prototype learning for zero-shot handwriting recognition. In: ICDAR, pp. 589–594 (2019)
2. Austin, J., Johnson, D.D., Ho, J., Tarlow, D., van den Berg, R.: Structured denoising diffusion models in discrete state-spaces. In: NeurIPS, vol. 34, pp. 17981–17993 (2021)
3. Brock, A., Donahue, J., Simonyan, K.: Large scale GAN training for high fidelity natural image synthesis. In: ICLR (2019)
4. Cao, Z., Lu, J., Cui, S., Zhang, C.: Zero-shot handwritten Chinese character recognition with hierarchical decomposition embedding. Pattern Recogn. **107**, 107488 (2020)
5. Chen, J., Li, B., Xue, X.: Zero-shot Chinese character recognition with stroke-level decomposition. In: IJCAI, pp. 615–621 (2021)
6. Chen, L., Wang, S., Fan, W., Sun, J., Naoi, S.: Beyond human recognition: a CNN-based framework for handwritten character recognition. In: ACPR, pp. 695–699 (2015)
7. Cireşan, D., Meier, U.: Multi-column deep neural networks for offline handwritten Chinese character classification. In: IJCNN, pp. 1–6 (2015)
8. Croitoru, F.A., Hondru, V., Ionescu, R.T., Shah, M.: Diffusion models in vision: a survey. CoRR abs/2209.04747 (2022)
9. Dhariwal, P., Nichol, A.: Diffusion models beat GANs on image synthesis. In: NeurIPS, vol. 34, pp. 8780–8794 (2021)
10. Diao, X., Shi, D., Tang, H., Wu, L., Li, Y., Xu, H.: REZCR: a zero-shot character recognition method via radical extraction. CoRR abs/2207.05842 (2022)
11. Gao, Y., Guo, Y., Lian, Z., Tang, Y., Xiao, J.: Artistic glyph image synthesis via one-stage few-shot learning. ACM TOG **38**(6), 1–12 (2019)
12. Graves, A.: Generating sequences with recurrent neural networks. arXiv preprint arXiv:1308.0850 (2013)
13. He, K., Zhang, X., Ren, S., Sun, J.: Deep residual learning for image recognition. In: CVPR, pp. 770–778 (2016)
14. Heusel, M., Ramsauer, H., Unterthiner, T., Nessler, B., Hochreiter, S.: GANs trained by a two time-scale update rule converge to a local nash equilibrium. In: NeurIPS, vol. 30, pp. 6626–6637 (2017)
15. Ho, J., Jain, A., Abbeel, P.: Denoising diffusion probabilistic models. In: NeurIPS. vol. 33, pp. 6840–6851 (2020)
16. Ho, J., Salimans, T.: Classifier-free diffusion guidance. In: NeurIPS 2021 Workshop DGMs Applications (2021)
17. Huang, G., Luo, X., Wang, S., Gu, T., Su, K.: Hippocampus-heuristic character recognition network for zero-shot learning in Chinese character recognition. Pattern Recogn. **130**, 108818 (2022)
18. Huang, Y., He, M., Jin, L., Wang, Y.: RD-GAN: few/zero-shot Chinese character style transfer via radical decomposition and rendering. In: ECCV, pp. 156–172 (2020)

19. Huang, Y., Jin, L., Peng, D.: Zero-shot Chinese text recognition via matching class embedding. In: ICDAR, pp. 127–141 (2021)
20. Li, Z., Teng, N., Jin, M., Lu, H.: Building efficient CNN architecture for offline handwritten Chinese character recognition. Int. J. Document Anal. Recog. **21**(4), 233–240 (2018)
21. Liu, C., Yang, C., Qin, H.B., Zhu, X., Liu, C.L., Yin, X.C.: Towards open-set text recognition via label-to-prototype learning. Pattern Recogn. **134**, 109109 (2022)
22. Liu, C., Yang, C., Yin, X.C.: Open-set text recognition via character-context decoupling. In: CVPR, pp. 4523–4532 (2022)
23. Liu, C.L., Yin, F., Wang, D.H., Wang, Q.F.: CASIA online and offline Chinese handwriting databases. In: ICDAR, pp. 37–41 (2011)
24. Liu, W., Liu, F., Ding, F., He, Q., Yi, Z.: XMP-Font: self-supervised cross-modality pre-training for few-shot font generation. In: CVPR, pp. 7905–7914 (2022)
25. Liu, Y., Lian, Z.: FontTransformer: few-shot high-resolution Chinese glyph image synthesis via stacked Transformers. CoRR abs/2210.06301 (2022)
26. Loshchilov, I., Hutter, F.: Decoupled weight decay regularization. In: ICLR (2019)
27. Luhman, T., Luhman, E.: Diffusion models for handwriting generation. CoRR abs/2011.06704 (2020)
28. Nichol, A., et al.: GLIDE: Towards photorealistic image generation and editing with text-guided diffusion models. In: ICML, vol. 162, pp. 16784–16804 (2022)
29. Nichol, A.Q., Dhariwal, P.: Improved denoising diffusion probabilistic models. In: ICML, pp. 8162–8171 (2021)
30. Pang, Y., Lin, J., Qin, T., Chen, Z.: Image-to-image translation: methods and applications. IEEE Trans. Multimedia **24**, 3859–3881 (2021)
31. Park, S., Chun, S., Cha, J., Lee, B., Shim, H.: Few-shot font generation with localized style representations and factorization. In: AAAI, pp. 2393–2402 (2021)
32. Perez, E., Strub, F., De Vries, H., Dumoulin, V., Courville, A.: FiLM: visual reasoning with a general conditioning layer. In: AAAI, pp. 3942–3951 (2018)
33. Ramesh, A., Dhariwal, P., Nichol, A., Chu, C., Chen, M.: Hierarchical text-conditional image generation with CLIP latents. CoRR abs/2204.06125 (2022)
34. Rombach, R., Blattmann, A., Lorenz, D., Esser, P., Ommer, B.: High-resolution image synthesis with latent diffusion models. In: CVPR, pp. 10684–10695 (2022)
35. Saharia, C., et al.: Palette: image-to-image diffusion models. In: ACM SIGGRAPH 2022 Conference Proceedings, pp. 1–10 (2022)
36. Saharia, C., et al.: Photorealistic text-to-image diffusion models with deep language understanding. CoRR abs/2205.11487 (2022)
37. Salimans, T., et al.: Improved techniques for training GANs. In: NeurIPS, vol. 29, pp. 2226–2234 (2016)
38. Sohl-Dickstein, J., Weiss, E., Maheswaranathan, N., Ganguli, S.: Deep unsupervised learning using nonequilibrium thermodynamics. In: ICML, pp. 2256–2265 (2015)
39. Song, J., Meng, C., Ermon, S.: Denoising diffusion implicit models. In: ICLR (2021)
40. Song, Y., Ermon, S.: Generative modeling by estimating gradients of the data distribution. In: NeurIPS, vol. 32, pp. 11895–11907 (2019)
41. Song, Y., Ermon, S.: Improved techniques for training score-based generative models. In: NeurIPS, vol. 33, pp. 12438–12448 (2020)
42. Song, Y., Sohl-Dickstein, J., Kingma, D.P., Kumar, A., Ermon, S., Poole, B.: Score-based generative modeling through stochastic differential equations. In: ICLR (2021)
43. Wang, T., et al.: Pretraining is all you need for image-to-image translation. CoRR abs/2205.12952 (2022)

44. Wang, T., Xie, Z., Li, Z., Jin, L., Chen, X.: Radical aggregation network for few-shot offline handwritten Chinese character recognition. Pattern Recogn. Lett. **125**, 821–827 (2019)
45. Wang, W., Zhang, J., Du, J., Wang, Z.R., Zhu, Y.: DenseRAN for offline handwritten Chinese character recognition. In: ICFHR, pp. 104–109 (2018)
46. Weng, L.: What are diffusion models? https://lilianweng.github.io/posts/2021-07-11-diffusion-models/, July 2021
47. Xie, Y., Chen, X., Sun, L., Lu, Y.: DG-Font: deformable generative networks for unsupervised font generation. In: CVPR, pp. 5130–5140 (2021)
48. Xue, M., Du, J., Zhang, J., Wang, Z.R., Wang, B., Ren, B.: Radical composition network for Chinese character generation. In: ICDAR, pp. 252–267 (2021)
49. Yang, L., et al.: Diffusion models: a comprehensive survey of methods and applications. CoRR abs/2209.00796 (2022)
50. Yin, F., Wang, Q.F., Zhang, X.Y., Liu, C.L.: ICDAR 2013 Chinese handwriting recognition competition. In: ICDAR, pp. 1464–1470 (2013)
51. Zhang, Y., Zhang, Y., Cai, W.: Separating style and content for generalized style transfer. In: CVPR, pp. 8447–8455 (2018)
52. Zhong, Z., Zhang, X.Y., Yin, F., Liu, C.L.: Handwritten Chinese character recognition with spatial Transformer and deep residual networks. In: ICPR, pp. 3440–3445 (2016)
53. Zhong, Z., Jin, L., Xie, Z.: High performance offline handwritten Chinese character recognition using GoogLeNet and directional feature maps. In: ICDAR, pp. 846–850 (2015)
54. Zhu, A., Lu, X., Bai, X., Uchida, S., Iwana, B.K., Xiong, S.: Few-shot text style transfer via deep feature similarity. IEEE Trans. Image Process. **29**, 6932–6946 (2020)

TBM-GAN: Synthetic Document Generation with Degraded Background

Arnab Poddar[1]($\boxtimes$), Soumyadeep Dey[2], Pratik Jawanpuria[2],
Jayanta Mukhopadhyay[1], and Prabir Kumar Biswas[1]

[1] Indian Institute of Technology Kharagpur, Kharagpur, India
arnabpoddar@iitkgp.ac.in, jay@cse.iitkgp.ac.in, pkb@ece.iitkgp.ac.in
[2] Microsoft, Hyderabad, India
{soumyadeep.dey,pratik.jawanpuria}@microsoft.com

Abstract. Deep document enhancement models often suffer in real world applications due to limited annotation and bias in training data. Moreover, generative models are often prone to spectral bias towards certain frequencies. The background (noisy) texture is usually harder to learn as it is composed from different frequency regions. In this work, we propose TBM-GAN, a generative adversarial network based framework to synthesise realistic handwritten documents with degraded background. In addition to the spatial information, TBM-GAN also incorporates the frequency information in its loss function to focus on complex noisy texture. Overall, we develop an automated pipeline for TBM-GAN and train it with artificially annotated data from publicly available resources. The pipeline provides both text-label and corresponding pixel-level annotation. We evaluate the quality of synthetic images in the downstream task of OCR. In text images with historical noisy background, we observe an 11% reduction in the character error rate when the OCR is trained with synthetic data from TBM-GAN.

Keywords: Synthetic Data · Data Augmentation · OCR · GAN · Focal Frequency Loss · Historical Document · Noisy Background

1 Introduction

The performance of document image understanding applications (DIUs) in real world are largely affected by both structural deformation of written content and background textural degradation of documents. Documents captured using handheld mobile devices inherently come with textural degradation [8,22,34] due to shadows, non-uniform illumination, warping in an uneven surface, and blurring for out-of-scale focus, to name a few. Typical examples of such structural and textural deformation in real-world document images are shown in Fig. 1.

Addressing real-world challenges like degraded texture and shape deformation altogether is relatively less explored due to the unavailability of appropriate training data with annotation. It requires text information and corresponding

G. A. Fink et al. (Eds.): ICDAR 2023, LNCS 14188, pp. 366–383, 2023.
https://doi.org/10.1007/978-3-031-41679-8_21

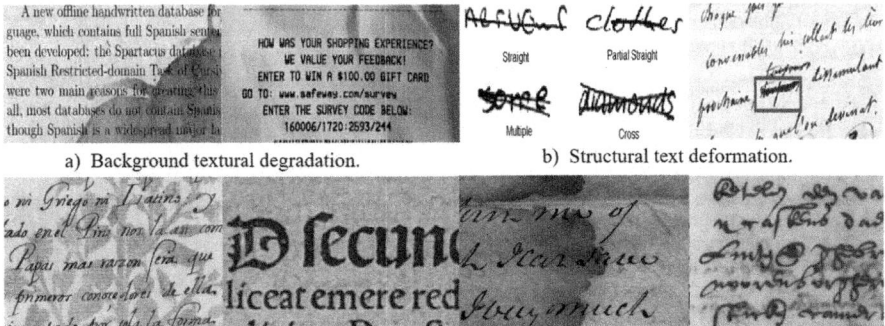

a) Background textural degradation.

b) Structural text deformation.

c) Historical documents from DIBCO/H-DIBCO databases with various degradation.

Fig. 1. Illustration of real text images with several natural degradation both in (a) textural aspects and in (b) structural aspects. It also depicts of diverse real background texture of historical documents in (c) DIBCO/H-DIBCO databases [38].

pixel-level ground-truth simultaneously. Moreover, the manual annotation process for the same is costly in terms of time, resource and man-power. In this regard, generative networks may contribute to augment synthetic text images with structural deformation and noisy background texture to compensate the scarcity of real world data. Generative approaches have proven beneficial for data augmentation in a wide range of computer vision applications including DIUs [20,21,41,42]. They have also been applied successfully for various document enhancement systems [3,10,28,29,34,37]. Generative framework is also explored to augment clean handwritten images with writer variability and text labels [17,18,39]. In particular, Vögtlin et al. [40] proposed a semi-supervised OCR constrained GAN model to synthesise labeled data with ground-truth.

In addition to the quality and quantity of training data, the performance of a generative model is also influenced by the information being captured by its loss function. For instance, while the popular spatial loss functions such as the ℓ_1-norm loss encourages less blurring [8,34]. However, the generative models with only spatial losses often suffer in capturing the image properties with a spectral bias towards certain frequency regions [25,31,36]. In the context of natural images synthesis, recent works [6,15] demonstrated the usefulness of using frequency based information in the loss function of generative models.

In this work, we propose a conditional GAN approach (TBM-GAN) for synthesising handwritten documents with variable noisy background and deformed texts. Given a text with clean background, we model the problem of generating a noisy handwritten document as an image to image translation problem. TBM-GAN incorporates both spatial and frequency information to learn a mapping function of the noisy background while keeping the shape of the given text-region unaltered. The proposed pipeline not only delivers texts with degraded background but also delivers annotation of text label and pixels both of the noisy

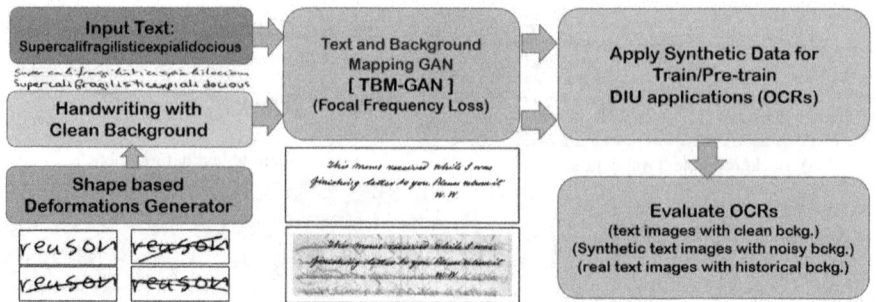

Fig. 2. Workflow diagram of the proposed framework for generation and evaluation of handwritten text images with structural and background degradation.

document. Hence the proposed pipeline can be applied simultaneously for data augmentation in multiple DIUs.

We also propose an artificial annotation process to prepare training data for TBM-GAN. Thus the proposed TBM-GAN is trained without any manual intervention. It significantly reduces the human effort, error rate, and time in the annotation process. The annotation process utilises widely used unlabeled data of other DIU applications like OCR, image binarisation, etc. to prepare training data of TBM-GAN. The method produces a trio of (a) text label, (b) text-pixel annotation, and (c) corresponding text with noisy background. The generated documents can be used as additional training data for various downstream DIU applications.

In this work, we consider one of the most popularly used DIU task i.e., OCR to evaluate the effectiveness of the synthetic images. The flow of work is diagrammatically outlined in Fig. 2. The synthetic images are used for improvement of OCR in real-world challenging condition with noisy historical documents. We evaluate the OCR systems in terms of character error rates (CER%) and words error rates (WER%) in three major evaluation condition i.e., with clean, synthetic and historical background from widely used benchmark databases. We observe significant improvement in performance of synthetic data augmented OCR over the same, trained with clean texts. The improvements are also consistent for realistic unseen synthetic images as well. Finally we summarise the major contributions of this work,

- We propose a framework to generate synthetic document images with both text and pixel-level annotation.
- We also study the effect of different frequency bands on noisy document images. This motivates us to incorporate frequency information in our loss function.
- We develop an automated pipeline to train the proposed TBM-GAN. Our pipeline utilises the existing datasets of various DIU applications for this purpose.

- Finally, we showcase the efficacy of the proposed data generation approach in OCR application. We observed a significant improvement in OCR performance when the OCR model is trained with synthetic data generated by the proposed TBM-GAN.

The rest of the work is presented as follows. Section 2 presents a brief discussion on related prior works. Section 3 deals with methodology and model of image synthesis module based on GAN and Focal Frequency Loss. The Sect. 4 discusses the synthetic image generation and artificial annotation process for training data of TBM-GAN. Subsequently, Sect. 5 deals with experiments on OCR, datasets, results and discussion. Finally, the work is concluded in Sect. 6.

2 Related Work

The deep learning models often suffer from limited amount of labeled data and its corresponding bias and cost implications. Commonly, strategies such as transfer learning, data augmentation, synthesizing artificial data, etc., are used to tackle such problems. To augment real documents with systematic degradation, Baird [2], Kieu et al. [19] and Seuret et al. [32] applied defect models in their work. Subsequently, tools like DocEmul [7] and DocCreator [16] target to synthesise documents with degraded background using a combination of user-specified lay-outs. Recently, Augraphy [23] is also introduced to augment noisy documents with user specified degradation in python. However, such approaches still costs human interventions in designing and directing appropriate set-up to synthesise realistic document images.

The work in [1] shows that deep autoencoders trained for reconstruction, may not be effective for task of synthesising document images from unlabeled data. However, GANs can potentially be used to synthesise realistic artificial images using unlabeled data [12,14]. The works in [13,18,30,37,40] have used GAN based approaches for document synthesis. The work in [5] proposed a generative model for synthetic data generation with given user-defined layouts and objects. The generative approaches are also found useful to augment clean handwritten images with writer variability and text labels [17,18,39]. The work in [40] proposed generative models to synthesise labeled data with ground truth with OCR constrained setting. This attempt also requires human effort and cost to prepare data for training.

The approaches primarily focus towards a particular DIU application like OCR, binarisation, shadow, etc. In the literature, there is hardly any work studying both text deformation and background degradation simultaneously. Subsequently, the existing approaches often require manual supervision for input condition. The approaches may require labeled data or paired images for training. Additionally, most of the generative deep models primarily focus only on the spatial distribution for loss function. However, the relevance of the frequency spectrum in loss function remained a potential area to be studied in DIUs.

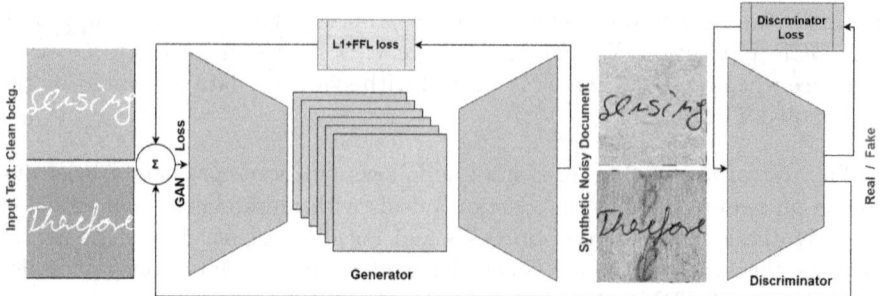

Fig. 3. TBM-GAN takes handwritten texts input with clean color background. It outputs the same text image with synthetic noisy texture. It uses focal frequency loss in addition to ℓ_1-norm loss to learn a text and background mapping function.

3   Text and Background Mapping GAN (TBM-GAN)

In this work, generating a noisy handwritten document from a given clean handwritten document is modeled as an image to image translation problem. In this regard, we explore a conditional GAN [14,26] based framework. While the usual (unconditional) GAN models the joint probability, the conditional GAN models the conditional probability. In particular, we learn a mapping function from conditional observed image domain and a randomized noise vector, to the targeted image domain.

We propose a conditional GAN, **T**ext and **B**ackground **M**apping **GAN** (TBM-GAN), to synthesise handwritten images with noisy background texture. Given a text image on clean color background, TBM-GAN generates suitable noisy background while keeping the tangibility of the input text pixels intact. The network is able to handle different languages, structural deformation, and writing styles. The model architecture of the TBM-GAN is shown in Fig. 3.

Existing GAN based architectures for document images enhancement [34, 37] have explored spatial loss functions such as ℓ_1-norm loss, ℓ_2-norm loss, and focal loss, to name a few. The ℓ_1-norm loss function, for instance, encourages less blurring [8,34]. However, GANs with only spatial loss function often suffer from a spectral bias towards low frequency region [25,31,36]. The low frequency components of the spectrum refers to less variability in spatial domain. Thus, the generative models with only spatial loss tends to avoid frequency components that are hard to synthesise. In context of noisy document image generation, this spectral bias makes it harder for the model to learn detailed texture belonging to various frequencies. In Fig. 4(a), we show synthetic images generated by TBM-GAN with only ℓ_1-norm loss. We observe that the generated image texture in Fig. 4(a) are visibly smooth and unable to capture the uneven and rough noisy texture resembling real world documents.

Recent works [6,15] have shown the benefits of employing frequency based information for generating more realistic natural images. However, to the best of our knowledge, incorporating frequency based information for document image

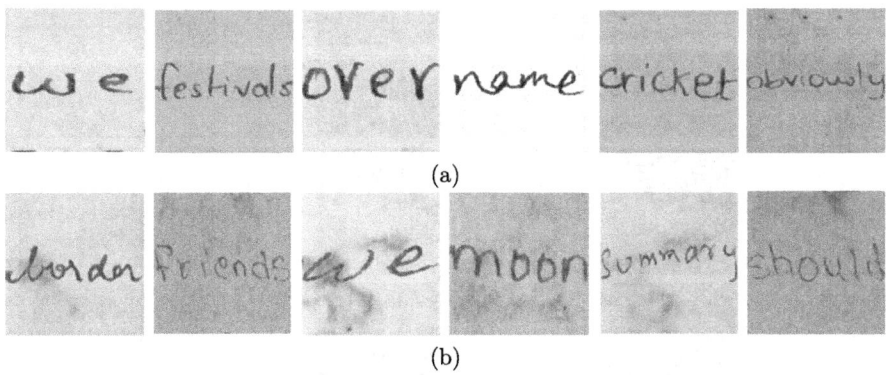

(a)

(b)

Fig. 4. (a) represents the images generated by TBM-GAN with only ℓ_1-norm loss. (b) represents the images generated with TBM-GAN using focal frequency loss and ℓ_1-norm loss

generation has not been explored. In Fig. 5, we observe that different bands in the frequency domain affects the features of both background and texts of document image. This motivates us to also use frequency representation of document images. In particular, we explore the usage of focal frequency loss (FFL) function [15] in TBM-GAN for generating document images.

Overall, we propose the following objective for TBM-GAN, which includes the spatial ℓ_1-norm loss $\mathcal{L}_{\ell_1}(\cdot)$ and FFL $\mathcal{L}_{FFL}(\cdot)$:

$$\hat{G} = \arg\min_G \max_D \mathcal{L}_{cGAN}(G, D) + \lambda\mathcal{L}_{\ell_1}(G) + \alpha\mathcal{L}_{FFL}(G). \tag{1}$$

The conditional GAN loss function function [14,26], $\mathcal{L}_{cGAN}(G, D)$, is defined as

$$\mathcal{L}_{cGAN}(G, D) = \mathbb{E}_{I_C, I_R}[\log D(I_R|I_C)] + \mathbb{E}_{I_C, z}[\log(1 - D(G(z|I_C)))], \tag{2}$$

where G is the generative model, D is the discriminative model, I_C represents the text images with clean background, I_R represents the (real) noisy text images, and z represents the random noise vector. The hyper-parameters $\lambda \geq 0$ and $\alpha \geq 0$ in (1) modulate the ratio of spatial ℓ_1-norm loss and FFL, respectively. The spatial ℓ_1-norm loss function is defined as

$$\mathcal{L}_{\ell_1}(G) = \mathbb{E}_{I_C, I_R, z}[\|I_R - G(z|I_C)\|_1]. \tag{3}$$

We now describe the FFL computation [15]. Let the frequency representation with 2D discrete Fourier transform of an image $\mathcal{I}$ of size $M \times N$ be represented by $\mathcal{H}$. Let (x, y) denotes the coordinate of an image pixel in the spatial domain and $\mathcal{I}(x, y)$ be the pixel value corresponding to (x, y). Let (u, v) represents the coordinate of a spatial frequency and $\mathcal{H}(u, v)$ represents complex frequency value. Let $\mathcal{H}_r(u, v)$ and $\mathcal{H}_f(u, v)$ be the spatial frequency value of real and fake images respectively.

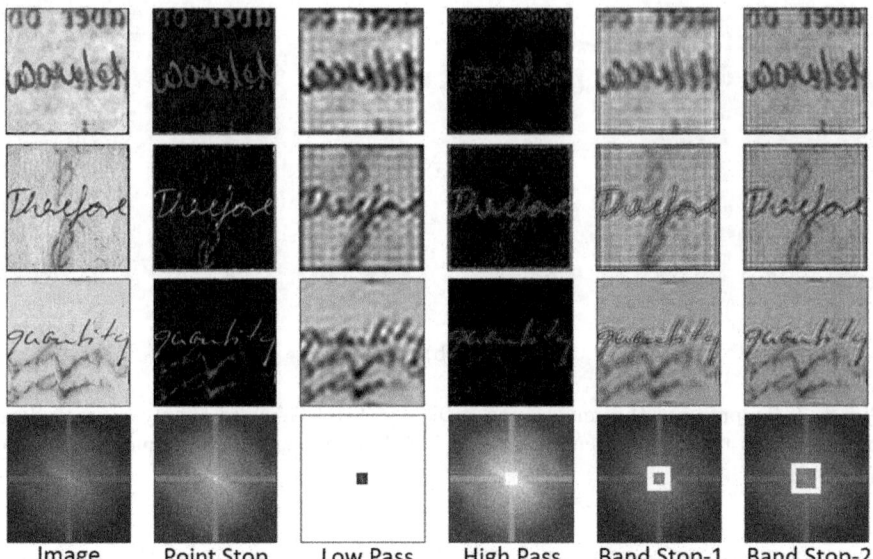

Fig. 5. Frequency spectrum (magnitude) of a document image with noisy background with standard band-limiting operation

Then, FFL is defined as:

$$\mathcal{L}_{FFL} = \frac{1}{MN} \sum_{u=0}^{M-1} \sum_{v=0}^{N-1} \omega(u,v)|\mathcal{H}_r(u,v) - \mathcal{H}_f(u,v)|^2 \tag{4}$$

In (4), $\omega(u,v)$ represents a spectrum weight matrix which dynamically weighs down the easy frequencies and helps in focusing more on hard frequencies. The matrix element $\omega(u,v)$ is defined as:

$$\omega(u,v) = |\mathcal{H}_r(u,v) - \mathcal{H}_f(u,v)|^\gamma \tag{5}$$

where $\gamma > 0$ is the scaling factor. In our experiments, we use $\gamma = 1$. The spectrum weight matrix is further normalized to the range $[0,1]$, where higher weights correspond to lost frequencies.

4   Synthetic Document Image Generation

We now discuss our overall pipeline to generate synthetic document images using the proposed TBM-GAN. As discussed in Sect. 3, TBM-GAN requires pairs of images (I_C, I_R), where I_R represents real world text image (with noisy background) and I_C represents the corresponding text image with clean background. Publicly available datasets which contain such pairs (I_C, I_R) are not able to capture the diversity of examples present in real world. Hence, we next propose a methodology to synthesise documents with structural and textural degradation.

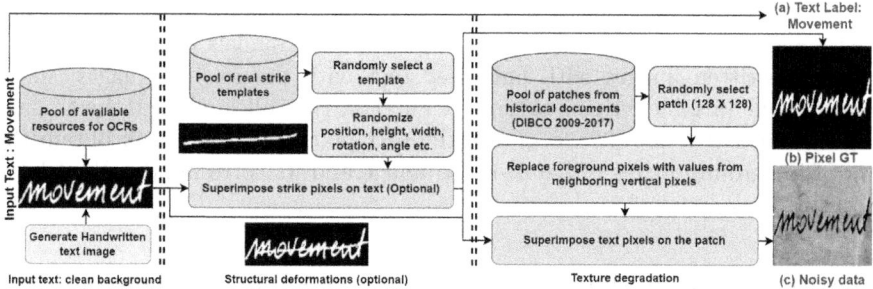

Fig. 6. Framework of artificial annotation process for training data of TBM-GAN. The artificial annotation process enables TBM-GAN to be trained without manual intervention. It is also used to prepare evaluation set of augmented OCR

4.1 Artificial Annotation of Background and Text in Documents

The artificial annotation process is designed to generate document images with structural deformation and noisy background. The procedure also generates both the text labels and associated pixel-level annotation of the text for these noisy document images. The proposed process consists of three major parts as shown in Fig. 6. The process uses datasets of available DIUs like OCR and image binarisation as the source of annotation.

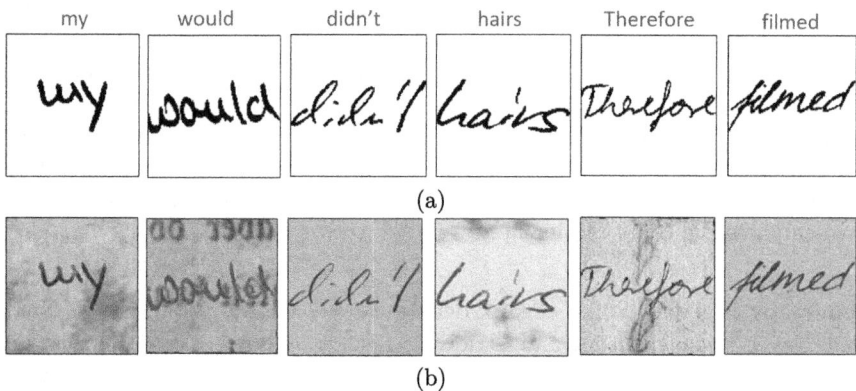

Fig. 7. Illustration of artificially prepared training data having both pixel level and text annotation. Figure (a) represents pixel-level ground-truth of text region. (b) represents corresponding images with real historical background

Clean Text Collection: The proposed annotation method takes a text label as input. The OCR databases can be used for clean handwritten images with a given text label. Alternatively, deep generative models [4,17,18] may also be used for generating clean handwritten images. Such models can generate realistic handwritten images with unlimited text with different languages, shape deformation, variable writing style, etc. Further, the collected clean text image is binarised with Otsu [27] to obtain the text pixels. Thus, we can obtain the pair of foreground pixels of handwritten image and its corresponding text label as in Fig. 6.

Structural Deformation: Existing works have studied incorporating shape deformation on clean handwritten images [28,29]. Strike-out and underline templates can be superimposed on clean texts with random movements, rotation, and spatial-deviations.

Background Texture: A patch is selected randomly from historical data and the foreground pixels are replaced with vertically neighboring pixels in the same column. Finally, the noisy patches are superimposed with the clean text images as shown in Fig. 6. Thus, the artificial annotation process is used to obtain noisy documents with (a) text labels (b) text pixels and (c) (real) noisy text image as in Fig. 6. Figure 7 illustrates few examples of training data generated by artificial annotation process.

4.2   TBM-GAN Model Architecture

As discussed, the proposed TBM-GAN model aims to learn a mapping from observed text image with clean background (I_C) and a random noise vector (z) to a text image with real (noisy) background (I_R). The generator network G of TBM-GAN is designed to synthesise text images with noisy texture background (I_S). The input pair (I_R, I_C) is generated using the approach described in Sect. 4.1. The generator and discriminator model architectures of TBM-GAN are adopted from Zhu et al. [43]. In the following, we detail the exact architectures employed in our experiments.

Generator Network: Let c7s1-k denote a 7×7 Convolution-InstanceNorm-ReLU layer with k filters and stride 1. Let dk denote a 3×3 layer with Convolution-InstanceNorm-ReLU and k filters with stride 2. Reflection padding is applied to diminish artifacts and glitches. Let Rk denote residual block which consists of two 3×3 convolutional layers with a same number of filters on both layers. Let uk denote a 3×3 fractional-strided-Convolution-InstanceNorm-ReLU layer with k filters and stride $\frac{1}{2}$. Nine residual blocks are used in the generator network to train images of size 128×128. Overall, the image generation network is as follows: c7s1-64, d128, d256, R256, R256, R256, R256, R256, R256, R256, R256, R256, u128, u64, c7s1-3. We include dropout layers as noise distribution on several layers of our generator architecture in both learning and testing phases.

Discriminator Network: In the discriminator model (D), we employ a 70×70 PatchGAN [11,14,37]. Let Ck denote a 4×4 Convolution-InstanceNorm-LeakyReLU layer with k filters and stride 2. At the end of the last layer, another layer of convolution is appended to produce an output of 1-dimension. The network does not apply InstanceNorm in the initial c64 layer. It uses leaky ReLUs with a slope of 0.2. Overall, the discriminator architecture is: C64, C128, C256, C512.

4.3   Dataset Creation

Figure 2 illustrates the proposed synthetic image generation framework with TBM-GAN. As discussed in Sect. 4.1, the proposed pipeline is trained with artificially annotated data. The real background are prepared from the DIBCO/H-DIBCO dataset series (2009–12, 2014, 2016–17) [37,38]. A total of 10 338 patches of size 128×128 are created from the DIBCO/H-DIBCO database. Among them, a random set of 7 000 real background patches are used to create the train set (I_R^{train}) and the remaining 3 338 patches are used to generate data the evaluation set (I_R^{eval}). The real words are adopted from publicly available IAM database [24] and has high variability in terms of writing style, gender, age, page-texture, writing ink, and stroke-width. We have simultaneously collected stroke images separately of various types like straight, slanted, cross, etc. A total of 2400 hand-drawn strokes are collected to introduce structural deformation such as struck-out strokes, underlines, etc.

We generate text images with real background using the artificial annotation framework described in Sect. 4.1. For training TBM-GAN, we create 76 412 pairs of images $(I_R^{train}, I_C^{train})$, where I_R^{train} contains text with real background and I_C^{train} contains text with clean background. We also generate 76 412 synthetic images I_S^{train} by applying TBM-GAN to I_C^{train}, each with a unique texture due to the random noise vector (z). Synthetic images provide more diversity and complexity of text and background appearance, which can improve the generalization performance of downstream applications, e.g. OCR. However, synthetic images may not capture all the real-world variations and challenges. Therefore, we use I_R^{train} to fine-tune our model. The background patches of I_R^{train} are selected from a pool of 7000 distinct real-world patches, which is much smaller than the number of text images. This may cause overfitting to the limited background variations and reduce the robustness of the model for real-world applications. We observe this empirically and discuss it in Sect. 5.2. For evaluation, we generate another set of 21 489 pairs of images (I_R^{eval}, I_C^{eval}), where I_R^{eval} contains text with real background and I_C^{eval} contains text with clean background. We also generate a synthetic evaluation set, I_S^{eval}, which consists of 21 489 synthetic images obtained by applying TBM-GAN to I_C^{eval}.

5   Results and Discussion

In this section, we discuss our experiments to showcase the effectiveness of synthetic data generated by our TBM-GAN based pipeline. We perform both qualitative and quantitative evaluations. For the latter, the synthetic image generated by TBM-GAN (I_S) can be used as additional training data for different DIU applications. In particular, we demonstrate the effectiveness of TBM-GAN on OCR application due to its wide applicability. In this regard, we have used a deep learning based benchmark OCR architecture [33,35].

5.1   Qualitative Results

In Fig. 4, we compare the synthetic document images generated by TBM-GAN with and without the FFL. We note that inclusion of FFL, Fig. 4 (b), enhances the variability of texture and prominence in text region pixels. On the other hand, TBM-GAN with only spatial loss, Fig. 4 (a), produces smooth background with fewer texture details.

Additional examples of synthetic document image generated by TBM-GAN are shown in Fig. 8. We observe that TBM-GAN is able to generate images with various background colors and degradation. Though TBM-GAN model is trained with clean handwritten images in English languages from IAM database, we note that it is also able to generate degraded documents with both text and numerals in non-English languages such as Telugu, Bengali, and Hindi. Figure 8 also shows that texts with shape deformation like struck-out strokes, underline, or cross are nicely captured by TBM-GAN.

5.2   Improving OCR Using TBM-GAN

We next discuss a downstream application of TBM-GAN in improving OCR with additional training data generated synthetically.

OCR architecture. We use state-of-the-art OCR model based on hybrid CNN-BiLSTM-CTC network [9,35] in our evaluation. The above OCR architecture includes five layers of convolution neural network (CNN), two layers of bi-directional LSTM, and a connectionist temporal classification (CTC) module at the end recognition.

Evaluation methodology. The OCR is trained in four different setups:

- OCR_C: the OCR is trained with only clean text images (I_C^{train}) from IAM database.
- OCR_R: the OCR is trained with real text images (I_R^{train}).
- OCR_S: the OCR is trained with the synthetically generated images (I_S^{train}) from TBM-GAN. It should be noted that the set I_S^{train} is the TBM-GAN's output corresponding to the input (I_R^{train}, I_C^{train}).
- OCR_{S+R}: the OCR is initially trained using I_S^{train} and subsequently fine tuned using I_R^{train}.

It should be noted that OCR_S and OCR_{S+R} are trained using TBM-GAN's synthetically generated output. In our experiments, we compare their performance against two baselines, OCR_C and OCR_R, to evaluate the influence of TBM-GAN on the downstream OCR application.

We compare the performance of the above discussed OCRs (OCR_C, OCR_R, OCR_S, OCR_{S+R}) on three evaluation datasets, each containing 21 000 samples:

Fig. 8. Output of synthesised images by TBM-GAN. The first four rows depict texts and numerals in English. Fifth and sixth rows show multiple languages such as Telugu, Bengali and Hindi (2 each) respectively. The seventh and eighth rows show texts with structural or shape deformation

- Clean text images (I_C^{eval})
- Synthetic images generated using TBM-GAN (I_S^{eval}). The set (I_S^{eval}) is the TBM-GAN's output corresponding to the input (I_R^{eval}, I_C^{eval}).
- Real images (I_R^{eval})

We report standard OCR evaluation metrics, character error rate (CER) and word error rates (WER%), in our experiments. The results are also broken down by the word length of the evaluation images, ranging from 2 to 6 or more characters.

Table 1. Evaluating OCRs on text images with clean background (I_C^{eval}) in terms of CER% (WER%). We observe that performance of OCRs trained with synthetic data from TBM-GAN (OCR$_S$ and OCR$_{S+R}$) are close to that of clean image baseline (OCR$_C$) when tested with clean documents (I_C^{eval}).

Word Length	OCR$_C$	OCR$_R$	OCR$_S$	OCR$_{S+R}$
2	8.36 (11.22)	19.73 (27.05)	10.61 (15.36)	9.41 (13.34)
3	7.73 (15.11)	26.08 (46.07)	10.32 (20.19)	8.16 (16.15)
4	10.62 (27.12)	40.02 (82.11)	13.88 (34.75)	11.33 (27.94)
5	11.32 (35.19)	43.62 (91.44)	14.67 (44.85)	11.84 (36.49)
>=6	14.11 (52.23)	34.62 (92.15)	18.73 (64.58)	14.72 (53.60)
All	11.91 (30.24)	34.14 (68.24)	15.71 (38.36)	12.53 (31.57)

Table 2. Evaluating OCRs on synthetic background text images (I_S^{eval}) in terms of CER% (WER%). We observe that the OCRs trained on TBM-GAN's generated output, OCR$_S$ and OCR$_{S+R}$, perform better than the baselines OCR$_C$ and OCR$_R$.

WordLength	OCR$_C$	OCR$_R$	OCR$_S$	OCR$_{S+R}$
2	19.41 (22.55)	29.48 (39.80)	12.83 (17.37)	**9.52 (12.88)**
3	15.38 (24.89)	30.10 (51.22)	11.75 (23.24)	**9.04 (16.49)**
4	17.34 (36.55)	35.44 (67.54)	15.15 (35.27)	**12.86 (29.22)**
5	18.28 (46.30)	36.43 (76.04)	16.51 (46.50)	**13.43 (37.54)**
>=6	23.26 (64.97)	46.62 (91.82)	23.17 (69.03)	**19.20 (59.01)**
All	20.37 (41.31)	40.08 (67.35)	18.77 (40.84)	**15.35 (33.57)**

Results: Table 1 reports the performance of different OCRs on the clean text images (I_C^{eval}). In this setting, the baseline OCR system OCR$_C$ obtains the best performance across all word lengths. This is because both the train set for OCR$_C$ and the test set are from the same domain (clean text images). On the other hand, the performance of the baseline OCR$_R$ deteriorates significantly.

Table 3. Evaluating OCRs on real background text images (I_R^{eval}) in terms of CER% (WER%). We observe that the OCRs trained on TBM-GAN's generated output, OCR$_S$ and OCR$_{S+R}$, outperform the baselines OCR$_C$ and OCR$_R$.

WordLength	OCR$_C$	OCR$_R$	OCR$_S$	OCR$_{S+R}$
2	41.36 (38.66)	24.82 (33.05)	30.69 (36.65)	**9.80 (13.40)**
3	25.16 (35.13)	28.38 (48.90)	18.42 (30.90)	**8.60 (16.28)**
4	22.04 (42.52)	43.64 (84.46)	17.83 (40.14)	**11.06 (26.64)**
5	21.78 (51.75)	49.93 (94.36)	18.44 (48.95)	**12.80 (36.82)**
>=6	23.67 (66.37)	44.58 (95.42)	23.08 (68.74)	**16.75 (54.94)**
All	24.76 (48.59)	41.28 (71.71)	21.66 (46.91)	**13.75 (31.84)**

This is because its train set, I_R^{train}, has a limited number of distinct real (noisy) background texture. Hence, OCR$_R$ gets conditioned to work well only on such background texture. We also observe that the performance of OCR systems trained using TBM-GAN's output, OCR$_S$ and OCR$_{S+R}$, are slightly worse than OCR$_C$. This is because the training sets of OCR$_S$ and OCR$_{S+R}$ has noisy background, while the test set has clean background. The performance of OCR$_{S+R}$ is better than OCR$_S$, which highlights that pre-training on synthetic background images helps to improve the generalization of the model to clean background images. The difference in CER% and WER% between OCR$_{S+R}$ and OCR$_C$ is less than 1% and 2%, respectively, for all word lengths greater than two. Overall, the results in Table 1 show that document images generated by TBM-GAN have reasonably good properties and help in training OCRs which are robust to change in test set distribution. It should also be emphasised that the setting of evaluation set I_C^{eval}, text images with clean background, is not common in real world scenarios. Real world images have noise, which is captured in the other two evaluation sets I_S^{eval} and I_R^{eval}.

Table 2 reports the performance of different OCRs on the synthetic text images with noisy background (I_S^{eval}). As discussed, I_S^{eval} is generated from TBM-GAN with I_C^{eval} as input. This is significantly more challenging evaluation set than I_C^{eval}. Hence, we observe a steep rise in the character error rates of OCR$_C$ and OCR$_R$ (6–11% and 4–12%, respectively) across all word lengths (when compared with results in Table 1). However, OCR$_S$ and OCR$_{S+R}$ perform much better than the baselines and are much more robust to the presence of background (noisy) texture. Compared to I_C^{eval} test set, the increase in the character error rates of OCR$_S$ and OCR$_{S+R}$ on I_S^{eval} is 1.5–4.5% and 0.11–4.5%, respectively. Overall, OCR$_{S+R}$ obtains the best performance across all word lengths and its average improvement over OCR$_C$ is 5% (over characters) and 7.7% (over words).

Table 3 compares the accuracy of different OCRs on text images with real background (I_R^{eval}), grouped by word length. The results show that the OCR trained on a combination of synthetic and real data (OCR$_{S+R}$) achieves the highest accuracy for all word lengths, demonstrating its superior generalization ability on real images (I_R^{eval}). This suggests that the synthetic data generated by TBM-GAN can complement the real data and enhance the robustness and accuracy of the model on real images. The model trained on real data (OCR_R) performs better than the model trained on clean data (OCR_C) on real images, but worse than the model trained on synthetic data (OCR_S). This suggests that the real data alone may not be sufficient or representative enough to train a reliable OCR model on real images and that the synthetic data can provide additional information and variation to the model.

The proposed TBM-GAN method can generate high-quality synthetic background text images that can be used to train OCR models that are robust and accurate across different types of backgrounds. Synthetic data from TBM-GAN improves OCR models' ability to recognize text images with different backgrounds. It preserves the text's sharpness, mimics the diversity and complexity of the backgrounds, and complements the real data. The best OCR model is the one trained on both synthetic and real data, which can achieve significant improvements over the baselines on all test sets, especially on real images. For example, on real images, the OCR_{S+R} model significantly reduces the CER and the WER by 44.5% and 66.7% and by 34.4% and 55.6% respectively, compared to the OCR_C and OCR_R models.

6   Conclusion

In this paper, we address the problem of generating realistic synthetic hand-written documents with structural and texture degradation, along with text and pixel-level annotation. We present TBM-GAN, a conditional GAN framework that uses FFL and a spatial loss to generate realistic document images with various background textures and degradation levels. Our approach maintains the spatial alignment and structure of the text and background, ensuring the readability and quality of the synthesized documents. Capturing information from document images with structural and texture degradation is an important problem in document-image understanding applications (DIUs). We show that the synthetic data generated by TBM-GAN can be used to enhance the training data for document-image understanding applications (DIUs), such as OCR. We evaluated the performance of TBM-GAN on three datasets of document images with clean, synthetic, and real background. We also trained and tested four OCR models with different combinations of data sources to assess the impact of TBM-GAN's generated images on OCR accuracy. Our experimental results showed that TBM-GAN can generate high-quality synthetic document images with diverse background degradation and that the OCR models trained on TBM-GAN's output can achieve better or comparable accuracy than the baselines on clean, and real background document images. We also present qualitative results of TBM-GAN's output on different languages and scripts, showing its ability to capture the variability and prominence of text and background regions. We believe that our framework can be a useful tool for creating large-scale synthetic datasets for various DIUs, especially for low-resource languages and scripts.

Acknowledgement. This work is partially supported by Microsoft Academic Partnership Grant (MAPG) 2022-2023 with grant number IIT/SRIC/CS/ADD/2022-2023/065.

References

1. Alberti, M., Seuret, M., Ingold, R., Liwicki, M.: A pitfall of unsupervised pretraining. arXiv preprint arXiv:1703.04332 (2017)
2. Baird, H.S., Bunke, H., Yamamoto, K.: Structured Document Image Analysis. Springer, Science & Business Media (2012). https://doi.org/10.1007/978-3-642-77281-8
3. Bhunia, A.K., Bhunia, A.K., Sain, A., Roy, P.P.: Improving document binarization via adversarial noise-texture augmentation. In: 2019 IEEE International Conference on Image Processing (ICIP), pp. 2721–2725. IEEE (2019)
4. Bhunia, A.K., Khan, S., Cholakkal, H., Anwer, R.M., Khan, F.S., Shah, M.: Handwriting transformers. In: Proceedings of the IEEE/CVF International Conference on Computer Vision (ICCV), pp. 1086–1094 (2021)
5. Biswas, S., Riba, P., Lladós, J., Pal, U.: DocSynth: a layout guided approach for controllable document image synthesis. In: Lladós, J., Lopresti, D., Uchida, S. (eds.) ICDAR 2021. LNCS, vol. 12823, pp. 555–568. Springer, Cham (2021). https://doi.org/10.1007/978-3-030-86334-0_36
6. Cai, M., Zhang, H., Huang, H., Geng, Q., Li, Y., Huang, G.: Frequency domain image translation: more photo-realistic, better identity-preserving. In: Proceedings of the IEEE/CVF International Conference on Computer Vision (ICCV), pp. 13930–13940 (2021)
7. Capobianco, S., Marinai, S.: Docemul: a toolkit to generate structured historical documents. In: 2017 14th IAPR International Conference on Document Analysis and Recognition (ICDAR), vol. 1, pp. 1186–1191. IEEE (2017)
8. Dey, S., Jawanpuria, P.: Light-weight document image cleanup using perceptual loss. In: Lladós, J., Lopresti, D., Uchida, S. (eds.) ICDAR 2021. LNCS, vol. 12823, pp. 238–253. Springer, Cham (2021). https://doi.org/10.1007/978-3-030-86334-0_16
9. Dutta, K., Krishnan, P., Mathew, M., Jawahar, C.: Improving CNN-RNN hybrid networks for handwriting recognition. In: 2018 16th International Conference on Frontiers in Handwriting Recognition (ICFHR), pp. 80–85. IEEE (2018)
10. Fogel, S., Averbuch-Elor, H., Cohen, S., Mazor, S., Litman, R.: Scrabblegan: semi-supervised varying length handwritten text generation. In: Proceedings of the IEEE/CVF Conference on Computer Vision and Pattern Recognition (CVPR), pp. 4324–4333 (2020)
11. Gatys, L.A., Ecker, A.S., Bethge, M.: Image style transfer using convolutional neural networks. In: Proceedings of the IEEE Conference on Computer Vision and Pattern Recognition (CVPR), pp. 2414–2423 (2016)
12. Goodfellow, I., et al.: Generative adversarial nets. In: Advances in Neural Information Processing Systems (NeurIPS), pp. 2672–2680 (2014)
13. Guan, M., Ding, H., Chen, K., Huo, Q.: Improving handwritten OCR with augmented text line images synthesized from online handwriting samples by style-conditioned GAN. In: 2020 17th International Conference on Frontiers in Handwriting Recognition (ICFHR), pp. 151–156. IEEE (2020)

14. Isola, P., Zhu, J.Y., Zhou, T., Efros, A.A.: Image-to-image translation with conditional adversarial networks. In: Proceedings of the IEEE Conference on Computer Vision and Pattern Recognition (CVPR), pp. 1125–1134 (2017)
15. Jiang, L., Dai, B., Wu, W., Loy, C.C.: Focal frequency loss for image reconstruction and synthesis. In: Proceedings of the IEEE/CVF International Conference on Computer Vision (ICCV), pp. 13919–13929 (2021)
16. Journet, N., Visani, M., Mansencal, B., Van-Cuong, K., Billy, A.: Doccreator: a new software for creating synthetic ground-truthed document images. J. Imaging **3**(4), 62 (2017)
17. Kang, L., Riba, P., Rusinol, M., Fornes, A., Villegas, M.: Content and style aware generation of text-line images for handwriting recognition. IEEE Trans. Pattern Anal. Mach. Intell. (T-PAMI) **44**(12), 8846–8860 (2021)
18. Kang, L., Riba, P., Wang, Y., Rusiñol, M., Fornés, A., Villegas, M.: GANwriting: content-conditioned generation of styled handwritten word images. In: Vedaldi, A., Bischof, H., Brox, T., Frahm, J.-M. (eds.) ECCV 2020. LNCS, vol. 12368, pp. 273–289. Springer, Cham (2020). https://doi.org/10.1007/978-3-030-58592-1_17
19. Kieu, V., Visani, M., Journet, N., Domenger, J.P., Mullot, R.: A character degradation model for grayscale ancient document images. In: Proceedings of the 21st International Conference on Pattern Recognition (ICPR), pp. 685–688. IEEE (2012)
20. Larson, S., Lim, G., Ai, Y., Kuang, D., Leach, K.: Evaluating out-of-distribution performance on document image classifiers. In: Thirty-sixth Conference on Neural Information Processing Systems Datasets and Benchmarks Track (NeurIPS) (2022)
21. Lee, Y., Hong, T., Kim, S.: Data augmentations for document images. In: SDU@ AAAI (2021)
22. Lin, Y.H., Chen, W.C., Chuang, Y.Y.: Bedsr-net: a deep shadow removal network from a single document image. In: Proceedings of the IEEE/CVF Conference on Computer Vision and Pattern Recognition (CVPR), pp. 12905–12914 (2020)
23. Maini, S., Groleau, A., Chee, K.W., Larson, S., Boarman, J.: Augraphy: A data augmentation library for document images. arXiv preprint arXiv:2208.14558 (2022)
24. Marti, U.V., Bunke, H.: The iam-database: an English sentence database for offline handwriting recognition. Int. J. Doc. Anal. Recogn. **5**(1), 39–46 (2002)
25. Mildenhall, B., Srinivasan, P.P., Tancik, M., Barron, J.T., Ramamoorthi, R., Ng, R.: Nerf: representing scenes as neural radiance fields for view synthesis. Commun. ACM **65**(1), 99–106 (2021)
26. Mirza, M., Osindero, S.: Conditional generative adversarial nets. arXiv preprint arXiv:1411.1784 (2014)
27. Otsu, N.: A threshold selection method from gray-level histograms. IEEE Trans. Syst. Man Cybern. **9**(1), 62–66 (1979)
28. Poddar, A., Chakraborty, A., Mukhopadhyay, J., Biswas, P.K.: Detection and localisation of struck-out-strokes in handwritten manuscripts. In: Barney Smith, E.H., Pal, U. (eds.) ICDAR 2021. LNCS, vol. 12917, pp. 98–112. Springer, Cham (2021). https://doi.org/10.1007/978-3-030-86159-9_7
29. Poddar, A., Chakraborty, A., Mukhopadhyay, J., Biswas, P.K.: Texrgan: a deep adversarial framework for text restoration from deformed handwritten documents. In: Proceedings of the Twelfth Indian Conference on Computer Vision, Graphics and Image Processing (ICVGIP), pp. 1–9 (2021)
30. Pondenkandath, V., Alberti, M., Diatta, M., Ingold, R., Liwicki, M.: Historical document synthesis with generative adversarial networks. In: 2019 International Conference on Document Analysis and Recognition Workshops (ICDARW), vol. 5, pp. 146–151. IEEE (2019)

31. Rahaman, N., et al.: On the spectral bias of neural networks. In: International Conference on Machine Learning (ICML), pp. 5301–5310. PMLR (2019)
32. Seuret, M., Chen, K., Eichenbergery, N., Liwicki, M., Ingold, R.: Gradient-domain degradations for improving historical documents images layout analysis. In: 2015 13th International Conference on Document Analysis and Recognition (ICDAR), pp. 1006–1010. IEEE (2015)
33. Shi, B., Bai, X., Yao, C.: An end-to-end trainable neural network for image-based sequence recognition and its application to scene text recognition. IEEE Trans. Pattern Anal. Mach. Intell. (T-PAMI) **39**(11), 2298–2304 (2016)
34. Souibgui, M.A., Kessentini, Y.: De-GAN: a conditional generative adversarial network for document enhancement. IEEE Trans. Patteren Anal. Mach. Intell. (T-PAMI) **44**(3), 1180–1191 (2020)
35. Strauß, T., Leifert, G., Labahn, R., Hodel, T., Mühlberger, G.: Icfhr 2018 competition on automated text recognition on a read dataset. In: 2018 16th International Conference on Frontiers in Handwriting Recognition (ICFHR), pp. 477–482. IEEE (2018)
36. Tancik, M., et al.: Fourier features let networks learn high frequency functions in low dimensional domains. Adv. Neural Inf. Process. Syst. (NeurIPS) **33**, 7537–7547 (2020)
37. Tensmeyer, C., Brodie, M., Saunders, D., Martinez, T.: Generating realistic binarization data with generative adversarial networks. In: 2019 International Conference on Document Analysis and Recognition (ICDAR), pp. 172–177. IEEE (2019)
38. Tensmeyer, C., Martinez, T.: Historical document image binarization: a review. SN Comput. Sci. **1**(3), 1–26 (2020)
39. Toshevska, M., Gievska, S.: A review of text style transfer using deep learning. IEEE Trans. Artif. Intell. (T-AI) **3**, 669–684 (2021)
40. Vögtlin, L., Drazyk, M., Pondenkandath, V., Alberti, M., Ingold, R.: Generating synthetic handwritten historical documents with OCR constrained GANs. In: Lladós, J., Lopresti, D., Uchida, S. (eds.) ICDAR 2021. LNCS, vol. 12823, pp. 610–625. Springer, Cham (2021). https://doi.org/10.1007/978-3-030-86334-0_40
41. Wigington, C., Stewart, S., Davis, B., Barrett, B., Price, B., Cohen, S.: Data augmentation for recognition of handwritten words and lines using a CNN-LSTM network. In: 2017 14th IAPR International Conference on Document Analysis and Recognition (ICDAR), vol. 1, pp. 639–645. IEEE (2017)
42. Zhong, Z., Zheng, L., Kang, G., Li, S., Yang, Y.: Random erasing data augmentation. In: Proceedings of the AAAI Conference on Artificial Intelligence (AAAI), vol. 34, pp. 13001–13008 (2020)
43. Zhu, J.Y., Park, T., Isola, P., Efros, A.A.: Unpaired image-to-image translation using cycle-consistent adversarial networks. In: Proceedings of the IEEE International Conference on Computer Vision (ICCV), pp. 2223–2232 (2017)

WordStylist: Styled Verbatim Handwritten Text Generation with Latent Diffusion Models

Konstantina Nikolaidou[1(✉)], George Retsinas[2], Vincent Christlein[3], Mathias Seuret[3], Giorgos Sfikas[4,5], Elisa Barney Smith[1], Hamam Mokayed[1], and Marcus Liwicki[1]

[1] Luleå University of Technology, Luleå, Sweden
`{konstantina.nikolaidou,elisa.barney,hamam.mokayed,marcus.liwicki}@ltu.se`
[2] National Technical University of Athens, Athens, Greece
`gretsinas@central.ntua.gr`
[3] Friedrich-Alexander-Universität, Erlangen, Germany
`{vincent.christlein,mathias.seuret}@fau.de`
[4] University of West Attica, Egaleo, Greece
`gsfikas@uniwa.gr`
[5] University of Ioannina, Ioannina, Greece

Abstract. Text-to-Image synthesis is the task of generating an image according to a specific text description. Generative Adversarial Networks have been considered the standard method for image synthesis virtually since their introduction. Denoising Diffusion Probabilistic Models are recently setting a new baseline, with remarkable results in Text-to-Image synthesis, among other fields. Aside its usefulness *per se*, it can also be particularly relevant as a tool for data augmentation to aid training models for other document image processing tasks. In this work, we present a latent diffusion-based method for styled text-to-text-content-image generation on word-level. Our proposed method is able to generate realistic word image samples from different writer styles, by using class index styles and text content prompts without the need of adversarial training, writer recognition, or text recognition. We gauge system performance with the Fréchet Inception Distance, writer recognition accuracy, and writer retrieval. We show that the proposed model produces samples that are aesthetically pleasing, help boosting text recognition performance, and get similar writer retrieval score as real data. Code is available at: https://github.com/koninik/WordStylist.

Keywords: Diffusion Models · Synthetic Image Generation · Text Content Generation · Handwriting Generation · Data Augmentation · Handwriting Text Recognition

1 Introduction

Image synthesis is a very challenging problem in Computer Vision, which has gained traction with the rekindling of interest in neural networks a decade prior,

G. A. Fink et al. (Eds.): ICDAR 2023, LNCS 14188, pp. 384–401, 2023.
https://doi.org/10.1007/978-3-031-41679-8_22

and especially the introduction of models and concepts such as Generative Adversarial Networks (GANs) [10], Variational Autoencoders (VAEs) [17] or Normalizing Flows (NFs) [18]. Apart from the utility of the generated image in itself, image synthesis has been employed as a tool to artificially augment training sets. This is an aspect that is critical when it comes to training Deep Learning models, which are notorious for typically requiring vast amounts of data to attain optimal performance. Annotating data is an expensive and time-consuming task that requires a lot of human effort and expertise. A particular variant of image synthesis is text-to-image synthesis, where the task is to generate an image given a text description. As stated in [9], a text description can indeed give more semantic and spatial information about the objects depicted in an image than a single label. Text-to-image synthesis has been established as a whole independent field as several applications have gained relative prominence.

Conditional Generative Adversarial Networks (cGANs) [26], the conditional variant of GANs, have further enabled the augmentation of existing datasets by generating data given a specific class or a specific input. With the advent of these models, adversarial training has been established as the standard for image generation, where a minimax game is "played" between two networks, aptly named Generator and Discriminator. The Generator is tasked with creating a sample – in the current context, the synthesized image – while the Discriminator is tasked with detecting instances that are outliers with respect to the training data. Unlike GANs, which do not explicitly define a data density, other state-of-the-art approaches have attempted to approach data generation as sampling from a probability density function (pdf). Variational Autoencoders cast the problem as one of estimating a latent representation for members of a given dataset, given the prior knowledge that latent embeddings are Gaussian-distributed. They are comprised of two network parts, named the Encoder and the Decoder. The Encoder produces (probabilistic) latent representations given a datum, while the Decoder is tasked with the inverse task, that is producing a sample given a latent representation. Normalizing flows also deal with estimating the pdf of a given set, and also assume the existence of a latent space that is to be estimated, like VAEs. Latent data are equidimensional to the image data, and training is performed by learning a series of non-linear mappings that gradually convert the data distribution from and to a Gaussian distribution. In VAEs as in NFs, once the model is trained, image generation can simply be performed by sampling from the latent space and applying the learned transformation back to the image/original space. The outburst of Diffusion Models, and in particular more recent variants such as Denoising Diffusion Probabilistic Models (DDPMs) or Latent Diffusion Models (LDM) have quickly begun to change the picture of the state of the art with achievements that can often be described to be no less than astonishing. The results of systems such as DALL·E-2 [30] and Imagen [34] have prompted many researchers to experiment with their use in different applications. Diffusion models [36] are based on a probabilistic framework like VAEs or NFs, but propose a different approach to the problem of image synthesis, cast in its standard form as density estimation followed by sampling. Like NFs, in their standard form the latent space dimensionality is defined to be equal to that of the original space, and learning is performed by estimating a series of non-linear transformations between latent space and original space. A "forward/diffusion" process gradually adds noise to inputs according to a predetermined schedule; with the "reverse" process the aim is to produce an estimate of an image given a latent, noisy sample.

In this work, instead of using text only as a description of the image contents, we also use it literally as image content, in the sense of generating handwriting. Thus, we address a task of Text-to-Text-Content-Image Synthesis. The main contributions of this work are the following:

1. We present a method based on a conditional Latent Diffusion Model, that takes as input a word string and a style class and generates a synthetic image containing that word.
2. We compare qualitative results of our method with other GAN-based generative model approaches.
3. We further evaluate our results by presenting qualitative and quantitative results for text recognition using the synthetic data. The synthetic data is used for data augmentation, resulting in boosting the performance of a state-of-the-art Handwriting Text Recognition (HTR) system.
4. And finally, we compare synthetic data and real handwritten paragraphs using a writer retrieval system. We show that data produced by our method show no significant difference in style to real data, and outperforms the other methods by a tremendous margin.

The paper is organized as follows. In Sect. 2, we present an overview of the related work. Our proposed method is introduced in Sect. 3, while Sect. 4 includes the evaluation process and results. Section 5 presents limitations and possible future directions. Finally, we discuss conclusions in Sect. 6.

2   Related Work

Text-to-Text-Content-Image Synthesis refers to the task of generating an image that depicts a specific text, whether it is on the character-, word-, sentence-, or page-level, given that text as the input condition. A field directly related to this task is Document Image Analysis and Recognition, notably one of the resource-constrained domains with respect to the availability of annotated data, at least compared to the current state in natural image-related tasks [7, 20].

Most existing works focus on conditioning on a string prompt and a writer style to generate images of realistic handwritten text using GAN-based approaches. GANwriting [15] creates realistic handwritten word images conditioned on text and writer style by guiding the generator. The method is able to produce out-of-vocabulary words. The authors extend this work in [14], generating realistic handwritten text-lines. SmartPatch [25] fixes artifact issues that GANwriting faces by deploying a patch discriminator loss. ScrabbleGAN [35] uses a semi-supervised method to generate long handwritten sentences of different style and content. A Transformer-based method is presented in [2], using a typical Transformer Encoder-Decoder architecture that takes as inputs style features of handwritten sentence images extracted by a CNN encoder and a query text in the decoding part. The model is trained with a four-part loss function, including an adversarial loss, a text recognition loss, a cycle loss, and a reconstruction loss.

Related to Historical Document Analysis [21,28], the work presented in [29] initially generates modern documents using LaTeX and then attempts to convert them into a historical style with the use of CycleGAN [40]. The work is further extended in [39], by adding text recognition to the framework and the loss function, which gives better readable text in the image synthesis.

3   Method

In this section, we present some general background information for the standard Diffusion and Latent Diffusion Models. We then illustrate in detail the proposed method that includes the forward process, model components, sampling and experimental setup for training and sampling from the model.

3.1   Diffusion Models Background

Denoising Diffusion Probabilistic Models (DDPM). Diffusion Models are a type of generative model that employ Markov chains to add noise and disrupt the structure of data. The models then learn to reverse this process and reconstruct the data. Inspired by Thermodynamics [36], Diffusion Models have gained popularity in the field of image synthesis due to their ability to generate high quality samples.

The Diffusion Model consists of two phases: the forward (diffusion) process and the reverse (denoising) process. In the forward process, a sample x_0 is initially drawn from a distribution $x_0 \sim q(x_0)$ corresponding to the observed data. This is subjected to Gaussian noise, which produces a latent variable x_1; noise is again added to x_1, giving latent variable x_2, and so on, until some predefined hyper-parameter T. This process forms a series of latent variables $x_1, x_2, \cdot, x_T$, Formally, we can write:

$$q(x_{1:T}|x_0) = \prod_{t=1}^{T} q(x_t|x_{t-1}), \quad q(x_t|x_{t-1}) = N(x_t; \sqrt{1-\beta_t}x_{t-1}, \beta_t I), \quad (1)$$

where we have $\beta_i \in [0,1], \forall i \in [1,T]$. Hyper-parameters $\beta_1, \beta_2, ..., \beta_T$ collectively form a noise variance schedule, used to control the amount of noise added at each timestep. In the final timestep, given large enough T and suitable noise schedule, we will have $q(x_T|x_0) = q(x_T) \approx N(0, I)$, i.e. the end result becomes practically a pure Gaussian noise sample with no structure. In the reverse (denoising) phase, a neural network learns to gradually remove the noise from the sampled by a stationary distribution until ending up with actual data. Hence, image synthesis will be performed according to an ancestral sampling scheme. This means that first we need to sample from $q(x_T)$, then we sample by the previous time-step conditioned on the sampled value of x_T, and so and so forth until we sample the required x_0.

The noise is gradually removed in reverse timesteps using the following transition:

$$p_\theta(x_{0:T}) = p(x_T) \prod_{t=1}^{T} p_\theta(x_{t-1}|x_t), \quad p_\theta(x_{t-1}|x_t) = N(x_{t-1}; \mu_\theta(x_t, t), \Sigma_\theta(x_t, t)). \quad (2)$$

The network is trained by optimizing the variational lower bound between the forward process posterior and the joint distribution of the reverse process p_θ. The training loss

$$L = \mathbb{E}_{x_0, t, \epsilon}[||\epsilon - \epsilon_\theta(x_t, t)||^2] \quad (3)$$

is calculated as the reconstruction error between the actual noise, ϵ, and the estimated noise, ϵ_θ. In the case of Latent Diffusion models the loss will be adapted to the latent representation z_t.

Latent Diffusion Models (LDM). Diffusion Models have demonstrated remarkable performance in image generation and transformation tasks [13,16, 19,27]. However, their computational cost is high due to the size of the input data and the use of cross-attention in images. To address this issue, Latent Diffusion Models were introduced in [32] to model the data distribution in a lower-dimensional latent representation space. This is accomplished by mapping the input images to a latent representation using an encoder, and then decoding the sampled latents back into an image using a decoder, both from a variational autoencoder architecture.

3.2   Proposed Approach

The goal of this work is to generate synthetic word-image samples given a word string and a style class as conditions from a known distribution. We approach this problem with the use of latent diffusion models to minimize training time and computational cost. To move to the latent space we use the pre-trained "stable-diffusion" VAE implementation from the Hugging Face repository[1]. Figure 1 presents the overall architecture of the proposed method.

Forward Process and Training. For the forward process, the VAE encoder V_E initially transforms an input image to a latent representation z. A diffusion model $p_\theta(x|Y, c_\tau)$ is learned on the style Y and text-condition c_τ pairs. Timesteps t are sampled from a uniform distribution and the latent representation z gets gradually corrupted by the diffusion process in every timestep. For the noise prediction, we use a U-Net architecture [33] with Residual Blocks [12] and intermediate

---

[1] https://huggingface.co/CompVis/stable-diffusion.

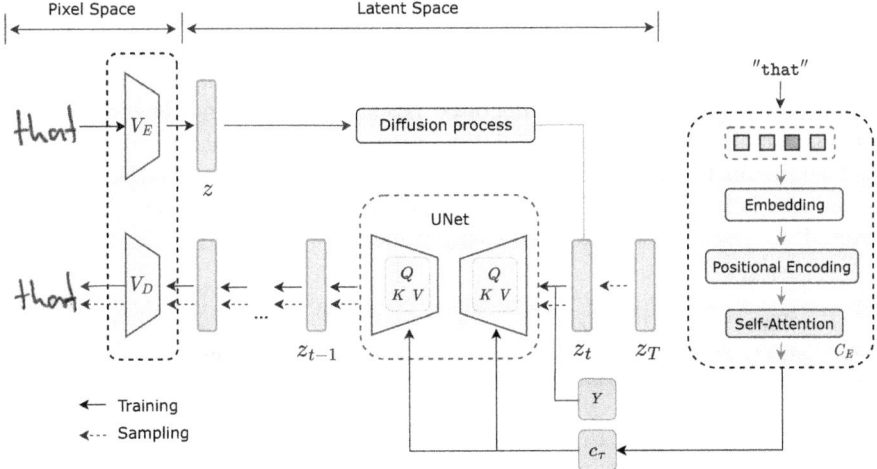

Fig. 1. The overall architecture. During training, an input image is fed to the encoder V_E to create a latent representation z, then noise is added to the latent. The noisy latent z_t is then fed to the U-Net noise predictor along with a style class index for the writer style and an encoded word as the content. The UNet predicts the noise of the noisy latent z_{t-1} where $t-1$ is the corresponding timestep. During sampling, a random noise latent z_T is given to predict its noise. Then the model uses the two noise predictions to reconstruct the latent of the image z_0 that is finally decoded by the decoder V_D that creates the synthetic image.

Transformer Blocks [38] to add the text condition to the model, as typically used by Ho et al. [13]. The network takes as input the noisy image latents, the corresponding timestep, and the desired conditions Y and τ. Timesteps are encoded using a sinusoidal position embedding, similar to [38] to inform the model about each particular timestep that is operating. The training objective is to minimize the reconstruction error between the network's noise prediction and the noise present in the image. For the diffusion process, a noise scheduler increases the amount of noise linearly from $\beta_1 = 10^{-4}$ to $\beta_T = 0.02$ for $T = 1000$ timesteps. While most works use multiple ResNet blocks within the U-Net components, in the context of the current problem we need to take into account that we must work with scarce data compared to other use-cases; larger models correspond to larger parameter spaces, which are exponentially harder to explore. Hence, we use 1 ResNet block in every module of the U-Net. To further reduce the parameters and complexity of the network we use an inner model dimension of 320 and 4 heads in the Multi-Headed Attention layers within the U-Net.

Sampling. We generate synthetic samples by deploying the reverse denoising process learned from the model. To this end, the noise of a random noisy sample z_T is predicted by the learned network p_θ and gradually removed in every timestep of the reversed process starting from T to $t = 0$. One of the main challenges associated with DDPM is the time required for sampling. Our experiments indicate that reducing the number of time steps from 1,000 to 600 does not compromise the quality of the generated samples. The final image is obtained in pixel space by decoding the denoised latent variable using decoder V_D. We demonstrate how the reduction of timesteps affects the quality of the generated sample in Fig. 2. The figure shows that below 500 timesteps the quality of the images is really affected, thus to make sure the generated samples are not affected dramatically we proceed with a value of 600.

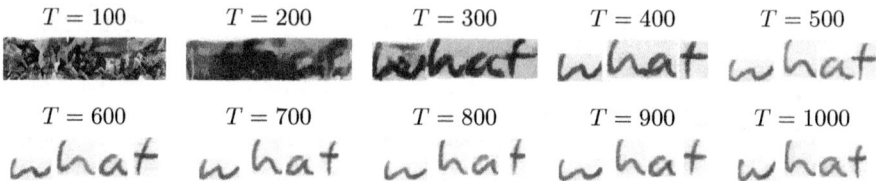

Fig. 2. Sampling outputs using various timesteps values in the reverse denoising process.

Style and Text Conditions. The input style condition Y is processed with an embedding layer and then added to the timestep embedding. For the text condition, a content encoder C_E is used to transform an input string τ into a meaningful context representation $c_\tau = C_E(\tau)$ for the model. Initially, the string is tokenized using a unique index for each letter and then passed through an embedding layer to transform it to an appropriate embedding dimension according to the vocabulary size which is the number of characters present in the training set. Then, positional encodings similar to [38] are used to inform the model about the character position in the sequence with the use of sine and cosine functions as $PE_{pos,i} = \sin(pos/1000^{i/emb\_dim})$ and $PE_{pos,i+1} = \sin(pos/1000^{i/emb\_dim})$, where pos is the position of each letter in the sequence and emb_dim is 320 as mentioned previously. Finally, to create the text input condition c_τ a dot-product attention layer is used, defined as $Att(Q, K, V) = softmax(\frac{QK^T}{\sqrt{d_k}})V$, to create a weighted sum of the character representations. To support the choice of the positional text encoding we present a few samples as ablation with and without the positional encoding and self-attention layers in Fig. 3.

3.3   Experimental Setup

We conducted extensive experiments using the IAM offline handwriting database on word-level [24]. Similar to [15,25], we used the Aachen split train set and included words of 2–7 characters to train the diffusion model. Thus, during training the model sees 339 writer styles and approximately 45K words. For consistency, all images were resized to a fixed height of 64 pixels, retaining their aspect ratio. To handle variations in width, images of width smaller than 256 pixels were center-padded, while larger ones were resized to the maximum width. Since the maximum number of characters is 7, this resizing did not cause significant distortions in the images. Moreover, these images were intended for training other models, which could eventually lead to resizing or modifications of the original images. AdamW [22] is used as the optimizer during training with a learning rate of 10^{-4}. To better understand the nature of the model, no augmentation is used on the images during training. Each model was trained for 1K epochs with a batch size of 224 on a single A100 SXM GPU.

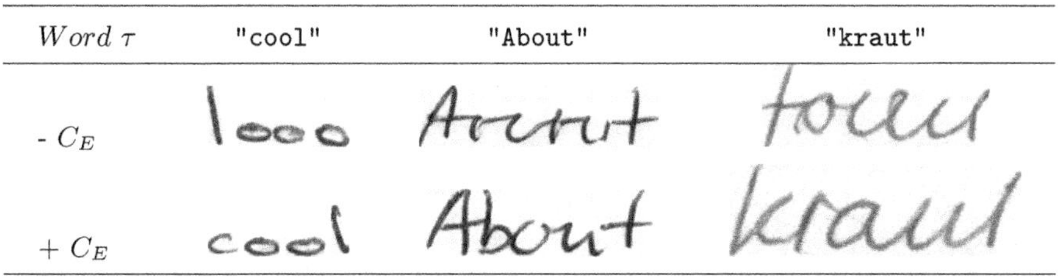

$Word\ \tau$	"cool"	"About"	"kraut"
$- C_E$			
$+ C_E$			

Fig. 3. Comparison of generated images with (top row) and without (bottom row) the positional encoding and self-attention layer of the context encoder C_E. All image pairs (top-bottom) share the same style condition.

4 Evaluation and Results

We evaluate the quality of the generated word-images using our method in three aspects: visual quality, text quality, and style quality. To assess visual quality, we compute the commonly used Fréchet Inception Distance (FID) score and provide examples of in-vocabulary and out-of-vocabulary words. Additionally, we demonstrate the results of blending two distinct writer styles through interpolation. To determine the effectiveness and text quality of our approach, we create a pseudo training set from the IAM database and conduct several experiments for handwriting text recognition (HTR). For comparison with other methods, we perform the same experiments using two GAN-based approaches, SmartPatch and GANwriting. Finally, we evaluate style quality in two ways. First, we train a standard Convolutional Neural Network (CNN) on the real IAM database for style classification and test it on the generated samples. Second, we apply a writer retrieval method and compare its performances using real or synthetic data. This enables us to measure the extent to which our method accurately captures the style of the original IAM database.

4.1 Qualitative Results

A comparative qualitative evaluation can be found in Fig. 4, where both Smart-Patch and GANwriting methods have been used to generate a set of word images. Specifically, the goal was to recreate the original images (further left column).

As we can see, all methods generate "readable" words without notable arti-facts/deformations. Nonetheless, SmartPatch has a smoother appearance com-pared to GANwriting, as it was designed to do, while the proposed Diffusion approach retains the original style to an outstanding degree.

Furthermore, to validate the variety of styles and the ability to generalize beyond already seen words, in Fig. 5 we present generated samples using our method of In-Vocabulary (IV) and Out-of-Vocabulary (OOV) words and random styles picked from the IAM training set. We can observe a notable variety over the writing style, indicating the good behavior of the proposed method, even for the case of OOV words that were never met in the training phase.

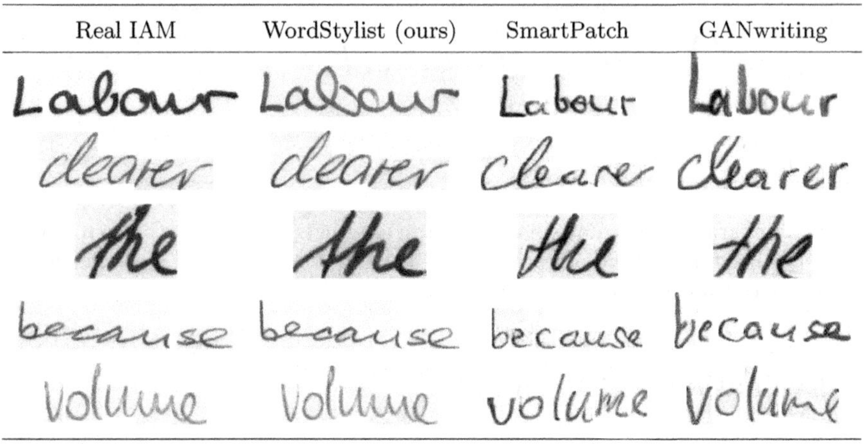

Real IAM	WordStylist (ours)	SmartPatch	GANwriting

Fig. 4. Comparison of (in-vocabulary) real word images and synthetic versions of these words.

Towards measuring the quality of the generation, a metric commonly used to evaluate generative models is the Fréchet Inception Distance (FID) score [8]. The metric computes the distance between two dataset feature vectors extracted by an InceptionV3 network [37] pre-trained on ImageNet [7]. Our approach achieves

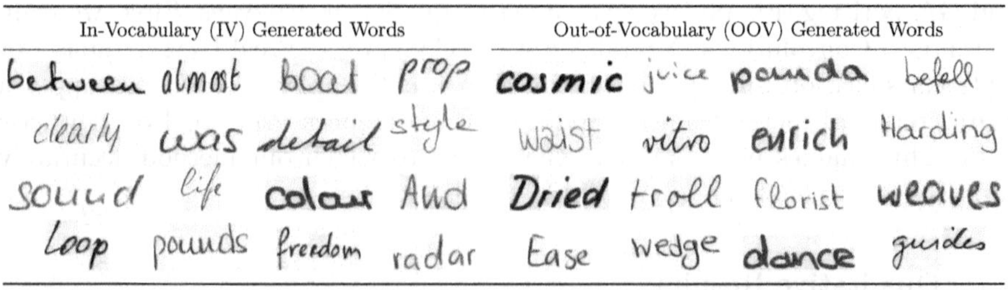

In-Vocabulary (IV) Generated Words	Out-of-Vocabulary (OOV) Generated Words

Fig. 5. Qualitative results from WordStylist of random writer styles from In-Vocabulary (IV) (left) and Out-of-Vocabulary (OOV) (right) word generation. All writer styles are randomly selected to produce each word meaning that the IV samples may not appear in the training set with the presented style-text combination.

an **FID** score of **22.74**, which is comparable to SmartPatch's score of **22.55**. GANwriting performs with an FID of **29.94**. While FID is a widely used metric for evaluating generative models, it may not be appropriate for tasks that do not involve natural images similar to those in ImageNet, on which the network was trained. In fact, this domain shift between natural images and handwritten documents lessens the fidelity of the evaluation protocol, but adapting this metric, by fine-tuning the FID network on document images, is out of the scope of this work. Despite this, the FID metric is still an indication of realistic images.

4.2 Latent Space Interpolation

Following the paradigm of GANwriting [15], we further interpolate between two writer styles Y_A and Y_B by a weight λ_{AB} to create mixed styles. Using a weighted average $Y_{AB} = (1 - \lambda_{AB})Y_A + \lambda_{AB}Y_B$, we interpolate between Y_A and Y_B for a fixed text condition. Figure 6 shows the results on fixed words with interpolation between two writing styles with various λ_{AB} values. One can observe the smooth progression between styles as the mix parameter λ_{AB} increases. This interpolation concept could be a useful tool for generating words of unseen/unknown style, especially if the goal is to create an augmented dataset for training document analysis methods.

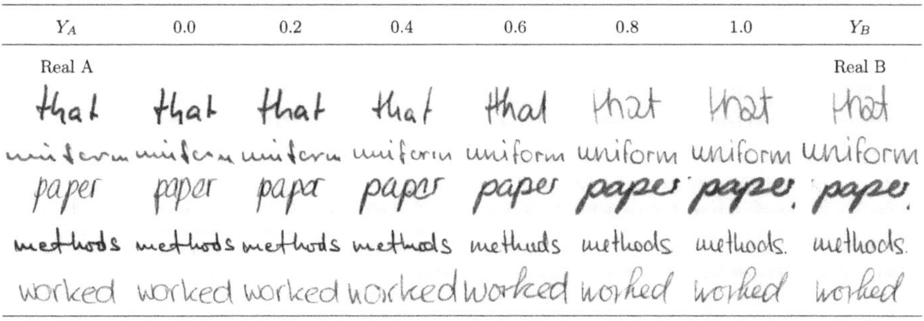

Y_A	0.0	0.2	0.4	0.6	0.8	1.0	Y_B

Fig. 6. Interpolation results between writer styles with various weights.

4.3 Handwriting Text Recognition (HTR)

We evaluate the generated data on the task of Handwriting Text Recognition and assess the usefulness of the data on a standard downstream task. We use the HTR system presented in [31]. Specifically, the used HTR system is a hybrid CNN-LSTM network with a ResNet-like CNN backbone followed by a 3-layers bi-directional LSTM head, trained with Connectionist Temporal Classification (CTC) loss [11]. We followed the modifications proposed in [31] and used a column-wise max-pooling operation between the CNN backbone and the recurrent head, as well as a CTC shortcut of a shallow 1D CNN head. This shortcut module, as described in the initial work, is discarded during testing and is used only for assisting the training procedure. Input word images have a fixed size of 64×256 by performing a padding operation (or resized if they exceed the pre-defined size).

For comparison with the related work on word-image generation, we further evaluate GANwriting and SmartPatch on the HTR task with the same model.

Using the generated images to train an HTR system and then evaluate the trained system on the original test set of real images aims to a multifaceted insight on the quality of the data; Achieving good results in the test set translates to "readable" words (at least in their majority), so that the system can understand the existing characters during training with CTC, as well as to a variability in writing styles, so that the training system could generalize well in the test set of unseen writing styles. The ideal generative model should abide to both these properties and thus can be used to train a well-performing HTR system.

Following the protocol of [25], for this recognition task, we discarded, both from the training and the test set, words containing non-alphanumeric characters, as well as words with more than 10 characters, since the generative models have been trained considering the same setup. We used the generative models to recreate the train set, both in text and in style. The results of this experiment are reported in Table 1, where we present the character error and word error rates (CER/WER) for the initial IAM train set, the recreated sets of the generative models (i.e., GANwriting, SmartPatch and our proposed Diffusion approach), as well as the combination of the original set with each one of the recreated (i.e., with ×2 training images, compared to the initial set). The reported results correspond to the mean value and the standard deviation over 3 different training/evaluation runs for each setup. The following observations can be made:

– The generated synthetic datasets under-perform with respect to the original IAM dataset. However, both GANwriting and SmartPatch approaches lead to a notable decrease in performance, indicating lack of writing style variability. On the other hand, the proposed method achieves considerably low error rates, but not on par with the real data.
– Combining the synthetic datasets with the real IAM train set, the performance is improved compared to training only on the original IAM set, with the exception of GANwriting and the CER metric, which is practically on par with the baseline model.
– SmartPatch, despite visually improving the results of GANwriting, does little to improve the HTR performance.
– The synthetic set, generated by our proposed method, along with the real set, considerably outperforms all other settings and is statistically significant with a p value of 0.035.

Table 1. HTR results, reporting the Character Error Rate (CER) and Word Error Rate (WER). For both metrics, the lower the better.

Training Data	CER (%) ↓	WER (%) ↓
Real IAM	4.86 ± 0.07	14.11 ± 0.12
GANwriting IAM	38.74 ± 0.57	68.47 ± 0.32
SmartPatch IAM	36.63 ± 0.71	65.25 ± 1.02
WordStylist IAM (Ours)	8.80 ± 0.12	21.93 ± 0.17
Real IAM + GANwriting IAM	4.87 ± 0.09	13.88 ± 0.10
Real IAM + SmartPatch IAM	4.83 ± 0.08	13.90 ± 0.22
Real IAM + WordStylist IAM (Ours)	**4.67 ± 0.08**	**13.28 ± 0.20**

4.4 Handwriting Style Evaluation

Qualitative results show that our proposed method is able to nicely capture the style of each writer present in the IAM database. In order to quantify this property, we employ an implicit evaluation via writer identification.

The most straightforward way to address this is via a writer classification formulation. Specifically, to evaluate the generated styles, we finetuned a ResNet18 CNN [12], pre-trained on ImageNet, on the IAM database for the task of writer classification. Then, we use the generated datasets from the three generative methods as test sets and present the obtained accuracy in Table 2. The network manages to successfully classify most of the generated samples from our proposed method with an accuracy of 70.67%, while it fails to recognize classes on samples from the other two methods. This result comes as no surprise since the proposed method learns explicitly the existing styles, while both the GAN-based approaches adapt the style based on a few-shot scheme. Furthermore, we use the features extracted by the model to plot t-SNE embeddings on the different datasets in Fig. 7. In more detail, we used the 512-dimensional feature vector extracted by the second-to-last layer, trying to simulate a style-based representation space. Again, the resulted projection of the data generated by our Diffusion approach appears to be much closer to the real data. On the contrary, the

Table 2. Classification accuracy of a ResNet18 trained for writer identification on real data.

Test Set	Accuracy (%)↑
GANwriting	4.81
SmartPatch	4.09
WordStylist (Ours)	**70.67**

GAN-based methods create "noisy" visualization with no distinct style neighborhoods. In fact, even the proposed method seems to have a similar noisy behavior (in the center of the plot) but to a much lesser extent. This phenomenon is in line with the HTR results, where the diffusion method provided results much closer to the real IAM, but not on par.

As an alternative to the straightforward implementation of writer classification, we also use a classic writer retrieval pipeline consisting of local feature extraction and computing a global feature representation [3, 4, 6]. While the local descriptors can also be trained in a self-supervised [5], we just use SIFT [23] descriptors extracted on SIFT keypoints. The descriptors are normalized using Hellinger normalization [1] (a. k. a. as RootSIFT) and are subsequently jointly whitened and dimensionality-reduced using PCA [4]. The global feature representation is computed using multi-VLAD [3], where the individual VLAD representations use generalized max-pooling [6].

This pipeline needs paragraphs as input in order to gather a sufficient amount of information. To produce synthetic text paragraphs, we paste randomly-selected synthetic words on a blank background, following a similar structure as the printed text of IAM: same number of lines, similar number of characters per line. Thus, no information from the handwritten text is used. Line spacing is constant, and a small randomness is added to word spacing.

We use a leave-one-image-out cross-validation, i. e., each sample is used as query and the results are averaged. As metrics, we give the top-1 accuracy and mean average precision (mAP). For our experiment, we use two paragraphs of 157 writers (IAM + IAM). In subsequent experiments, we replace the second paragraph by the synthesizers (GANWriting, SmartPatch, WordStylist). In this way, the query sample is either an original sample and the closest match should be the synthetic one or vice-versa.

The results, given in Table 3, show little difference between real data (IAM + IAM) and data produced by our method (IAM + WordStylist). Thus, our method produces persistent writing styles that are nearly indistinguishable for the writer retrieval pipeline. It is able to imitate handwriting much better than GANwriting and SmartPatch, which both achieve significantly lower scores in this experiment.

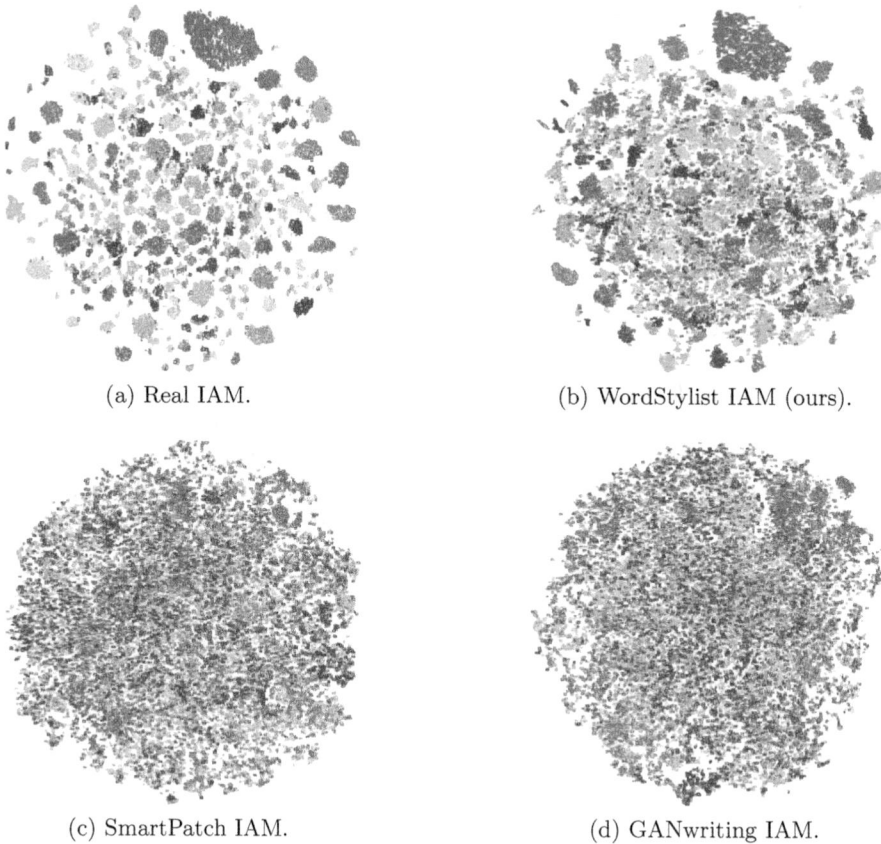

(a) Real IAM.   (b) WordStylist IAM (ours).

(c) SmartPatch IAM.   (d) GANwriting IAM.

Fig. 7. T-SNE projections of the feature vector produced by the ResNet18 trained for writer identification, as described in Sect. 4.4.

5   Limitations and Future Work

Here, we address the limitations of the proposed method that pave the way towards future directions. We identified two main limitations in the proposed method:

– *Style Adaptation:* Our proposed method, contrary to the compared GAN-based methods [15,25], explicitly takes the writing style as an input embedding. This way, the model can learn to recreate such styles accurately. Nonetheless, adaptation to new styles is not straightforward with this pipeline. The interpolation concept is a work-around to generate "new" styles, and can be extended to even interpolating K different styles. Nonetheless, even such ideas do not provide the ability to adapt to a specific given style via few word examples, as done in [15]. An interesting future direction is the projection of different style embeddings to a common style representation

Table 3. Writer retrieval results using a 157 writers subset of IAM.

	Top-1 [%] ↑	mAP [%] ↑
IAM + IAM	97.45	97.61
IAM + GANwriting	3.18	7.23
IAM + SmartPatch	3.18	7.72
IAM + WordStylist (Ours)	**97.13**	**97.84**

space, using deep features extracted by a writer classification model as done in Sect. 4.

- *Sampling Complexity:* Generating realistic examples requires many iterations (timesteps) in the sampling process. To generate a single image requires ∼ 12 sec, when using $T = 600$, making the creation of large-scale datasets impractical. We aim to explore ways to assist the generation of quality images in fewer steps, while also utilizing a more lightweight network to further reduce the time requirements of a single step.
- *Fixed Image Size:* As the generation process is initiated by sampling a latent Gaussian noise of a fixed shape, our proposed method currently generates images of a fixed shape. Generating text images of arbitrary shapes is a possible future direction to explore.

6   Conclusion

We presented WordStylist, a latent diffusion-based system for styled text-to-text-content image generation on word-level. Our model manages to capture the style and content without the use of any adversarial training, text recognition and writer identification. Qualitative and quantitative evaluation results show that our method produces high quality images which outperform significantly the state-of-art systems used for comparison. Also we show that our synthetic word images can be used as extra training data to improve HTR accuracy. The verisimilitude of the synthetic handwriting styles is proven by two experiments. Using a CNN for writer identification, we obtain a classification accuracy of 70% with our synthetic data, while the other generative methods used for comparison do not get higher than 5%. Also, t-SNE projections of the features learned by the CNN exhibit structures very similar to real data in the case of the proposed method only. Moreover, we showed that using a recognized writer retrieval pipeline, there is no significant difference between results on our synthetic data and real data, both having a mAP slightly below 98%. The other generative methods do not perform as well, obtaining mAP below 8%. For future work, we aim to investigate the parameters and sampling of the model. We further plan on extending this work for sentences and whole pages, focusing also on the layout of the document.

References

1. Arandjelović, R., Zisserman, A.: Three things everyone should know to improve object retrieval. In: 2012 IEEE Conference on Computer Vision and Pattern Recognition (CVPR), pp. 2911–2918. Providence, June 2012
2. Bhunia, A.K., Khan, S., Cholakkal, H., Anwer, R.M., Khan, F.S., Shah, M.: Handwriting transformers. In: Proceedings of the IEEE/CVF International Conference on Computer Vision (ICCV), pp. 1086–1094, October 2021
3. Christlein, V., Bernecker, D., Angelopoulou, E.: Writer identification using vlad encoded contour-zernike moments. In: 2015 13th International Conference on Document Analysis and Recognition (ICDAR), pp. 906–910. Nancy, August 2015
4. Christlein, V., Bernecker, D., Hönig, F., Maier, A., Angelopoulou, E.: Writer identification using GMM supervectors and exemplar-svms. Pattern Recogn. **63**, 258–267 (2017)
5. Christlein, V., Gropp, M., Fiel, S., Maier, A.: Unsupervised feature learning for writer identification and writer retrieval. In: 2017 14th International Conference on Document Analysis and Recognition, vol. 01, pp. 991–997. Kyoto (2017)
6. Christlein, V., Maier, A.: Encoding CNN activations for writer recognition. In: 13th IAPR International Workshop on Document Analysis Systems, pp. 169–174. Vienna (2018)
7. Deng, J., Dong, W., Socher, R., Li, L.J., Li, K., Fei-Fei, L.: ImageNet: a large-scale hierarchical image database. In: 2009 IEEE Conference on Computer Vision and Pattern Recognition, pp. 248–255 (2009). https://doi.org/10.1109/CVPR.2009.5206848
8. Dowson, D., Landau, B.: The Fréchet distance between multivariate normal distributions. J. Multivar. Anal. **12**(3), 450–455 (1982)
9. Frolov, S., Hinz, T., Raue, F., Hees, J., Dengel, A.: Adversarial text-to-image synthesis: a review. Neural Netw. **144**, 187–209 (2021)
10. Goodfellow, I., et al.: Generative adversarial nets. In: Ghahramani, Z., Welling, M., Cortes, C., Lawrence, N., Weinberger, K. (eds.) Advances in Neural Information Processing Systems, vol. 27. Curran Associates, Inc. (2014). https://proceedings.neurips.cc/paper_files/paper/2014/file/5ca3e9b122f61f8f06494c97b1afccf3-Paper.pdf
11. Graves, A., Liwicki, M., Fernández, S., Bertolami, R., Bunke, H., Schmidhuber, J.: A novel connectionist system for unconstrained handwriting recognition. IEEE Trans. Pattern Anal. Mach. Intell. **31**(5), 855–868 (2008)
12. He, K., Zhang, X., Ren, S., Sun, J.: Deep residual learning for image recognition. In: Proceedings of the IEEE Conference on Computer Vision and Pattern Recognition, pp. 770–778 (2016)
13. Ho, J., Jain, A., Abbeel, P.: Denoising diffusion probabilistic models. Adv. Neural Inf. Process. Syst. **33**, 6840–6851 (2020)
14. Kang, L., Riba, P., Rusinol, M., Fornés, A., Villegas, M.: Content and style aware generation of text-line images for handwriting recognition. IEEE Trans. Pattern Anal. Mach. Intell. **44**(12), 8846–8860 (2021)
15. Kang, L., Riba, P., Wang, Y., Rusiñol, M., Fornés, A., Villegas, M.: GANwriting: content-conditioned generation of styled handwritten word images. In: Vedaldi, A., Bischof, H., Brox, T., Frahm, J.-M. (eds.) ECCV 2020. LNCS, vol. 12368, pp. 273–289. Springer, Cham (2020). https://doi.org/10.1007/978-3-030-58592-1_17
16. Kingma, D., Salimans, T., Poole, B., Ho, J.: Variational diffusion models. Adv. Neural Inf. Process. Syst. **34**, 21696–21707 (2021)

17. Kingma, D.P., Welling, M.: Auto-encoding variational bayes. In: Bengio, Y., LeCun, Y. (eds.) 2nd International Conference on Learning Representations, ICLR 2014, Banff, AB, Canada, 14–16 April 2014, Conference Track Proceedings (2014)

18. Kingma, D.P., Dhariwal, P.: Glow: generative flow with invertible 1x1 convolutions. Adv. Neural Inf. Process. Syst. **31** (2018)

19. Kong, Z., Ping, W., Huang, J., Zhao, K., Catanzaro, B.: DiffWave: a versatile diffusion model for audio synthesis. In: International Conference on Learning Representations (2020)

20. Lin, T.Y., et al.: Microsoft COCO: common objects in context. In: European Conference on Computer Vision (2014)

21. Lombardi, F., Marinai, S.: Deep learning for historical document analysis and recognition-a survey. J. Imaging **6**(10), 110 (2020)

22. Loshchilov, I., Hutter, F.: Decoupled weight decay regularization. In: International Conference on Learning Representations (2017)

23. Lowe, D.G.: Distinctive image features from scale-invariant keypoints. Int. J. Comput. Vis. **60**(2), 91–110 (2004)

24. Marti, U.V., Bunke, H.: The IAM-database: an English sentence database for offline handwriting recognition. Int. J. Doc. Anal. Recogn. **5**, 39–46 (2002)

25. Mattick, A., Mayr, M., Seuret, M., Maier, A., Christlein, V.: SmartPatch: improving handwritten word imitation with patch discriminators. In: Lladós, J., Lopresti, D., Uchida, S. (eds.) ICDAR 2021. LNCS, vol. 12821, pp. 268–283. Springer, Cham (2021). https://doi.org/10.1007/978-3-030-86549-8_18

26. Mirza, M., Osindero, S.: Conditional generative adversarial nets. arXiv preprint arXiv:1411.1784 (2014)

27. Mittal, G., Engel, J.H., Hawthorne, C., Simon, I.: Symbolic music generation with diffusion models. In: Proceedings of the 22nd International Society for Music Information Retrieval Conference, ISMIR 2021, 7–12 November 2021, pp. 468–475 (2021). https://archives.ismir.net/ismir2021/paper/000058.pdf

28. Nikolaidou, K., Seuret, M., Mokayed, H., Liwicki, M.: A survey of historical document image datasets. Int. J. Doc. Anal. Recogn. (IJDAR) **25**, 305–338 (2022)

29. Pondenkandath, V., Alberti, M., Diatta, M., Ingold, R., Liwicki, M.: Historical document synthesis with generative adversarial networks. In: 2019 International Conference on Document Analysis and Recognition Workshops (ICDARW), vol. 5, pp. 146–151 (2019). https://doi.org/10.1109/ICDARW.2019.40096

30. Ramesh, A., Dhariwal, P., Nichol, A., Chu, C., Chen, M.: Hierarchical text-conditional image generation with CLIP Latents. ArXiv abs/2204.06125 (2022)

31. Retsinas, G., Sfikas, G., Gatos, B., Nikou, C.: Best practices for a handwritten text recognition system. In: Uchida, S., Barney, E., Eglin, V. (eds.) Document Analysis Systems, pp. 247–259. Springer International Publishing, Cham (2022). https://doi.org/10.1007/978-3-031-06555-2_17

32. Rombach, R., Blattmann, A., Lorenz, D., Esser, P., Ommer, B.: High-resolution image synthesis with latent diffusion models. In: Proceedings of the IEEE/CVF Conference on Computer Vision and Pattern Recognition, pp. 10684–10695 (2022)

33. Ronneberger, O., Fischer, P., Brox, T.: U-net: convolutional networks for biomedical image segmentation. In: Navab, N., Hornegger, J., Wells, W.M., Frangi, A.F. (eds.) MICCAI 2015. LNCS, vol. 9351, pp. 234–241. Springer, Cham (2015). https://doi.org/10.1007/978-3-319-24574-4_28

34. Saharia, C., et al.: Photorealistic text-to-image diffusion models with deep language understanding. Adv. Neural Inf. Process. Syst. **35**, 36479–36494 (2022)

35. Fogel, S., Averbuch-Elor, H., Cohen, S., Mazor, S., Litman, R.: ScrabbleGAN: semi-supervised varying length handwritten text generation. In: 2020 IEEE/CVF Conference on Computer Vision and Pattern Recognition (CVPR), pp. 4323–4332 (2020)
36. Sohl-Dickstein, J., Weiss, E., Maheswaranathan, N., Ganguli, S.: Deep unsupervised learning using nonequilibrium thermodynamics. In: International Conference on Machine Learning, pp. 2256–2265. PMLR (2015)
37. Szegedy, C., Vanhoucke, V., Ioffe, S., Shlens, J., Wojna, Z.: Rethinking the inception architecture for computer vision. In: 2016 IEEE Conference on Computer Vision and Pattern Recognition (CVPR), pp. 2818–2826 (2015)
38. Vaswani, A., et al.: Attention is all you need. Adv. Neural Inf. Process. Syst. **30** (2017)
39. Vögtlin, L., Drazyk, M., Pondenkandath, V., Alberti, M., Ingold, R.: Generating synthetic handwritten historical documents with OCR constrained GANs. In: Lladós, J., Lopresti, D., Uchida, S. (eds.) ICDAR 2021. LNCS, vol. 12823, pp. 610–625. Springer, Cham (2021). https://doi.org/10.1007/978-3-030-86334-0_40
40. Zhu, J.Y., Park, T., Isola, P., Efros, A.A.: Unpaired image-to-image translation using cycle-consistent adversarial networks. In: Proceedings of the IEEE International Conference on Computer Vision, pp. 2223–2232 (2017)

Competition

ICDAR 2023 Competition on Video Text Reading for Dense and Small Text

Weijia Wu[1]([✉])([iD]), Yuzhong Zhao[2], Zhuang Li[3], Jiahong Li[3], Mike Zheng Shou[4],
Umapada Pal[5], Dimosthenis Karatzas[6], and Xiang Bai[7]

[1] Zhejiang University, Hangzhou, China
weijiawu@zju.edu.cn
[2] University of Chinese Academy of Sciences, Beijing, China
[3] Kuaishou Technology, Beijing, China
[4] National University of Singapore, Singapore, Singapore
[5] Computer Vision and Pattern Recognition Unit, Indian Statistical Institute,
Chennai, India
[6] Computer Vision Centre, Universitat Autónoma de Barcelona, Barcelona, Spain
[7] Huazhong University of Science and Technology, Wuhan, China

Abstract. Recently, video text detection, tracking and recognition in natural scenes are becoming very popular in the computer vision community. However, most existing algorithms and benchmarks focus on common text cases (*e.g.,* normal size, density) and single scenario, while ignore extreme video texts challenges, *i.e.,* dense and small text in various scenarios. In this competition report, we establish a video text reading benchmark, named DSText, which focuses on dense and small text reading challenge in the video with various scenarios. Compared with the previous datasets, the proposed dataset mainly include three new challenges: 1) Dense video texts, new challenge for video text spotter. 2) High-proportioned small texts. 3) Various new scenarios, *e.g.,* 'Game', 'Sports', etc. The proposed DSText includes 100 video clips from 12 open scenarios, supporting two tasks (*i.e.,* video text tracking (Task 1) and end-to-end video text spotting (Task2)). During the competition period (opened on 15th February, 2023 and closed on 20th March, 2023), a total of 24 teams participated in the three proposed tasks with around 30 valid submissions, respectively. In this article, we describe detailed statistical information of the dataset, tasks, evaluation protocols and the results summaries of the ICDAR 2023 on DSText competition. Moreover, we hope the benchmark will promise the video text research in the community.

Keywords: Video Text Spotting · Small Text · Text Tracking · Dense Text

1 Introduction

Video text spotting [1] has received increasing attention due to its numerous applications in computer vision, *e.g.,* video understanding [2], video retrieval [3], video text translation, and license plate recognition [4], etc. There already exist

© The Author(s), under exclusive license to Springer Nature Switzerland AG 2023
G. A. Fink et al. (Eds.): ICDAR 2023, LNCS 14188, pp. 405–419, 2023.
https://doi.org/10.1007/978-3-031-41679-8_23

Fig. 1. Visualization of DSText. Different from previous benchmarks, DSText focuses on dense and small text challenges.

some video text spotting benchmarks, which focus on easy cases, *e.g.,* normal text size, density in single scenario. ICDAR2015 (Text in Videos) [5], as the most popular benchmark, was introduced during the ICDAR Robust Reading Competition in 2015 focus on wild scenarios: walking outdoors, searching for a shop in a shopping street, etc. YouTube Video Text (YVT) [6] contains 30 videos from YouTube. The text category mainly includes overlay text (caption) and scene text (*e.g.,* driving signs, business signs). RoadText-1K [7] provide 1,000 driving videos, which promote driver assistance and self-driving systems. LSVTD [8] proposes 100 text videos, 13 indoor (*e.g.,* bookstore, shopping mall) and 9 outdoor (*e.g.,* highway, city road) scenarios, and support two languages, *i.e.,* English and Chinese. BOVText [9] establishes a large-scale, bilingual video text benchmark, including abundant text types, *i.e.,* title, caption or scene text.

However, the above benchmarks still suffer from some limitations: 1) Most text instances present normal text size without challenge, *e.g.,* ICDAR2015(video) YVT, BOVText. 2) Sparse text density in single scenario, *e.g.,* RoadText-1k and YVT, which can not evaluate the small and dense text robustness of algorithm effectively. 3) Except for ICDAR2015(video), most benchmarks present unsatisfactory maintenance. YVT, RoadText-1k and BOVText all do not launch a corresponding competition and release open-source evaluation script. Besides, the download links of YVT even have become invalid. The poor maintenance plan is not helpful to the development of video text tasks in the community. To break these limitations, we establish one new benchmark, which focus on dense and small texts in various scenarios, as shown in Fig. 1. The benchmark mainly supports two tasks, *i.e., video text tracking,* and end to end *video text spotting* tasks, includes 100 videos with 56k frames and 671k text instances.

Therefore, we organize the ICDAR 2023 Video Text Reading competitive for dense and small text, which generates a large-scale video text database, and proposes video text tracking, spotting tasks, and corresponding evaluation methods. This competition can serve as a standard benchmark for assessing the robustness of algorithms that are designed for video text spotting in complex natural scenes, which is more challenging. The proposed competition and dataset will enhance the related direction (Video OCR) of the ICDAR community from two main aspects:

Table 1. Statistical Comparison. 'Box Type', 'Text Area' denote detection box annotation type and average area of text while the shorter side of image is 720 pixels. 'Text Density' refers to the average text number per frame. The proposed DSText presents more small and dense texts.

Dataset	Video	Frame	Text	Box Type	Text Area (# pixels)	Text Density	Supported Scenario (Domain)
YVT [6]	30	13k	16k	Upright	8,664	1.15	Cartoon, Outdoor(supermarket, shopping street, driving...)
ICD15 VT [10]	51	27k	144k	Oriented	5,013	5.33	Driving, Supermarket, Shopping street...
RoadText-1K [7]	1k	300k	1.2m	Upright	2,141	0.75	Driving
LSVTD [8]	100	66k	569k	Oriented	2,254	5.52	Shopping mall, Supermarket, Hotel...
BOVText [9]	2k	1.7m	8.8m	Oriented	10,309	5.12	Cartoon, Vlog, Travel, Game, Sport, News ...
DSText	100	56k	671k	Oriented	**1,984**	**23.5**	Driving, Activity, Vlog, Street View (indoor), Street View (outdoor), Travel, News, Movie, Cooking

- Compared to the current existing video text reading datasets, the proposed DSText has some special features and challenges, including 1) Abundant scenarios, 2) higher proportion of small text, 3) dense text distribution. Table 1, Figs. 2, 3 and 5 present detailed statistical comparison and analysis.
- The competition supports two tasks: video text tracking and end-to-end video text spotting. And we provide comprehensive evaluation metrics, including ID_P, ID_R, ID_{F1} [11], MOTA, and MOTP. These metrics are widely used on previous video text benchmarks, such as ICDAR2015 [10,12]. We are proud to report the successful completion of the competition, which has garnered over 25 submissions and attracted wide interest. The submissions have inspired new insights, ideas, and approaches, which promise to advance the state of the art in video text analysis.

2   Competition Organization

ICDAR 2023 video text reading competition for dense and small text is organized by a joint team, including Zhejiang University, University of Chinese Academy of Sciences, Kuaishou Technology, National University of Singapore, Computer Vision and Pattern Recognition Unit, Computer Vision Centre, Universitat Autónoma de Barcelona, and Huazhong University of Science and Technology. And we organize the competition on the Robust Reading Competition Website[1], where provide corresponding download links of the datasets, and user interfaces for participants and submission page for their results.

---

[1] https://rrc.cvc.uab.es/?ch=22&com=introduction.

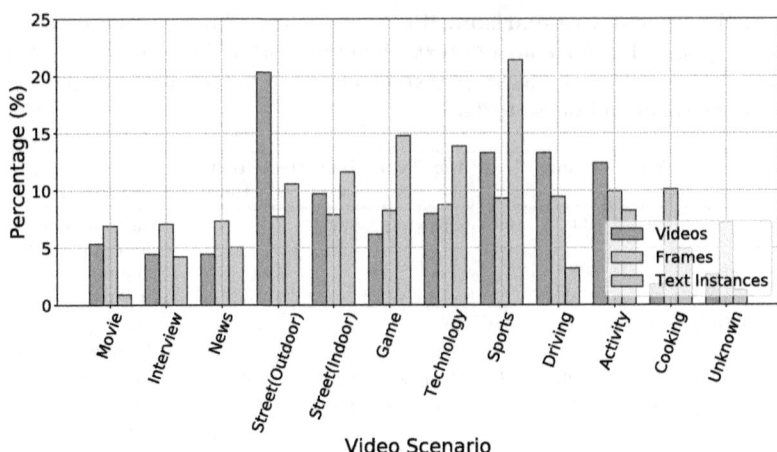

Fig. 2. The Data Distribution for 12 Open Scenarios. "%" denotes the percentage of each scenario data over the whole data.

After removing duplicate entries, we received 30 valid submissions for the three tasks. These came from a total of 24 teams, encompassing both research communities and industry sectors.

3   Dataset

3.1   Dataset and Annotations

Dataset Source. The videos in DSText are collected from three parts: 1) 30 videos sampled from the large-scale video text dataset BOVText [9]. BOVText, as the largest video text dataset with various scenarios, includes a mass of small and dense text videos. We select the top 30 videos with small and dense texts via the average text area of the video and the average number of text per frame. 2) 10 videos for driving scenario are collected from RoadText-1k [7]. As shown in Fig. 1, RoadText-1k contains abundant small texts, thus we also select 10 videos to enrich the driving scenario. 3) 60 videos for street view scenes are collected from YouTube. Except for BOVText and RoadText-1k, we also obtain 60 videos with dense and small texts from YouTube, which mainly cover street view scenarios. Therefore, we obtain 100 videos with 56k video frames, as shown in Table 1. Then the dataset is divided into two parts: the training set with 29,107 frames from 50 videos, and the testing set with 27,234 frames from 50 videos.

Annotation. For these videos from BOVText, we just adopt the original annotation, which includes four kinds of description information: the rotated bounding box of detection, the tracking identification(ID) of the same text, the content of the text for recognition, the category of text, *i.e.*, caption, title, scene text, or others. As for others from RoadText-1k and YouTube, we hire a professional

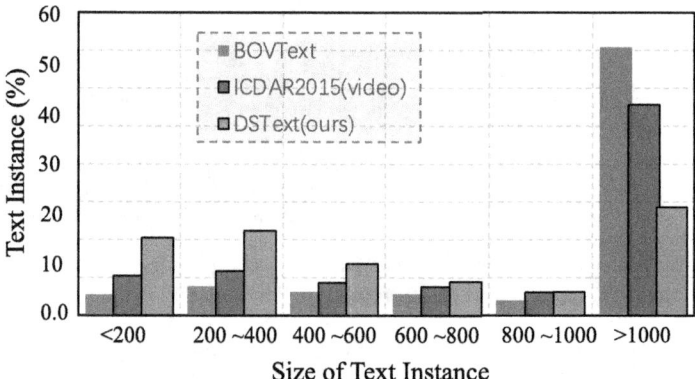

Fig. 3. The distribution of different text size range on different datasets "%"
denotes the percentage of text size region over the whole data. Text area (# pixels) is
calculated while the shorter side of the image is 720 pixels.

annotation team to label each text for each frame. The annotation format is the
same as BOVText. One mentionable point is that the videos from RoadText-
1k only provides the upright bounding box (two points), thus we abandon the
original annotation and annotate these videos with the oriented bounding box.
Due to the structure of the video source, it is not allowed to use BOVText and
RoadText-1k as extra data for training in this competition. As a *labor-intensive*
job, the whole labeling process takes **30** men in one month, *i.e.,* around **4,800**
man-hours, to complete the 70 video frame annotations. As shown in Fig. 6, it
is quite time-consuming and expensive to annotate a mass of text instances at
each frame.

3.2 Dataset Comparison and Analysis

The statistical comparison and analysis are presented by three figures and one
table. Table 2 presents an overall comparison for the basic information, *e.g.,*
number of video, frame, text, and supported scenarios. In comparison with previous works, the proposed DSText shows the denser text instances density per
frame (*i.e.,* average 23.5 texts per frame) and smaller text size (*i.e.,* average
1,984 pixels area of texts).

Video Scenario Attribute. As shown in Fig. 2, we present the distribution
of video, frame, and frame of 11 open scenarios and an "Unknown" scenario on
DSText. 'Street View (Outdoor)' and 'Sport' scenarios present most video and
text numbers, respectively. And the frame number of each scenario is almost the
same. We also present more visualizations for 'Game', 'Driving', 'Sports' and
'Street View' in Fig. 4.

Fig. 4. More Qualitative Video Text Visualization of DSText. DSText covers small and dense texts in various scenarios, which is more challenging.

Higher Proportion of Small Text. Figure 3 presents the proportion of different text areas. The proportion of big text (more than $1,000$ pixel area) on our DSText is less than that of BOVText and ICDAR2015(video) with at least 20%. Meanwhile, DSText presents a higher proportion for small texts (less 400 pixels) with around 22%. As shown in Table 1, RoadText-1k [7] and LSVTD [8] also show low average text area, but their text density is quite sparse (only 0.75 texts and 5.12 per frame), and RoadText-1k only focuses on the driving domain, which limits the evaluation of other scenarios.

Dense Text Distribution. Figure 5 presents the distribution of text density at each frame. The frame with more than 15 text instances occupies 42% in our dataset, at least 30% improvement than the previous work, which presents more dense text scenarios. Besides, the proportion of the frame with less 5 text instances is just half of the previous benchmarks, *i.e.,* BOVText, and ICDAR2015(video). Therefore, the proposed DSText shows the challenge of dense text tracking and recognition. More visualization can be found in Fig. 4 (Visualization for various scenarios) and Fig. 6 (Representative case with around 200 texts per frame).

WordCloud. We also visualize the word cloud for text content in Fig 5. All words from annotation must contain at least 3 characters, we consider the words less four characters usually are insignificant, *e.g.,* 'is'.

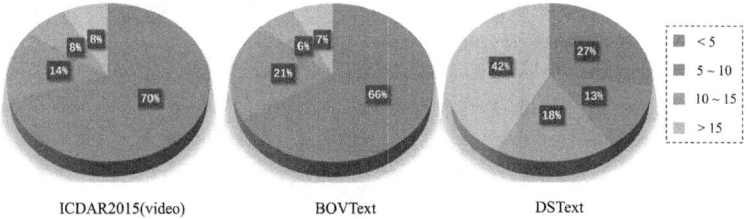

Fig. 5. Comparison for frame percentage of different text numbers. "%" denotes the percentage of the corresponding frame over the whole data.

4   Tasks and Evaluation Protocols

The competition include two tasks: 1) the video text tracking, where the objective is to localize and track all words in the video sequences. and 2) the end-to-end video text spotting: where the objective is to localize, track and recognize all words in the video sequence.

Task 1: Video Text Tracking. In this task, all the videos (50 train videos and 50 test videos) will be provided as MP4 files. Similar to ICDAR2015 Video [10], the ground truth will be provided as a single XML file per video. A single compressed (zip or rar) file should be submitted containing all the result files for all the videos of the test set.

The task requires one network to detect and track text over the video sequence simultaneously. Given an input video, the network should produce two results: a rotated detection box, and tracking ids of the same text. For simplicity, we adopt the evaluation method from the ICDAR 2015 Video Robust Reading competition [10] for the task. The evaluation is based on an adaptation of the MOTChallenge [11] for multiple object tracking. For each method, MOTChallenge provides three different metrics: the Multiple Object Tracking Precision (MOTP), the Multiple Object Tracking Accuracy (MOTA), and the IDF1. See the 2013 competition report [12] and MOTChallenge [11] for details about these metrics. In our competition, we reuse the evaluation scripts from the 2015 video text reading competition [10], and transfer the format of annotation to the same as that of the ICDAR2015 Video (Fig. 7).

Task 2: End-to-End Video Text Spotting. Video Text Spotting (VTS) task that requires simultaneously detecting, tracking, and recognizing text in the video. The word recognition performance is evaluated by simply whether a word recognition result is completely correct. And the word recognition evaluation is case-insensitive and accent-insensitive. All non-alphanumeric characters are not taken into account, including decimal points, such as '1.9' will be transferred to '19' in our GT. Similarly, the evaluation method (*i.e.,* ID_{F1}, MOTA and MOTP) from the ICDAR 2015 Robust Reading competition is also adopted for the task. In the training set, we provide the detection coordinates, tracking id, and transcription result.

Fig. 6. One Case for around 200 Texts per Frames. DSText includes huge amounts of small and dense text scenarios, which is a new challenge.

Fig. 7. Wordcloud visualizations for DSText.

Note: From 2020, the ICDAR 2015 Robust Reading competition online evaluation[2] has updated the evaluation method, and added one new metric, *i.e.,* ID metrics (ID_{F1}) [13,14]. Similarly, we adopted the updated metric for the two tasks.

5   Baseline: TransDETR

To help participants more easily engage in our competition, we have also provided corresponding baseline algorithms in the competition website[3], *i.e.,* TransDETR [15]. where including the corresponding training and inference code for the competition. TransDETR is a novel, simple and end to end video text DEtection, Tracking, and Recognition framework (TransDETR), which views the video text spotting task as a direct long-sequence temporal modeling problem.

---

[2] https://rrc.cvc.uab.es/?ch=3&com=evaluation&task=1.

[3] https://rrc.cvc.uab.es/?ch=22&com=downloads.

Table 2. Task 1: Video Text Tracking Results. - denotes missing descriptions in affiliations. 'M-M', 'P-M', 'M-L' denotes 'Mostly Matched', 'Partially Matched' and 'Mostly Lost', respectively.

User ID	Rank	MOTA	MOTP	ID_{F1}/%	M-M	P-M	M-L	Affiliations
TencentOCR	1	62.56%	79.88%	75.87%	8114	1800	2663	Tencent
DA	2	50.52%	78.33%	70.99%	7121	2405	3051	Guangzhou Shiyuan Electronic Technology Company Limited
Tianyu Zhang	3	43.52%	78.15%	62.27%	4980	2264	5333	AI Lab, Du Xiaoman Financial
Liu Hongen	4	36.87%	79.24%	48.99%	2123	3625	6829	Tianjin Univeristy
Yu Hao	5	31.01%	78.00%	50.39%	2361	1767	8449	-
Hu Jijin	6	28.92%	78.46%	43.96%	1385	1186	10006	Beijing University of Posts & Telecommunications
CccJ	7	27.55%	78.40%	44.28%	1583	1103	9891	CQUT
MiniDragon	8	25.75%	74.03%	50.22%	3302	2806	6469	-
FanZhengDuo	9	23.41%	75.54%	49.66%	5216	3578	3783	-
zjb	10	19.85%	71.98%	39.87%	2815	3354	6408	-
dunaichao	11	19.84%	73.82%	31.18%	924	1765	9888	China Mobile Communications Research Institute
JiangQing	12	13.83%	75.75%	58.41%	6924	2622	3031	South China University of Technology; Shanghai AI Laboratory; KingSoft Office CV&D Department
Kebin Liu	13	7.49%	75.62%	45.68%	5403	3835	3339	Beijing University of Posts and Telecommunications
TungLX	14	0%	0%	0%	0	0	0	-

6 Submissions

The result for Task 1 and Task 2 are presented on Table 2 and Table 3, respectively.

6.1 Top 3 Submissions in Task 1

Tencent-OCR team. The top 1 solution method follows the framework of Cascade Mask R-CNN [16]. Multiple backbones including HRNet [17] and Intern-Image [18] are used to enhance the performance. On the text tracking task, the team designed four different metrics to compare the matching similarity between the current frame detection box and the existing text trajectory, *i.e.*, box IoU, text content similarity, box size proximity and text geometric neighborhood relationship measurement. These matching confidence scores are used as a weighted sum for the matching cost between the currently detected box and tracklet. When there is a time difference between the current detection box and

Table 3. Task2: End-to-End Video Text Spotting Results. - denotes missing descriptions in affiliations. 'M-M', 'P-M', 'M-L' denotes 'Mostly Matched', 'Partially Matched' and 'Mostly Lost', respectively.

User ID	Rank	MOTA	MOTP	ID_{F1}/%	M-M	P-M	M-L	Affiliations
TencentOCR	1	22.44%	80.82%	56.45%	5062	1075	6440	Tencent
DA	2	10.51%	78.97%	53.45%	4629	1392	6556	Guangzhou Shiyuan Electronic Technology Company Limited
dunaichao	3	5.54%	74.61%	24.25%	528	946	11103	China Mobile Communications Research Institute
cnn_lin	4	0%	0%	0%	0	0	0	South China Agricultural University
Hu Jijin	5	0%	0%	0%	0	0	0	Beijing University of Posts and Telecommunications
XUE CHUHUI	6	0%	0%	0%	0	0	0	–
MiniDragon	7	-25.09%	74.95%	26.38%	1388	1127	10062	–
JiangQing	8	−27.47%	76.59%	43.61%	4090	1471	7016	South China University of Technology; Shanghai AI Laboratory; KingSoft Office CV&D Department
Tianyu Zhang	9	−28.58%	80.36%	26.20%	1556	543	10478	AI Lab, Du Xiaoman Financial

the last appearance of the tracklet, the IoU and box size metrics are divided by the corresponding frame number difference to prioritize matching with the latest detection box in the trajectory set. They construct a cost matrix for each detected box and existing trajectory in each frame, where the Kuhn-Munkres algorithm is used to obtain matching pairs. When the metrics are less than a certain threshold, their corresponding costs are set to 0. Finally, they perform grid search to find better hyperparameters. Referring to ByteTrack [19], boxes with high detection/recognition scores are prioritized for matching, followed by boxes with lower detection/recognition scores. Each box that is not linked to an existing trajectory is only considered as a starting point for a new trajectory when its detection/recognition score is high enough. Finally, we removed low-quality trajectories with low text confidence scores and noise trajectories with only one detection box.

Strong data augmentation strategies are adopted such as photometric distortions, random motion blur, random rotation, random crop, and random horizontal flip. And IC13 [12], IC15 [5], IC15 Video [5] and Synth800k [20] are involved during the training phase. Furthermore, they treat non-alphanumeric characters as negative samples and regard text instances that are labeled '##DONT#CARE##' as ignored ones during the training phase. In the inference phase, they use multiple resolutions of 600, 800, 1000, 1333, 1666, and 2000.

Guangzhou Shiyuan Technology Team. The team utilized Mask R-CNN [21] and DBNet [22] as their base architectures. These were trained separately, and their prediction polygons were fused through non-maximum suppression. For the tracking stage, VideoTextSCM [23] was adopted, with Bot-SORT [24] replacing the tracker in the VideoTextSCM model. Bot-SORT is an enhanced multi-object tracker that leverages MOT bag-of-tricks to achieve robust association. It combines the strengths of motion and appearance information, and also incorporates camera-motion compensation and a more accurate Kalman filter state vector. COCO-Text [25], RCTW17 [26], ArT [27], LSVT [28], LSVTD [8], as the public datasets, are used in the training stage. RandomHorizontalFlip, RandomRotate, ColorJitter, MotionBlur and GaussNoise were used for data augmentation.

AI Lab, Du Xiaoman Financial. The team selected TransDETR [15] as the baseline and employed the public datasets COCO-Text V2.0 [25] and SynthText [20] as the pre-training data. To enhance the model's capacity to detect small texts, additional small texts were added to the SynthText images. Furthermore, the HRNet [17] was employed as the new backbone, which demonstrated superiority in identifying faint text objects. The team modified the original hyper-parameters of TransDETR to detect more texts from a single frame. When loading the training data, the maximum number of text instance queries of the Transformer module is set to 400.

6.2 Top 3 Submissions in Task 2

Tencent-OCR Team. To enable end-to-end video text spotting, two methods, namely Parseq [29] and ABINet [30], were utilized in the recognition stage. Both methods were trained on a dataset of 20 million samples, extracted from various open-source datasets, including ICDAR-2013 [12], ICDAR-2015 [5], COCO-Text [25], SynthText [20], among others.

During the end-to-end text spotting stage, different recognition methods are applied to predict all the detected boxes of a trajectory. The final text result corresponding to the trajectory is selected based on confidence and character length. Trajectories with low-quality text results, indicated by low scores or containing only one character, are removed.

Guangzhou Shiyuan Technology Team. The text tracking task was addressed in a similar manner to Task 1. To recognize text, the PARSeq method was employed, which involves learning an ensemble of internal autoregressive (AR) language models with shared weights using Permutation Language Modeling. This approach unifies context-free non-AR and context-aware AR inference, along with iterative refinement using bidirectional context. The recognition model was trained using several extra public datasets, including COCO-Text [25], RCTW17 [26], ArT [27], LSVT [28], and LSVTD [8].

China Mobile Communications Research Institute Team. The team used CoText [31] as the baseline and utilized ABINet [30] to enhance the recognition head. The Cotext model was trained using the ICDAR2015 [5], ICDAR2015 Video [5], and ICDAR2023 DSText datasets for text detection and tracking. The recognition part employed a pretrained ABINet model based on the MJSynth and SynthText [20] datasets.

7 Discussion

Text tracking Task. In this task, most participants firstly employ the powerful backbone to enhance the performance, *e.g.,* HRNet, Res2Net and SENet. With multiple backbones, TencentOCR team achieves the best score in three main metrics, *i.e.,* MOTA, MOTP, and $ID_{F1}/\%$, as shown in Table 2. For the text tracking, based on ByteTrack, the team designed four different metrics to compare the matching similarity between the current frame detection box and the existing text trajectory, i.e., box IoU, text content similarity, box size proximity and text geometric neighborhood relationship. To further enhance the performance of result, most participants use the various data augmentations, *e.g.,* random motion blur, random rotation, random crop. Besides, various public datasets, *e.g.,* COCO-Text [25], RCTW17 [26], ArT [27], LSVT [28], and LSVTD [8] are used for joint training.

End-to-End Video Text Spotting Task. To enhance the end-to-end text spotting, most participants adopted the advanced recognition models, *i.e.,* Parseq [29] and ABINet [30]. Large synthetic datasets (*e.g.,* SynthText [20]), firstly are used to pretrain the model, and then further finetuned on the released training dataset (DSText). With various data augmentation, large public datasets, powerful network backbones and model ensembles, TencentOCR team achieves the best score in three main metrics, as shown in Table 3.

Overall, while many participants implemented various improvement techniques such as using extra datasets and data augmentation, the majority of their results were unsatisfactory, with MOTA scores below 25% and ID_{F1} scores below 70%. As a result, there is still a significant amount of room for improvement in this benchmark and many technical challenges to overcome. It is worth mentioning that many of the top ranking methods utilize an ensemble of multiple models and large public datasets to enhance their performance. However, these pipelines tend to be complex and the corresponding inference speeds are slow. Simplifying the pipeline and accelerating inference are also important considerations for the video text spotting task. Additionally, it is noteworthy that many of the submitted methods adopted different ideas and strategies, providing the community with new insights and potential solutions. We expect that more innovative approaches will be proposed following this competition.

8 Conclusion

Here, we present a new video text reading benchmark, which focuses on dense and small video text. Compare with the previous datasets, the proposed dataset mainly includes two new challenges for dense and small video text spotting. High-proportioned small texts are a new challenge for the existing video text methods. Meanwhile, we also organize the corresponding competition on Robust Reading

Competition Website[4], where we received around 30 valid submissions from 24 teams. These submissions will provide the community with new insights and potential solutions. Overall, we believe and hope the benchmark, as one standard benchmark, develops and improve the video text tasks in the community.

9   Potential Negative Societal Impacts and Solution

Similar to BOVText [9], we blur the human faces in DSText with two steps. Firstly, detecting human faces in each frame with *face recognition*[5] - an easy-to-use face recognition open source project with complete development documents and application cases. Secondly, after obtaining the detection box, we blur the face with Gaussian Blur operation in OpenCV[6].

10   Competition Organizers

The benchmark is mainly done by Weijia Wu and Yuzhong Zhao, while they are research interns at Kuaishou Technology. The establishment of the benchmark is supported by the annotation team of Kuaishou Technology. Prof. Xiang Bai at Huazhong University of Science and Technology, Prof. Dimosthenis Karatzas at the Universitat Autónoma de Barcelona, Prof. Umapada Pal at Indian Statistical Institute, and Asst Prof. Mike Shou at the National University of Singapore, as the four main supervisors, provide many valuable suggestions and comments, *e.g.,* annotation format suggestion from Prof. Xiang Bai, competition schedule plan from Prof. Dimosthenis Karatzas, submission plan and suggestions for the proposal from Prof.Umapada Pal, and statistical analysis from Asst Prof. Mike Shou. Therefore, our team mainly includes eight people from seven institutions.

Acknowledgements. This competition is supported by the National Natural Science Foundation (NSFC#62225603).

References

1. Yin, X.-C., Zuo, Z.-Y., Tian, S., Liu, C.-L.: Text detection, tracking and recognition in video: a comprehensive survey. IEEE Trans. Image Process. **25**(6), 2752–2773 (2016)
2. Srivastava, N., Mansimov, E., Salakhudinov, R.: Unsupervised learning of video representations using lstms. In: International Conference on Machine Learning, pp. 843–852 (2015)
3. Dong, J., et al.: Dual encoding for video retrieval by text. IEEE Trans. Pattern Anal. Mach. Intell. **44**(8), 4065–4080 (2021)

---

[4] https://rrc.cvc.uab.es/?ch=22&com=introduction.
[5] https://github.com/ageitgey/face_recognition.
[6] https://www.tutorialspoint.com/opencv/opencv_gaussian_blur.htm.

4. Anagnostopoulos, C.-N.E., Anagnostopoulos, I.E., Psoroulas, I.D., Loumos, V., Kayafas, E.: License plate recognition from still images and video sequences: a survey. IEEE Trans. Intell. Transp. Syst. **9**(3), 377–391 (2008)
5. Karatzas, D., et al.: Competition on robust reading. IEEE Int. Conf. Doc. Anal. Recogn. **2015**, 1156–1160 (2015)
6. Nguyen, P.X., Wang, K., Belongie, S.: Video text detection and recognition: dataset and benchmark. In: IEEE Winter Conference on Applications of Computer Vision, pp. 776–783 (2014)
7. Reddy, S., Mathew, M., Gomez, L., Rusinol, M., Karatzas, D., Jawahar, C.: Roadtext-1k: text detection & recognition dataset for driving videos. In: IEEE International Conference on Robotics and Automation, pp. 11 074–11 080 (2020)
8. Cheng, Z., Lu, J., Niu, Y., Pu, S., Wu, F., Zhou, S.: You only recognize once: towards fast video text spotting. In: ACM International Conference on Multimedia, pp. 855–863 (2019)
9. Wu, W., et al.: A bilingual, Openworld video text dataset and end-to-end video text spotter with transformer. In: Thirty-fifth Conference on Neural Information Processing Systems Datasets and Benchmarks Track (Round 2) (2021)
10. Zhou, X., Zhou, S., Yao, C., Cao, Z., Yin, Q.: Icdar 2015 text reading in the wild competition, arXiv preprintarXiv:1506.03184 (2015)
11. Dendorfer, P., et al.: Cvpr19 tracking and detection challenge: how crowded can it get? arXiv preprintarXiv:1906.04567 (2019)
12. Karatzas, D., et al.: Icdar,: robust reading competition. In: 2013 12th International Conference on Document Analysis and Recognition, vol. 2013, pp. 1484–1493. IEEE (2013)
13. Li, Y., Huang, C., Nevatia, R.: Learning to associate: Hybridboosted multi-target tracker for crowded scene. In: IEEE Conference on Computer Vision and Pattern Recognition, vol. 2009, pp. 2953–2960. IEEE (2009)
14. Ristani, E., Solera, F., Zou, R., Cucchiara, R., Tomasi, C.: Performance measures and a data set for multi-target, multi-camera tracking. In: Workshops of European Conference on Computer Vision, pp. 17–35 (2016)
15. Wu, W., et al.: End-to-end video text spotting with transformer, arXiv preprintarXiv:2203.10539, (2022)
16. Cai, Z., Vasconcelos, N.: Cascade r-cnn: delving into high quality object detection. In: Proceedings of the IEEE Conference on Computer Vision and Pattern Recognition, pp. 6154–6162 (2018)
17. Wang, J., et al.: Deep high-resolution representation learning for visual recognition. IEEE Trans. Pattern Anal. Mach. Intell. **43**(10), 3349–3364 (2020)
18. Wang, W., et al.: Internimage: exploring large-scale vision foundation models with deformable convolutions, arXiv preprintarXiv:2211.05778 (2022)
19. Zhang, Y., et al.: Bytetrack: multi-object tracking by associating every detection box. In: Computer Vision-ECCV,: 17th European Conference, Tel Aviv, Israel, 23–27 October 2022, Proceedings, Part XXII. Springer vol. 2022, pp. 1–21 (2022). https://doi.org/10.1007/978-3-031-20047-2_1
20. Gupta, A., Vedaldi, A., Zisserman, A.: Synthetic data for text localisation in natural images. In: IEEE Conference on Computer Vision and Pattern Recognition, pp. 2315–2324 (2016)
21. He, K., Gkioxari, G., Dollár, P., Girshick, R.: Mask r-cnn. In: Proceedings of the IEEE International Conference on Computer Vision, pp. 2961–2969 (2017)
22. Liao, M., Wan, Z., Yao, C., Chen, K., Bai, X.: Real-time scene text detection with differentiable binarization. In: Proceedings of the AAAI Conference on Artificial Intelligence, vol. 34, no. 07, pp. 11 474–11 481 (2020)

23. Gao, Y., et al.: Video text tracking with a spatio-temporal complementary model. IEEE Trans. Image Process. **30**, 9321–9331 (2021)
24. Aharon, N., Orfaig, R., Bobrovsky, B.-Z.: Bot-sort: robust associations multi-pedestrian tracking, arXiv preprintarXiv:2206.14651 (2022)
25. Veit, A., Matera, T., Neumann, L., Matas, J., Belongie, S.: Coco-text: dataset and benchmark for text detection and recognition in natural images, arXiv preprintarXiv:1601.07140 (2016)
26. Shi, B., et al.: Icdar2017 competition on reading Chinese text in the wild (rctw-17). In: 14th IAPR International Conference on Document Analysis and Recognition (ICDAR), vol. 1, pp. 1429–1434. IEEE (2017)
27. Chng, C.K., et al.: Icdar2019 robust reading challenge on arbitrary-shaped text-RRC-art. In: 2019 International Conference on Document Analysis and Recognition (ICDAR), pp. 1571–1576. IEEE (2019)
28. Sun, Y., Liu, J., Liu, W., Han, J., Ding, E., Liu, J.: Chinese street view text: large-scale Chinese text reading with partially supervised learning. In: Proceedings of the IEEE/CVF International Conference on Computer Vision, pp. 9086–9095 (2019)
29. autista, D., Atienza, R.: Scene text recognition with permuted autoregressive sequence models. In: Avidan, S., Brostow, G., Cissé, M., Farinella, G.M., Hassner, T. (eds.) Computer Vision - ECCV 2022. ECCV 2022. LNCS, vol. 13688, pp. 178–196. Springer, Cham (2022).
30. Fang, S., Xie, H., Wang, Y., Mao, Z., Zhang, Y.: Read like humans: autonomous, bidirectional and iterative language modeling for scene text recognition. In: Proceedings of the IEEE/CVF Conference on Computer Vision and Pattern Recognition, pp. 7098–7107 (2021)
31. Wu, W., et al.: Real-time end-to-end video text spotter with contrastive representation learning, arXiv preprintarXiv:2207.08417 (2022)

ICDAR 2023 Competition on Document UnderstanDing of Everything (DUDE)

Jordy Van Landeghem[1,2(✉)], Rubèn Tito[5], Łukasz Borchmann[3], Michał Pietruszka[3,6], Dawid Jurkiewicz[3,7], Rafał Powalski[8], Paweł Józiak[3,4], Sanket Biswas[5], Mickaël Coustaty[9], and Tomasz Stanisławek[3,4]

[1] KU Leuven, Leuven, Belgium
[2] Contract.fit, Brussels, Belgium
jordy@contract.fit
[3] Snowflake, Bozeman, USA
tomasz.stanislawek@snowflake.com
[4] Warsaw University of Technology, Warsaw, Poland
[5] Computer Vision Center, Universitat Autónoma de Barcelona, Barcelona, Spain
[6] Jagiellonian University, Kraków, Poland
[7] Adam Mickiewicz University, Poznań, Poland
[8] Instabase, San Francisco, USA
[9] University of La Rochelle, La Rochelle, France

Abstract. This paper presents the results of the ICDAR 2023 competition on Document UnderstanDing of Everything. DUDE introduces a new dataset comprising 5 K visually-rich documents (VRDs) with 40 K questions with novelties related to types of questions, answers, and document layouts based on **multi-industry**, **multi-domain**, and **multi-page** VRDs of various origins and dates. The competition was structured as a single task with a multi-phased evaluation protocol that assesses the few-shot capabilities of models by testing generalization to previously unseen questions and domains, a condition essential to business use cases prevailing in the field. A new and independent diagnostic test set is additionally constructed for fine-grained performance analysis. A thorough analysis of results from different participant methods is presented. Under the newly studied settings, current state-of-the-art models show a significant performance gap, even when improving visual evidence and handling multi-page documents. We conclude that the DUDE dataset proposed in this competition will be an essential, long-standing benchmark to further explore for achieving improved generalization and adaptation under low-resource fine-tuning, as desired in the real world.

1 Introduction

Document UnderstanDing of Everything (DUDE) is a concept rooted in both machine learning and philosophy, seeking to *expand* the boundaries of document AI systems by creating highly challenging datasets that encompass a diverse range of topics, disciplines, and complexities. Inspired by the philosophical 'Theory of Everything', which aims to provide a comprehensive explanation of the nature of reality, DUDE endeavors to stimulate the development of AI models that can effectively comprehend, analyze, and respond to *any* question on *any* complex document.

G. A. Fink et al. (Eds.): ICDAR 2023, LNCS 14188, pp. 420–434, 2023.
https://doi.org/10.1007/978-3-031-41679-8_24

Incorporating philosophical perspectives into DUDE enriches the approach by engaging with fundamental questions about knowledge, understanding, and the nature of documents. By addressing these dimensions, researchers can develop AI systems that not only exhibit advanced problem-solving skills but also demonstrate a deeper understanding of the context, nuances, and implications of the information they process.

Over the past few years, the field of Document Analysis and Recognition (DAR) has embraced multi-modality with contributions from both Natural Language Processing (NLP) and Computer Vision (CV). This has given rise to Document Understanding (DU) as the all-encompassing solution [1, 10, 23] for handling Visually Rich Documents (VRDs), where layout and visual information is decisive in understanding a document.

This umbrella term subsumes multiple subtasks ranging from key-value information extraction (KIE) [12, 28], document layout analysis (DLA) [36], visual question answering (VQA) [20, 32], table recognition [13, 24], and so on. For each of these subtasks, influential challenges have been proposed, e.g., the ICDAR 2019 Scene Text VQA [2, 3] and ICDAR 2021 Document VQA (DocVQA) [21, 32] challenges, which in turn have generated novel ideas that have impacted the new wave of architectures that are currently transforming the DAR field.

Nevertheless, we argue that the DAR community must encompass the future challenges (multi-domain, multi-task, multi-page, low-resource settings) that naturally juxtapose the previous competitions with pragmatic feedback attained via its business-driven applications.

Challenge Objectives. We aim to support the emergence of models with strong multi-domain layout reasoning abilities by adopting a diversified setting where multiple document types with different properties are present (Fig. 2). Moreover, a low-resource setting (number of samples) is assumed for every domain provided, which formulated as a DocVQA competition allows us to measure progress with regard to the desired generalization (Sect. 2). Additionally, we strive for the development of confidence estimation methods that can not only improve predictive performance but also adjust the calibration of model outputs, leading to more practical and reliable DU solutions.

We believe that DUDE's emphasis on task adaptation and the capability of handling a wide range of document types, layouts, and complexities will encourage researchers to push the boundaries of current DU techniques, fostering innovation in areas such as multi-modal learning, transfer learning, and zero-shot generalization.

Challenge Contributions. DUDE answers the call for measuring improvements closer to the real-world applicability of DU models. By design of the dataset and competition, participants were forced to make novel contributions in order to make a significant impact on the DU task. Competitors showcased intriguing model extensions, such as combining models that learn strong document representations with the strengths of recent large language or vision-language models (ChatGPT [4] and BLIP2 [16,17]) to better understand questions and extract information from a document context more effectively. HiVT5 + modules extended Hi-VT5 [31] with token/object embeddings for various DU subtasks, while MMT5 employed a two-stage pre-training process and multiple objectives to enhance performance. These innovative extensions highlight the ingenuity in addressing the complex challenges of document understanding.

2   Motivation and Scope

We posit that progress in DU is determined not only by the improvements in each of its related predecessor fields (CV, NLP) but even more by the factors connecting to document intelligence, as explicitly understood in business settings. To improve the real-world applicability of DU models, one must consider (i) the availability and variety of types of documents in a dataset, as well as (ii) the problem-framing methods.

Currently, publicly available datasets avoid **multi-page** documents, are not concerned with **multi-task** settings, nor provide **multi-domain** documents of sufficiently different types. These limitations hinder real-world DU systems, given the ever-increasing number of document types occurring in various business scenarios. This problem is often bypassed by building systems based on private datasets, which leads to a situation where datasets cannot be shared, documents of interest are not covered in benchmarks, and published methods cannot be compared objectively. DUDE counters these limitations by explicitly incorporating a large variety of multi-page documents and document types (see e.g., Fig. 2). Furthermore, the adaptability of DU to the real world is slowed down by a low-resource setting, since only a limited number of training examples can be provided, involving unpleasant manual labor, and subsequently costly model development. Anytime a new dataset is produced in the scientific or commercial context, a new model must be specifically designed and trained on it to achieve satisfactory performance. At the same time, transfer learning is the most promising solution for rapid model improvements, while zero- and few-shot performance still needs to be addressed in evaluation benchmarks.

Bearing in mind the characteristics outlined above, we formulated the DUDE dataset as an instance of *DocVQA* to evaluate how well current solutions can simultaneously handle the complexity and variety of real-world documents and all subtasks that can be expected. Optimally, a DU model should understand layout in a way that allows for zero-shot performance through attaining "desired generalization", i.e., generalization to *any documents* (e.g., drawn from previously unseen distributions of layouts, domains, and types) and *any questions* (e.g., regarding document elements, their properties, and compositions). Therefore, we incorporated these criteria while designing our dataset, which may stand as a common starting point and a cooperative path toward progress in this emerging area.

Desired Generalization. The challenge presented by DUDE is an instance of a Multi-Domain Long-Tailed Recognition (*MDLT*) problem [34].

Definition 1 (Multi-Domain Long-Tailed Recognition). **MDLT** *focuses on learning from multi-domain imbalanced data whilst addressing label imbalance, divergent label distributions across domains, and potential train-test domain shift. This framework naturally motivates targeting estimators that generalize to all domain-label pairs.*

A *domain* $D = \{(x_i, y_i)\}_{i=1}^N$ is composed of data sampled from a distribution P_{XY}, where $\mathcal{X}$ denotes an input space (documents) and $\mathcal{Y}$ the output space (QA pairs). Each $x \in \mathcal{X}$ represents a document, forming a tuple of (v, l, t), expressing a complex composition of visual, layout and textual elements. For simplicity, consider that each

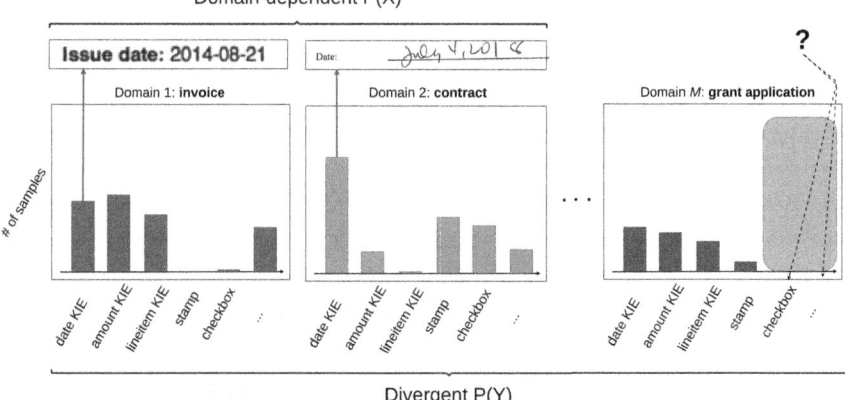

Fig. 1. Illustration of MDLT as applicable to the DUDE problem setting. The y-axis aggregates skills related to specific KIE or reasoning tasks over document elements (checkbox, signature, logo, footnote, ...). The x-axis denotes the obtained samples (QA pairs) per task. Each domain has a different label distribution $P(Y)$, typically relating to within-domain document properties $P(X)$. This training data exhibits label distribution shifts across domains, often requiring zero-shot generalization (marked red).

'label' $y \in \mathcal{Y}$ represents a question-answer pair, relating to implicit tasks to be completed (such as date KIE in *What is the document date?*). Due to the potentially compositional nature of QA, the label distribution is evidently *long-tailed*. During training, we are given MM domains (*document types*) on which we expect a solution to generalize (Fig. 1), both within (different number of samples for each unique task) and across domains (even without examples of a task in a given domain).

What sets apart domains is any difference in their joint distributions $P_{XY}^j \neq P_{XY}^k$. For example, an invoice is less similar (in terms of language use, visual appearance, and layout) to a contract than to a receipt or credit note. Yet, a credit note naturally contains a stamp stating information such as "invoice paid", whereas receipts rarely contain stamps. This might require a system to transfer 'stamp detection' learned within another domain, say on notary deeds.

Notably, it will be 'organic' to obtain more examples of certain questions (*tasks*) in a given domain. This should also encourage models to learn a certain skill in the domains where they have more training examples. Put plainly, it is better to learn checkbox detection on contracts than on invoices, which rarely contain any. This MDLT framework allows us to create a lasting, challenging benchmark that can be easily extended in the future with more tasks (formulated as QA pairs) and domains (relating to document types). In the first iteration of the DUDE competition, we have targeted specific skills by guiding annotators with focused instructions, which we share for future extensions.

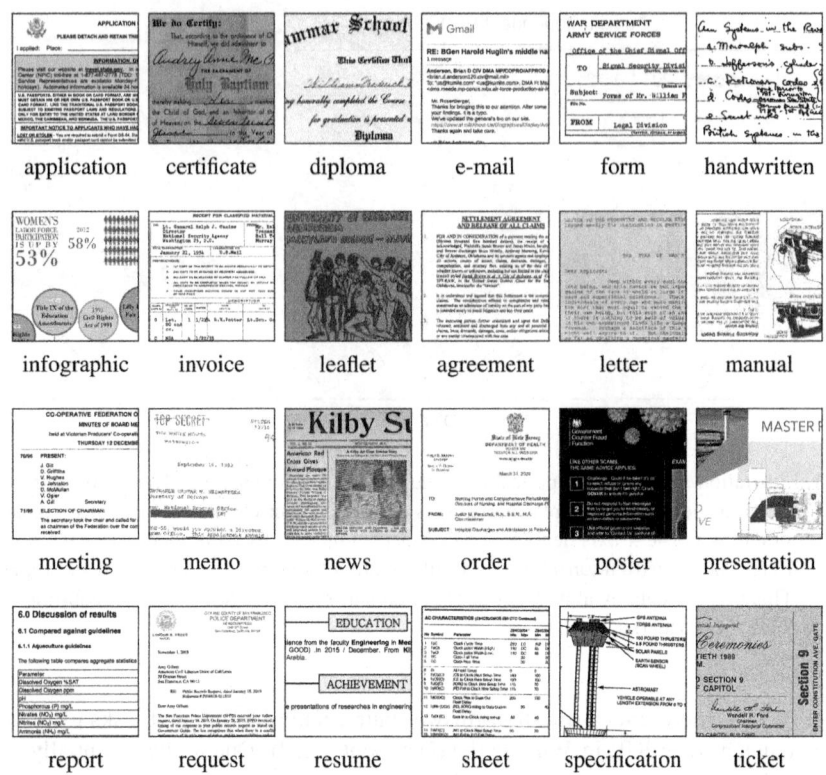

Fig. 2. Excerpts from DUDE documents (one visualized per type). Note that it is not an exhaustive list of document types collected.

3   DUDE Dataset

As part of the ICDAR 2023 DUDE competition, the authors constructed a novel dataset from scratch. A separate publication [33] describes the dataset in more detail, together with how it is different from related VQA datasets with an analysis of baseline methods. As part of the report, we will summarize the most important statistics and provide more insight into how the dataset and diagnostic subset were annotated and controlled for data quality.

The DUDE dataset is diverse, covering a wide range of document types (±200), sources, dates (1900–2023), and industries (±15). It contains documents with varying layouts and font styles, targeting diverse questions that require comprehension beyond document content. It includes abstractive and extractive questions, covering various answer types like textual, numerical, dates, yes/no, lists, or 'no answer'.

Annotation Process. To create the dataset, diverse documents were manually collected from websites such as Archive, Wikimedia Commons, and DocumentCloud. The selection ensured that the documents were visually distinct and free from controversial content, privacy, or legal concerns. A total of 5,000 multi-page English documents were gathered.

The annotation process involved in-house annotators and Amazon Mechanical Turk freelancers. The process consisted of four stages: generating candidate QA pairs, verifying QA pairs, selecting the best answers, and an optional review by Qualified Linguists for test set annotations. The total cost of annotation was estimated at $20,000.

Our multi-stage annotation process started with freelancers and in-house annotators proposing QA pairs, which were semi-automatically filtered for length, non-typical character combinations, and type-specific criteria. This was followed by the stage in which freelancers answered the accepted questions. Cases with an inter-answer agreement (ANLS) above 0.8 were added to the final dataset; otherwise, they were directed to further investigation. This stage employed freelancers with the highest historic quality score, who evaluated document, question, and answer variants, making corrections when necessary. Outliers were assessed by Qualified Linguists and corrected if needed (see Van Landeghem et al. [33] for the detailed description).

Future Extensions. To extend the dataset, one could follow the document collection and annotation process outlined in the original description [33]. This involves manually gathering diverse documents from various sources, ensuring they meet the dataset's criteria, and then following the multi-phase annotation process to generate and verify new QA pairs.

4 DUDE Competition Protocol

The ICDAR 2023 competition on Document UnderstanDing of Everything took place from February to May of 2023. A *training-validation* set with 30 K QA annotations on 3.7 K documents was given to participants at the beginning of February. The 11.4 K questions on 12.1 K documents for the *test set* were only made accessible for a window between March and May. Participants were asked to submit results obtained on the public, blind test set documents rather than deliver model executables, although they were encouraged to open-source their implementations. We relied on the scientific integrity of the participants to adhere to the competition's guidelines specified on The Robust Reading Competition (RRC) portal[1].

Task Formulation. Given an input consisting of a PDF with multiple pages and a natural language question, the objective is to provide a natural language answer together with an assessment of the answer confidence (a float value scaled between 0 and 1). Each unique document is annotated with multiple questions of different types, including extractive, abstractive, list, and non-answerable. Annotated QA pairs are not restricted to the answer being explicitly present in the document. Instead, any question on aspect, form, or visual/layout appearance relative to the document under review is allowed.

Additionally, competitors were allowed to submit results for only a specific answer type (provided in annotations) such that, for example for extractive questions, encoder-only architectures could compete in DUDE. Another important subtask is to obtain a *calibrated* and *selective* DocVQA system, which lowers answer confidence when

---

[1] https://rrc.cvc.uab.es/?ch=23.

unsure about its answers and does not hallucinate in case of non-answerable questions. Regardless of the number of answers (zero in the case of non-answerable or multiple in list-questions), we expect a single confidence estimate for the whole answer to guarantee consistency in calibration evaluation. To promote fair competition, we provided for each document three OCR versions obtained from one open-source (Tesseract) and two commercial engines (Azure, AWS).

Evaluation Protocol. The first evaluation phase assumes only independently and identically distributed (iid) data containing a similar mixture of document and question-answer types for the train-validation-test splits. To support scoring all possible answer types, the evaluation metric is the Average Normalized Levenshtein Similarity (ANLS) metric, modified for non-answerable questions (0/1 loss) and made invariant to the order of provided answers for list answers (ANLSL [30]). To assess the calibration and ranking of answer confidence, we applied two metrics, Expected Calibration Error (ECE) [8,22] ($\ell2$ norm, equal-mass binning with 100 bins) and Area-Under-Risk-Coverage-Curve (AURC) [7,11,15], respectively.

The (implicit) second evaluation phase created a mixture of seen and unseen domain test data. This was launched jointly with the first evaluation phase, as otherwise, one would be able to already detect the novel unseen domain test samples. To score how gracefully a system deals with unseen domain data, the evaluation metric is AUROC [18], which roughly corresponds to the probability that a positive example (in-domain) is assigned a higher detection score than a negative example (out-of-domain). A system is expected to either lower its confidence or abstain from giving an answer.

There is a strict difference between a non-answerable question and an unseen domain question. For the former, the document is from a domain that was included during training, yet the question cannot be solved with the document content, e.g., asking about who signed the document without any signatures present. For the latter, the question is apt for the document content, yet the document is from a domain that was not included during training and validation, which we would expect the system to pick up on.

For an in-depth explanation of these metrics and design choices, we refer the reader to [33, Appendix B.4.]. All metric implementations and evaluation scripts are made available as a standalone repository to allow participants to evaluate close to official blind test evaluations[2].

All submitted predictions are automatically evaluated, and the competition site provides ranking tables and visualization tools newly adapted to PDF inputs to examine the results. After the formal competition period, it will serve as an open archive of results. The main competition winner will be decided based on the aggregate high scores for ANLS, AURC, and AUROC.

To ensure proper validation and interpretability of competitor method results, we have created a diagnostic hold-out test set, where each instance is expert-annotated with specific metadata (QA type, document category, expected answer form and type, visual evidence) or operations (counting, normalization, arithmetic) required to answer). Fur-

---

[2] https://github.com/Jordy-VL/DUDEeval.

thermore, we sourced an independent human expert baseline on this diagnostic subset (see [33, Section 3.4]) to further perform a ceiling analysis on the submitted methods.

5 Results and Analysis

Together with the creation of the DUDE dataset, we did a preliminary study with some reference baseline methods [33]. These will not be covered in the competition report, unless relevant for comparison or analysis.

Submitted Methods. Overall, 6 methods from 3 different participants were submitted for the proposed tasks in the DUDE competition. To avoid cherry-picking from considering all submissions of individual participants, we consider only the last submission (accentuated) for the final ranking. All the methods followed an encoder-decoder architecture, which is a standard choice for VQA when abstractive questions are involved. Specifically, the submitted methods are mostly based on T5-base [25] as the decoder. For this reason, we include the *T5-base* baseline to compare how the participant methods improved on it. A short description of each method can be found in Table 1.

Two very recent state-of-the-art architectures, UDOP and HiVT5, have been extensively leveraged by participants. The former is geared toward improved document page representations, while the latter targets multi-page document representations. In their method reports, the UDOP-based models by LENOVO RESEARCH mention calculating confidence by multiplying the maximum softmax score of decoded output tokens with two additional post-processing rules: a) predicted not-answerable questions confidence is set to 1, b) when abstaining, confidence is set to 0.

Performance Analysis. Table 2 reports the competition results ranking comparing the submitted methods' performance on the test set. Higher ANLS and AUROC values indicate better performance, while lower ECE and AURC values signify improved calibration and confidence ranking. According to the findings, the UDOP+BLIP2+GPT approach attains the highest ANLS score (50.02), achieving the best calibration and OOD (out-of-distribution) detection performance. In a direct comparison of the MMT5 and HiVT5+modules methods, the former shows a higher ANLS score, yet did not provide any confidence estimates.

Thus, the overall winner is UDOP+BLIP2+GPT by LENOVO RESEARCH. Their submitted methods (ranked by highest ANLS) also differentiate themselves by their additional attention to confidence estimation. Based on the numbers in the table, several interesting observations can be made to support the suggested future directions and propose additional experiments:

- **ANLS.** The integration of UDOP, BLIP2, and ChatGPT contributes to the method's superior overall performance in answering different question types.
- **ECE, AURC.** Integrating UDOP, BLIP2 visual encoder, and ChatGPT for question decomposition contributes to the method's performance in handling uncertainty across various question types.
- **Abstractive.** The top performance of UDOP+BLIP2+GPT in abstractive questions reveals the potential of combining the UDOP ensemble, BLIP2 visual encoder, and ChatGPT to enable abstract reasoning and synthesis of information beyond simple extraction.

Table 1. Short descriptions of the methods participating to the DUDE competition, in order of submission. The last submitted method is considered for the final ranking.

Method	Description
T5-base (ours)	T5-base [25] fine-tuned on DUDE (AWS OCR), with a delimiter combining list answers into a single string, and replacing not-answerable questions with 'none'.
LENOVO RESEARCH	
UDOP(M)	Ensemble (M=10) of UDOP [29] (794M each) models without self-supervised pre-training, only fine-tuned in two stages: 1) SP-DocVQA [32] and MP-DocVQA [31], and 2) DUDE (switching between Azure and AWS OCR).
UDOP +BLIP2	UDOP(M=1) with integrated BLIP2 [16] predictions to optimize the image encoder and additional page number features.
UDOP +BLIP2+GPT	UDOP(M=1) and BLIP2 visual encoder with ChatGPT to generate Python-like modular programs to decompose questions for improved predictions [6,9].
UPSTAGE AI	
MMT5	Multimodal T5 pre-trained in two stages: single-page (ScienceQA [27], VQAonBD2023 [26], HotpotQA [35], SP-DocVQA) with objectives (masked language modeling (MLM) and next sentence prediction (NSP)), multi-page (MP-DocVQA and DUDE) with three objectives (MLM, NSP, page order matching). Fine-tuning on DUDE with answers per page combined for final output.
INFRRD.AI	
HiVT5	Hi-VT5 [31] with 20 <PAGE> tokens pre-trained with private document collection (*no information provided*) using span masking objective [14]. Fine-tuned with MP-DocVQA and DUDE.
HiVT5 +modules	Hi-VT5 extended with token/object embeddings for a variety of modular document understanding subtasks (detection: table structure, signatures, logo, stamp, checkbox; KIE: generic named entities; classification: font style)

- **List.** The performance of UDOP+BLIP2+GPT in list-based questions suggests that incorporating page number features can enhance the model's capability to process and generate list information, which might be spread across pages.

Figure 3 visualizes an overview of the performance of each submitted method respective to diagnostic subset samples matching a certain diagnostic category. The models generally struggle with operations involving *counting, arithmetic, normalization*, and *comparisons*. As expected, models have higher performance when dealing with simpler questions (*complexity simple*) compared to more complex questions (*complexity multi-hop, complexity other hard*, and *complexity meta*). Models tend to perform better when handling evidence in the form of plain text (*evidence plain*) compared to other forms of evidence, such as visual charts, maps, or signatures. Performance across models is notably lower for tasks involving lists compared to other question types. Models show varying performance when dealing with different types of forms (e.g., *date, numeric, other, proper*).

Figure 5 studies the ability of the competitors' methods to answer questions respective to increasingly longer documents. We observe a significant drop in ANLS when

Table 2. Summary of Method performance on the DUDE test set. Average ANLS results per question/answer type are abbreviated as (Abs)tractive, (Ex)tractive, (N)ot-(A)nswerable, (Li)st. (*) All scalars are scaled between 0 and 100 for readability.

	Answer	**Calibration**		**OOD Detection**	ANLS/answer type			
Method	ANLS ↑	ECE ↓	AURC ↓	AUROC ↑	*Ex*	*Abs*	*Li*	*NA*
UDOP+BLIP+GPT	**50.02**	**22.40**	**42.10**	**87.44**	**51.86**	**48.32**	**28.22**	**62.04**
MMT5	37.90	59.31	59.31	50.00	41.55	40.24	20.21	34.67
HiVT5+modules	35.59	28.03	46.03	51.24	30.95	35.15	11.76	52.50

aggregating scores over gradually longer documents. This is expected as the longer the document is, the more probable that the answer will either be located on a later page or rely on a long-range dependency between the tokens (e.g., a multi-hop question). Strikingly, all methods' scores, except Hi-VT5+modules, drop significantly for questions on 2-page documents. This is likely to have the root cause in the standard input size of T5-based methods equal to 512 tokens, covering roughly 1 page.

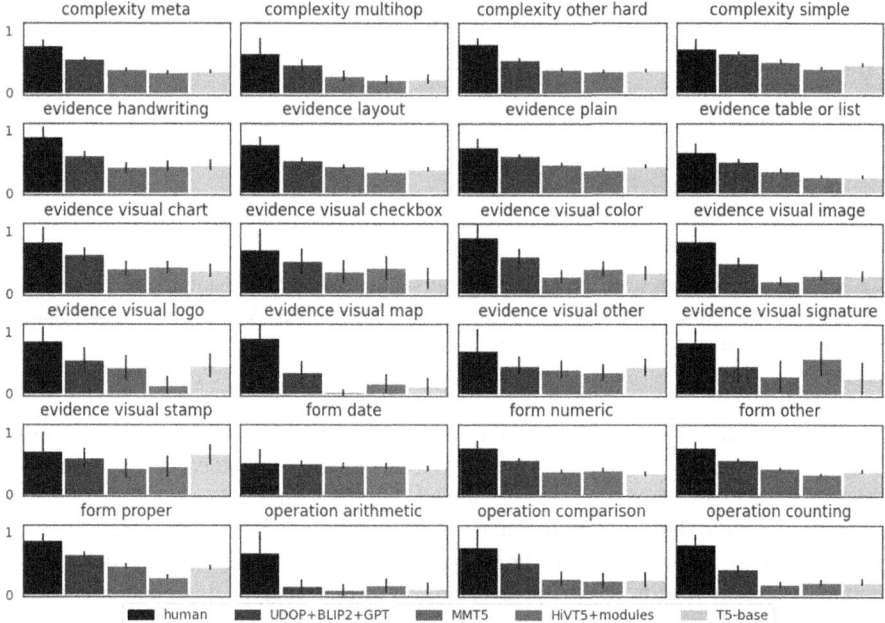

Fig. 3. We report the average ANLS per diagnostic category for each of the submitted methods vs. **human** and a baseline method T5-base. Since the diagnostic dataset contains a different number of samples per diagnostic category, we added error bars representing 95% confidence intervals. This helps visually determine statistically significant differences.

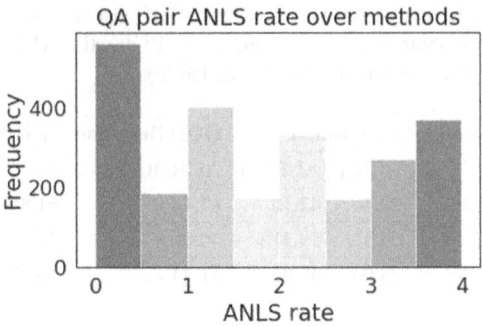

Fig. 4. A histogram (bins=8, matching ANLS-threshold of 0.5) of the average ANLS rate per QA pair when summing ANLS scores over competitor methods.

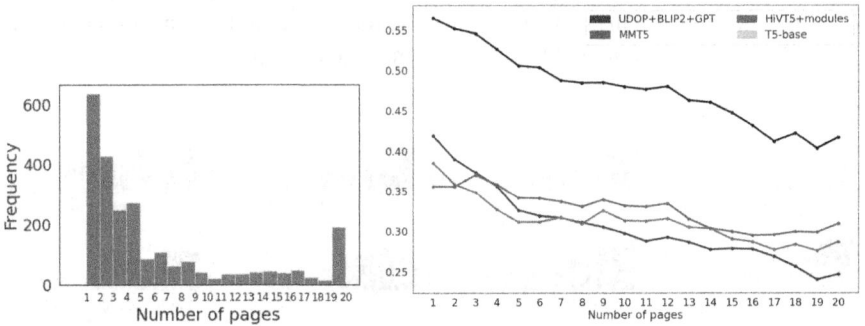

Fig. 5. Left: A histogram over the number of questions relative to the number of pages in the document (limited to 20 pages). Right: A line plot of the average ANLS score per QA pair: – documents of length *at least* (x-axis) pages.

Figure 4 analyzes the correlation of errors over competitor methods. A large portion of QA pairs is predicted completely wrong (ANLS-rate = 0) by all competitor methods. This can have many plausible causes: a) by all sharing a similar decoder (T5), methods suffer from similar deficiencies, b) some QA pairs are too complex for current state-of-the-art competitor methods, particularly questions requiring more complex reasoning or unique document-specific layout processing. To further analyze this phenomenon, we will sample qualitative examples with different ANLS rates.

5.1   Qualitative Examples

We provide some interesting, hand-picked test set examples with predictions from the submitted competition methods.

Low Complexity. *Who is the president and vice-chancellor?* Despite the question's relatively straightforward nature, some systems struggle with providing the appropriate answer. One can hypothesize it is the result of limited context (the answer is located at the end of the document), i.e., models either hallucinate a value or provide a name found earlier within the document.

Source	Answer	ANLS	Conf.
Ground truth	Jack N. Lightstone		
Human	Jack N. Lightstone	1.0	—
T5-base	James L. Turk	0.0	0.0
MMT5	james l. turk	0.0	1.0
UDOP+BLIP2+GPT	jack n. lightstone	1.0	0.9
HiVT5+modules	Jack N. Whiteside	0.6	0.6

Requires Graphical Comprehension. *Which is the basis for jurisdiction?* To provide a valid answer, the model needs to comprehend the meaning of the form field and recognize the selected checkbox. None of the participating systems was able to spot the answer correctly.

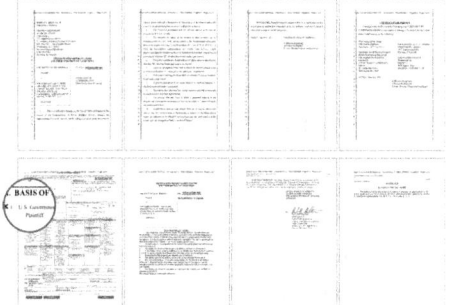

Source	Answer	ANLS	Conf.
Ground truth	U.S. Goverment Plaintiff		
Human	U.S. Goverment Plaintiff	1.0	—
T5-base	Declaration of taking	0.0	0.1
MMT5	united states district court	0.0	1.0
HiVT5+modules		0.0	1.0
UDOP+BLIP2+GPT	public purpose	0.0	0.4

Requires Comparison. *In which year does the Net Requirement exceed 25,000?* The question requires comprehending a multi-page table and spotting if any values fulfill the posed condition. Some of the models resort to plausible answers (one of the three dates that the document covers), whereas others correctly decide there is no value exceeding the provided amount.

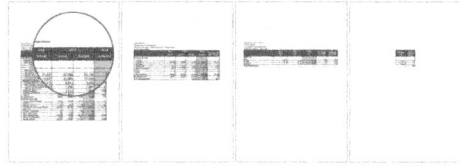

Source	Answer	ANLS	Conf.
Ground truth	[*Unanswerable*]		
Human	[*Unanswerable*]	1.0	—
T5-base	[*Unanswerable*]	1.0	0.2
MMT5	2018	0.0	1.0
UDOP+BLIP2+GPT	[*Unanswerable*]	1.0	1.0
HiVT5+modules	2017	0.0	0.8

Requires Arithmetic. *What is the difference between how much Operator II and Operator III make per hour?* The question requires table comprehension, determining relevant values, and dividing extracted integers. None of the participating models was able to fulfill this requirement.

Source	Answer	ANLS	Conf.
Ground truth	$5		
Human	$5	1.0	—
T5-base	$0.00	0.0	0.0
MMT5	65%	0.0	1.0
UDOP+BLIP2+GPT	-1.5 mile	0.0	0.0
HiVT5+modules	$5,700.00	0.0	0.4

Requires Counting and List Output. *What are the first two behavioral and intellectual disabilities of people with FASDs?* It seems most of the models correctly recognized that this type of question requires a list answer but either failed to comprehend the question or provided a list with incorrect length (incomplete or with too many values).

Source	Answer	ANLS	Conf.
Ground truth	Learning disabilities I Hyperactivity		
Human	learning disabilities	0.5	—
T5-base	Early embryo brain development I External Genitals	0.0	0.0
MMT5	heart beats I difficulty with attention I lung function I hyperactivity I problem with judgment I speech and language delays	0.2	1.0
UDOP+BLIP2+GPT	hyperactivity I speech and language delays	0.5	0.2
HiVT5+modules	HIV/AIDS	0.0	0.6

6   Conclusion and Future Work

As a core contribution of DUDE, we wanted to emphasize the importance of evaluation beyond mere predictive performance. DUDE offers an interesting and varied test bed for the evaluation of novel calibration and selective QA approaches (e.g., [5, 19]). While this was not explicitly attempted in this iteration of the competition, we hope that future work will consider testing their methods against DUDE.

Future of the Shared Task. As the competition evolves, we hope that DUDE will serve as an essential platform for pushing the frontiers of research and driving innovation in the DU field. Currently, our competition focuses on English language documents, which means we miss out on the potential of incorporating *multilingual* data. An ideal extension for future iterations of the shared task would be to introduce multilingualism, which our framework can accommodate, provided that source documents are readily available. However, this would also require specifying language qualifications for annotation experts. Moreover, one could automate part of the data collection process and annotation process by allowing the best-performing competition system to validate the aptitude and complexity of human-proposed QA pairs.

Acknowledgment. Jordy Van Landeghem acknowledges the financial support of VLAIO (Flemish Innovation & Entrepreneurship) through the Baekeland Ph.D. mandate (HBC.2019.2604). The Smart Growth Operational Programme partially supported this research under projects no. POIR.01.01.01-00-1624/20 (*Hiper-OCR - an innovative solution for information extraction from scanned documents*) and POIR.01.01.01-00-0605/19 (*Disruptive adoption of Neural Language Modelling for automation of text-intensive work*).

References

1. Appalaraju, S., Jasani, B., Kota, B.U., Xie, Y., Manmatha, R.: DocFormer: end-to-end transformer for document understanding. In: Proceedings of the IEEE/CVF International Conference on Computer Vision, pp. 993–1003 (2021)
2. Biten, A.F., et al.: ICDAR 2019 competition on scene text visual question answering. In: 2019 International Conference on Document Analysis and Recognition (ICDAR), pp. 1563–1570. IEEE (2019)

3. Biten, A.F., et al.: Scene text visual question answering. In: Proceedings of the IEEE/CVF International Conference on Computer Vision (2019)
4. Brown, T., et al.: Language models are few-shot learners. In: Advances in Neural Information Processing Systems, vol. 33, pp. 1877–1901 (2020)
5. Dhuliawala, S., Adolphs, L., Das, R., Sachan, M.: Calibration of machine reading systems at scale. In: Findings of the Association for Computational Linguistics: ACL 2022. Association for Computational Linguistics, Dublin, Ireland, pp. 1682–1693 (2022). https://doi.org/10.18653/v1/2022.findings-acl.133, https://aclanthology.org/2022.findings-acl.133
6. Dídac, S., Menon, S., Vondrick, C.: ViperGPT: visual inference via python execution for reasoning. arXiv preprint: arXiv:2303.08128 (2023)
7. Geifman, Y., El-Yaniv, R.: Selective classification for deep neural networks. In: Advances in neural information processing systems, vol. 30 (2017)
8. Guo, C., Pleiss, G., Sun, Y., Weinberger, K.Q.: On calibration of modern neural networks. In: Proceedings of the 34th International Conference on Machine Learning, ICML'17, vol. 70, pp. 1321–1330 (2017)
9. Gupta, T., Kembhavi, A.: Visual programming: compositional visual reasoning without training. arXiv preprint: arXiv:2211.11559 (2022)
10. Huang, Y., Lv, T., Cui, L., Lu, Y., Wei, F.: LayoutLMv3: pre-training for document AI with unified text and image masking, MM '22, pp. 4083–4091. Association for Computing Machinery, New York (2022). https://doi.org/10.1145/3503161.3548112
11. Jaeger, P.F., Lüth, C.T., Klein, L., Bungert, T.J.: A call to reflect on evaluation practices for failure detection in image classification. In: International Conference on Learning Representations (2023). https://openreview.net/forum?id=YnkGMIh0gvX
12. Jaume, G., Ekenel, H.K., Thiran, J.P.: FUNSD: a dataset for form understanding in noisy scanned documents. In: 2019 International Conference on Document Analysis and Recognition Workshops (ICDARW), vol. 2, pp. 1–6. IEEE (2019)
13. Jimeno Yepes, A., Zhong, P., Burdick, D.: ICDAR 2021 competition on scientific literature parsing. In: Lladós, J., Lopresti, D., Uchida, S. (eds.) ICDAR 2021. LNCS, vol. 12824, pp. 605–617. Springer, Cham (2021). https://doi.org/10.1007/978-3-030-86337-1_40
14. Joshi, M., Chen, D., Liu, Y., Weld, D.S., Zettlemoyer, L., Levy, O.: SpanBERT: improving pre-training by representing and predicting spans. Trans. Assoc. Comput. Linguist. **8**, 64–77 (2020)
15. Kamath, A., Jia, R., Liang, P.: Selective question answering under domain shift. In: Proceedings of the 58th Annual Meeting of the Association for Computational Linguistics, pp. 5684–5696 (2020)
16. Li, J., Li, D., Savarese, S., Hoi, S.: BLIP-2: bootstrapping language-image pre-training with frozen image encoders and large language models. arXiv preprint: arXiv:2301.12597 (2023)
17. Li, J., Li, D., Xiong, C., Hoi, S.: BLIP: bootstrapping language-image pre-training for unified vision-language understanding and generation. In: International Conference on Machine Learning, pp. 12888–12900. PMLR (2022)
18. Liang, S., Li, Y., Srikant, R.: Enhancing the reliability of out-of-distribution image detection in neural networks. In: International Conference on Learning Representations (2018). https://openreview.net/forum?id=H1VGkIxRZ
19. Lin, S., Hilton, J., Evans, O.: Teaching models to express their uncertainty in words. Trans. Mach. Learn. Res. (2022). https://openreview.net/forum?id=8s8K2UZGTZ
20. Mathew, M., Bagal, V., Tito, R., Karatzas, D., Valveny, E., Jawahar, C.: InfographicVQA. In: Proceedings of the IEEE/CVF Winter Conference on Applications of Computer Vision, pp. 1697–1706 (2022)
21. Mathew, M., Tito, R., Karatzas, D., Manmatha, R., Jawahar, C.: Document visual question answering challenge 2020. arXiv preprint: arXiv:2008.08899 (2020)

22. Naeini, M.P., Cooper, G., Hauskrecht, M.: Obtaining well calibrated probabilities using Bayesian binning. In: Proceedings of the AAAI Conference on Artificial Intelligence, vol. 29 (2015)
23. Powalski, R., Borchmann, Ł., Jurkiewicz, D., Dwojak, T., Pietruszka, M., Pałka, G.: Going full-TILT boogie on document understanding with text-image-layout transformer. In: Lladós, J., Lopresti, D., Uchida, S. (eds.) ICDAR 2021. LNCS, vol. 12822, pp. 732–747. Springer, Cham (2021). https://doi.org/10.1007/978-3-030-86331-9_47
24. Qiao, L., et al.: LGPMA: complicated table structure recognition with local and global pyramid mask alignment. In: Lladós, J., Lopresti, D., Uchida, S. (eds.) ICDAR 2021. LNCS, vol. 12821, pp. 99–114. Springer, Cham (2021). https://doi.org/10.1007/978-3-030-86549-8_7
25. Raffel, C., et al.: Exploring the limits of transfer learning with a unified text-to-text transformer. J. Mach. Learn. Res. 21(140), 1–67 (2020)
26. Raja, S., Mondal, A., Jawahar, C.: ICDAR 2023 competition on visual question answering on business document images (2023)
27. Saikh, T., Ghosal, T., Mittal, A., Ekbal, A., Bhattacharyya, P.: ScienceQA: a novel resource for question answering on scholarly articles. Int. J. Digit. Libr. 23(3), 289–301 (2022)
28. Stanisławek, T., et al.: Kleister: key information extraction datasets involving long documents with complex layouts. In: Lladós, J., Lopresti, D., Uchida, S. (eds.) ICDAR 2021. LNCS, vol. 12821, pp. 564–579. Springer, Cham (2021). https://doi.org/10.1007/978-3-030-86549-8_36
29. Tang, Z., et al.: Unifying vision, text, and layout for universal document processing. arXiv preprint: arXiv:2212.02623 (2022)
30. Tito, R., Karatzas, D., Valveny, E.: Document collection visual question answering. In: Lladós, J., Lopresti, D., Uchida, S. (eds.) ICDAR 2021. LNCS, vol. 12822, pp. 778–792. Springer, Cham (2021). https://doi.org/10.1007/978-3-030-86331-9_50
31. Tito, R., Karatzas, D., Valveny, E.: Hierarchical multimodal transformers for multi-page DocVQA. arXiv preprint: arXiv:2212.05935 (2022)
32. Tito, R., Mathew, M., Jawahar, C.V., Valveny, E., Karatzas, D.: ICDAR 2021 competition on document visual question answering. In: Lladós, J., Lopresti, D., Uchida, S. (eds.) ICDAR 2021. LNCS, vol. 12824, pp. 635–649. Springer, Cham (2021). https://doi.org/10.1007/978-3-030-86337-1_42
33. Van Landeghem, J., et al.: Document understanding dataset and evaluation (DUDE). In: International Conference on Computer Vision (2023)
34. Yang, Y., Wang, H., Katabi, D.: On multi-domain long-tailed recognition, imbalanced domain generalization and beyond. In: Computer Vision - ECCV 2022: 17th European Conference, Proceedings, Part XX, Tel Aviv, Israel, 23–27 October 2022, pp. 57–75. Springer-Verlag, Berlin, Heidelberg (2022). https://doi.org/10.1007/978-3-031-20044-1_4
35. Yang, Z., Qi, P., et al.: HotpotQA: a dataset for diverse, explainable multi-hop question answering. In: Proceedings of the 2018 Conference on Empirical Methods in Natural Language Processing, pp. 2369–2380. Association for Computational Linguistics, Brussels (2018). https://doi.org/10.18653/v1/D18-1259, https://aclanthology.org/D18-1259
36. Zhong, X., Tang, J., Yepes, A.J.: PubLayNet: largest dataset ever for document layout analysis. In: 2019 International Conference on Document Analysis and Recognition (ICDAR), pp. 1015–1022. IEEE (2019)

ICDAR 2023 Competition on Indic Handwriting Text Recognition

Ajoy Mondal[✉][iD] and C. V. Jawahar[iD]

International Institute of Information Technology, Hyderabad, India
{ajoy.mondal,jawahar}@iiit.ac.in

Abstract. This paper presents the competition report on Indic Handwriting Text Recognition (IHTR) held at the 17th International Conference on Document Analysis and Recognition (ICDAR 2023 IHTR). Handwriting text recognition is an essential component for analyzing handwritten documents. Several good recognizers are available for English handwriting text in the literature. In the case of Indic languages, limited work is available due to several challenging factors. (i) Two or more characters are often combined to form conjunct characters, (ii) most Indic scripts have around 100 unique Unicode characters, (iii) diversity in handwriting styles, (iv) varying ink density around the words, (v) challenging layouts with overlap between words and natural unstructured writing, and (vi) datasets with only a limited number of writers and examples.
With this competition, we motivate the researchers to continue researching Indic handwriting text recognition tasks to prevent the risk of vanishing Indic scripts/languages. In this competition, we use a training set of an existing benchmark dataset [3,4,6]. We create a new manually annotated validation set and test set for validation and testing purposes. A total of eighteen different teams around the world registered for this competition. Among them, only six teams submitted the results along with algorithm details. The winning team Upstage KR achieves an average 95.94% Character Recognition Rate (CRR) and 88.31% Word Recognition Rate (WRR) over ten languages.

Keywords: OCR · Indic handwriting text recognition · Indic language · Indic script · conjunct characters

1 Introduction

Optical Character Recognition (OCR) is the process of converting printed or handwritten images into machine-readable formats. OCR is an essential component in document image analysis. The OCR system usually consists of two main modules (i) text detection module and (ii) text recognition module. The text detection module locates all text blocks within an image, either at the word or line level. The text recognition module attempts to interpret the text image content and translate the visual signals into natural language tokens. Handwriting text recognition is more challenging than printed text because of the imbalanced

G. A. Fink et al. (Eds.): ICDAR 2023, LNCS 14188, pp. 435–453, 2023.
https://doi.org/10.1007/978-3-031-41679-8_25

differences in handwriting styles, content, and time. An individual's handwriting is always unique, and this property creates motivation and interest among researchers to work in this demanding and challenging area.

Many languages worldwide are disappearing due to their limited usage. We can easily use OCR and natural language processing techniques to stop the extermination of languages worldwide. Among 7000 languages[1], the handwritten OCR is available only for a few languages: English [7,10,19], Chinese [18,22,23], Arabic [8,13], and Japanese [12,16]. Several Indian scripts and languages are at risk due to insufficient research efforts. So, there is an immense need for research on text recognition for Indic scripts and languages.

Among 22 languages in India, few are used only for communication purposes. Hindi, Bengali, and Telugu are the most spoken languages [9]. In most Indic scripts, two or more characters often combine to form conjunct characters [1]. These inherent features of Indic scripts make Handwriting Text Recognition (HWR) more challenging than Latin scripts. Compared to the 52 unique (upper case and lower case) characters in English, most Indic scripts have over 100 unique basic Unicode characters [17].

We organize a challenge on Indic handwriting text recognition tasks in ten different scripts: Bengali, Devanagari, Gujarati, Gurumukhi, Kannada, Malayalam, Odia, Tamil, Telugu, and Urdu. The competition is hosted on http://cvit.iiit.ac.in/ihtr2022/. This challenge motivates researchers to develop/design new algorithms for the above scripts. This challenge suggests a significant new direction for the community to recognize handwritten text in Indic scripts.

The paper is organized as follows. In Sect. 3, we give details about the dataset used for the competition. The submitted methods are discussed in Sect. 4. Section 5 shows the results of the competition. The conclusive remark is drawn in Sect. 6.

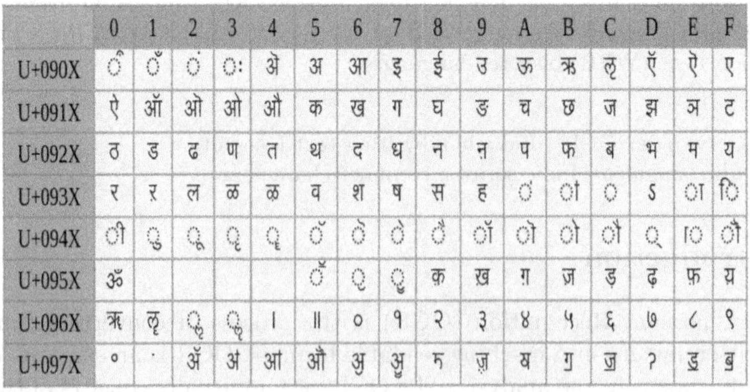

	0	1	2	3	4	5	6	7	8	9	A	B	C	D	E	F
U+090X	◌ऀ	◌ँ	◌ं	◌ः	ऄ	अ	आ	इ	ई	उ	ऊ	ऋ	ऌ	ऍ	ऎ	ए
U+091X	ऐ	ऑ	ऒ	ओ	औ	क	ख	ग	घ	ङ	च	छ	ज	झ	ञ	ट
U+092X	ठ	ड	ढ	ण	त	थ	द	ध	न	ऩ	प	फ	ब	भ	म	य
U+093X	र	ऱ	ल	ळ	ऴ	व	श	ष	स	ह	◌ऺ	◌ऻ	◌़	ऽ	◌ा	◌ि
U+094X	◌ी	◌ु	◌ू	◌ृ	◌ॄ	◌ॅ	◌ॆ	◌े	◌ै	◌ॉ	◌ॊ	◌ो	◌ौ	◌्	◌ॎ	◌ॏ
U+095X	ॐ					◌ॕ	◌ॖ	◌ॗ	क़	ख़	ग़	ज़	ड़	ढ़	फ़	य़
U+096X	ॠ	ॡ	◌ॢ	◌ॣ	।	॥	०	१	२	३	४	५	६	७	८	९
U+097X	॰	ॱ	ॲ	ॳ	ॴ	ॵ	ॶ	ॷ	ॸ	ॹ	ॺ	ॻ	ॼ	ॽ	ॾ	ॿ

Fig. 1. Shows a list of characters and corresponding Unicode characters in the Devanagari script.

[1] https://www.ethnologue.com/guides/how-many-languages.

2   Indic Languages

There are 22 official languages in India, Assamese, Bengali, Bodo, Dogri, Gujarati, Hindi, Kannada, Kashmiri, Konkani, Malayalam, Manipuri, Marathi, Maithili, Nepali, Odia, Punjabi, Sanskrit, Santali, Sindhi, Tamil, Telugu, and Urdu. Each language has its script. Multiple languages are also derived from a single script. Assamese and Bengali languages use Bengali script. Bodo, Dogri, Hindi, Kashmiri, Konkani, Marathi, Maithili, Nepali, Sanskrit, and Sindhi use the Devanagari script. Gujarati, Kannada, Malayalam, Odia, Punjabi, Tamil, Telugu, and Urdu languages use Gujarati, Kannada, Malayalam, Odia, Gurumukhi, Tamil, Telugu, and Urdu scripts, respectively. While Manipuri uses Bengali/Meitei script, and Santali uses Ol Chiki/Bengali/Odia scripts. Twenty-two languages are derived from 12 scripts. Each script has a set of unique characters and digits. Figure 1 and Fig. 2 show the groups of unique characters and corresponding Unicode characters in Bengali and Devanagari scripts, respectively.

	0	1	2	3	4	5	6	7	8	9	A	B	C	D	E	F
U+098X	ী	ঁ	ং	ঃ		অ	আ	ই	ঈ	উ	ঊ	ঋ				এ
U+099X	ঐ			ও	ঔ	ক	খ	গ	ঘ	ঙ	চ	ছ	জ	ঝ	ঞ	ট
U+09AX	ঠ	ড	ঢ	ণ	ত	থ	দ	ধ	ন		প	ফ	ব	ভ	ম	য
U+09BX	র		ল				শ	ষ	স	হ			়	ঽ	া	ি
U+09CX	ী	ু	ূ	ৃ	ৄ			ে	ৈ			ো	ৌ	্	ৎ	
U+09DX								ৗ					ড়	ঢ়		য়
U+09EX	ৠ	ৡ	ৢ	ৣ			০	১	২	৩	৪	৫	৬	৭	৮	৯
U+09FX	ৰ	ৱ	৲	৳	৴	৵	৶	৷	৸	৹	৺	৻	ৼ	৽	৾	

Fig. 2. Shows a list of characters and corresponding Unicode characters in Bengali script.

Compared to English, most Indic scripts consist of over 100 Unicode characters. Two or more characters are often combined to form a conjunct character. Figure 3 shows a few examples of conjunct characters in Bengali and Hindi. It also shows how the unique characters are combined to create conjunct characters. It also presents a sequence of Unicode code characters corresponding to the unique characters to create conjunct characters. From Fig. 3, for the Bengali language, we observe that 1st and 3rd conjunct characters are obtained by combining three unique characters. At the same time, 2nd and 4th conjunct characters are created by combining five unique characters. In the case of the Hindi language, all four conjunct characters are obtained by combining three unique characters.

Words in Indic languages are created by combining characters from the corresponding script. Figure 4 shows a few examples of words in Bengali and Hindi languages. It also shows how these (e.g., Bengali and Hindi) words are formed using unique characters from respective scripts (Bengali and Devanagari). Form Fig. 4, we observe 1st and 4th words have 'Upper Matra', 1st and 2nd words have 'Lower Matra'. In the case of Hindi words, all four words have 'Upper Matra', and 1st and 2nd words have both 'Upper Matra and Lower Matra'. The conjunct characters, 'Upper Matra' and 'Lower Matra' in Indic languages, differentiate them from Latin scripts/languages. Due to these special characteristics, the recognition of Indic words is more complex than Latin words.

3   Dataset

We use publicly available benchmark Indic handwriting text recognition datasets [3, 4, 6] for this competition. We use the training set mentioned in [3, 4, 6]. For evaluation purposes, we create a new validation set (VAL-IHW-ICDAR-2023) and test set (TEST-IHW-ICDAR-2023), which are manually annotated. For each of the ten languages, we manually annotate 6000 word images from handwriting pages written by 100 writers. Table 1 presents the statistics of training, validation, and test sets for this competition. It contains word-level images of ten Indic languages: Bengali, Devanagari, Gujarati, Gurumukhi, Kannada, Odia, Malayalam, Tamil, Telugu, and Urdu. We also create a list of unique words from the training set of a language which indicates the lexicon of this language. The lexicon can be used as a language model for post OCR error correction to improve OCR performance. Table 1 also shows the statistic of words in the dataset. The table shows that Malayalam has the most significant average word length among

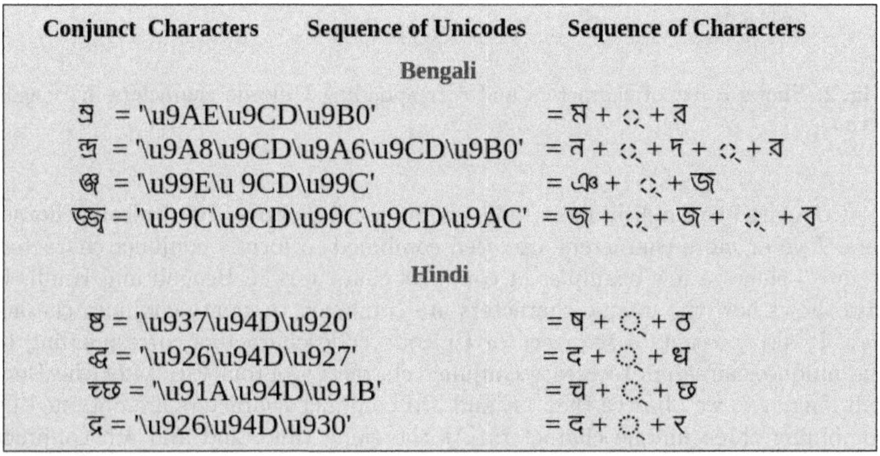

Conjunct Characters	Sequence of Unicodes	Sequence of Characters
	Bengali	
ম্র = '\u9AE\u9CD\u9B0'		= ম + ্ + র
ন্দ্র = '\u9A8\u9CD\u9A6\u9CD\u9B0'		= ন + ্ + দ + ্ + র
ঞ্জ = '\u99E\u 9CD\u99C'		= ঞ + ্ + জ
জ্জ্ব = '\u99C\u9CD\u99C\u9CD\u9AC'		= জ + ্ + জ + ্ + ব
	Hindi	
ष्ठ = '\u937\u94D\u920'		= ष + ् + ठ
द्ध = '\u926\u94D\u927'		= द + ् + ध
च्छ = '\u91A\u94D\u91B'		= च + ् + छ
द्र = '\u926\u94D\u930'		= द + ् + र

Fig. 3. Shows a few samples of conjunct characters in both Bengali and Hindi languages.

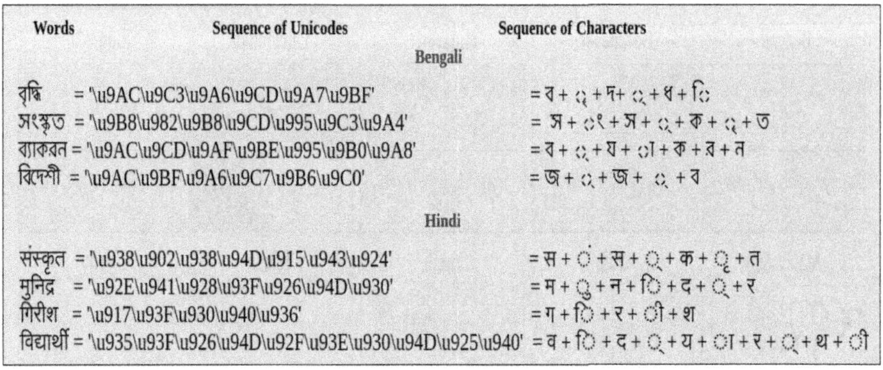

Words	Sequence of Unicodes	Sequence of Characters
	Bengali	
বৃদ্ধি	= '\u9AC\u9C3\u9A6\u9CD\u9A7\u9BF'	= ব + ৃ + দ্ + ্ + ধ + ি
সংস্কৃত	= '\u9B8\u982\u9B8\u9CD\u995\u9C3\u9A4'	= স + ং + স + ্ + ক + ৃ + ত
ব্যাকরন	= '\u9AC\u9CD\u9AF\u9BE\u995\u9B0\u9A8'	= ব + ্ + য + া + ক + র + ন
বিদেশী	= '\u9AC\u9BF\u9A6\u9C7\u9B6\u9C0'	= জ + ি + জ + ে + ব
	Hindi	
संस्कृत	= '\u938\u902\u938\u94D\u915\u943\u924'	= स + ं + स + ् + क + ृ + त
मुनिद्र	= '\u92E\u941\u928\u93F\u926\u94D\u930'	= म + ु + न + ि + द + ् + र
गिरीश	= '\u917\u93F\u930\u940\u936'	= ग + ि + र + ी + श
विद्यार्थी	= '\u935\u93F\u926\u94D\u92F\u93E\u930\u94D\u925\u940'	= व + ि + द + ् + य + ा + र + ् + थ + ी

Fig. 4. Shows a few sample words from both Bengali and Hindi languages.

all other languages. Figure 5 shows sample word images from the test set. The users in the competition are allowed to use additional real and/or synthetic data for training purposes. But they need to provide useful information about the additional dataset for training.

Table 1. Division of dataset into training, validation, and test sets. **#Image:** indicates the number of word-level images. **#Lexicon:** shows the number of unique words from the training set of respective languages.

Script	Training Set				Validation Set				Test Set				#Lexicon
	#Image	Word Length			#Image	Word Length			#Image	Word Length			
		Max.	Min.	Avg.		Max.	Min.	Avg.		Max.	Min.	Avg.	
Bengali	82554	29	1	7	1000	13	1	5	5000	19	1	5	11295
Devanagari	69853	23	1	5	1000	13	1	4	5000	15	1	4	11030
Gujarati	82563	22	1	6	1000	15	1	5	5000	20	2	4	10963
Gurumukhi	81042	22	1	5	1000	11	1	4	5000	12	1	4	11093
Kannada	73517	32	1	9	1000	17	1	6	5000	23	1	7	11766
Malayalam	85270	41	1	11	1000	21	1	7	5000	24	1	8	13401
Odia	73400	29	1	7	1000	15	1	5	5000	18	3	5	13314
Tamil	75736	31	1	9	1000	24	1	7	5000	24	1	7	13292
Telugu	80693	27	1	8	1000	20	1	6	5000	21	1	7	12945
Urdu	69212	14	3	5	1000	11	3	5	5000	9	1	3	11936

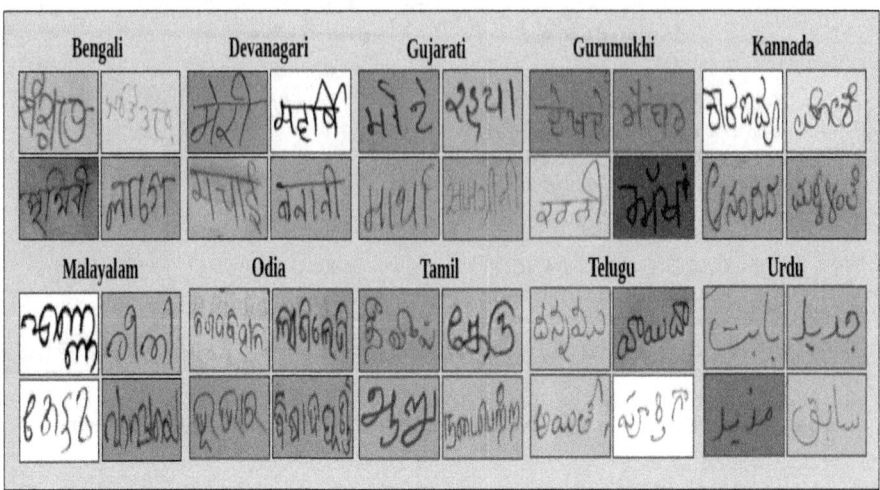

Fig. 5. Shows sample word images from the test set.

4   Methods

This section discusses each submitted method, including the baseline, in detail. Eighteen participants around the world registered for the competition. However, we got submissions from eight of them. These eight teams are (i) EE-Noobies, Indian Institute of Technology, Bombay, India, (ii) Upstage KR, Upstage KR, (iii) huanxiteam, (iv) GG-GradientGurus, Indian Institute of Technology, Delhi, India, (v) light, CCB Financial Technology Co. Ltd, China, (vi) PERO, Faculty of Information Technology, Brno University of Technology, (vii) SRUKR, Samsung R&D Institute Ukraine, and (viii) LEAP-OCR, Indian Institute of Technology, Bombay, India. However, the teams huanxiteam and GG-GradientGurus are not interested to submitted algorithm details. Due to this, these teams are not included in the draft.

4.1   Baseline

We evaluated the method proposed by Gongidi and Jawahar [6] as our baseline. The baseline network consists of four modules: Transformation Network (TN), Feature Extractor (FE), Sequence Modeling (SM), and finally, Predictive Modeling (PM). The transformation network has six plain convolutional layers with 16, 32, 64, 128, 128, and 128 channels. Each layer has a filter size, stride, and padding size of 3, 1, and 1, followed by a 2×2 max-pooling layer with a stride of 2. The feature extractor module consists of ResNet architecture. The sequence modeling component consists of a 2 layer Bidirectional LSTM (BLSTM) architecture with 256 hidden neurons in each layer. The predictive modeling layer consists of Connectionist Temporal Classification (CTC) to decode and recognize the characters by aligning the feature and target character sequences. We resize input images into 96×256. We use the Adadelta optimizer with Stochastic Gradient Descent (SGD) for all the experiments. We set the learning rate to 1.0, batch size to 64, and momentum to 0.09. For more details, refer to [6][2].

4.2 EE-Noobies

Convolutional Recurrent Neural Networks (CRNN) have emerged as a robust end-to-end trainable architecture for various image-based sequence recognition problems, particularly scene text recognition. A detailed overview of the CRNN implementation pipeline is presented here, highlighting the key components and steps involved in training and inference.

CRNN Implementation Pipeline: The following section describes the key components and steps involved in the CRNN implementation pipeline for handwritten text recognition.

- **Pre-processing:** Input images are pre-processed by resizing them to a fixed height while maintaining their aspect ratios. This step ensures that the input images have a consistent format and size suitable for the CRNN model.
- **Convolutional Layers:** The pre-processed images are passed through convolutional layers, extracting feature maps from the input images. Pooling layers follow these layers to reduce spatial dimensions and capture local information. Batch normalization and activation functions, such as ReLU, are applied to improve training stability and performance.
- **Recurrent Layers:** The feature maps produced by the convolutional layers are fed into a series of recurrent layers, typically implemented using Long Short-Term Memory (LSTM) or Gated Recurrent Unit (GRU) cells. These layers capture the sequential information in the feature maps and model long-range dependencies. To convert the feature maps into a sequence format, a "Map-to- Sequence" operation is performed, which involves reshaping the feature maps by collapsing the spatial dimensions, resulting in a sequence of feature vectors.
- **Bidirectional RNN:** A bidirectional RNN often captures both forward and backward context information in the sequence. It is achieved using two RNNs, one processing the sequence in the forward direction and the other in the reverse order. The outputs of these RNNs are then concatenated at each time step.
- **Transcription Layer:** The output from the recurrent layers is passed through a transcription layer that employs the Connectionist Temporal Classification (CTC) loss. The CTC layer allows the model to align the input sequences with the target label sequences without explicit character segmentation. It also helps the model recognize variable-length text by collapsing repeated characters and inserting blank tokens between characters.

---

[2] The code is available at https://github.com/sanny26/indic-htr.

- **Training:** The CRNN model is optimized during training using gradient-based optimization methods, such as Stochastic Gradient Descent (SGD) or Adam. The CTC loss is used as the objective function, and the gradients are back-propagated through the entire network, updating the weights of both the convolutional and recurrent layers.
- **Inference:** The CRNN model takes an input image and generates an output sequence for inference. The CTC layer outputs a probability distribution over the possible label sequences. A decoding algorithm, such as the best path decoding or beam search decoding, is applied to find the most likely label sequence to obtain the final prediction.

4.3   Upstage UK

To solve the Indic Handwritten Text Recognition task, the Upstage UK team uses the PARSeq framework [2], changing the vision encoder to SwinV2 [11]. The training details are as follows.

- The Upstage UK team uses an in-house synthetic data generator derived from the open-source SynthTiger [24] to generate English synthetic data. It uses Pango Text Renderer to generate 100k Indic synthetic data for each language for pre-training purposes.
- The team gathers a large pool of real-world in-house Korean scene text datasets, English handwriting dataset (IAM) [14], and various Indic handwriting datasets on Kaggle[3] to perform real-world data pre-training.
- The team used the competition data for ten languages to train a multilingual model from the second pre-training stage. Lastly, they train single language models for each of the ten languages respectively, using only the corresponding competition data for that language, starting from the multilingual model in the third stage. For the final submission, they use an ensemble of multiple models per language, whose combination is determined by utilizing the validation accuracy to pick the best ensemble candidates.
  In the processing stage, they cropped the text image to remove the empty area of the image using the Otsu threshold algorithm. In the post-processing stage, they remove the ',' character from Urdu words.

4.4   Light

The team light participated in the Indic Handwriting Text Recognition Competition using the TrOCR algorithm. In terms of data, they used GAN networks to perform data augmentation for each language category, increasing the robustness of the models.

---

[3] https://www.kaggle.com/competitions/bengaliai-cv19/.

Text Recognition Model TrOCR: TrOCR [10] is an end-to-end OCR model based on Transformer. Unlike existing methods, TrOCR is simple and efficient, does not use CNN as the backbone network but instead splits the input text image into image slices and then inputs them into the image Transformer. The encoder and decoder of TrOCR use standard Transformer structures and self-attention mechanisms, with the decoder generating word pieces as the recognition text of the input image. To train the TrOCR model more effectively, the researchers used pre-training models in ViT and BERT modes to initialize the encoder and decoder.

Data Augmentation Method: To increase the diversity of training data and improve the robustness of the model, the team utilized various techniques such as image compression, image warping, stretching, random cropping, and grid mask perspective overlay on the input images.

Additionally, the team attempted to use the ScrabbleGAN [5] generative network for data generation. The team generated 100,000 training data for each language class during model training. However, after comparing the accuracy of the training data with and without data augmentation, the team observed a 1.6% reduction in accuracy. This decrease in accuracy can be attributed to the limited simulation of the generated data. Consequently, the team discarded the data augmentation method.

Training Implementation: First, the team initializes the model to a custom training dataset and generates a character set lexicon file for the current data. Then, the team initializes the custom model weights based on the lexicon file and pre-trained model weights. Finally, the team loads the training data and initializes the custom model weights for model training.

4.5   PERO

The OCR system consists of an optical model (OM) and a language model (LM). For each script, the individual OM and LM are trained. Besides the provided competition datasets, the team also used the original IIIT-INDIC-HW-WORDS validation and test datasets, which differ from the corresponding competition datasets, as additional training data.

The OM is a CRNN-based neural network comprising convolutional blocks and LSTM layers, and it is trained using the CTC loss function. Each OM is trained for 100k iterations with an initial learning rate of 2×10^{-4} and was halved after 80k iterations. After every 2k iterations of training, the model is evaluated on the validation dataset. As the final model, the team selected the one with the lowest character error rate on the validation dataset. The team used data augmentations during training, including color changing, binarization, affine transformations, adding noise, blurring, and masking.

The LM is an LSTM network trained to predict the next character in a sequence. As training data, the team used vocabulary scraped from internet sources (i.e., each word type occurring once in the training data). The team used the transcriptions of the training data from the challenge to select the best model by perplexity, exploring different model widths and dropout rates. The language model is combined with the optical model during prefix beam search over the OM outputs.

4.6   SRUKR

This report describes the approach used for offline handwritten text recognition of Indic languages. It is based on Convolutional Recurrent Neural Network (CRNN) with Connectionist Temporal Classification (CTC) objective function, which allows the processing of image-based sequence-to-sequence data and is widely used in handwriting recognition [20, 21, 26]. There is no need for character segmentation or horizontal scale normalization.

The developed solution is trained on the IHTR2022 and ICDAR2023 datasets. Training scripts were implemented using PyTorch. An essential step of offline handwriting recognition is pre-processing because images can have different sizes, and writing styles differ in the characters' skew, slant, height, and thickness. With pre-processing, the recognition rate is significantly higher. Pre-processing algorithms [21] used in the current recognition solution is described below.

The first pre-processing step is image binarization. For this, we are taking grey-scale images and using Otsu's method. After that, we perform skew and slant correction, then crop and normalize the image height to 64 px.

Each handwriting has its structure of the character's shape with different thicknesses of writing trajectory. To process such samples, we extract the skeleton of these binary images to a skeletal remnant using skeletonization. For the Urdu language, the team flips images to account for right-to-left writing direction. Also, the team performed a set of experiments with blurring and data augmentation to extend the training dataset.

The architecture of the CRNN consists of three components: convolutional layers (CNN), recurrent layers (RNN), and a transcription layer. The CNN extracts feature from the image based on MobileNet-V3 and EfficientNet-V2. The team modifies strides to reduce the sequence length by the X-axis 8 times. It provides a sufficient number of frames necessary for recognition. EfficientNet-V2 gives better results, and at the same time, it has more weight and works longer. The team uses a 1-layer bidirectional GRU with 128 cells as the RNN. Detailed configuration of every layer for trained recognition CRNN is given below. The number of CNN parameters is 5326200, and the number of RNN parameters is 198144.

Configuration of trained CRNN:
Type Configuration
Input width x 64×1
CNN (EfficientNet-V2) channels, kernel, stride, layers

Conv2d 24, 3 × 3, (2, 2), 1
Fused-MBConv 24, 3 × 3, (1, 1), 2
Fused-MBConv 48, 3 × 3, (2, 2), 4
Fused-MBConv 64, 3 × 3, (2, 2), 4
MBConv 128, 3 × 3, (2, 1), 6
MBConv 160, 3 × 3, (2, 1), 9
Reshape the image to a sequence
Conv1d in: 320, out: 128, kernel: 1, stride: 1
RNN
Bidirectional GRU hidden units: 128
Transcription
Linear (256 + 1(bias)) x (chars + 1 (blank))
Output width / 8 x (chars + 1 (blank))

In the recognition phase for decoding, the team uses an adapted version of the token passing algorithm [25]. It allows the utilization of a language model and improves the recognition rates. An adapted token-passing algorithm finds the most probable sequence of complete words, given the class-based 3-gram language model. The dictionary contains about 150 thousands of words.

4.7 LEAP-OCR

Handwriting recognition is one of the most widely researched problems. Handwritten characters have many variations and are available in many scripts and languages, making this problem more challenging. Furthermore, handwritten text in Indic languages also has concerned complexities of conjunct consonants, which further adds to the challenges. Deep learning is widely used to recognize handwriting. The team proposed a Convolutional Recurrent Neural Network (CRNN) based text recognition model as CRNNs have proven beneficial for image-based sequential identification tasks.

Method: The team uses a CRNN [20] based recognition model for Indic Language Handwritten Text Recognition as proposed in the DocTR [15] framework. The competition dataset is pre-processed in the format required by DocTR, which involves creating a JSON file for image names and corresponding ground truths. The image pre-processing also includes making all of them equal sizes (maintaining aspect ratio) by necessary padding, color inversion for a few images, and adding random noise to make the model more robust. The processed images from the competition dataset for the selected language and the ground are fed to the randomly initialized CRNN model. Similarly, the team trained the CRNN models language-wise, each composed of a distinct set of vocabulary. Once the concerned epochs are completed, we use the learned models for handwriting recognition. Along with the trained model, the language of the handwritten data is also given as input for the inference stage to carry out proper predictions in inference stage.

5   Evaluation

5.1   Evaluation Metrics

Two popular evaluation metrics such as Character Recognition Rate (CRR) (alternatively Character Error Rate, CER) and Word Recognition Rate (WRR) (alternatively Word Error Rate, WER) are used to evaluate the performance of recognizers. Error Rate (ER) is defined as

$$ER = \frac{S + D + I}{N}, \tag{1}$$

where S indicates the number of substitutions, D indicates the number of deletions, I indicates the number of insertions, and N number of instances in reference text. In the case of CER, Eq. (1) operates on the character level, and in the case of WER, Eq. (1) operates on word level. Recognition Rate (RR) is defined as

$$RR = 1 - ER. \tag{2}$$

In case of CRR, Eq. (2) operates on the character level and in the case of WRR, Eq. (2) operates on word level.

We assign a rank to a team based on WRR for each language. Each team will get points based on their rank on each script: rank 1 gets 5 points, rank 2 gets 4 points, rank 3 gets 3 points, rank 4 gets 2 points, and rank 5 and below get 1 point for participation. Not submitting results for a language gets a point 0. Since the competition has ten languages, each team may get different ranks and corresponding points for ten languages. Therefore, finding the winner and runner-up team from the list of groups for ten languages is tricky. To solve this issue, we calculate the final point for a team by summing all points corresponding to the languages of that team. We declare the winner and runner-up teams based on the final points.

5.2   Competition Results

Among eighteen registered participants, six teams submitted the results of ten languages (except EE-Noobies submitted results corresponding to three languages: Devanagari, Tamil, and Urdu. Tables 2, 3, 4, 5, 6, 7, 8, 9, 10, 11 show obtained results by different methods for Bengali, Devanagari, Gujarati, Gurumukhi, Kannada, Malayalam, Odia, Tamil, Telugu, and Urdu, respectively. While Table 12 shows the average CRR and WRR over ten languages for all methods, including the baseline.

For Bengali, only five teams submitted results. Upstage KR obtains the best CRR (98.99%) and WRR (96.10%) among the teams. It is because of using additional synthetic and real datasets and multi-stage training strategies. The second highest scoring team, PERO, achieves CRR (98.11%) and WRR (92.02%), which is significantly closer to the best CRR (98.99%) and WRR (96.10%). The team PERO also used additional real data for training and language models to correct prediction. The team light obtains CRR (97.54%) and WRR (91.62%) and is in the third position. During training, this team also used additional synthetic data (10M word images, 1M word images per language). LEAP-OCR is the minor performer. From the results of Bengali, we observed that using additional training data and language models improves CRR and WRR.

For Devanagari, the team Upstage KR achieves the best CRR (98.02%) and WRR (93.16%). The other three groups, PERO (CRR 97.64%, WRR 91.16%), light (CRR 97.24%, WRR 91.16%), and SRUKR (CRR 96.97%, WRR 91.98%), obtain significantly closer CRR and WRR. All these four teams used additional data for training. However, multi-stage training helps Upstage KR to achieve the best CRR and WRR. On the other hand, teams LEAP-OCR and EE-Noobies obtained the least CRR and WRR as they have not used any additional data during training.

Table 2. Shows comparison of results obtained by several methods on **Bengali**.

Method	CRR	WRR	Rank (Point)
Baseline [6]	93.46	75.34	–
Upstage KR	**98.99**	**96.10**	**1 (5)**
PERO	98.11	92.02	2 (4)
LEAP-OCR	74.58	40.98	5 (1)
SRUKR	96.01	88.06	4 (2)
light	97.54	91.62	3 (3)
EE-Noobies	0.0	0.0	0

Table 3. Shows comparison of results obtained by several methods on **Devanagari**.

Method	CRR	WRR	Rank (Point)
Baseline [6]	92.99	74.72	–
Upstage KR	**98.02**	**93.16**	**1 (5)**
PERO	97.64	91.16	3 (3)
LEAP-OCR	80.68	58.16	4 (2)
SRUKR	96.97	91.98	2 (4)
light	97.24	91.16	3 (3)
EE-Noobies	76.04	41.30	5 (1)

Table 4. Shows comparison of results obtained by several methods on **Gujarati**.

Method	CRR	WRR	Rank (Point)
Baseline [6]	54.70	23.75	–
Upstage KR	83.88	61.4	4 (2)
PERO	**84.38**	61.96	3 (3)
LEAP-OCR	50.12	21.25	5 (1)
SRUKR	82.82	62.38	2 (4)
light	84.08	**62.80**	1 (5)
EE-Noobies	0.0	0.0	0

Table 5. Shows comparison of results obtained by several methods on **Gurumukhi**.

Method	CRR	WRR	Rank (Point)
Baseline [6]	90.88	71.42	–
Upstage KR	98.52	**95.28**	1 (5)
PERO	**98.62**	95.0	3 (3)
LEAP-OCR	83.97	56.66	5 (1)
SRUKR	97.06	90.22	4 (2)
light	98.44	95.16	2 (4)
EE-Noobies	0.0	0.0	0

Table 6. Shows comparison of results obtained by several methods on **Kannada**.

Method	CRR	WRR	Rank (Point)
Baseline [6]	89.16	56.08	–
Upstage KR	98.8	93.62	2 (4)
PERO	**99.06**	**94.54**	1 (5)
LEAP-OCR	85.72	50.46	5 (1)
SRUKR	97.77	91.36	4 (2)
light	98.6	92.58	3 (3)
EE-Noobies	0.0	0.0	0

For Gujarati, Gurumukhi, Kannada, Odia, and Tamil languages, the four teams, Upstage KR, PERO, SRUKR, and light, obtained much closer CRR and WRR. For these languages, additional synthetic and real data help the models to predict higher and closer CRR and WRR.

In the case of the Malayalam language, using additional real data for the teams Upstage KR, PERO, light, and SRUKR help the corresponding models achieve good performance. Using synthetic and real data makes the Upstage KR team achieve the highest CRR and WRR.

For Telugu, two teams, Upstage KR (CRR 98.53, WRR 91.18) and PERO (CRR 98.63, WRR 91.44), obtain significantly closer output. While the other two

Table 7. Shows comparison of results obtained by several methods on **Malayalam**.

Method	CRR	WRR	Rank (Point)
Baseline [6]	95.79	77.20	–
Upstage KR	**99.47**	**97.16**	1 (5)
PERO	99.19	94.88	2 (4)
LEAP-OCR	73.78	43.24	5 (1)
SRUKR	97.26	87.22	4 (2)
light	98.98	94.46	3 (3)
EE-Noobies	0.0	0.0	0

Table 8. Shows comparison of results obtained by several methods on **Odia**.

Method	CRR	WRR	Rank (Point)
Baseline [6]	89.83	66.34	–
Upstage KR	94.16	81.0	4 (2)
PERO	**94.77**	**83.38**	1 (5)
LEAP-OCR	82.07	46.68	5 (1)
SRUKR	94.04	81.48	3 (3)
light	94.37	82.96	2 (4)
EE-Noobies	0.0	0.0	0

Table 9. Shows comparison of results obtained by several methods on **Tamil**.

Method	CRR	WRR	Rank (Point)
Baseline [6]	97.88	88.48	–
Upstage KR	99.61	97.92	2 (4)
PERO	99.59	97.54	3 (3)
LEAP-OCR	90.63	60.90	6 (1)
SRUKR	99.15	95.62	4 (2)
light	**99.63**	**98.08**	1 (5)
EE-Noobies	92.24	66.92	5 (1)

teams SRUKR (CRR 97.5, WRR 88.84) and light (CRR 97.61, WRR 86.62), also get substantially closer results. The additional synthetic and real data helps the model better recognize all these four teams. The teams LEAP-OCR and EE-Noobies are inferior CRR and WRR because of no other synthetic and real data for training.

For Urdu, the writing sequence is different from other languages. It is written from right-to-left, while the word is written from left-to-right for other languages. All four teams, Upstage KR, PERO, light, and SRUKR, used additional data for training, reasonable CRR, and WRR obtained. As the words are written from

Table 10. Shows comparison of results obtained by several methods on **Telugu**.

Method	CRR	WRR	Rank (Point)
Baseline [6]	95.82	76.53	–
Upstage KR	98.53	91.18	2 (4)
PERO	**98.63**	**91.44**	**1 (5)**
LEAP-OCR	81.07	36.12	6 (1)
SRUKR	97.5	88.84	3 (3)
light	97.61	86.62	4 (2)
EE-Noobies	84.2	40.48	5 (1)

Table 11. Shows comparison of results obtained by several methods on **Urdu**.

Method	CRR	WRR	Rank (Point)
Baseline [6]	76.23	36.78	–
Upstage KR	89.38	76.3	2 (4)
PERO	89.79	73.34	3 (3)
LEAP-OCR	59.1	22.42	5 (1)
SRUKR	**91.68**	**82.12**	**1 (5)**
light	78.32	48.5	4 (2)
EE-Noobies	0.0	0.0	0

right-to-left, the results of the teams Upstage KR, PERO, and light are lesser than that of SRUKR. The team SRUKR obtained the best CRR (91.68%) and WRR (82.12%). The team SRUKR flips word images to account for right-to-left writing direction for Urdu. Because of this strategy, the SRUKR obtains the best CRR and WRR and 6% margin in WRR than the 2nd best team Upstage KR.

From the experiments, we observed that additional real and synthetic data helped to learn the representation of word images for recognition. Multi-stage

Table 12. Shows comparison of average results over ten languages obtained by several methods.

Method	CRR	WRR	Points
Baseline [6]	87.67	64.66	–
Upstage KR	95.94	**88.31**	40
PERO	**95.98**	87.53	38
light	94.48	84.39	34
SRUKR	95.03	85.93	29
LEAP-OCR	76.17	43.69	11
EE-Noobies	84.16	49.57	3

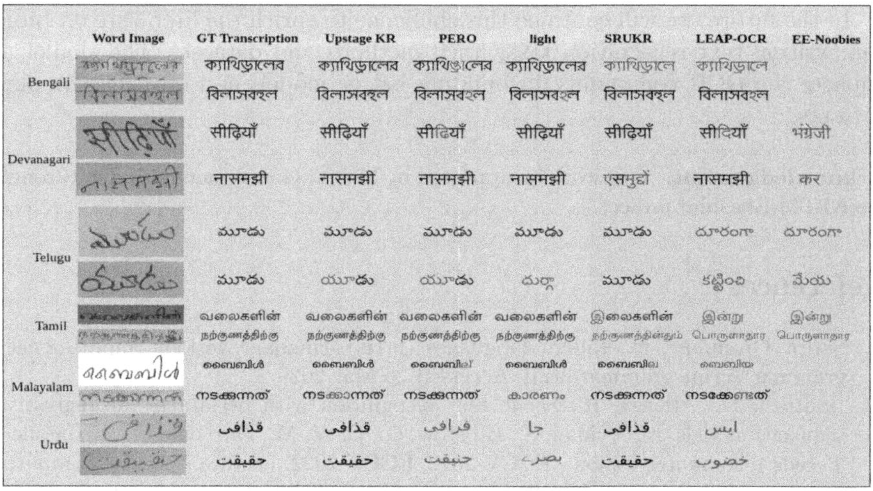

Fig. 6. Shows sample word images recognized by different techniques.

learning further improves recognition accuracy. The language model helps to rectify and correct wrongly recognized words. Though four methods, Upstage KR, PERO, light, and SRUKR, achieve more than 84% WRR over ten languages on average, several wrongly recognized words still need to be corrected. Figure 6 shows a few sample word images recognized by the submitted methods.

6   Conclusion

This competition motivates the researchers to continue researching Indic handwritten text recognition tasks to prevent the risk of vanishing a few Indic scripts/languages. Eighteen teams registered for this competition. Among them, only six teams submitted results along with algorithm details. The Upstage KR team won the competition and achieved an average CRR of 95.94% and WRR of 88.31% over all languages. At the same time, team PERO (average CRR 95.98% and WRR 87.53% over ten languages) won the runner-up position in this competition. Four of the six teams, Upstage KR, PERO, light, and SRUKR, obtain much closer recognition results. The following factors (i) additional synthetic and/or real training data, (ii) pre-processing of the training data, and (iii) language model for recognition error correction help these teams to achieve high recognition scores. In the case of the Urdu language, words are written from right-to-left direction, totally different from the writing direction from left-to-right for other Indic languages. For Urdu, flipping word images helps the SRUKR team to achieve the best performance (CRR 91.68% and WRR 82.12%). All these four methods set a new benchmark for the Indic handwriting text recognition tasks. These methods open a direction for solving Indic handwriting recognition tasks.

In the future, we will continue this challenge to enrich the literature on Indic handwriting text recognition tasks with methods and datasets. The challenge impacts the OCR community in building better models and creating complex datasets.

Acknowledgement. This work is supported by MeitY, Government of India, through the NLTM-Bhashini project.

References

1. Script Grammar. for Indian languages. http://language.worldofcomputing.net/grammar/script-grammar.html. Accessed 26 Mar 2020
2. Bautista, D., Atienza, R.: Scene text recognition with permuted autoregressive sequence models. In: Avidan, S., Brostow, G., Cisse, M., Farinella, G.M., Hassner, T. (eds.) Computer Vision - ECCV 2022. ECCV 2022. Lecture Notes in Computer Science, vol. 13688, pp. 178–196. Springer, Cham (2022). https://doi.org/10.1007/978-3-031-19815-1_11
3. Dutta, K., Krishnan, P., Mathew, M., Jawahar, C.V.: Towards spotting and recognition of handwritten words in Indic scripts. In: ICFHR, pp. 32–37 (2018)
4. Dutta, K., Krishnan, P., Mathew, M., Jawahar, C.: Offline handwriting recognition on Devanagari using a new benchmark dataset. In: DAS, pp. 25–30 (2018)
5. Fogel, S., Averbuch-Elor, H., Cohen, S., Mazor, S., Litman, R.: ScrabbleGAN: semi-supervised varying length handwritten text generation. In: IEEE/CVF Conference on Computer Vision and Pattern Recognition, pp. 4324–4333 (2020)
6. Gongidi, S., Jawahar, C.V.: IIIT-INDIC-HW-WORDS: a dataset for Indic handwritten text recognition. In: Lladós, J., Lopresti, D., Uchida, S. (eds.) ICDAR 2021. LNCS, vol. 12824, pp. 444–459. Springer, Cham (2021). https://doi.org/10.1007/978-3-030-86337-1_30
7. Graves, A., Schmidhuber, J.: Offline handwriting recognition with multidimensional recurrent neural networks. In: NIPS (2008)
8. Jemni, S.K., Ammar, S., Kessentini, Y.: Domain and writer adaptation of offline Arabic handwriting recognition using deep neural networks. Neural Comput. Appl. **34**, 2055–2071 (2022)
9. Krishnan, P., Jawahar, C.V.: HWNet v2: an efficient word image representation for handwritten documents. IJDAR (2019)
10. Li, M., et al.: TrOCR: transformer-based optical character recognition with pre-trained models. arXiv (2021)
11. Liu, Z., et al.: Swin transformer v2: scaling up capacity and resolution. In: IEEE/CVF Conference on Computer Vision and Pattern Recognition, pp. 12009–12019 (2022)
12. Ly, N.T., Nguyen, C.T., Nakagawa, M.: Training an end-to-end model for offline handwritten Japanese text recognition by generated synthetic patterns. In: ICFHR (2018)
13. Maalej, R., Kherallah, M.: Improving the DBLSTM for on-line Arabic handwriting recognition. Multimedia Tools Appl. **79**, 17969–17990 (2020)
14. Marti, U.V., Bunke, H.: The IAM-database: an English sentence database for offline handwriting recognition. Int. J. Doc. Anal. Recogn. **5**, 39–46 (2002)
15. Mindee: docTR: document text recognition (2021). https://github.com/mindee/doctr

16. Nguyen, K.C., Nguyen, C.T., Nakagawa, M.: A semantic segmentation-based method for handwritten Japanese text recognition. In: ICFHR (2020)
17. Pal, U., Chaudhuri, B.: Indian script character recognition: a survey. Pattern Recogn. **37**, 1887–1889 (2004)
18. Peng, D., et al.: Recognition of handwritten Chinese text by segmentation: a segment-annotation-free approach. IEEE Trans. Multimedia (2022)
19. Pham, V., Bluche, T., Kermorvant, C., Louradour, J.: Dropout improves recurrent neural networks for handwriting recognition. In: ICFHR (2014)
20. Shi, B., Bai, X., Yao, C.: An end-to-end trainable neural network for image-based sequence recognition and its application to scene text recognition. IEEE Trans. Pattern Anal. Mach. Intell. **39**(11), 2298–2304 (2016)
21. Viatchaninov, O., Dziubliuk, V., Radyvonenko, O., Yakishyn, Y., Zlotnyk, M.: CalliScan: on-device privacy-preserving image-based handwritten text recognition with visual hints. In: The Adjunct Publication of the 32nd Annual ACM Symposium on User Interface Software and Technology, pp. 72–74 (2019)
22. Wu, Y.C., Yin, F., Chen, Z., Liu, C.L.: Handwritten Chinese text recognition using separable multi-dimensional recurrent neural network. In: ICDAR (2017)
23. Xie, Z., Sun, Z., Jin, L., Feng, Z., Zhang, S.: Fully convolutional recurrent network for handwritten Chinese text recognition. In: ICPR (2016)
24. Yim, M., Kim, Y., Cho, H.-C., Park, S.: SynthTIGER: synthetic text image GEneratoR towards better text recognition models. In: Lladós, J., Lopresti, D., Uchida, S. (eds.) ICDAR 2021. LNCS, vol. 12824, pp. 109–124. Springer, Cham (2021). https://doi.org/10.1007/978-3-030-86337-1_8
25. Young, S.J., Russell, N., Thornton, J.: Token passing: a simple conceptual model for connected speech recognition systems. Citeseer (1989)
26. Zhelezniakov, D., Zaytsev, V., Radyvonenko, O.: Acceleration of online recognition of 2D sequences using deep bidirectional LSTM and dynamic programming. In: Rojas, I., Joya, G., Catala, A. (eds.) IWANN 2019. LNCS, vol. 11507, pp. 438–449. Springer, Cham (2019). https://doi.org/10.1007/978-3-030-20518-8_37

ICDAR 2023 Competition on Visual Question Answering on Business Document Images

Sachin Raja, Ajoy Mondal$^{(\boxtimes)}$, and C. V. Jawahar

International Institute of Information Technology, Hyderabad, India
sachin.raja@research.iiit.ac.in, {ajoy.mondal,jawahar}@iiit.ac.in

Abstract. This paper presents the competition report on Visual Question Answering (VQA) on Business Document Images (VQAonBD) held at the 17th International Conference on Document Analysis and Recognition (ICDAR 2023). Understanding business documents is a crucial step toward making an important financial decision. It remains a manual process in most industrial applications. Given the requirement for a large-scale solution to this problem, it has recently seen a surge in interest from the document image research community. Credit underwriters and business analysts often look for answers to a particular set of questions to reach a decisive conclusion. This competition is designed to encourage research in this broader area to find answers to questions with minimal human supervision. Some problem-specific challenges include an accurate understanding of the questions/queries, figuring out cross-document questions and answers, the automatic building of domain-specific ontology, accurate syntactic parsing, calculating aggregates for complex queries, and so on. Further, despite having the same accounting fundamentals, the terminologies and ontologies used across different organizations and geographic locations may vary significantly. This makes the problem of generic VQA on such documents only more challenging. Since this is the first iteration of the competition, it was restricted in terms of some of the challenges listed; however, the further iterations of this competition aim to include many additional sub-tasks with the larger vision of accurate semantic understanding of business documents as images.

Eleven different teams around the world registered for this competition. Five teams out of those submitted methods spanning multiple approaches, among which Team Upstage KR won the competition with a weighted average score of 95.9%. The runner-up team, NII-TablQA obtained a weighted average score of 90.1%

Keywords: Optical Character Recognition (OCR) · Visual Question Answering (VQA) · Business Documents · Table Structure Recognition (TSR)

1 Introduction

Visual question-answering generally aims to answer a query described in natural language, taking cues from the document image as the only input. As a part of

G. A. Fink et al. (Eds.): ICDAR 2023, LNCS 14188, pp. 454–470, 2023.
https://doi.org/10.1007/978-3-031-41679-8_26

this competition, we propose a visual question-answering dataset and baseline model from business document images. While a lot of work has already been done in the broader VQA space [1–11], the questions from business documents present many niche challenges that may require cross-document referencing, additional numeric computations over the simple search query to reach the final solution, and so on. Further, since most business documents are usually presented in a tabular format, leveraging this structural conformity to answer more challenging queries may be non-trivial. Given the unique nature of the problem, its tremendous prospect in the industry, layers of challenges to be tackled, and the recent surge of interest in visual question answering, we believe that there would be a surge in the research interest in this area in the near future (Fig. 1).

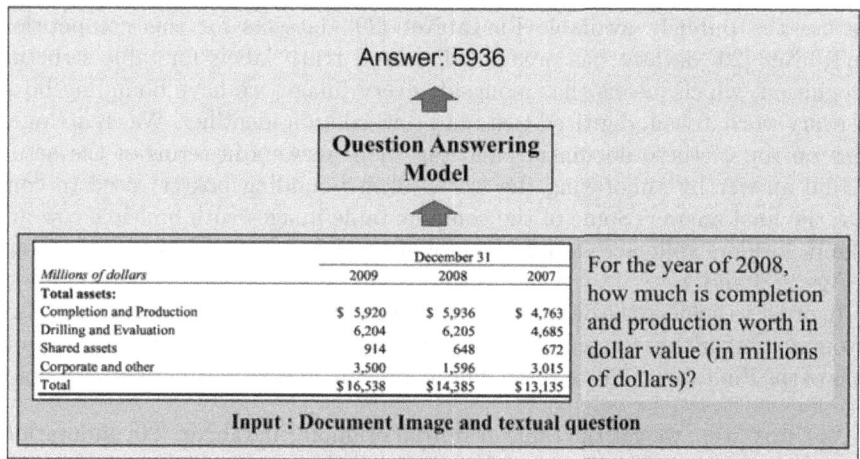

Fig. 1. Given a document image and questions, the task of the competition is to produce answers corresponding to the questions.

The recent works in the broader problem of visual question answering on generic scene images demonstrate the ability of deep-learning models to understand the context of the scene at hand. While at first glance, the problem of document VQA, particularly VQA on tabular images seems quite similar, the reality is quite different. Tabular data often presents highly dense data compressed in a structured format, with limited linguistic contexts. This is usually because most of the data present is in numeric format, which is more complex to digitize and understand in the broader context than standard documents containing sentences and paragraphs.

Another possible way to approach the problem may be through a more pipeline-driven methodology which would need table detection and table reconstruction as precursors. Though this process involves multiple stages, it would result in easy explainability with respect to question-answering. Moreover, the success of recent methods in table reconstruction space [12–19] make this approach as a reasonable prospective.

The problem at hand has an immense utility, primarily in banking and insurance verticals where analysts manually digitize the incoming financial reports (including but not limited to balance sheets, income statements and cash flow statements). As a next step, subject matter experts, such as credit underwriters, peruse these reports to extract answers for a specific set of queries to make a decision. This competition aims to pose this problem as a cognitive machine learning task to answer the queries at hand, given only the table image along with queries as inputs.

The paper is organized as follows. In Sect. 2, we give details about the dataset used for the competition. The submitted methods are discussed in Sect. 3. Section 4 shows the results of the competition. The conclusive remark is drawn in Sect. 5.

2   Dataset

We use the publicly available FinTabNet [20] datasets for this competition. FinTabNet [20] dataset has predefined ground truth labels for table structure recognition, which means that alongside every image, we have bounding boxes for every word/token, digitized text, and row/column identifier. We create questions on top of these documents and tag their answers in terms of the actual textual answer by annotating the word/token bounding box(es) used to compute the final answer. Some of the complex table images with multiple row and column headers split across different columns and rows respectively are shown in Figs. 2, 3 and 4.

In order to achieve the desired scale of the dataset, we employ a heuristic-based automated algorithm to create the dataset using original table structure annotations of the FinTabNet [20] dataset. The algorithm, in brief, is described as follows:

- As a first step, we get the table grid from original FinTabNet [20] annotations that allow us to identify all table cells, including those which are empty.
- The next step is to identify the data-type of each cell based on its content. The data-types include string, integer, floating-point, empty, percentage-value, year, month, date, special chars and ranges, to name a few.
- Once the datatype of every cell is identified, we employ heuristics to identify row headers and column headers depending on data-types and whether the cell spans multiple rows and/or columns.
- In case the image contains multiple tables, as shown in Figs. 2, 3 and 4, we use the header information to split the tables horizontally and/or vertically.
- At this point, all the information and metadata (row headers, column headers, cell data-types) of the table are extracted.
- Most business report document tables can often be represented in a tree-like structure where certain rows add up corresponding to a row below in the table in a recursive manner. We extract this tree structure for every table in the dataset to identify inter-row relationships.
- Lastly, we generate questions of varying difficulty levels using all the table-level and cell-level metadata collected as described above.

| | Amount of Gain / (Loss) Recognized in OCI | | | | Amount of Gain / (Loss) Reclassified from OCI | | |
| | Year Ended December 31, | | | | Year Ended December 31, | | |
Derivative Instrument	2013	2012	2011	Location on Statement of Earnings	2013	2012	2011
Foreign exchange forward contracts	$63.9	$16.3	$(34.9)	Cost of products sold	$ 8.0	$(12.0)	$(32.9)
Foreign exchange options	(0.3)	(1.1)	(0.2)	Cost of products sold	(0.2)	(0.4)	–
Cross-currency interest rate swaps	–	–	0.2	Interest expense	–	0.2	(8.3)
	$63.6	$15.2	$(34.9)		$ 7.8	$(12.2)	$(41.2)

Fig. 2. Example of a complex table image that has row headers split across different columns.

The different categories of questions imply varying difficulty levels of the questions as described below:

| (In millions) | Zions Bank | | | Amegy | | | CB&T | | |
	2017	2016	2015	2017	2016	2015	2017	2016	2015
SELECTED INCOME STATEMENT DATA									
Net interest income	$ 650	$ 624	$ 544	$ 483	$ 460	$ 387	$ 476	$ 434	$ 377
Provision for loan losses	19	(22)	(28)	25	163	91	(5)	(9)	(4)
Net interest income after provision for loan losses	631	646	572	458	297	296	481	443	381
Noninterest income	151	149	133	118	123	121	75	67	63
Noninterest expense	436	424	430	336	326	373	299	290	294
Income before income taxes	$ 346	$ 371	$ 275	$ 240	$ 94	$ 44	$ 257	$ 220	$ 150
SELECTED AVERAGE BALANCE SHEET DATA									
Total average loans	$ 12,481	$ 12,538	$ 12,118	$ 11,021	$ 10,595	$ 10,148	$ 9,539	$ 9,211	$ 8,556
Total average deposits	15,986	15,991	15,688	11,096	11,130	11,495	11,030	10,827	10,063
(In millions)	NBAZ			NSB			Vectra		
	2017	2016	2015	2017	2016	2015	2017	2016	2015
SELECTED INCOME STATEMENT DATA									
Net interest income	$ 206	$ 190	$ 152	$ 134	$ 122	$ 94	$ 126	$ 120	$ 101
Provision for loan losses	(8)	(3)	8	(11)	(28)	(28)	1	(8)	5
Net interest income after provision for loan losses	214	193	144	145	150	122	125	128	96
Noninterest income	40	40	36	40	39	36	25	23	21
Noninterest expense	148	144	133	139	137	131	101	97	98
Income before income taxes	$ 106	$ 89	$ 47	$ 46	$ 52	$ 27	$ 49	$ 54	$ 19
SELECTED AVERAGE BALANCE SHEET DATA									
Total average loans	$ 4,267	$ 4,086	$ 3,811	$ 2,357	$ 2,284	$ 2,344	$ 2,644	$ 2,469	$ 2,400
Total average deposits	4,762	4,576	4,311	4,254	4,137	3,891	2,756	2,720	2,792
(In millions)	TCBW			Other			Consolidated Company		
	2017	2016	2015	2017	2016	2015	2017	2016	2015
SELECTED INCOME STATEMENT DATA									
Net interest income	$ 46	$ 38	$ 28	$ (56)	$ (121)	$ 32	$ 2,065	$ 1,867	$ 1,715
Provision for loan losses	2	—	(3)	1	—	(1)	24	93	40
Net interest income after provision for loan losses	44	38	31	(57)	(121)	33	2,041	1,774	1,675
Noninterest income	5	5	4	90	70	(57)	544	516	357
Noninterest expense	20	19	17	170	148	105	1,649	1,585	1,581
Income (loss) before income taxes	$ 29	$ 24	$ 18	$ (137)	$ (199)	$ (129)	$ 936	$ 705	$ 451
SELECTED AVERAGE BALANCE SHEET DATA									
Total average loans	$ 926	$ 791	$ 707	$ 266	$ 88	$ 87	$ 43,501	$ 42,062	$ 40,171
Total average deposits	1,107	1,007	879	1,209	207	(481)	52,200	50,595	48,638

Fig. 3. Example of a complex table image that have column headers split across different rows.

- To generate category 1 questions, which are simple extraction queries, we define multiple question templates and depending on the cell data-type and metadata, we curate the question accordingly.
- For the questions of category type 2, we compute ratios of cells that belong to the same row but across two different columns. The question is then curated according to the pre-defined multi-paraphrased templates by populating the corresponding values of the row header and the two-column headers.
- For the questions of category type 3, we compute ratios of cells across two different rows. The question is then curated according to the pre-defined multi-paraphrased templates by populating the corresponding values of the row and column headers.
- For the questions of category type 4, we compute aggregation functions (among minimum, maximum, mean, median and cumulative) across cells with the same row header but belonging to different years or months of the report. The question is then curated according to the pre-defined multi-paraphrased templates by populating the corresponding values of the row and column headers (years).
- For the questions of category type 5, we make use of the recursive inter-rows relationships to compute aggregation (among minimum, maximum, mean, median and cumulative) across a group. The questions around these groups are generated from the same column header and group row headers of the report. The question is then curated according to the pre-defined multi-paraphrased templates by populating the corresponding values of the row and column headers.

(In thousands)	Net unrealized gains (losses) on investment securities	Net unrealized gains (losses) on derivatives and other	Pension and post-retirement	Total
2015				
Balance at December 31, 2014	$ (91,921)	$ 2,226	$ (38,346)	$ (128,041)
Other comprehensive income (loss) before reclassifications, net of tax	(12,471)	4,903	(3,161)	(10,729)
Amounts reclassified from AOCI, net of tax	86,023	(5,583)	3,718	84,158
Other comprehensive income (loss)	73,552	(680)	557	73,429
Balance at December 31, 2015	$ (18,369)	$ 1,546	$ (37,789)	$ (54,612)
Income tax expense (benefit) included in other comprehensive income (loss)	$ 48,422	$ (331)	$ 374	$ 48,465
2014				
Balance at December 31, 2013	$ (168,805)	$ 1,556	$ (24,852)	$ (192,101)
Other comprehensive income (loss) before reclassifications, net of tax	82,204	2,275	(15,284)	69,195
Amounts reclassified from AOCI, net of tax	(5,320)	(1,605)	1,790	(5,135)
Other comprehensive income (loss)	76,884	670	(13,494)	64,060
Balance at December 31, 2014	$ (91,921)	$ 2,226	$ (38,346)	$ (128,041)
Income tax expense (benefit) included in other comprehensive income (loss)	$ 60,795	$ 467	$ (8,764)	$ 52,498

Fig. 4. Example of a complex table image that have column headers (year of the table) split across different rows.

During the training phase, the dataset is divided into two categories - training and validation sets containing 39,999 and 4535 table images respectively. Ground truth corresponding to each table image consists of the following: Table Structure Annotation: Each cell is annotated with information about its bounding box, digitised content, and cell spans in terms of start-row, start-column, end-row and end-column indices. Difficulty-Wise Sample Questions and Answers: Corresponding to every table image, a few sets of questions along with their answers are annotated in the JSON file. The questions are organised into five categories in increasing order of difficulty. The question types primarily include extraction type query, ratio calculations and aggregations across rows and/or columns. Further, answer types are classified as text or numeric. While text answers will be evaluated according to edit-distance-based measures, for numeric-type answers, the absolute difference between the ground-truth and predicted value will also be taken into account. Ideally, to answer all the questions correctly, both syntactic along with a semantic understanding of the business document would be required. Each table image would have annotations for a maximum of 50 questions and corresponding answers for training and validation. Depending on the format and content of the table, the total number of questions from each category within a single table will be in the following range:

- Category 1 : 0–25
- Category 2 : 0–10
- Category 3 : 0–3
- Category 4 : 0–7
- Category 5 : 0–5

Every training annotation is in the form of a json file that contains two primary keys:

- Table Structure (table_structure): Each key within this object is represented by an integer value, cell_id. The object corresponding to this cell_id has information about its bounding box, start_row, start_col, end_row, end_col and content.
- Questions and Answers (questions_answers): The keys within this object denote the category of questions (category_1, etc.). Further, the object corresponding to each category is again a dictionary with a key corresponding to the question_id and a value corresponding to the question_object containing the question as the string, its answer and answer type.

During the evaluation, the predictions are expected in a similar JSON format such that the key at the first level is the category_id. Within each category is a nested dictionary such that its key is the question_id and the corresponding value is the predicted answer.

The statistics of the dataset are as shown below:

Table 1. Division of dataset into training, validation, and test sets. #: indicates counts.

Dataset-Type	#Images	#Total Questions	#Numeric Questions	#Text Questions
Training	39,999	1,254,165	1,197,358	56,807
Validation	4,535	141,465	134,651	6,814
Test	4,361	135,825	129,861	5,964

3   Methods

In this section, we discuss each of the submitted methods including the baseline in detail. Eleven teams registered for the competition. However, we obtained complete submissions from five of them, which include results, submission reports, trained model(s) and inference codes. One team did submit the results but did not submit other details to test for reproducibility and hence, won't be included in the leaderboard. These five final participants are:

Table 2. Category-wise distribution of questions in training, validation and test datasets. #: indicates counts.

Question	Training Dataset		Validation Dataset		Test Dataset	
Category	#Numeric	#Text	#Numeric	#Text	#Numeric	#Text
Category 1	632,037	56,807	69,458	6,814	68,439	5,964
Category 2	137,395	0	15,396	0	14,705	0
Category 3	107,712	0	12,471	0	11,863	0
Category 4	187,844	0	21,696	0	20,609	0
Category 5	132,370	0	15,630	0	14,245	0

- **Upstage KR**, affiliation: Upstage
- **NII-TABIQA**, affiliation: National Institute of Informatics, Japan
- **DEEPSE-X-UPSTAGE-HK**, affiliation: DeepSE x Upstage HK
- **BD-VQA**, affiliation: Apple Inc.
- **SFANC57**, affiliation: OneConnect FinTech

3.1   Baseline

We evaluated the method proposed by Xu and Li. [21,22] for our baseline. In their work, they proposed the model called **LayoutLM**, which jointly models interactions between text and layout information across scanned document images. This becomes beneficial for a great number of real-world document image understanding tasks such as information extraction from scanned documents.

To add to this, authors also leverage image features to incorporate words' visual information into LayoutLM. Their architecture extends the well-known Bert [23] model by adding two types of input embeddings: (i)a 2-D position embedding that denotes the relative position of a token within a document; and (ii) an image embedding for scanned token images within a document. The proposed 2-D position embedding captures the relationship among tokens within a document, meanwhile, the image embedding captures visual characteristics including but not limited to fonts, font-styles, colors, etc. In addition, authors employ a multi-task learning objective for LayoutLM [21,22], which includes a Masked Visual-Language Model (MVLM) loss and a Multi-label Document Classification (MDC) loss. The two losses combined allow for joint pre-training of text and layout collectively. It is important to note that in order to extract the token, authors use an OCR tool as a precursor to the joint training. The pre-trained model was then finetuned on form understanding, receipt understanding and document image classification as the downstream tasks. The implementation that we have employed is in the form of an API available on Hugging-Face, which has been further finetuned on both the SQuAD2.0 [24] and DocVQA [6] datasets. This makes it a go to choice for our baseline[1]

3.2   Upstage KR

Participants use three models named CPRQ (Component Prediction from Raw Question), CPEQ (Component Prediction from Extracted Questions), and CPEQ Pseudo. After the prediction of each model, they generate the final result using weighted hard voting (Table 1).

CPRQ. Component Prediction from Raw Question (CPRQ) attempts to train the generative model (Donut [25]) to predict the values of the components needed to answer the original raw question. Taking the ratio-type questions as an example, instead of training the model to predict the final ratio answer, it was trained to output the values present within the table that are needed to solve the ratio question. After successfully extracting of the necessary component values, subsequent mathematical operations (e.g. ratio) could be applied in the post-processing step. To obtain the component values corresponding to the different mathematical operation questions, both rule-based algorithms and external generative model API were used. For the external generative model API, ChatGPT 3.5 [26] API to be specific, only the training dataset was used to find the component values and train their model (Table 2).

CPEQ. Component Prediction from Extracted Questions (CPEQ) attempts to train the generative model (Donut [25]) to predict component value from extracted questions.

---

[1] The code is available at https://huggingface.co/impira/layoutlm-document-qa.

First, a raw question is divided into multiple extractive questions similar to those in category 1 by pre-defined rules. For example, "What is the ratio of the value of due after 10 years for the year 2018 to the year 2017?" is divided into two extractive questions such as "What is the ratio of the value of due after 10 years for the year 2018?" and "What is the ratio of the value of due after 10 years for the year 2017?". Participants defined some dividing patterns that can cover all questions.

Second, a trained model using only category 1 data as training data predict both category 1 questions and extracted questions in categories 2–5. Lastly, predictions from extracted questions in categories 2–5 are post-processed to generate the final result by operation (e.g. maximum, minimum, ratio).

CPEQ Pseudo. CPEQ is trained using only category 1 data. For data augmentation, pseudo question-answer pair is generated by the CPEQ algorithm, and the trained CPEQ model is fine-tuned on pseudo data. The resulting model is CPEQ Pseudo.

3.3   NII-TablQA

The team introduces TabIQA, a system designed for question-answering using table images in business documents, as illustrated in Fig. 5. Given a table image of a business document and a question about the image, the system utilizes the table recognition module to extract table structure information and the text content of each table cell and convert them into HTML format. Subsequently, the high-level table structure is extracted to identify the headers, data cells, and hierarchical structure with the post-structure extraction module. Once the table is structured, it is converted to a data frame for further processing. The question-answering module processes the input question and the table data frame with an encoder and generates the final answer from a decoder.

Table Recognition. This module aims to predict the table structure information and the text content of each table cell from a table image and represent them in a machine-readable format (HTML). Specifically, this module consists of one shared encoder, one shared decoder, and three separate decoders for three sub-tasks of table recognition: table structure recognition, cell detection, and cell-content recognition. Participants trained this model on the training set of VQAonBD 2023 and validated it on the validation set of VQAonBD 2023 for model selection and choosing the hyperparameters.

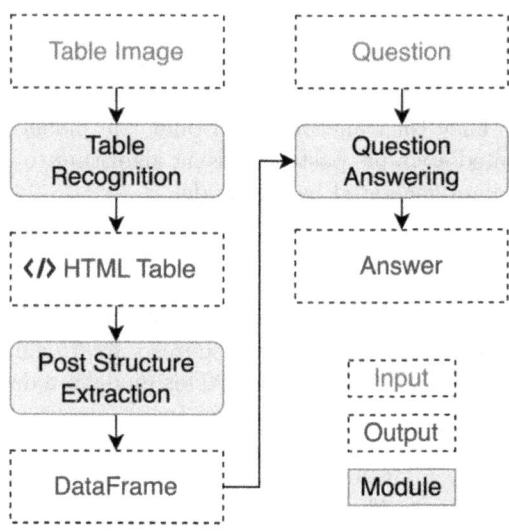

Fig. 5. Architectural diagram of the team NII-TablQA.

Post-structure Extraction. The TabIQA system classifies table headers and data rows from HTML tables using a set of heuristics. Specifically, the system identifies headers as some of the first table rows with column spans, nan cells, or duplicate values in the same rows. The system designates the first row as the table header if no header is found. The system then classifies the remaining rows as data cells. The system identifies hierarchical rows by focusing on data cells with column spans for entire rows. Once the system has identified the structured table, it generates a table data frame by concatenating the values of the header rows to form a one-row header and concatenating the value of each hierarchical row to the lower-level cell values to improve the interpretation of each cell value and provide a more accurate representation of the table data in the data frame.

Question Answering. This module is built on the state-of-the-art table-based question-answering model, OmniTab [27]. The team fine-tuned the OmniTab [27] large pre-trained models using the VQAonBD 2023 training set.

3.4 DeepSE-x-Upstage-HK

Their method, Donut-EAMA (Extract Answer Merge Answer), is based on the end-to-end OCR-free document understanding model - Donut [25] (https:// github.com/clovaai/donut). To apply it on the VQAonBD task, they first pre-trained the model on the training set with the text-reading task. Then considering the model had no training involving arithmetic calculations, they believed that asking it to answer the questions directly would probably not work well. Therefore, the team developed a rule-based algorithm that extracts relevant cell values based on the question and the provided table annotations for the training set and uses those extracted values as labels to reformulate the task into an extractive one. They then finetune the Donut [25] model on this extractive task and implemented a simple post-processing algorithm to calculate the final answer from the values generated by the model.

3.5 BD-VQA

As part of this challenge, the team has used Donut [25] VQA (Visual Question Answering) pre-trained model open-sourced by Hugging face (https://huggingface.co/naver-clova-ix/donut-base). This model is a deep learning model that is designed to answer questions about images of donuts.

Before feeding the image and question list as inputs into the Donut VQA system [25], they performed data pre-processing, handling questions from different categories in distinct ways. They left Category 1 questions as they were, while for Category 2 and Category 3 questions, they split them into two independent questions and subsequently computed the ratio of the two values in the table. This was done because they noticed that these questions relied on the ratio of two values.

For Category 4 and 5 questions that involved operations such as median, maximum, minimum, cumulative, and average, were found to rely on the final aggregate output of three values in the table. Hence, the team split them into three separate questions. Using Donut VQA [25], they predicted the value of each question, and then computed the corresponding operator value to obtain the final result.

3.6 SFANC57

For the system used for VQAonBD, the team has chosen the OCR-free VDU model Donut [25]. For category 1 questions: most answers can be directly selected from the original table content; thus we generate the answer from the Donut-VQA model. For category 2–5 questions, they developed a simple query parsing script to split the logic into content selection and aggregation calculation.

4   Evaluation

4.1   Evaluation Metrics

During the evaluation, a model is expected to take only the document image and question as the input to produce the output. This output is then compared against the ground truth answer to obtain a quantitative evaluation score computed over the entire evaluation dataset.

In most cases, the expected answers to questions from business documents are single numeric token ones. It makes classical accuracy a good prospect for evaluating this task. While for a more generic assignment of visual question answering, there may be some subjectivity in the answers (e.g., white, off-white, and cream may all be correct answers), the solutions for the proposed task are primarily objective and absolute. It makes evaluation relatively straightforward. Hence, we use standard accuracy as the primary criterion for evaluation. Further, we also employ averaged absolute deviation as one of the criteria for numeric-type answers. If the absolute difference between the ground truth and the predicted value is more than 100%, we give a score of 0. In the other case, the score is defined by:

$$Deviation\ Score\ =\ 1\ -\ \frac{absolute\ distance}{ground\ truth\ value} \tag{1}$$

However, since the input to the model will only be by the document image to answer a specific query, penalizing the VQA model word/token detection and recognition is not fair. Therefore, we also employ Averaged Normalized Levenshtein Similarity (ANLS) as proposed in [28,29], which responds softly to answer mismatches due to OCR imperfections. ANLS is given by Eq. 2, where N is the total number of questions, M are possible ground truth answers per question, $i\ =\ 0...N, j\ =\ 0...M$ and o_{q_i} is the answer to the ith question q_i.

$$ANLS = \frac{1}{N} \sum_{i=0}^{N} \left(\max_j\ s(a_{ij}, o_{q_i}) \right)$$

$$s(a_{ij}, o_{q_i}) = \begin{cases} 1 - NL(a_{ij}, o_{q_i}), & \text{if } NL(a_{ij}, o_{q_i}) < \tau. \\ 0, & \text{otherwise.} \end{cases} \tag{2}$$

where $NL(a_{ij}, o_{q_i})$ is the normalized Levenshtein distance (ranges between 0 and 1) between the strings a_{ij} and o_{q_i}. The value of τ can be set to add softness toward recognition errors. If the normalized edit distance exceeds τ, it is assumed that the error is because of an incorrectly located answer rather than an OCR mistake.

The final score is an L2 norm of the deviation score and the ANLS score, both of which range between 0 and 1 for the numeric values. For text answers, the final score is the same as the ANLS score.

4.2   Results

Out of eleven registered participants, we received submissions from a total of six teams. Five of them submitted their results along with a brief description of their method, trained model(s) and inference codes. The final leaderboard consists of those five submissions. Furthermore, we have executed the inference codes for each of the submissions to ensure that the submission score could be replicated within a +−1% score. Also, we received multiple submissions from each team. To ensure there was no cherry-picking of the best-performing submission, we only considered the most recent submission by the team within the deadline window.

Table 3. Final Scores corresponding to the latest submissions of all the participating teams. Categories 1 through 5 indicate the average scores corresponding to questions of each category, All Avg indicates the average scores and Weighted average indicates the weighted average score, based on which the final ranking was decided.

Team	Category 1	Category 2	Category 3	Category 4	Category 5	All Avg	Weighted Avg
Baseline	0.281	0.091	0.096	0.200	0.169	0.168	0.163
UPSTAGE KR	0.963	0.942	0.953	0.974	0.956	0.957	0.959
NII-TABIQA	0.932	0.876	0.855	0.895	0.931	0.898	0.901
DEEPSE-X-UPSTAGE-HK	0.939	0.874	0.859	0.902	0.858	0.886	0.879
BD-VQA	0.799	0.794	0.729	0.736	0.422	0.696	0.640
SFANC57	0.648	0.119	0.132	0.463	0.418	0.356	0.359

Table 4. Final exact match accuracy scores corresponding to the latest submissions of all the participating teams. Categories 1 through 5 indicate the average exact match scores corresponding to questions of each category, All Avg indicates the average exact match scores and Weighted average indicates the weighted average exact match score.

Team	Category 1	Category 2	Category 3	Category 4	Category 5	All Avg	Weighted Avg
BASELINE	0.085	0.000	0.000	0.015	0.012	0.023	0.015
UPSTAGE KR	0.933	0.907	0.925	0.957	0.924	0.929	0.931
DEEPSE-X-UPSTAGE-HK	0.872	0.799	0.784	0.791	0.734	0.796	0.778
BD-VQA	0.586	0.630	0.533	0.501	0.110	0.472	0.397
NII-TABIQA	0.874	0.554	0.451	0.215	0.259	0.470	0.374
SFANC57	0.111	0.001	0.002	0.090	0.140	0.069	0.082

Table 5. Evaluation based on answer data types.

Team	Numeric Score	Text Score	Micro-Average Score	Numeric Exact Match Score	Text Exact Match Score	Micro Average Exact Match Score
BASELINE	0.214	0.359	0.220	0.051	0.020	0.050
UPSTAGE	0.962	0.929	0.960	0.934	0.880	0.932
NII-TABIQA	0.924	0.674	0.913	0.645	0.470	0.637
DEEPSE-X-UPSTAGE-HK	0.912	0.870	0.910	0.833	0.750	0.829
BD-VQA	0.753	0.522	0.743	0.545	0.051	0.523
SFANC57	0.494	0.470	0.493	0.091	0.057	0.089

From Tables 3 and 4, it is evident that the team **UPSTAGE KR** won the competition by a significant margin of 5.8% average weighted final score across all the categories of questions as compared to the runner-up team, which obtained a score of 90.1%. There are many interesting conclusions that can be drawn from these results. If we only consider the simple extractive questions, which belong to category 1, we observe that the results obtained by the top three teams are within a close range of 3% scores. Among the participants, we observe three very distinct approaches toward the solution. The first team follows a weighted ensemble-driven approach where they train three different generative models using the architecture of Donut [25] and ChatGPT 3.5 [26] API to answer the questions. The second team, on the other hand, follows a more pipeline-driven approach where they perform table recognition as a precursor step for post-structure data extraction using heuristics to extract row and column headers. On top of the structured information extracted, they use the OmniTab [27] model to generate answers. The third team used the Donut [25] model but reformulated the task into an extractive task instead of a text reading task. The fourth and fifth teams used Donut [25] model to extract answers to the questions. The fourth team developed parsers to break down complex questions into simple ones, while the fifth standing team did not fine-tune or developed any query parsers but used the standard Donut [25] model API available on hugging-face to generate answers.

The numbers clearly indicate that the fine-tuning of the pre-trained generative models like Donut [25] is imperative to obtain any meaningful results in the first place because of completely different dataset distributions. The difference between the scores of the third and fourth teams also clearly indicates the significance of training a problem-specific downstream task for a generative model instead of using it right out of the box. Further, a difference of almost 19% score between the BD-VQA and SFANC57 teams indicates that developing complex question parsers and transforming those into simple extractive queries can significantly aid generative models; however, such models fail to perform well directly on the aggregation and ratio-type complex questions.

Further, Table 5 compares the performance of each submission on text and numeric-type questions. The non-trivial difference between the proposed evaluation score and exact match accuracy scores clearly demonstrates that there is some error induced because of OCR mistakes. The difference however is particularly stark for the team NII-TABIQA. Our qualitative analysis suggests that the difference is primarily in the least significant bits of the numeric values. The significant difference for the same submission for text-based questions further signifies that OCR does not seem to be as accurate as compared to the other submissions.

As discussed above, we draw many interesting conclusions from various submissions of this competition. In this first iteration of the competition, we only requested for the answers of every question put forward in front of the model and did not ask for where the relevant information was picked up from in order to answer the query. This makes it hard for us to thoroughly investigate the errors made by the OCR tool in extracting tokens. In the next version, we would definitely ask for the coordinates of the relevant tokens which would allow us to thoroughly investigate the submissions from the OCR dimension as well.

5   Conclusion

This competition aims to bridge the gap between the document research community in the academia and the industry. Through this competition, we have seen two primary distinct ways in which researchers go about tackling this problem -

(i) through direct VQA on images as a black box; and (ii) a more pipeline-driven approach using table structure recognition and OCR as precursors to answering the query. The high-performing quantitative results show both approaches as promising directions of research in this space.

Since this was the first version of this competition and in turn the dataset, the questions were generated primarily using keywords from the underlying ground-truth tokens of the document itself. Furthermore, the aggregation queries by themselves contained many cues using which it was not so difficult to break them down into simpler questions to answer (as we have seen in most of the submissions). The reasonable number of participants and submissions in this challenge motivates us to take this further and build upon the dataset to make it all the more challenging. Some of the ways in which we plan to do this are to (i) increase the scope of the documents (including invoices, receipts, etc.); (ii) add cross-document questions; (iii) add additional sub-tasks (such as table-specific tokens detection and recognition, table structure recognition, key-value pair detection); and (iv) by building domain specific taxonomy and ontology which would make the questions independent of the absolute keywords seen in the document thereby making them generic for multiple similar style of documents. We believe that in the future, our competition would play a vital role in getting towards a rather "Grand Challenge" in the document research space at large.

In conclusion, we hope that this competition would continue to bridge the gap between the document research community in academia and the industry. We also hope that models presented in this competition will eventually lead to the building of state-of-the-art artificially intelligent methods that could solve the real-world problem efficiently at a large scale.

Acknowledgement. This work is supported by MeitY, Government of India.

References

1. Antol, S., et al.: VQA: visual question answering. In: Proceedings of the IEEE International Conference on Computer Vision, pp. 2425–2433 (2015)
2. Selvaraju, R.R., Cogswell, M., Das, A., Vedantam, R., Parikh, D., Batra, D.: Grad-CAM: visual explanations from deep networks via gradient-based localization. In: Proceedings of the IEEE International Conference on Computer Vision, pp. 618–626 (2017)
3. Anderson, P., et al.: Bottom-up and top-down attention for image captioning and visual question answering. In: Proceedings of the IEEE Conference on Computer Vision and Pattern Recognition, pp. 6077–6086 (2018)
4. Zhou, L., Palangi, H., Zhang, L., Houdong, H., Corso, J., Gao, J.: Unified vision-language pre-training for image captioning and VQA. In: Proceedings of the AAAI Conference on Artificial Intelligence, vol. 34, pp. 13041–13049 (2020)
5. Changpinyo, S., Kukliansky, D., Szpektor, I., Chen, X., Ding, N., Soricut, R.: All you may need for VQA are image captions. arXiv preprint: arXiv:2205.01883 (2022)
6. Mathew, M., Karatzas, D., Jawahar, C.V.: DocVQA: a dataset for VQA on document images. In: Proceedings of the IEEE/CVF Winter Conference on Applications of Computer Vision, pp. 2200–2209 (2021)

7. Mishra, A., Shekhar, S., Singh, A.K., Chakraborty, A.: OCR-VQA: visual question answering by reading text in images. In: 2019 International Conference on Document Analysis and Recognition (ICDAR), pp. 947–952. IEEE (2019)
8. Yusuf, A.A., Chong, F., Xianling, M.: An analysis of graph convolutional networks and recent datasets for visual question answering. Artif. Intell. Rev. **55**, 1–24 (2022)
9. Lu, J., Yang, J., Batra, D., Parikh, D.: Hierarchical question-image co-attention for visual question answering. In: Advances in Neural Information Processing Systems, vol. 29 (2016)
10. Jang, Y., Song, Y., Yu, Y., Kim, Y., Kim, G.: TGIF-QA: toward Spatio-temporal reasoning in visual question answering. In: Proceedings of the IEEE Conference on Computer Vision and Pattern Recognition, pp. 2758–2766 (2017)
11. Gokhale, T., Banerjee, P., Baral, C., Yang, Y.: VQA-LOL: visual question answering under the lens of logic. In: Vedaldi, A., Bischof, H., Brox, T., Frahm, J.-M. (eds.) ECCV 2020. LNCS, vol. 12366, pp. 379–396. Springer, Cham (2020). https://doi.org/10.1007/978-3-030-58589-1_23
12. Tensmeyer, C., Morariu, V.I., Price, B., Cohen, S., Martinez, T.: Deep splitting and merging for table structure decomposition. In: ICDAR (2019)
13. Qasim, S.R., Mahmood, H., Shafait, F.: Rethinking table parsing using graph neural networks. In: ICDAR (2019)
14. Qiao, L., et al.: LGPMA: complicated table structure recognition with local and global pyramid mask alignment. In: Lladós, J., Lopresti, D., Uchida, S. (eds.) ICDAR 2021. LNCS, vol. 12821, pp. 99–114. Springer, Cham (2021). https://doi.org/10.1007/978-3-030-86549-8_7
15. Zhang, Z., Zhang, J., Jun, D., Wang, F.: Split, embed and merge: an accurate table structure recognizer. Pattern Recogn. **126**, 108565 (2022)
16. Lin, W., et al.: TSRFormer: table structure recognition with transformers. arXiv preprint: arXiv:2208.04921 (2022)
17. Long, R., et al.: Parsing table structures in the wild. In: Proceedings of the IEEE/CVF International Conference on Computer Vision, pp. 944–952 (2021)
18. Raja, S., Mondal, A., Jawahar, C.V.: Table structure recognition using top-down and bottom-up cues. In: Vedaldi, A., Bischof, H., Brox, T., Frahm, J.-M. (eds.) ECCV 2020. LNCS, vol. 12373, pp. 70–86. Springer, Cham (2020). https://doi.org/10.1007/978-3-030-58604-1_5
19. Raja, S., Mondal, A., Jawahar, C.V.: Visual understanding of complex table structures from document images. In: Proceedings of the IEEE/CVF Winter Conference on Applications of Computer Vision, pp. 2299–2308 (2022)
20. Zheng, X., Burdick, D., Popa, L., Zhong, X., Wang, N.X.R.: Global table extractor (GTE): a framework for joint table identification and cell structure recognition using visual context. In: Proceedings of the IEEE/CVF Winter Conference on Applications of Computer Vision, pp. 697–706 (2021)
21. Xu, Y., Li, M., Cui, L., Huang, S., Wei, F., Zhou, M.: LayoutLM: pre-training of text and layout for document image understanding. In: Proceedings of the 26th ACM SIGKDD International Conference on Knowledge Discovery & Data Mining, pp. 1192–1200 (2020)
22. Xu, Y., et al.: LayoutLMv2: multi-modal pre-training for visually-rich document understanding. arXiv preprint: arXiv:2012.14740 (2020)
23. Devlin, J., Chang, M.W., Lee, K., Toutanova, K.: BERT: pre-training of deep bidirectional transformers for language understanding. arXiv preprint: arXiv:1810.04805 (2018)
24. Rajpurkar, P., Jia, R., Liang, P.: Know what you don't know: unanswerable questions for SQuAD. arXiv preprint: arXiv:1806.03822 (2018)

25. Kim, G., et al.: OCR-free document understanding transformer. In: Avidan, S., Brostow, G., Cisse, M., Farinella, G.M., Hassner, T. (eds.) Computer Vision - ECCV 2022. ECCV 2022. Lecture Notes in Computer Science, vol. 13688, pp. 498–517. Springer, Cham (2022). https://doi.org/10.1007/978-3-031-19815-1_29
26. Hagendorff, T., Fabi, S., Kosinski, M.: Machine intuition: uncovering human-like intuitive decision-making in GPT-3.5. arXiv preprint: arXiv:2212.05206 (2022)
27. Jiang, Z., Mao, Y., He, P., Neubig, G., Chen, W.: OmniTab: pretraining with natural and synthetic data for few-shot table-based question answering. arXiv preprint: arXiv:2207.03637 (2022)
28. Biten, A. F., et al.: ICDAR 2019 competition on scene text visual question answering. In: 2019 International Conference on Document Analysis and Recognition (ICDAR), pp. 1563–1570. IEEE (2019)
29. Tito, R., Mathew, M., Jawahar, C.V., Valveny, E., Karatzas, D.: ICDAR 2021 competition on document visual question answering. In: Lladós, J., Lopresti, D., Uchida, S. (eds.) ICDAR 2021. LNCS, vol. 12824, pp. 635–649. Springer, Cham (2021). https://doi.org/10.1007/978-3-030-86337-1_42

ICDAR 2023 Competition on Robust Layout Segmentation in Corporate Documents

Christoph Auer[(✉)], Ahmed Nassar, Maksym Lysak, Michele Dolfi, Nikolaos Livathinos, and Peter Staar

IBM Research, Zurich, Switzerland
{cau,ahn,mly,dol,nli,taa}@zurich.ibm.com

Abstract. Transforming documents into machine-processable representations is a challenging task due to their complex structures and variability in formats. Recovering the layout structure and content from PDF files or scanned material has remained a key problem for decades. ICDAR has a long tradition in hosting competitions to benchmark the state-of-the-art and encourage the development of novel solutions to document layout understanding. In this report, we present the results of our *ICDAR 2023 Competition on Robust Layout Segmentation in Corporate Documents*, which posed the challenge to accurately segment the page layout in a broad range of document styles and domains, including corporate reports, technical literature and patents. To raise the bar over previous competitions, we engineered a hard competition dataset and proposed the recent DocLayNet dataset for training. We recorded 45 team registrations and received official submissions from 21 teams. In the presented solutions, we recognize interesting combinations of recent computer vision models, data augmentation strategies and ensemble methods to achieve remarkable accuracy in the task we posed. A clear trend towards adoption of vision-transformer based methods is evident. The results demonstrate substantial progress towards achieving robust and highly generalizing methods for document layout understanding.

Keywords: Document Layout Analysis · Machine Learning · Computer Vision · Object Detection · ICDAR Competition

1 Introduction

Document understanding is a key business process in the data-driven economy since documents are central to knowledge discovery and business insights. Converting documents into a machine-processable format is a particular challenge due to their huge variability in formats and complex structure. Recovering the layout structure and content from either PDF files or scanned material has remained a key problem since decades, and is as relevant-as-ever today. One can find vast amounts of approaches and solutions to this task [7,11,14–16,18],

G. A. Fink et al. (Eds.): ICDAR 2023, LNCS 14188, pp. 471–482, 2023.
https://doi.org/10.1007/978-3-031-41679-8_27

all of which are constrained to different degrees in the domains and document styles that they can perform well on. A highly generalising model for structure and layout understanding has yet to be achieved.

ICDAR has organized various competitions in the past to benchmark the state-of-the-art and encourage the development of novel approaches and solutions to layout segmentation problems in documents [2–4,21]. In this report, we present the results of our *ICDAR 2023 Competition on Robust Layout Segmentation in Corporate Documents*, which posed the challenge to accurately segment the layout of a broad range of document styles and domains, including corporate reports, technical literature and patents. Participants were challenged to develop a method that could identify layout components in document pages as bounding boxes. These components include paragraphs, (sub)titles, tables, figures, lists, mathematical formulas, and several more. The performance of submissions was evaluated using the commonplace mean average precision metric (mAP) used in the COCO object detection competition [13]. To raise the bar over previous competitions, we proposed to use our recently published DocLayNet dataset [16] for model training, and engineered a challenging, multi-modal competition dataset with a unique distribution of new page samples.

Below, we present a detailed overview of this competition, including its datasets, evaluation metrics, participation, and results.

2   Datasets

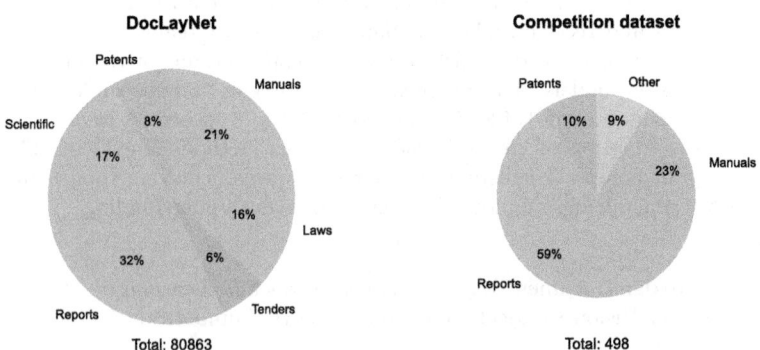

Fig. 1. Dataset statistics of DocLayNet and the competition dataset.

2.1   Related Work

Layout segmentation datasets published in the recent past, such as Pub-LayNet [25] or DocBank [12], have enabled a big leap forward for ML-driven document understanding approaches due to their huge ground-truth size compared to earlier work. However, these datasets still remain limited to a narrow domain of predominantly scientific documents, which is owed to their automatic ground-truth generation approach from mostly uniform XML or LATEX sources. Despite exposing many different publisher layouts, all documents strongly share common traits and general structure. This has led to a saturation of ML model accuracy baselines at a very high level, with little room for improvement [21]. Yet, all publicly proposed ML models trained on these datasets generalize rather poorly to out-of-domain document samples, such as those found in the corporate world. For example, tables in invoices or manuals are difficult to detect correctly with models trained on scientific literature or books.

2.2   DocLayNet Dataset

The DocLayNet dataset [16] addresses these known limitations by providing 80,863 page samples from a broad range of document styles and domains, which are fully layout annotated by human experts to a high-quality standard. DocLayNet is the first large-scale dataset covering a wide range of layout styles and domains, which includes Financial reports, Patents, Manuals, Laws, Tenders, and Technical Papers. It defines 11 class labels for rectangular bounding-box annotations, namely *Caption, Footnote, Formula, List-item, Page-footer, Page-header, Picture, Section-header, Table, Text* and *Title*. Detailed instructions and guidance on how to consistently annotate the layout of DocLayNet pages were published in the accompanying layout annotation guideline.

Fig. 2. Select samples in the competition dataset (*Other* category) which fall outside of the layout distribution in DocLayNet.

Additionally, DocLayNet provides a JSON representation of each page with the original text tokens and coordinates from the programmatic PDF code. This opens the opportunity for new multi-modal ML approaches to the layout segmentation problem.

2.3 Competition Dataset

To assess the layout segmentation performance of each team's submissions, we engineered a competition dataset of 498 new pages in the same representation as the original DocLayNet dataset, which was provided to the participants without any annotation ground-truth. This competition dataset includes a mix of corporate document samples as shown in Fig. 2. Samples in the new *Other* category expose layouts which fall outside of the DocLayNet layout space.

3 Task

We designed the competition objective as a straightforward object detection task, since this is well-understood in the computer-vision community and fits the representation format of our DocLayNet dataset. Participants of our competition were challenged to develop methods that can identify layout components in document pages as rectangular bounding boxes, labelled with one of the 11 classes defined in the DocLayNet dataset (see Fig. 3). The performance of each team's approach was evaluated on our competition dataset using the well established COCO mAP metric.

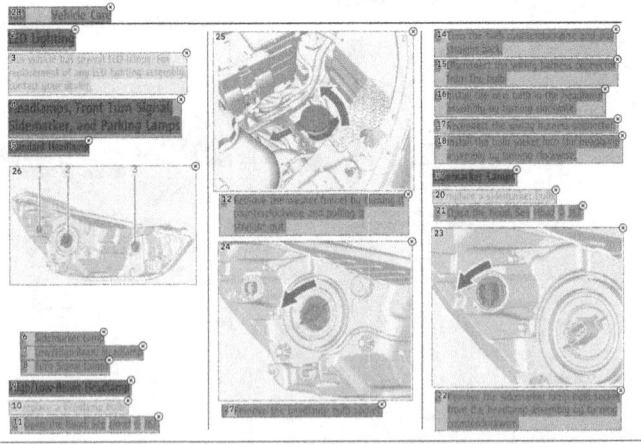

○ Text ○ Caption ● List-Item ○ Formula ○ Table ○ Picture ● Section-Header ○ Page-Header ○ Page-Footer ○ Title

Fig. 3. Example page with bounding-box annotations.

Submission Format: Since the COCO dataset format [13] and tooling is well established in the object detection community, we provided a standard COCO dataset file as part of our competition dataset, which includes the definition of class labels and image identifiers, but no ground-truth annotation data. Submissions were expected in the format of a JSON file complying with the commonly used *COCO results* schema, including complete bounding-box predictions for each page sample, matching to the identifiers defined in our provided dataset file.

Evaluation Metric: All submissions were evaluated using the Mean Average Precision (mAP) @ Intersection-over-Union (IoU) [0.50:0.95] metric, as used in the COCO object detection competition. In detail, we calculate the average precision for a sequence of IoU thresholds ranging from 0.50 to 0.95 with a step size of 0.05, using the standard *pycocotools* library[1]. This metric was computed for every document category in the competition dataset separately. Then, the mean of the mAPs across all categories was computed with equal weights per category. The final ranking of every team's submissions was based on the overall mAP.

4   Competition

Schedule: Our competition was officially announced on December 19th, 2022 and ended on April 3rd, 2023. The regular competition phase ended on March 26th, 2023 and the final week was run as a dedicated extension phase. Results of both phases are reflected in Sect. 5.

Setup: We launched a competition website[2] to provide task descriptions, instructions, resources and news updates for the competition. For submission management, automatic online evaluation and tracking team submissions on a leader board, we relied on the free-to-use *EvalAI* platform [19]. To ensure fair conditions and prevent reverse engineering of our ground-truth, each team was originally granted 10 submission attempts on the evaluation platform. We increased this limit by 5 attempts for the extension phase. The feature in EvalAI to declare submissions private or public allowed teams to create multiple private submissions and check how they perform in evaluation before deciding to re-submit one of them as an official entry. The test-score for each submission was provided directly after submission. The latter has advantages and disadvantages. On the one hand, teams have a direct feedback on the quality of their results and can explore different strategies, which is one of the main motivations of this competition. On the other hand, it can also be used to overfit the model. For this explicit reason, we limited the number of submissions of each team to 10 (with extension 15). To set a baseline for the leader board, the competition organizers created an initial submission entry, which was visible to all teams.

---

[1] pypi.org/project/pycocotools.
[2] ds4sd.github.io/icdar23-doclaynet.

5   Results

5.1   Overview

After the competition ended, we counted 45 team registrations, which altogether created 374 private or public submissions. Out of these, 21 team decided to make at least one public submission which counts towards the final ranking. Table 1 shows the results achieved by the participating teams for the regular submission phase and the extension phase of the competition. More detailed analysis and descriptions of selected methods from the participants are presented below.

Table 1. Leaderboard of our competition with all teams ranking above our baseline (rank 19). Ranks are shown separately for the regular phase (*reg*) and the extension phase (*ext*)

Team	mAPs after extension					Ranking		
	Overall	Rep	Man	Pat	Other	reg.	ext.	Diff
docdog	**0.70**	**0.66**	**0.69**	0.84	**0.62**	1	1	↔
BOE_AIoT_CTO	0.64	0.54	0.67	0.84	0.52	2	2	↔
INNERCONV	0.63	0.57	0.63	**0.85**	0.48	3	3	↔
LC-OCR	0.63	0.61	0.65	0.77	0.48	5	4	∧1
DXM-DI-AI-CV-TEAM	0.63	0.54	0.63	0.82	0.51	4	5	∨1
alexsue	0.61	0.53	0.63	0.81	0.49	6	6	↔
PIX	0.61	0.52	0.63	0.82	0.46	-	7	*
Acodis	0.60	0.53	0.62	0.80	0.46	15	8	∧7
Linkus	0.59	0.49	0.64	0.77	0.48	7	9	∨2
TTW	0.58	0.49	0.61	0.80	0.42	8	10	∨2
amdoc	0.58	0.47	0.62	0.80	0.42	12	11	∧1
CVC-DAG	0.58	0.49	0.61	0.77	0.44	9	12	∨3
SPDB LAB	0.57	0.48	0.57	0.80	0.44	10	13	∨3
Alphastream.ai	0.57	0.47	0.57	0.79	0.45	11	14	∨3
Hisign	0.57	0.48	0.62	0.79	0.39	13	15	∨2
DLVC	0.55	0.53	0.57	0.74	0.38	14	16	∨2
Vamshikancharla	0.49	0.36	0.48	0.76	0.37	16	17	∨1
Azure	0.49	0.44	0.55	0.59	0.38	17	18	∨1
ICDAR23 DocLayNet **organizers (Baseline)**	0.49	0.38	0.52	0.70	0.35	18	19	∨1

5.2   General Analysis

Layout Segmentation Performance: It is apparent that the top-ranking team (docdog) has presented a solution that is performing notably superior compared to the remainder of the field, as evidenced by their 6% lead in total mAP score over the second best submission. This result is achieved through outperforming every other team in the *Reports* category (5% lead) and the particularly difficult *Others* category (10% lead). From the second rank down, we observe a very competitive field with many teams achieving similar levels of mAP performance, ranging from 0.64 (rank 2) to 0.55 (rank 16). Two more teams ranked just slightly above our baseline mAP of 0.49 (see Fig. 4).

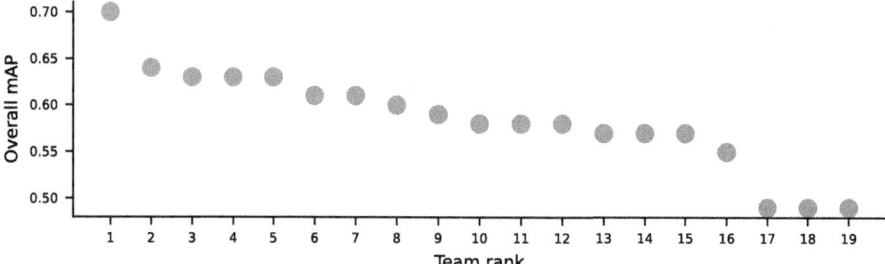

Fig. 4. Distribution of overall mAP achieved by teams. Numbers and ranks refer to extension phase.

Throughout the extension phase, we observed mostly small improvements of overall mAP within a 1–2% range, with few exceptions such as team *docdog* and team *Acodis*, which managed to improve by 4% and 6% over their result from the regular competition phase, respectively. Team *PIX* joined as a new entrant in the extension phase only.

The highest, and also most consistent performance across submissions, is observed in the *Patent* category, with 12 teams achieving an mAP of 0.79 or better. This is consistent with our expectations, since Patent document layouts are the most uniform and structured. The diverse, free-style layouts in the *Reports* and *Others* categories posed a considerably bigger challenge, with mAPs generally ranging in the low 60%s and 50%s respectively.

Two interesting observations can be made. On one hand, we find significantly lower mAPs in the submissions across our competition set categories than those which were achieved for example on PubMed Central papers in the ICDAR 2021 Competition on Scientific Literature Parsing [21]. This can be attributed both to the more challenging layouts and higher class count of the DocLayNet dataset, as well as to the distribution bias and hard samples we engineered the competition dataset to expose. On the other hand, we see a significant spread of mAPs across the final submissions, with almost all teams exceeding the baseline by a significant margin. This delivers evidence that the participating teams have created solutions that differentiate themselves significantly from previous off-the-shelf object detection methods (see baseline). It also shows that the investment to develop sophisticated methods is beneficial to obtain superior performance on this dataset.

Models and Strategies: In the solutions presented by the top five teams, we were pleased to see novel and interesting combinations of recent computer vision models, data augmentation strategies and ensemble methods applied to solve the layout segmentation task with high accuracy. All top-ranking solutions adopt, to different degrees, the recently emerging deep-learning models based on vision transformer methods and self-supervised pre-training, such as the generic DINO [22] and MaskDINO [10] models, or the document-understanding focused DiT [11] and LayoutLMv3 [7] models. The DocLayNet dataset was used for fine-tuning in this context. Several solutions combine these new-generation vision models with more traditional, CNN-based object detectors such as YOLO [8] through model ensemble, for example through *Weighted Boxes Fusion* [17]. Data augmentation strategies used by the teams include multi-scale and mosaic methods, as well as deriving synthetic datasets from DocLayNet. Two of the top five teams reported that they include the additional text cell layer provided by DocLayNet and the competition dataset in their approach. No teams stated to create private ground-truth data that was not derived from DocLayNet.

5.3   Method Descriptions

Below we summarize the methods reported by the top five teams for comparison reasons to our best understanding. We would like to extend our thanks to all competition teams who took the time to provide us with a comprehensive description of their methods.

Team Docdog (Tencent WeChat AI)

The team created a synthetic image dataset of 300,000 samples based on the training dataset. For the task of layout prediction, the team used two models, YOLOv8 [8] and DINO [22]. An extra classification model was trained to categorize the samples of the competition dataset into the document categories. YOLOv8 models with different network sizes (medium, large, x-large) were trained, each with different input resolutions, for ensemble and optimization of the detection performance. For the DINO model, the team applied a carefully designed augmentation strategy and integrated focal modulation networks [20] in the backbone for improved performance. Separate models were trained per category, both with and without synthetic data. Model hyper-parameters were optimized using a Tree-Structured Parzen Estimator (TPE) [1] to find the best weights. Prediction results from the individual models were combined using Weighted Boxes Fusion (WBF) [17] and fine-tuned using text cell coordinates from the JSON representation of the samples in the competition dataset. Further detail on the approach is provided in the team's *WeLayout* paper [23].

Team BOE_AIoT_CTO

The team relied exclusively on the DocLayNet dataset for training, and applied scale and mosaic methods for image augmentation. For the task of layout prediction, the team trained two object detection models, YOLOv5 [9] and

YOLOv8 [8]. Training was conducted over 150 epochs using BCELoss with Focal-Loss, and mosaic augmentation was cancelled for the final 20 epochs. Additionally, a DiT model [11] (dit-large) was fine-tuned using the DocLayNet dataset. To improve vertical text detection, the team added multi-scale image training. Predictions for the final submission were ensembled from three detectors to achieve superior performance.

Team INNERCONV

For the task of layout prediction, the team uses the MaskDINO model [10]. MaskDINO is derivative of DINO [22] which introduces a mask prediction branch in parallel to the box prediction branch of DINO. It achieves better alignment of features between detection and segmentation. In training, only the image representation of the DocLayNet dataset is used. In inference, the team applied the Weighted Boxes Fusion (WBF) technique [17] to ensemble the predictions on multiple scales of the same input image.

Team LC-OCR (CVTE)

For the task of layout prediction, the team applied two models, VSR [24] and LayoutLMv3 [7], which use pre-trained weights. Prediction results from both models are merged in inference. Detections for the classes *Footnote*, *Picture*, *Table* and *Title* were taken from LayoutLMv3, the remainder of classes from VSR. In VSR, the team included the text cell information provided in the JSON representation of DocLayNet.

Team DXM-DI-AI-CV-TEAM (Du Xiaoman Financial)

For the task of layout prediction, the team trained different versions of Cascade Mask R-CNN [6] models, based on a DiT [11] backbone (DiT-large), and fuse prediction results using different models.

Baseline of ICDAR 2023 DocLayNet Organizers

To set a comparison baseline for the competition, the organizers used a YOLOv5 model (medium size), and trained it solely on the DocLayNet training dataset, with images re-scaled to square 1024 by 1024 pixels. The model was trained from scratch with default settings for 80 epochs. We applied standard augmentation techniques such as mosaic, scale, flipping, rotation, mix-up and image levels.

6   Conclusions

We believe that this ICDAR competition served its purpose well to benchmark the state-of-the-art solutions to the layout segmentation task in documents, and again encouraged the development of unique new approaches. Our new competition dataset was designed to raise the bar over previous competitions by providing diverse, challenging page layouts, paired with multi-modal representation. This enabled participants to test the generalization power of the latest computer-vision methods, especially with recently emerging models based on self-supervised pre-training and visual transformers.

We were pleasantly surprised by the high level of engagement in this competition, with 45 teams registering, out of which 21 teams created an official final submission. The budget of 15 total submissions was fully used by the majority of the contestants. Overall, the level of sophistication demonstrated in the approaches went well beyond our anticipation. One core take-away is the importance of data augmentation and ensemble techniques to improve the layout prediction performance beyond the level of what any single end-to-end model currently delivers. It was also interesting to observe how the various techniques applied by the different teams in many cases yielded similar results in overall accuracy. The remarkable progress demonstrated by the top-performing teams in this competition will be valuable for future research on highly capable document understanding models.

We are also glad to see this competition spark wider interest in the community, as it prompted some members to build and share fully runnable example codes and publish blog articles on training and inference with DocLayNet and pre-trained models [5]. To support these community efforts, we made DocLayNet available on the HuggingFace datasets hub[3]. As such, we believe that this ICDAR competition has also helped to establish the DocLayNet dataset as a well known asset for document understanding research and applications.

Acknowledgments. We would like to thank all participants for their remarkable efforts and contributions to this competition, and the Competitions Chairs for providing the opportunity to host this competition in ICDAR 2023.

References

1. Bergstra, J., Bardenet, R., Bengio, Y., Kégl, B.: Algorithms for hyper-parameter optimization. In: Shawe-Taylor, J., Zemel, R., Bartlett, P., Pereira, F., Weinberger, K. (eds.) Advances in Neural Information Processing Systems, vol. 24. Curran Associates, Inc. (2011)
2. Clausner, C., Antonacopoulos, A., Pletschacher, S.: Icdar 2017 competition on recognition of documents with complex layouts - rdcl2017. In: 2017 14th IAPR International Conference on Document Analysis and Recognition (ICDAR), vol. 01, pp. 1404–1410 (2017). https://doi.org/10.1109/ICDAR.2017.229
3. Déjean, H., et al.: ICDAR 2019 competition on table detection and recognition (cTDaR), April 2019. https://doi.org/10.5281/zenodo.2649217
4. Göbel, M., Hassan, T., Oro, E., Orsi, G.: Icdar 2013 table competition. In: 2013 12th International Conference on Document Analysis and Recognition, pp. 1449–1453 (2013). https://doi.org/10.1109/ICDAR.2013.292

---

[3] https://huggingface.co/datasets/ds4sd/DocLayNet.

5. Guillou, P.: Document AI — processing of doclaynet dataset to be used by layout models of the hugging face hub (finetuning, inference), January 2023. https://medium.com/@pierre_guillou/308d8bd81cdb

6. He, K., Gkioxari, G., Dollár, P., Girshick, R.: Mask r-cnn (2018)

7. Huang, Y., Lv, T., Cui, L., Lu, Y., Wei, F.: Layoutlmv3: pre-training for document AI with unified text and image masking (2022)

8. Jocher, G., Chaurasia, A., Qiu, J.: YOLO by Ultralytics, January 2023. https://github.com/ultralytics/ultralytics

9. Jocher, G., et al.: ultralytics/yolov5: v7.0 - YOLOv5 SOTA realtime instance segmentation, November 2022. https://doi.org/10.5281/zenodo.7347926

10. Li, F., et al.: Mask dino: towards a unified transformer-based framework for object detection and segmentation (2022)

11. Li, J., Xu, Y., Lv, T., Cui, L., Zhang, C., Wei, F.: Dit: self-supervised pre-training for document image transformer. In: ACM Multimedia 2022, October 2022

12. Li, M., et al.: Docbank: a benchmark dataset for document layout analysis. In: Proceedings of the 28th International Conference on Computational Linguistics, pp. 949–960. COLING, International Committee on Computational Linguistics, December 2020. https://doi.org/10.18653/v1/2020.coling-main.82, https://aclanthology.org/2020.coling-main.82

13. Lin, T.Y., et al.: Microsoft coco: common objects in context (2015)

14. Livathinos, N., et al.: Robust pdf document conversion using recurrent neural networks. In: Proceedings of the AAAI Conference on Artificial Intelligence, vol. 35, no. 17, pp. 15137–15145, May 2021. https://ojs.aaai.org/index.php/AAAI/article/view/17777

15. Nassar, A., Livathinos, N., Lysak, M., Staar, P.: Tableformer: table structure understanding with transformers. In: Proceedings of the IEEE/CVF Conference on Computer Vision and Pattern Recognition (CVPR), pp. 4614–4623, June 2022

16. Pfitzmann, B., Auer, C., Dolfi, M., Nassar, A.S., Staar, P.W.J.: Doclaynet: a large human-annotated dataset for document-layout segmentation. In: Zhang, A., Rangwala, H. (eds.) KDD 2022: The 28th ACM SIGKDD Conference on Knowledge Discovery and Data Mining, Washington, DC, USA, 14–18 August 2022, pp. 3743–3751. ACM (2022). https://doi.org/10.1145/3534678.3539043

17. Solovyev, R., Wang, W., Gabruseva, T.: Weighted boxes fusion: Ensembling boxes from different object detection models. Image Vis. Comput. **107**, 104117 (2021). https://doi.org/10.1016/j.imavis.2021.104117

18. Staar, P.W.J., Dolfi, M., Auer, C., Bekas, C.: Corpus conversion service: a machine learning platform to ingest documents at scale. In: Proceedings of the 24th ACM SIGKDD International Conference on Knowledge Discovery and Data Mining, KDD 2018, pp. 774–782. Association for Computing Machinery, New York, NY, USA (2018). https://doi.org/10.1145/3219819.3219834

19. Yadav, D., et al.: Evalai: towards better evaluation systems for AI agents (2019)

20. Yang, J., Li, C., Dai, X., Gao, J.: Focal modulation networks. In: Koyejo, S., Mohamed, S., Agarwal, A., Belgrave, D., Cho, K., Oh, A. (eds.) Advances in Neural Information Processing Systems, vol. 35, pp. 4203–4217. Curran Associates, Inc. (2022)

21. Jimeno Yepes, A., Zhong, P., Burdick, D.: ICDAR 2021 competition on scientific literature parsing. In: Lladós, J., Lopresti, D., Uchida, S. (eds.) ICDAR 2021. LNCS, vol. 12824, pp. 605–617. Springer, Cham (2021). https://doi.org/10.1007/978-3-030-86337-1_40

22. Zhang, H., et al.: Dino: Detr with improved denoising anchor boxes for end-to-end object detection (2022)

23. Zhang, M., Cao, Z., Liu, J., Niu, L., Meng, F., Zhou, J.: Welayout: Wechat layout analysis system for the ICDAR 2023 competition on robust layout segmentation in corporate documents (2023). https://arxiv.org/abs/2305.06553
24. Zhang, P., et al.: Vsr: a unified framework for document layout analysis combining vision, semantics and relations (2021)
25. Zhong, X., Tang, J., Yepes, A.J.: Publaynet: largest dataset ever for document layout analysis. In: 2019 International Conference on Document Analysis and Recognition (ICDAR), pp. 1015–1022. IEEE, September 2019. https://doi.org/10.1109/ICDAR.2019.00166

ICDAR 2023 Competition on Hierarchical Text Detection and Recognition

Shangbang Long[✉], Siyang Qin, Dmitry Panteleev, Alessandro Bissacco, Yasuhisa Fujii, and Michalis Raptis

Google Research, Mountain View, CA, USA
{longshangbang,qinb,dpantele,bissacco,yasuhisaf,mraptis}@google.com

Abstract. We organize a competition on hierarchical text detection and recognition. The competition is aimed to promote research into deep learning models and systems that can jointly perform text detection and recognition and geometric layout analysis. We present details of the proposed competition organization, including tasks, datasets, evaluations, and schedule. During the competition period (from January 2nd 2023 to April 1st 2023), at least 50 submissions from more than 20 teams were made in the 2 proposed tasks. Considering the number of teams and submissions, we conclude that the HierText competition has been successfully held. In this report, we will also present the competition results and insights from them.

Keywords: OCR · Text Detection and Recognition · Layout Analysis

1 Introduction

Text detection and recognition systems [10] and geometric layout analysis techniques [11,12] have long been developed separately as independent tasks. Research on text detection and recognition [13–16] has mainly focused on the domain of natural images and aimed at single level text spotting (mostly, word-level). Conversely, research on geometric layout analysis [11,12,17,18], which is targeted at parsing text paragraphs and forming text clusters, has assumed document images as input and taken OCR results as fixed and given by independent systems. The synergy between the two tasks remains largely under-explored.

Recently, the Unified Detector work by Long et al. [19] shows that the unification of line-level detection of text and geometric layout analysis benefits both tasks significantly. StructuralLM [20] and LayoutLMv3 [26] show that text line grouping signals are beneficial to the downstream task of document understanding and are superior to word-level bounding box signals. These initial studies demonstrate that the unification of OCR and layout analysis, which we term as *Hierarchical Text Detection and Recognition (HTDR)*, can be mutually beneficial to OCR, layout analysis, and downstream tasks.

Given the promising potential benefits, we propose the **ICDAR 2023 Competition on Hierarchical Text Detection and Recognition**. In this competition, candidate systems are expected to perform the unified task of text detection and recognition and geometric layout analysis. Specifically, we define the

G. A. Fink et al. (Eds.): ICDAR 2023, LNCS 14188, pp. 483–497, 2023.
https://doi.org/10.1007/978-3-031-41679-8_28

unified task as producing a hierarchical text representation, including word-level bounding boxes and text transcriptions, as well as line-level and paragraph-level clustering of these word-level text entities. We defer the rigorous definitions of word/line/paragraph later to the dataset section. Figure 1 illustrates our notion of the unified task.

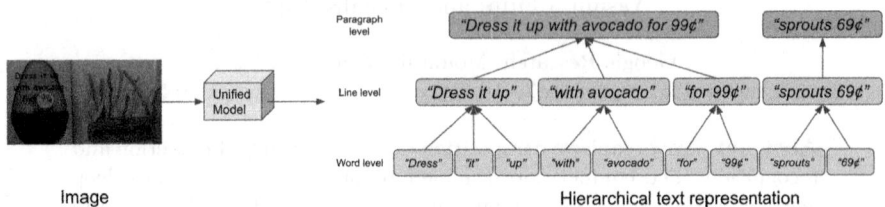

Fig. 1. Illustration for the proposed unified task: **Hierarchical Text Detection and Recognition (HTDR)**. Given an input image, the unified model is expected to produce a hierarchical text representation, which resembles the form of a forest. Each tree in the forest represents one paragraph and has three layers, representing the clustering of words into lines and then paragraphs.

We believe this competition will have profound and long-term impact on the whole image-based text understanding field by unifying the efforts of text detection and recognition and geometric layout analysis, and furthering providing new signals for downstream tasks.

The competition started on January 2nd 2023, received more than 50 submissions in 2 tasks in total, and closed on April 1st 2023. This report provides details into the motivation, preparation, and results of the competition. We believe the success of this competition greatly promotes the development of this research field. Furthermore, the dataset except the test set annotation and evaluation script are made publicly available. The competition website[1] remains open to submission and provides evaluation on the test set.

2   Competition Protocols

2.1   Dataset

The competition is based on the HierText dataset [19]. Images in HierText are collected from the Open Images v6 dataset [27], by first applying the *Google Cloud Platform (GCP) Text Detection API*[2] and then filtering out inappropriate images, for example those with too few text or non-English text. In total, 11639 images are obtained. In this competition, we follow the original split of 8281/1724/1634 for *train, validation, test* sets. Images and annotations of the train and validation set are released publicly. The test set annotation is kept private and will remain so even after the end of the competition.

---

[1] https://rrc.cvc.uab.es/?ch=18.

[2] https://cloud.google.com/vision/docs/ocr.

As noted in the original paper [19], we check the cross-dataset overlap rates with the two other OCR datasets that are based on Open Images. We find that 1.5% of the 11639 images we have are also in TextOCR [28] and 3.6% in Intel OCR [29]. Our splits ensure that our training images are not in the validation or test set of Text OCR and Intel OCR, and vice versa.

```
paragraph 0:                                paragraph 5:                                paragraph 9:
  line 0:                                     line 0:                                     line 0:
    word 0                                      word 0                                      word 0
      text: MOZART                                text: COCONUT                                text: FREE
      polygon: [[330, 56], [741, 208], [697, 327], [285, 175]]    polygon: [[902, 434], [984, 434], [984, 453], [901, 452]]    polygon: [[1023, 407], [1143, 401], [1146, 445], [1025, 451]]
  line 1:                                     line 1:                                     line 1:
    word 0                                      word 0                                      word 0
      text: MIKE                                  text: PREMIUM                                text: SHIPPING
      polygon: [[467, 40], [671, 113], [650, 172], [445, 99]]      polygon: [[903, 409], [982, 410], [982, 428], [902, 426]]    polygon: [[1026, 455], [1148, 450], [1149, 473], [1026, 478]]
  line 2:                                     line 2:                                     line 2:
    word 0                                      word 0                                      word 0
      text: natural                               text: WATER                                  text: @TARGET.COM
      polygon: [[794, 215], [897, 210], [899, 234], [795, 239]]    polygon: [[901, 460], [961, 460], [961, 478], [900, 477]]    polygon: [[1028, 483], [1151, 479], [1152, 494], [1028, 498]]
paragraph 1:                                  line 3:                                   paragraph 10:
  line 0:                                       word 0                                     line 0:
    word 0                                        text: PURE                                  word 0
      text: natural                                 polygon: [[902, 382], [965, 382], [965, 402], [902, 402]]    text: 14
      polygon: [[810, 17], [886, 15], [887, 36], [810, 37]]  paragraph 6:                                polygon: [[782, 653], [775, 853], [774, 868], [760, 867]]
paragraph 2:                                    line 0:                                     word 1
  line 0:                                         word 0                                      text: FL
    word 0                                          text: from                                 polygon: [[782, 855], [799, 857], [797, 873], [779, 870]]
      text: TODAY                                    polygon: [[900, 488], [930, 488], [930, 501], [900, 501]]    word 2
      polygon: [[1003, 71], [1077, 65], [1079, 84], [1004, 90]]    word 1                                       text: 0Z
    word 1                                          text: concentred                            polygon: [[806, 860], [826, 861], [825, 877], [804, 875]]
      text: &                                         polygon: [[932, 491], [989, 491], [989, 501], [932, 501]]    word 3
      polygon: [[1086, 69], [1095, 68], [1097, 78], [1087, 79]]  paragraph 7:                                  text: (414
  line 1:                                       line 0:                                       polygon: [[833, 862], [870, 866], [868, 886], [830, 881]]
    word 0                                        word 0                                      word 4
      text: EVERY                                    text: natura                                text: mL)
      polygon: [[1006, 94], [1064, 90], [1065, 105], [1007, 109]]    polygon: [[899, 535], [951, 535], [951, 549], [899, 549]]    polygon: [[878, 865], [906, 868], [906, 885], [875, 881]]
    word 1                                        line 1:                                   paragraph 11:
      text: DAY                                       word 0                                     line 0:
      polygon: [[1068, 88], [1103, 85], [1105, 102], [1069, 105]]    text: flavor                                word 0
  line 2:                                           polygon: [[898, 553], [937, 553], [937, 567], [898, 567]]    text: 7
    word 0                                        line 2:                                       polygon: [[924, 830], [935, 829], [937, 847], [925, 848]]
      text: 5%                                        word 0                                   paragraph 12:
      polygon: [[1003, 0], [1099, 0], [1099, 66], [1003, 66]]      text: with                                 line 0:
    word 1                                            polygon: [[899, 520], [927, 519], [928, 531], [899, 532]]    word 0
      text: OFF                                   paragraph 8:                                  text: XTRA
      polygon: [[1051, 44], [1098, 40], [1100, 58], [1052, 63]]    line 0:                                      polygon: [[1143, 890], [1201, 889], [1202, 909], [1143, 910]]
paragraph 4:                                      word 0                                     line 1:
  line 0:                                           text: zlco                                 word 0
    word 0                                            polygon: [[729, 753], [731, 295], [894, 296], [891, 754]]    text: AYS
      text: natural                                                                              polygon: [[1152, 916], [1203, 914], [1205, 941], [1153, 943]]
      polygon: [[904, 360], [948, 360], [948, 373], [904, 373]]                                 line 2:
                                                                                                word 0
                                                                                                  text: RNS
                                                                                                  polygon: [[1162, 948], [1208, 948], [1208, 960], [1162, 960]]
```

Fig. 2. Example of hierarchical annotation format of the dataset.

The images are annotated in a hierarchical way of *word*-to-*line*-to-*paragraph*, as shown in Fig. 2. *Words* are defined as a sequence of textual characters not interrupted by *spaces*. *Lines* are then defined as *space*-separated clusters of *words* that are logically connected and aligned in spatial proximity. Finally, *paragraphs* are composed of *lines* that belong to the same semantic topic and are geometrically coherent. Figure 3 illustrates some annotated samples. Words are annotated with polygons, with 4 vertices for straight text and more for curved text depending on the shape. Then, words are transcribed regardless of the scripts and languages, as long as they are legible. Note that we do not limit the character sets, so the annotation could contain case-sensitive characters, digits, punctuation, as well as non-Latin characters such as Cyrillic and Greek. After word-level annotation, we group words into lines and then group lines into paragraphs. In this way, we obtain a hierarchical annotation that resembles a forest structure of the text in an image.

Sample 1

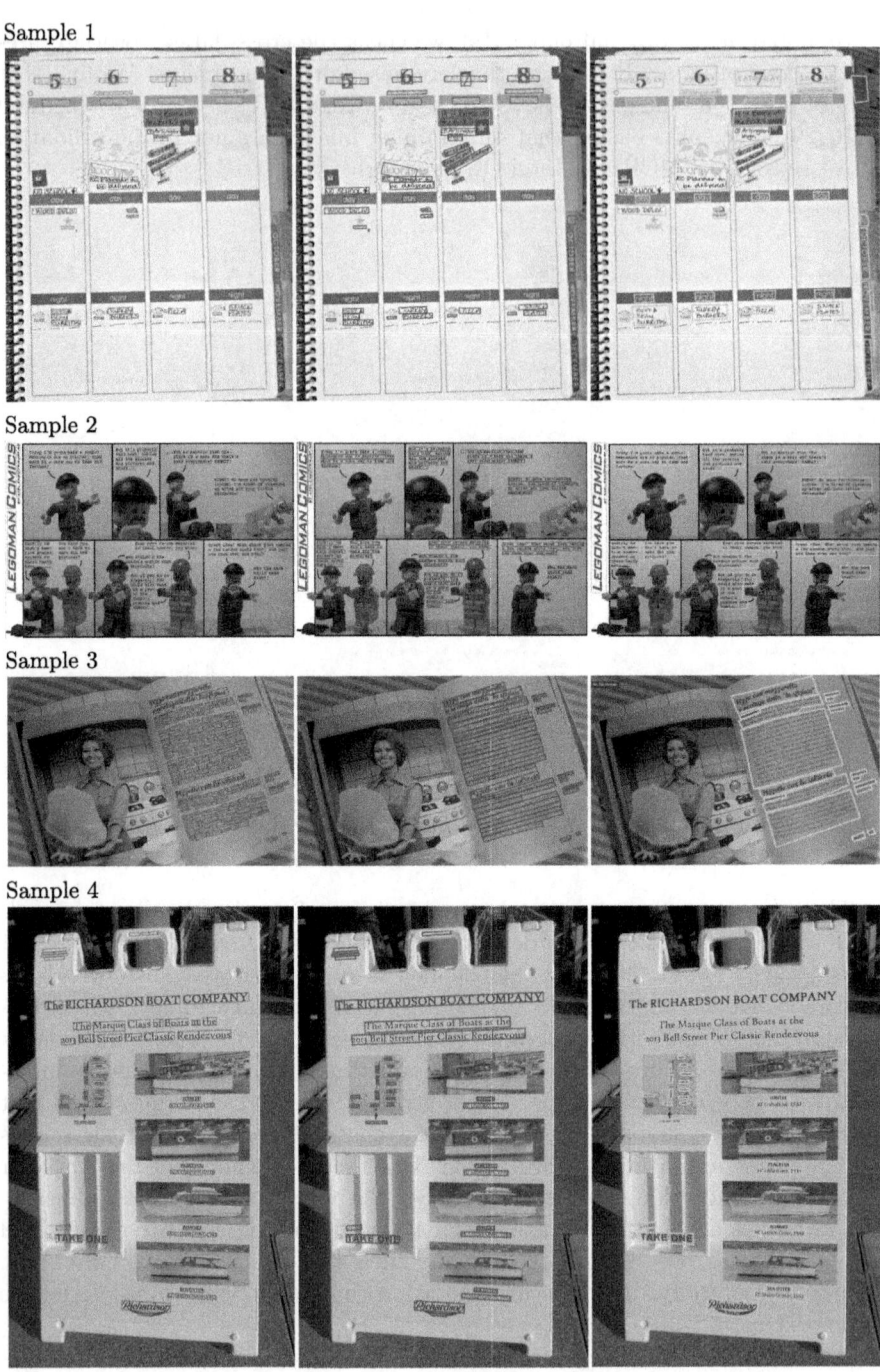

Fig. 3. Illustration for the hierarchical annotation of text in images. From **left** to **right**: **word**, **line**, **paragraph** level annotations. Words (blue) are annotated with polygons. Lines (green) and paragraphs (yellow) are annotated as hierarchical clusters and visualized as polygons. Images are taken from the train split.

2.2   Tasks

Our challenge consists of 2 competition tracks, **Hierarchical Text Detection** and **Word-Level End-to-End Text Detection and Recognition**. In the future, we plan to merge them into a single unified Hierarchical Text Spotting task that requires participants to give a unified representation of text with layout.

Task 1: Hierarchical Text Detection. This task itself is formulated as a combination of 3 tasks: word detection, text line detection, and paragraph detection, where lines and paragraphs are represented as clusters of words hierarchically.

In this task, participants are provided with images and expected to produce the hierarchical text detection results. Specifically, the results are composed of **word-level bounding polygons** and **line and paragraph clusters** on top of words. The clusters are represented as forests, as in Fig. 1, where each paragraph is a tree and words are leaves. For this task, participants do not need to provide text recognition results.

Fig. 4. Illustration of how hierarchical text detection can be evaluated as 3 instance segmentation sub-tasks. The coloring of each column indicates the instance segmentation for each sub-task.

As illustrated in Fig. 4, we evaluate this task as 3 instance segmentation sub-tasks for word, line, and paragraph respectively. For word level, each word is one instance. For line level, we take the union of each line's children words as one instance. For paragraph level, we aggregate each paragraph's children lines, and take that as one instance. With this formulation, all the 3 sub-tasks will be evaluated with the PQ metric [30] designed for instance segmentation, as specified in [19]:

$$PQ = \frac{\sum_{(p,g) \in TP} IoU(p,g)}{|TP| + \frac{1}{2}|FP| + \frac{1}{2}|FN|} \tag{1}$$

where TP, FP, FN represent true positives, false positives, and false negatives respectively. We use an IoU threshold of 0.5 to count true positives. Note that the PQ metric is mathematically equal to the product of the *Tightness* score, which is defined as the average IoU scores of all TP pairs, and the *F1*, score which is commonly used in previous OCR benchmarks. Previous OCR evaluation protocols only report F1 scores which do not fully reflect the detection quality. We argue that tightness is very important in evaluating hierarchical detection. It gives an accurate measurement of how well detections match ground-truths. For words, a detection needs to enclose all its characters and not overlap with other words, so that the recognition can be correct. The tightness score can penalize missing characters and oversized boxes. For lines and paragraphs, they are represented as clusters of words, and are evaluated as unions of masks. Wrong clustering of words can also be reflected in the IoU scores for lines and paragraphs. In this way, using the PQ score is an ideal way to accurately evaluate the hierarchical detection task.

Each submission has 3 PQ scores for word, line, and paragraph respectively. There are 3 rankings for these 3 sub-tasks respectively. For the final ranking of the whole task, we compute the final score as a harmonic mean of the 3 PQ scores (dubbed *H-PQ*) and rank accordingly.

Task 2: Word-Level End-to-End Text Detection and Recognition. For this task, images are provided and participants are expected to produce word-level text detection and recognition results, i.e. a set of word bounding polygons and transcriptions for each image. Line and paragraph clustering is not required. This is a challenging task, as the dataset has the most dense images, with more than 100 words per image on average, 3 times as many as the second dense dataset TextOCR [28]. It also features a large number of recognizable characters. In the training set alone, there are more than 960 different character classes, as shown in Fig. 5, while most previous OCR benchmarks limit the tasks to recognize only digits and case-insensitive English characters. These factors make this task challenging.

For evaluation, we use the F1 measure, which is a harmonic mean of word-level prediction and recall. A word result is considered true positive if the IoU with ground-truth polygon is greater or equal to 0.5 and the transcription is the same as the ground-truth. The transcription comparison considers all characters and will be case-sensitive. Note that, some words in the dataset are marked as illegible words. Detection with high overlap with these words (IoU larger than 0.5) will be removed in the evaluation process, and ground-truths marked as illegible do not count as false negative even if they are not matched.

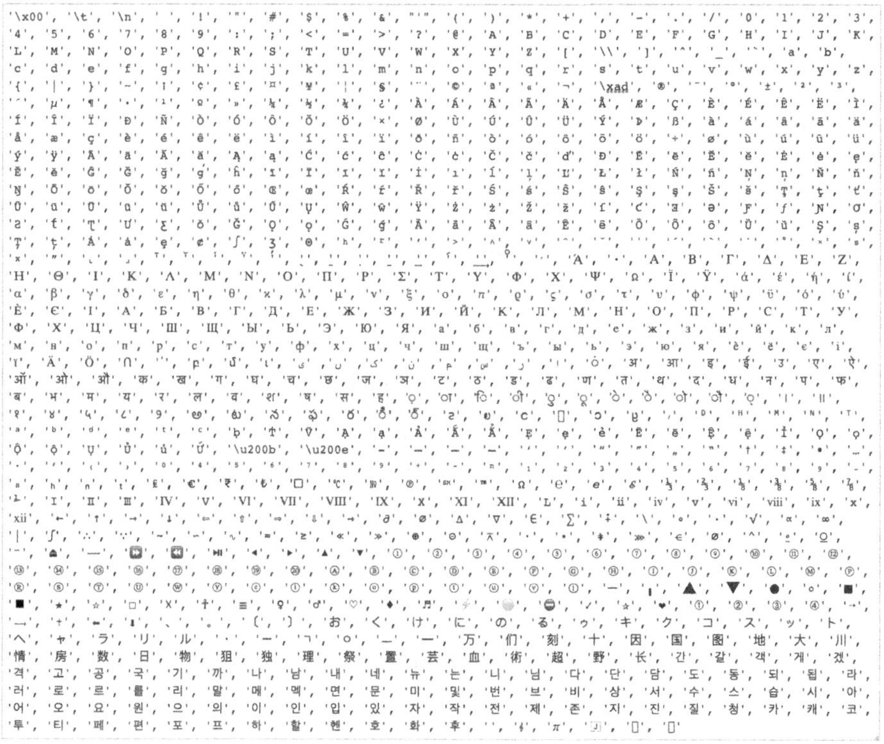

Fig. 5. Character set in the training split.

2.3 Evaluation and Competition Website

We host the competition on the widely recognized Robust Reading Competition (RRC) website[3] and set up our own competition page. The RRC website has been the hub of scene text and document understanding research for a long time and hosted numerous prestigious competitions. It provides easy-to-use infrastructure to set up competition, tasks, and carry out evaluation. It also supports running the competition continuously, making it an ideal candidate.

---

[3] https://rrc.cvc.uab.es/.

2.4   Competition Schedule

We propose and execute the following competition schedule, in accordance with the conference timeline:

- **January 2nd, 2023**: Start of the competition; submissions of results were enabled on the website.
- **April 1st, 2023**: Deadline for competition submissions.
- **April 15th, 2023**: Announcement of results.

2.5   Other Competition Rules

In addition to the aforementioned competition specifications, we also apply the following rules:

- **Regarding the usage of other publicly available datasets**: HierText is the only allowed annotated OCR dataset. However, participants are also allowed to do self-labeling on other public OCR datasets as long as they don't use their ground-truth labels. In other words, they can use the images of other public datasets, but not their labels. They can also use non-OCR datasets, whether labeled or not, to pretrain their models. We believe they are important techniques that can benefit this field.
- **Usage of synthetic datasets** Synthetic data has been an important part of OCR recently [21–25]. Participants can use any synthetic datasets, whether they are public or private, but are expected to reveal how they are synthesized and some basic statistics of the synthetic datasets if they are private.
- Participants should not use the validation split in training their models.
- Participants can make as many submissions as desired before the deadline, but we only archive the latest one submission of each participant in the final competition ranking.

2.6   Organizer Profiles

Authors are all members of the OCR team at Google Research. In addition to academic publications, authors have years of experience in building industrial OCR systems that are accurate and efficient for a diversity of image types and computation platforms.

3   Competition Results

In total, the competition received 30 submissions in Task 1 and 20 submissions in Task 2. Note that, we encourage participants to submit multiple entries using different methods, for example, to understand the effect of applying different techniques such as pretraining and synthetic data. To produce the final leaderboard in compliance with the ICDAR competition protocols, we only keep the latest 1 submission from each participants. The final deduplicated competition results are summarized in Table 1/Fig. 6 and Table 2/Fig. 7. In total, the competition received 11 unique submissions in Task 1 and 7 in Task 2.

Table 1. Results for Task 1. F/P/R/T/PQ stand for *F1-score, Precision, Recall, Tightness,* and *Panoptic Quality* respectively. The submissions are ranked by the *H-PQ* score. H-PQ can be interpreted as *Hierarchical-PQ* or *Harmonic-PQ*. H-PQ is calculated as the harmonic means of the PQ scores of the 3 hierarchies: word, line, and paragraph. It represents the comprehensive ability of a method to detect the text hierarchy in image. We omit the % for all these numbers for simplicity.

User	Method	Rank	Task 1 metric	Word					Line					Paragraph				
			H-PQ	PQ	F	P	R	T	PQ	F	P	R	T	PQ	F	P	R	T
YunSu Kim	Upstage KR	1	76.85	79.80	91.88	94.73	89.20	86.85	76.40	88.34	91.32	85.56	86.48	74.54	86.15	87.40	84.94	86.52
DeepSE x Upstage	DeepSE hierarchical detection model	2	70.96	75.30	88.49	93.50	83.99	85.10	69.43	82.43	82.65	82.21	84.23	68.51	81.39	81.69	81.10	84.17
zhm	hiertext_submit_0401 curve_199_v2	3	70.31	76.71	88.18	92.71	84.08	86.99	71.43	83.32	89.32	78.07	85.73	63.97	74.83	81.25	69.35	85.48
Mike Ranzinger	NVTextSpotter	4	68.82	73.69	87.07	95.10	80.29	84.63	67.76	80.42	93.87	70.35	84.25	65.51	78.04	81.82	74.60	83.94
ssm	Ensemble of three task-specific Clova DEER detection	5	68.72	71.54	92.03	93.82	90.31	77.74	69.64	89.04	91.75	86.49	78.21	65.29	83.70	84.17	83.23	78.01
xswl	Global and local instance segmentations for hierarchical text detection	6	68.62	76.16	90.72	93.45	88.16	83.95	68.50	82.22	80.24	84.31	83.31	62.55	75.11	74.00	76.25	83.28
Asaf Gendler	Hierarchical Transformers for Text Detection	7	67.59	70.44	86.09	88.47	83.83	81.82	69.30	85.23	87.83	82.78	81.31	63.46	78.40	77.84	78.97	80.94
JiangQing	SCUT-HUAWEI	8	62.68	70.08	89.58	89.79	89.37	78.23	67.70	86.20	90.46	82.33	78.53	53.14	69.06	74.03	64.72	76.96
Jiawei Wang	DQ-DETR	9	27.81	61.01	77.27	80.64	74.17	78.96	26.96	35.91	26.81	54.39	75.07	18.38	24.72	15.99	54.41	74.36
ZiqianShao	test	10	21.94	27.45	41.75	51.82	34.95	65.76	25.61	39.04	51.50	31.43	65.59	16.32	24.52	35.61	18.70	66.57
Yichuan Cheng	a	11	0.00	0.00	0.00	0.24	0.00	53.62	0.01	0.01	0.25	0.01	51.29	0.01	0.02	0.21	0.01	50.89

Task 1 results

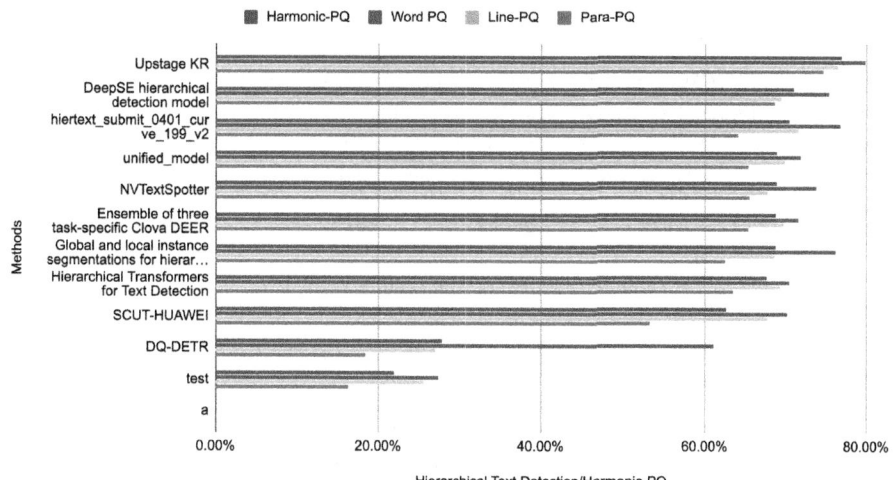

Fig. 6. Figure for the results of task 1.

Table 2. Results for Task 2. F/P/R/T/PQ stand for *F1-score, Precision, Recall, Tightness*, and *Panoptic Quality* respectively. The submissions are ranked by the F1 score. We omit the % for all these numbers for simplicity.

User	Method	Rank	Word				
			PQ	F	P	R	T
YunSu Kim	Upstage KR	1	70.00	79.58	82.05	77.25	87.97
DeepSE x Upstage	DeepSE End-to-End Text Detection and Recognition Model	2	67.46	77.93	88.05	69.89	86.57
ssm	Ensemble of three task-specific Clova DEER	3	59.84	76.15	77.63	74.73	78.59
Mike Ranzinger	NVTextSpotter	4	63.57	74.10	80.94	68.34	85.78
JiangQing	SCUT-HUAWEI	5	58.12	73.41	74.38	72.46	79.17
kuli.cyd	DBNet++ and SATRN	6	51.62	71.64	82.76	63.15	72.06
LGS	keba	7	44.87	54.30	68.37	45.03	82.64

Task 2 results

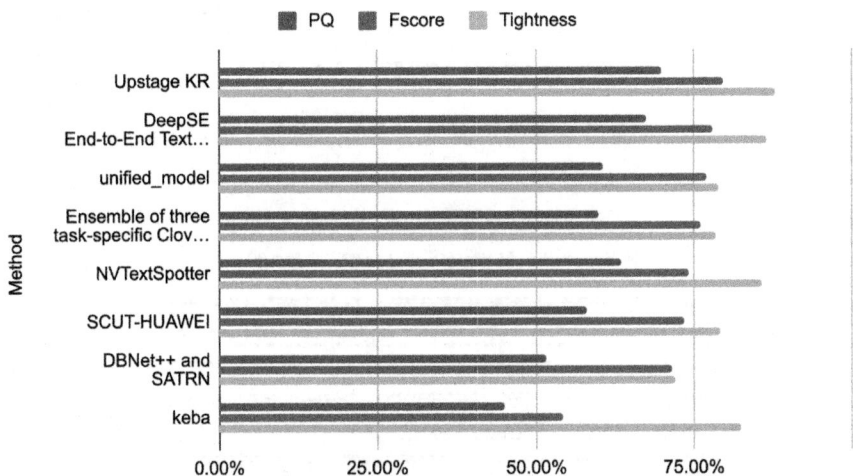

Fig. 7. Figure for the results of task 2.

3.1   Submission Validation

In the final leaderboard, each participant is only allowed to have one submission. We validate each submission and examine the number of submissions from each team. If a team has more than one submission, we keep the latest one and remove the rest from the leaderboard. Note that these removed submissions will remain on the RRC portal for reference, since they also provide important aspects into this research field. We adopt the following rules to determine the authorship of each submission:

- **user_id**: If two submissions have the same user_id field, it means they are submitted by the same RRC user account and thus should be from the same team.
- **method description**: Participants are asked to provide descriptive information of their submissions, including authors, method details, etc. If two submissions have strictly almost identical author list and method description, we consider them to be from the same team.

3.2 Task 1 Methodology

Task 1 in our competition, i.e. Hierarchical Text Detection, is a novel task in the research field. There are no existing methods that participants can refer to. Even the previous work Unified Detector [19] can only produce line and paragraph outputs but no word-level results. Among the 8 submissions in Task 1 which have disclosed their methods, we observed that 5 of them develop '*multi-head plus postprocessing*' systems. These methods treat words, lines, and paragraphs as generic objects, and train detection or segmentation models to localize these three levels of text entities in parallel with separate prediction branches for each level. In the post-processing, they use IoU-based rules to build the hierarchy in the post-processing step, i.e. assigning words to lines and lines to paragraphs. The most of the top ranking solutions belong to this type of methods. One submission (from the SCUT-HUAWEI team) adopts a cascade pipeline, by first detecting words and then applying LayoutLMv3 [26] to cluster words into lines and paragraphs. The *Hierarchical Transformers for Text Detection* method develops a unified detector similar to [19] for line detection and paragraph grouping and also a line-to-word detection model that produces bounding boxes for words. Here we briefly introduce the top 2 methods in this task:

Upstage KR team ranks 1st place in Task 1, achieving an H-PQ metric of 76.85%. It beats the second place by almost 6% in the H-PQ metric. They implemented a two-step approach to address hierarchical text detection. First, they performed multi-class semantic segmentation where classes were word, line, and paragraph regions. Then, they used the predicted probability map to extract and organize these entities hierarchically. Specifically, an ensemble of UNets with ImageNet-pretrained EfficientNetB7 [8]/MitB4 [7] backbones was utilized to extract class masks. Connected components were identified in the predicted mask to separate words from each other, same for lines and paragraphs. Then, a word was assigned as a child of a line if the line had the highest IoU with the word compared to all other lines. This process was similarly applied to lines and paragraphs. For training, they eroded target entities and dilated predicted entities. Also, they ensured that target entities maintained a gap between them. They used symmetric Lovasz loss [9] and pre-trained their models on the SynthText dataset [24].

DeepSE X Upstage HK team ranks 2nd in the leaderboard. They fundamentally used DBNet [6] as the scene text detector, and leveraged the oCLIP [5] pretrained Swin Transformer-Base [4] model as the backbone to make direct predictions at three different levels. Following DBNet, they employed Balanced

Cross-Entropy for binary map and L1 loss for threshold map. The authors also further fine-tuned the model with lovasz loss [9] for finer localization.

3.3 Task 2 Methodology

Task 2, i.e. Word-Level End-to-End Text Detection and Recognition, is a more widely studied task. Recent research [2,15] focuses on building end-to-end trainable OCR models, as opposed to separately trained detection and recognition models. It's widely believed that end-to-end models enjoy shared feature extraction which leads to better accuracy. However, the results of our competition say otherwise. The top 2 methods by the **Upstage KR team** and **DeepSE End-to-End Text Detection and Recognition Model team** are all separately trained models. There are two end-to-end submissions. The **unified_model team** applies a deformable attention decoder based text recognizer and ranks 3th place. Here we briefly introduce the top 2 methods in this task:

Upstage KR team uses the same task 1 method for detecting words. For word-level text recognition, they use the ParSeq [1] model but replace the visual feature extractor with SwinV2 [3]. The text recognizer is pretrained with synthetic data before fine-tuning it on the HierText dataset. They use an in-house synthetic data generator derived from the open source SynthTiger [25] to generate word images using English and Korean corpus. Notably, they generate 5M English/Korean word images with vertical layout, in addition to 10M English/Korean word images with horizontal layout. For the final submission, they use an ensemble of three text recognizers for strong and stable performance.

DeepSE End-to-End Text Detection and Recognition Model team also uses the ParSeq [1] model as their recognizer. They point out that, in order to make the data domain consistent between the training and inference stages, they run their detector on training data, and then crop words using detected boxes. This step is important int adapting the training domain to the inference domain. This trick essentially improves their model's performance.

4 Discussion

In the Hierarchical Text Detection task, the original Unified Detector [19] can only achieve PQ scores of 48.21%, 62.23%, 53.60% on the words, lines, and paragraphs respectively. The H-PQ score for Unified Detector is only 54.08%, ranking at 10th place if put in the competition leaderboard. The winning solution exceeds Unified Detector by more than 20%. These submissions greatly push the envelope of state-of-the-art Hierarchical Text Detection method. However, current methods are still not satisfactory. As shown in Fig. 6, we can easily notice that for all methods, word PQ scores are much higher than line PQ scores, and line PQ scores are again much higher than paragraph PQ scores. It indicates that, line and paragraph level detections are still more difficult than word detection. Additionally, Fig. 8 shows that layout analysis performance is only marginally correlated with word detection performance, especially when outliers are ignored.

We believe there's still hidden challenges and chances for improvement in layout analysis. Furthermore, winning solutions in our competition rely on postprocessing which can be potentially complicated and error-prone. It's also important to improve end-to-end methods.

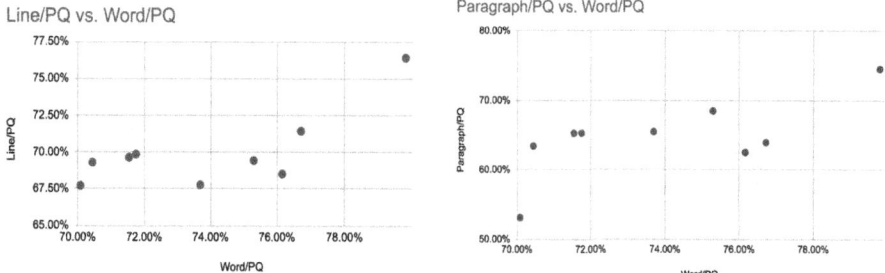

Fig. 8. Correlation between text levels. Each dot is a submission in the Task 1. **Left**: Correlation between word PQ and line PQ. **Right**: Correlation between word PQ and paragraph PQ.

The task 2 of our challenge is a standard yet unique end-to-end detection and recognition task. While it inherits the basic setting of an end-to-end task, it is based on a diversity of images which has high word density, and it has an unlimited character set. For this task, we see most of the submissions are two-stage methods, where the detection and recognition models are trained separately, and there's no feature sharing. These two-stage methods achieve much better performances than end-to-end submissions. This contrasts with the trend in research paper that favors end-to-end trainable approaches with feature sharing between the two stage. Therefore, we believe the HierText dataset can be a very useful benchmark in end-to-end OCR research. Another interesting observation for Task 2 is that, while most submissions achieve a tightness score of around 80%, the correlation between tightness scores and F1 scores and very low, with a correlation coefficient of 0.06. It could indicate that recognition is less sensitive to the accuracy of bounding boxes after it surpasses some threshold. This would mean that the mainstream training objective of maximizing bounding box IoU might not be the optimal target. For example, a slightly oversized bounding box is better than a small one which might miss some characters. With that said, a precise bounding box is still useful itself, which indicates the localization. Another potential reason is that bounding box annotation is not always accurate – it's always oversized because text are not strictly rectangular.

5   Conclusion

This paper summarizes the organization and results of ICDAR 2023 Competition on Hierarchical Text Detection and Recognition. We share details of competition motivation, dataset collection, competition organization, and result analysis. In total, we have 18 valid and unique competition entries, showing great interest from both research communities and industries. We keep the competition submission site open to promote research into this field. We also plan to extend and improve this competition, for example, adding multilingual data.

References

1. Bautista, D., Atienza, R.: Scene text recognition with permuted autoregressive sequence models. In: Computer Vision-ECCV,: 17th European Conference, Tel Aviv, Israel, 23–27 October 2022, Proceedings, Part XXVIII, p. 2022. Springer, Cham (2022). https://doi.org/10.1007/978-3-031-19815-1_11
2. Ye, M., et al.: DeepSolo: let transformer decoder with explicit points solo for text spotting. arXiv preprint arXiv:2211.10772 (2022)
3. Liu, Z., et al.: Swin transformer v2: scaling up capacity and resolution. In: Proceedings of the IEEE/CVF Conference on Computer Vision and Pattern Recognition (2022)
4. Liu, Z., et al.: Swin transformer: hierarchical vision transformer using shifted windows. In: Proceedings of the IEEE/CVF International Conference on Computer Vision (2021)
5. Xue, C.: Language matters: a weakly supervised vision-language pre-training approach for scene text detection and spotting. In: Computer Vision-ECCV, et al.: 17th European Conference, Tel Aviv, Israel, 23–27 October 2022, Proceedings, Part XXVIII, p. 2022. Springer, Cham (2022). https://doi.org/10.1007/978-3-031-19815-1_17
6. Liao, M., et al.: Real-time scene text detection with differentiable binarization. In: Proceedings of the AAAI Conference on Artificial Intelligence, vol. 34, no. 07 (2020)
7. Xie, E., et al.: SegFormer: simple and efficient design for semantic segmentation with transformers. Adv. Neural Inf. Process. Syst. **34**, 12077–12090 (2021)
8. Tan, M., Le, Q.: Efficientnet: rethinking model scaling for convolutional neural networks. In: International Conference on Machine Learning, PMLR (2019)
9. Berman, M., Amal, R.T., Matthew, B.B.: The lovász-softmax loss: a tractable surrogate for the optimization of the intersection-over-union measure in neural networks. In: Proceedings of the IEEE Conference on Computer Vision and Pattern Recognition (2018)
10. Long, S., He, X., Yao, C.: Scene text detection and recognition: the deep learning era. Int. J. Comput. Vis. **129**(1), 161–184 (2021)
11. Lee, J., et al.: Page segmentation using a convolutional neural network with trainable co-occurrence features. In: 2019 International Conference on Document Analysis and Recognition (ICDAR), IEEE (2019)
12. Yang, X., et al.: Learning to extract semantic structure from documents using multimodal fully convolutional neural networks. In: Proceedings of the IEEE Conference on Computer Vision and Pattern Recognition (2017)
13. Ronen, R., et al.: GLASS: global to local attention for scene-text spotting. arXiv preprint arXiv:2208.03364 (2022)
14. Long, S., et al.: Textsnake: a flexible representation for detecting text of arbitrary shapes. In: Proceedings of the European Conference on Computer Vision (ECCV) (2018)

15. Qin, S., et al.: Towards unconstrained end-to-end text spotting. In: Proceedings of the IEEE/CVF International Conference on Computer Vision (2019)
16. Kittenplon, Y., et al.: Towards weakly-supervised text spotting using a multi-task transformer. In: Proceedings of the IEEE/CVF Conference on Computer Vision and Pattern Recognition (2022)
17. Liu, S., et al.: Unified line and paragraph detection by graph convolutional networks. In: Uchida, S., Barney, E., Eglin, V. (eds.) Document Analysis Systems. DAS 2022. LNCS, vol. 13237, pp. 33–47. Springer, Cham (2022). https://doi.org/10.1007/978-3-031-06555-2_3
18. Wang, R., Yasuhisa, F., Ashok, C.P.: Post-ocr paragraph recognition by graph convolutional networks. In: Proceedings of the IEEE/CVF Winter Conference on Applications of Computer Vision (2022)
19. Long, S., et al.: Towards end-to-end unified scene text detection and layout analysis. In: Proceedings of the IEEE/CVF Conference on Computer Vision and Pattern Recognition (2022)
20. Li, C., et al.: StructuralLM: structural pre-training for form understanding. arXiv preprint arXiv:2105.11210 (2021)
21. Long, S., Cong, Y.: Unrealtext: synthesizing realistic scene text images from the unreal world. arXiv preprint arXiv:2003.10608 (2020)
22. Liao, M., et al.: SynthText3D: synthesizing scene text images from 3D virtual worlds. Sci. China Inf. Sci. **63**, 1–14 (2020)
23. Jaderberg, M., et al.: Synthetic data and artificial neural networks for natural scene text recognition. arXiv preprint arXiv:1406.2227 (2014)
24. Gupta, A., Andrea, V., Andrew, Z.: Synthetic data for text localisation in natural images. In: Proceedings of the IEEE Conference on Computer Vision and Pattern Recognition (2016)
25. Yim, M., Kim, Y., Cho, H.-C., Park, S.: SynthTIGER: synthetic text image GEneratoR towards better text recognition models. In: Lladós, J., Lopresti, D., Uchida, S. (eds.) ICDAR 2021, Part IV. LNCS, vol. 12824, pp. 109–124. Springer, Cham (2021). https://doi.org/10.1007/978-3-030-86337-1_8
26. Huang, Y., et al.: LayoutLMv3: pre-training for document AI with unified text and image masking. arXiv preprint arXiv:2204.08387 (2022)
27. Kuznetsova, A., et al.: The open images dataset v4. Int. J. Comput. Vis. **128**(7), 1956–1981 (2020)
28. Singh, A., et al.: TextOCR: towards large-scale end-to-end reasoning for arbitrary-shaped scene text. In: Proceedings of the IEEE/CVF Conference on Computer Vision and Pattern Recognition (2021)
29. Krylov, I., Sergei, N., Vladislav, S.: Open images v5 text annotation and yet another mask text spotter. In: Asian Conference on Machine Learning, PMLR (2021)
30. Kirillov, A., et al.: Panoptic segmentation. In: Proceedings of the IEEE/CVF Conference on Computer Vision and Pattern Recognition (2019)

ICDAR 2023 Competition on Detection and Recognition of Greek Letters on Papyri

Mathias Seuret[4]([✉]), Isabelle Marthot-Santaniello[1], Stephen A. White[2,3], Olga Serbaeva Saraogi[1], Selaudin Agolli[1], Guillaume Carrière[5], Dalia Rodriguez-Salas[4], and Vincent Christlein[4]

[1] Departement Altertumswissenschaften, Universität Basel, Basel, Switzerland
{i.marthot-santaniello,olga.serbaevasaraogi,s.agolli}@unibas.ch
[2] International University of Venice, Digital Humanities Software Engineer/Consultant, Venice, Italy
[3] Library Institute of Asian and Oriental Studies, University of Zurich, Zürich, Switzerland
[4] Pattern Recognition Lab, Friedrich-Alexander-Universität Erlangen-Nürnberg, Erlangen, Germany
{mathias.seuret,dalia.rodriguez-Salas,Vincent.Christlein}@fau.de
[5] École Pour l'Informatique et les Techniques Avancées, Le Kremlin-Bicêtre, France
guillaume.carriere@epita.fr

Abstract. This competition investigates the performance of glyph detection and recognition on a very challenging type of historical document: Greek papyri. The detection and recognition of Greek letters on papyri is a preliminary step for computational analysis of handwriting that can lead to major steps forward in our understanding of this major source of information on Antiquity. It can be done manually by trained papyrologists. It is however a time-consuming task that would need automatising. We provide two different tasks: localization and classification. The document images are provided by several institutions and are representative of the diversity of book hands on papyri (a millennium time span, various script styles, provenance, states of preservation, means of digitization and resolution).

Keywords: Document analysis · papyrus · character spotting

1 Introduction

Greek papyri are a unique source of information on Antiquity. In particular, they allowed rediscovering lost pieces of ancient literature that had not survived through the medieval manuscript tradition. Literary papyri are usually written in a careful hand close to calligraphy with isolated (i.e. not connected) letters. They are however heavily damaged and often broken into fragments, such as the ones shown in Fig. 1, as papyrus becomes very brittle as it ages and dries.

© The Author(s), under exclusive license to Springer Nature Switzerland AG 2023
G. A. Fink et al. (Eds.): ICDAR 2023, LNCS 14188, pp. 498–507, 2023.
https://doi.org/10.1007/978-3-031-41679-8_29

Collections are often composed of disparate fragments, without necessarily information about which ones come from the same document. Therefore comparing shape similarities at the character level is a promising approach to improve the assignment of dates, provenance, and possibly schools of writers, even to find fragments that used to belong together. Correctly detecting and recognizing the individual letters is thus a key preliminary step before applying methods based on similarity measurement and clustering or classification.

2   Data

The dataset we produced is formed by 194 high-resolution (jpeg) images of papyri bearing the text of the best-attested book of Antiquity, Homer's *Iliad*. Images have been collected from online catalogs that allow their use for scientific and teaching purposes or have been requested from the owning institutions along with their copyright authorization. Each image has been annotated by specialists at the character level by drawing bounding boxes around each letter using the Research Environment for Ancient Documents (READ) platform[1].

Besides the transcription (which of the 24 Greek letters is in the bounding box), an evaluation of the state of preservation of the letter has been made using the following criteria:

- State 1: complete or almost complete character (the pattern is complete)
- State 2: damaged but recognizable character (at least one important component of the pattern is missing but what remains can unequivocally be recognized as one class of letter)
- State 3: damaged letter that requires the context to be identified (what remains can be interpreted as more than one class of letter).
- Stage 4: incomplete letter that ambiguously looks like another class than the one given by the context (due to a discriminative element now damaged).

Isolated blots of ink on the border of holes have not been annotated. Two signs of punctuation have been annotated and transcribed ("" ' for apostrophe, "." for period) but will not be taken into account in the evaluation. All other remains of ink that are neither letters nor punctuation (accents, marks indicating paragraph or text section) have been annotated and assigned the "?" sign and won't be part of the evaluation.

The images have been carefully split into training and test subsets (respectively 160 and 34 images).

Additionally, we provide a baseline method[2] consisting to the PyTorch tutorial on finetuning for object detection[3], based on Fast-RCNN [3].

3   Evaluation

The evaluation is based on the mean Average Precision (AP) defined by COCO[4], with intersection over union (IoU) going from 50% to 95%, with steps of 5%.

---

[1] https://github.com/readsoftware/read.

[2] https://github.com/daliarodriguez/icdar2023-papyri.

[3] https://pytorch.org/tutorials/intermediate/torchvision_tutorial.html.

[4] https://cocodataset.org/#detection-eval.

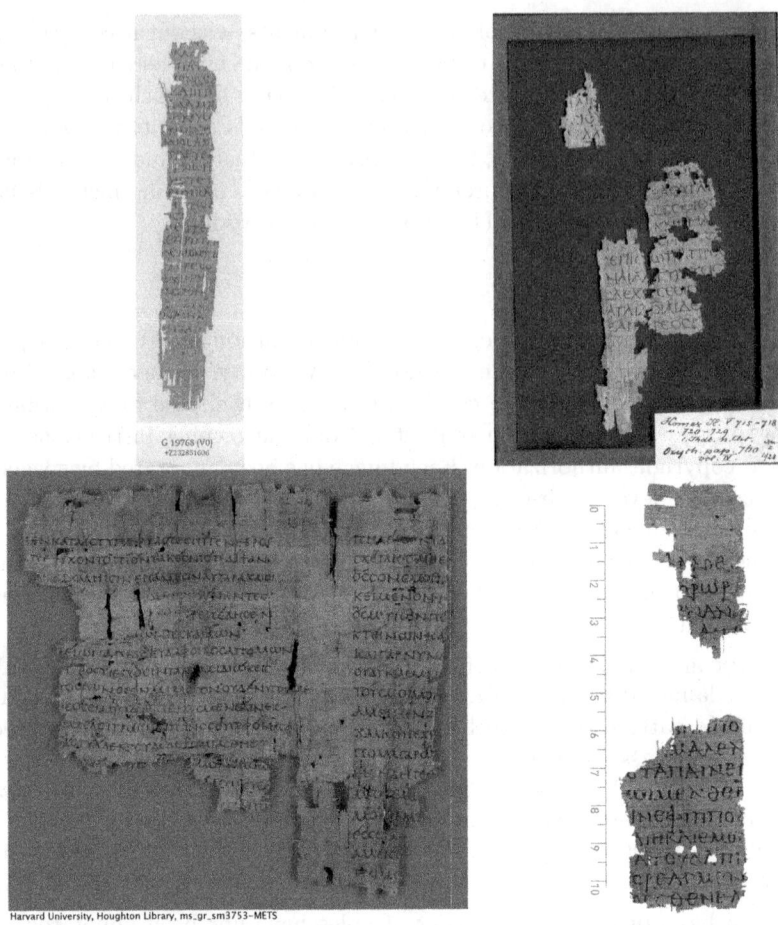

Fig. 1. Illustration of 4 random fragments from the test set.

This metric, which code is publicly available[5], has however to be tuned to our task for the reason detailed below. The detected bounding boxes have to be provided together with a confidence score. The function which computes the AP with a specific IoU, e.g., $AP^{IoU=.75}$, has a parameter to indicate how many of the bounding boxes, from the highest confidence to the lowest, have to be taken into account for the evaluation. However, this parameter cannot be set when computing the mean AP, and the default value of 100 is used[6]. Computing the mean AP requires computing AP with 10 different IoU thresholds, thus we believe this implementation choice is motivated by the computation time.

---

[5] https://github.com/cocodataset/cocoapi/blob/master/PythonAPI/pycocotools/cocoeval.py.

[6] Lines 460 & 427 in the GitHub link, version of the 25th of December 2019.

We created an unrestricted version of the `summarize` class and attributed it to `COCOeval`, which solved the issue.

We distinguish the detection and the recognition tasks as follows. For the detection, we want the system to output bounding boxes around characters, regardless of the characters' classes. This is done by setting `cocoEval.params.useCats` to false – in practice, it is equivalent to assuming that all detection predictions have the correct class label. The recognition task is done by using the class labels, i.e., in addition to having the bounding box correctly located, it must have the correct label. Providing the participants with crops to recognize, and thus fully decouple this metric from the detection, would imply sending to the participants the solution for the detection task. Punctuation symbols are not taken into account.

Moreover, state 3 characters (damaged beyond readability) are considered as a special case. We run the evaluation for the detection task once ignoring them, and once taking them into account, and keep the highest score. While they are heavily damaged (or even missing, in some cases), a method able to spot faint ink remains should benefit from it.

4   Competition Platform: CodaLab

To manage user evaluations, we used CodaLab [8], a web platform designed for running competitions. CodaLab conveniently offers to run custom evaluation scripts when participants submit results and manages a leaderboard automatically.

We intend to re-open the competition, and thus allow researchers to continue comparing their results on the leaderboard; any new submission will however not be considered as taking part to the competition.

We have to mention that while CodaLab works very smoothly once fully set up, some problems in the administration panel forced us to create a second competition page and close the first one. Some leaderboard settings, such as which variables to use, can be selected only when uploading the initial Yaml file; the administration controls have no effect on these.

5   Proposed Methods

The descriptions of methods provided by the participants are given below, sorted in alphabetical order.

5.1   Carson Brown

Images are passed through an RCNN (using a sliding window to crop the images to a standard size) to detect bounding boxes around the glyphs without classes. Once the set of glyphs is cleaned and duplicates are removed, the detected bounding boxes are collected into lines using horizontal and vertical proximity. Each line is then fed into a recurrent CNN that classifies each glyph, with an LSTM layer [4] acting as a language model to increase classification accuracy. Done as a part of a Masters Thesis for the University of St Andrews, Scotland, with the supervision of Dr Mark-Jan Nederhof.

5.2 KittyDetection

Proposed by Martin Leipert, from TH Deggendorf (Deggendorf Institute of Technology). He used Faster-RCNN [9] and plugged in FocalNet-T [14] (pre-trained on ImageNet) with small receptive field as a backbone. He sliced the image into overlapping patches and used them for training. Additionally, he did on-the-fly augmentation with Albumentations [1]. He used class weighting to compensate for class imbalance.

5.3 Nara Information

Proposed by Dr. Joon Mo Ahn, Tae Hong Jang, and Dr. Sojung Lucia KIM, from Nara Information, Nara Labs in South Korea. They first apply some preprocessing steps to the input images in order to reduce the noise, especially on the background, and enhance the edges. They then apply an HRNet [13] for detection and classification, as this popular model allows to process high-resolution data.

5.4 Shi et al.

They trained a YOLOv8 detection model. Instead of feeding it entire images, they pre-processed the training set and divided each image into lines of 10 letters based on the K-means clustering of the central points of each box. For the prediction, they used the model first to predict the letters on the entire image. Central points of the detected boxes were used to determine the line alignment and also the possible letters that the model did not detect (based on intervals). Then, the test images were cut into lines of 10 letters, which went through the detection model again for local detection and classification.

5.5 Turnbull and Mannix

Robert Turnbull and Evelyn Mannix, from the Melbourne Data Analytics Platform, University of Melbourne divided the dataset into five partitions, such that all the images from any particular manuscript were in the same partition. They took YOLOv8 models [5] for object detection and classification, at sizes medium, large, and extra large and trained them using the competition dataset, using cross-validation from the five partitions. They then downloaded roughly 4,000 published images of the Oxyrhynchus Papyri and used the YOLOv8 models to detect and classify characters from these images. These predictions were used to fine-tune the YOLOv8 models. The resulting models for each partition and model size were used to make predictions on the test dataset. These predictions on the test set were ensembled using Weighted Boxes Fusion [10]. To enhance performance, we trained image classifiers for detected characters. They were interested to see if we could improve the classification performance using additional data, namely the AL-PUB dataset [11] and the predictions on the Oxyrhynchus Papyri. In the first instance, they trained a SimCLR ResNet50 self-supervised model on all of these data and used the available annotations and the AL-PUB labels to fine-tune a classifier [2]. Then, They looked into using transfer learning with transformers, following the DeiT approach [12]. These approaches achieved similar performance, so they followed the above approach and ensembled them with the YOLOv8 classifier. In the instance these three approaches agreed, they kept the same label, but if they disagreed we created a new box with the same dimensions that contained the other possible candidate characters.

5.6   ENCyclops

The approach proposed by Carolina Macedo, Malamatenia Vlachou Efstathiou, Violette Saïag, Noé Leroy, and Chahan Vidal-Gorene, from the Ecole Nationale des Chartes, is based on YOLOv5 [6], trained on the detection task (localization + classification), mixed with a custom quite-intensive data-augmentation. The objective was to experiment with an approach different from Faster-RCNN, which performs detection in two stages. YOLO has already been tested on historical documents (YaltAI), but never, to our knowledge, for character detection. In detail, the task being complex, they used the pre-trained model *yolov5l6u.pt*, which they specialized with Greek characters. The size of the images is fixed at 1536 pixels. YOLO is usually very permissive, with very low detection thresholds, and default configuration leads to inaccurate results. In order to limit the number of false positives caused by partial localization of information (default IoU 0.45), they have drastically increased the thresholds to bring them to 0.9 of IoU in localization and 0.65 of confidence for classification (during the training step). In addition, the dataset presenting a disparity in the distribution of each class, they developed a significant data augmentation, combining the following methods: mixup (100%), mosaic (50%), pixel dropout (50%), image inversion (25%), multiplicative noise (15%), blur (5%), random contrast and brightness (5%) and elastic transform (5%). A combination of these methods is randomly applied during each step of the training. A general dropout of 0.1 is applied, and we have also implemented a scheduler for the learning rate (by step, every 200 epochs). Due to a lack of computing power, they restricted the training to a batch of 2 and image size of 1536 pixels. They are convinced that doubling the batch size and increasing the images to at least 2000 pixels would significantly improve the results. Training is quite easy and the results converge quickly.

5.7   Vu and Aimar

Manh Tu Vu and Marie Beurton Aimar, from the Laboratoire Bordelais de Recherche en Informatique, propose a new deep learning network called

Table 1. Detection mean average precision

Rank	Team	mean AP (%)
1	Vu & Aimar	51.83
2	Turnbull & Mannix	51.42
3	ENCyclops	48.82
4	Shi et al.	44.09
5	Nara Information	39.25
6	Carson Brown	38.42
7	KittyDetection	28.81

PapyTwin which consists of two subnetworks, the first and second twin, that cooperate together to address the challenge of detecting Greek letters on ancient papyri. While the first twin network aims to uniform the letter size across the images, the second twin network predicts letter bounding boxes based on these letter-uniformed images. Both of these two subnetworks have the same network architecture which is Faster RCNN pretrained on COCO database [7].

For the first twin network, they resized the images to the size of 800×800 pixels. During training, random image translation and random colour jitter were applied. They train the network in 80 epochs using Adam optimizer with a learning rate of Lr×BatchSize/256 with a base Lr $= 4 \cdot 10^{-3}$, and batches of 5 images. The output bounding box predictions of this network are used to compute the average letter height $l(x)$ of the image. Then they compute the scaling factor of the image by $f(x) = s/l(x)$, where s is a reference size hyperparameter. They found that their network works best with $s = 96$.

For the second twin network, they resize the images based on the scaling factor that they have computed using the first twin network. Then, they train the network using a patch-based training strategy. Rather than training an entire large image, they split the image into a grid of overlapping patches and treat each patch as an individual training sample. In the inference phase, they also split the large image into a grid of overlapping patches with a fixed size of 800×800 pixels and predict each patch separately. Then, they merge the prediction results back together based the location of the patch in the original image.

Finally, to obtain the best prediction, they train several PapyTwin models and merge their predictions by averaging the coordinate of bounding boxes that have their IoU > 0.7. The most dominant label of these bounding boxes is chosen to be the label of the merged bounding box.

6   Results

The competition results for the detection and recognition tasks are given in Tables 1 and 2. We can see that the is very stable, with only the first two positions being swapped in the tables.

Table 2. Recognition mean average precision

Rank	Team	mean AP (%)
1	Turnbull & Mannix	42.16
2	Vu & Aimar	41.73
3	ENCyclops	38.60
4	Shi et al.	31.07
5	Nara Information	26.33
6	Carson Brown	25.51
7	KittyDetection	18.90

The winning team for the detection task is Mahn Tu Vu and Marie Beurton Aimar, from the Laboratoire Bordelais de Recherche en Informatique, with a mean AP of 51.83%. For the recognition task, the winning team is Robert Turnbull and Evelyn Mannix, from the Melbourne Data Analytics Platform, University of Melbourne, with a mean AP of 42.16%.

Table 3. Additional details for the AP

Team	Detection (%)		Recognition (%)	
	$AP^{IoU=.5}$	$AP^{IoU=.75}$	$AP^{IoU=.5}$	$AP^{IoU=.75}$
Turnbull & Mannix	93.2	49.9	74.7	41.2
Vu & Aimar	92.5	52.7	72.1	43.9
ENCyclops	90.6	46.1	69.9	37.9
Shi et al	84.9	39.3	57.9	28.3
Nara Information	77.9	33.5	49.3	25.1
Carson Brown	85.9	24.5	53.9	17.5
KittyDetection	68.2	17.5	41.2	12.9

More detailed results given in Tables 3 and 4 give some insight into the different methods' performances. While the three best detection mean AP are at around 50%, the AP with an IoU threshold of 0.5 are significantly higher, ranging from 68.2% to 93.2% for the detection, and 41.2% to 74.7% for the recognition. High AP could be linked to outputting few but accurate bounding boxes; for this reason, we also measured the average recall (AR). Table 4 shows that the recall is actually higher than the precision. Thus, the main source of error is false positives, which decrease the AP without impacting the AR.

Another aspect of the results which arises from these measures is that the AP and AR drop significantly when the IoU threshold increases. While lower values are not surprising, such a drop implies that while most characters are correctly detected, the bounding boxes do not match well the ground truth.

Table 4. Additional details for the average recall AR

Team	Detection (%)		Recognition (%)	
	$AR^{IoU=.5}$	$AR^{IoU=.75}$	$AR^{IoU=.5}$	$AR^{IoU=.75}$
Turnbull & Mannix	98.6	81.2	69.9	56.4
Vu & Aimar	98.5	76.2	69.7	55.5
ENCyclops	97.5	75.1	64.9	51.8
Shi et al	90.9	62.5	56.3	39.8
Nara Information	84.7	53.7	50.4	34.8
Carson Brown	93.6	59.2	47.1	31.7

7   Conclusion

Overall the two highest-ranked teams for both tasks have very similar results, but obtained with approaches following two very different philosophies. While Turnbull & Mannix used a large amount of external papyri data for training their method, Vu & Aimar developed a system using two subnetworks pre-trained on external non-papyri data. We can note that ENCyclops, at the third rank, did use heavy data augmentation. This highlights very well the importance of data diversity for this task.

In general, the detection results were better than what we expected, which is a pleasant surprise. The recognition, however, still leaves some room for improvement, and the diversity of the approaches explored by the participants, such as preprocessing, using external data, subnetworks, or language models predicts exciting results in computer-assisted papyrology.

References

1. Buslaev, A., Parinov, E., Khvedchenya, V.I., Iglovikov, Kalinin, A.A.: Albumentations: fast and flexible image augmentations. ArXiv e-prints (2018)
2. Chen, T., Kornblith, S., Swersky, K., Norouzi, M., Hinton, G.E.: Big self-supervised models are strong semi-supervised learners. Adv. Neural Inf. Process. Syst. **33**, 22243–22255 (2020)
3. Girshick, R.: Fast r-cnn. In: Proceedings of the IEEE International Conference on Computer Vision, pp. 1440–1448 (2015)
4. Hochreiter, S., Schmidhuber, J.: Long short-term memory. Neural Comput. **9**(8), 1735–1780 (1997)
5. Jocher, G., Chaurasia, A., Qiu, J.: YOLO by Ultralytics, January 2023. https://github.com/ultralytics/ultralytics
6. Jocher, G., et al.: ultralytics/yolov5: v7.0 - YOLOv5 SOTA realtime Instance Segmentation, November 2022. https://doi.org/10.5281/zenodo.7347926
7. Lin, T.Y., et al.: Microsoft coco: common objects in context, pp. 740–755 (2014)
8. Pavao, A., et al.: Codalab competitions: An open source platform to organize scientific challenges. Technical report (2022). https://hal.inria.fr/hal-03629462v1

9. Ren, S., He, K., Girshick, R., Sun, J.: Faster r-cnn: towards real-time object detection with region proposal networks. Adv. Neural Inf. Process. Syst. **28** (2015)
10. Solovyev, R., Wang, W., Gabruseva, T.: Weighted boxes fusion: Ensembling boxes from different object detection models. Image Vis. Comput. **107**, 1–6 (2021)
11. Swindall, M.I., et al.: Exploring learning approaches for ancient greekcharacter recognition with citizen science data. In: 2021 17th International Conference on eScience (eScience), pp. 128–137. IEEE (2021)
12. Touvron, H., Cord, M., Douze, M., Massa, F., Sablayrolles, A., Jégou, H.: Training data-efficient image transformers & distillation through attention. In: International Conference on Machine Learning, pp. 10347–10357. PMLR (2021)
13. Wang, J., et al.: Deep high-resolution representation learning for visual recognition. IEEE Trans. Pattern Anal. Mach. Intell. **43**(10), 3349–3364 (2020)
14. Yang, J., Li, C., Dai, X., Gao, J.: Focal modulation networks. Adv. Neural Inf. Process. Syst. **35**, 4203–4217 (2022)

ICDAR 2023 Competition on Born Digital Video Text Question Answering

Zhibo Yang[1,3]($\boxtimes$) ![ID], Xiaoge Song[2] ![ID], Sibo Song[1] ![ID], Tong Lu[2] ![ID], Xiang Bai[3] ![ID], Cheng-Lin Liu[4] ![ID], Fei Huang[1] ![ID], and Cong Yao[1] ![ID]

[1] Alibaba Group, Hangzhou, China
yangzhibo450@gmail.com
[2] Nanjing University, Nanjing, China
dz1833021@nju.edu.cn , lutong@nju.edu.cn
[3] Huazhong University of Science and Technology, Wuhan, China
[4] Institute of Automation of Chinese Academy of Sciences, Beijing, China

Abstract. This paper presents the final results of the ICDAR 2023 Competition on Born Digital Video Text Question Answering (i.e., BDVT-QA) which contains two major task tracks: 1) End-to-End Video Text Spotting, and 2) Video Text Question Answering. BDVT-QA aims to spot texts and answer questions from born-digital videos. The proposed competition introduces a brand new dataset consisting of 1,000 video clips fully annotated with manually-designed question/answer pairs, where the answers are based on the text captions presented in the video clips. A total of 23 final submissions were received for this competition. The top-3 performances of each track are as follows: 1)T1.1 - 57.53%, T1.2 - 53.3%, T1.3 - 52.35%, and 2) T2.1 - 31.2%, T2.2 - 28.84%, T2.3 - 21.19%. We summarize the submitted methods and give a deep analysis. Besides, this paper also includes dataset descriptions, task definitions and evaluation protocols. The dataset and the final ranking of submissions are publicly available on the challenge's official website: https://tianchi.aliyun.com/specials/promotion/ICDAR_2023_Competition_on_Born_Digital_Video_Text_QA.

Keywords: Born digital video · Video Text Spotting · Text-based Video Question Answering · Video Text Understanding

1 Introduction

Textual content plays an important role in video understanding, as text instances are either direct indicators of scenes or lingual cues about ongoing stories. They can provide explicit and high-level semantic information which not available in other form in the scene. Reading textual content in the videos such as captions, product descriptions and brand names in man-made environments is important

---

Z. Yang, X. Song and S. Song—Equal Contribution.

for downstream applications like navigation in advanced driver assistance system, assistive shopping on the live stream, and conversation understanding in drama.

Though numerous works have been proposed for related tasks such as video text recognition [5,14,17] and image text QA [2,3,13], less effort are put on born-digital video question answering. In BDVT-QA, we would like to go one step forward to explore the video text QA problem, which requires a holistic, precise and in-depth understanding of text information across space and time over video frames. Though widely used, video text has been rarely explored because of its challenging factors such as arbitrarily-shaped text trajectories, animation of text's presentation and long-term text language processing.

Fig. 1. Challenging factors in BDVT-QA. Row1: Animation of texts. Row2: Circular trajectory of text caused by rotating object.

Currently, there are already several comprehensive benchmarks for text-based VQA (ST-VQA [3], TextVQA [13], etc). Nevertheless, these works mainly focus on image level and lack benchmarks on video text-based tasks. In the proposed competition, we make a significant step further and present a new VTQA (short for Video Text QA) benchmark. There are two steps in building the proposed dataset: collection and labelling. In collection, we crawled more than 50 thousand videos from public website, filtered out those videos without texts or with too much texts, discarded those with stationary texts, and finally got 1000 qualified videos. In labelling, detailed information are labelled, including period of the target question, location and trajectory of text instances, transcription of the text lines and corresponding question-answer pairs. In BDVT-QA, the questions mainly involve inferring topics from videos and understanding the temporal context between descriptive texts. This means that potential answers are presented progressively alongside the video.

Two tasks are proposed within this challenge: **(1) End-to-End Video Text Spotting**, which requires simultaneously detecting, tracking and recognizing text instances in a video sequence, and **(2) Video Text Question Answering**, encourages participating teams to contribute state-of-the-art works that understands question and video texts for answering specified questions.

In the End-to-End Video Text Spotting task, as shown in Fig. 1, texts are frequently accompanied by animated elements, special effects, and visual imagery, which can present considerable impediments to the accurate recognition of textual content. The majority of participants attain text spotting results through a consecutive process consisting of a text detector, a text recognition component, and a tracker. The winner is from team *DA* and gets a 57.53% NED score [12] in their final submission, where a multistage solution is adopted. We believe that the integration of advanced sub-methods led to a comparatively high score.

The Video Text Question Answering task focuses on questions about the text content in videos. However, unlike previous image or frame-level QA tasks that rely on a single frame, this task includes lots of questions that require long-term frames. Exemplar questions are "How many items are presented?" and "What is the second step?", as shwon in Fig. 2. We appreciate that Top 2 participants both use large language models, such as T5 and GPT3.5. The winner uses T5-large as base model, trains it with resistance training and exponential sliding average training skills, and gets a final 31.2% ANLS score [12].

2   Competition Organization

ICDAR 2023 Competition on Born Digital Video Text Question Answering is organized by a joint team of Alibaba Damo Academy, Nanjing University, Huazhong University of Science and Technology and Chinese Academy of Sciences. The competition started and the training data was fully released on 1st March, while the test set videos and questions were only made available for a 15 day period between 15th March and 30th March. The participants were requested to submit results over the test set videos and not executable of their systems. At all times we relied on the scientific integrity of the authors to follow the established rules of the challenge.

The competition is hosted on the TIANCHI plartform provided by Alibaba Group. Participants can register, check information, download datasets, and submit results on TIANCHI. All submitted results are evaluated automatically, and the platform updated the leader board 10 AM from 15th March 30th March.

The complete schedule of the proposed competition is as follows:

(1) 1st March 2023: Registration is started for competition participants. Training datasets of two tasks are available for download.
(2) 15th March 2023: Submissions of both tasks are open for participants. Test data (without ground-truth information) is released.
(3) 1st April 2023: Registration and submission deadline for all the tasks for participants.
(4) 15th April 2023: Technical report submission deadline for participants.

Overall, we received more than 300 submissions, but half of which were invalid due to factors such as false formatting, shortcuts in JSON, and illegal characters. On average, each team managed to submit 5.5 valid entries. There are 27 final submissions on the leader board, submitted by 23 teams. We have discussed over-fitting issues with the competition chairs, given that the evaluation system grants scores to teams that post on the leader board. In fact, the test set is extensive and the video QA problems are complex, making it extremely challenging for participants to overfit the test set. However, we have recognized that there were drawbacks of the way that used in the evaluation. It is possible for some participating teams to pick up a best model by submitting multiple times. We appreciate the feedback from the competition chairs and will strive to improve our organization of scientific contests in the future. Taking into account the reasonable number of teams and the thought-provoking concepts presented in the submissions, it can be inferred that the deductions and performance evaluation are useful.

Question: "What is the second step to put on your swim cap?"
Answer: "Stretch outwards and backwards. "

Question: "What does Treasure Race refer to? "
Answer: "The hunt for the treasure of Gold Roger"

Fig. 2. Exemplar QA pairs in BDVT-QA

3   Dataset

The proposed dataset is deliberately collected from online sources and public websites and consists of 1,000 video clips (ranging from 3 s to 3 min) from 3 common born-digital video scenes such as advertising, short videos of product introductions and movies. There are multi-lingual texts with irregular shape and movement in these videos. Thus, text instances in this proposed dataset are only partially visible or recognizable through the whole video due to animation, view changes and blurry scenes. Statistics of the datasets are shown in Table 1. Distributions of dataset frames are shown in Fig. 5.

Dataset Split. We divided the dataset into training set and testing set, comprising 750 and 250 videos respectively, with no overlap between the two different splits. Both the training set and testing set contain videos from all scene categories, and have similar distributions of videos and annotations. Distribution of words in the trainset is shown in Fig. 3.

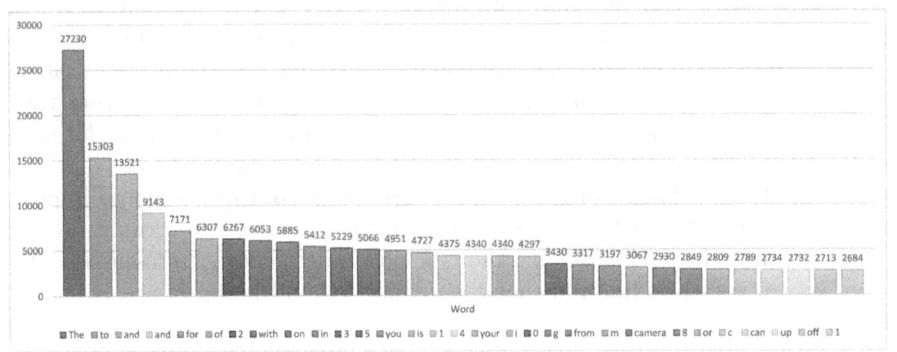

Fig. 3. High frequency words in the training set and their distribution

Annotations. To meet the demand of practical application and better evaluation, the proposed dataset has different transcription and bounding boxes of ground truths from existing video text datasets. Text instances in our dataset are annotated with (a) polygon ground truth per frame (in a similar way to ICDAR2019 Robust Reading Challenge on Arbitrary-Shaped Text [6]) with text instance id, and (b) transcription with text id. For the fully annotated video frames, horizontal, multi-oriented, and vertical text instances are labelled with quadrilateral bounding boxes. The points in the polygons are arranged in a clockwise sequence, starting from the reading direction. The transcriptions of every text instance are annotated and encoded in UTF-8. The illegibility regions in images are labeled with **visibility** as **False**, which are ignored to calculate final scores.

These annotations mentioned above cater for (1) the End-to-End Video Text Spotting task. For (2) Video Text Question Answering task, we provide large-scale video-question-answer triplets, in which questions are mainly focused on the temporal context between video texts. Some question/answer pair examples are shown in Fig. 2. The proposed dataset also provides multiple alternative human answers to each question in order to closely match the natural language style. The common questions and their frequency are shown in Fig. 4.

Table 1. Number details of the proposed dataset

Original Dataset	Videos	Frames	Texts	Average Text Length
Trainset	750	380323	1530184	2.6
Testset	250	98944	417901	2.3

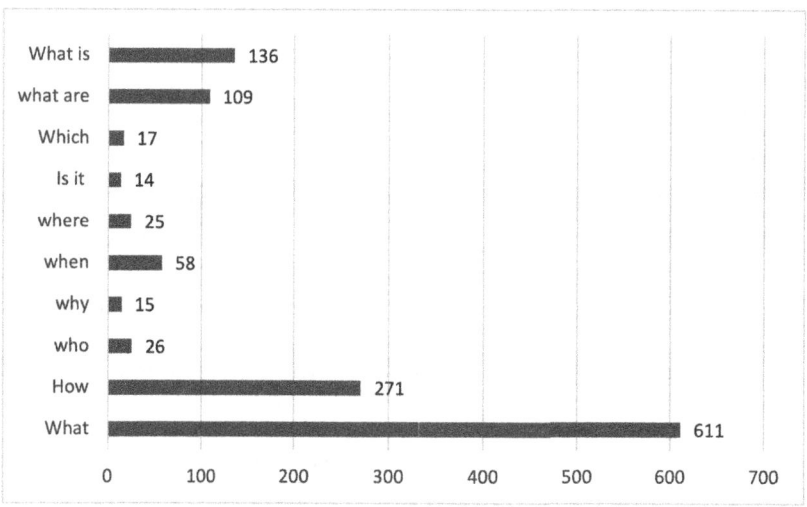

Fig. 4. High frequency questions of training set

4   Tasks

The proposed competition has two major tasks: (1) End-to-End Video Text Spotting, and (2) Video Text Question Answering. For the first task, alphanumeric and Chinese text instances are considered to be evaluated. For the second task, only alphanumeric text instances are considered to be evaluated.

4.1   Task 1 - End-to-End Video Text Spotting

The objective of this task is to assess end-to-end system performance of video text spotting. It requires models to localize, track and recognize words simultaneously. Following the previous competition settings [6,11], sequence-level metrics like Recall, Precision, F-score, and Normalized Edit Distance measures are adopted for the end-to-end evaluation. In particular, we use Recall, Precision and F-score to evaluate text detection performance. A predicted text instance is considered as a true-positive if and only if it has the maximum IoU with a groud-truth text and the IoU ≥ 0.5.

Prediction results of this track are required to be submitted as a single compressed (zip or rar) file that contains all the result files and the videos of the test set. Participants are required to automatically localize and recognize text lines in all the frames of one specific video and output prediction results in a single JSON file. If the submitted model fails to produce any prediction results for a particular video, then the compressed file only includes the corresponding JSON of that video. The naming of all the submitted JSON files follows the format: res_[video_id].json. For example, the JSON result file corresponding to the test video "vid_6371842732441621.mp4" should be "res_vid_6371842732441621.json".

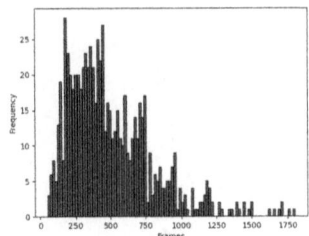
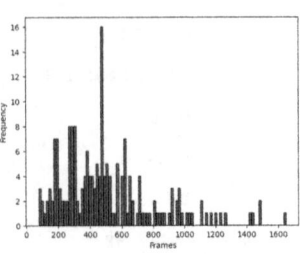

(a) Distribution of trainset frames   (b) Distribution of testset frames

Fig. 5. Video frames distribution of trainset and testset

4.2   Task 2 - Video Text Question Answering

This task is the more generic and challenging one since it requires the participants to combine video text spotting and video text question-answering technologies. The submitted methods for this task should be able to provide correct answers for the given questions by reading, tracking and comprehending all text instances in videos. In this task, the evaluation metric will be the Average Normalized Levenshtein Similarity (ANLS), the same as the ST-VQA competition [3].

The prediction results of this track are expected to be submitted as a single JSON file. The JSON file contains a list of dictionaries and each dictionary should have two keys: "question_id" and"answer". The "question_id" key refers to the unique id of the question and the "answer" corresponds to the model's prediction.

4.3   Evaluation Metric

For task 1, we use Recall, Precision and F-score to evaluate text detection performance, and Normalized Edit Distance metric (1-N.E.D specifically) as evaluation protocol of text recognition over frames. In task1, only the Normalized Edit Distance is treated as the official ranking metric while the results of other metrics are published for reference only. The calculation of Recall, Precision and F-score are as:

$$Recall = \frac{TP}{TP + FN} \tag{1}$$

$$Precision = \frac{TP}{TP + FN} \tag{2}$$

$$F = \frac{2 \times Precision \times Recall}{Precision + Recall} \tag{3}$$

where TP, FP, FN and F denote true positive, false positive, false negative and F-score, respectively. For text recognition, we adopt Normalized Edit Distance metric (1-N.E.D specifically) as evaluation protocol, which is commonly used in previous competitions [12]. Normalized Edit Distance (N.E.D) is formulated as:

$$N.E.D = 1 - \frac{1}{N} \sum_{i=1}^{N} D\left(s_i, \bar{s}_i\right) / max\left(s_i, \bar{s}_i\right) \tag{4}$$

where D represents the Levenshtein Distance, and $s_i, \bar{s}_i$, denote the predicted text and its corresponding ground-truth, respectively.

The conventional QA evaluation metrics based on standard accuracy criteria pose a challenge in determining the correctness of answers in the proposed competition. This is because the answers to the questions comprise text instances recognized from video frames and not specific textual responses. Therefore, we require an evaluation metric that can account for recognition errors and mismatches in answers resulting from imperfections in OCR. Such a metric should be able to respond in a more flexible manner to such errors and mismatches without rigidly judging the correctness of answers.

For task 2, we adopt Average Normalized Levenshtein Distance (ANLS), following the previous competitions [12]. Given N questions and M ground-truth answers per question, ANLS can be formulated as:

$$ANLS = \frac{1}{N} \sum_{i=0}^{N} max_j s\left(a_{ij}, o_{q_i}\right) \tag{5}$$

where $a_{ij}(i = 1..., N, \ j = 1..., M)$ is the ground-truth answers, and o_{q_i} represents the predicted answer for i_{th} question q_i. Note that, N is the maximum number of "paired" ground truths and detected regions, which include singletons, e.g., the ground truth regions that were not matched with any detection (paired with an empty string) and detected regions that were not matched with any ground truth region (paired with an empty string).

For both task 1 and task 2, text regions with visibility of False are excluded and the detected results matched to such regions do not contribute to the final score. To avoid the ambiguity in annotations, pre-processing are utilized before comparing two strings: 1) The evaluation for English text recognition is case insensitive; 2) The Chinese traditional and simplified characters are considered to be the same categories; 3) The blank spaces and symbols, e.g., comma and dots, etc., are ignored in distance calculation.

5   Submissions

Overall, we received 150 valid submissions from both research communities and industries, among which 27 submissions are final-classified. Table 2 and Table 3 summarize top valid submitted results of the two tasks, respectively.

5.1   Top 3 Submissions in Task 1

For task 1, 14 submissions from different participants are final verified. In this section we make brief introductions of the top 3 methods. The first place of this task is **DA** by She *et al.* from Guangzhou Shiyuan Electronic Technology Company Limited, with the wining 1-N.E.D score of 57.53% and the best detection performance of 77.04% in terms of H-mean.

#1 - DA Team divides this task into three sub-tasks: text detection, text line recognition ,and text tracker post-processing. In detection module, they fuse the Cascade R-CNN [4] and DBNet [10], to get the better detection results. Their detection models are implemented using MMOCR [9]. The backbone for Cascade Mask R-CNN is ResNeXt101 [15] with official pretraining weights. The network structures and parameter settings for Cascade R-CNN follow the default configuration in MMOCR. As for DBNet, they use the ResNet50 [8] pretrained on SynthText [7] using oclip [16] from MMOCR as the backbone.

To recognize text, the PARSeq [1] method is employed, which involves learning an ensemble of internal auto-regressive (AR) language models with shared weights using Permutation Language Modeling.

To alleviate the low-quality text problems such as blurring, perspective distortion, rotation and camera motion, they also adopt BoT-SORT, an advanced multi-object tracker with MOT bag-of-tricks for a robust association. For each text trajectory, they get the top 3 results according to the number of occurrences. Then, the longest text is selected as the best text in the top 3 results. Every text in the trajectory is replaced by the best text.

Table 2. Top 10 results of task 1 - End-To-End Text Spotting. Note that * denotes missing descriptions in affiliations.

Team	Rank	Precision	Recall	H-mean	1-N.E.D	Affiliations
DA	1	85.32%	**70.23%**	**77.04%**	**57.53%**	Guangzhou Shiyuan Electronic Technology Company Limited
MFC	2	77.54%	62.60%	69.28%	53.30%	*
Winforever	3	**88.27%**	61.53%	72.52%	52.35%	Shandong University of Science and Technology
Lin *et al.*	4	87.76%	61.79%	72.52%	51.33%	Xi'an Jiaotong University
OCR-ing	5	87.77%	61.81%	72.54%	51.32%	Peking University
Xie *et al.*	6	87.35%	61.86%	72.42%	50.25%	Harbin Institute of Technology
Tang *et al.*	7	87.32%	60.97%	71.80%	50.25%	Xi'an Jiaotong University
Liu *et al.*	8	81.88%	62.98%	71.20%	49.54%	*
Qian *et al.*	9	83.45%	61.52%	70.16%	46.36%	*
MatrixQ	10	58.96%	62.29%	60.58%	37.85%	East China Normal University

#2 - MFC Team use the PP-OCRv3 version of the PaddleOCR framework relative model. Specifically, the main model of text detection is DBNet network and the main model for text recognition is SVTR. This team adopts the following scheme to prepare data and predict end-to-end results. 1) Frame cutting of the videos in the training set and test set. 2) Cutting out the text detection box given in the training set, save the picture and corresponding text, and form a new training set of text recognition, 3) Using PaddleOCR to train the text recognition model of the above data set, 4) For the video test set, the PaddleOCR pre-trained text detection model and the text recognition model trained by the team were used to get the final result.

#3 - Winforever Team conducts OCR directly on a frame-by-frame basis to see the results. Considering that the dataset has both Chinese and English text, they choose to use paddle_ch_PP-OCRv3 as the pre-training model, and then convert the trained model to the familiar Pytorch model before processing, which involves a PaddlePaddle model to Pytorch model process here. And a confidence threshold is set to remove the results with lower confidence.

5.2 Top 3 Submissions in Task 2

In total, 10 submissions are received from different participants for Task 2. The first place in this task was taken by MFC Team which achieves the highest ANLS score of 31.2%. We conclude top5 results of task 2 in Table 3. Here is a brief introduction to the top 3 methods:

#1 - MFC Team tackles this task by fine-tuning a T5-large model after processing video text data. In order to address the issue of repeated text across multiple frames, they propose to select key frames as the input frames. Furthermore, for some bilingual videos, they apply a language translation model to translate Chinese into English to improve the QA performance.

#2 - Lin's Team designs a system which consists of three sub-modules, namely the frame-level OCR module, the video text tracking module, and the text question answering module. The frame-level OCR module performs OCR on a frame-by-frame basis. Then, the video text tracking module is responsible for detecting, extracting and cleaning text from the video, through measuring both content and location similarities. Finally, they employ prompt engineering with Large Language Models (LLMs) to build the QA module as these models are capable of generating coherent and precise responses to questions.

#3 - Qian's Team propose to use Bert-large and Roberta-large for Span QA to handle the task. Owing to the question types presented in the training set, they employ an architecture that incorporates one backbone, such as BERT, along with two heads, namely the answer type head and the span QA head. The answer type head is a 3-class classifier implemented with a linear projection layer. The three answer types correspond to "yes" "no" and "span" The input for the answer type head is derived from the [CLS] token after processing through BERT. The span QA head is responsible for predicting the start and end logits for each token. During the training phase, the gradients for the span prediction branch are disregarded if no span answer exists. To effectively address the challenge

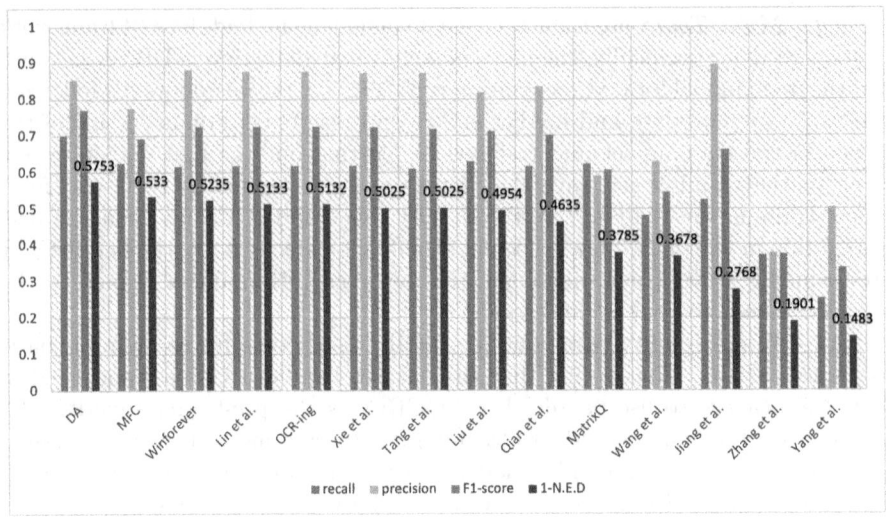

Fig. 6. Detection and recognition performance of all submissions in Task 1

posed by duplicate video text, they obtain the context by setting a threshold of edit distance. During the training stage, they incorporate a stratified K-fold strategy to address the issue of imbalanced distribution between yes/no questions and span questions.

6   Performance Analysis

6.1   Performance Analysis on Task 1

For the task 1, most of the submitted methods adopt a multi-stage based pipeline for end-to-end text recognition. They either divide the main task into two stages (i.e., text detection and text recognition) or three stages (i.e., text detection, text recognition and text tracking). Figure 6 shows the detection and recognition performance of all submissions, with the 1-N.E.D scores labelled on the image.

In the detection part, most of the competitors employ DBNet and Cascade R-CNN for training and prediction. The Winforever Team ranks first in terms of precision, while DA team has top performances in terms of recall and H-mean. The detection performance of all submissions varies from 33.64% to 77.04% in terms of H-mean, and the average H-mean is 64.41%. Most top-ranking methods have an H-mean above 72% while the best detection performance is 77.04%. Compared to current benchmarks for video text, text samples obtained from born-digital videos exhibit greater discrepancies in font, size, shapes, and trajectories, which poses a persisting challenge.

In the recognition part, the PARSeq and SVTR are mostly used. As shown in Table 2, the performance of text recognition highly relies on the detection performance. The DA team adopts an additional text tracking module to refine its text recognition performance, and it has the best 1-N.E.D result which is 4.23% higher than the second place. The MFC team and Winforever team share a similar pipeline by adopting PaddleOCR. Though the former has a sub-optimal detection performance in terms of H-mean compared to the latter one, it achieves slightly higher recognition results by optimizing the recognition module and data processing procedure.

6.2   Performance Analysis on Task 2

In this task, many top-ranking methods generate context input for QA through carefully designed text-processing strategies. For example, the MFC Team adopts key frame selection and Lin's Team propose text tracking methods to merge and generate text tracklets across frames. Some teams use Edit Distance to eliminate duplicate text sequences from their data processing. However, this approach may face limitations due to inconsistent OCR results across frames. Such inconsistencies can be attributed to variations in the backgrounds of the frames, which ultimately affect the final performance. It has been demonstrated that text processing strategies play a crucial role in preparing a concise and effective context input to the QA module, therefore, they are important for achieving optimal results.

In terms of the QA module, these teams employ different models including span-based approaches such as BERT and Roberta, and generative models like T5 or even LLMs like ChatGPT. We also note that several top-ranking methods choose generative models for answering various types of questions such as counting problems, yes-or-no questions or extractive questions. The employment of the generative and unified framework has emerged as a prominent research direction in diverse fields. The rationale behind the adoption is that the unified approach can be applied to answer multiple types of questions rather than formulating specific solutions or pipelines for each type of question.

Table 3. Top5 results of Task2 - Video Text Question Answering. Note that * denotes missing descriptions in affiliations.

Team Name	Rank	ANLS	Affiliations
MFC	1	31.20%	*
Lin *et al.*	2	28.84%	Xi'an Jiaotong University
Qian *et al.*	3	21.19%	East China Normal University
Winforever	4	18.95%	Shandong University of Science and Technology
Shen *et al.*	5	18.03%	*

7   Conclusions and Future Directions

We organized the first Born Digital Video Text Question Answering ICDAR BDVTQA 2023 competition. A novel video dataset featured with born-digital text was proposed. The competition lasted for one month, which has received large attention and participation from diversified teams, showing that the video text QA problem is a popular topic in the community. The report suggests that the performance for both tasks was below 60% and 40%, respectively, indicating that there is room for improvement in this area. Through our inspection, we summarize some reasons that may affect the results: a) For task 1, animation and effects of motion are key factors that result in false text recognition and tracking; b) For task 2, QA with long-term context is the biggest challenge. Besides, the way to handle the gap between OCR and QA is also important for end-to-end QA. In the future, We plan to reserve the website and the evaluation program for continuous research. We also plan to extend born digital video text QA to multi-modal version which focus both text and info-graphics in the video.

Acknowledgments. The authors express their gratitude to the Competition Chairs for their valuable input in organizing the competition and for their critical review of the competition report. This challenge is sponsored by Alibaba Group. This work is also supported by NSFC (62225603), NSFC (61672273) and NSFC (61832008).

References

1. Bautista, D., Atienza, R.: Scene text recognition with permuted autoregressive sequence models. In: Avidan, S., Brostow, G., Cissé, M., Farinella, G.M., Hassner, T. (eds.) ECCV 2022. LNCS, vol. 13688, pp. 178–196. Springer, Cham (2022). https://doi.org/10.1007/978-3-031-19815-1_11
2. Biten, A.F., et al.: ICDAR 2019 competition on scene text visual question answering. In: 2019 International Conference on Document Analysis and Recognition (ICDAR), pp. 1563–1570. IEEE (2019)
3. Biten, A.F., et al.: Scene text visual question answering. In: 2019 IEEE/CVF International Conference on Computer Vision (ICCV), pp. 4290–4300 (2019)
4. Cai, Z., Vasconcelos, N.: Cascade R-CNN: delving into high quality object detection. In: 2018 IEEE Conference on Computer Vision and Pattern Recognition, CVPR 2018, Salt Lake City, UT, USA, 18–22 June 2018, pp. 6154–6162. Computer Vision Foundation/IEEE Computer Society (2018)
5. Cheng, Z., Lu, J., Niu, Y., Pu, S., Wu, F., Zhou, S.: You only recognize once: towards fast video text spotting. In: Proceedings of the 27th ACM International Conference on Multimedia, pp. 855–863 (2019)
6. Chng, C.K., et al.: ICDAR2019 robust reading challenge on arbitrary-shaped text - RRC-art. In: 2019 International Conference on Document Analysis and Recognition, ICDAR 2019, Sydney, Australia, 20–25 September 2019, pp. 1571–1576. IEEE (2019)
7. Gupta, A., Vedaldi, A., Zisserman, A.: Synthetic data for text localisation in natural images. In: 2016 IEEE Conference on Computer Vision and Pattern Recognition, CVPR 2016, Las Vegas, NV, USA, 27–30 June 2016, pp. 2315–2324. IEEE Computer Society (2016)
8. He, K., Zhang, X., Ren, S., Sun, J.: Deep residual learning for image recognition. In: 2016 IEEE Conference on Computer Vision and Pattern Recognition, CVPR 2016, Las Vegas, NV, USA, 27–30 June 2016, pp. 770–778. IEEE Computer Society (2016)

9. Kuang, Z., et al.: MMOCR: a comprehensive toolbox for text detection, recognition and understanding. In: MM 2021: ACM Multimedia Conference, Virtual Event, China, 20–24 October 2021, pp. 3791–3794. ACM (2021)
10. Liao, M., Wan, Z., Yao, C., Chen, K., Bai, X.: Real-time scene text detection with differentiable binarization. In: Proceedings of the AAAI Conference on Artificial Intelligence, vol. 34, pp. 11474–11481 (2020)
11. Nayef, N., et al.: ICDAR2019 robust reading challenge on multi-lingual scene text detection and recognition - RRC-MLT-2019. In: 2019 International Conference on Document Analysis and Recognition, ICDAR 2019, Sydney, Australia, 20–25 September 2019, pp. 1582–1587. IEEE (2019)
12. Reddy, S., Mathew, M., Gómez, L., Rusiñol, M., Karatzas, D., Jawahar, C.V.: Roadtext-1k: text detection & recognition dataset for driving videos. In: 2020 IEEE International Conference on Robotics and Automation, ICRA 2020, Paris, France, 31 May–31 August 2020, pp. 11074–11080. IEEE (2020)
13. Singh, A., et al.: Towards VQA models that can read. In: 2019 IEEE/CVF Conference on Computer Vision and Pattern Recognition (CVPR), pp. 8309–8318 (2019)
14. Tian, S., Pei, W.Y., Zuo, Z.Y., Yin, X.C.: Scene text detection in video by learning locally and globally. In: International Joint Conference on Artificial Intelligence, IJCAI, pp. 2647–2653 (2016)
15. Xie, S., Girshick, R.B., Dollár, P., Tu, Z., He, K.: Aggregated residual transformations for deep neural networks. In: 2017 IEEE Conference on Computer Vision and Pattern Recognition, CVPR 2017, Honolulu, HI, USA, 21–26 July 2017, pp. 5987–5995. IEEE Computer Society (2017)
16. Xue, C., Zhang, W., Hao, Y., Lu, S., Torr, P.H.S., Bai, S.: Language matters: a weakly supervised vision-language pre-training approach for scene text detection and spotting. In: Avidan, S., Brostow, G., Cissé, M., Farinella, G.M., Hassner, T. (eds.) ECCV 2022. LNCS, vol. 13688, pp. 284–302. Springer, Cham (2022)
17. Yang, X.H., He, W., Yin, F., Liu, C.L.: A unified video text detection method with network flow. In: 2017 14th IAPR International Conference on Document Analysis and Recognition (ICDAR), vol. 1, pp. 331–336 (2017)

ICDAR 2023 Competition on Reading the Seal Title

Wenwen Yu[1], Mingyu Liu[1], Mingrui Chen[1], Ning Lu[2], Yinlong Wen[3], Yuliang Liu[1], Dimosthenis Karatzas[4], and Xiang Bai[1(✉)]

[1] Huazhong University of Science and Technology, Wuhan, China
{wenwenyu,mingyuliu,charmier,ylliu,xbai}@hust.edu.cn
[2] Huawei Technologies Ltd., Shenzhen, China
luning12@huawei.com
[3] Sichuan Optical Character Technology Co., Ltd., Chengdu, China
ylwen@chineseocr.com
[4] Computer Vision Centre, Universitat Autónoma de Barcelona, Bellaterra, Spain
dimos@cvc.uab.es

Abstract. Reading seal title text is a challenging task due to the variable shapes of seals, curved text, background noise, and overlapped text. However, this important element is commonly found in official and financial scenarios, and has not received the attention it deserves in the field of OCR technology. To promote research in this area, we organized ICDAR 2023 competition on reading the seal title (ReST), which included two tasks: seal title text detection (Task 1) and end-to-end seal title recognition (Task 2). We constructed a dataset of 10,000 real seal data, covering the most common classes of seals, and labeled all seal title texts with text polygons and text contents. The competition opened on 30th December, 2022 and closed on 20th March, 2023. The competition attracted 53 participants and received 135 submissions from academia and industry, including 28 participants and 72 submissions for Task 1, and 25 participants and 63 submissions for Task 2, which demonstrated significant interest in this challenging task. In this report, we present an overview of the competition, including the organization, challenges, and results. We describe the dataset and tasks, and summarize the submissions and evaluation results. The results show that significant progress has been made in the field of seal title text reading, and we hope that this competition will inspire further research and development in this important area of OCR technology.

1 Introduction

Based on the flourish of deep learning method, we have witnessed the maturity of regular and general OCR technology, including scene text detection and recognition. However, as a common element which can be seen everywhere in official and financial scenarios, seal title text has not gain its attention. And the task of reading seal title text is also faced with many challenges, such as variable shapes of seal (for example, circle, ellipse, triangle and rectangle), curved text, background noise and overlapped text, as shown in Figs. 1, 2 and 3. In order to promote the research of seal text, we propose the competition on reading the seal title.

Considering there are no existing datasets for seal title text reading. We construct a dataset including 10,000 real seal data, which covers the most common classes of seal. In the dataset, all seal title texts are labeled with text polygons and text contents. Besides, two tasks are presents for this competition: (1) Seal title text detection; (2) End-to-end seal title recognition. We hope that the dataset and tasks could greatly promote the research in seal text reading.

Fig. 1. Different shapes of seals samples in the ReST.

Fig. 2. Seals with curved texts in the ReST.

Fig. 3. Seals with overlapped texts in the ReST.

1.1 Competition Organization

ICDAR 2023 competition on reading the seal title is organized by a joint team, including Huazhong University of Science and Technology and Universitat Autónoma de Barcelona.

We organize the competition on the Robust Reading Competition (RRC) website[1], where provide corresponding download links of the datasets, and user interfaces for participants and submission page for their results. Great support has been received from the RRC web team. The online evaluation server[2] will remain available for future usage of this benchmark.

2 Dataset and Annotations

We name our dataset ReST, as it focuses on Reading Seal Title text. It totally includes 10,000 images collected from real scene. The data is mainly in Chinese, with English data accounting for 1%.

The datasets cover the most common classes of seals:

- **Circle/Ellipse shapes**: This type of seals are commonly existing in official seals, invoice seals, contract seals, and bank seals.
- **Rectangle shapes**: This type of seals are commonly seen in driving licenses, corporate seals, and medical bills.
- **Triangle shapes**: This type of seals are seen in bank receipts and financial occasions. This type is uncommon seal and has a small amount of data.

The dataset is split half into a training set and a test set. Every image in the dataset is annotated with text line locations and the labels. Locations are annotated in terms of polygons, which are in clockwise order. Transcripts are UTF-8 encoded strings. Annotations for an image are stored in a json file with the identical file name, following the naming convention: gt_[image_id], where image_id refers to the index of the image in the dataset.

In the JSON file, each gt_[image_id] corresponds to a list, where each line in the list correspond to one text instance in the image and gives its bounding box coordinates and transcription, in the following format:

{
"gt_1": [
"points": [[x1, y1], [x2, y2], ..., [xn, yn]], "transcription" : "trans1"],
"gt_2": [
"points": [[x1, y1], [x2, y2], ..., [xn, yn]] , "transcription" : "trans3"], ...
}

where x1, y1, x2, y2, ..., xn, yn in "points" are the coordinates of the polygon bounding boxes,. The "transcription" denotes the text of each text line.

Note: There may be some inaccurate annotations in the training set, which can measure the robustness of the algorithm, and participants may filter this part of the data as appropriate. The test set is manually corrected and the annotations are accurate.

---

[1] https://rrc.cvc.uab.es/?ch=20.

[2] https://rrc.cvc.uab.es/?ch=20&com=mymethods&task=1.

3   Competition Tasks and Evaluation Protocols

The competition include two tasks: 1) seal title text detection, where the objective is to localize the title text in seal image. and 2) the end-to-end seal title recognition, where the main objective of this task is to extract the title of a seal.

3.1   Task 1: Seal Title Text Detection

The aim of this task is to localize the title text in seal image, The input examples are shown in Fig. 4.

Fig. 4. Example images of the Seal Text dataset. Green color binding lines are formed with polygon ground truth format. (Color figure online)

Submission Format. Participants will be asked to submit a JSON file containing results for all test images. The results format is:

```
{
"res_1": [
"points": [[x1, y1], [x2, y2], ..., [xn, yn]], "confidence" : c],
"res_2": [
"points": [[x1, y1], [x2, y2], ..., [xn, yn]] , "confidence" : c ],
......
}
```

where the key of JSON file should adhere to the format of res_[image_id]. Also, n is the total number of vertices (could be unfixed, varied among different predicted text instance), and c is the confidence score of the prediction and the range is 0-1.

Evaluation Protocol. For Task 1, we adopt IoU-based evaluation protocol by following CTW1500 [3,10]. IoU is a threshold-based evaluation protocol, with 0.5 set as the default threshold. We will report results on 0.5 and 0.7 thresholds but only H-Mean under 0.7 will be treated as the final score for each submitted model, and to be used as submission ranking purpose. To ensure fairness, the competitors are required to submit confidence score for each detection, and thus we can iterate all confidence thresholds to find the best H-Mean score. Meanwhile, in the case of multiple matches, we only consider the detection region with the highest IOU, the rest of the matches will be counted as False Positive. The calculation of Precision, Recall, and F-score are as follows:

$$\text{Precision} = \frac{TP}{TP + FP},$$
$$\text{Recall} = \frac{TP}{TP + FN}, \quad (1)$$
$$F = \frac{2 * \text{Precision} * \text{Recall}}{\text{Precision} + \text{Recall}}$$

where TP, FP, FN and F denote true positive, false positive, false negative and H-Mean, respectively.

3.2 Task 2: End-to-End Seal Title Recognition

The main objective of this task is to extract the title of a seal, as shown in Fig. 5, the input is a whole seal image and the output is the seal's title.

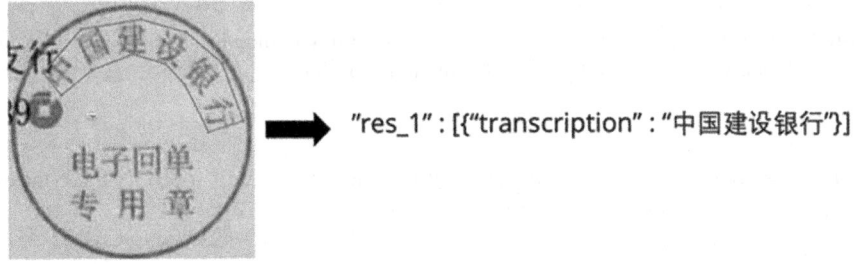

Fig. 5. Example of the task2 input-output.

Submission Format. For Task 2, participants are required to submit the predicted titles for all the images in a single JSON file:

```
{
"res_1": [ "transcription" : "title1"],
"res_2": [ "transcription" : "title2"],
"res_3": [ "transcription" : "title3"],
......
}
```

where the key of JSON file should adhere to the format of res_[image_id].

Evaluation Protocol. Metrics for this task is case-insensitive word accuracy. We will compute the ratio of correctly predicted titles and the total titles.

4 Submissions and Results

By the submission deadline, we received 135 submission from 53 participants in total, including 72 submissions from 28 participants for Task 1, and 63 submissions from 25 participants for Task 2.

After the submission deadlines, we collected all submissions and evaluate their performance through automated process with scripts developed by the RRC web team. Participants did not receive feedback during the submission process, and for those who made multiple submissions, only the last submission prior to the final deadline was considered for ranking purposes. The winners are determined for each task based on the score achieved by the corresponding primary metric. The complete leaderboard can be accessed on the official competition website[3] for all tasks. The following table presents the top 10 results due to limited space.

4.1 Task 1 Seal Title Text Detection

The result for Task 1 is presented on Table 1.

Table 1. Task-1: Seal Title Text Detection Results.

Rank	Method Name	Team Members	Insititute	Precision-0.7	Recall-0.7	Hmean-0.7
1	Dao Xianghu light of TianQuan	Kai Yang, Ye Wang, Bin Wang, Wentao Liu, Xiaolu Ding, Jun Zhu, Ming Chen, Peng Yao, Zhixin Qiu	CCB Financial Technology Co. Ltd, China	99.06%	99.06%	99.06%
2	det314_4	Huajian Zhou	China Mobile Cloud Centre	98.18%	98.18%	98.18%
3	INTIME_OCR	Wei Wang, Chengxiang Ran, Jin Wei, Xinye Yang, Tianjiao Cao, Fangmin Zhao	Institute of Information Engineering, Chinese Academy of Sciences; Mashang Consumer Finance Co., Ltd	98.14%	98.06%	98.10%
4	AntFin-UperNet	Yangkun Lin, Tao Xu	Ant Group	97.72%	97.70%	97.71%
5	SPDB LAB	Jie Li, Wei Wang, Yuqi Zhang, Ruixue Zhang, Yiru Zhao, Danya Zhou, Di Wang, Dong Xiang, Hui Wang, Min Xu, Pengyu Chen, Bin Zhang, Chao Li, Shiyu Hu, Songtao Li, Yunxin Yang	Shanghai Pudong Development Bank	97.60%	97.60%	97.60%
6	Aaaaa_v3	Wudao, Liaoming	cmb	97.34%	97.32%	97.33%
7	PAN_ReST_4	Yuchen Su, Yongkun Du, Tianlun Zheng, Yi Gan, Zhineng Chen	Fudan University, Paddle OCR	96.86%	96.86%	96.86%
8	DB with SegFormer	Sehwan Joo, Wonho Song	Upstage AI	98.11%	95.42%	96.75%
9	AppAI for Seal	Chuanjian Liu, Miao Rang, Zhenni Bi, Zhicheng Liu, Wenhui Dong, Yuyang Li, Dehua Zheng, Hailin Wu, Kai Han, Yunhe Wang	Noah	96.00%	96.00%	96.00%
10	ratio_4.0	sunyifan	SY_007	95.96%	95.96%	95.96%

---

[3] https://rrc.cvc.uab.es/?ch=20&com=evaluation&task=1.

The methods used by the top 3 submissions for Task 1 are presented below.

1st Ranking Method. The team of "CCB Financial Technology Co. Ltd, China" are elaborated in detail from the following three perspectives:

- Data Analysis: This competition provided 5000 pieces of training data officially. Upon analyzing the data, they found that it can be classified into four categories: round, oval, square, and triangular, with the round and oval categories being the primary ones. The training set contains various conditions, including multi-directional rotations, uneven colors, overlapping seals, and indistinct seal patterns.
- Data Processing: When it comes to data analysis, they began by re-annotating the training set images and enlarging them to squares. They then rotated the data and produced a total of 15,000 images. Data generation was carried out on difficult samples, including those with overlapping or blurry stamps. Prior to generating the seal data, they gathered a large number of company and organization names from the internet. Then, they generated the rotation angle and position of each individual character based on its length and merged them into the seal's background image. Moreover, they output the coordinates of the outer edge points of the text. To create a more realistic representation of seals in the generated data, they incorporated various colors, fonts, backgrounds, and textures. The base image for each seal was created by randomly cropping backgrounds, and they used RGBA format during data generation to allow for control over the color depth of the seal by adding a transparency channel. They also included two types of seal borders: solid and fragmented.
- Model Introduction: In this segmentation task, they employed a "voting ensemble" method to detect the content of the seal title. Five models are utilized in the method, namely Mask R-CNN [5], K-Net [17], Segformer [12], Segmenter [11], and UperNet [16]. Each model generates a mask. And they utilize a majority vote to derive the final mask, which allows them to identify the seal title area on the mask.

2nd Ranking Method. "China Mobile Cloud Centre" team are elaborated in detail from the following two perspectives:

- Regarding the synthesized data, the team generated a dataset comprising 7000 seals, including circular, elliptical, rectangular, and triangular seals. Additionally, the team addressed the issue of redundant annotation data. Specifically, the annotation process often resulted in unnecessary parts being included on the sides of elliptical and circular seal text. This redundancy could potentially impact the segmentation and subsequent text recognition. To mitigate this, the team developed a correction program that automatically removes the redundant parts by leveraging the geometric properties of the elliptical ring.
- The team utilized a single detection model, specifically the VitDet detection part from EVA's (Exploring the Limits of Masked Visual Representation

Learning at Scale) framework. The backbone network employed VIT-Giant, while the network head employed Cascade Mask-Rcnn. Considering the small size of the seals, the network input size was adjusted to 320 * 320. To ensure a smooth mask output, the network head's mask output size was increased from 28 * 28 to 56 * 56.

3rd Ranking Method. "Institute of Information Engineering, Chinese Academy of Sciences; Mashang Consumer Finance Co., Ltd" team's competition solution is based on the TPSNet detection model [14]. To better adapt the seals, the team has modified the regression branch to regress the bezier control points [9]. Due to some seal titles being too long for a single-stage model to regress, the team has designed a regression-merging post-process where the regressed curves belonging to the same title are merged, weighted by the distance between the feature location and text boundary. The backbone of the team's model is ResNet50 with DCN, pretrained on ImageNet.

In terms of data, the team has designed a script to convert the original polygon annotation to two long curves for every title, allowing training of the regression-based model. Inaccurate or wrong annotations have been re-annotated to ensure data quality. To increase the size of the training set, the team has implemented a seal synthesis pipeline based on Synthtext [4]. They have modified the character layout to create various seals, used the WTW document dataset images[4] as background images, and employed the Company-Name-Corpus[5] as the corpus of seal titles. This effort has resulted in the generation of 10,000 synthetic seals that have been added to the training set.

During training and testing, both real and synthetic seals are trained together. The team has applied ColorJitter, Random Rotate, and Random Resize as training augmentations. For testing, an input scale of 448×448 is used without any augmentation.

4.2   Task 2 End-to-End Seal Title Recognition

The result for Task 2 is presented on Table 2.

The methods used by the top 3 submissions for Task 1 are presented below.

1st Ranking Method. "Shanghai Pudong Development Bank" team's method can be described in detail from the following two perspectives:

- Circle seals and Ellipse seals: Based on the results of the circle and ellipse seals title detection in task1, PCA technology was used to correct the rotated seal, the image processing technology was used to separate the seal title, and finally the curved text was sent to the recognition model for recognition. The recognition model was selected by Trocr [6], and the training data includes the provided training data and synthetic data.

---

[4] https://github.com/wangwen-whu/WTW-Dataset.
[5] https://github.com/wainshine/Company-Names-Corpus.

- Rectangle seals and Triangle seals: Rectangle seals and triangle seals were not based on the task1 detection model, but train a text line detection mode [15]l. the image processing technology was used to separate the seal title. The recognition model was selected by Trocr [6], and the training data includes the provided by synthetic data.

Table 2. Task-2: End-to-end Seal Title Recognition Results.

Rank	Method Name	Team Members	Insititute	Accuracy
1	SPDB LAB	Jie Li, Wei Wang, Yuqi Zhang, Ruixue Zhang, Yiru Zhao, Danya Zhou, Di Wang, Dong Xiang, Hui Wang, Min Xu, Pengyu Chen, Bin Zhang, Chao Li, Shiyu Hu, Songtao Li, Yunxin Yang	Shanghai Pudong Development Bank	91.88%
2	rec320_3	Huajian Zhou	China Mobile Cloud Centre	91.74%
3	Dao Xianghu light of TianQuan	Kai Yang, Ye Wang, Bin Wang, Wentao Liu, Xiaolu Ding, Jun Zhu, Ming Chen, Peng Yao, Zhixin Qiu	CCB Financial Technology Co. Ltd, China	91.22%
4	AppAI for Seal	Chuanjian Liu, Miao Rang, Zhenni Bi, Zhicheng Liu, Wenhui Dong, Yuyang Li, Dehua Zheng, Hailin Wu, Kai Han, Yunhe Wang	Noah	90.20%
5	ensemble	xubo	-	90.08%
6	task2_test_submit2	jgj aksbob	pa	88.90%
7	INTIME_OCR(e2e)	Wei Wang, Jin Wei, Chengxiang Ran, Xinye Yang, Tianjiao Cao, Fangmin Zhao	Institute of Information Engineering, Chinese Academy of Sciences; Mashang Consumer Finance Co., Ltd	84.24%
8	task2 result	DH	-	84.22%
9	Seal Recognize	Shente Zhou, Tianyi Zhu, Weihua Cao, Mingchao Fang, Xiaogang Ouyang	Shizai Intellect	83.02%
10	SealRecognizor	Qiao Liang	Zhejiang University	83.00%

2nd Ranking Method. "China Mobile Cloud Centre" team's method can be described in detail from the following perspectives:

- Text Detection and Segmentation Module:
  Data:
  1. Synthesize data: Synthesize 7000 seals (including circular, elliptical, rectangular, and triangular seals).
  2. Correction of annotation data: In the annotation process, there are redundant parts at two sides of elliptical and circular seal text, which can affect text recognition. The correct program removes the redundant parts automatically by utilizing the geometric properties of the elliptical ring.

  Method:
  The text detection and segmentation module follows the following approach:
  1. Detection Model: Only one detection model is used, using EVA's (Exploring the Limits of Masked Visual Representation Learning at Scale) VitDet detection part.
  2. Backbone Network: The backbone network uses VIT-Giant.
  3. Network Head: The network head uses Cascade Mask-Rcnn.
  4. Adjustment for Seal Size: Considering the small size of the seals, the network input size is adjusted to 320×320.
  5. Mask Output Enhancement: The network head's mask output size is adjusted from 28×28 to 56×56 to achieve a smooth mask output.

- Text Rectification Module:
  1. For triangular and rectangular texts, the text is rectified to horizontal text by using the direction classification model combined with affine transformation.
  2. For elliptical and circular text, the least squares method is used to obtain the upper and lower curve equations of the text. Based on the curve equations, the curve region is divided into several small regions, and affine transformations are performed on these regions. Then, they are concatenated to get the horizontal text.

- Text Recognition Model:
  Data:
  1. Synthetic data: Extracting millions of lines of corpus from the open-source THUCNews, News2016zh, and wiki_zh corpus datasets, and using this data to synthesize horizontal and curved text images.
  2. Rectification data: Rectifying or cropping official training images to obtain text images.

  Method:
  1. Recognition model 1: Using SVTR-Small (Scene Text Recognition with a Single Visual Model), with the network input size adjusted to 48×320.
  2. Recognition model 2: Using DIG (Reading and Writing: Discriminative and Generative Modeling for Self-Supervised Text Recognition), with the network input size adjusted to 48×288.

3. Recognition correction: If the results of the two models are different, the correction program uses the open-source Chinese administrative division dataset for correction.

3rd Ranking Method. "CCB Financial Technology Co. Ltd, China" team finds that the difficulties of recognition mainly focus on multi-directional recognition, overlapping interference from handwritten or printed characters, fuzzy and blurred images, and multiple reading orders, after data exploration and analysis. Based on the analysis, they build the following solution. First, they make a seal title segmentation that masks out the non-title area, and removes the interference of irrelevant regions. Then, they train a TrOCR [6] model using over 6 million data from the training set, open dataset, and synthetic dataset. Finally, in the post process, place names correction is implemented.

In the seal title segmentation, they adopt an ensemble strategy with five segmentation models to vote for the title segmentation, laying a good foundation for the recognition. Since the training set only has 5000 images, it is far from enough for the recognition task. They use the official chars.txt dictionary and collect the corpus of company names and organization names on the Internet, and generate a large number of seals by codes. To simulate the real situation, they use various fonts, colors, backgrounds, and textures to synthesize the images, and they perform kinds of data augmentation strategies for improving generalization including rotation, gaussian blur, stretching, perspective transformation, contour expansion or contraction and so on. In addition, they use 10k seals from Baidu public dataset.

At the early stage of the competition, they use the public dataset and the synthesized dataset as the training set and the original training set of the competition as the test set. They continuously synthesize kinds of data to improve the accuracy of the test set. To further improve the accuracy, they design a classifier to separate circular seals (Circle/Ellipse shapes) and non-circular seals (Rectangle/Triangle shapes). They generate nearly 400k non-circular seals. And they compare the single recognition model solution with the solution of classifying then recognizing with multiple models. And they verify that the former solution is better.

When analyzing bad cases, they find that smudging and character overlapping often lead to recognition errors. So they design place names based postprocessing strategy to correct some of these errors.

5   Discussion

Task 1 Seal Title Text Detection. Many teams use data augmentation techniques such as random scaling, flipping, rotation, cropping, and synthesis of hard samples with various shapes, colors, fonts, and backgrounds to improve the generalization ability of their models. Additionally, they employ powerful methods such as DBNet++ [7], ResNet, Segformer, Unet, and PANNet to enhance performance. Some teams also use a "voting ensemble" strategy to achieve better results. Certain teams take into account the four main shapes of seal images during training and data generation. Task 1, which involves general text detection, has produced excellent results, with 12 teams achieving an Hmean of over 95%. However, in scenarios that require stricter standards or zero error rates, even the best method with a 99.06% Hmean in Task 1 cannot meet the required performance. Therefore, there is still room for improvement, and further efforts and exploration are needed.

Task 2 End-to-End Seal Title Recognition. Extending from Task 1, the task of end-to-end seal title recognition poses greater challenges, requiring flexible adjustments and further refinement based on the findings from Task 1. To address the issues of curved or overlapped text, some teams have designed image processing technologies such as PCA or post-processing strategies to correct these errors. Diverse model ensembles continue to be utilized for improved results. For more accurate recognition, powerful models such as Parseq [1] and TrOCR are employed. To further enhance accuracy, many teams have designed a novel classification network to differentiate circular seals (Circle/Ellipse shapes) and non-circular seals (Rectangle/Triangle shapes). Additionally, pre-trained models and fine-tuning strategies with augmentations and label smoothing based on joint datasets, including training sets, open datasets such as Synthtext-Chinese, ReCTs [8], LSVT [13], ArT [2], and synthetic datasets, are used. However, as shown in Table 2, only five teams achieved accuracy of over 90.00%, with the top-1 accuracy being 91.88%. Therefore, we can conclude that end-to-end seal title recognition remains a challenging task, with most methods submitted using different ideas and approaches. We look forward to seeing more innovative approaches proposed following this competition.

6   Conclusion

We organized the ReST competition, with a focus on reading challenging seal title text, an area that has not received sufficient attention in the field of document analysis. To this end, we constructed new datasets, comprising 10,000 real seal data labeled with text polygons and transcripts. Strong interest from both academia and industry was evident, with a large number of submissions showcasing novel ideas and approaches for the competition tasks. Reading seal titles holds significant potential for numerous document analysis applications, making it a rewarding task. However, despite the top-performing team achieving a remarkable performance of approximately 99% in Task 1, the task still warrants continued research and exploration, particularly in strict scenarios with zero error tolerance rates. Additionally, Task 2 proved to be a challenging task, with only five teams achieving an accuracy above 90%, and the top-1 accuracy reaching 91.88%. However, the excellent submissions by these teams provide valuable insights for other researchers. Future competitions could expand on this topic with more challenging datasets and applications, thus attracting researchers from the fields of computer vision and advancing the state-of-the-art in document analysis.

Acknowledgements. This competition is supported by the National Natural Science Foundation of China (No. 62225603, No. 62206103, No. 62206104). The organizers thank Sergi Robles and the RRC web team for their tremendous support on the registration, submission and evaluation jobs.

References

1. Bautista, D., Atienza, R.: Scene text recognition with permuted autoregressive sequence models. In: Avidan, S., Brostow, G., Cissé, M., Farinella, G.M., Hassner, T. (eds.) ECCV 2022. LNCS, vol. 13688, pp. 178–196. Springer, Cham (2022). https://doi.org/10.1007/978-3-031-19815-1_11
2. Chng, C.K., et al.: ICDAR 2019 robust reading challenge on arbitrary-shaped text - RRC-art. In: 2019 International Conference on Document Analysis and Recognition (ICDAR), pp. 1571–1576 (2019)
3. Chng, C.K., et al.: ICDAR 2019 robust reading challenge on arbitrary-shaped text-RRC-art. In: 2019 International Conference on Document Analysis and Recognition (ICDAR), pp. 1571–1576. IEEE (2019)
4. Gupta, A., Vedaldi, A., Zisserman, A.: Synthetic data for text localisation in natural images. In: IEEE Conference on Computer Vision and Pattern Recognition, pp. 2315–2324 (2016)
5. He, K., Gkioxari, G., Dollár, P., Girshick, R.B.: Mask R-CNN. IEEE Trans. Pattern Anal. Mach. Intell. **42**, 386–397 (2017)
6. Li, M., et al.: TrOCR: transformer-based optical character recognition with pretrained models. arXiv abs/2109.10282 (2021)
7. Liao, M., Zou, Z., Wan, Z., Yao, C., Bai, X.: Real-time scene text detection with differentiable binarization and adaptive scale fusion. IEEE Trans. Pattern Anal. Mach. Intell. **45**(1), 919–931 (2022)
8. Liu, X., et al.: ICDAR 2019 robust reading challenge on reading Chinese text on signboard. In: 2019 International Conference on Document Analysis and Recognition (ICDAR), pp. 1577–1581 (2019)
9. Liu, Y., Chen, H., Shen, C., He, T., Jin, L., Wang, L.: ABCNet: real-time scene text spotting with adaptive bezier-curve network. In: 2020 IEEE/CVF Conference on Computer Vision and Pattern Recognition (CVPR), pp. 9806–9815 (2020)
10. Liu, Y., Jin, L., Zhang, S., Luo, C., Zhang, S.: Curved scene text detection via transverse and longitudinal sequence connection. Pattern Recogn. **90**, 337–345 (2019)
11. Strudel, R., Garcia, R., Laptev, I., Schmid, C.: Segmenter: transformer for semantic segmentation. In: Proceedings of the IEEE/CVF International Conference on Computer Vision, pp. 7262–7272 (2021)
12. Strudel, R., Pinel, R.G., Laptev, I., Schmid, C.: Segmenter: transformer for semantic segmentation. In: 2021 IEEE/CVF International Conference on Computer Vision (ICCV), pp. 7242–7252 (2021)
13. Sun, Y., et al.: ICDAR 2019 competition on large-scale street view text with partial labeling - RRC-LSVT. In: 2019 International Conference on Document Analysis and Recognition (ICDAR), pp. 1557–1562 (2019)
14. Wang, W., et al.: Tpsnet: reverse thinking of thin plate splines for arbitrary shape scene text representation. In: Proceedings of the 30th ACM International Conference on Multimedia (2021)

15. Wang, W., et al.: Efficient and accurate arbitrary-shaped text detection with pixel aggregation network. In: 2019 IEEE/CVF International Conference on Computer Vision (ICCV), pp. 8439–8448 (2019)
16. Xiao, T., Liu, Y., Zhou, B., Jiang, Y., Sun, J.: Unified perceptual parsing for scene understanding. In: Proceedings of the European Conference on Computer Vision (ECCV), pp. 418–434 (2018)
17. Zhang, W., Pang, J., Chen, K., Loy, C.C.: K-Net: towards unified image segmentation. In: NeurIPS (2021)

ICDAR 2023 Competition on Structured Text Extraction from Visually-Rich Document Images

Wenwen Yu[1], Chengquan Zhang[2], Haoyu Cao[3], Wei Hua[1], Bohan Li[2],
Huang Chen[3], Mingyu Liu[1], Mingrui Chen[1], Jianfeng Kuang[1],
Mengjun Cheng[5], Yuning Du[2], Shikun Feng[2], Xiaoguang Hu[2], Pengyuan Lyu[2],
Kun Yao[2], Yuechen Yu[2], Yuliang Liu[1], Wanxiang Che[6], Errui Ding[2],
Cheng-Lin Liu[7], Jiebo Luo[8], Shuicheng Yan[9], Min Zhang[6],
Dimosthenis Karatzas[4], Xing Sun[3], Jingdong Wang[2], and Xiang Bai[1(✉)]

[1] Huazhong University of Science and Technology, Wuhan, China
{wenwenyu,xbai}@hust.edu.cn
[2] Baidu Inc., Beijing, China
zhangchengquan@baidu.com
[3] Tencent YouTu Lab, Shanghai, China
{rechycao,huaangchen,winfredsun}@tencent.com
[4] Universitat Autónoma de Barcelona, Bellaterra, Spain
[5] Peking University, Beijing, China
[6] Harbin Institute of Technology, Harbin, China
[7] CAS Institute of Automation, Beijing, China
[8] University of Rochester, Rochester, USA
[9] Sea AI Lab, Singapore, Singapore

Abstract. Structured text extraction is one of the most valuable and challenging application directions in the field of Document AI. However, the scenarios of past benchmarks are limited, and the corresponding evaluation protocols usually focus on the submodules of the structured text extraction scheme. In order to eliminate these problems, we organized the ICDAR 2023 competition on Structured text extraction from Visually-Rich Document images (SVRD). We set up two tracks for SVRD including Track 1: **HUST-CELL** and Track 2: **Baidu-FEST**, where **HUST-CELL** aims to evaluate the end-to-end performance of **C**omplex **E**ntity **L**inking and **L**abeling, and **Baidu-FEST** focuses on evaluating the performance and generalization of Zero-shot/Few-shot Structured Text extraction from an end-to-end perspective. Compared to the current document benchmarks, our two tracks of competition benchmark enriches the scenarios greatly and contains more than 50 types of visually-rich document images (mainly from the actual enterprise applications). The competition opened on 30th December, 2022 and closed on 24th March, 2023. There are 35 participants and 91 valid submissions received for Track 1, and 15 participants and 26 valid submissions received for Track 2. In this report we will presents the motivation, competition datasets,

---

W. Yu, C. Zhang and H. Cao—Contributed equally.

G. A. Fink et al. (Eds.): ICDAR 2023, LNCS 14188, pp. 536–552, 2023.
https://doi.org/10.1007/978-3-031-41679-8_32

task definition, evaluation protocol, and submission summaries. According to the performance of the submissions, we believe there is still a large gap on the expected information extraction performance for complex and zero-shot scenarios. It is hoped that this competition will attract many researchers in the field of CV and NLP, and bring some new thoughts to the field of Document AI.

1 Introduction

In recent years, the domain of Document AI has gradually become a hot research topic. As one of the most concerned Document AI technologies, structured text extraction aims to capture text fields with specified semantic attributes from complex visually-rich documents (VRDs). It is widely used in many applications and services, such as customs information inspection, accounting in the financial field, office automation, and so on.

In the past, several benchmarks have been established in the community, such as FUNSD [7], CORD [12], XFUND [16], EPHOIE [14], etc., to measure relative technical efforts. However, there are many imperfections in these benchmarks and the corresponding evaluation protocols. The obvious shortcomings are as follows: 1) The performance of structured text extraction is not evaluated from the end-to-end perspective, but is disassembled into three independent functional modules, namely text detection [11,19,22], text recognition [4,13,17] and entity labelling or linking) [1,10,15,18], for respective evaluation, which is not intuitive for downstream applications. 2) The scenarios covered by the above benchmarks are relatively few, and only focus on a certain receipt or form scenario, which is difficult to guide objectively evaluate the effectiveness and robustness of the model.

Therefore, we propose a new structured text extraction competition benchmark including two tracks, which covers the most abundant visually-rich document images of scenarios and types as far as we know. The whole benchmark will contain more than 50 document types and more than 100 semantic attributes of text fields. In order to evaluate the performance of structured information extraction from an end-to-end perspective, we will set up two tracks: (1) end-to-end Complex Entity Linking and Labeling created by Huanzhong University of Science and Technology (**HUST-CELL**), where we have designed entity linking and labeling tasks. (2) end-to-end Few-shot Structured Text extraction created by **Baidu** Company (**Baidu-FEST**) to explore the generalization and robustness of the submitted models, where we have newly designed zero-shot and few-shot structured text extraction tasks. The motivation and relevance to ICDAR community including:

– The intelligent analysis of visually-rich document images has always been an important domain of concern for ICDAR community. Its core technologies include text detection, recognition and named-entity recognition. Our proposed competition is the first time to guide the evaluation of the effect and generalization of structured text extraction scheme from an end-to-end perspective, which is more valuable but more challenging.

– This competition aims to further connect researchers from both the document image understanding and NLP communities to bring more inspiration.

1.1 Competition Organization

ICDAR 2023 competition on SVRD is organized by a joint team, including Huazhong University of Science and Technology, Baidu Inc., Tencent YouTu Lab, Universitat Autónoma de Barcelona, Peking University, Harbin Institute of Technology, CAS Institute of Automation, University of Rochester, and Sea AI Lab.

We organize the SVRD competition on the Robust Reading Competition (RRC) website[1], where provide corresponding download links of the datasets, and user interfaces for participants and submission page for their results[2]. Great support has been received from the RRC web team.

2 Related Works

This section discusses most of the well-received visually-rich document benchmarks as following:

In 2019, the Robust Reading Competition (RRC) web portal introduced a new challenge, known as Scanned Receipts OCR and Information Extraction reading competition which also commonly named as SROIE [6]. The main feature of SROIE is that all the images are collected from scanned receipts. It contains of 626 receipts for training and 347 receipts for testing, and each receipt only contains four predefined values: company, date, address, and total.

Meanwhile, EATEN [5] constructs a dataset of 1,900 real images in train ticket scenarios, and it only has entity values annotation without OCR annotation. CORD [12] consists of 1,000 Indonesian receipts, which contains images and box/text annotations for OCR, and multi-level semantic labels for parsing, which can be used to address various OCR and entity extraction tasks.

FUNSD [7] dataset was presented at the ICDAR workshop in 2019. FUNSD is a form understanding benchmark with 199 real, fully annotated, scanned form images, such as marketing, advertising, and scientific reports, which is split into 149 training samples and 50 testing samples. FUNSD dataset is suitable for a variety of tasks including text detection, recognition, entity labelling, etc. And its entity has only three types of semantic attributes including question, answer and header.

XFUND [16] is an extended version of FUNSD that was proposed in 2021. It is launched to introduce the multi-script structured text extraction problem.

---

[1] https://rrc.cvc.uab.es/?ch=21.
[2] Scores achieved using the ChatGPT large model interface during the competition are temporarily excluded from the leaderboard.

It consists of human-labeled forms with key-value pairs in 7 languages (Chinese, Japanese, Spanish, French, Italian, German, Portuguese). Each language includes 199 forms, where the training set includes 149 forms, and the test set includes 50 forms.

EPHOIE [14] was also released in 2021, and contains 1,494 images which are collected and scanned from real examination papers of various schools in China. There are 10 text entity types including Subject, Test Time, Name, School, Examination Number, Seat Number, Class, Student Number, Grade, and Score. As an overview, Table 1 lists the details of above datasets.

Table 1. Existing Visually-rich Document Images Datasets.

Dataset	Image number(Train/Test)	Language	Granularity	Document type	Year
SROIE	973(626/347)	English	Word	Receipt	2019
CORD	1,000(800/200)	English	Word	Receipt	2019
EATEN	1,900(1,500/400)	Chinese	w/o OCR	Train Ticket	2019
FUNSD	199(149/50)	English	Word	Form	2019
XFUND	1,393(1,043/350)	7 languages	Word/Line	Form	2021
EPHOIE	1,494(1,183/311)	Chinese	Character	Paper	2021

3 Benchmark Description

3.1 Track 1: HUST-CELL

Our proposed HUST-CELL complexity goes over and above previous datasets in four distinct aspects. First, we provide 30 categories of documents with more than 4k documents, 2 times larger than the existing English and Chinese datasets including SROIE (973), CORD (1,000), EATEN (1,900), FUNSD (199), XFUNSD (1,393), and EPHOIE (1,494). Second, HUST-CELL contains 400+ diverse keys and values. Third, HUST-CELL covers complex keys more challenging than others, for instance, nested keys, fine-grained key-value pairs, multi-line keys/values, long-tailed key-value pairs, as shown in Fig. 1. Current state-of-the-art Key Information Extraction (KIE) techniques [2,3,15,18,20] fail to deal with such situations that are essential for a robust KIE system in the real world. Fourth, our dataset comprises real-world documents reflecting real-life diversity of content and the complexity of the background, e.g. different fonts, noise, blur, seal.

In this regard, under the consideration of the importance and huge application value of KIE, we propose to set up track 1 competition on complex entity linking and labeling.

Our proposed HUST-CELL were collected from public websites and cover a variety of scenarios, e.g., receipt, certificate, and license of various industries. The language of the documents is mainly Chinese, along with a small portion of English. The number of images collected for each specific scenario varies, ranging from 10 to 300, with a long-tail distribution, which can avoid introducing any bias towards specific real application scenarios. Due to the complexity of the data source, the diversity of this dataset can be guaranteed. To be able to use publicly, this data is collected from open websites, and we delete images that contain private information for privacy protection. Some examples are shown in Fig. 1.

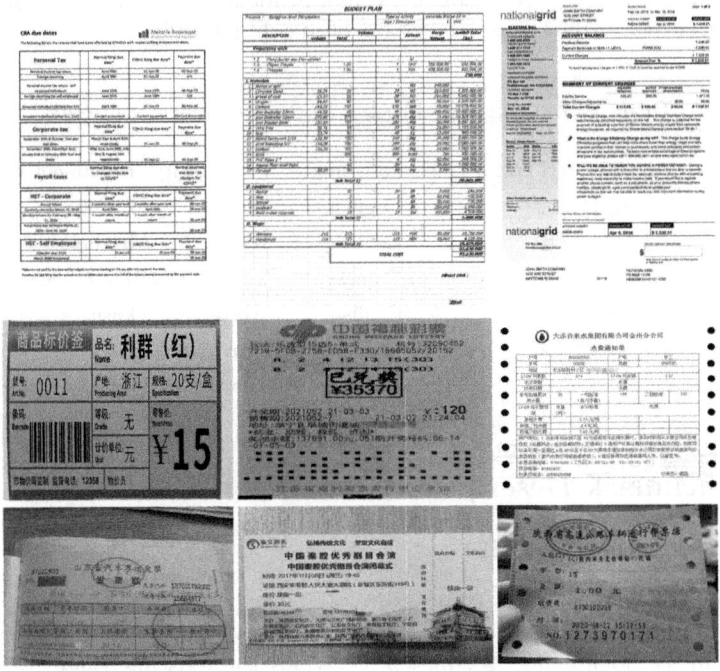

Fig. 1. Samples of HUST-CELL collected from various scenarios.

The dataset is split into the training set and test set. The training set consists of 2,000 images, which will be available to the participants along with OCR and KIE annotations. The test set consists of 2,000 images, whose OCR annotations and KIE annotations will not be released, but the online evaluation server[3] will remain available for future usage of this benchmark.

3.2　Track 2: Baidu-FEST

Our proposed Baidu-FEST benchmark comes from the practical scenarios, mainly including finance, insurance, logistics, customs inspection, and other fields. Different applications have different requirements for text fields of interest. In addition, the data collection methods in different scenarios may be affected by

---

different cameras and environments, thus the benchmark is relatively rich and challenging.

Specifically, the benchmark contains about 11 kinds of synthetic business documents for training, and 10 types of real visually-rich document images for testing. The format of documents major consists of cards, receipts, and forms. Each type of document provides about 60 images.

Each image in the dataset is annotated with text-field bounding boxes (bbox) and the transcript, entity caption, entity id of each text bbox. Locations are annotated as rectangles with four vertices, which are in clockwise order starting from the top. Some examples of images and the corresponding annotations are shown in Fig. 2.

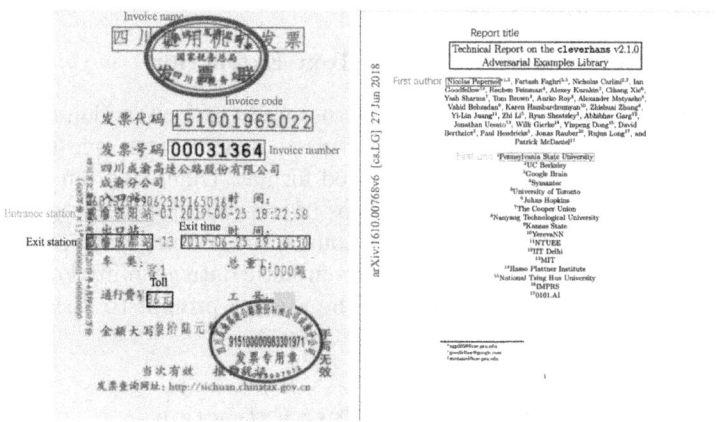

Fig. 2. Some visually-rich document samples of Baidu-FEST.

4 Competition Tasks and Evaluation Protocols

Our proposed competition has two tracks totaling four main tasks.

4.1 Track 1: HUST-CELL

Task-1: E2E Complex Entity Linking

- **Task Description**: This task aims to extract key-value pairs (entity linking) from given images only, then save the key-value pairs of each image into a JSON file. For the train set, both KIE annotation files for training and human-checked OCR annotation files are provided. So the OCR annotation is clean and can be used as the ground truth of the OCR task. The test set of Task 1 will only provide images without any annotation including OCR and KIE. It requires the method to accomplish both OCR and KIE tasks in an end-to-end manner.

Task-2: E2E Complex Entity Labeling

- **Task Description**: The end-to-end complex entity labeling is to extract texts of a number of predefined key text fields from given images (entity labeling), and save the texts for each image in a JSON file with required format. Task 2 has 13 predefined entities. For the train set, both KIE annotation files for training and human-checked OCR annotation files are provided. So the OCR annotation is clean and can be used as the ground truth of the OCR task. The test set of Task 2 will only provide images without any annotation including OCR and KIE. It requires the method to accomplish both OCR and KIE tasks in an end-to-end manner.

4.2   Track 2: Baidu-FEST

Task-3: E2E Zero-Shot Structured Text Extraction

- **Task Description**: The zero-shot structured text extraction is to extract texts of a number of key fields from given images, and save the texts for each image in a JSON file with required format. Different from Task2, there is no intersection between the scenarios of the provided training-set and the scenarios of the provided test-set. Of course, the training data consists of the real data provided by Track 1 and the synthetic data generated officially. The caption_en and caption_ch in GT can be used as prompt to assist extraction but it is not allowed to be modified.

Task-4: E2E Few-Shot Structured Text Extraction

- **Task Description**: The few-shot structured text extraction is to extract texts of a number of key fields from given images, and save the texts for each image in a JSON file with required format. Different from Task-3, the localization information and transcript will be provided, but the total number of the provided training-set will no more than five images for each scenario of the provided test-set. The caption_en and caption_ch in GT can be used as prompt to assist extraction but it is not allowed to be modified.

4.3   Evaluation Protocol

Task 1 Evaluation. For Task 1, the evaluation metrics include two parts:

Normalized Edit Distance. For each predicted kv-pair (key-value pair), if it matched with GT kv-pair in the given image, the normalized edit distance (NED) between the predicted kv-pair s1 and ground-truth kv-pair s2 will be calculated as following:

$$NED(s1, s2) = \left(\frac{ed\,(s1\_k, s2\_k)}{\max\,(\operatorname{len}\,(s1\_k)\,,\operatorname{len}\,(s2\_k))} + \frac{ed\,(s1\_v, s2\_v)}{\max\,(\operatorname{len}\,(s1\_v)\,,\operatorname{len}\,(s2\_v))} \right) /2 \quad (1)$$

$$\text{score}\,1 = 1 - \sum_{i=1}^{n} \frac{NED\,(s_{i1}, s_{i2})}{n} \tag{2}$$

where n denotes the number of matched kv-pairs (both the edit distance of key and value are larger than a threshold simultaneously. $ed()$ denotes the edit distance function. The calculated details refer to the following Matching Protocol.). s1_k/s2_k, s1_v/s2_v indicate the content of key and value of the kv-pair s1/s2, respectively. Note that for predicted kv-pairs that do not matched in the GT of the given image, the edit distance will be calculated between predicted kv-pairs and empty string.

Matching Protocol: Given the predicted kv-pair s1 and ground-truth kv-pair s2. The matching protocol is calculated as following:

$$\text{Match}(s1,\ s2) = \begin{cases} True, & ed(s1\_k, s2\_k) \le th\_k \ and \ ed(s1\_v, s2\_v) \le th\_v \\ False, & other \end{cases} \tag{3}$$

$$th\_k = max(factor\_k * min(len(s1\_k), len(s2\_k)), 0) \tag{4}$$

$$th\_v = max(factor\_v * min(len(s1\_v), len(s2\_v)), 0) \tag{5}$$

where $ed()$ denotes the edit distance function. the factor_k and factor_v are set to 0.15, 0.15, respectively.

F-score. Considering all the predicted kv-pairs and all GT kv-pairs, the F-score will be calculated as following:

$$\text{Precision} = \frac{N3}{N2} \tag{6}$$

$$\text{Recall} = \frac{N3}{N1} \tag{7}$$

$$\text{score2} = \frac{2 * \text{Precision} * \text{Recall}}{\text{Precision} + \text{Recall}} \tag{8}$$

where N1 denotes the number of kv-pairs that exists in the given image, N2 denotes the number of predicted kv-pairs, N3 denotes the number of perfectly matched kv-pairs (both the edit distance of key and value are larger than a threshold simultaneously. Specifically, the factor_k and factor_v in matching protocol are set to 0.). The final score is the weighted score of score1 and score2:

$$\text{score} = 0.5 * \text{score1} + 0.5 * \text{score2} \tag{9}$$

The final weighted score will be used as submission ranking purpose for Task 1.

Task 2–4 Evaluation. For Task 2, Task 3, and Task 4, the evaluation metrics include two parts:

Normalized Edit Distance. For each predefined key text field, if it exists in the given image, the normalized edit distance (NED) between predicted text s1 and ground-truth text s2 will be calculated as following:

$$NED(s1, s2) = \frac{edit\_distance\ (s1, s2)}{max(\ length\ (s1), length(s2))} \tag{10}$$

where n denotes the number of perfectly matched key text fields (both entity_id and text are predicted correctly). Note that for predicted key text fields that do not exist in the given image, the edit distance will be calculated between predicted text and empty string. Then the *score1* can be calculated by Eq. 2.

F-score. Considering all the predicted key text fields and all the predefined key text fields, the *score2* (F-score) can also be calculated as Eq. 8. In this scenarios, N1 denotes the number of key text fields that exists in the given image, N2 denotes the number of predicted key text fields, N3 denotes the number of perfectly matched key text fields (both entity_id and text are predicted correctly). The final score is the weighted score of score1 and score2:

$$score = 0.8 * score1 + 0.2 * score2 \tag{11}$$

The final weighted score will be used as submission ranking purpose for Task 2, Task 3, and Task 4.

5 Submissions and Results

The competition attracted 50 participants and 117 submissions from academia and industry, including 19 participants and 53 submissions for Task 1, 16 participants and 38 submissions for Task 2, 7 participants and 15 submissions for Task 3 and 8 participants and 11 submissions for Task 4, which demonstrated significant interest in this challenging task.

After the submission deadlines, we collected all submissions and evaluate their performance through automated process with scripts developed by the RRC web team. No feedback was given to the participants during the submission process. If participants have multiple submissions, we pick the last submission made before the final submission deadline for ranking. The winners are determined for each task based on the score achieved by the corresponding primary metric. The complete leaderboard is available on the online website[4] for all tasks. However, due to limited space, the results table presented below showcases a maximum of 10 top performers.

5.1 Task 1 Performance and Ranking

The result for Task 1 is presented on Table 2.

The methods used by the top 3 submissions for Task 1 are presented below.

---

[4] https://rrc.cvc.uab.es/?ch=21&com=evaluation&task=1.

Table 2. Task-1: E2E Complex Entity Linking Results. * means the reproducible script has been submitted by participants and verified by organizers.

Rank	Method Name	Team Members	Insititute	Score1(NED)	Score2(F-score)	Score(Total)
1	Super_KVer*	Lele Xie, Zuming Huang, Boqian Xia, Yu Wang, Yadong Li, Hongbin Wang, Jingdong Chen	Ant Group	49.93%	62.97%	56.45%
2	End-to-end document relationship extraction	Huiyan Wu, Pengfei Li, Can Li	University of Chinese Academy of Sciences	43.55%	57.90%	50.73%
3	sample-3*	Zhenrong Zhang, Lei Jiang, Youhui Guo, Jianshu Zhang, Jun Du	University of Science and Technology of China, iFLYTEK AI Research	42.52%	56.68%	49.60%
4	Pre-trained model based fullpipe pair extraction (opti_v3, no inf_aug)*	Zening Lin, Teng Li, Wenhui Liao, Jiapeng Wang, Songxuan Lai, Lianwen Jin	South China University of Technology; Huawei Cloud	42.17%	55.63%	48.90%
5	Meituan OCR V4*	Jianqiang Liu, Kai Zhou, Chen Duan, Shuaishuai Chang, Ran Wei, Shan Guo	Meituan	41.10%	54.55%	47.83%
6	submit-trainall	hsy	-	40.65%	52.98%	46.82%
7	f2	Zhi Zhang	cocopark	41.07%	50.82%	45.94%
8	LayoutLM & STrucText Based Method	Wumin Hui, Mei Jiang	PKU & BUPT	33.09%	45.92%	39.51%
9	Layoutlmv3	Li Jie, Wang Wei, Li Songtao, Yang Yunxin, Chen Pengyu, Zhou Danya, Li Chao, Hu Shiyu, Zhang Yuqi, Xu Min, Zhao Yiru, Zhang Bin, Zhang Ruixue, Wang Di, Wang Hui, Xiang Dong	SPDB LAB	29.81%	41.45%	35.63%
10	Data Relation2	-	-	23.26%	35.07%	29.16%

1st Ranking Method. "Ant Group" team apply an ensemble of both discriminative and generative models. The former is a multimodal method which utilizes text, layout and image, and they train this model with two different sequence lengths, 2048 and 512 respectively. The texts and boxes are generated by independent OCR models. The latter model is an end-to-end method which directly generates K-V pairs for an input image.

2nd Ranking Method. "University of Chinese Academy of Sciences" team realized end-to-end information extraction through OCR, NER and RE technologies. Text information extracted by OCR and image information are jointly transmitted to NER to identify key and value entities. RE module extracts entity pair relationships through multi-classification. The training dataset is Hust-Cell.

3rd Ranking Method. "University of Science and Technology of China (USTC), iFLYTEK AI Research" firstly perform key-value-background triplet classification for each OCR bounding box using a PretrainedLM called GraphDoc [21] which utilizes text, layout, and visual information simultaneously. Then they use a detection model (DBNet [11]) to detect all the table cells in input images and split images into table-images and non-table images. For table images, they merge ocr boxes into table cells and then group all the left and

Table 3. Task-2: E2E Complex Entity Labeling Results. * means the reproducible script has been submitted by participants and verified by organizers.

Rank	Method Name	Team Members	Insititute	Score1(NED)	Score2(F-score)	Score(Total)
1	LayoutLMV3 &StrucText*	Minhui Wu, Mei Jiang, Chen Li, Jing Lv, Qingxiang Lin, Fan Yang	TencentOCR	57.78%	55.32%	57.29%
2	sample-3*	Zhenrong Zhang, Lei Jiang, Youhui Guo, Jianshu Zhang, Jun Du	University of Science and Technology of China, iFLYTEK AI Research	47.15%	41.91%	46.10%
3	task 1 transfer learning LiLT + task3 transfer learning LiLT + LilLT + Layoutlmv3 ensemble*	Hengguang Zhou, Zeyin Lin, Xingjian Zhao, Yue Zhang, Dahyun Kim, Sehwan Joo, Minsoo Khang, Teakgyu Hong	Deep SE x Upstage HK	45.70%	40.20%	44.60%
4	LayoutMask-v3*	Yi Tu	Ant Group	44.79%	42.53%	44.34%
5	Pre-trained model based entity extraction (ro)*	Zening Lin, Teng Li, Wenhui Liao, Jiapeng Wang, Songxuan Lai, Lianwen Jin	South China University of Technology, Huawei Cloud	44.98%	40.06%	43.99%
6	EXO-brain for KIE	Boqian Xia, Yu Wang, Yadong Li, Zuming Huang, Lele Xie, Jingdong Chen, Hongbin Wang	Ant Group	44.02%	39.63%	43.14%
7	multi-modal based KIE through model fusion	Jie Li, Wei Wang, Min Xu, Yiru Zhao, Bin Zhang, Pengyu Chen, Danya Zhou, Yuqi Zhang, Ruixue Zhang, Di Wang, Hui Wang, Chao Li, Shiyu Hu, Dong Xiang, Songtao Li, Yunxin Yang	SPDB LAB	42.42%	37.97%	41.53%
8	Aaaa	Li Rihong, Zheng Bowen	Shenzhen Runnable Information Technology Co., Ltd.	42.03%	37.14%	41.05%
9	donut	zy	-	41.64%	37.65%	40.84%
10	Ant-FinCV	Tao Huang, Jie Wang, Tao Xu	Ant Group	41.61%	35.98%	40.48%

top keys for each value table cell as its corresponding key content. For non-table images (including all text in non-table images and text outside tabel cells in table images), they directly use a MLP model to predict all keys for each value box.

5.2   Task 2 Performance and Ranking

The result for Task 2 is presented on Table 3.

The methods used by the top 3 submissions for Task 2 are presented below.

1st Ranking Method. "TencentOCR" team are mainly based on LayoutLMv3 and StrucTextv1 model architecture. All training models are finetuned on large pretrained models of LayoutLM and StrucText. During training and testing, they did some preprocessings to merge and split some badly detected boxes. Since entity label of kv-pair boxes are ignored, they used model trained on task1 images to predict kv relations of text boxes in Task 2 training/testing images. Thus they added additional 2 classes of labels (question/answer) and mapped original labels to new labels (other → question/answer) to ease the difficulty of training. Similarly, During testing, they used kv-prediction model to filter those text boxes with kv relations and used model trained on Task 2 to predict entity label of the lefted boxes. In addition, they combined predicted results of different models based on scores and rules and did some postprocessings to merge texts with same entity label and generated final output.

2nd Ranking Method. "University of Science and Technology of China (USTC), iFLYTEK AI Research" team uses the GraphDoc [21] to perform bounding box classification, which utilizes text, layout, and visual information simultaneously.

3rd Ranking Method. "Deep SE x Upstage HK" team, for the OCR, uses a cascade approach where the pipeline is broken up into text detection and text recognition. For text detection, they use the CRAFT architecture with the backbone changed to EfficientUNet-b3. For text recognition, they use the ParSeq architecture with the visual feature extractor changed to SwinV2. Regarding the parsing models, they trained both the LiLT and LayoutLMv3 models on the Task2 dataset. For LiLT, they also conducted transfer learning on either task1 or task3 before fine-tuning on Task 2 dataset. Finally, they take an ensemble of these four models to get the final predictions.

5.3   Task 3 Performance and Ranking

The result for Task 3 is presented on Table 4.

Table 4. Task-3: E2E Zero-shot Structured Text Extraction Results. * means the reproducible script has been submitted by participants and verified by organizers.

Rank	Method Name	Team Members	Insititute	Score1(NED)	Score2(F-score)	Score(Total)
1	sample-1*	Zhenrong Zhang, Lei Jiang, Youhui Guo, Jianshu Zhang, Jun Du	University of Science and Technology of China, iFLYTEK AI Research	82.07%	65.27%	78.71%
2	LayoutLMv3*	Minhui Wu, Mei Jiang, Chen Li, Jing Lv, Huiwen Shi	TencentOCR	80.01%	66.71%	77.35%
3	KIE-Brain3*	Boqian Xia, Yu Wang, Yadong Li, Ruyi Zhao, Zuming Huang, Lele Xie, Jingdong Chen, Hongbin Wang	Ant Group	74.90%	57.59%	71.44%
4	zero-shot-qa	-	-	74.24%	56.81%	70.75%
5	task3-2	chengl	CMSS	65.52%	50.85%	62.59%

The methods used by the top 3 submissions for Task 3 are presented below.

1st Ranking Method. "University of Science and Technology of China (USTC), iFLYTEK AI Research" team first use OCR models (DBNet-det + SVTR-rec) to get each bounding box coordinate and it's text content of input images, then sort boxes with manual rules, and concatenate all text content to a string according to the box sequence. Result string of step 1 is fed into a seq2seq model to directly predict target text content, which consists of 8 open-source bert models, including chinese-roberta, chinese-lilt-roberta, chinese-pert, chinese-lilt-pert, chinese-lert, chinese-lilt-lert, chinese-macbert and chinese-lilt-macbert, and these models are trained using the UniLM[5] toolbox. Data augmentation including random text replacing and erasing, box random scaling and shifting is used. As for english doc images, they directly use DocPrompt[6] outputs as final result

2nd Ranking Method. "TencentOCR" team's method is based on LayoutLMv3 and StrucTextv1 model architecture. All training models are finetuned on large pretrained models of LayoutLM and StrucText. During training and testing, we did some preprocessings to merge and split some badly detected boxes. They used models trained on task 1 images to predict key-value pairs in test images and models trained on task 2 images to predict entity labels(title, date, etc.) for text boxes. Besides, they applied rule based post processing methods and assembled results of different models to generate final outputs.

3rd Ranking Method. "Ant Group" team apply an ensemble of multi-task end-to-end information extraction models. The document question answering task and the document information extraction task are jointly realized, and the model performance is improved. At the same time, this solution is an end-to-end information extraction method and does not rely on external OCR.

5.4   Task 4 Performance and Ranking

The result for Task 4 is presented on Table 5.

The methods used by the top 3 submissions for Task 4 are presented below.

1st Ranking Method. "TencentOCR" team's methods are mainly based on LayoutLMv3 and StrucTextv1 model architecture. All training models are finetuned on large pretrained models of LayoutLM and StrucText. During training and testing, they did some preprocessings to merge and split some badly detected boxes. They also trained our own ocr models including dc-convnet based detection and ctc/eda based recognition. They applied merging methods to merge ocr results from different sources. Based on predicted results of task 3, they also introduced self-supervising training to train segment-based classification models for folder 9 & folder 10 to predict entity labels. Similar to task 3, they also did rule based post processing methods and assembled results of different models to generate final outputs.

---

[5] https://github.com/microsoft/unilm/blob/master/s2s-ft/.
[6] https://github.com/PaddlePaddle/PaddleNLP/blob/develop/model_zoo/ernie-layout.

Table 5. Task-4: E2E Few-shot Structured Text Extraction Results. * means the reproducible script has been submitted by participants and verified by organizers.

Rank	Method Name	Team Members	Insititute	Score1(NED)	Score2(F-score)	Score(Total)
1	LayoutLMv3& StrucText*	Mei Jiang, Minhui Wu, Chen Li, Jing Lv, Haoxi Li, Lifu Wang, Sicong Liu	TencentOCR	87.14%	73.59%	84.43%
2	sample-1*	Zhenrong Zhang, Lei Jiang, Youhui Guo, Jianshu Zhang, Jun Du	University of Science and Technology of China, iFLYTEK AI Research	85.24%	69.68%	82.13%
3	task4-base	chengl	CMSS	78.57%	60.21%	74.90%
4	Fewshot-brain_v1*	Boqian Xia, Yu Wang, Yadong Li, Hongbin Wang	Ant Group	77.81%	60.71%	74.39%
5	Dao Xianghu light of TianQuan	Kai Yang, Tingmao Lin, Ye Wang, Shuqiang Lin, Jian Xie, Bin Wang, Wentao Liu, Xiaolu Ding, Jun Zhu, Hongyan Pan, Jia Lv	CCB Financial Technology Co. Ltd, China	71.48%	55.03%	68.19%
6	GRGBanking	Liu Kaihang, Yue Xuyao, Xu Tianshi, Zhang Huajun, Liang Tiankai	GRGBanking	45.44%	35.83%	43.52%

2nd Ranking Method. "University of Science and Technology of China (USTC), iFLYTEK AI Research" team first use OCR models (DBNet-det + SVTR-rec) to get each bounding box coordinate and it's text content of input images, then sort boxes with manual rules, and concatenate all text content to a string according to the box sequence. Result string of step 1 is fed into a seq2seq model to directly predict target text content, which consists of 8 open-source bert models, including chinese-roberta, chinese-lilt-roberta, chinese-pert, chinese-lilt-pert, chinese-lert, chinese-lilt-lert, chinese-macbert and chinese-lilt-macbert, and these models are trained using the UniLM[7] toolbox. Data augmentation including random text replacing and erasing, box random scaling and shifting is used. As for english doc images, they directly use DocPrompt[8] outputs as final result.

3rd Ranking Method. "CMSS" team's method firstly perform data cleaning, data synthesis, and random augmentation based on the given data. For blurry images, they utilize a variety of cascaded models to optimize the text detection results. To further improve the accuracy of text recognition algorithm, they use a neural network structure and remove special characters. Particularly, they have found that the position of the text plays a significant role in the structure, especially for structured text with key and value. To address this issue, they optimize the model parameters based on the UIE-X basic model (Unified Structure Generation for Universal Information Extraction) and optimize the data pre-processing part to enhance the relationship features between text lines.

---

[7] https://github.com/microsoft/unilm/blob/master/s2s-ft/.
[8] https://github.com/PaddlePaddle/PaddleNLP/blob/develop/model_zoo/ernie-layout.

Finally, they merge the model inference results by designing rules to achieve the best possible outcome.

6   Discussion

The competition saw a considerable number of submissions from both academic and industrial participants, indicating a notable level of interest in the topic of complex structured text extraction. The competition focused on two tracks: Track 1 involved end-to-end complex entity linking and labeling, while Track 2 introduced new zero/few-shot scenario tasks for structured text extraction.

The leading approaches in Track 1 of the competition utilized ensemble models that integrated multiple modalities, including text, layout, and image information. These models were enhanced by incorporating techniques such as OCR, NER, and RE. The prominence of multimodal models and ensemble techniques highlights their significance. Nevertheless, there exists a notable performance gap in achieving the necessary accuracy for information extraction in complex scenarios for large-scale practical applications. The observed performance gap is notable when comparing the top-performing F1 score of 55.32% for Task 2 to the state-of-the-art end-to-end method on other datasets. Specifically, Kuang et al. [9] achieved an 85.87% F1 score on the SROIE dataset and Donut [8] achieved an 84.1% F1 score on the CORD dataset in end-to-end evaluations.

Track 2 of the SVRD competition focused on new zero/few-shot scenario tasks for structured text extraction. Specifically, Task 3 required participants to extract structured text from images that were not present in the training set, whereas Task 4 allowed participants to perform structured text extraction using a small number of provided labeled examples. The methods used for zero-shot and few-shot learning involved various deep learning techniques, such as pre-trained large multimodal model and model ensemble. The top-performing methods achieved 78.71% and 84.43% for Task 3 and Task 4, respectively. Despite the commendable average performance of Task 3 and Task 4 across all 10 scenarios, the individual performance analysis of the champion method reveals significantly low effectiveness in more challenging scenarios. Specifically, the champion method attained a score of 53.25% in the Letter/Email scenario and 46.2% in the Technical Report scenario for Task 3. In Task 4, the champion method achieved a score of 59.23% in the Car Ticket scenario. These results underscore the need for further research to enhance the performance of these methods in structured text extraction for such challenging scenarios.

Despite the top-performing methods showcased promising results, a significant performance gap remains for challenging scenarios in structured text extraction. Advancements in developing robust techniques for structured text extraction are crucial for the progress of Document AI. The SVRD competition provided a platform for researchers in CV and NLP to collaborate and showcase their expertise. The results highlight the potential of structured text extraction for various document analysis applications, emphasizing the need for continued research.

7 Conclusion

We hosted the SVRD competition, focused on key information extraction from visually-rich document images, and provided new datasets, including HUST-CELL and Baidu-FEST, and designed new evaluation protocols for our tasks. The submissions indicated strong interest from academia and industry, but also revealed significant challenges in achieving high performance for complex and zero-shot scenarios. Key information extraction remains a difficult task with potential for numerous document analysis applications. Future competitions could expand on this topic with more challenging datasets and applications, attracting researchers in CV and NLP and advancing the field of Document AI.

Acknowledgements. This competition is supported by the National Natural Science Foundation of China (No. 62225603, No. 62206103, No. 62206104). The organizers thank Sergi Robles and the RRC web team for their tremendous support on the registration, submission and evaluation jobs.

References

1. Appalaraju, S., Jasani, B., Kota, B.U., Xie, Y., Manmatha, R.: Docformer: end-to-end transformer for document understanding. In: Proceedings of the IEEE/CVF International Conference on Computer Vision, pp. 993–1003 (2021)
2. Cao, H., et al.: Query-driven generative network for document information extraction in the wild. In: Proceedings of the 30th ACM International Conference on Multimedia (2022)
3. Cao, H., et al.: GMN: generative multi-modal network for practical document information extraction. In: North American Chapter of the Association for Computational Linguistics (2022)
4. Fang, S., Xie, H., Wang, Y., Mao, Z., Zhang, Y.: Read like humans: autonomous, bidirectional and iterative language modeling for scene text recognition. In: Proceedings of the IEEE/CVF Conference on Computer Vision and Pattern Recognition, pp. 7098–7107 (2021)
5. Guo, H., Qin, X., Liu, J., Han, J., Liu, J., Ding, E.: Eaten: entity-aware attention for single shot visual text extraction. In: 2019 International Conference on Document Analysis and Recognition (ICDAR), pp. 254–259. IEEE (2019)
6. Huang, Z., et al.: ICDAR 2019 competition on scanned receipt OCR and information extraction. In: 2019 International Conference on Document Analysis and Recognition (ICDAR), pp. 1516–1520. IEEE (2019)
7. Jaume, G., Ekenel, H.K., Thiran, J.P.: FUNSD: a dataset for form understanding in noisy scanned documents. In: ICDARW, vol. 2, pp. 1–6 (2019)
8. Kim, G., et al.: OCR-free document understanding transformer. In: Avidan, S., Brostow, G., Cissé, M., Farinella, G.M., Hassner, T. (eds.) ECCV 2022. LNCS, vol. 13688, pp. 498–517. Springer, Cham (2022). https://doi.org/10.1007/978-3-031-19815-1_29
9. Kuang, J., et al.: Visual information extraction in the wild: practical dataset and end-to-end solution. In: ICDAR (2023)
10. Li, Y., et al.: Structext: structured text understanding with multi-modal transformers. In: Proceedings of the 29th ACM International Conference on Multimedia, pp. 1912–1920 (2021)

11. Liao, M., Wan, Z., Yao, C., Chen, K., Bai, X.: Real-time scene text detection with differentiable binarization. In: Proceedings of the AAAI Conference on Artificial Intelligence, vol. 34, pp. 11474–11481 (2020)

12. Park, S., et al.: Cord: a consolidated receipt dataset for post-OCR parsing. In: Workshop on Document Intelligence at NeurIPS 2019 (2019)

13. Shi, B., Bai, X., Yao, C.: An end-to-end trainable neural network for image-based sequence recognition and its application to scene text recognition. IEEE Trans. Pattern Anal. Mach. Intell. **39**(11), 2298–2304 (2016)

14. Wang, J., et al.: Towards robust visual information extraction in real world: new dataset and novel solution. In: Proceedings of the AAAI Conference on Artificial Intelligence, vol. 35, pp. 2738–2745 (2021)

15. Xu, Y., Li, M., Cui, L., Huang, S., Wei, F., Zhou, M.: Layoutlm: -training of text and layout for document image understanding. In: Proceedings of the 26th ACM SIGKDD International Conference on Knowledge Discovery & Data Mining, pp. 1192–1200 (2020)

16. Xu, Y., et al.: XFUND: a benchmark dataset for multilingual visually rich form understanding. In: Findings of the Association for Computational Linguistics: ACL 2022, pp. 3214–3224 (2022)

17. Yu, D., et al.: Towards accurate scene text recognition with semantic reasoning networks. In: Proceedings of the IEEE/CVF Conference on Computer Vision and Pattern Recognition, pp. 12113–12122 (2020)

18. Yu, W., Lu, N., Qi, X., Gong, P., Xiao, R.: Pick: processing key information extraction from documents using improved graph learning-convolutional networks. In: 2020 25th International Conference on Pattern Recognition (ICPR), pp. 4363–4370. IEEE (2021)

19. Zhang, C., et al.: Look more than once: An accurate detector for text of arbitrary shapes. In: Proceedings of the IEEE/CVF Conference on Computer Vision and Pattern Recognition, pp. 10552–10561 (2019)

20. Zhang, P., et al.: Trie: end-to-end text reading and information extraction for document understanding. In: Proceedings of the 28th ACM International Conference on Multimedia, pp. 1413–1422 (2020)

21. Zhang, Z., Ma, J., Du, J., Wang, L., Zhang, J.: Multimodal pre-training based on graph attention network for document understanding. IEEE Trans. Multimedia (2022)

22. Zhou, X., et al.: East: an efficient and accurate scene text detector. In: Proceedings of the IEEE Conference on Computer Vision and Pattern Recognition, pp. 5551–5560 (2017)

ICDAR 2023 CROHME: Competition on Recognition of Handwritten Mathematical Expressions

Yejing Xie[1], Harold Mouchère[1(✉)], Foteini Simistira Liwicki[2], Sumit Rakesh[2], Rajkumar Saini[2], Masaki Nakagawa[3], Cuong Tuan Nguyen[4], and Thanh-Nghia Truong[3]

[1] Nantes Université, École Centrale Nantes, CNRS, LS2N, UMR 6004, 44000 Nantes, France
{chrome2023,harold.mouchere}@univ-nantes.fr
[2] Lulea University of Technology, Luleå, Sweden
[3] Tokyo University of Agriculture and Technology, Fuchu, Japan
[4] FPT University, Hanoi, Vietnam

Abstract. This paper overviews the 7th edition of the Competition on Recognition of Handwritten Mathematical Expressions. ICDAR 2023 CROHME proposes three tasks with three different modalities: on-line, off-line and bimodal. 3905 new handwritten equations have been collected to propose new training, validation and test sets for the two modalities. The complete training set includes previous CROHME training set extented with complementary off-line (from OffRaSHME competition) and on-line samples (generated). The evaluation is conducted using the same protocol as the previous CROHME, allowing a fair comparison with previous results. This competition allows for the first time the comparison of the on-line and off-line systems on the same test set. Six participating teams have been evaluated. Finally the same team won all 3 tasks with more than 80% of expression recognition rate.

Keywords: mathematical expression recognition · handwriting recognition · bimodal · evaluation · dataset

1 Introduction

Handwritten mathematical recognition is an important and challenging task with numerous real-world applications. The CROHME (Competition on Recognition of On-line Handwritten Mathematical Expressions) competition has been organized since 2011 to foster research and development in this field. The competition provides a platform for researchers to compare and evaluate their methods for recognizing on-line and off-line handwritten mathematical expressions. CROHME has contributed to significant advances in this area by promoting the development of new approaches and datasets, as well as facilitating the exchange of ideas among researchers.

G. A. Fink et al. (Eds.): ICDAR 2023, LNCS 14188, pp. 553–565, 2023.
https://doi.org/10.1007/978-3-031-41679-8_33

At the very beginning CROHME (2011 [5], 2012 [6], 2013 [8], 2014 [9], 2016 [7]) focused only on-line content. Recently, CROHME 2019 [4] has considered also off-line content using the rendering of the on-line signal to produce a perfect image of the handwritten expressions. This was a first step towards the off-line domain. After that, other competitions and dataset as OffSRAHME [15] appeared focusing on the off-line domain with more realistic content. This time the competition includes bimodal expressions that combine both on-line and off-line data.

In this paper, we provide an overview of the ICDAR 2023 CROHME competition, describing the three tasks, the corresponding datasets and the evaluation metrics. We also provide the results of the participating systems with a short description. We conclude by discussing the main trends and challenges in handwritten mathematical recognition.

2   Tasks

Task 1. On-line Handwritten Formula Recognition. As the main task in the previous CROHME, participants utilise InkML format data, which has series of handwritten strokes collected by a tablet or similar device (as in Fig. 1), and should convert it to Symbol Label Graph (SymLG), which is a specific Symbol Layout Tree since CROHME 2019. Note that the participants were advised to use strictly only the on-line information and not the off-line information.

Task 2. Off-line Handwritten Formula Recognition. Real scanned images (as in Fig. 3) or rendered images from InkML (as in Fig. 2) are used and should be converted to SymLG. In the same manner as for Task 1, the participants were advised to use strictly only the off-line information and not the on-line information.

Task 3. Bi-modal Handwritten Formula Recognition. Both the on-line strokes and off-line images which come from the same acquisition were used for the bi-modal system. While similar with the 2 tasks mentioned above, the outputs of bi-modal system are also SymLG. Note that the participants in this task could use both the on-line and off-line information.

All three tasks involved in the above system are ranked according to the rate of completely correct recognition, but each task is ranked separately.

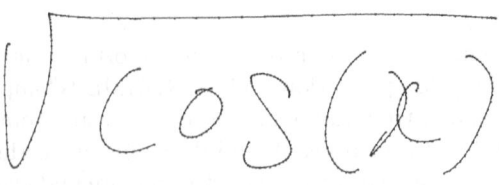

Fig. 1. On-line handwritten formula. Strokes are sequence of points (in Inkml file format).

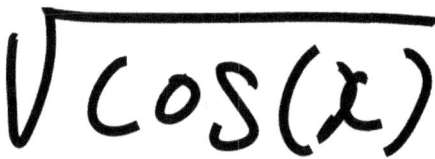

Fig. 2. Off-line handwritten formula rendered from Inkml (PNG file format).

Fig. 3. Off-line handwritten formula scanned from real paper (400 dpi, PNG file format).

3 Datasets and Formula Encodings

In this section, we describe the data sources, data collection, data format and data encoding that we used in this competition. As indicated in Table 1, CROHME 2023 collects new bimodal data both from off-line real images and on-line strokes, as well as merges data from the previous CROHME. On this basis, it is supplemented by the off-line dataset OffRaSHME and large scale of artificial on-line data.

3.1 Handwritten Formulas Data Format

In general, CROHME 2023 provides 2 types of input handwritten formula data, which are InkML for on-line and PNG format image for off-line, and the bi-modal is combining these 2 formats of one same handwritten formula. All 3 tasks used Symbol Label Graph (SymLG) as the ground truth for system training and also as the final output of the system for further performance evaluation.

InkML: Strokes are defined by lists of (x, y) coordinates, representing sampled points which are collected by on-line input equipment. Figure 1 gives a example of on-line version of a expression, each point is visible and connected to the next point to see the strokes. The specific strokes compose the symbols, and these symbols compose the handwritten formulas. Each InkML (except test set), in addition to giving information on the position of the strokes, also provides stroke-level annotation information to indicate the relationship between strokes with the assistance of MathML (XML-based representation) in presentation mode. Thus the symbol segmentations and labels are available in this type of data. Similarly, the Label Graph (LG), which is a CSV-based representation for Stroke-level Label Graph used in previous CROHME is also provided

Table 1. CROHME 2023 Data Sets. On-line data are only inkml files, off-line are images, bimodal are inkml and image for each equation.

Tasks	Train	#	Validation	#	Test	#
Task 1						
On-line	Train 2019	9 993	Test 2016	1147		
	Validation 2019	986				
	artificial	145 108				
	new samples	1045	new samples	555	new samples	2 300
	Total = 157 132		**Total = 1 702**		**Total = 2 300**	
Task 2						
Off-line	rendered	9975	rendered	1147		
	real	1604				
	OffRaSHME	10 000				
	new samples	1045	new samples	555	new samples	2 300
	Total = 20 979		**Total = 1 702**		**Total = 2 300**	
Task 3:						
Bimodal	InkML + rendered	9 975	InkML + rendered	1147		
	InkML + real	1 604				
	new samples	1045	new samples	555	new samples	2 300
	Total = 10 979		**Total = 1 702**		**Total = 2 300**	

Image: Due to the different sources of data, there are two different images types. Indeed, in first competitions several types of acquisition devices have been used: Annoto pens (which produce on-line and off-line signal), tablet-pc with pen based sensitive screen, and numeric white board. So a number of early CROHME data, which are only available as on-line data, lack real off-line images, so these were rendered automatically from the on-line data with size 1000×1000 pixels and 5 pixels of edge padding. At the same time, as OffSRaHME [15], we scanned all the formulas that existed in real writing on paper; these real off-line images are not gray-scale and have no fixed size due to the different sizes of handwritten formulas scale.

Symbol Label Graph: With the development of end-to-end handwritten mathematical recognition system, the stroke level segmentation of each symbol is no longer produced. Therefore, CROHME 2023 does not demand stroke level Label Graph (LG) files as the final output of the system, only Symbol Label Graph (SymLG) is required as formula structure representation for evaluation. SymLG and the related metrics are detailed in Sect. 4.

3.2   Formula Data Selection and Collection

Firstly a corpus of expressions is built then these expressions are written by volunteers.

Formula Data Selection. The selection of the mathematical expressions is an important step because it selects the language domain of the competition. To build this list of expressions, we have followed the same process as in the past contests, as described in [10]. Using a public set of LaTeX sources of scientific papers [1] all math expressions have been extracted and filtered using the grammar IV defined in previous context. Then the expressions are selected in this list to keep the term frequency and expressions length frequency comparable to the training set. 4000 expressions have been pre-selected.

Handwritten Formula Collection. The 2023 new data collection was done over 150 of participants in three sites: France, Japan and Sweden with the Wacom Intuio device, which is allowed to record the pen trajectory while writing on a standard paper. Each A4 size paper has 5 or 8 formulas written on it. Most of the formulas need to be written in rectangular boxes in order to easily separate the formulas on the same sheet of paper. We also tried to let the participants write down a series of formulas freely, without any limitation of boxes.

After the collection, each expression have been individually checked. A part of them has been discarded because of issues in the acquisition (miss matching of on-line and off-line), connecting symbols (the same stroke shared by 2 symbols), or scratching... leading to a new set of 3900 handwritten expressions.

3.3 Generated Artificial HME

Two pattern generation sets are provided for Task 1. The first set consists of syntactic patterns generated by parsing the structure of handwritten mathematical expressions (HMEs) to identify the syntactic role of each component within an HME [12]. After that, each HME is decomposed to identify all valid sub-HMEs based on their syntactic role. These sub-HMEs are considered as new meaningful HMEs. Moreover, we also interchange sub-HMEs that have the same syntactic role inside each HME to get new HME patterns. The generated syntactic patterns differ from the original patterns, and we have provided approximately 76,224 syntactic patterns in CROHME 2023.

The second set consists of synthetic patterns generated from LaTeX sequences. We create an XML layout from each LaTeX sequence and update the layout with the handwritten symbols extracted from CROHME 2019 training set. We generated approximately 68,884 HME patterns from LaTeX sequences of the CROHME 2019 Wikipedia corpus and collected LaTeX sequences.

The Table 1 describes clearly the usage of the different data sets in the competition.

4   Evaluation Metrics

4.1   Symbol Layout Graph: SymLG

As CROHME 2023 concentrates on formula recognition for a variety of input formats, the evaluation metric of formula recognition is always the same symbol-level evaluation for different tasks. The unified SymLG representation forgets which strokes belong to which symbol and allows all systems, whether they produce stroke-level or symbol-level results, to have an identical standard that can be compared directly. The Fig. 4(a) gives an example of a SymLG for an expression containing 4 symbols. This graph can be generated from the LATEX string, from the mathML representation (available in InkML files) or from Stroke level Label Graphs.

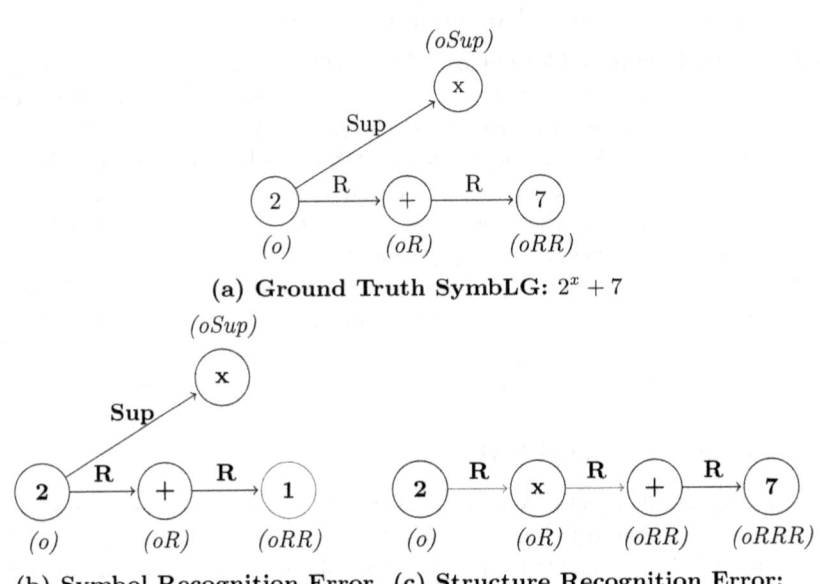

(a) **Ground Truth SymbLG:** $2^x + 7$

(b) **Symbol Recognition Error** $2^x + 1$ (c) **Structure Recognition Error:** $2x + 7$

Fig. 4. A visualisation example for formula recognition evaluation with SymLG. Compared with the Ground Truth (a) '$2^x + 7$', (b) '$2^x + 1$' and (c) '$2x + 7$' are 2 possible mistakes of system recognition. (b) has a symbol recognition error, which means that '1' and '7' are mistaken, while (c) has two structure recognition errors, which means that the relationship between '2' and 'x', '2' and '+' are misunderstood, which leads to new errors of mismatching nodes oRR and $oRRR$.

4.2   Metrics

As previous CROHME, initially described in [10], updated in [4], we consider expression and structure recognition rates as the most important metrics.

For each system, we generated a table giving the percentage of expressions with matching MathML trees, and with at most 3 incorrect symbols or relations. And the completely correct recognition rates were used to rank systems for all Tasks. It should be noticed that the evaluation metric applied on SymLG looses a part of its advantages.

Indeed, originally designed for stroke level comparison, the tool lgeval first aligned the node identifiers. For one expression, stroke identifiers are unique and the same whatever the recognized output. In Symbol level LG, the node identifiers are based on the position of the symbol in the relation tree, starting from the origin. This identification leads to add new errors during the evaluation in the case of relation errors.

For example, in Fig. 4(c) the symbols x, $+$ and 7 are correctly recognized, but because of the relation error at the beginning of the expression, all following nodes are renamed and miss-matching with the ground-truth. So, in case of perfect recognition, every nodes are matching and no errors will be produced. In case of only symbol recognition errors (without structure errors), the matching is pertinent and the provided errors are meaning full. In case of a relation recognition error, all the symbols from the concerned sub-expression will be in error.

5   Participating Methods

As clearly shown in Table 4, there are in total 6 teams joined in 3 tasks. Sunia, YP_OCR and TUAT participated all the 3 tasks, while DPRL_RIT only participated in task 1, PERO and UIT@AIClub_Tensor only participated in task 2. We need to mention that, probably because of our technique problem, the system description of team UIT@AIClub_Tensor is missed, thus the descriptions of the 5 team are shown as follow.

Sunia PTE.LTD[1]. A standard encoder-decoder models are used to translate a bitmap image or a sequence of offsets and pen-up flags to a sequence of tokens in a compact language where each mathematical expression has a unique representation. The encoder consists of stacked CNN/BLSTM layers, and the decoder relies on an attention mechanism that can model convergence and estimate the location of each symbol. For on-line recognition, preprocessing steps such as reordering (for some of the combined models which were trained on spatially sorted strokes and thus stroke-order free), resampling, and normalization are performed to ensure that our system is not too sensitive to stroke order, device, location, and size. Model ensembling and LL(1) grammar parsing are also employed during beam search to boost the accuracy and avoid illegal output. We have augmented the training set by assembling synthetic samples from the subexpressions in the official training set and applying local and global distortions. For data usage, based on-line official data, more synthetic samples were generated with stroke-level annotations by official labels of these artificial samples. And for off-line and bimodal task, rendered version of on-line samples are also used in additional to the official train set.

YP_OCR, CVTE Research: The system utilized an attention-based encoder-decoder structure, where the encoder adopted the DenseNet architecture. In the decoding stage, both a bidirectional tree decoder and a regular decoder are

---

[1] https://patentscope.wipo.int/search/zh/detail.jsf?docId=DE375155214.

applied, and two different models are obtained. For Task 1, the InkML files were rendered uniformly before feeding them into the models with different decoders. Additionally, a multi-stage curriculum learning training strategy is employed to train the models. For Task 2, the models is trained by only using the official training dataset and alleviated the problem of limited training data by using data augmentation methods. Finally, the trained models are fused in different stages to obtain better results and integrated the results of both decoders through beam search during the inference stage.

PERO System: The system consists of an optical model (OM) and a language model (LM). The OM is based on a CRNN architecture trained using the CTC loss function. It operates on images with a normalized height of 128 pixels and arbitrary length. The LM is an LSTM network trained to predict the next LaTeX token given its predecessors. To train the OM, the provided off-line dataset and also the provided synthetic data are used. As training data for the LM, the ground truth provided in the datasets and also publicly available equations from Wikipedia1 are used. The OM is trained to produce a LaTeX code. The LM decodes logits generated by the OM using prefix search decoding and the hypothesis representing a valid LaTeX code with the highest probability is taken as a result. Finally, LaTeX codes are converted into SymLG using the provided tools.

Team DPRL RIT. QD-GGA [3] system is extended, a visual parser that (1) creates a line-of-sight (LOS) graph over strokes, (2) scores stroke segmentation, symbol class, and relationship hypotheses with a multi-task CNN, (3) segments and classifies symbols, and (4) selects relationships using a maximum spanning tree. We have improved visual features and speed. Input images are now 64×64 for all stroke and formula window images, with formula windows centered on the target/parent stroke. The 2 nearest neighbors [14] of strokes define context windows. A modified graph attention network is used with cropped strokes and their two cropped nearest neighbors as additional context features. Spatial pyramidal pooling (SPP) [2] was added to avoid spatial information loss. These features have improved symbol segmentation and classification rates, while relationships require more work. Memory optimizations provide inference times of 25.9 ms/formula on a desktop system with two Nvidia GeForce GTX 1080 Ti GPUs (12 GB each).

Team TUAT. Two models and their combination are provided for on-line, off-line, and bi-modal recognition tasks. For the on-line recognition model, a deep BLSTM network is used for jointly classifying symbols and relations. Then, a 2D-CFG is used to parse the symbols into mathematical structure [11]. For the off-line recognition model, an end-to-end deep neural network that is trained using weakly supervised learning is used, and a symbol classifier is added to the encoder-decoder model to improve the localization and classification of the CNN features. More details of this method can be found in the paper [13]. For the combination of on-line and off-line models, firstly several candidates with their probability scores during the Beam Search process are generated for each online and off-line recognition model. Then, their prediction scores are combined using a trained weight. The weight is chosen by the best combination recognition rate on the validation set. For the data, the systems are trained using the given dataset as provided by the competition.

6   Results

In this section, we summarize all of the results in CROHME 2023. Table 2 and 3 present respectively results on the new test set 2023 and on the previous test set 2019. There are in total 6 teams participating in 3 tasks, 3 of them participate in all of these 3 tasks, while the other 3 teams only selected one of on-line or off-line task.

Table 2. Formula Recognition Results (Test set 2023)

	Structure + Symbol Labels			Structure
	Correct	$1 \leq$ s.err	$2 \leq$ s.err	Correct rate
Task 1: On-line				
Sunia*	**82.34**	**90.26**	**92.47**	**92.41**
YP_OCR	72.55	83.57	86.22	86.60
TUAT	41.10	54.52	60.04	56.85
DPRL_RIT	38.19	53.39	58.39	59.98
Task 2: Off-line				
Sunia*	**70.81**	**81.74**	**86.13**	**86.95**
YP_OCR	67.86	80.86	85.13	85.99
PERO	58.37	71.22	75.57	75.55
TUAT	51.02	63.17	67.48	64.59
UIT@AIClub_Tensor	14.83	22.87	28.57	32.19
Task 3: Bi-modal				
Sunia*	**84.12**	**91.43**	**93.70**	**94.13**
YP_OCR	72.55	83.57	86.22	86.60
TUAT	53.76	68.83	74.22	70.81

Sunia team obtained all the highest recognition rate in on-line, off-line and bi-modal tasks, and is ahead of **YP_OCR** who also took part in all of 3 tasks.

Compared to the Test set 2019, Test set 2023 had lower recognition rates for the same participants. AS this is a global observation, we can conclude that the new test set is harder than the previous one. However, Sunia, which won CROHME 2023, had a better recognition rate than the systems that participated in CROHME 2019 on the same data. It is a significant indication of the continuous developments in Handwritten Mathematical Expression Recognition over recent years.

Table 3. Formula Recognition Results (Testset 2019)

	Structure + Symbol Labels			Structure
	Correct	1 ≤ s.err	2 ≤ s.err	Correct rate
Task 1: On-line				
Sunia*	**88.24**	**93.08**	**93.99**	**94.25**
YP_OCR	84.74	90.99	92.33	92.66
TUAT	56.88	71.89	76.31	70.73
DPRL_RIT	40.70	59.47	67.06	70.23
2019 winner	*80.73*	*88.99*	*90.74*	*91.49*
Task 2: Off-line				
Sunia*	**77.73**	**87.41**	**90.08**	**90.58**
YP_OCR	73.39	85.74	88.49	89.16
PERO	67.39	77.23	81.07	78.98
TUAT	50.13	63.55	66.89	63.30
UIT@AIClub_Tensor	38.28	52.29	58.80	59.13
2019 winner	*77.15*	*86.82*	*88.99*	*89.49*
Task 3: Bi-modal				
Sunia*	**86.91**	**92.16**	**93.58**	**93.74**
YP_OCR	84.74	90.99	92.33	92.66
TUAT	57.88	73.81	82.82	72.14

Since the 3 tasks in this competition use exactly the same evaluation metrics, it is interesting to conduct a comparison between different tasks. In general, bi-modal systems have the best performance, which has a more enriched data dimension therefore fits the common sense. In most cases, on-line systems have better capabilities than off-line systems, except on-line and off-line systems from TUAT. This illustrates that on-line data with the combination of spatial and temporal information is more conducive than off-line data which only consider the spatial information. However, we do not exclude that generated artificial on-line HME greatly increases the sample size of the on-line dataset, and improving the performance of the deep learning system. Therefore, it is also necessary to further explore the usage of artificial data for off-line datasets, which can greatly increase the data volume of off-line datasets and save manual annotation costs.

A summary of confusion histograms that tabulate errors in symbols and symbol pairs as defined in [10] is presented in Table 4. It is important to note that the most common errors are confusion with absent symbols which are not shown in the table. Indeed, the confusion with absent symbols is always caused by the errors in structure, different structures can not be matched symbol by symbol, that raises errors in the position and classification of the symbols, as explained in Sect. 4.2. The statistics in the table show that symbols 1, 2, and − are the most difficult to recognize, which have commonalities across systems, maybe because these symbols are the most frequent. For symbol pairs, most of systems are hard to identify -1, $\frac{1}{\ }$, and $\frac{\ }{2}$, there is also specificity in the different systems.

Table 4. Most frequent symbol and sub-structure recognition errors (Test set 2023). **E** columns give the GT symbol or sub-structure which have been mis-recognized. **#** columns give the number of occurrences of this errors.

| | Symbols | | | | | | Symbol Pairs | | | | | |
| | 1st | | 2nd | | 3rd | | 1st | | 2nd | | 3rd | |
	E	#	E	#	E	#	E	#	E	#	E	#
Task1: On-line												
Sunia	1	84	)	71	2	68	-1	34	$--$	27	$\frac{\ }{2}$	26
YP_OCR	1	193	2	152	-	132	-1	39	$(x$	26	$\frac{\ }{2}$	26
TUAT	1	405	2	380	-	362	$\frac{1}{\ }$	70	-1	70	$--$	59
DPRL_RIT	1	599	2	558	-	540	$\frac{\ }{2}$	112	-1	109	$\frac{1}{\ }$	104
Task2: Off-line												
Sunia	1	181	2	157	-	136	-1	27	$--$	26	$\frac{\ }{2}$	21
YP_OCR	1	213	x	172	2	168	$x)$	38	-1	36	$(x$	33
PERO	1	331	-	225	2	214	-1	57	$=1$	52	$\frac{1}{\ }$	37
TUAT	1	291	-	213	2	211	$\frac{1}{\ }$	51	$=1$	50	$--$	48
UIT@AIClub_Tensor	-	986	1	982	2	978	$\frac{\ }{2}$	169	$\frac{1}{\ }$	165	-1	148
Task3: Bi-modal												
Sunia	1	70	2	56	x	46	$x)$	14	ab	14	dx	12
YP_OCR	1	193	2	152	-	132	-1	39	$(x$	26	$\frac{\ }{2}$	26
TUAT	1	337	2	254	-	245	$\frac{1}{\ }$	58	$=1$	55	$--$	55

7 Conclusion

CROHME 2023 provides 1,045 new expressions in train set, 555 in validation set and 2,300 in test set of handwritten formulas in bi-modal with manual annotations, as well as large-scale of artificial data. Compared to the CROHME 2019 results, the CROHME 2023 winner had a superior performance, especially in on-line task. But the winner's improvements are not as impressive as those in 2019. CROHME 2023 participants all adopted the deep learning approaches, training models with large amounts of data. However, several system continue to use structural constraints (as LL(1) or CFG grammars). Furthermore, statistical language models are now integrated by some participating systems. It is worth noting that, the winner team Sunia, used data augmentation strategy to increase the amount of provided official data. In addition, most on-line systems have better performance than off-line systems, there is still a significant improvement potential for Handwritten Mathematical Expression Recognition systems, especially off-line.

On the basis of this topic, we have more in-depth ongoing discussions scheduled for the future. Nowadays, deep learning models, especially those with encoder-decoder structure, have become the dominant solution for Handwritten Mathematical Expression Recognition. These systems are sensitive to samples quantity and quality of data sets, therefore, it will be interesting to extend the amount of existing data set by exploring more cost-effective annotation methods as well as gaining more data with the help of synthetic algorithms. Exploring non-supervised training strategy would also be benefit. In addition, more benchmarking experiments are possible to explore the highlights and weaknesses of each system, in order to provide more guidance for them.

Acknowledgments. The organization team of CROHME 2023 would like to thank all the writers who participated in the input of the thousands of equations, the colleagues who helped with the annotation, the participants for their constructive feedback. We particularly thank Da-Han Wang and the OffRASHME team for sharing their data.

References

1. Gehrke, J., Ginsparg, P., Kleinberg, J.: Overview of the 2003 KDD cup. ACM SIGKDD Explorations Newsl. **5**(2), 149–151 (2003)
2. He, K., Zhang, X., Ren, S., Sun, J.: Spatial pyramid pooling in deep convolutional networks for visual recognition. IEEE Trans. Pattern Anal. Mach. Intell. **37**(9), 1904–1916 (2015)
3. Mahdavi, M., Zanibbi, R.: Visual parsing with query-driven global graph attention (QD-GGA): preliminary results for handwritten math formula recognition. In: Proceedings of the IEEE/CVF Conference on Computer Vision and Pattern Recognition Workshops, pp. 570–571 (2020)
4. Mahdavi, M., Zanibbi, R., Mouchère, H., Viard-Gaudin, C., Garain, U.: ICDAR 2019 CROHME + TFD: competition on recognition of handwritten mathematical expressions and typeset formula detection. In: 15th IAPR International Conference on Document Analysis and Recognition (ICDAR 2019), Sydney, Australia (2019)
5. Mouchère, H., Viard-Gaudin, C., Kim, D.H., Kim, J.H., Utpal, G.: CROHME 2011: competition on recognition of online handwritten mathematical expressions. In: ICDAR, Beijing, China (2011)
6. Mouchère, H., Viard-Gaudin, C., Kim, D.H., Kim, J.H., Utpal, G.: ICFHR 2012 - competition on recognition of on-line mathematical expressions (CROHME 2012). In: ICFHR, Bari, Italy (2012)
7. Mouchère, H., Viard-Gaudin, C., Zanibbi, R., Garain, U.: ICFHR 2016 competition on recognition of handwritten mathematical expressions (CROHME 2016). In: ICFHR, Shenzhen, Chine (2016)
8. Mouchère, H., Viard-Gaudin, C., Zanibbi, R., Garain, U., Kim, D.H., Kim, J.H.: ICDAR 2013 CROHME: third international competition on recognition of online handwritten mathematical expressions. In: ICDAR, Washington, DC, USA (2013)
9. Mouchère, H., Viard-Gaudin, C., Zanibbi, R., Utpal, G.: ICFHR 2014 - competition on recognition of on-line mathematical expressions (CROHME 2014). In: ICFHR, Crete, Greece (2014)

10. Mouchère, H., Zanibbi, R., Garain, U., Viard-Gaudin, C.: Advancing the state of the art for handwritten math recognition: the CROHME competitions, 2011–2014. Int. J. Doc. Anal. Recognit. (IJDAR) **19**(2), 173–189 (2016). https://doi.org/10.1007/s10032-016-0263-5

11. Nguyen, C.T., Truong, T.-N., Nguyen, H.T., Nakagawa, M.: Global context for improving recognition of online handwritten mathematical expressions. In: Lladós, J., Lopresti, D., Uchida, S. (eds.) ICDAR 2021. LNCS, vol. 12822, pp. 617–631. Springer, Cham (2021). https://doi.org/10.1007/978-3-030-86331-9_40

12. Truong, T.N., Nguyen, C.T., Nakagawa, M.: Syntactic data generation for handwritten mathematical expression recognition. Pattern Recogn. Lett. **153**, 83–91 (2022)

13. Truong, T.N., Nguyen, C.T., Phan, K.M., Nakagawa, M.: Improvement of end-to-end offline handwritten mathematical expression recognition by weakly supervised learning. In: 2020 17th International Conference on Frontiers in Handwriting Recognition (ICFHR), pp. 181–186. IEEE (2020)

14. Veličković, P., Cucurull, G., Casanova, A., Romero, A., Lio, P., Bengio, Y.: Graph attention networks. arXiv preprint arXiv:1710.10903 (2017)

15. Wang, D.H., et al.: ICFHR 2020 competition on offline recognition and spotting of handwritten mathematical expressions-offrashme. In: 2020 17th International Conference on Frontiers in Handwriting Recognition (ICFHR), pp. 211–215. IEEE (2020)

ICDAR 2023 Competition on Recognition of Multi-line Handwritten Mathematical Expressions

Chenyang Gao[1], Yuliang Liu[1], Shiyu Yao[2], Jinfeng Bai[2], Xiang Bai[1(✉)], Lianwen Jin[3], and Cheng-Lin Liu[4]

[1] Huazhong University of Science and Technology, Wuhan, China
{m202172425,ylliu,xbai}@hust.edu.cn
[2] Tomorrow Advancing Life Education Group, Beijing, China
yaoshiyu@tal.com, jfbai.bit@gmail.com
[3] South China University of Technology, Guangzhou, China
eelwjin@scut.edu.cn
[4] Institute of Automation, Chinese Academy of Sciences, Beijing, China
liucl@nlpr.ia.ac.cn

Abstract. Mathematical expressions play an essential role in scientific documents and are critical for describing problems and theories in various fields, such as mathematics and physics. Consequently, the automatic recognition of handwritten mathematical expressions in images has received significant attention. While existing datasets have primarily focused on single-line mathematical expressions, multi-line mathematical expressions also appear frequently in our daily lives and are important in the field of handwritten mathematical expression recognition. Additionally, the structure of multi-line mathematical expressions is more complex, making this task even more challenging. Despite this, no benchmarks or methods for multi-line handwritten mathematical expressions have been explored. To address this issue, we present a new challenge dataset that contains multi-line handwritten mathematical expressions, along with a challenging task: recognition of multi-line handwritten mathematical expressions (MLHMER). The competition was held from January 10, 2023 to March 26, 2023 with 16 valid submissions. In this report, we describe the details of this new dataset, the task, the evaluation protocols, and the summaries of the results.

Keywords: Handwritten mathematical expression recognition · Multi-line handwriting recognition

1 Introduction

Mathematical expressions are a crucial component of scientific documents, providing a concise and precise way of describing complex problems and theories in various fields, such as mathematics, physics, engineering, and economics. Consequently, the automatic recognition of handwritten mathematical expressions is

G. A. Fink et al. (Eds.): ICDAR 2023, LNCS 14188, pp. 566–576, 2023.
https://doi.org/10.1007/978-3-031-41679-8_34

an essential task as it can facilitate the digitization and analysis of mathematical texts, assist visually impaired students in learning mathematics, and enable natural handwriting interfaces for mathematical notation.

In recent years, the field of handwritten mathematical expression recognition has received significant attention, with various algorithms and techniques proposed for recognizing single-line mathematical expressions. However, multi-line mathematical expressions are equally important and challenging, appearing frequently in our daily lives and in various scientific documents. More specifically, multi-line expressions are used to represent complex mathematical concepts, such as equations, matrices, systems of equations, and proofs. Moreover, the structure of multi-line expressions is more complex, making their recognition a challenging task.

Despite the importance of recognizing multi-line handwritten mathematical expressions, no benchmarks or methods for multi-line expressions have been explored. Therefore, we present a new challenge dataset containing multi-line handwritten mathematical expressions and a task of recognizing them. As shown in Fig. 1, we give a simple comparison between multi-line mathematical expressions in our MLHME-38K dataset and single-line mathematical expressions in CROHME [10] and HME100K [11] datasets. Our goal is to promote research in handwritten mathematical expression recognition and to address the gap in the availability of datasets and methods for multi-line expressions.

Recognizing multi-line handwritten mathematical expressions is a challenging task due to several factors. First, multi-line expressions can be of varying lengths and have complex structures, with nested subexpressions and dependencies between lines. Second, the handwriting styles and fonts used in multi-line expressions can vary significantly, making it difficult to generalize across different writers and domains. Third, the recognition of multi-line expressions requires not only the recognition of individual characters and symbols but also the identification of their spatial and structural relationships.

Considering the importance of recognizing multi-line handwritten mathematical expressions and the challenges it faces, we host the ICDAR 2023 Competition on Recognition of Multi-line Handwritten Mathematical Expressions. We hope this competition can attract more researchers to pay attention to this field and promote research in this area.

2 Competition Organization

ICDAR 2023 Competition on Recognition of Multi-line Handwritten Mathematical Expressions is organized by a joint team of Tomorrow Advancing Life Education Group, Huazhong University of Science and Technology, South China University of Technology, and Institute of Automation, Chinese Academy of Sciences.

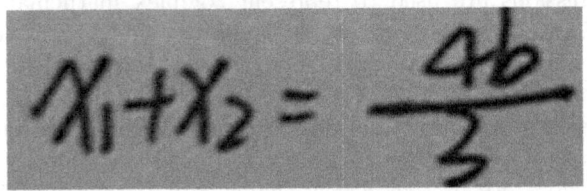

(a) A single-line mathematical expression from CROHME dataset.

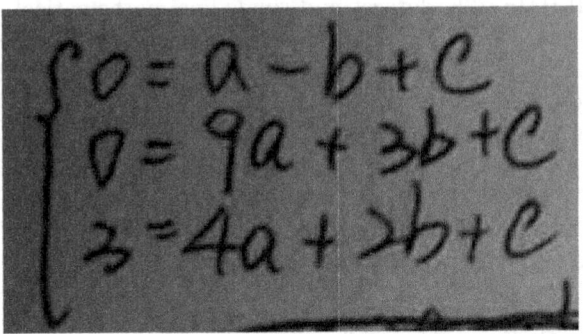

(b) A single-line mathematical expression from HME100K dataset.

(c) A multi-line mathematical expression from our MLHME-38K dataset.

Fig. 1. Comparison between multi-line mathematical expressions in our MLHME-38K dataset and single-line mathematical expressions in CROHME and HME100K dataset.

3   Dataset

We name our dataset MLHME-38K, as it focuses on **M**ulti-**L**ine **H**andwritten **M**athematical **E**xpressions. It totally includes 38,000 labeled images, of which 9,971 images are multi-line mathematical expressions and 28,029 images are single-line. All these images were uploaded by users from real-world scenarios. Consequently, our dataset MLHME-38K becomes more authentic and realistic with variations in color, blur, complicated background, twist, illumination, longer length, and complicated structure. Some examples of our dataset are shown in Fig. 2.

The dataset is divided into a training set, a test_A set, and a test_B set. The training set consists of 30,000 images which will be available to the participants

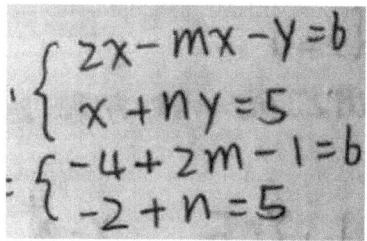

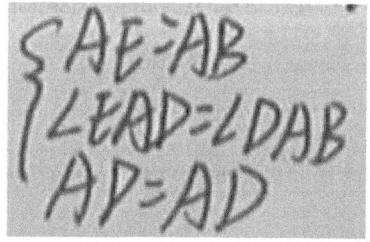

```
\begin{matrix} \left\{
\begin{matrix} 2 x - m x - y = 6
\\ x + n y = 5 \end{matrix}
\right. \\ \left\{ \begin{matrix}
- 4 + 2 m - 1 = 6 \\ - 2 + n = 5
\end{matrix} \right. \end{matrix}
```

```
\left\{ \begin{matrix} A E = A B
\\ \angle E A D = \angle D A B \\
A D = A D \end{matrix} \right.
```

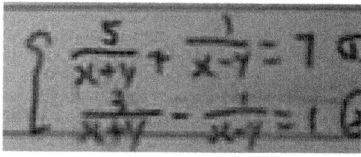

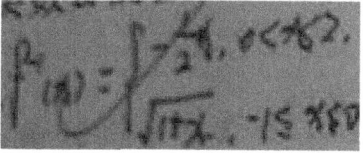

```
\left\{ \begin{matrix} \frac { 5 }
{ x + y } + \frac { 1 } { x - y }
= 7 \textcircled { 1 } \\ \frac {
3 } { x + y } - \frac { 1 } { x -
y } = 1 \textcircled { 2 }
\end{matrix} \right.
```

```
f ^ { - 1 } ( x ) = \left\{
\begin{matrix} - \frac { 1 } { 2 }
x , 0 < x < 2 , \\ \sqrt { 1 + x }
, - 1 \leq x \leq 0 \end{matrix}
\right.
```

Fig. 2. Some examples from our MLHME-38K dataset

along with their annotations. The test_A set consists of 3,000 images and the test_B set consists of 5,000 images, whose annotations will not be released. In the competition stage, only images are offered. The participants are required to submit their results on the test_A set and test_B set with the specific format. Dividing these two test sets can prevent the model from overfitting on a single test set. During the competition, test_A set will be open first and the top ranked teams on the leaderboard will be selected for the evaluation of test_B set. The final ranking is based on test_B set. Every image in the dataset is annotated with a string of LaTeX sequence denoting the mathematical expression. Annotations for images are stored in a txt file with the format shown in Table 1.

Table 1. The annotation format of our MLHME-38K dataset.

File Name	LaTeX String
train_0.jpg	\begin{matrix} 9 x − y = 3 \\ 2 x + y = 5 \end{matrix}
...	...
train_26.jpg	\begin{matrix} n + 2 m = a \\ 2 n − m = 3 a \end{matrix}

4   Tasks

The competition has only one task: recognition of multi-line handwritten mathematical expressions. The aim of this task is to recognize the multi-line handwritten mathematical expressions in images and output them in LaTeX format.

4.1   Evaluation Protocol

In this task, expression recall and character recall are utilized to evaluate the performance.

- Expression recall: The percentage of predicted LaTeX formula sequences matching ground truth (ignore space).

$$S_{\text{recall}} = \frac{S_{\text{right}}}{S_{\text{sum}}} \tag{1}$$

- Character recall: C_{diff} is the sum of edit distances for all images and C_{sum} is the number of characters for all labels.

$$C_{\text{recall}} = 1 - \frac{C_{\text{diff}}}{C_{\text{sum}}} \tag{2}$$

For example, Suppose there are two predictions: "1+1==2" and "a−b=11", and the corresponding labels are "1+1=2" and "a−b=11". As for the expression recall, S_{sum} is 2 since there are two predictions, and S_{right} is 1 since only "a−b=11" match the target. So the expression recall is 0.5 in this case. For the character recall, C_{sum} is 11 (5+6) since the label "1+1=2" contains 5 characters and "a−b=11" contains 6 characters. C_{diff} is 1 (1+0) because the edit distance between "1+1==2" and "1+1=2" is 1. The edit distance between "a−b=11" and "a−b=11" is 0. So the character recall is 0.909 in this case.

4.2   Evaluation Details

The ranking results on the leaderboard of both test_A and test_B set is based on the expression recall. During the competition, test_A set will be open first and the top-10 teams on the leaderboard will be selected for the evaluation of test_B set. After the submission of test_B set due, the final ranking will be given based on the expression recall (higher priority) and character recall:

$$\text{Better} = \begin{cases} S_{r1}, \text{ if } S_{r1} \geq S_{r2} + 0.001 \\ S_{r1}, \text{ if } |S_{r1} - S_{r2}| < 0.001 \text{ and } 0.9\,(S_{r1} - S_{r2}) + 0.1\,(C_{r1} - C_{r2}) > 0 \\ S_{r1}, \text{ if } S_{r1} > S_{r2} \text{ and } 0.9\,(S_{r1} - S_{r2}) + 0.1\,(C_{r1} - C_{r2}) = 0 \\ S_{r2}, \text{ otherwise} \end{cases}$$

where S_{r1} and C_{r1} denotes the expression recall and the character recall from one team. While S_{r2} and C_{r2} denote the expression recall and the character recall from another team.

The final ranking mainly depends on the expression recall. When the expression recall difference between the two teams does not exceed 0.001, we will utilize character recall as an additional measure. Since tiny expression recall differences indicate that the performance of the two methods is very close. In this case, the character recall can provide a better comparison since it focuses on individual characters. We also give different weights for the expression recall and the character recall. The higher weight is given to the expression recall because it is more commonly used in recognition of handwritten mathematical expressions. The chosen hyperparameters mainly rely on previous experience in the similar competition[1].

5   Submissions

Overall, we received valid submissions from 16 teams from both research communities and industries. The submission results of test_A and test_B set are shown in Table 2 and 3. The final ranking is based on the results of test_B set (Table 3).

5.1   Top 5 Submissions

iFLYTEK-OCR team uses an encoder-decoder architecture that formulates HMER as an image-to-sequence translation problem. Specifically, the Conv2Former [2] is employed as the image encoder, and a bi-directional trained Transformer decoder [13] with Attention Refinement Module [12] is utilized as the latex sequence decoder. A Beam Search Ensemble is proposed to ensemble the models trained with different sizes of characters. Specifically, at each decoding step, probability distributions produced by all member models are averaged by certain weights, and the top-k candidate characters to be output are decided by the averaged probability distribution. As for the data augmentation, blur, random, color jitter, scale [6], and TIA Transform [9] are applied to improve the generalization ability of the model.

100-Gan Car University team utilizes CoMER [12] as the baseline model. To efficiently establish time-series information in the encoder, they use two conformer blocks [1] to extract the sequential information in the image feature map

---

[1] https://www.heywhale.com/home/competition/5f703ac023f41e002c3ed5e4/content.

Table 2. Test_A set results and rankings on recognition of multi-line handwritten mathematical expressions.

Rank	Team Name	Exp. Recall	Team Members	Affiliations
1	iFLYTEK-OCR	0.6697	Hao Wu, Mingjun Chen, Xuejing Niu, Changpeng Pi	iFLYTEK
2	BOE_AIOT_AIBD	0.6573	Zhanfu An, Guangwei Huang, Ruijiao Shi, Rui Zheng	Boe Technology Group Co., Ltd.
3	MACARON	0.6557	Yamato Okamoto, Baek Youngmin, Ichimura Shuta, Nakao Ryota, Nakagome Yu	LINE Corporation
4	Just do your best	0.6463	Yingnan Fu, Tiandi Ye	East China Normal University
5	TianyuAI	0.6383	Yu Yan	Wuhan Tianyu Information Industry Co., Ltd.
6	100-Gan Car University	0.6330	Zhuoyan Luo, Yinghao Wu, Zihang Xu, Qi Jing, Hui Xue	Southeast University
7	HMEfly	0.6300	Chenyu Liu	University of Science and Technology of China
8	zyyjyvision	0.6217	Hui Zheng, Xiahan Yang, Qianwen Jia	None
9	sysu	0.6210	Qiqiang Lin, Haiyang Xiao	Sun Yat-sen University
10	FM	0.5993	Xiaoyu Zhang, Maojin Xia, Wenhui Dong	Shanghai Yichuang Information Technology Co., Ltd., Huawei Technologies Co., Ltd.
11	C_OCR	0.5937	Haoyang Shen	Guangzhou Shiyuan Electronics Co., Ltd
12	None	0.5687	Dan Luo	None
13	Not ready yet	0.5937	Weiwei Yi	Vivo Communication Technology Co. Ltd.
14	AYYTeam	0.4363	Zhe Wang, Yifan Bian, Mingyi Ma, Zeyu Chen	None
15	BeatMER	0.2717	Jing Xian, Xingran Zhao	Ant Group Co., Ltd.
16	atd_doc_cits	0.2347	Nobukiyo Watanabe, Tadahito Yao	Canon IT Solutions Inc.

from the horizontal and vertical axes, respectively. Then a fusion operation is conducted on the reshaped feature maps from two different perspectives to get the final image representations. Apart from the image encoder, the other parts are the same as CoMER. The scale augmentation [6], distortion, sketch and perspective are employed to augment the images. Moreover, a simple voting scheme is utilized to ensemble the models trained with different settings.

Table 3. Test_B set results and rankings on recognition of multi-line handwritten mathematical expressions. Note that * denotes character recall.

Rank	Team Name	Exp. Recall	Team Members	Affiliations
1	iFLYTEK-OCR	0.6790 (0.9695*)	Hao Wu, Mingjun Chen, Xuejing Niu, Changpeng Pi	iFLYTEK
2	100-Gan Car University	0.6300 (0.9603*)	Zhuoyan Luo, Yinghao Wu, Zihang Xu, Qi Jing, Hui Xue	Southeast University
3	BOE_AIOT_AIBD	0.6244 (0.9618*)	Zhanfu An, Guangwei Huang, Ruijiao Shi, Rui Zheng	Boe Technology Group Co., Ltd.
4	TianyuAI	0.6186 (0.9583*)	Yu Yan	Wuhan Tianyu Information Industry Co., Ltd.
5	MACARON	0.6166 (0.9552*)	Yamato Okamoto, Baek Youngmin, Ichimura Shuta, Nakao Ryota, Nakagome Yu	LINE Corporation
6	Just do your best	0.6036 (0.9492*)	Yingnan Fu, Tiandi Ye	East China Normal University
7	sysu	0.5950 (0.9589*)	Qiqiang Lin, Haiyang Xiao	Sun Yat-sen University
8	zyyjyvision	0.5950 (0.9463*)	Hui Zheng, Xiahan Yang, Qianwen Jia	None
9	FM	0.5456 (0.9381*)	Xiaoyu Zhang, Maojin Xia, Wenhui Dong	Shanghai Yichuang Information Technology Co., Ltd., Huawei Technologies Co., Ltd.
10	HMEfly	0.2272 (0.1714*)	Chenyu Liu	University of Science and Technology of China

BOE_AIOT_AIBD team adopts three steps to solve the problem. First, an adaptive image adjustment is proposed to determine the number of lines of the expression in an image. Second, to solve the problem of different image resolutions, an Image Super Resolution (ISR) module is added to CAN [5] and CoMER [12], resulting in six models with different training strategies. The simple voting scheme is utilized to ensemble these models. Finally, they judge the latex format of the fusion results. If the latex format is incorrect, the SAN network is utilized to further correct these results. As for the data augment methods, color enhancement (adjusting gamma value, adding Gaussian blur, adjusting hue, saturation, and value), scale augmentation (scale ratio value range $[0.7, 1.4]$), and rotation (rotation angle value range $[-5, 5]$) are applied to improve the robustness of the model.

TianyuAI team adopts CAN [5] and CoMER [12] as the baseline models. To enhance the prediction of single-line mathematical expressions, they utilize a CTC branch during the training phase. They also introduce focal loss [7] to address the unbalanced character distribution in the dataset. Similar to most teams, the model ensemble strategy is adopted to improve the performance of their methods. Specifically, a simple voting scheme is utilized to ensemble the CAN and CoMER models trained with different settings. As for the data augmentation, blur, random sharpness, random contrast, color jitter, rotate and stretch are utilized to improve the performance of the models. Moreover, they use Test Time Augmentation (TTA) during the testing phase.

MACARON team use Donut [4] as the baseline framework, which is a method of document understanding that utilizes an OCR-free end-to-end Transformer model. The encoder and decoder of Donut are SwinTransformer [8] and Multilingual BART. They utilize RGB shift, random brightness, random contrast, hue saturation value, channel shuffle, CLAHE, random sun flare, sharpen, gaussian blur, optical distortion, coarse dropout, and rotate as the data augment methods. Different from most teams, no model ensemble strategy is applied to further improve the performance of their method.

5.2 Discussion

In this task, most participants utilize an encoder-decode framework that models the recognition process in a sequence-to-sequence manner. They first employ a powerful backbone as the image encoder to enhance the performance. DenseNet [3], SwinTransformer [8] and Conv2Former [2] are the commonly used image encoder. Additionally, the conformer block [1] and counting sub-task [5] are utilized to reinforce the extraction of image features. As for the LaTeX sequence decoder, counting-based GRU decoder [5], Transformer decoder [13] with Attention Refinement Module [12], and syntax-aware decoder [11] are frequently utilized.

To improve the generalization ability of the model, most participants use various data augmentations, e.g., blur, color jitter, rotation, distortion, sketch, perspective, and so on. Apart from these commonly used data augmentations, the scale [6] and TIA Transform [9] can also effectively improve the performance of different methods. The scale [6] augmentation can address the recognition difficulty caused by symbols of various sizes. The TIA Transform [9] is a learnable geometric augmentation that can bridge the isolated processes of data augmentation and model training.

The model ensemble strategy is also utilized by most teams to improve the performance of their methods. The typical process is first to train a predefined number of models with different training settings of the same method or with different methods. During the inference stage, probability distributions produced by all member models are ensembled to produce the final results at each decoding step. The voting scheme and weighted average are the most widely used ensemble strategies. However, the model ensemble strategy greatly increases the inference time, limiting its application in real life.

Since our MLHME-38K dataset contains not only multi-line mathematical expressions but also single-line mathematical expressions. More specifically, the test B set contains 1,880 multi-line mathematical expressions and 3,120 single-line mathematical expressions. In order to see the performance of different methods on the multi-line mathematical expressions, we conduct experiments on the multi-line subset of the test B set. As shown in Table 4, the iFLYTEK-OCR team still ranks first. However, the MACARON team rises from 5th to 2nd which indicates that document understanding frameworks may have more potential for multi-line mathematical expressions.

From Table 4, it can be observed that the scores for the multi-line subset of Test B are higher than the scores for the single-line subset. The major reason is that the diversity of the multi-line expressions is much smaller than that of single-line expressions in our MLHME-38K dataset. The image sources of our dataset are mainly the test questions and solutions from middle school and high school, which leads to the fact that the multi-line expressions in our dataset are mostly equation sets. We will try to incorporate more diverse data (e.g. university test questions) into our dataset in the future.

Table 4. The results on the multi-line subset and single-line subset of the test_B set. The rankings are based on the multi-line subset.

Rank	Team Name	Multi-line Exp. Recall	Single-line Exp. Recall
1	iFLYTEK-OCR	0.7532	0.6343
2	MACARON	0.7197	0.5545
3	100-Gan Car University	0.7160	0.5782
4	BOE_AIOT_AIBD	0.6968	0.5792
5	Just do your best	0.6952	0.5484
6	zyyjyvision	0.6952	0.5311
7	TianyuAI	0.6947	0.5715
8	sysu	0.6814	0.5429
9	FM	0.6542	0.4801
10	HMEfly	0.3069	0.1792

6   Conclusion

This paper summarizes the organization and results of ICDAR 2023 Competition on Recognition of Multi-line Handwritten Mathematical Expressions. In this competition, we present a new dataset named MLHME-38K, which focuses on multi-Line handwritten mathematical expressions. Compared with single-line mathematical expressions, the structure of multi-line mathematical expressions is more complicated which makes this task more challenging. We received valid submissions from 16 teams from both research communities and industries. These submissions will provide this research area with new insights and potential solutions. We also believe that our dataset will contribute to advancing the field of handwritten mathematical expression recognition.

Acknowledgments. This competition is supported by the National Natural Science Foundation (NSFC#62225603).

References

1. Gulati, A., et al.: Conformer: convolution-augmented transformer for speech recognition. arXiv preprint arXiv:2005.08100 (2020)
2. Hou, Q., Lu, C.Z., Cheng, M.M., Feng, J.: Conv2former: a simple transformer-style convnet for visual recognition. arXiv preprint arXiv:2211.11943 (2022)
3. Huang, G., Liu, Z., Van Der Maaten, L., Weinberger, K.Q.: Densely connected convolutional networks. In: Proceedings of the IEEE/CVF Conference on Computer Vision and Pattern Recognition, pp. 4700–4708 (2017)
4. Kim, G., et al.: OCR-free document understanding transformer. In: Avidan, S., Brostow, G., Cissé, M., Farinella, G.M., Hassner, T. (eds.) ECCV 2022. LNCS, vol. 13688, pp. 498–517. Springer, Cham (2022). https://doi.org/10.1007/978-3-031-19815-1_29
5. Li, B., et al.: When counting meets HMER: counting-aware network for handwritten mathematical expression recognition. In: Avidan, S., Brostow, G., Cissé, M., Farinella, G.M., Hassner, T. (eds.) ECCV 2022. LNCS, vol. 13688, pp. 197–214. Springer, Cham (2022). https://doi.org/10.1007/978-3-031-19815-1_12
6. Li, Z., Jin, L., Lai, S., Zhu, Y.: Improving attention-based handwritten mathematical expression recognition with scale augmentation and drop attention. In: 17th International Conference on Frontiers in Handwriting Recognition, pp. 175–180. IEEE (2020)
7. Lin, T.Y., Goyal, P., Girshick, R., He, K., Dollár, P.: Focal loss for dense object detection. In: Proceedings of the IEEE/CVF International Conference on Computer Vision, pp. 2980–2988 (2017)
8. Liu, Z., et al.: Swin transformer: hierarchical vision transformer using shifted windows. In: Proceedings of the IEEE/CVF International Conference on Computer Vision, pp. 10012–10022 (2021)
9. Luo, C., Zhu, Y., Jin, L., Wang, Y.: Learn to augment: joint data augmentation and network optimization for text recognition. In: Proceedings of the IEEE/CVF Conference on Computer Vision and Pattern Recognition, pp. 13746–13755 (2020)
10. Mouchere, H., Viard-Gaudin, C., Zanibbi, R., Garain, U.: ICFHR 2014 competition on recognition of on-line handwritten mathematical expressions (CROHME 2014). In: 14th International Conference on Frontiers in Handwriting Recognition, pp. 791–796. IEEE (2014)
11. Yuan, Y., et al.: Syntax-aware network for handwritten mathematical expression recognition. In: Proceedings of the IEEE/CVF Conference on Computer Vision and Pattern Recognition, pp. 4553–4562 (2022)
12. Zhao, W., Gao, L.: Comer: modeling coverage for transformer-based handwritten mathematical expression recognition. In: Avidan, S., Brostow, G., Cissé, M., Farinella, G.M., Hassner, T. (eds.) ECCV 2022. LNCS, vol. 13688, pp. 392–408. Springer, Cham (2022). https://doi.org/10.1007/978-3-031-19815-1_23
13. Zhao, W., Gao, L., Yan, Z., Peng, S., Du, L., Zhang, Z.: Handwritten mathematical expression recognition with bidirectionally trained transformer. In: Lladós, J., Lopresti, D., Uchida, S. (eds.) ICDAR 2021. LNCS, vol. 12822, pp. 570–584. Springer, Cham (2021). https://doi.org/10.1007/978-3-030-86331-9_37

ICDAR 2023 Competition on RoadText Video Text Detection, Tracking and Recognition

George Tom[1][✉][ID], Minesh Mathew[1][ID], Sergi Garcia-Bordils[2,3][ID],
Dimosthenis Karatzas[2][ID], and C.V. Jawahar[1][ID]

[1] Center for Visual Information Technology (CVIT), IIIT Hyderabad, Hyderabad,
India
{george.tom,minesh.mathew}@research.iiit.ac.in, jawahar@iiit.ac.in
[2] Computer Vision Center (CVC), UAB, Barcelona, Spain
{sergi.garcia,dimos}@cvc.uab.cat
[3] AllRead Machine Learning Technologies, Barcelona, Spain

Abstract. In this report, we present the final results of the ICDAR 2023
Competition on RoadText Video Text Detection, Tracking and Recognition. The RoadText challenge is based on the RoadText-1K dataset and
aims to assess and enhance current methods for scene text detection,
recognition, and tracking in videos. The RoadText-1K dataset contains
1000 dash cam videos with annotations for text bounding boxes and
transcriptions in every frame. The competition features an end-to-end
task, requiring systems to accurately detect, track, and recognize text
in dash cam videos. The paper presents a comprehensive review of the
submitted methods along with a detailed analysis of the results obtained
by the methods. The analysis provides valuable insights into the current capabilities and limitations of video text detection, tracking, and
recognition systems for dashcam videos.

Keywords: Scene text · Tracking · Recognition

1 Introduction

Text detection and recognition in videos have traditionally been explored by
the document analysis community. The last text-tracking competition was held
nearly a decade ago and introduced the Text in Videos [9] dataset, which comprises 51 egocentric videos encompassing indoor and outdoor scenarios. Othe
rpopular datasets that deal with text in videos are USTB-VidTEXT [18] and
YouTube Video Text(YVT) [14]. They contain videos sourced from YouTube.
The USTB-VidTEXT dataset primarily consists of text in the form of overlaid
captions, whereas the YVT includes both born-digital text and scene text. These
datasets contain videos with text that are incidental and widely dispersed across
the scene.

© The Author(s), under exclusive license to Springer Nature Switzerland AG 2023
G. A. Fink et al. (Eds.): ICDAR 2023, LNCS 14188, pp. 577–586, 2023.
https://doi.org/10.1007/978-3-031-41679-8_35

Compared to the ICDAR 2013–15 Text-in-Videos Challenge that used a dataset containing 50 videos, our challenge uses the RoadText1K [15] dataset having much larger and diverse set of videos. The text objects in driving videos typically have short lifetimes, which require models tolerant to occlusions, able to handle tiny text instances, and robust to motion blur and significant perspective distortions. Additionally, text instances may not be fully readable in any single frame, necessitating the combination of detections across various frames to transcribe them successfully. Furthermore, camera movement during driving introduces distortions, such as motion blur. As a result, the approaches developed for existing video text datasets tend to be challenging to adapt to real-world applications, such as driver assistance and self-driving systems (Fig. 1).

Fig. 1. Sample frames from RoadText-1K illustrating the various challenges and artefacts like glare, raindrops, out-of-focus, low contrast, and motion blur often encountered in driving videos.

2   Competition Protocol

The competition took place between December 2022 and March 2023. The training and validation data were made available at the end of December 2022, while the test data was released in mid-February 2023. Submissions were accepted between March 1st and March 27th. The expectation was that participating authors would adhere to the established rules of the challenge, to which they had agreed when registering at the Robust Reading Competition (RRC) portal[1], as a means of ensuring scientific integrity throughout the competition.

The RRC portal serves as the host platform for the challenge. Submissions are assessed through automated methods, and the outcomes reported in this report represent the state of submissions at the conclusion of the challenge. However, the challenge will remain open to accept new submissions. But the submissions made after the official challenge period are not considered as an official challenge entry (Table 1).

3   The RoadText-1K Dataset

The RoadText-1K [15] dataset comprises 10-second video clips extracted from the BDD100K [21] dataset. The videos are 720p and 30 fps, and capture diverse

---

[1] https://rrc.cvc.uab.es/?ch=25.

Table 1. Comparison of RoadText-1K with existing text video datasets.

Dataset	Text in Videos [9]	USTB-VidTEXT [18]	YouTube Video Text [14]	RoadText-1K [15]
Source	Egocentric	Youtube	Youtube	car-mounted
Size (Videos)	51	5	30	**1000**
Length (Seconds)	varying	varying	15	10
Resolution	720 × 480	480 × 320	1280 × 720	1280 × 720
Annotated Frames	27,824	27,670	13,500	**300,000**
Total Text Instances	143,588	41,932	16,620	**1,280,613**
Text type	Scene Text	Digital (captions)	Scene Text and Digital	Scene Text
Unique Words	3,563	306	224	8,263
Avg. text frequency per frame	5.1	1.5	1.23	4.2
Avg. Text Track length	46	161	72	48

locations, weather conditions (such as sunny, overcast, and rainy), as well as different times of day. To identify videos with a significant number of text instances, an off-the-shelf text detector was utilised to scan through the frames of the videos in BDD100K. The dataset was randomly partitioned into train, validation and test sets of 500, 200 and 300 videos, respectively.

The bounding boxes and their transcriptions are provided at line level for all the frames in the dataset. The tracks are classified into English, Non-English, and Illegible. Ground truth transcriptions are provided only for text instances of the English category. In contrast to most scene text datasets, text lines rather than individual "words" (separated by spaces) were annotated to expedite annotation and avoid ambiguity in cases involving numbers or abbreviations (Fig. 2).

4   RoadText-1K Challenge

4.1   Evaluation Metrics

The evaluation is based on an adaptation of the CLEAR-MOT [3, 13] and ID [16] frameworks, designed for tracking multiple objects. Each submission is evaluated using three different metrics, namely Multiple Object Tracking Precision (MOTP), Multiple Object Tracking Accuracy (MOTA), and IDF1 score. The number of objects tracked for at least 80 per cent of their lifespan are considered as "Mostly Matched". Those objects that are tracked between 20 and 80 per

Fig. 2. These are sample frames from clips in RoadText-1K, and they have annotations indicating the location and transcription of the text overlaid on them. The boxes that are colored green indicate English text, the ones in blue represent non-English text, and the red boxes represent text that is illegible. (Color figure online)

Table 2. Affiliations and the methods of the competition participants.

Method	Affiliation
ClusterFlow	Google
TH-DL	Tsinghua University
TencentOCR	TencentOCR
TransDETR	ByteDance Inc
RoadText DRTE	KLE Technological University
SCUT-MMOCR-KS	South China University of Technology, Shanghai AI Laboratory and KingSoft Office CV R&D Department

cent of their lifespan fall under "Partially Matched", and those tracked for less than 20 per cent of their lifespan are categorised as "Mostly Lost". For ranking the submissions, MOTA is used. During the evaluation process, a predicted word is classified as a true positive if its intersection over union with a ground-truth word is greater than 0.5 and the predicted transcription matches the ground truth transcription. The assessment of transcription is case insensitive and it is only done for English category tracks. Leading and trailing spaces are disregarded, and instances of two or more spaces are treated as a single space. The recognition of punctuation marks at the start or end of a ground truth word is discretionary and does not influence the evaluation. The evaluation process does not consider areas that contain illegible or non-English legible text. As a result, if a method fails to detect such words, it will not be penalised. Similarly, a method that is successful in detecting such words does not receive a higher score. Even though we only have a single end-to-end task, we also provide results of detection and tracking without taking recognition into account.

4.2   Submitted Methods

The challenge received a total of 16 submissions, out of which 6 were unique and had fulfilled all the competition criteria. The contestants were permitted to submit multiple entries, but they were required to select a single submission as their official entry for the competition. This selection had to be made blindly, without access to the evaluation scores for the submissions. Table 2 presents the names of the submitted methods and affiliations. A brief description of the 6 submitted methods is provided below:

ClusterFlow - ClusterFlow benefits from merging multiple algorithms, including optical character recognition (OCR), optical flow, clustering, and decision trees. The approach involves using a cloud API to extract OCR results at the line level for every image frame of each video, followed by calculating a dense optical flow field using a modern RAFT implementation. The optical flow field is then used to temporally extend the OCR line results to generate tubes or tracklets of lines, which are then grouped into clusters across the entire video using an unsupervised clustering algorithm. To achieve this, the algorithm searches for the optimal distance metric between tracklets, clustering algorithm, and hyper-parameters using the training dataset. Once the tracklets are clustered, the algorithm selects geometry and text from the tracklet to create tracked lines that appear at most once within any video frame. This is accomplished by generating a set of features from each line appearance, tracklet, and cluster, which are then inputted into a classification algorithm. The classification algorithm is trained to select the appearances of the cluster that match the ground truth in the training set. During inference, the classification probabilities are used to choose the most suitable line text appearance within a cluster at any video frame.

TH-DL - It uses an integrated approach for text detection, recognition, and tracking in driving videos. For text detection and recognition, the algorithm adopts TESTR [22] based on Transformer and finetunes the pre-trained TESTR model on the training set of the Roadtext Challenge. For multi-object tracking, ByteTrack [23] is employed, which uses similarities with tracklets to recover true objects from low score detection boxes. A post-processing module is included to filter duplicate instances of text detection and recognition.

TencentOCR - It integrates the detection results of DBNet [10] and Cascade MaskRCNN [4], built with multiple backbone architectures, with the Parseq [2] English recognition model for recognition and further improves the end-to-end tracking with OCSort [5]. The result is end-to-end tracking and trajectory recognition.

TransDETR - The method used in this submission is TransDETR [19]. The approach involves pre-training the network weights on the ICDAR2015 video [9] and fine-tuning the network on the RoadText-3K [7] and BOVText [20] datasets for 20 epochs each. Finally, the network is fine-tuned on the RoadText-1K dataset for 20 epochs.

RoadText DRTE - EasyOCR [8] is used to perform the subtasks of detection and recognition on the RoadText-1K [15] dataset. The algorithm uses the CRAFT [1] algorithm for detection and the CRNN [17] model for recognition. Once the video is processed frame by frame, the algorithm performs the tracking subtask by assigning a unique ID to each unique transcription in the video. Instances of the same unique transcription are assigned the same ID throughout the video.

SCUT-MMOCR-KS - This submission utilizes DBNet++ [11] for text detection, which is first pre-trained on a collection of TextOCR, HierText [12], DSText, YVT [14], ICDAR2015-Video [9], and Minetto before being fine-tuned on DSText. For text recognition, a ViT-based [6] recognizer is used, which is pre-trained on 10M unlabeled real STR images and fine-tuned on 4M labelled real STR images. CoText tracking module is used for text tracking (Fig. 3).

Table 3. Results of RoadText video text detection, tracking

Method	MOTA	MOTP	IDF1	Mostly Matched	Partially Matched	Mostly Lost
TransDETR	37.53	74.18%	60.27%	1665	1762	1563
ClusterFlow	36.01	70.29%	61.19%	1757	1194	2029
TH-DL	31.07	75.20%	62.35%	2180	1495	1317
TencentOCR	16.40	66.59%	42.58%	746	894	3231
SCUT-MMOCR-KS	−10.27	71.84%	56.91%	2354	1660	978
RoadText DRTE	−27.61	70.46%	17.42%	1083	1692	2214

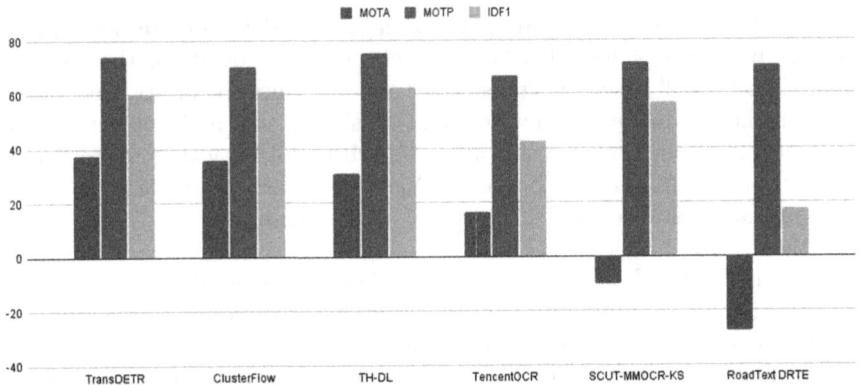

Fig. 3. The chart illustrates the results for text detection and tracking, with MOTA, MOTP, and IDF1 represented by blue, red, and yellow bars, respectively. (Color figure online)

Table 4. Results of RoadText video text detection, tracking and recognition

Method	MOTA	MOTP	IDF1	Mostly Matched	Partially Matched	Mostly Lost
ClusterFlow	11.09	69.04%	48.07%	1392	920	2668
TH-DL	−23.10	72.83%	37.34%	1235	737	3020
TencentOCR	−23.87	56.19%	19.71%	315	454	4102
TransDETR	−28.50	68.74%	26.87%	660	741	3589
RoadText DRTE	−61.39	65.47%	12.08%	146	823	4020
SCUT-MMOCR-KS	−77.1	67.83%	29.6%	1196	918	2878

The participants could use any dataset for training their methods, except the RoadText-1K test set.

4.3   Analysis

The results of the evaluation are presented in Table 3 and Table 4, with the first one focusing on text detection and tracking and the second one displaying text tracking results with recognition. In the absence of recognition, the method with the highest MOTA score was TransDETR, while TH-DL achieved the highest MOTP score and IDF1 score for text tracking. However, in the presence of recognition, ClusterFlow is the winner of the competition and the only method

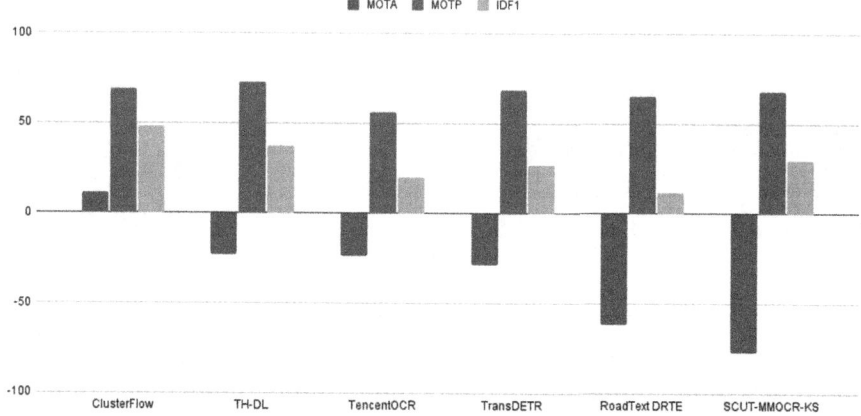

Fig. 4. The chart illustrates the results for text detection, tracking and recognition, with MOTA, MOTP, and IDF1 represented by blue, red, and yellow bars, respectively. (Color figure online)

with a positive MOTA value and also achieved the highest IDF1 score, while TH-DL maintained its position for the highest MOTP value. The commercial Google OCR performs better in comparison to TH-DL, which utilizes TESTR, SCUT-MMOCR-KS, employing ViT-based, and TencentOCR, which relies on Parseq methods for recognition. In the evaluation process, predicted words are only considered true positives when they match the ground truth. This means that if the recognition fails to identify a word, the corresponding track will be considered a false positive, leading to negative MOTA values. Text that appears in a frontal or head-on position is relatively easy to detect. However, text detection methods appear to struggle when presented with text instances such as fancy shop signage or text situated on distant portions of the road beyond the driver's lane (Figs. 4 and 5).

The participants utilized various approaches and strategies to enhance the effectiveness of their methods. These include pre-training and fine-tuning models on diverse datasets, implementing post-processing steps like filtering out repeated text detection and recognition instances to improve outcomes, and merging multiple algorithms and methods. Despite these efforts, the detection, tracking, and recognition still have significant room for improvement, particularly recognition in challenging scenarios presented by the dataset.

(i) Ground Truth

(ii) ClusterFlow

(iii) TH-DL

(iii) TencentOCR

Fig. 5. Sample visualisation of the detected text and the recognition are shown for the ground truth and the top three methods. Green bounding boxes are drawn over detected text, and the recognised text is displayed over the bounding box. (Color figure online)

5   Conclusions and Future Work

The text detection, tracking and recognition challenge introduces a robust benchmark based on driving videos. The challenge is based on the already existing RoadText-1K dataset and has received a total of 16 submissions from multiple teams. In this report, we have summarised the unique features of the RoadText-1K dataset, which make it particularly challenging and different from previous datasets. The report also details a concise overview and an analysis of the submissions. The RoadText challenge will remain open for new submissions in the future, thereby providing a platform for researchers to benchmark and showcase their methods. Looking ahead, we plan to expand the RoadText challenge further by gaining deeper insights into the results and incorporating additional tasks that encompass multilingual settings.

Acknowledgement. This work has been supported by IHub-Data at IIIT-Hyderabad, and MeitY, and grants PDC2021-121512-I00, and PID2020-116298GB-I00 funded by MCIN/AEI/10.13039/501100011033 and the European Union NextGenerationEU/PRTR.

References

1. Baek, Y., Lee, B., Han, D., Yun, S., Lee, H.: Character region awareness for text detection. In: Proceedings of the IEEE Conference on Computer Vision and Pattern Recognition, pp. 9365–9374 (2019)
2. Bautista, D., Atienza, R.: Scene text recognition with permuted autoregressive sequence models. In: Avidan, S., Brostow, G., Cissé, M., Farinella, G.M., Hassner, T. (eds.) Computer Vision - ECCV 2022. ECCV 2022. LNCS, vol. 13688, pp. 178–196. Springer, Cham (2022). https://doi.org/10.1007/978-3-031-19815-1_11
3. Bernardin, K., Stiefelhagen, R.: Evaluating multiple object tracking performance: the clear mot metrics. EURASIP J. Image Video Process. **2008**, 1–10 (2008)
4. Cai, Z., Vasconcelos, N.: Cascade r-cnn: high quality object detection and instance segmentation. IEEE Trans. Pattern Anal. Mach. Intell. **43**(5), 1483–1498 (2019)
5. Cao, J., Weng, X., Khirodkar, R., Pang, J., Kitani, K.: Observation-centric sort: rethinking sort for robust multi-object tracking. arXiv preprint arXiv:2203.14360 (2022)
6. Dosovitskiy, A., et al.: An image is worth 16x16 words: transformers for image recognition at scale. arXiv preprint arXiv:2010.11929 (2020)
7. Garcia-Bordils, S., et al.: Read while you drive - multilingual text tracking on the road. In: Uchida, S., Barney, E., Eglin, V. (eds.) Document Analysis Systems. DAS 2022. LNCS, vol. 13237, pp. 756–770. Springer, Cham (2022). https://doi.org/10.1007/978-3-031-06555-2_51
8. JaidedAI: Easyocr. https://github.com/JaidedAI/EasyOCR
9. Karatzas, D., et al.: ICDAR 2013 robust reading competition. In: ICDAR (2013)
10. Liao, M., Wan, Z., Yao, C., Chen, K., Bai, X.: Real-time scene text detection with differentiable binarization. In: Proceedings of the AAAI Conference on Artificial Intelligence, vol. 34, pp. 11474–11481 (2020)
11. Liao, M., Zou, Z., Wan, Z., Yao, C., Bai, X.: Real-time scene text detection with differentiable binarization and adaptive scale fusion. IEEE Trans. Pattern Anal. Mach. Intell. **45**(1), 919–931 (2022)

12. Long, S., Qin, S., Panteleev, D., Bissacco, A., Fujii, Y., Raptis, M.: Towards end-to-end unified scene text detection and layout analysis. In: Proceedings of the IEEE/CVF Conference on Computer Vision and Pattern Recognition (2022)

13. Milan, A., Leal-Taixé, L., Reid, I., Roth, S., Schindler, K.: Mot16: a benchmark for multi-object tracking. arXiv preprint arXiv:1603.00831 (2016)

14. Nguyen, P.X., Wang, K., Belongie, S.J.: Video text detection and recognition: dataset and benchmark. In: WACV (2014)

15. Reddy, S., Mathew, M., Gomez, L., Rusinol, M., Karatzas, D., Jawahar, C.: Roadtext-1k: text detection & recognition dataset for driving videos. In: 2020 IEEE International Conference on Robotics and Automation (ICRA), pp. 11074–11080. IEEE (2020)

16. Ristani, E., Solera, F., Zou, R., Cucchiara, R., Tomasi, C.: Performance measures and a data set for multi-target, multi-camera tracking. In: Hua, G., Jégou, H. (eds.) ECCV 2016, Part II. LNCS, vol. 9914, pp. 17–35. Springer, Cham (2016). https://doi.org/10.1007/978-3-319-48881-3_2

17. Shi, B., Bai, X., Yao, C.: An end-to-end trainable neural network for image-based sequence recognition and its application to scene text recognition. IEEE Trans. Pattern Anal. Mach. Intell. **39**(11), 2298–2304 (2016)

18. Tian, S., Yin, X., Su, Y., Hao, H.W.: A unified framework for tracking based text detection and recognition from web videos. In: TPAMI (2018)

19. Wu, W., Shen, C., et al.: End-to-end video text spotting with transformer. arxiv (2022)

20. Wu, W., et al.: Bovtext: a large-scale, multidimensional multilingual dataset for video text spotting. Organization (2021)

21. Yu, F., et al.: Bdd100k: a diverse driving dataset for heterogeneous multitask learning. In: Proceedings of the IEEE/CVF Conference on Computer Vision and Pattern Recognition, pp. 2636–2645 (2020)

22. Zhang, X., Su, Y., Tripathi, S., Tu, Z.: Text spotting transformers. In: Proceedings of the IEEE/CVF Conference on Computer Vision and Pattern Recognition, pp. 9519–9528 (2022)

23. Zhang, Y., et al.: Bytetrack: multi-object tracking by associating every detection box. In: Avidan, S., Brostow, G., Cissé, M., Farinella, G.M., Hassner, T. (eds.) Computer Vision - ECCV 2022. ECCV 2022, Part XXII, LNCS, vol. 13682, pp. 1–21. Springer, Cham (2022). https://doi.org/10.1007/978-3-031-20047-2_1

ICDAR 2023 Competition on Detecting Tampered Text in Images

Dongliang Luo[1], Yu Zhou[1], Rui Yang[2], Yuliang Liu[1], Xianjin Liu[2],
Jishen Zeng[2], Enming Zhang[1], Biao Yang[1], Ziming Huang[1], Lianwen Jin[3],
and Xiang Bai[1(✉)]

[1] Huazhong University of Science and Technology, Wuhan, China
{ldl,yuzhou,ylliu,emzhang,hust_byang,zmhuang,xbai}@hust.edu.cn
[2] Alibaba Group, Hangzhou, China
{duming.yr,xianjinliu.lxj,jishen.zjs}@alibaba-inc.com
[3] South China University of Technology, Guangzhou, China
eelwjin@scut.edu.cn

Abstract. Document analysis and recognition techniques are evolving quickly and have been widely used in real-world applications. However, detecting tampered text in images is rarely studied. Existing image forensics research mainly focuses on detecting tampered objects in natural images. Text manipulation in images exhibits different characteristics, *e.g.*,, text consistency, imperceptibility, etc., which bring new challenges for image forensics. Therefore, We organized the ICDAR 2023 Competition on Detecting Tampered Text in Images (DTTI) and established a new dataset named TII, which consists of 11,385 images. 5,500 images are tampered using various manipulation techniques and annotated by pixel-level masks. Two tasks are set up: text manipulation classification and text manipulation detection. The contest started on 15th February, 2023 and ended on 20th March, 2023. Over a thousand teams were registered for the competition, with 277 and 176 valid submissions in each task. In this competition report, we describe the details of the proposed dataset, tasks, evaluation protocols and the results summaries. We hope that this competition could promote the research of text manipulation detection in images.

Keywords: Images forensics · Tampered text detection · Image manipulation detection · Document forgery detection

1 Introduction

Texts in images efficiently deliver a wealth of important information and have become one of the most common mediums for various applications, such as digital finance, electronic commerce, security audit, and qualification review. Therefore, it is crucial that we can prevent text from tampering. A slight change in a sentence might significantly twist the whole carried semantic information. However, most of the previous studies in Document Analysis and Recognition

G. A. Fink et al. (Eds.): ICDAR 2023, LNCS 14188, pp. 587–600, 2023.
https://doi.org/10.1007/978-3-031-41679-8_36

focus on detecting and understanding the content of texts, such as text detection [22,23,30], recognition [31,32,41,42] and spotting [21,26]. The authenticity of them is rarely discussed, raising growing concerns about information security in daily life.

In recent years, image forensics have received increasing attention from both academia and industry aiming to defend against malicious image manipulation. Most of the studies focus on natural images, such as Columbia [28], CASIA [8], COVER [38], NIST16 [10] and RTD [19], in which the tampered subjects are usually objects, *e.g.,*, a human or a car. The individual characteristics of texts bring new challenges, such as various fonts, lengths, and shapes, which might cause limitations when using previous image-based manipulation detection methods for detecting tampered texts. In comparison, tampered text detection is more challenging due to the unstructured presentation of the texts. For example, the tampered areas can be very small (*e.g.,* a character in a paragraph); the contrast between tampered regions and surroundings can be very low.

However, most of the previous works [2,3,6,7] towards tampered text detection are based on private datasets, which restricts the advance of this area. In addition, most works [2,4,6] define the task as image classification, while localizing the forgery region is also of great importance. Recently, [36] and [37] propose synthetic tempered text datasets for document and scene text, respectively. In real scenes, manual text manipulation is one of the most common approaches, and the detection difficulty of manual tampering is often greater than that of synthetic tampering. Nevertheless, manual text tampering requires professional skills, and the cost of text manipulation and pixel-level annotation is expensive.

Therefore, we organized the ICDAR 2023 Competition on Detecting Tampered Text in Images. We build a new tampered text dataset, namely Tampered Text in Images (TTI), which simulates the common cases in real-world electronic commerce scenarios. The images are captured via several sources for diversity, especially some are highly related to actual applications. It contains 11,385 text images in total, and 5,500 images are manipulated using several types of manipulation techniques, including manual and automatic approaches. Each tampered image is annotated with a binary mask indicating the location of tampered regions. Correspondingly, two tasks are present for this competition: (1) text manipulation classification, and (2) text manipulation detection. We hope the dataset and competition could facilitate the research community and promote the research in tampered text detection in images.

2   Competition Organization

ICDAR2023 Competition on Detecting Tampered Text in Images is organized by a joint team, including Huazhong University of Science and Technology, Alibaba Group, and the South China University of Technology. We organize the competition through the Aliyun Tianchi platform online. The website[1,2] for each task

---

[1] https://tianchi.aliyun.com/competition/entrance/532048/introduction.

[2] https://tianchi.aliyun.com/competition/entrance/532052/introduction.

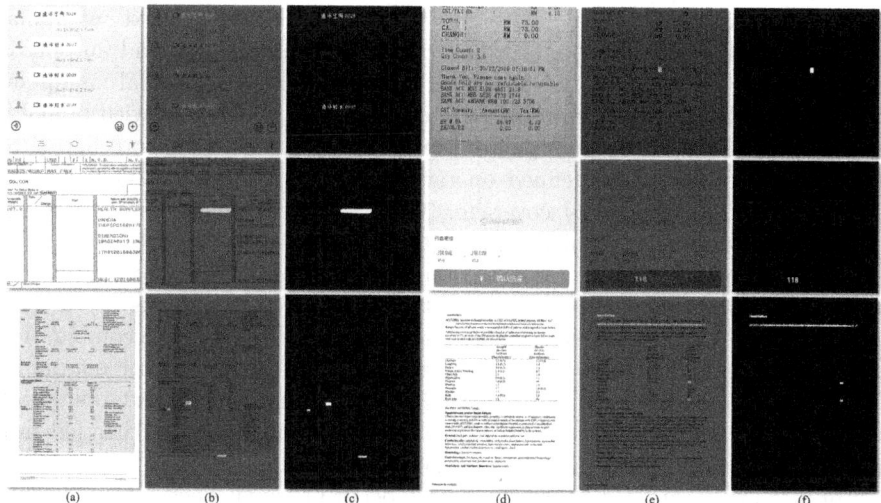

Fig. 1. Samples from TTI dataset. (a), (d) are the tampered images, (b), (e) indicate the locations of the tampered regions, and (c), (f) are corresponding binary masks. Some are cropped from the full-size images for better visualization.

provides user interfaces for participants, including corresponding download links of the dataset, descriptions of the evaluation protocol, and a submission page. Participants are welcome to choose one of the tasks or both.

Schedule. The competition started on 15th February 2023 and ended on 20th March 2023. The code was submitted and evaluated from 15th March to 20th March.

Awards. To encourage the participants, we have established prize awards for the top-performing teams. The prize for the Competition on Detecting Tampered Text in Images is 87,000 CNY in total, sponsored by Alibaba Group.

Overall, the competition attracted 1267 and 1156 teams from both research communities and industries in each task, receiving 277 and 176 valid submissions, respectively. Note that duplicate submissions are removed.

3   Dataset

3.1   Dataset Construction

Source of Images. The images are captured from three sources: (1) the text images from actual e-commerce applications provided by Alibaba Group after data desensitization, which includes certifications, contracts and screenshots, etc.; (2) the text images photographed by 10 volunteers from daily life, including books, commodity packages and signboards; (3) the text images selected from the open-source datasets, specifically, receipt images from SROIE [17], scan document images from FUNSD [18] and TNCR [1].

Text Manipulation. To simulate the practical situations of text manipulation in the real world, we perform both manual manipulations and automatic manipulations on the collected text images. Specifically, the manual manipulations are conducted by more than 20 professional image editors with different editing preferences. They manipulate the images using Adobe Photoshop (PS) following six types of pre-defined operations: (1) copy-move, (2) splicing, (3) insertion, (4) inpainting, (5) coverage, (6) replacement. These operations are summarized based on the observation of the most commonly used manipulation techniques in practical electronic commerce scenarios. The main purpose of operations (1)-(3) is to add texts, while the intention of operations (4) and (5) is removing texts. The replacement operation often represents the combination of text removal and text addition in the same region. The automatic manipulations are performed through PS scripts that mimic the realistic text tampering process mentioned above. Moreover, all tampered images are further fed into a series of post-processing, including random crop, random resize, and random compression to create distortions that often occur in actual electronic commerce scenarios.

Ground Truth. Images in the dataset are first categorized into tampered and untampered. As shown in Fig. 1, each tampered image in TTI is annotated with a binary mask, where the manipulated pixels are labeled as 1 (white) and the other pixels are labeled as 0 (black). The binary mask is annotated in parallel with the manual manipulation process. The annotations of automatically manipulated images are generated under the same criteria by comparing the differences between the images before and after tampering.

As a result, TTI contains multiple types of tampering. Figure 1 visualizes some typical cases. The first row is text addition and replacement. The tampered text is highly similar to adjacent text. The second row shows the situation of text removal. The tampered regions and the surrounding background are well integrated. The bottom row demonstrates the cases of automatic manipulation. Although they might not be as stealthy as manual tampering, they increase the diversity of tampering and are also likely to occur in real applications. More data examples from TII can be found in Appendix A.

3.2 Dataset Statistics and Analysis

Statistics. There are 11,385 images in TTI dataset. 5,500 of them are tampered images and the rest are untampered images. The dataset is split into a training set and a test set to evaluate the performance of the two tasks. The training set consists of 8,285 images (4,000 tampered and 4,285 untampered). The test set contains 3,100 images, whose ground truth will be privately served for online evaluation purposes. Figure 2 shows the statistics of the dataset. The side lengths of each image are between 512 to 2000. The differences in image size between the training set and the test set also introduce challenges. Most of the tampered regions are smaller than 5‰. The proportion of images with more than 30‰ tampered pixels is below 5%.

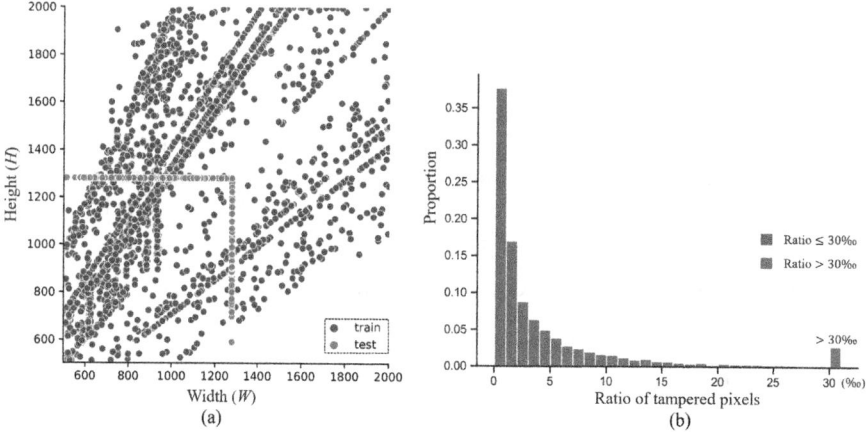

Fig. 2. Illustration of the statistics of TTI dataset. (a) is the distribution of image size and aspect ratio; (b) is the distribution of the tampered pixel ratio in each forged image.

Challenges. We briefly summarize the challenges of TTI into two parts, namely diversity and imperceptibility, according to the visualization and statistical analysis.

1) Diversity. As mentioned, the diversity of text images in the real world increases the difficulty of tampered text detection. Therefore, we try to simulate it from the following aspects When building TTI dataset. Firstly, the source image comes from various sources from daily life. The texts in the images differ in colors, fonts, perspectives, sizes and languages. Secondly, the size of images varies, bringing challenges to tracing subtle tampering cues. As shown in Fig. 2(a), the resolution of images in TTI dataset reflects an approximate decentralized distribution from 512 to 2000 with different aspect ratios. Thirdly, the manipulation methods are diverse. In addition, the shape of tampered regions is also various. As shown in Fig. 1, the shape of tampered regions might be regular shapes like a box or circle or irregular shapes. The size of the manipulated text region varies from 0.001% to 50% as shown in Fig. 2(b). The diversity of TTI challenges the generalization of detection methods.

2) Imperceptibility. Compared to natural image manipulation datasets, TTI dataset is more difficult since the tampered texts leave fewer visual traces and are well integrated with the surrounding context. Firstly, the tampered regions are relatively small. As shown in Fig. 2(b), the size of most tampered text regions is below 1% of the image. The smallness of tampered regions also means subtle visual inconsistency. Secondly, the local contrast between tampered and authentic regions is low, increasing the difficulty of distinguishing tampered from authentic areas. The distortions including random resize and random compression further weaken the visual traces. The imperceptibility challenges the discovery ability of artifacts and the robustness of solutions.

4   Tasks and Evaluation Protocols

The competition consists of two tasks: 1) text manipulation classification, and 2) text manipulation detection. The former aims to determine whether the input image contains tampered text, while the latter requires further localize the tampered region. In real applications, the detection methods are expected to achieve a higher recall within a range of tolerable false alarms. Based on such consideration, the evaluation protocols are designed as follows.

Task 1: Text Manipulation Classification.

Submission: Probability scores that an image is predicted as tampered.

Evaluation Protocols: we use the average true positive rate (TPR) under a fixed range of true negative rate (TNR) to evaluate the performance of submitted methods. TPR and TNR under a certain threshold are formulated as:

$$TPR = \frac{TP}{TP + FN} \tag{1}$$

$$TNR = \frac{TN}{TN + FP} \tag{2}$$

Similar to threshold-free metrics like Area Under Curve (AUC), the average true positive rate mPN is formulated as:

$$mPN_{[0.9:0.99]} = \frac{TPR_{TNR=0.90} + TPR_{TNR=0.91} + \ldots + TPR_{TNR=0.99}}{10} \tag{3}$$

The range of tolerable TNR is set from 0.9 to 0.99.

Task 2: Text Manipulation Detection.

Submission: The probability maps for all test images are saved as 8-bit gray-scale images, where the value of predicted tampered and authentic pixels is closer to 255 and 0, respectively. The shape of each map is equal to the corresponding input image. The gray-scale images will be normalized before evaluation.

Evaluation Protocols: We use the maximum IOU under the fixed range of image classification false positive rate as the evaluation index. Firstly, thresholds are determined according to the predicted tampered image mask submitted by the contestant, where the maximum value of each mask M_i is regarded as its image-level classification score:

$$Score_i = max(M_i) \tag{4}$$

Accordingly, the minimum threshold T under the lowest tolerable TNR is calculated based on the proportion of the number of correctly predicted authentic images. Specifically, classification scores for authentic test samples $\{Score_i^A\}$ are sorted, and TNR is formulated as:

$$TNR = \frac{\sum_i \mathbb{I}(Score_i^A < T)}{\sum_i \mathbb{I}(Score_i^A \geq T)} \tag{5}$$

For each TNR, the corresponding threshold is determined. Finally, we calculate the sum of maximum IOU ($MIOU$) of the tampered image within the threshold range, which is formulated as:

$$IOU_{[0.9:0.99]} = MIOU_{TNR=0.90} + MIOU_{TNR=0.91} + \ldots + MIOU_{TNR=0.99} \tag{6}$$

The range of tolerable TNR is set from 0.9 to 0.99 following Task 1.

5   Submissions

The top-5 performing results for Task 1 and Task 2 are presented in Table 1, The complete ranking can be viewed on the competition website[3,4]. We introduce the representative public submissions.

5.1   Top Submissions in Task 1

Team DouyinCV. Multiple backbones including ResNet [16], EfficientNet [33] and Swin Transformer V2 [24] are leveraged to enhance the performance. Various manipulation trace extracting methods are used as data pre-processing, such as Error Level Analysis (ELA), SRM filter, and pre-trained MantraNet [39]. Data augmentations include random resize, random compression, random aspect ratio and random crop. As some images are tampered from the same picture in the training and test set, they design a post-processing method called local detection to fully utilize the authentic prior. For each image in the test set, they retrieve the most similar image in the training set based on the cosine similarity of features extracted by ResNet18 and align them using ECC [9]. Both the image-level and pixel-level labels of the reference image are taken into account when scoring the test image.

Team Xixihaha. The team selects DeepLabv3+ [5] as the basic architecture with DiNAT [14] as its backbone, followed by a classification head and segmentation head. Multi-task learning can optimize the segmentation performance by regarding classification probabilities as prior. Furthermore, A improved DCT Volume stream based on CAT-Net [20] is paralleled to extract JPEG artifacts from the DCT domain. Inspired by TransForensics [13], they replace the original segmentation decoder in DeepLabv3+ [5] with an attention decoder to enhance the relation between different positions and levels. In addition, random resize and random compression are employed during training to balance the distribution. Focal loss and dice loss are leveraged to alleviate the data imbalance problem.

---

[3] https://tianchi.aliyun.com/competition/entrance/532048/rankingList.
[4] https://tianchi.aliyun.com/competition/entrance/532052/rankingList.

Table 1. Results summary for the top-5 submissions of the competition

	Team Name	Affiliation	Score ⇑	Rank
Task 1	DouyinCV	ByteDance	88.06	1
	Xixihaha	Xi'an Jiaotong University	87.28	2
	Lykken	Ant Group Co., Ltd.	86.33	3
	Overfit Again	IntSig Information Co., Ltd.	86.33	4
	Heiha	Shenzhen University	86.33	5
Task 2	Duiduidui	IntSig Information Co., Ltd.	3.890	1
	Xixihaha	Xi'an Jiaotong University	3.419	2
	Heiha	Shenzhen University	3.163	3
	Easy Electronics LLL	Shenzhen University	3.087	4
	ustc-imcc	University of Science and Technology of China	3.087	5

Team Lykken. Like previous teams, they adopt model ensemble using YOLO [29], Mask-CNN [15] and HRNet [35]. Multi-task learning of classification and segmentation is utilized. Furthermore, they simulate our tampering operations, such as text replacement, copy-move and removal to expand the training data. Test Time Augmentation (TTA) is employed by taking three scales of the test image as the input (480×480, 640×640 and 960×960) separately. Similar to Team DouyinCV, they perform multi-mode retrieval based on image features and text features. The discrepancy between the test image and its homologous template in the database is fused with the prediction of the detection network.

Team Overfit Again. The team selects SORIE [17], TNCR [1], FUNSD [18] and other two tampered detection competition data as pre-training data to enhance the performance on receipts images and scan documents. To obtain more robust results, five models are ensembled, including MobileViT-v1, MobileViT-v3 [27], GhostNet [12], EfficientNet-B5 [33] and EfficientNetV2-M [34]. In addition, they perform automatic text tampering on the authentic images in our dataset and additional data to expand the tampered samples during training. During inference, the result of different models are fused and multiple steps of post-processing are designed to refine the predictions.

5.2   Top Submissions in Task 2

Team Duiduidui. To combine the advantages of different architectures, five models, namely ConvNeXt [25], HRNet [35], SegNeXt [11], DeepLabV3+ [5] and SegFormer [40], are ensembled. Various data augmentation strategies are adopted including random crop, random compression, photometric distortions, random motion blur, random rotation and random horizontal flip. Furthermore, they synthesize tampered text using untampered images in the training set to expand positive samples. To address the issue of data imbalance, they apply Online Hard Example Mining (OHEM) and weighted loss functions. The combination of cross-entropy loss, lovasz loss and dice loss also contributes to the performance. To comprehensively explore the local and global information, multi-scale input during inference.

Team Xixihaha. Same as in Sect. 5.1.

Team Easy Electronics LLL. The team leverages ConvNeXt [25] and Seg-Former [40] as their basic segmentation models. Dice loss and cross-entropy loss were used to remedy the unbalance problem of the positive and negative samples. In addition, they introduce multi-task learning by adding an auxiliary classification head to alleviate false positives. To solve the problem of scale diversity, they adopt the curriculum learning paradigm by continuously migrating from small size to large size during training. Random flip, random brightness and contrast transform, random compression and random Gaussian noise were also used for data augmentation. TTA is applied to enhance the performance during inference.

5.3   Discussion

In this competition, task 1 is the foundation of task 2. The submissions point out some common difficulties, such as data imbalance, resolution diversity, small targets, etc. Most teams design their methods based on semantic segmentation architectures. Model ensemble strategy is also utilized by most teams as an effective approach to improve performance, including CNN-based and ViT-based architectures. Some participants introduce the manipulation trace extractors from image manipulation detection methods to enhance the tamper detection capacity through another domain, such as noise, ELA and frequency domain, while others take the RGB image as the only cue as the trace may not be evident as expected.

To overcome the data imbalance problem, various data augmentations and OHEM are leveraged. In addition, synthetic tampered samples are generated by some teams to further increase positive data. The design of the loss function is also useful, participants mainly choose focal loss, dice loss and lovasz loss.

To boost the performance of models on images with different resolution, multi-scale training and TTA is adopted by most contestants. By fusing predictions from multi-scale inputs, the model is capable of discovering the tampered regions from both local and global cues. As our evaluation protocols impose great penalties for false positives, which is one of the major reasons that many teams obtained poor scores, most teams introduce multi-task learning to alleviate it. Some make further efforts to mine the complementarity of two tasks and fuse their predictions.

Based on the observation of the existence of homologous data, Some teams propose to exploit the untampered images by retrieving the most similar template. The discrepancy map between them is regarded as an additional cue and fused into the prediction. They provide a potential aspect of treating the tampered regions as anomalies if the authentic template exists. However, their generalization and flexibility when facing text images with brand-new layouts are of concern.

In summary, the submissions provide valuable insights into our task and propose pertinent approaches against some of the challenges. However, the characteristics and challenges of tampered text detection are not comprehensively studied. Some solutions might also have limitations. Thus this field still waits for further exploration.

6   Potential Negative Societal Impacts and Solution

To avoid the malicious use of the proposed dataset, TTI is constructed after data desensitization by cropping off the sensitive information. TTI is only available for participating in the challenge.

7   Conclusion

We organize the Competition on Detecting Tampered Text in Images (DTTI). TTI dataset of 5,500 tampered and 5,785 untampered images is released alone with two tasks. During the challenge, more than a thousand teams are registered and submitted 277 and 176 results in each task, demonstrating the broad interest in the community. These submissions will provide the community with new insights and potential solutions. We hope that our dataset will contribute to advancing the field of tampered text detection in images.

Acknowledgments. This competition is supported by the National Natural Science Foundation (NSFC#62225603).

Appendix

A   More Examples from TTI

Figure 3 gives more examples from TII dataset, including tampered documents and their ground truth.

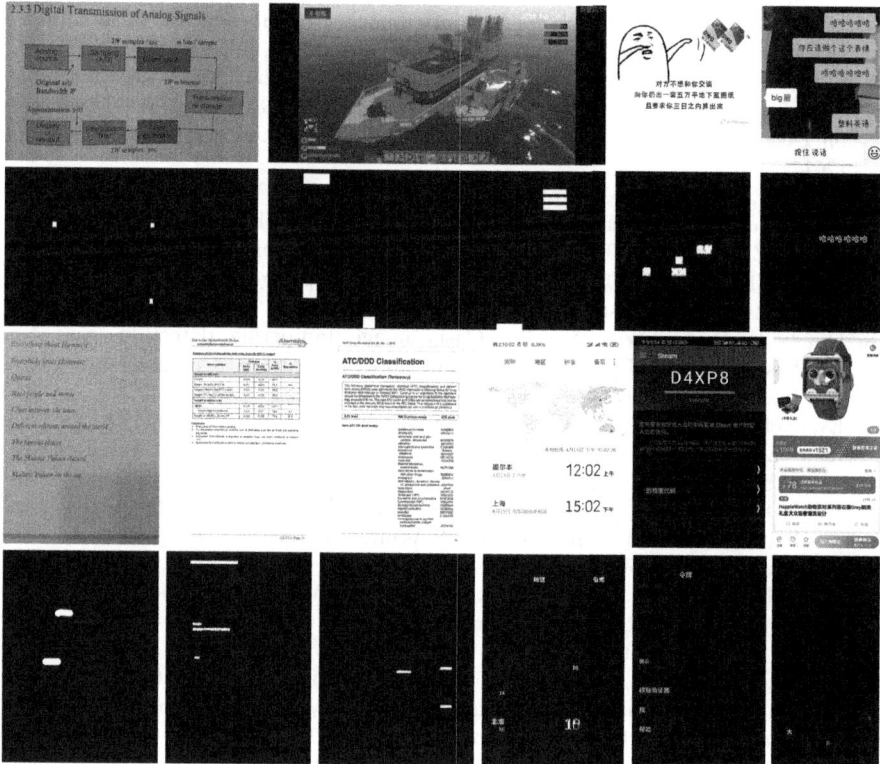

Fig. 3. More examples from TTI dataset. The forged images and their ground truth are presented in the odd rows and even rows, respectively.

References

1. Abdallah, A., Berendeyev, A., Nuradin, I., Nurseitov, D.: TNCR: table net detection and classification dataset. Neurocomputing **473**, 79–97 (2022)
2. Bertrand, R., Gomez-Krämer, P., Terrades, O.R., Franco, P., Ogier, J.M.: A system based on intrinsic features for fraudulent document detection. In: 2013 12th International Conference on Document Analysis and Recognition, pp. 106–110. IEEE (2013)

3. van Beusekom, J., Stahl, A., Shafait, F.: Lessons learned from automatic forgery detection in over 100,000 invoices. In: Garain, U., Shafait, F. (eds.) IWCF 2012/2014. LNCS, vol. 8915, pp. 130–142. Springer, Cham (2015). https://doi.org/10.1007/978-3-319-20125-2_12
4. Bibi, M., Hamid, A., Moetesum, M., Siddiqi, I.: Document forgery detection using printer source identification-a text-independent approach. In: 2019 International Conference on Document Analysis and Recognition Workshops (ICDARW), vol. 8, pp. 7–12. IEEE (2019)
5. Chen, L.C., Zhu, Y., Papandreou, G., Schroff, F., Adam, H.: Encoder-decoder with atrous separable convolution for semantic image segmentation. In: Proceedings of the European Conference on Computer Vision (ECCV), pp. 801–818 (2018)
6. Cruz, F., Sidere, N., Coustaty, M., d'Andecy, V.P., Ogier, J.M.: Local binary patterns for document forgery detection. In: 2017 14th IAPR International Conference on Document Analysis and Recognition (ICDAR), vol. 1, pp. 1223–1228. IEEE (2017)
7. Cruz, F., Sidère, N., Coustaty, M., Poulain d'Andecy, V., Ogier, J.-M.: Categorization of document image tampering techniques and how to identify them. In: Zhang, Z., Suter, D., Tian, Y., Branzan Albu, A., Sidère, N., Jair Escalante, H. (eds.) ICPR 2018. LNCS, vol. 11188, pp. 117–124. Springer, Cham (2019). https://doi.org/10.1007/978-3-030-05792-3_11
8. Dong, J., Wang, W., Tan, T.: Casia image tampering detection evaluation database. In: 2013 IEEE China Summit and International Conference on Signal and Information Processing, pp. 422–426. IEEE (2013)
9. Evangelidis, G.D., Psarakis, E.Z.: Parametric image alignment using enhanced correlation coefficient maximization. IEEE Trans. Pattern Anal. Mach. Intell. **30**(10), 1858–1865 (2008)
10. Guan, H., et al.: MFC datasets: large-scale benchmark datasets for media forensic challenge evaluation. In: 2019 IEEE Winter Applications of Computer Vision Workshops (WACVW), pp. 63–72. IEEE (2019)
11. Guo, M.H., Lu, C.Z., Hou, Q., Liu, Z., Cheng, M.M., Hu, S.M.: Segnext: rethinking convolutional attention design for semantic segmentation. arXiv preprint arXiv:2209.08575 (2022)
12. Han, K., Wang, Y., Tian, Q., Guo, J., Xu, C., Xu, C.: Ghostnet: more features from cheap operations. In: Proceedings of the IEEE/CVF Conference on Computer Vision and Pattern Recognition, pp. 1580–1589 (2020)
13. Hao, J., Zhang, Z., Yang, S., Xie, D., Pu, S.: Transforensics: image forgery localization with dense self-attention. In: Proceedings of the IEEE/CVF International Conference on Computer Vision, pp. 15055–15064 (2021)
14. Hassani, A., Shi, H.: Dilated neighborhood attention transformer. arXiv preprint arXiv:2209.15001 (2022)
15. He, K., Gkioxari, G., Dollár, P., Girshick, R.: Mask R-CNN. In: Proceedings of the IEEE International Conference on Computer Vision, pp. 2961–2969 (2017)
16. He, K., Zhang, X., Ren, S., Sun, J.: Deep residual learning for image recognition. In: Proceedings of the IEEE Conference on Computer Vision and Pattern Recognition, pp. 770–778 (2016)
17. Huang, Z., et al.: ICDAR 2019 competition on scanned receipt OCR and information extraction. In: 2019 International Conference on Document Analysis and Recognition (ICDAR), pp. 1516–1520. IEEE (2019)
18. Jaume, G., Ekenel, H.K., Thiran, J.P.: FUNSD: a dataset for form understanding in noisy scanned documents. In: 2019 International Conference on Document Analysis and Recognition Workshops (ICDARW), vol. 2, pp. 1–6. IEEE (2019)

19. Korus, P., Huang, J.: Evaluation of random field models in multi-modal unsupervised tampering localization. In: 2016 IEEE International Workshop on Information Forensics and Security (WIFS), pp. 1–6. IEEE (2016)
20. Kwon, M.J., Yu, I.J., Nam, S.H., Lee, H.K.: Cat-net: compression artifact tracing network for detection and localization of image splicing. In: Proceedings of the IEEE/CVF Winter Conference on Applications of Computer Vision, pp. 375–384 (2021)
21. Liao, M., Lyu, P., He, M., Yao, C., Wu, W., Bai, X.: Mask textspotter: an end-to-end trainable neural network for spotting text with arbitrary shapes. IEEE Trans. Pattern Anal. Mach. Intell. 43(2), 532–548 (2019)
22. Liao, M., Shi, B., Bai, X., Wang, X., Liu, W.: Textboxes: a fast text detector with a single deep neural network. In: Thirty-First AAAI Conference on Artificial Intelligence (2017)
23. Liao, M., Wan, Z., Yao, C., Chen, K., Bai, X.: Real-time scene text detection with differentiable binarization. In: Proceedings of the AAAI Conference on Artificial Intelligence, vol. 34, pp. 11474–11481 (2020)
24. Liu, Z., et al.: Swin transformer V2: scaling up capacity and resolution. In: Proceedings of the IEEE/CVF Conference on Computer Vision and Pattern Recognition, pp. 12009–12019 (2022)
25. Liu, Z., Mao, H., Wu, C.Y., Feichtenhofer, C., Darrell, T., Xie, S.: A convnet for the 2020s. In: Proceedings of the IEEE/CVF Conference on Computer Vision and Pattern Recognition, pp. 11976–11986 (2022)
26. Lu, P., Wang, H., Zhu, S., Wang, J., Bai, X., Liu, W.: Boundary textspotter: toward arbitrary-shaped scene text spotting. IEEE Trans. Image Process. 31, 6200–6212 (2022)
27. Mehta, S., Rastegari, M.: Mobilevit: light-weight, general-purpose, and mobile-friendly vision transformer. arXiv preprint arXiv:2110.02178 (2021)
28. Ng, T.T., Hsu, J., Chang, S.F.: Columbia image splicing detection evaluation dataset. DVMM lab. Columbia Univ CalPhotos Digit Libr (2009)
29. Redmon, J., Divvala, S., Girshick, R., Farhadi, A.: You only look once: unified, real-time object detection. In: Proceedings of the IEEE Conference on Computer Vision and Pattern Recognition, pp. 779–788 (2016)
30. Shi, B., Bai, X., Belongie, S.: Detecting oriented text in natural images by linking segments. In: Proceedings of the IEEE Conference on Computer Vision and Pattern Recognition, pp. 2550–2558 (2017)
31. Shi, B., Bai, X., Yao, C.: An end-to-end trainable neural network for image-based sequence recognition and its application to scene text recognition. IEEE Trans. Pattern Anal. Mach. Intell. 39(11), 2298–2304 (2016)
32. Shi, B., Yang, M., Wang, X., Lyu, P., Yao, C., Bai, X.: Aster: an attentional scene text recognizer with flexible rectification. IEEE Trans. Pattern Anal. Mach. Intell. 41(9), 2035–2048 (2018)
33. Tan, M., Le, Q.: Efficientnet: rethinking model scaling for convolutional neural networks. In: International Conference on Machine Learning, pp. 6105–6114. PMLR (2019)
34. Tan, M., Le, Q.: EfficientNetV2: smaller models and faster training. In: International Conference on Machine Learning, pp. 10096–10106. PMLR (2021)
35. Wang, J., et al.: Deep high-resolution representation learning for visual recognition. IEEE Trans. Pattern Anal. Mach. Intell. 43(10), 3349–3364 (2020)
36. Wang, Y., Zhang, B., Xie, H., Zhang, Y.: Tampered text detection via RGB and frequency relationship modeling. Chin. J. Netw. Inf. Secur. 8(3), 29–40 (2022)

37. Wang, Y., Xie, H., Xing, M., Wang, J., Zhu, S., Zhang, Y.: Detecting tampered scene text in the wild. In: Avidan, S., Brostow, G., Cissé, M., Farinella, G.M., Hassner, T. (eds.) ECCV 2022. LNCS, vol. 13688, pp. 215–232. Springer, Cham (2022). https://doi.org/10.1007/978-3-031-19815-1_13

38. Wen, B., Zhu, Y., Subramanian, R., Ng, T.T., Shen, X., Winkler, S.: Coverage- a novel database for copy-move forgery detection. In: 2016 IEEE International Conference on Image Processing (ICIP), pp. 161–165. IEEE (2016)

39. Wu, Y., AbdAlmageed, W., Natarajan, P.: Mantra-net: manipulation tracing network for detection and localization of image forgeries with anomalous features. In: Proceedings of the IEEE/CVF Conference on Computer Vision and Pattern Recognition, pp. 9543–9552 (2019)

40. Xie, E., Wang, W., Yu, Z., Anandkumar, A., Alvarez, J.M., Luo, P.: Segformer: simple and efficient design for semantic segmentation with transformers. Adv. Neural. Inf. Process. Syst. **34**, 12077–12090 (2021)

41. Xie, X., Fu, L., Zhang, Z., Wang, Z., Bai, X.: Toward understanding wordart: corner-guided transformer for scene text recognition. In: Avidan, S., Brostow, G., Cissé, M., Farinella, G.M., Hassner, T. (eds.) ECCV 2022. LNCS, vol. 13688, pp. 303–321. Springer, Cham (2022). https://doi.org/10.1007/978-3-031-19815-1_18

42. Yang, M., et al.: Reading and writing: discriminative and generative modeling for self-supervised text recognition. In: Proceedings of the 30th ACM International Conference on Multimedia, pp. 4214–4223 (2022)

Author Index

Printed in the United States
by Baker & Taylor Publisher Services